Completely Regular Codes in Distance Regular Graphs

The concept of completely regular codes was introduced by Delsarte in his celebrated 1973 thesis, which created the field of Algebraic Combinatorics. This notion was extended by several authors from classical codes over finite fields to codes in distance-regular graphs. Half a century later, there was no book dedicated uniquely to this notion. Most of Delsarte examples were in the Hamming and Johnson graphs. In recent years, many examples were constructed in other distance regular graphs including q-analogues of the previous and the Doob graph.

Completely Regular Codes in Distance Regular Graphs provides, for the first time, a definitive source for the main theoretical notions underpinning this fascinating area of study. It also supplies several useful surveys of constructions using coding theory, design theory, and finite geometry in various families of distance regular graphs of large diameters.

Features

- Written by pioneering experts in the domain
- Suitable as a research reference at the master's level
- Includes extensive tables of completely regular codes in the Hamming graph
- Features a collection of up-to-date surveys.

Minjia Shi received the Ph.D. degree from the Institute of Computer Network Systems, Hefei University of Technology, China, in 2010. From August 2012 to August 2013, he was a Visiting Researcher with the School of Physical and Mathematical Sciences, Nanyang Technological University, Singapore. From July 2016 to August 2016, he was a Visiting Researcher with Telecom Paris Tech, Paris, France. Later, he visited the Sobolev Institute of Mathematics for half a month in 2020. He has been the title of a Professor of the fourth grade from 2017 to 2020 in Anhui University. He has been the title of a Professor of the third grade from 2020 to 2023 in Anhui University. He has been the title of a Professor of the second grade since 2024 in Anhui University. He is the author of more than 160 journal articles and three books. His current research interests include algebraic coding theory and cryptography.

Patrick Solé received the Ingénieur and Docteur-Ingénieur degrees both from Ecole Nationale Supérieure des Télécommunications, Paris, France, in 1984 and 1987, respectively, and the habilitation à diriger des recherches from Université de Nice-Sophia Antipolis, Sophia Antipolis, France, in 1993.

He has held visiting positions at Syracuse University, Syracuse, NY, from 1987 to 1989, Macquarie University, Sydney, Australia, from 1994 to 1996, and Lille University, Lille, France, from 1999 to 2000.

Since 1989, he has been a permanent member of the CNRS and became Directeur de Recherche in 1996. He is currently a member of the CNRS lab I2M, Marseilles, France.

His research interests include coding theory (codes over rings, quasi-cyclic codes), interconnection networks (graph spectra, expanders), vector quantization (lattices), and cryptography (boolean functions, secret sharing schemes). He is the author of over 300 journal articles and three books.

Monographs and Research Notes in Mathematics

This series is designed to capture new developments and summarize what is known over the entire field of mathematics, both pure and applied. It will include a broad range of monographs and research notes on current and developing topics that will appeal to academics, graduate students, and practitioners. Interdisciplinary books appealing not only to the mathematical community but also to engineers, physicists, and computer scientists are encouraged.

This series will maintain the highest editorial standards, publishing well-developed monographs as well as research notes on new topics that are final, but not yet refined into a formal monograph. The notes are meant to be a rapid means of publication for current material where the style of exposition reflects a developing topic.

Direct and Projective Limits of Geometric Banach Structures
Patrick Cabau and Fernand Pelletier

Classical Clifford Algebras
Operator-Algebraic and Free-Probabilistic Approaches
Ilwoo Cho

Perturbed Functional Iterations
A Matrix Free Large-Scale Nonlinear System Solver in Applied Science with an Introduction to D-Mapping
Suhrit Dey

Separate and Joint Continuity
Jiling Cao and Warren B. Moors

Fractional Integrals, Potentials, and Radon Transforms, Second Edition
Boris Rubin

Variational-Hemivariational Inequalities with Applications, Second Edition
Mircea Sofonea and Stanisław Migórski

Classification of Lipschitz Mappings, Second Edition
Torrey M. Gallagher, Víctor Pérez-García, and Łukasz Piasecki

Completely Regular Codes in Distance Regular Graphs
Patrick Solé and Minjia Shi

For more information about this series please visit:
https://www.routledge.com/Chapman--HallCRC-Monographs-and-Research-Notes-in-Mathematics/book-series/CRCMONRESNOT

Completely Regular Codes in Distance Regular Graphs

Edited by

Minjia Shi and Patrick Solé

CRC Press
Taylor & Francis Group
Boca Raton London New York

CRC Press is an imprint of the
Taylor & Francis Group, an **informa** business
A CHAPMAN & HALL BOOK

Front cover image: Denis Krotov. A completely regular code of size 4 in the Desargues graph.

First edition published 2025
by CRC Press
2385 NW Executive Center Drive, Suite 320, Boca Raton FL 33431

and by CRC Press
4 Park Square, Milton Park, Abingdon, Oxon, OX14 4RN

CRC Press is an imprint of Taylor & Francis Group, LLC

ISBN: 978-1-032-49444-9 (hbk)
ISBN: 978-1-032-49458-6 (pbk)
ISBN: 978-1-003-39393-1 (ebk)

DOI: 10.1201/9781003393931

Typeset in Minion
by codeMantra

This book is supported in part by National Natural Science Foundation of China No.12471490

Contents

List of Contributors

Evgeny A. Bespalov
Sobolev Institute of Mathematics
Novosibirsk, Russia

Alexander Gavrilyuk
Department of Mathematical Sciences
University of Memphis
Memphis, Tennessee

Daniel R. Hawtin
Faculty of Mathematics
University of Rijeka
Rijeka, Croatia

Jack H. Koolen
School of Mathematical Sciences
University of Science and Technology of China
Hefei, China

Denis S. Krotov
Sobolev Institute of Mathematics
Novosibirsk, Russia

William J. Martin
Department of Mathematical Sciences
Worcester Polytechnic Institute
Worcester, Massachusetts

Vladimir N. Potapov
Faculty of Engineering and Natural Science
Sabancı University
Istanbul, Turkey

Cheryl E. Praeger
Centre for the Mathematics of Symmetry and Computation
University of Western Australia
Perth, Australia

Minjia Shi
School of Mathematical Sciences
Anhui University
Hefei, China

Victor A. Zinoviev
A.A. Kharkevich Institute for Problems of Information Transmission
Russian Academy of Sciences
Moscow, Russia

Introduction

In the early 1970s, the trend in combinatorial coding theory was the characterization of families of highly regular codes: perfect codes [4, Chapter 6, Section 10] and uniformly packed codes [6]. In this context, Philippe Delsarte in his Ph.D. dissertation [2], introduced the concept of completely regular codes (shortly CR codes), that is to say, codes whose coset weight distributions solely depend on the coset weight. The motivation was partly to generalize the two said families. A more important application, from the standpoint of Algebraic Combinatorics, was the fact that the Hamming graph, when quotiented by a linear CR code yielded a distance regular graph. The main examples (Golay codes, Kasami codes, and so on) are described in Ref. [1, Chapter 11]. A more general situation is that of a distance regular graph with an equitable partition. As observed by Neumaier [5], CR codes in distance regular graphs are equivalent to the datum of equitable partitions that are distance partitions with respect to a code. A very important technique in constructing CR codes is through the action of a group. A notion of a completely transitive code based on coset weight distributions was introduced in Ref. [7]. A more general notion of completely transitive codes was studied by Giudici and Praeger [3].

The aims of this collection of surveys are to recall the theory of CR codes in distance regular graphs and to describe the main known constructions. The two graphs of choice are the Hamming and Johnson graphs, due to their organic relations with codes and combinatorial designs, respectively. Chapter 1, written by D. S. Krotov and V. N. Potapov, lays the foundations for the other chapters. Chapter 2, authored by V. A. Zinoviev, collects the main constructions of codes over finite fields. Chapter 4, written by M. Shi, collects constructions based on codes over rings and modules. Chapter 3, written by A. Gavrilyuk and V. A. Zinoviev, describes completely regular designs, or, equivalently designs in the Johnson graphs. Chapter 5, written by C. E. Praeger and D. R. Hawtin, gives constructions and characterizations of completely transitive codes and their generalizations (neighbor transitive and others). Chapter 6, written by D. S. Krotov and E. A. Bespalov, is dedicated to the Doob graph. Chapter 7 contains tables of unrestricted CR codes in the Hamming graph, compiled by J. H. Koolen, D. S. Krotov, and W. J. Martin.

REFERENCES

[1] A. E. Brouwer, A. M. Cohen, A. Neumaier, *Distance-Regular Graphs*, Springer Verlag, Berlin (1989).

[2] P. Delsarte, An algebraic approach to the association schemes of Coding Theory, *Philips Res. Rep. Supp.* 10 (1973).

[3] M. Giudici and C. E. Praeger, Completely transitive codes in Hamming graphs, *Eur. J. Comb.* 20 (1999), 647–662.

[4] F. J. MacWilliams, N. J. A. Sloane, *The Theory of Error-Correcting Codes.* North-Holland, Amsterdam (1977).

[5] A. Neumaier, Completely regular codes, *Discrete Math.* 106/107 (1992), 353–360.

[6] N. V. Semakov, V. A. Zinoviev, G. V. Zaitsev, Uniformly packed codes, Probl. Inform. Trans. 7(1) (1971), 30–39.

[7] P. Solé, Completely regular codes and completely transitive codes, *Discrete Math.* 81 (1990), 193–201.

CHAPTER 1

Completely Regular Codes and Equitable Partitions

Denis S. Krotov and Vladimir N. Potapov

THIS CHAPTER CONTAINS SOME background material on equitable partitions (perfect colorings) and completely regular codes, including a survey of known results and some original results. Section 1.1 contains the main definitions related to the theory of completely regular codes and important correspondences between defined objects. Section 1.2 describes known properties of equitable partitions. In Section 1.3, we consider some very general ways to construct equitable partitions. In Section 1.4, we consider classes of combinatorial objects (mostly, classes of objects that are optimal in terms of a certain bound on their parameters) that are equivalent to equitable partitions, i.e., can be alternatively defined as a cell of an equitable partition with special parameters. Section 1.5 contains a small survey of results on completely regular codes in distance-regular graphs different from Hamming, Johnson, Grassmann, and Doob graphs. The following results of this section are new, up to our knowledge: Theorem 1.1 (generalization of Delsarte's definition of completely regular codes to non-distance-regular graphs), Lemma 1.4 (generalized correlation-immunity bound for T-designs), Theorems 1.2 and 1.3 (a new necessary condition on the existence of equitable partitions: a triangle inequality for eigenvectors of the quotient matrix), Lemma 1.7 (a sufficient condition for the strict monotonicity of an intersection array), Theorem 1.4 (strong distance invariance of some combinatorial designs), Theorem 1.6 (generalized Bierbrauer–Friedman bound), the complete list of parameters of completely-regular codes in cubic distance-regular graphs in Section 1.5 (it was known earlier only for some of these graphs). Some other results, e.g., Propositions 1.33 and 1.58, are more-or-less straightforward consequences of the general theory and might be known in the folklore. The content of this chapter is limited by the knowledge and expertize of its authors, and some deep results on completely regular codes related to the theory of association schemes (for example, most of the results of [62]) are not mentioned here.

DOI: 10.1201/9781003393931-1

1.1 DEFINITIONS AND BASIC FACTS

1.1.1 Graphs and Related Concepts

We will assume that, unless otherwise stated, every graph is finite and every group is finite.

Definition 1.1 (graph, simple graph, regular graph, subgraph, induced subgraph). *Formally, a graph is an ordered pair (V, E) from some set V, whose elements are called* vertices, *and a multiset E of 1- and 2- subsets of V, called* edges *(1-subsets are called* loops*). A* simple graph *is a graph without loops and multiple edges (the edge set of a simple graph can be considered as an ordinary set). If, in a graph, $\{x, y\}$ is an edge (x and y are not necessarily distinct), then the vertices x and y are called* neighbors *of each other or* adjacent *to each other. An edge and a vertex in it are called* incident *to each other. Unless otherwise stated, all considered graphs are finite, i.e., the vertex set is finite. The number of vertices in a graph is called its* order*; the number of edges containing a given vertex is called its* degree *(in particular, every loop scores 1 to the degree of its vertex) or* valency*. A graph is called* regular *(k-*regular*) if all its vertices have the same degree (equal to k). A graph (V', E') is a* subgraph *of a graph (V, E) if $V' \subseteq V$ and $E' \subseteq E$; the subgraph is* induced *(by V') if $e \in E$ and $e \subseteq V'$ implies that e belongs to E' and has there the same multiplicity as in E.*

Definition 1.2 (path, connectivity, distance, metric space, distance-i graph $\Gamma^{(i)}$). *In a graph, a* path *of length s is a sequence $(x_0, x_1, \ldots, x_s)$ of $s+1$ vertices in which any two subsequent vertices x_{i-1}, x_i, $i \in \{1, \ldots, s\}$ are adjacent. The relation "to be connected by a path" is an equivalence on the vertex set of a graph, and the subgraphs induced by the corresponding equivalence classes are called* connected components *of the graph. A graph is* connected *if it has only one connected component. The* distance *$d(x, y) = d_\Gamma(x, y)$ between two vertices x and y in a connected graph Γ is the length of the shortest path between these vertices. So, the vertex set of a connected graph with the corresponding distance form a metric space, which usually has a name related to the corresponding graph, for example, the Hamming graph $H(n, q)$ is associated to the Hamming (metric) space, which is also denoted $H(n, q)$. Given a metric space, we also imply that related natural concepts are defined, such as diameter (of a vertex set or of the graph), the distance between a vertex and a set or between two sets, the ball of a given radius centered in a given vertex, and so on. In particular, the diameter* $\mathrm{diam}(\Gamma)$ *of a connected graph Γ is the maximum distance between two vertices of Γ. Given a graph Γ, we define the* distance-i graph $\Gamma^{(i)}$ *on the same vertex set as follows: two vertices are adjacent (with multiplicity 1 of the edge) in $\Gamma^{(i)}$ if and only if the distance between them in Γ is i. In particular, $\Gamma^{(0)}$ has one loop at each vertex and no other edges; $\Gamma^{(i)}$, $i \in \{1, \ldots, \mathrm{diam}(\Gamma)\}$, is a simple graph; for any other i, $\Gamma^{(i)}$ is edge-free.*

Definition 1.3 (bipartite graph, halved graph $\frac{1}{2}\Gamma$). *A graph is* bipartite *if its vertex set can be partitioned into two parts such that every edge intersects with both parts* *(in particular, there are no loops). If a simple connected graph Γ is bipartite, then its distance-2 graph $\Gamma^{(2)}$ has exactly two connected components, called a* halved Γ *and denoted* $\frac{1}{2}\Gamma$.

Definition 1.4 (antipodal graph, folded graph $\widetilde{\Gamma}$). *A connected graph Γ is called* antipodal *if the relation "to be at distance* $\mathrm{diam}(\Gamma)$*" on the vertices is an equivalence (equivalently, the graph $\Gamma^{(\mathrm{diam}(\Gamma))}$ is the direct sum of complete graphs, see Definitions 1.8 and 1.11 below). The graph on the corresponding equivalence classes (also called* antipodal classes*), where two classes are adjacent if and only if they contain adjacent vertices, is called the* folded Γ *and denoted* $\widetilde{\Gamma}$.

Definition 1.5 (walk). *Formally, a* walk *in a graph $\Gamma = (V, E)$ can be defined as a path where each pair of consequent vertices x_i, x_{i+1} is assigned a label from* 1 *to $\mu_{x_i,x_{i+1}}$, where $\mu_{x_i,x_{i+1}}$ is the multiplicity of the edge $\{x_i, x_{i+1}\}$. So, each path $(x_0, x_1, \ldots, x_s)$ corresponds to $\mu_{x_0,x_1}\mu_{x_1,x_2}\cdots\mu_{x_{s-1},x_s}$ walks. We mention walks when we discuss the number of ways to reach one vertex from another respecting the edge multiplicities, i.e., distinguishing the $\mu_{x,y}$ occurrences of the same edge $\{x, y\}$ in the multiset E.*

Definition 1.6 (automorphism, automorphism group, isomorphism). *An* automorphism *of a graph is a permutation of the vertices that respects adjacency, i.e., sends every edge to an edge. The set of all automorphisms of a graph Γ forms the* automorphism group *of Γ, denoted* $\mathrm{Aut}(\Gamma)$*, with composition as the group operation. Two graphs are called* isomorphic *if there is a bijection (*isomorphism*) between their vertex sets that respects adjacency.*

Definition 1.7 (retract, i-retract). *Given a graph Γ and a smaller graph Δ, whose isomorphic copies occur as induced subgraphs of Γ, by a* Δ-retract *of a set of vertices C of Γ, we will imply the intersection of C with the set V_0 of vertices of such isomorphic copy, treated as a set of vertices of Δ (i.e., $\pi(C \cap V_0)$, where π is the corresponding isomorphism). Similarly, a* Δ-retract *of a function f on the set of vertices of Γ is defined as the function $f_\Delta(\cdot) \equiv f(\pi^{-1}(\cdot))$. Considering retracts, we are usually interested in its properties that are invariant under equivalence and hence independent of the choice of a concrete isomorphism. For the Hamming graph $H(n, q)$,* i-retract *is a synonym of $H(i, q)$-retract.*

Definition 1.8 (sum of graphs, complement graph). *The* sum *of two graphs $\Gamma_1 = (V_1, E_1)$ and $\Gamma_2 = (V_2, E_2)$ is the graph $\Gamma_1 + \Gamma_2 = (V_1 \cup V_2, E_1 \uplus E_2)$ (the multiset union $\uplus$ means that the multiplicity of elements are summarized); the sum is called* direct *if $V_1 \cap V_2 = \emptyset$. The* complement $\overline{\Gamma}$ *of a simple graph $\Gamma = (V, E)$ is the graph on the same vertex set V with the edge set equal the complement of E in the set of all 2-subsets of V.*

Definition 1.9 (Cartesian product). *The* Cartesian product $\Gamma_1\Box\ldots\Box\Gamma_n$ *of n graphs $\Gamma_1 = (V_1, E_1)$, ..., $\Gamma_n = (V_n, E_n)$ is the graph whose vertex set is the set of tuples (words) $(x_1, \ldots, x_n)$, where $x_i \in V_i$, and two vertices $(x_1, \ldots, x_n)$ and*

$(y_1, \ldots, y_n)$ are adjacent if and only if $\{x_i, y_i\} \in E_i$ for some i in $\{1, \ldots, n\}$ and $x_i = y_i$ for all other i in $\{1, \ldots, n\}$.

Definition 1.10 (Cayley graph). *The* Cayley graph $\mathrm{Cay}(G, S)$ *on a group G (written in the multiplicative notation) with* connecting *multiset S of elements of G, $S = S^{-1}$, is the graph whose vertices are the elements of G and whose edge multiset E is $\{\{x, ax\} : x \in G, a \in S\}$ (equivalently, $\{x, y\} \in E$ if and only if $y \cdot x^{-1} \in S$). The Cayley graph is simple if and only if $\mathrm{Id} \notin S$ and S is a simple set.*

Proposition 1.1. *The right multiplication $G \to Gc$ by a constant c in a group G is an automorphism of any Cayley graph $\mathrm{Cay}(G, S)$.*

Definition 1.11 (complete graph, Hamming graph, n-cube, hypercube, k-face). *The* complete graph K_q *is the simple graph of order q where every two different vertices are adjacent. The* complete k-partite graph $K_{q_1,\ldots,q_k}$ *is a simple graph of order $q_1 + \cdots + q_k$ where the vertex set is partitioned into k* parts *of sizes $q_1, \ldots, q_k$ and two vertices are adjacent if and only if they belong to different parts. The* Hamming graph $H(n, q)$ *is the Cartesian product of n copies of K_q. According to the definition of the Cartesian product, the vertex set of $H(n, q)$ consists of all n-words in some alphabet of order q, which, if q is a prime power, is often associated with the elements of the Galois field $\mathbb{F}_q = \mathrm{GF}(q)$ of order q. In this case, the vertex set of $H(n, q)$ forms an n-dimensional vector space $\mathbb{F}_q^n$ over $\mathbb{F}_q$, and $H(n, q)$ is a Cayley graph on its additive group $\mathbb{F}_q^{n+}$. The binary ($q = 2$) Hamming graph $H(n, 2)$ is also called the* n-cube, *or a* hypercube. *A* k-face *of the Hamming graph $H(n, q)$ is a set of q^k vertices that induces a subgraph isomorphic to $H(k, q)$.*

Definition 1.12 (clique, coclique = independent set). *In a graph, a set of vertices that are mutually adjacent (mutually non-adjacent) to each other is called a* clique *(*coclique *or* independent set*) or, more specifically,* k-clique *(*k-coclique*) where k is the cardinality of this set.*

Definition 1.13 (multineighborhood operator). *The* multineighborhood $\Gamma(x)$ *of a vertex x of a graph $\Gamma = (V, E)$ is the multiset $\{y : \{x, y\} \in E\}$ of all vertices adjacent to x, where the multiplicity of a neighbor y in $\Gamma(x)$ equals the multiplicity of $\{x, y\}$ in E. The* multineighborhood $\Gamma(X)$ *of a set or multiset X of vertices is the multiset sum of the multineighborhoods of all its elements: $\Gamma(X) = \biguplus_{x \in X} \Gamma(x)$. We will also naturally define the powers of the multineighborhood operator as $\Gamma^0(X) = X$, $\Gamma^{i+1}(X) = \Gamma(\Gamma^i(X))$ and extend the action of $\Gamma(\cdot)$ to the real-valued functions $f : V \to \mathbb{R}$ by linearity: $\Gamma(f)(x) = \sum_{y \in \Gamma(x)} f(y)$.*

The action of the multineighborhood operator on a multiset X of vertices of a graph Γ can be represented by the action of the adjacency matrix of Γ, defined below, on the multiplicity vector of X (for an ordinary set X the multiplicity vector is also known as the characteristic vector, or indicator).

Definition 1.14 (adjacency matrix, eigenfunction, eigenvalue). *Given a graph* $\Gamma = (V, E)$*, the* adjacency matrix *of* Γ *is the* $|V| \times |V|$ *matrix* $A = (A_{i,j})_{i,j \in V}$ *whose rows and columns are indexed by elements of* V *such that* $A_{i,j}$ *is the multiplicity of* $\{i, j\}$ *in* E*. An* eigenfunction *of* Γ *is a real-valued function* $f : V \to \mathbb{R}$ *on its vertices that is essentially an eigenvector of the adjacency matrix, that is,* $f \not\equiv 0$ *and*

$$\sum_{y \in \Gamma(x)} f(y) = \theta f(x) \qquad \text{for all } x \in V,$$

where θ *is the corresponding* eigenvalue *(for short, we will say that* f *is a* θ*-*eigenfunction *of* Γ*, or a* θ*-*eigenvector *of its adjacency matrix). All* θ*-eigenfunctions together with the all-zero function form a vector space called the* θ*-*eigenspace *of* Γ*; its dimension is called the* multiplicity *of the eigenvalue* θ*. The multiset of all eigenvalues of a graph with the corresponding multiplicities is called its* eigenspectrum*. The degree of a regular graph is always its largest eigenvalue (see Proposition 1.2 below), called the* main eigenvalue.

It is well known that

Proposition 1.2. *The largest eigenvalue of a* k*-regular graph* Γ *is* k*, whose multiplicity equals the number of connected components. The smallest eigenvalue of a* k*-regular graph* Γ *is not smaller than* $-k$*, with equality if and only if the graph has a bipartite connected component.*

Proof. Let f be a θ-eigenfunction of a k-regular graph $\Gamma = (V, E)$ and x a vertex of Γ such that $|f(x)| = \max_{y \in V} |f(y)|$. In particular, $-|f(x)| \leq f(y) \leq |f(x)|$ for every vertex y. Therefore, from the k-regularity, we have

$$-k|f(x)| \leq \sum_{y \in \Gamma(x)} f(y) = \theta f(x) \leq k|f(x)|.$$

So, $-k \leq \theta \leq k$.

If $\theta = -k$, then $f(y) = -f(x)$ for all neighbors of x. Then, the same argument can be applied to the neighbors of x, then to the neighbors of neighbors, and so on. Finally, $f(y) = -f(z)$ holds for all neighbor y and z from the same connected component as x, and we see that this connected component is bipartite.

Similarly, for $\theta = k$, we can show that f is constant on each connected component of Γ. Therefore, the dimension of the k-eigenspace equals the number of connected components. □

Definition 1.15 (polynomials in graphs, Γ^i, $\Pi(\Gamma)$). *For a graph* Γ *with adjacency matrix* A*, by* Γ^i*,* $i = 0, 1, 2, \ldots$*, we denote the graph with adjacency matrix* A^i*; by* $\Pi(\Gamma)$*, where* Π *is a polynomial, we denote the graph with adjacency matrix* $\Pi(A)$ *if this matrix is nonnegative. In particular,* $\Gamma + a$ *is the graph obtained from* Γ *by attaching* a *loops to each vertex,* $a = 0, 1, 2, \ldots$.

Remark 1.1. *The definition of Γ^i is in agree with the notation $\Gamma^i(X)$ in Definition 1.13. Indeed, the ith degree of the multineighborhood $\Gamma(X)$ trivially coincides with the multineighborhood $\Delta(X)$ where Δ is the ith degree of Γ, $\Delta = \Gamma^i$. It can also be noted that the multiplicity of an edge $\{x, y\}$ in Γ^i is the number of walks of length i starting in x and ending in y in Γ.*

1.1.2 Perfect Colorings, Completely Regular Codes, and Related Concepts

Definition 1.16 (coloring, spectrum). *An arbitrary surjective function from the vertex set of a graph Γ onto a finite set Σ (of* colors*) of cardinality k is called a (vertex)* coloring, *or k-*coloring, *of Γ. For a multiset X of vertices of a graph Γ and a coloring f of Γ, the f-*spectrum, *$\bar{f}(X)$, or just the* spectrum *(when f is clear from the context), of X is the multiset*

$$\bar{f}(X) = \{f(x) : x \in X\}$$

of colors of the vertices from X. The spectrum, as a multiset, will be identified with its multiplicity vector $(\mu(s))_{s\in\Sigma}$, where $\mu(s)$ is the multiplicity of the color s in the spectrum. In this context, it would be nice to imply that $\Sigma = \{0, 1, \ldots, k-1\}$; however, this is not always convenient, and the reader should not be confused if the elements of a tuple or rows/columns of a matrix are indexed by elements of some other set.

A coloring f of a bipartite graph with parts V', V'' and the corresponding partition of the vertex set into color classes are called bipartite *if the spectra $\bar{f}(V')$ and $\bar{f}(V'')$ are disjoint. A coloring f of a connected graph Γ and the corresponding partition of the vertex set into color classes are called* self-antipodal *if the colors of any two vertices at distance* $\mathrm{diam}(\Gamma)$ *coincide.*

Definition 1.17 (perfect coloring, equitable partition, quotient matrix). *A k-coloring of a graph Γ is called* perfect *if there is a k-by-k matrix $(S_{i,j})_{i,j\in\Sigma}$ (the* quotient *matrix), whose rows and columns are indexed by colors, such that for every vertex of color i, its multineighborhood (for simple graph, neighborhood) contains exactly $S_{i,j}$ vertices of color j. An ordered partition $(C_0, \ldots, C_{k-1})$ of the vertex set of Γ into nonempty sets C_0, $\ldots$, C_{k-1} is called* equitable, *or an* equitable k-partition *if the vertex coloring f defined as $f(x) = i$ if $x \in C_i$ is perfect. We will also use the shorter forms "S-perfect coloring" for "perfect coloring with quotient matrix S" and "equitable S-partition" for "equitable partition with quotient matrix S".*

Proposition 1.3 (an operator definition of perfect coloring). *A coloring f of the vertices of a graph Γ is perfect with quotient matrix S if and only if*

$$\bar{f}(\Gamma(X)) = \bar{f}(X) \cdot S \tag{1.1}$$

for every set X of vertices of Γ.

Proof. For singleton sets, (1.1) just repeats the definition of perfect coloring. Because both the spectrum $f(\cdot)$ and the multineighborhood $\Gamma(\cdot)$ are linear with respect to the multiplicity vectors, it follows that (1.1) holds for any vertex set as well. □

The same claim rewritten in the matrix language looks as follows.

Proposition 1.4 (matrix definition of perfect coloring, see e.g. [40, Chapter 5, Lemma 2.1]). *A surjective function $f : V \to \Sigma$ is a perfect coloring of a graph $\Gamma = (V, E)$ with quotient $|\Sigma| \times |\Sigma|$ matrix S if and only if*

$$A \cdot F = F \cdot S \tag{1.2}$$

where A is the $|V|\times|V|$ adjacency matrix of Γ and F is the incidence (characteristic) $|V| \times |\Sigma|$ $\{0,1\}$-matrix of f.

Proof. Trivially, the action of $\Gamma(\cdot)$ on a multiset X of vertices is the same as the right action of A on the multiplicity vector $\bar{\mu}_X$ of this multiset: $\bar{\mu}_{\Gamma(X)} = \bar{\mu}_X \cdot A$. Hence,

$$\bar{\mu}_X \cdot A \cdot F = \bar{\mu}_X \cdot F \cdot S \tag{1.3}$$

is equivalent to (1.1). Applying (1.3) to each singleton set X, we get (1.2). □

Remark 1.2. *A method of constructing all A and F for a given S under condition (1.2), where F is a $\{0,1\}$-matrix with all row sums equal to 1, is known as voltage assignment [46, Theorem 2].*

Definition 1.18 (twin colors). *Two colors of a perfect k-coloring (the corresponding two cells of the corresponding equitable partition) are called* twin colors *(twin cells) or just* twins *if identifying them results in a perfect $(k-1)$-coloring.*

Clearly, the two colors of a perfect 2-coloring of a regular graph are always twins; this shows that twin cells can have different cardinalities.

In the theory of equitable partitions, equation (1.2) plays an important role in more general form, where A is not necessarily symmetric and F is not necessarily $\{0,1\}$-valued, see in Section 1.2.3. For this reason, we introduce a more general concept as follows.

Definition 1.19 (perfect structure). *If* (1.2) *holds for some real-valued square matrices A, S, and rectangular matrix F, then we will say that F is a* perfect structure *over A (or over Γ if A is the adjacency matrix of a graph Γ) with quotient matrix S, or an S-*perfect structure.

A very important special case of perfect structures are eigenfunctions of a graph, which are essentially perfect structures over the adjacency matrix with a 1×1 quotient matrix. General properties of perfect structures in application to the theory of equitable partitions were studied in [68,118].

As we will see in Section 1.2.3, some invariants of perfect colorings can also be represented as perfect structures. Also, equitable partitions can be generalized as follows (see also Proposition 1.41 for an example of a meaningful statement about equitable families).

Definition 1.20 (equitable family). *A collection $\bar{C} = (C_0, \dots, C_{k-1})$ of sets of vertices of a graph Γ with adjacency matrix A is called an* equitable family *(of sets) with quotient $k \times k$ matrix S if the incidence $\{0,1\}$-matrix F of the family satisfies* (1.2).

Definition 1.21 (covering radius, completely regular (CR) code, CR partition). *Given a nonempty set C of vertices of a graph $\Gamma = (V, E)$, the* distance partition *with respect to C is the partition $(C^{(0)}, C^{(1)}, \dots, C^{(\rho)})$ of V with respect to the distance from C: $C^{(i)} = \{x \in V : d(x, C) = i\}$. The number $\rho = \rho(C)$ of non-empty cells different from C in the distance partition is called the* covering radius *of C. A nonempty set C (code) of vertices of a graph is called* completely regular *(*CR*) (CR set or CR code) if the partition of the vertex set with respect to the distance from C is equitable, or, equivalently, if the coloring $f(x) = d(x, C)$ of the vertex set with respect to the distance from C is perfect. The quotient matrix $(S_{i,j})_{i,j=0}^{\rho}$ corresponding to a CR code is always tridiagonal, and for regular graphs it is often written in the form of the* intersection array

$$\{S_{0,1}, S_{1,2}, \dots, S_{\rho-1,\rho}; S_{1,0}, S_{2,1}, \dots, S_{\rho,\rho-1}\}$$

(the diagonal elements of the quotient matrix are uniquely determined from the intersection array and the degree of the graph). An equitable partition whose cells are CR codes with the same quotient matrix is called a completely regular (CR) partition.

We will use the abbreviation "A-CR" for "completely regular with intersection array A" or "completely regular with quotient matrix A", depending on the context, and the abbreviation "CR-ρ" for "completely regular with covering radius ρ".

A very important special case of perfect colorings is graph coverings.

Definition 1.22 (graph covering). *Assume that f is a perfect coloring of a graph Γ with symmetric quotient matrix S and $\overline{P}$ is the equitable partition corresponding to f. Then S can be considered as the adjacency matrix of some graph, denoted $\Gamma/\overline{P}$, on the set of colors (if S is a $\{0,1\}$-matrix with zero diagonal, then $\Gamma/\overline{P}$ is a simple graph), and we say that the coloring f, as well as the corresponding partition $\overline{P}$, is a* covering *of the* target *graph $\Gamma/\overline{P}$ by a* cover *graph Γ. In other words, a covering is a locally bijective (bijective on the set of edges incident to a vertex) surjective homomorphism from Γ to $\Gamma/\overline{P}$.*

Lemma 1.1. *A surjective homomorphism φ from a group G onto a group $\varphi(G)$ is a covering of* $\mathrm{Cay}(\varphi(G), \varphi(S))$ *by* $\mathrm{Cay}(G, S)$*, for any multiset S of elements of G such that $S^{-1} = S$.*

Remark 1.3. *In the literature, other names for equitable partitions (perfect colorings) can be found, such as regular partitions [13, 11.1.B], partition designs [18]. Some names are rarely used or not used anymore: distributive colorings [134], feasible colourings [27], coherent partitions [52]. Other names are in some special* *graphs for special parameters, especially for equitable 2-partitions (CR-1 codes). For example, CR-1 codes are known as* intriguing sets *[29]; in $J_q(n,2)$ (the line graph of a projective space), CR-1 codes with non-minimum eigenvalues are known as Cameron–Liebler line classes, while CR-1 codes with minimum eigenvalue are multifold spreads.*

1.1.3 Distance-Regular Graphs and Related Concepts

Definition 1.23 (amply regular graph, strongly regular graph). *A simple graph $\Gamma = (V, E)$ that is neither complete nor edge-free is said to be* amply regular *with parameters (v, k, λ, μ) if Γ is a regular graph of degree k on v vertices, any pair of adjacent vertices has exactly λ common neighbors, and any pair of nonadjacent vertices has exactly μ or 0 common neighbors. By removing the words "or 0", we obtain the definition of a* strongly regular graph.

It is known that strongly regular graphs different from the disjoint sum of complete graphs are characterized as connected regular graphs with exactly three distinct eigenvalues [39, Lemma 10.2.1] and as distance-regular graphs (defined below) of diameter 2. The complement of an $\mathrm{srg}(v, k, \lambda, \mu)$ is an $\mathrm{srg}(v, v-k-1, v-2-2k+\mu, v-2k+\lambda)$. The eigenvalues of $\mathrm{srg}(v, k, \lambda, \mu)$ are (see, e.g. [15, Theorem 9.1.3])

$$k, \quad \frac{\lambda-\mu \pm \sqrt{(\lambda-\mu)^2+4(k-\mu)}}{2}, \tag{1.4}$$

with multiplicities, respectively,

$$1, \quad \frac{1}{2}\left(v-1 \mp \frac{2k+(v-1)(\lambda-\mu)}{\sqrt{(\lambda-\mu)^2+4(k-\mu)}}\right). \tag{1.5}$$

Definition 1.24 (distance-regular multigraph, distance-regular graph). *A connected regular graph is called a* distance-regular multigraph *with* intersection array $I = \{b_0, \ldots, b_{r-1}; c_1, \ldots, c_r\}$ *if every one-vertex set is I-CR. If, in addition, the graph is simple (without loops and multiple edges), then it is called a* distance-regular graph.

So, according to the standard terminology, distance-regular graphs are always simple. The non-simple version of distance-regular graphs, distance-regular multigraphs, are quite rarely mentioned in the literature. The reason is that they are first-degree polynomials from (simple) distance-regular graphs, see the trivial proposition below. However, we find this concept convenient to generalize one known result (see Proposition 1.8).

Proposition 1.5. *A graph is a distance-regular multigraph with intersection array*

$$\{b_0, \ldots, b_{r-1}; c_1, \ldots, c_r\}$$

if and only if it is obtained from a distance-regular graph with intersection array

$$\{b_0/c_1, \ldots, b_{r-1}/c_1; 1 = c_1/c_1, \ldots, c_r/c_1\}$$

by setting multiplicity c_1 *to every edge and attaching a constant number of loops to every vertex.*

Proposition 1.6 ([40, Lemma 2.1]). *Let* Γ *be a simple connected graph with adjacency matrix* A*, and let* A_i *denote the adjacency matrix of the distance-i graph* $\Gamma^{(i)}$ *of* Γ, $i = 0, 1, \dots, \mathrm{diam}(\Gamma) + 1$. *Then* Γ *is distance-regular if and only if for* $i = 1, \dots, \mathrm{diam}(\Gamma)$ *the product* AA_i *is a linear combination of* A_{i-1}, A_i, A_{i+1}.

Corollary 1.1. *A connected simple graph* Γ *is distance-regular if and only if there are polynomials* $K^{(i)}$ *of degree* i, $i = 0, \dots, \mathrm{diam}(\Gamma)$, *such that* $\Gamma^{(i)} = K^{(i)}(\Gamma)$.

Definition 1.25 (Q-polynomial ordering, Q-polynomial graph). *The ordering* S_0, S_1, …, S_d *of the eigenspaces (and the corresponding ordering of the eigenvalues) of a graph* Γ *is* Q-polynomial *if*

- *the pointwise (Hadamard) product of any two functions in* S_i *and* S_j *belongs to* $S_{|i-j|} \oplus \dots \oplus S_{\min\{i+j,d\}}$ *and*
- *for each* i *and* j *with* $i+j \le d$, *there is some function in* S_i *and some function in* S_j *whose pointwise product is not orthogonal to* S_{i+j}.

If such an ordering exists, the graph is called Q-polynomial.

Example 1.1. *The hypercube* $H(n, 2)$ *is Q-polynomial with respect to the natural, descending, ordering* θ_0, …, θ_n *of eigenvalues,* $\theta_i = n - 2i$. *It is sufficient to check this for basis eigenfunctions of form* $\chi_y(x) = (-1)^{(x,y)}$ *(for each vertex* y, χ_y *is a* $\theta_{\mathrm{wt}(y)}$*-eigenfunction of* $H(n, 2)$*). Similarly, it is not difficult to check that for even* n, *the ordering* θ'_0, …, θ'_n *of eigenvalues,* $\theta'_i = (-1)^i(n - 2i)$, *is also Q-polynomial.*

For odd n, *consider the eigenvalue ordering* η_0, …, $\eta_{\frac{n-1}{2}}$, $\xi_{\frac{n+1}{2}}$, …, ξ_n, *where* $\eta_i = (-1)^i(n - 2i)$ *and* $\xi_{\frac{n+1}{2}+i} = -(-1)^i(n - 2i)$. *This ordering satisfies the first condition of the definition, but the second condition fails if* $i, j < \frac{n}{2} < i + j$.

An example of a Q-polynomial distance-regular graph where the natural, descending, ordering of the eigenvalues is not Q-polynomial is the Odd graph, see the definition below.

Definition 1.26 (distance-transitive graph). *A connected simple graph* $\Gamma = (V, E)$ *is called* distance-transitive *if for every* i, $\mathrm{Aut}(\Gamma)$ *acts transitively on the ordered pairs* (x, y) *of vertices at distance* i *from each other, i.e., for every* x, y, x', y' *in* V *such that* $d(x, y) = d(x', y')$ *there exists an automorphism* π *in* $\mathrm{Aut}(\Gamma)$ *such that* $\pi(x) = x'$ *and* $\pi(y) = y'$.

It is straightforward that every distance-transitive graph is distance-regular. Below, we define several families of distance-transitive graph. We have already defined Hamming graphs $H(n, q)$. For $\Gamma = H(n, q)$, the degree-w polynomial $K^{(w)}$ such that $\Gamma^{(w)} = K^{(w)}(\Gamma)$, is

$K^{(w)}(\cdot) = P_w(P_1^{-1}(\cdot)),$

$$\text{where } P_w(x) = P_w(x; n, q) = \sum_{j=0}^{w} (-1)^j (q-1)^{w-j} \binom{x}{j} \binom{n-x}{w-j}$$

is the *Krawtchouk polynomial*; $P_1(x) = (q-1)n - qx$, $P_1^{-1}(y) = \frac{(q-1)-y}{q}$.

Definition 1.27 (Johnson graph $J(n,k)$). *The* Johnson graph $J(n,k)$ *has the vertex set consisting of all w-subsets of a finite set of cardinality n. Two k-subsets are adjacent in $J(n,k)$ if they intersect in $k-1$ elements. Alternatively, $J(n,k)$ can be defined as the subgraph of $H(n,2)^{(2)}$ induced by the set of vertices of weight k.*

For $\Gamma = J(n,k)$, the degree-w polynomial $K^{(w)}$ such that $\Gamma^{(w)} = K^{(w)}(\Gamma)$, is defined by

$$K^{(w)}(\cdot) = E_w(E_1^{-1}(\cdot)),$$

where

$$E_w(x) = E_w(x; n, k) = \sum_{j=0}^{w} (-1)^{w-j} \binom{k-j}{w-j} \binom{k-x}{j} \binom{n-k+j-x}{j}$$
$$= \sum_{j=0}^{w} (-1)^j \binom{x}{j} \binom{k-x}{w-j} \binom{n-k-x}{w-j}$$

is the *Eberlein polynomial* [31, 4.2.1] of degree $2w$; $E_1(x) = k(n-k) - x(n+1-x)$ (it is easy to see that

$$\binom{k-x}{j}\binom{n-k+j-x}{j} = \frac{1}{j!^2}\sum_{s=0}^{j-1}\big((-x)+(k-s)\big)\cdot\big((n+1-x)-(k-s)\big)$$
$$= \frac{1}{j!^2}\sum_{s=0}^{j-1}\big(-\underline{x(n+1-x)} + (n+1-k+s)(k-s)\big)$$

is a polynomial in $x(n-1-x)$ and hence in $E_1(x)$; so, $E_w(E_1^{-1}(\cdot))$ is well defined).

Definition 1.28 (odd graph, doubled odd graph). *The distance-k graph of the Johnson graph $J(2k+1,k)$ is called the* Odd graph O_{k+1} *(the subscript gives the valency). The subgraph of the hypercube $H(2k+1,2)$ induced by the vertices of weights k and $k+1$ is called a* doubled Odd graph DO_{k+1}.

Definition 1.29 (Grassmann graph $J_q(n,k)$). *The* Grassmann graph $J_q(n,k)$ *(also known as* q-ary Johnson graph*) has the vertex set consisting of all k-dimensional subspaces of an n-dimensional vector space over the finite field $\mathbb{F}_q$ of order q. Two such subspaces are adjacent in $J_q(n,k)$ if they intersect in a $(k-1)$-dimensional subspace.*

Definition 1.30 (bilinear forms graph $B_q(m,n)$). *The vertices of the* bilinear forms graph $B_q(m,n)$ *are $m\times n$ matrices over $\mathbb{F}_q$. Two such matrices are adjacent is their difference has rank 1. Codes in $B_q(m,n)$ are called* rank-metric codes *in coding theory.*

1.1.4 Codes and Related Concepts; Coset Graph

Definition 1.31 (code, code distance, automorphism group of a code, equivalent codes). *A* code *in a graph* Γ *is an arbitrary nonempty set of vertices of* Γ. *Vertices that belong to the code under consideration are referred to as* code vertices *or* codewords. *The* distance *(code distance, minimum distance) of a code is the minimum distance between distinct code vertices.*[1] *A code of cardinality* M *and distance* d *in* $H(n,q)$ *is called an* $(n,M,d)_q$ *code, sometimes with omitted* q *if it is clear from the context. The* automorphism group $\mathrm{Aut}(C)$ *of a code* C *in a graph* Γ *is the setwise stabilizer of* C *in* $\mathrm{Aut}(\Gamma)$, *that is, the subgroup of all automorphisms of* Γ *that map codewords to codewords. Two codes* C *and* D *in the same graph* Γ *are called* equivalent *if* $\varphi(C) = D$ *for some automorphism* φ *of* Γ.

Definition 1.32 (perfect codes). *A code in a graph is called* r-perfect *(*perfect*) if every vertex is at distance at most* r *from exactly one code vertex. In other words, the radius-*r *balls centered in the elements of an* r*-perfect code partition the vertex set of the graph.*

It is straightforward that a 1-perfect code in a regular graph is CR-1, a 2-perfect code in an amply regular graph is CR-2, an r-perfect code in a distance-regular graph is CR-r for any r.

Definition 1.33 (subgroup codes, additive codes, linear codes, generator matrix, check matrix). *A code in a Cayley graph over a group* G *is called a* subgroup code *if it is closed under the group operation, i.e., it is a subgroup of* G. *If* G *is the additive group of a module over a commutative ring, then subgroup codes are also called* additive, *while a* linear code *is any submodule (subspace, if the ring is a field). If* q *is a prime power, then* $\mathbb{F}_q$ *denotes the Galois field of order* q, *and for a linear code in* $H(n,q)$ *(understood as a Cayley graph on* $\mathbb{F}_q^n$*), the notation* $[n,k,d]_q$ *means the same as* $(n,q^k,d)_q$. *A linear* $[n,k,d]_q$ *code* C *can be defined by a* generator matrix *whose rows span* C *or by a* check matrix H *such that* C *is the nullspace of* H: $C = \{c \in \mathbb{F}_q^n : Hc^{\mathrm{T}} = \bar{0}\}$.

Definition 1.34 (transitive, r-neighbor-transitive, completely transitive, and propelinear codes). *A code* C *in a graph* Γ *is called* transitive *if the automorphism group* $\mathrm{Aut}(C)$ *of the code acts transitively on the codewords, i.e., for every two code vertices* x *and* y, *there is* $\pi \in \mathrm{Aut}(C)$ *such that* $\pi(x) = y$. *A code* C *in a graph* Γ *is called* r-neighbor-transitive *if for* $i = 0,1,\ldots,r$, $\mathrm{Aut}(C)$ *acts transitively on the set* $C^{(i)}$ *of vertices at distance* i *from* C. *A code is* completely transitive *if it is* r*-neighbor-transitive for every* r. *A code* C *is called* propelinear *if some subgroup* G *of* $\mathrm{Aut}(C)$ *acts regularly on* C, *i.e., for every two code vertices* x *and* y, *there is exactly one* $\pi \in G$ *such that* $\pi(x) = y$.

Definition 1.35 (weight, weight distribution, distance distribution, distance invariance). *The concept of weight implies that in the graph* Γ *we have some initial vertex* $\mathbf{0}$. *In Cayley graphs, it is by default the identity element of the group;*

however, sometimes it is convenient to define the concepts below with respect to an arbitrary vertex. The weight $\mathrm{wt}(x)$ *of a vertex* x *is the distance from* x *to* $\mathbf{0}$. *The* weight distribution *of a code (or, in general, multiset of vertices)* C *with respect to a vertex* o *(by default,* $\mathbf{0}$*) or a non-empty set of vertices* o *is the multiset* $\{d(x,o) : x \in C\}$ *of distances from codewords to* o *(in some literature, this is called the outer distance distribution of* C *with respect to* o*). The weight distribution is often written in the form of the multiplicity vector*

$$(A_0, \ldots, A_{\mathrm{diam}(\Gamma)}), \qquad \textit{where } A_i = |\{x \in C : d(x,o) = i\}| = |\Gamma^{(i)}(o) \cap C|,$$

or

$$\{\langle 0, A_0\rangle, \langle 1, A_1\rangle, \ldots\} \qquad \textit{or } \{0^{A_0}, 1^{A_1}, \ldots\}$$

(in the last two cases, the entries with $A_i = 0$ *are usually skipped). The* distance distribution *of* C *is the average of the weight distributions with respect to all codewords. It equals the multiset of all distances between codewords divided by* $|C|$. *A code is called* distance-invariant *if its weight distributions with respect to all codewords coincide (and hence equal to the distance distribution).*

Any linear code is obviously additive; in its turn, any additive code is subgroup; any subgroup code is propelinear; any propelinear code is transitive; any transitive code is distance-invariant. Inverse inclusions can take place for some special graphs, but not in general.

Proposition 1.7 (on the left-cosets partition). *The partition* $\overline{L}$ *of a finite group* G *into left cosets of a subgroup* C *is equitable in* $\mathrm{Cay}(G,S)$*; moreover, it is a covering.*

Proof. By Proposition 1.1, the action of an element of C on G by right multiplication is an automorphism of $\mathrm{Cay}(G,S)$. Hence, with this action, C indices a group of automorphisms of $\mathrm{Cay}(G,S)$, a subgroup of $\mathrm{Aut}(\mathrm{Cay}(G,S))$. The partition $\overline{L}$ of G into left cosets of C is the orbit partition with respect to this group. The orbit partition with respect to a group of automorphisms of a graph is trivially equitable. Since the cardinalities of the orbits are the same, the quotient matrix is symmetric and we have a covering of the graph $\mathrm{Cay}(G,S)/\overline{L}$. □

Definition 1.36 (coset graph). *If* $\Gamma = \mathrm{Cay}(G,S)$ *is a Cayley graph on a group* G *and* C *is a subgroup code, then the target graph* $\Gamma/\overline{L}$ *of the covering in Proposition 1.7 is called the* coset graph Γ/C *of* C. *The vertices are the left cosets of* C *and the number of edges between cosets* xC *and* yC *in* Γ/C *equals the number of edges between* x *and elements of* yC *in* Γ *(by Proposition 1.7, this number does not depend on the choice of the representative* x *in* xC*). The coset graph* Γ/C *is simple if and only if* Γ *is simple and the code distance of* C *is at least* 3.

According to [31], one of the important parameters of a code is the following.

Definition 1.37 (external distance). *Let* C *be a set of vertices of a distance-regular graph* Γ *and* χ_C *its characteristic function. The number of nonzero projections of* χ_C *onto eigenspaces of* Γ *minus one is called the* external distance *of* C *(in some literature, the* dual degree *of* C*).*

For a linear or additive code C with a properly defined duality, the external distance coincides with the number of non-zero distances of the code dual to C.

1.1.5 Connection between CR Partitions and Distance-Regular Graphs

The following statement is a variant of [13, Theorem 11.1.6]; the difference is that we treat the same result in terms of multigraphs, when the quotient graph respects the multiplicity of edges.

Proposition 1.8. *A partition $\overline{P}$ of the vertex set of a connected graph Γ is CR if and only if it is a covering of a distance-regular multigraph. The intersection arrays of the CR codes from $\overline{P}$ and the distance-regular multigraph $\Gamma/\overline{P}$ coincide.*

Note that the code distance of the CR codes from $\overline{P}$ is at least 3 if and only if $\Gamma/\overline{P}$ is a (simple) distance-regular graph.

Proof. Only if. Let $\overline{P} = (P_0, \ldots, P_{k-1})$ be a CR partition of a graph Γ with quotient matrix S, where each cell is a $\{b_0, \ldots, b_{\rho-1}; c_1, \ldots, c_\rho\}$-CR code. If different cells P_i, P_j of $\overline{P}$ contain adjacent elements $p \in P_i$, $p' \in P_j$, then $p' \in P_i^{(1)}$ and p' has b_0 neighbors in P_i. It follows that $S_{i,j} = b_0$. Similarly, $S_{i,j} = S_{j,i} \in \{0, b_0\}$ for every i and j, $0 \le i < j \le k-1$. Since, by the definition, P_0, $\ldots$, P_{k-1} are CR codes with the same quotient matrix, say S', all diagonal elements of S equal $S'_{0,0}$. So, $\overline{P}$ is a covering of a graph, denoted $\Gamma/\overline{P}$, having the constant multiplicity b_0 of all non-loop edges and the constant number $S'_{0,0}$ of loops at every vertex.

It remains to show that $\Gamma/\overline{P}$ is a distance-regular multigraph. We need the following simple but useful fact.

(*) *The distance in $\Gamma/\overline{P}$ between two cells C, D of $\overline{P}$ is l if and only if $D \subseteq C^{(l)}$ (similarly, $C \subseteq D^{(l)}$).*

To show (*), we will prove by induction that

$$C^{(l)} = \{D \in \overline{P} : \; d_{\Gamma/\overline{P}}(C, D) = l\}. \tag{1.6}$$

For $l = 0$, it is trivial. Assume that $l > 0$ and (1.6) holds for all smaller values of l. Consider $D \in \overline{P}$ such that $d_{\Gamma/\overline{P}}(C, D) = l$ and a vertex x in P. There is a cell E of $\overline{P}$ such that $d_{\Gamma/\overline{P}}(C, E) = l - 1$ and $d_{\Gamma/\overline{P}}(E, D) = 1$. There is a vertex y in E adjacent to x (indeed, E and D are neighbor cells of the equitable partition $\overline{P}$). By the induction hypothesis, $d(y, C) = l - 1$. Hence, $d(x, C) \le l$. Moreover, from the induction hypothesis we get $d(x, C) \notin \{0, \ldots, l-1\}$. Hence, $d(x, C) = l$. We conclude that $D \subseteq C^{(l)}$, and the set in the right part of (1.6) is included in the left-part set. To show the inverse inclusion, consider an arbitrary x in $C^{(l)}$. It belongs to some $D \in \overline{P}$. There is y in $C^{(l-1)}$ adjacent to x. By the induction hypothesis, y belongs to some $E \in \overline{P}$ such that $d_{\Gamma/\overline{P}}(C, E) = l - 1$. Since E and D are adjacent in $\Gamma/\overline{P}$, we have $d_{\Gamma/\overline{P}}(C, D) \le l$. On the other hand, from the induction hypothesis we get

$d_{\Gamma/\overline{P}}(C,D) \notin \{0,\ldots,l-1\}$; hence $d_{\Gamma/\overline{P}}(C,D) = l$. This completes the proof of (1.6); (*) follows.

Now, consider the cells C and D at distance l (in $\Gamma/\overline{P}$) from each other. By (*), all cells of $\overline{P}$ that are vertices from $(\Gamma/\overline{P})^{(l-1)}(C)$ form a partition of $C^{(l-1)}$. By the definition of $\Gamma/\overline{P}$, the number of edges between D and $E \in (\Gamma/\overline{P})^{(l-1)}(C)$ in the graph $\Gamma/\overline{P}$ equals the number of edges between any element of D and E in Γ. It follows that the total number of vertices between D and $(\Gamma/\overline{P})^{(l-1)}(C)$ in $\Gamma/\overline{P}$ equals the total number of vertices between any element x of D and $C^{(l-1)}$. By (*), we have $x \in C^{(l)}$, and hence this number is c_l from the intersection array of the CR code C. Similarly, the total number of vertices between D and $(\Gamma/\overline{P})^{(l+1)}(C)$ is b_l. By the definition, $\Gamma/\overline{P}$ is a distance-regular multigraph with intersection array $\{b_0,\ldots,b_{\rho-1}; c_1,\ldots,c_\rho\}$.

The "*if*" statement is proved similarly, as the arguments are invertible. □

Corollary 1.2. *Consider a group G, a multiset S of elements of G such that $S^{-1} = S$, and a subgroup $C < G$ such that $SC = CS$. The subgroup code C in* $\mathrm{Cay}(G,S)$ *is CR if and only if the partition $\overline{R}$ of G into right cosets of C is a covering of a distance-regular multigraph (with the same intersection array as C).*

Note that if C is a non-normal subgroup, then the right-cosets partition $\overline{R}$ is different from the left-cosets partition $\overline{L}$ discussed in Proposition 1.7. If C is a normal subgroup of G (e.g., if G is abelian, then every subgroup is normal), then the condition $SC = CS$ is satisfied for any $S \subseteq G$.

Proof. By Proposition 1.7, $\overline{R}$ is a CR partition if and only Γ/C is a distance-regular multigraph (with the same intersection array). It remains to prove that

- *C is a CR code if and only if $\overline{R}$ is a CR partition.*

The "if" is trivial; let us show the "only if". If C is a CR code, then by Proposition 1.1 all sets Ca, $a \in G$, are equivalent to C and hence are CR codes too. It remains to show that $\overline{R}$ is an equitable partition.

Consider arbitrary right cosets Ca and Cb, $a, b \in G$. For any xa in Ca (i.e., for any x in C), its multineighborhood $\Gamma(xa)$ is Sxa and

$$|\Gamma(xa) \cap Cb| = |Sxa \cap Cb| = |\{s \in S : sxa \in Cb\}| \overset{(S=S^{-1})}{=} |\{s \in S : xa \in sCb\}|.$$

The last is the multiplicity of xa in SCb, the multiplicity of xa in CSb (by the hypothesis, $SC = CS$), and the multiplicity of a in CSb (because $x^{-1}C = C$), i.e., does not depend on x. So, the number of edges between an element xa of Ca and Cb does not depend on the choice of xa. By the definition, the right-cosets partition $\overline{R}$ is equitable. □

The following statement characterizes all linear distance-1 and distance-2 CR codes in Hamming graphs in terms of linear CR codes with larger code distance.

Corollary 1.3 (see [126] for $q = 2$ and [63, Lemma 3.15] in general). *Up to equivalence, a parity check matrix of every linear* $\{b_0, \dots, b_{\rho-1}; c_1, \dots, c_\rho\}$*-CR code* C *in* $H(n, q)$ *(*q *is a prime power) is obtained from a parity check matrix of some linear* $\{b_0/c_1, \dots, b_{\rho-1}/c_1; 1 = c_1/c_1, \dots, c_\rho/c_1\}$*-CR code with code distance at least 3 in* $H((n-\lambda)/c_1, q)$ *by repeating every column* c_1 *times and adding* λ *all-zero columns, where* $\lambda = n - b_0/(q-1)$.

Proof. Let M be a full-rank check matrix of C; denote by λ the number of its all-zero columns. The cosets C_s of C can be indexed by column vectors s, *syndromes*, such that

$$C_s = \{x \in \mathbb{F}_q^n : Mx^{\mathrm{T}} = s\}.$$

Two different cosets S_s and $S_{s'}$ are adjacent in $H(n, q)/C$ if $S_s = S_{s'} + e$ for some weight-1 vertex e of $H(n, q)$. It follows that $He^{\mathrm{T}} = s - s'$, and hence $s - s'$ is collinear to some nonzero column of M. Moreover, the multiplicity of the edge $\{S_s, S_{s'}\}$ in $H(n, q)/C$ equals the number of nonzero columns collinear to $s - s'$ in M. By Corollary 1.2, $H(n, q)/C$ is a distance-regular multigraph, and by Proposition 1.5 all its non-loop edges have the same multiplicity. It is easy to see that this multiplicity is c_1. Now, denote by M' an matrix obtained from M by keeping only one column from each group of mutually collinear nonzero columns; by C', we denote the nullspace of M'. Easily, $H(n', q)/C'$ (where $n' = (n-\lambda)/c_1 + \lambda$ is the corresponding length) is the graph obtained from $H(n, q)/C$ by setting the multiplicities of all non-loop edges to 1. Hence, it is a distance-regular multigraph (it can still have loops). Its intersection array coefficients are smaller than those of $H(n, q)/C$ by factor of c_1.

Now consider the loops. The number of loops at C'_s is the number of weight-1 vectors e in $H(n', q)$ such that $C'_s + e = C'_s$, i.e., $He^{\mathrm{T}} = \bar{0}$. It is easy to see that the number of such vectors is $\lambda(q-1)$ and does not depend on s. By removing the λ all-zero columns of M' we obtain a code C'' whose coset graph is the distance-regular graph obtained from the previous one by removing all loops. Since the check matrix has no zero columns and collinear columns, the code distance of the code is neither 1 nor 2. □

1.1.6 Walk Distribution, an Alternative Definition of CR Codes

Weight and distance distributions are very powerful tools in the study of codes and designs. In particular, based on these distributions, several invariants were derived for equitable partitions in distance regular graphs. Some of them can be partially generalized to arbitrary graphs, using the concept introduced in the next definition.

Definition 1.38 (walk distribution). *Given a graph* $\Gamma = (V, E)$*, a multiset* C *of vertices of* Γ *and a vertex* $x \in V$*, the* walk distribution *of* C *with respect to* x *is the infinite sequence* $(W_0, W_1, \dots)$*, where* W_i *is the number of length-*i *paths (walks) beginning in* x *and ending in* C*, respecting the multiplicities, i.e., a path* $(x = x_0, x_1, \dots, x_i \in C)$ *is counted* $\mu(\{x_0, x_1\}) \cdot \mu(\{x_1, x_2\}) \cdot \ldots \cdot \mu(\{x_{i-1}, x_i\}) \cdot \mu(x_i)$

times, where $\mu(\{x_{j-1},x_j\})$ *is the multiplicity of the edge* $\{x_{j-1},x_j\}$ *in* E *and* $\mu(x_i)$ *is the multiplicity of the vertex* x_i *in* C.

It is known that in distance-regular graphs, the walk distributions (see, e.g. the combinatorial matrix A in [18]) are closely related with distance distributions. For arbitrary graphs in general, it is not so, and the alternative definition of CR codes in terms of distance distributions below (Theorem 1.1) cannot be treated in terms of distance distributions in general, without additional assumption on the distance regularity of the graph.

By induction, it is easy to see that Γ^i is the graph where the multiplicity of a vertex pair $\{x,y\}$ in the edge set equals the number of walks of length i between x and y. This implies the first claim of the following lemma (the second claim is straightforward).

Lemma 1.2. *The walk distribution of* C *with respect to* x *in* Γ *equals* $(|\Gamma^i(x) \Cap C|)_{i=0,1,2,\ldots}$, *where the multiset intersection* $X \Cap Y$ *denotes the multiset where the multiplicity of each element equals the product of its multiplicities in* X *and* Y. *The weight distribution of* C *with respect to* x *in* Γ *equals* $(|\Gamma^{(i)}(x) \Cap C|)_{i=0,1,\ldots,\mathrm{diam}(\Gamma)}$.

Theorem 1.1 (an alternative definition of CR codes). *A non-empty set* C *of vertices of a connected graph is CR if and only if its walk distribution with respect to a vertex* x *depends only on the distance from* x *to* C *(i.e., if there are exactly* $\rho(C)+1$ *different walk distributions).*

Proof. Consider the distance partition $(C^{(0)},\ldots,C^{(\rho)})$ of the vertex set of a connected graph Γ with respect to a code C in Γ.

If. Assume that the walk distribution

$$(W_0^{(i)},W_1^{(i)},\ldots)$$

of C with respect to a vertex x in $C^{(i)}$ depends only on i and does not depend on the choice of x in $C^{(i)}$. Denote by $S_{i,j}$ the number of vertices from $\Gamma(x)$ that belong to $C^{(j)}$ (we are going to show that $S_{i,j}$ does not depend on x, but at this moment this is not implied). Obviously, $S_{i,j}=0$ if $|i-j|>1$. The number of walks of length l from x to C such that the first edge goes from x to $C^{(j)}$ is obviously $S_{i,j}\cdot W_{l-1}^{(j)}$, so we have

$$W_l^{(i)} = S_{i,i-1}\cdot W_{l-1}^{(i-1)} + S_{i,i}\cdot W_{l-1}^{(i)} + S_{i,i+1}\cdot W_{l-1}^{(i+1)} \tag{1.7}$$

(for convenience, we agree that if some index j is out of $\{0,\ldots,\rho\}$ then $S_{i,j} = W_{l-1}^{(j)}=0$). Applying (1.7) for $l=i$, $l=i+1$, and $l=i+2$ and taking into account that $W_{i-1}^{(i)}=W_{i-1}^{(i+1)}=W_i^{(i+1)}=0$, we get

$$\begin{aligned} W_i^{(i)} &= S_{i,i-1}\cdot W_{i-1}^{(i-1)}, \\ W_{i+1}^{(i)} &= S_{i,i-1}\cdot W_i^{(i-1)} + S_{i,i}\cdot W_i^{(i)}, \\ W_{i+2}^{(i)} &= S_{i,i-1}\cdot W_{i+1}^{(i-1)} + S_{i,i}\cdot W_{i+1}^{(i)} + S_{i,i+1}\cdot W_{i+1}^{(i+1)}. \end{aligned}$$

We then can find consequently $S_{i,i-1}$, $S_{i,i}$, and $S_{i,i+1}$. We see that $S_{i,j}$ are determined from the walk distributions (actually, only from the components $W_j^{(j)}$, $W_{j+1}^{(j)}$, $W_{j+2}^{(j)}$) and hence do not depend on the choice of x. By the definition, $(C^{(0)}, \ldots, C^{(\rho)})$ is an equitable partition and C is CR.

Only if. If the partition $(C^{(0)}, \ldots, C^{(\rho)})$ is equitable with quotient matrix $(S_{i,j})$, then from (1.7) we can see by induction on l that the elements $W_l(i)$ of the walk distribution depend only on i and l and do not depend on the choice on the initial vertex in $C^{(i)}$ (the induction base is $l = 0$: $W_0(0) = 1$, $W_0(i) = 1$ for $i > 0$). □

Neumaier [97] was the first who defined CR codes in a manner similar to Definition 1.21 and proved the equivalence of that definition to the original definition by Delsarte [31, Section 5.2.3, p. 67] in the case of distance-regular graphs (a similar equivalence was proved independently in [18, Corollary 3.1], formally for Hamming graphs, but essentially the proof works for any distance-regular graph).

Corollary 1.4 (Delsarte's definition of CR codes). *A non-empty set C of vertices of a distance-regular graph Γ is CR if and only if its weight distribution with respect to a vertex x depends only on the distance from x to C (i.e., if there are exactly $\rho(C) + 1$ different such distributions).*

Proof. From Lemma 1.2 and Corollary 1.1 (see also Propositios 2.1 and 2.2 in [18]), we see that in a distance-regular graph the coefficients A_i and W_i of the weight and walk distributions, respectively (with respect to the same initial vertex), can be expressed from each other: A_i is a linear combination of W_0, ..., W_i and vice versa. So, we can replace walk distributions by weight distributions in the claim of Theorem 1.1 if the graph is distance-regular. □

We can consider Theorem 1.1 as a generalization of Corollary 1.4 and also of Delsarte's definition of CR from distance-regular to arbitrary graphs.

1.1.7 Orthogonal Arrays, Algebraic t-Designs

Definition 1.39 (orthogonal arrays, algebraic t-designs). *A nonempty multiset C of vertices of $H(n, q)$ is called an* orthogonal array *(OA) of strength t, or $OA(|C|, n, q, t)$ if every $H(n-t, q)$-subgraph of $H(n, q)$ contains exactly $\lambda = \frac{|C|}{q^t}$ elements of C. An orthogonal array is* simple *if it is a usual set (without multiplicities larger than 1). More generally (see Proposition 1.9 below), in an arbitrary regular graph $\Gamma = (V, E)$, a nonempty multiset C of vertices is called an* algebraic t-design *if $\varphi_1 = \varphi_2 = \cdots = \varphi_t = 0$ in the decomposition*

$$f = \varphi_0 + \varphi_1 + \cdots + \varphi_D \tag{1.8}$$

of the multiplicity function f of C into eigenfunctions of Γ, where φ_i corresponds to the ith (starting from 0) largest eigenvalue of Γ. More generally, a real-valued t-design *is defined as an arbitrary function $f : V \to \mathbb{R}$, $f \not\equiv 0$, that satisfy the property above without necessarily being the multiplicity function of some multiset.*

Remark 1.4. *In classical definitions of an orthogonal array (see, e.g., [49]), it is often defined as a* $|C| \times n$ *or* $n \times |C|$ *array whose rows (respectively, columns) are codewords of* C *from our definition. In connection to applications, the elements of* C *are referred to as (experimental)* runs, *coordinate positions from* 1 *to* n *as* factors, *the elements of the alphabet as* levels. *The value* λ *is known as the* index *of the orthogonal array.*

Remark 1.5. *We defined algebraic t-designs with respect to the natural, descending, ordering of eigenvalues. This is not the only way, and in the literature one can find valuable theorems about t-designs with respect to a Q-polynomial ordering, not necessarily descending. Delsarte [31] defined more general* T*-designs, where* T *is the list of the numbers of components to be vanished in* (1.8).

Proposition 1.9 (Theorem 4.4 in [31])**.** *In the Hamming graph* $H(n, q)$, *a multiset* C *is an* $OA(|C|, n, q, t)$ *if and only if it is an algebraic t-design.*

As we will see in Section 1.2.1.1, cells of equitable partitions are algebraic designs, whose strength can be found from the eigenspectrum of the quotient matrix.

1.1.8 Induced Colorings and Colorings of Hypergraphs

Definition 1.40 (induced coloring)**.** *Let* Γ *be a bipartite graph with parts* V' *and* V''. *Let* f *be a coloring of* V', *and let, according to Definition 1.16,* $\bar{f}(X)$ *denote the* f*-spectrum of a set* $X \subseteq V''$. *The vertex coloring*

$$\begin{cases} f(x), & \text{if } x \in V' \\ \bar{f}(\Gamma(x)), & \text{if } x \in V'' \end{cases}$$

is called the coloring induced by f.

The following easy fact can be helpful in the characterization of perfect colorings of a bipartite graph.

Lemma 1.3 ([82])**.** *Let* Γ *be a bipartite graph with parts* V' *and* V''. *Let* f *be a coloring of* V'. *If there is a perfect coloring* f'' *of* Γ *whose retract on* V' *is* f, *then the coloring* f' *of* Γ *induced by* f *is also perfect and, moreover, either* f'' *coincides with* f' *up to renaming of colors, of* f'' *is a refinement of* f' *with twin colors.*

Induced colorings can be productively used to construct CR codes, see e.g. [6,30,89,93,95,95]. With the concept of induced colorings, we can generalize perfect colorings from graphs to hypergraphs.

Definition 1.41 (hypergraph, uniform hypergraph)**.** *A hypergraph is an ordered pair* (V, E) *from some set* V, *whose elements are called* vertices, *and a set* E *of subsets of* V, *called* hyperedges. *A hypergraph is called* k-uniform, *or just* uniform, *if all its hyperedges have cardinality* k.

Definition 1.42 (incidence graph). *For a simple graph or hypergraph* (V, E)*, its* incidence graph *(also known as the* Levi graph*) is the bipartite graph* $(V \cup E, \{\{v, e\} : e \in E,\, v \in e\}$.

Proposition 1.10. *A vertex coloring* f *of a simple graph* Γ *is perfect if and only if in the incidence graph of* Γ *the coloring induced by* f *is perfect.*

The proof is straightforward. In view of Proposition 1.10, we can define perfect colorings of hypergraphs in such a way that perfect colorings of 2-uniform hypergraphs are exactly the same as perfect colorings of simple graphs.

Definition 1.43 (perfect colorings of hypergraphs [102,117]). *A vertex coloring* f *of a hypergraph* (V, E)*, is called a* perfect coloring *if in the incidence graph of* $(V \cup E, \{\{v, e\} : e \in E,\, v \in e\}$ *the coloring induced by* f *is perfect.*

Remark 1.6. *It is not difficult, with properly defined incidence graph, to treat Proposition 1.10 and Definition 1.43 for graphs and hypergraphs with multiple (hyper)edges. However, loops in graphs cannot be properly treated. In particular, the role of hyperedges of cardinality* 1 *in the definition of perfect colorings of hypergraphs is different from the role of loops for perfect colorings of graphs.*

An example of using hypergraphs in the theory of perfect colorings can be found in Section 1.4.6.

1.2 PROPERTIES OF EQUITABLE PARTITIONS

1.2.1 Algebraic Designs and other Spectral Restrictions

1.2.1.1 Equitable Partitions Are Algebraic Designs

Proposition 1.11 (algebraic-design property of equitable partitions). *Each cell of an equitable partition in a regular graph is an algebraic* t*-design, where the second largest eigenvalue* θ *of the quotient matrix is the* $(t + 1)$*th largest non-main eigenvalue of* Γ.

Proof. Let $(C_0, \ldots, C_{k-1})$ be an equitable S-partition of Γ, and let $F = (F_{i,j})$ be its incidence $\{0, 1\}$-matrix, where $F_{i,j} = 1$ if and only if $i \in C_j$. Consider the jth cell C_j of an equitable k-partition and denote by $\bar{e}_j$ the column vector with 1 in the jth place and 0 in the others. Then $F \cdot \bar{e}_j$, the jth column of F, is the characteristic vector of C_j. Let

$$\bar{e}_j = \sum_s \bar{\psi}_s$$

be the decomposition of $\bar{e}_j$ into eigenvectors of S corresponding to eigenvalues β_s of S. Then we have

$$A \cdot F \cdot \bar{\psi}_s = F \cdot S \cdot \bar{\psi}_s = F \cdot \beta_s \bar{\psi}_s = \beta_s F \cdot \bar{\psi}_s, \tag{1.9}$$

i.e., $F \cdot \bar{\psi}_s$ is a β_s-eigenvector of A, and hence

$$F \cdot \bar{e}_j = F \cdot \sum_s \bar{\psi}_s$$

is the decomposition of $F \cdot \bar{e}_j$ into the sum of eigenfunctions of A. Since, by the hypothesis, the first t non-main eigenvalues of Γ do not happen among β_s, the claim follows. □

Since the eigenvalues of $H(n,q)$ are $\theta_i = (q-1)n - qi$, $i = 0, \dots, n$, by Propositions 1.9 and 1.11 we have the following corollary (which is also a special case of Proposition 1.26, see Remark 1.14).

Corollary 1.5. *Each cell of an equitable partition of $H(n,q)$ is an orthogonal array of strength $t = \frac{(q-1)n-\theta}{q} - 1$, where θ is the second largest eigenvalue of the quotient matrix.*

1.2.1.2 Generalized Correlation-Immunity Bound

In this section, we discuss the following bound, which was firstly proved in [35] for the case of simple binary orthogonal arrays. Simple binary orthogonal arrays of strength t are essentially (via characteristic functions) tth order correlation-immune Boolean functions; this motivates the name of the bound. Together with [80, Corollary 6], which rejects some boundary cases, this bound is the only known nontrivial necessary condition for the existence of CR-1 codes in hypercubes $H(n,2)$.

Proposition 1.12 (correlation-immunity bound [35,58])**.** *For an orthogonal array $OA(N,n,2,t)$, $N < 2^{n-1}$ implies $t \le \frac{2n}{3} - 1$. Moreover, if $N = 2^{n-1}$ and $t > \frac{2n}{3} - 1$, then the array is simple.*

To prove this fact in more general settings, we need to define two auxiliary concepts.

Definition 1.44 (triangle triples of eigenvalues)**.** *A triple $(\theta', \theta'', \theta)$ of (not necessarily distinct) eigenvalues of a graph Γ is called a* triangle triple *if there are θ'-, θ''-, and θ-eigenfunctions ϕ', ϕ'', and ϕ, respectively, such that the pointwise product $\phi' \bullet \phi''$ is not orthogonal to ϕ.*

Remark 1.7. *For a distance-regular graph, triangle triples $(\theta_j, \theta_l, \theta_h)$ correspond to nonzero Krein parameters $q^h_{j,l}$ (dual intersection numbers, see, e.g., [13, Section 2.3]).*

Definition 1.45 (Delsarte's T-design [31])**.** *Let T be a subset of the set of eigenvalues of a graph Γ. A multiset C of vertices of Γ is called a T-design if for each θ in T, the multiplicity function χ_C of C is orthogonal to the θ-eigenspace of Γ.*

Lemma 1.4. *Let C be a T-design in a k-regular graph $\Gamma = (V,E)$, where T is a set of eigenvalues of Γ. If all triangle triples of eigenvalues of Γ intersect with $\{k\} \cup T$, then*

i. $|C| \geq \frac{|V|}{2}$;

ii. $|C| = \frac{|V|}{2}$ *implies that* C *is a simple set.*

Proof. Let χ_C be the multiplicity function of C and a its average value over V, $a = |C|/|V|$. For each eigenvalue θ, denote by S_θ the corresponding eigenspace of Γ. Denote $g = \chi_C - a$. We claim that (recall that $\bullet$ denotes the poinwise product)

$$g \bullet g \in \bigoplus_{\theta \in \{k\} \cup T} S_\theta \tag{1.10}$$

Proof of (1.10). Indeed, by hypothesis, $g = \sum_{i=1}^{s} \Psi_i$ for some eigenfunctions $\Psi_i \in S_{\psi_i}$, $\psi_i \notin \{k\} \cup T$, $i = 1, \dots, s$. For each $i, j \in \{1, \dots, s\}$, any triangle with ψ_i and ψ_j must have the third component in $\{k\} \cup T$. It follows that $\Psi_i \bullet \Psi_j \in \bigoplus_{\theta \in \{k\} \cup T} S_\theta$. Finally, we have $g \bullet g = \sum_{i,j=1}^{s} \Psi_i \bullet \Psi_j \in \bigoplus_{\theta \in \{k\} \cup T} S_\theta$.

From (1.10), we see that $g \bullet g$ is orthogonal to g, i.e.,

$$\sum_{x \in V} (g(x))^3 = 0. \tag{1.11}$$

Also, by the definition of g, we have

$$\sum_{x \in V} g(x) = 0. \tag{1.12}$$

(i) If, seeking a contradiction, we have $|C| < \frac{|V|}{2}$, then $a < \frac{1}{2}$. It follows that all negative values of g are between $-\frac{1}{2}$ and 0, while all positive values of g are larger than $\frac{1}{2}$. It follows that (1.12) implies $\sum_{x \in V} (g(x))^3 > 0$, contradicting (1.11).

(ii) If $|C| = \frac{|V|}{2}$, i.e., $a = \frac{1}{2}$, then the negative values of g are all $-\frac{1}{2}$, while the positive are at least $\frac{1}{2}$. Clearly, in this case (1.11) and (1.12) can hold together only if $g(x) \in \{-\frac{1}{2}, \frac{1}{2}\}$, i.e., $\chi_C(x) \in \{0, 1\}$, for all $x \in V$. □

Proof of Proposition 1.12. Any orthogonal array $\mathrm{OA}(N, n, 2, t)$ is a T-design in $H(n, 2)$, where $T = \{\theta_1, \dots, \theta_t\}$, $\theta_i = n - 2i$. Taking into account Lemma 1.4, it remains to show that if $t > \frac{2n}{3} - 1$ then every triangle triple of eigenvalues intersects with $T_0 = \{\theta_0, \dots, \theta_t\}$. Indeed, if we take θ_i- and θ_j-eigenfunctions of form $\chi_y(x) = (-1)^{(x,y)}$ and $\chi_z(x) = (-1)^{(x,z)}$, where $\mathrm{wt}(y) = i$, $\mathrm{wt}(z) = j$, and $i, j > \frac{2n}{3}$, then $\chi_y \bullet \chi_z(x) = (-1)^{(x,y+z)} = \chi_{y+z}$ is a θ_h-eigenfunction, where $h = \mathrm{wt}(y+z) = \mathrm{wt}(\bar{1}+y+\bar{1}+z) \leq \mathrm{wt}(\bar{1}+y) + \mathrm{wt}(\bar{1}+y) < \frac{n}{3} + \frac{n}{3} = \frac{2n}{3}$ and hence $\theta_h \in T_0$. Since any θ_i-eigenfunction is a linear combination of the function of form χ_y, where $\mathrm{wt}(y) = i$ (similarly, for θ_j), the pointwise product of a θ_i-eigenfunction and a θ_j-eigenfunction is a linear combination of θ_h-eigenfunctions with eigenvalues θ_h in T_0. □

1.2.1.3 Q-Polynomial Ordering and Triangle Inequality

The following lemma rejected many putative intersection arrays while generating the tables of small parameters for CR codes in Hamming graphs (Chapter 7).

Theorem 1.2 (triangle inequality). *Let Γ be a connected regular Q-polynomial graph with eigenvalues $\theta_0, \theta_1, \ldots, \theta_d$ in a Q-polynomial ordering, where θ_0 is the degree of Γ but the ordering is not required to be descending. Let $\bar{P} = (P_0, \ldots, P_k)$ be an equitable $(k+1)$-partition with quotient matrix S having a complete set of eigenvectors $\phi_0, \phi_1, \ldots, \phi_k$, where ϕ_j is a θ_{i_j}-eigenvector, $0 = i_0 < i_1 \le \ldots \le i_k$. For j and l in $\{0, \ldots, k\}$, let $\beta^h_{j,l}$, $h = 0, \ldots, k$, denote the coefficients in the unique decomposition of the pointwise product $\phi_j \bullet \phi_l$ as a linear combination of the eigenvectors:*

$$\phi_j \bullet \phi_l = \sum_{h=0}^{k} \beta^h_{j,l} \phi_h. \tag{1.13}$$

Then for the nonzero coefficients $\beta^h_{j,l}$, the indices i_j, i_l, i_h satisfy the triangle inequality

$$|i_j - i_l| \le i_h \le i_j + i_l. \tag{1.14}$$

In other words, if, for a putative quotient matrix S (implying that it satisfies Lloyd's condition), some *nonzero* coefficient $\beta^h_{j,l}$ in the decomposition (1.13) *does not* satisfy (1.14), then equitable S-partitions of Γ do *not exist*.

Proof. Let A be the adjacency matrix of Γ and F the incidence matrix of the partition $\bar{P}$. Denote $\Phi_j = F\phi_j$ and $\Phi = F\phi$. From $AF = FS$ we see that Φ_j is a θ_{i_j}-eigenvector of A, $j = 0, \ldots, k$. Since F is a coloring matrix, it holds

$$F(\phi_j \bullet \phi_l) = (F\phi_j) \bullet (F\phi_l) = \Phi_j \bullet \Phi_l.$$

Therefore, (1.13) is equivalent to

$$\Phi_j \bullet \Phi_l = \sum_{h=0}^{k} \beta^h_{j,l} \Phi_h,$$

and by the definition of a Q-polynomial order, $\beta^h_{j,l} \neq 0$ implies (1.14). □

Remark 1.8. *By linear algebra, the conclusion of Theorem 1.2 is equivalent to say that* (1.13) *holds for all j, l, h such that $(\phi_j \bullet \phi_l) D \phi_h \neq 0$, where D is the diagonal matrix with the diagonal elements $|P_0|, \ldots, |P_k|$. Indeed, the decomposition of $\phi_j \bullet \phi_l$ into a linear combination of ϕ_h, $h = 0, \ldots, k$, is the same as the decomposition of $D^{\frac{1}{2}}(\phi_j \bullet \phi_l)$ into a linear combination of $D^{\frac{1}{2}}\phi_h$, which are eigenvectors of the symmetric matrix $D^{\frac{1}{2}} S D^{-\frac{1}{2}}$ and hence mutually orthogonal (well, this is not necessarily so if S has multiple eigenvalues, but we actually check orthogonality to the concrete eigenspace, not to particularly chosen basis eigenvectors).*

Theorem 1.2 effectively rejects putative quotient matrices even in the case $j = l = 1$.

Example 1.2. *Consider a putative CR-4 code in $H(16,3)$ with quotient matrix*

$$\begin{pmatrix} 0 & 32 & 0 & 0 & 0 \\ 1 & 1 & 30 & 0 & 0 \\ 0 & 2 & 2 & 28 & 0 \\ 0 & 0 & 6 & 9 & 17 \\ 0 & 0 & 0 & 10 & 22 \end{pmatrix}.$$

It has cardinality 3^8, code distance 6, and formally self-dual weight distribution

$$[1,0,0,0,0,0,224,0,0,2720,0,0,3360,0,0,256,0].$$

The eigenvalues are $\theta_0 = 32$, $\theta_6 = 14$, $\theta_9 = 5$, $\theta_{12} = -4$, $\theta_{15} = -13$; the corresponding indices are $i_0 = 0$, $i_1 = 6$, $i_2 = 9$, $i_3 = 12$, $i_4 = 15$. Sadly, such a nice code does not exist by Theorem 1.2. Indeed, the corresponding eigenvectors are $\phi_0 = (1,1,1,1,1)$, $\phi_1 = (1, \frac{7}{16}, \frac{5}{32}, \frac{1}{28}, -\frac{5}{112})$, $\phi_2 = (1, \frac{5}{32}, -\frac{1}{80}, -\frac{1}{80}, \frac{1}{136})$, $\phi_3 = (1, -\frac{1}{8}, -\frac{1}{80}, \frac{13}{1120}, -\frac{1}{224})$, $\phi_4 = (1, -\frac{13}{32}, \frac{5}{32}, -\frac{7}{128}, \frac{1}{64})$. We have

$$\phi_1 \bullet \phi_1 = \sum_{h=0}^{k} \beta_{1,1}^h \phi_h = \frac{1}{224}\phi_0 + \frac{121}{672}\phi_1 + \frac{85}{112}\phi_2 + \frac{5}{112}\phi_3 + \frac{1}{84}\phi_4$$

and see that $\beta_{1,1}^4 \neq 0$ (which, according to Remark 1.8, can also be established by checking that $(\phi_1 \bullet \phi_1) \cdot (D \bullet \phi_4) \neq 0$, where $D = 3^8 \cdot (1,32,160,2240,3808)$ is the vector of cell cardinalities, found from S), while $i_4 = 15 \not\leq 6+6 = i_1 + i_1$, which contradicts (1.14). Moreover, the eigenvectors and the indices $i_0, \ldots, i_4$ do not change if we add a constant to the diagonal elements of S; hence, the same argument continues to work and $\{32,30,28,17;1,2,6,10\}$-CR codes do not exist in $H(n,3)$ for any n.

Remark 1.9. *The triangle inequality (1.14) is not the only possible condition for nonzero coefficients $\beta_{j,l}^h$, see, e.g., the next theorem. Even for a non-Q-polynomial graph, $\beta_{j,l}^h$ cannot be nonzero if the Krein parameter $q_{i_j,i_l}^{i_h}$ (a dual intersection number, see, e.g., [13, Section 2.3]) of the graph is zero. It remains unclear if some efficient necessary conditions for quotient matrices can be derived from concrete values of nonzero Krein parameters of the graph.*

Theorem 1.3 (additional restrictions for $H(n,2)$). *If, under the hypothesis and notation of Theorem 1.2, we have $\Gamma = H(n,2)$ and $\theta_i = n - 2i$, $i = 0,\ldots,n$, then for each nonzero coefficient $\beta_{j,l}^h$, the indices i_j, i_l, i_h satisfy*

$$|i_j - i_l| \leq i_h \leq \min\{i_j + i_l, 2n - i_j - i_l\} \quad \text{and } i_j + i_l + i_h \equiv 0 \bmod 2. \tag{1.15}$$

Proof. Similarly to the proof of Theorem 1.2, it is sufficient to show

$$|a-b| \le c \le \min\{a+b, 2n-a-b\} \quad \text{and } a+b+c \equiv 0 \bmod 2 \tag{1.16}$$

for nonzero components Φ_c in the decomposition

$$\Phi' \bullet \Phi'' = \sum_{c=0}^{n} \Phi_c,$$

of the pointwise product of θ_a- and θ_b-eigenfunctions Φ' and Φ'' of $H(n,2)$ into the sum of its projections Φ_c on the θ_c-eigenspaces S_c of $H(n,2)$.

To prove this, will use the basis of the space of functions on $H(n,2)$ consisting of eigenfunctions $\chi_y(x) = (-1)^{(y,x)}$, where $((y_1,\ldots,y_n),(x_1,\ldots,x_n)) = x_1y_1 + \cdots + x_ny_n$. For each vertex y, $\chi_y(x)$ is a $\theta_{\mathrm{wt}(y)}$-eigenfunction of $H(n,2)$. For y' and y'' of weight a and b, we have

$$\chi_{y'} \bullet \chi_{y''}(x) = (-1)^{(y',x)} \cdot (-1)^{(y'',x)} = (-1)^{(y'+y'',x)} = \chi_{y'+y''}(x).$$

We know that $\chi_{y'+y''}$ is a $\theta_{\mathrm{wt}(y'+y'')}$-eigenfunction of $H(n,2)$; moreover, it is easy to see that $\mathrm{wt}(y'+y'')$ satisfies

$$|a-b| \le \mathrm{wt}(y'+y'') \le \min\{a+b, 2n-a-b\}, \quad (-1)^{a+b+\mathrm{wt}(y'+y'')} = 1.$$

Therefore, representing Φ' and Φ'' as linear combinations of basis θ_a- and θ_b-eigenfunctions, respectively, we get that their pointwise product is decomposed into the sum of θ_c-eigenfunctions, where the indices c satisfy (1.16). □

Sometimes, we can see that the triangle inequality is not satisfied if we know only the eigenvalues of the quotient matrix. The following fact was proved in [62, Lemma 2.6] for CR codes, but holds also for arbitrary partitions, with essentially the same arguments as in [62].

Lemma 1.5. *If, under the hypothesis and notation of Theorem 1.2, all $k+1$ entries of some θ_{i_1}-eigenvector ϕ of S are pairwise different (as noted in [62, Corollary 2.7], this condition is automatically satisfied if $\theta_0 > \theta_1 > \ldots > \theta_d$ and S is tridiagonal), then*

$$i_{j+1} - i_j \le i_1, \qquad j = 1,\ldots,k-1.$$

Proof. Since all entries of ϕ are distinct, its entry-wise (Hadamard) degrees $\phi^{\bullet 0}$, $\phi^{\bullet 1}$, ..., $\phi^{\bullet k}$ form a basis of the $(k+1)$-dimensional space (indeed, they are the rows of a $(k+1)\times(k+1)$ Vandermonde matrix). We can assume without loss of generality that $\phi_1 = \phi$.

Now, let A be the adjacency matrix of Γ and F the incidence matrix of the partition $\bar{P}$. Denote $\Phi_j = F\phi_j$ and $\Phi = F\phi$. From $AF = FS$ we see that Φ_j is an θ_{i_j}-eigenvector of A, $j = 0,\ldots,k$. Since F is a coloring matrix, we have $F(\phi^{\bullet j}) = (F\phi)^{\bullet j} = \Phi^{\bullet j}$.

So, we have

$$\langle \Phi^{\bullet 0}, \ldots, \Phi^{\bullet k} \rangle = \langle \Phi_0, \ldots, \Phi_k \rangle,$$

where $\langle \ldots \rangle$ denotes the linear span. For $j \in \{1, \ldots, k-1\}$, let s be a number such that $\Phi^{\bullet s}$ is a linear combination of $\Phi_0, \ldots, \Phi_j$, say $\Phi^{\bullet s} = \sum_{t=0}^{j} \alpha_t \Phi_t$, but $\Phi^{\bullet s+1}$ is not a linear combination of $\Phi_0, \ldots, \Phi_j$. This means that $\Phi^{\bullet s+1}$ is a linear combination of $\Phi_0, \ldots, \Phi_k$ involving *at least one* of $\Phi_{j+1}, \ldots, \Phi_k$. Since

$$\Phi^{\bullet s+1} = \Phi \bullet \Phi^{\bullet s} = \Phi_1 \bullet \sum_{t=0}^{j} \alpha_t \Phi_t = \sum_{t=0}^{j} \alpha_t (\Phi_1 \bullet \Phi_t),$$

from the definition of Q-polynomiality, we have $i_1 + i_j \geq i_{j+1}$, i.e., $i_{j+1} - i_j \leq i_1$. □

1.2.2 Quotient Matrices, Invariant Subspaces, Projectors

1.2.2.1 *Properties of Quotient Matrices*

Some basic properties of quotient matrices of equitable partitions can be derived from the definition and equation (1.2), we repeat it for convenience:

$$A \cdot F = F \cdot S. \tag{1.17}$$

We will call a rectangular $\{0,1\}$-matrix F a *coloring matrix* if it does not have all-zero columns and each or its rows contains exactly one 1. A square matrix A is called *symmetrizable* if there exists an invertible diagonal matrix D such that $A = DB$ where B is a symmetric matrix. Although some statements below are more general, we will pay a special attention to symmetrizable non-negative integer (*s.n.i.*) matrices.

If square matrices A and S satisfy (1.17) for some coloring matrix F, then we will write $S \preceq A$ and say that A is *S-regular* (as well as the graph with adjacency matrix A if any) if there exists a coloring matrix F such that $AF = FS$. Some properties of S-regular matrices is proved in [25]. It is easy to see that the relation $\preceq$ is a partial order on the set of square matrices (see the proof of Proposition 1.36).

Proposition 1.13. *If $S \preceq A$, then*

a. *the set of row sums of A coincides with the set of row sums of S;*

b. *if A is a non-negative matrix, then S is non-negative too;*

c. *if A is an integer matrix, then S is integer too;*

d. *if A is a symmetrizable matrix, then S is also symmetrizable and, in particular, the zero entries of S are placed symmetrically with respect to the main diagonal;*

e. ([99], see also [27, Theorem 4.5], [40, Lemma 2.2(c)]) *the eigenspectrum of S is included (respecting the multiplicities) in the eigenspectrum of A;*

f. *if A is a non-negative symmetrizable matrix, then there is a basis from eigenvectors of A and the same is true for S.*

Proof. Let F be a coloring matrix such that $AF = FS$. We first note that

(*) *the sets of row vectors of the matrices FS and S coincide.*

(a) Denote by $\bar{1}_i$ the all-one column of height i. Trivially, for any $i' \times i$ matrix M, the column $M\bar{1}_i$ consists of row sums of M. In particular, $F\bar{1}_k = \bar{1}_N$, where N and k are the height and the width of F, respectively. We have

$$A\bar{1}_N = AF\bar{1}_k = FS\bar{1}_k.$$

By (*) the last is a column containing all (and only) row sums of S.

(b) If A is non-negative, then AF is non-negative, FS is non-negative (from $AF = FS$), and S is non-negative, by (*).

(c) is proved similarly to (b).

(d) If A is symmetrizable, then DA is symmetric for some invertible diagonal D. Multiplying both parts of $AF = FS$ by $F^{\mathrm{T}}D$, we get

$$F^{\mathrm{T}}DAF = F^{\mathrm{T}}DFS,$$

where $F^{\mathrm{T}}DAF$ is symmetric and $F^{\mathrm{T}}DF$ is diagonal (indeed, for different columns $\bar{a}$ and $\bar{b}$ of F, the vectors $\bar{a}$ and $D\bar{b}$ are orthogonal) and invertible.

(e) From $AF = FS$, we see that every θ-eigenvector $\bar{u}$ of S corresponds to the θ-eigenvector $F\bar{u}$ of A:

$$AF\bar{u} = FS\bar{u} = F\theta\bar{u} = \theta F\bar{u} \tag{1.18}$$

(actually, this is true for any F of rank k, not only for coloring matrices).

(f) Assume A is non-negative and symmetrizable. By (b) and (d), S is also non-negative and symmetrizable.

Since A is symmetrizable, we have $A = DB$ for symmetric B and invertible diagonal D. Since A is non-negative, we can assume that D and B are both non-negative (if it is not so, then removing the signs from all entries of D and B keeps the equality $A = DB$). Then there exists an invertible diagonal matrix T such that $T^2 = D$. From $A = T^2B$ we get $T^{-1}AT = TBT$. Readily, TBT is a symmetric matrix. Hence, there is a basis $\bar{e}_1, \ldots, \bar{e}_n$ from eigenvectors of TBT. Then $T\bar{e}_1, \ldots, T\bar{e}_n$ is a basis from eigenvectors of A. □

In application to equitable partitions, claim (e) of Proposition 1.13 is known as Lloyd's necessary condition of existence; it was firstly proved it [83] for perfect codes, in another form, which have different generalizations, not only for equitable partitions. We summarize the basic properties of quotient matrices in the following corollary.

Corollary 1.6. *If there is an equitable S-partition of a graph Γ, then S is an s.n.i. matrix with complete eigenspectrum, which is included (as a multiset) in the eigenspectrum of Γ, and the set of row sums of S coincides with the set of vertex degrees of Γ.*

A partial inversion of the fact that the quotient matrix of an equitable partition is s.n.i. is the following.

Proposition 1.14 ([59] or [25, Proposition 2.3]). *If S is an s.n.i. matrix, then there is an S-regular simple graph, i.e., there is an equitable partition of a simple graph with quotient matrix S.*

The next result says that we can construct graphs admitting two equitable partitions with given quotient matrices (actually, by induction, the same is true for any number of quotient matrices satisfying the hypothesis).

Proposition 1.15 ([81]). *Let A_1 and A_2 be s.n.i. matrices. If there exists an s.n.i. matrix S such that $S \preceq A_1$ and $S \preceq A_2$, then there exists an s.n.i. matrix B such that $A_1 \preceq B$ and $A_2 \preceq B$.*

Corollary 1.7 ([81]). *If Γ_1 and Γ_2 are S-regular graphs, then there exists a graph Γ covering both Γ_1 and Γ_2.*

In particular [3], if Γ_1 and Γ_2 are r-regular graphs, then there exists an r-regular graph Γ covering both Γ_1 and Γ_2.

1.2.2.2 *Invariant Subspaces and Projectors*

Let F be an $N \times k$ matrix of rank k. Denote by V_F the subspace generated by the columns of F. Denote by P_F the $N \times N$ symmetric matrix $F(F^{\mathrm{T}}F)^{-1}F^{\mathrm{T}}$. We have

a. $P_F^2 = P_F$;

b. $P_F F = F$ and hence $P_F\bar{v} = \bar{v}$ for every $\bar{v} \in V_F$;

c. $F^{\mathrm{T}}\bar{v} = \bar{0}$ implies $P_F\bar{v} = \bar{0}$, and hence the space $V_F^{\perp}$ dual to V_F is the nullspace of P_F.

Properties (a)–(c) imply that P_F is the orthogonal projection from the N-dimensional space onto V_F.

For an $N \times N$ matrix A, the subspace V_F is called A-*invariant* (invariant under the action of A) if $AV_F \subseteq V_F$, which is obviously equivalent to $AF = FS$ for some $k \times k$ matrix S. An invariant space of a symmetrizable non-negative matrix A has a basis consisting of eigenvectors of A:

Proposition 1.16. *If $AF = FS$, where A is a symmetrizable non-negative matrix and the rank of F coincides with its width, then the invariant subspace V_F is generated by eigenvectors $F\bar{u}$, where $\bar{u}$ are eigenvectors of S.*

Proof. It is easy to see (1.18) that for any eigenvector $\bar{u}$ of S, $F\bar{u}$ is an eigenvector of A. Since there is a basis $(\bar{u}_1, \ldots, \bar{u}_k)$ from eigenvectors of S (Proposition 1.13(f)), a required basis is formed by $F\bar{u}_1, \ldots, F\bar{u}_k$. □

Given the adjacency matrix A of a graph Γ and the incidence matrix F of a vertex partition of Γ, the following proposition (proved in a more general form) provides an algebraic criterion on the partition to be equitable and a formula [39, Lemma 9.3.1] for its quotient matrix.

Proposition 1.17. *Suppose A is a symmetric $N \times N$ matrix, F is an $N \times k$ matrix of rank k, and $P_F = F(F^{\mathrm{T}}F)^{-1}F^{\mathrm{T}}$. The following assertions are equivalent:*

i. *$AP_F = P_F A$ and $S = (F^{\mathrm{T}}F)^{-1}F^{\mathrm{T}}AF$;*

ii. *$AF = FS$.*

Proof. (i)⇒(ii). From (i) and $P_F F = F$ (property (b) of the projector), we derive $FS = F(F^{\mathrm{T}}F)^{-1}F^{\mathrm{T}}AF = P_F AF = AP_F F = AF$.

(ii)⇒(i). From $AF = FS$ and $P_F F = F$ we find $AF = P_F FS = P_F AF$. Multiplying both parts by $(F^{\mathrm{T}}F)^{-1}F^{\mathrm{T}}$ in the right, we find $AP_F = P_F AP_F$. Since P_F and A are symmetric, we get $P_F A = P_F AP_F = AP_F$.

Finally, multiplying both parts of $AF = FS$ by $(F^{\mathrm{T}}F)^{-1}F^{\mathrm{T}}$ in the left, we get $(F^{\mathrm{T}}F)^{-1}F^{\mathrm{T}}AF = (F^{\mathrm{T}}F)^{-1}F^{\mathrm{T}}FS = S$. □

Corollary 1.8. *If A is the adjacency matrix of a graph Γ and F is a coloring matrix, i.e., the incidence matrix of some partition $\overline{C}$ of the vertex set of Γ, then $\overline{C}$ is an equitable partition if and only if $AP_F = P_F A$, in which case the quotient matrix is $(F^{\mathrm{T}}F)^{-1}F^{\mathrm{T}}AF$.*

1.2.2.3 Description of Equitable Partitions in Terms of Eigenfunctions

Sometimes, we can represent every perfect k-coloring with a given quotient matrix by a k-valued eigenfunction with given eigenvalue and reals values $c_0, c_1, \ldots, c_{k-1}$. The fact below is trivial, but it helps easily expand some theorems from eigenfunctions to perfect colorings.

Proposition 1.18. *Assume S is an s.n.i. $k \times k$ matrix with a θ-eigenvector $\bar{v} = (a_0, \ldots, a_{k-1})^{\mathrm{T}}$ whose entries a_i, $i = 0, \ldots, k-1$, are all distinct. Then by renaming the colors $i \to a_i$, every S-perfect coloring $f : V \to \{0, \ldots, k-1\}$ of any graph $\Gamma = (V, E)$ can be turned into a perfect k-coloring which is at the same time a θ-eigenfunction of Γ.*

Proof. If F if the corresponding coloring matrix, then $F\bar{v}$ is a θ-eigenvector of the adjacency matrix (see (1.9)) and at the same time a k-valued function on V corresponding to the same equitable k-partition as f. □

Note that apart of the case $k = 2$ (see Corollary 1.9 below), the mapping above from perfect k-colorings to $(a_0, \ldots, a_{k-1})$-valued eigenfunctions is not surjective in general. As we will see in the next proposition, for arbitrary k, we can represent perfect k-colorings by invariant subspaces spanned by k eigenvectors.

Recall that for real-valued functions f and g on the same set, $f \bullet g$ denotes their pointwise (*Hadamard*) product: $(f \bullet g)(x) = f(x)g(x)$ for all x. We also denote by $f^{\bullet i}$ the ith degree of f with respect to the Hadamard product, i.e., $f^{\bullet i}(\cdot) \equiv (f(\cdot))^i$. A subspace V of the space of real-valued functions on the vertex set of a graph is called *closed with respect to the Hadamard product* if $f \bullet g \in V$ for any $f, g \in V$.

Proposition 1.19. (a) *A partition $\overline{C}$ of the vertex set of a graph of Γ is equitable if and only if the linear span of characteristic functions of cells of $\overline{C}$ has a basis from eigenfunctions of Γ.*

(b) *Let f be a real-valued k-coloring of a graph Γ with adjacency matrix A. The coloring f is perfect if and only if $\langle f^{\bullet 0}, f^{\bullet 1}, \ldots, f^{\bullet k-1}\rangle$ is an invariant k-subspace of A, where $f^{\bullet i}$ denotes the i'th degree of f with respect to the pointwise (Hadamard) product.*

(c) [88, Proposition 2.1.4] *Let Γ be a regular graph with adjacency matrix A. An A-invariant subspace V is closed with respect to the Hadamard product if and only if it is generated by the characteristic functions of the cells of an equitable partition of Γ.*

Proof. (a) "*Only if*" is a special case of Proposition 1.16.

If. Let us have a coloring $N \times k$ matrix F such that V_F is generated by k eigenfunctions of an $N \times N$ matrix A. Since every eigenfunction, by the definition, generates an A-invariant 1-dimensional space, the sum of such spaces is also A-invariant. Therefore, V_F is A-invariant and $AF = FS$ for some S.

Applying the argument above to the adjacency matrix A of the graph Γ and the incidence matrix F of the partition $\overline{C}$, we conclude that $\overline{C}$ is an equitable partition.

(b) Let f possess k distinct real values c_i, $i = 0, \ldots, k-1$. The coloring f is perfect if and only if the characteristic functions of the cells $f^{-1}(c_i)$, $i = 0, \ldots, k-1$, generate an invariant space of the adjacency matrix of $A(\Gamma)$. This invariant subspace is equal to the subspace generated by $f^{\bullet 0}$, $f^{\bullet 1}$, ..., $f^{\bullet k-1}$ because the Vandermonde determinant of the matrix $(c_i^j)_{i,j=1}^{k-1}$ is not zero.

(c) is proved similarly. □

The following well-known special case is easy to prove independently.

Corollary 1.9. *A 2-partition (C_0, C_1) of a vertex set of a regular graph is equitable (i.e., each of C_0, C_1 is CR-1) if and only if C_0 and C_1 are the preimages of the values of a two-valued eigenfunction of the graph.*

For two partitions B and C of the same set V, the partition of V of the largest cardinality such that both B and C are its refinements is denoted by $B \vee C$ (so, $\vee$ is the least upper bound in the natural lattice of partitions). Since the intersection of two invariant subspaces is an invariant subspace, the following proposition is true.

Proposition 1.20. *Let $C = (C_0, \ldots, C_{k-1})$ and $B = (B_0, \ldots, B_{m-1})$ be equitable partitions of a graph Γ. Then*

$$B \vee C$$

is an equitable partition of Γ.

1.2.2.4 Monotonicity and other Combinatorial Properties

In [88], Martin conjectured that the intersection array $\{b_0, \ldots, b_{\rho-1}; c_1, \ldots, c_\rho\}$ of any CR code in a distance-regular graph is monotone, i.e., $b_0 \geq b_1 \geq \ldots \geq b_{\rho-1}$ and $c_1 \leq c_2 \leq \ldots \leq c_\rho$. Koolen [61] found counterexamples in the dodecahedron (a CR-$\{1,2,1;1,2,1\}$), the Biggs–Smith graph (a CR-$\{1,1,2;2,1,1\}$), Odd graphs and doubled Odd graphs and proved the conjecture for many families of distance-regular (and not only) graphs. The key fact proved in [61] is the following:

Lemma 1.6 ([61, Lemma 9]). *If for two $\{b_0, \ldots, b_{\rho-1}; c_1, \ldots, c_\rho\}$-CR codes C and D in the same graph $s = \max\{j : C^{(j)} \cap D \neq \emptyset\}$, then $b_i \geq b_{i+s}$ and $c_i \leq c_{i+s}$, $i = 0, \ldots, \rho - s$.*

Corollary 1.10 ([61, Lemma 10]). *Let C be a there is a $\{b_0, \ldots, b_{\rho-1}; c_1, \ldots, c_\rho\}$-CR code a graph Γ. If Γ has an automorphism ϕ such that $\phi(C) \neq C$ and $d(\phi(c), C) \leq s$ for each $c \in C$, then $b_i \geq b_{i+s}$ and $c_i \leq c_{i+s}$, $i = 0, \ldots, \rho - s$.*

In particular, the hypothesis of Lemma 1.10 is always true if Γ is a translation graph (a Cayley graph on an abelian group). For some graphs, we can prove that the intersection arrays of CR codes are strictly monotone. In Chapter 7, this property is shown for Hamming graphs, which actually satisfy the sufficient conditions in the following lemma.

Lemma 1.7. *Let $\Gamma = (V, E)$ be a connected graph such that*

i. *every two vertices at distance 2 have exactly μ common neighbors, where $\mu > 1$ is some constant, and*

ii. *for every vertices v, x, y such that $d(v,x) = d(v,y) = d(x,y) = 1$ there are at most $\mu - 1$ vertices z such that $d(v,z) > d(x,z) = d(y,z) = 1$.*

Then the intersection array $\{b_0, \ldots, b_{\rho-1}; c_1, \ldots, c_\rho\}$ of any CR code is strictly monotone:

$$b_0 > b_1 > \ldots > b_{\rho-1}, \qquad c_1 < c_2 < \ldots < c_\rho.$$

Proof. Consider the distance partition $C^{(0)}, C^{(1)}, C^{(2)}, \ldots$ of V with respect to an arbitrary set C of vertices. Take adjacent $v \in C^{(i)}$ and $x \in C^{(i+1)}$. Denote by Z the set of neighbors of x in $C^{(i+2)}$ and denote $|Z|$ by b_{i+1}; denote by Y the set of neighbors of v in $C^{(i+1)} \setminus \{x\}$ and denote $|Y| + 1$ by b_i (both b_i and b_{i+1} are in agree with the corresponding intersection numbers if C is CR). Now, we count the number of edges $\{z, y\}$ such that $z \in Z$ and $y \in Y$. From one side, it is $|Z| \cdot (\mu - 1)$, by (i). From the other side, it is not greater than $|Y| \cdot (\mu - 1)$, by (i) (for $y \in Y$ not adjacent to x) and (ii) (for $y \in Y$ adjacent to x). We conclude that $|Z| \leq |Y|$, which is $b_{i+1} \leq b_i - 1$. For c_i, the proof is similar. □

Ending Section 1.2.2, which considers properties of quotient matrices, we mention some simple observations that can reject some matrices for concrete graphs.

Lemma 1.8 ([116]). *Let f be an S-perfect coloring of a graph Γ. Let x and y be two vertices of Γ with $f(x) = i$ and $f(y) = i'$. Then the number of common neighbors of x and y cannot be larger than*

$$\sum_j \min\{S_{i,j}, S_{i',j}\}. \tag{1.19}$$

Corollary 1.11. *Let f be an S-perfect coloring of an amply regular Γ with parameters (v, k, λ, μ). Then*

- *for any i and i' such that $S_{i,i'} \neq 0$, the value of (1.19) is not less than λ;*
- *for any different i and i' such that $S_{i,i'} = 0$ but $S_{i,i''} \neq 0$ and $S_{i',i''} \neq 0$ for some i'', the value of (1.19) is not less than μ.*

1.2.3 Distance Invariance, Mutual Distribution, and Orthogonality of Perfect Colorings

The following fact is a straightforward generalization of the famous Lloyd–Shapiro–Slotnick [83,106] theorem, which is actually a special case saying that perfect codes in Hamming graphs are distance invariant. For CR codes, it was shown in [97, Theorem 3.2].

Proposition 1.21 ([88], distance invariance of equitable partitions). *1. In a distance-regular graph Γ, the distance distribution of each cell of an equitable partition $\bar{P}$ depends only on the number of this cell, the number of the cell containing the initial vertex, the quotient matrix of $\bar{P}$ and intersection array of Γ and does* not *depend on the choice of the initial vertex (in the same cell), the choice of the partition (with the same quotient matrix), and even on the choice of the graph with the same intersection array.*

2. In other words, an equitable S-partition $\bar{P}$ of a distance regular graph Γ is necessarily an equitable partition of the distance-i graph $\Gamma^{(i)}$. Moreover, the quotient matrix of $\bar{P}$ in $\Gamma^{(i)}$ is $K^{(i)}(S)$, where $K^{(i)}$ is the degree-i polynomial (see Proposition 1.1) such that $\Gamma^{(i)} = K^{(i)}(\Gamma)$.

A modern proof of this fact is very easy (see e.g. [68,88]). Moreover, Proposition 1.21 is a special case $\Pi = K^{(i)}$ of the following simple fact.

Proposition 1.22. *If Γ and Δ are graphs such that $\Delta = \Pi(\Gamma)$ for some polynomial Π (i.e., the adjacency matrix of Δ is $\Pi(A)$, where A is the adjacency matrix of Γ), then any S-perfect structure F over Γ is a $\Pi(S)$-perfect structure over Δ.*

Proof. From $A \cdot F = F \cdot S$, see (1.2), it follows that $A^i \cdot F = A^{i-1} \cdot F \cdot S = \ldots = F \cdot S^i$ for every nonnegative integer degree i and hence $\Pi(A) \cdot F = F \cdot \Pi(S)$ for every polynomial Π. □

Corollary 1.12. *The second paragraph of Proposition 1.21 remains valid if we replace the partition $\bar{P}$ by any perfect structure over Γ.*

In the following subsections, we discuss different corollaries and generalizations of this fact.

1.2.3.1 Minimum Supports of Eigenfunctions and Difference between Perfect Colorings

For eigenfunctions, as a special case of perfect structures, Proposition 1.22 is just a well-known obvious fact that a θ-eigenfunction of a matrix A is a $K(\theta)$-eigenfunction of $K(A)$, for any polynomial K. From Corollary 1.12, we can deduce the following bound on the minimum cardinality of the support of an eigenfunction of a distance-regular graph (see also the survey [109] on eigenfunctions with minimum support).

Proposition 1.23 (the weight-distribution (w.d.) bound on the support of an eigenfunction [75, Corollary 1]). *A θ-eigenfunction f of a distance regular graph Γ has nonzero values in at least*

$$\sum_{i=0}^{\mathrm{diam}(\Gamma)} \lceil |K^{(i)}(\theta)| \rceil \tag{1.20}$$

vertices, where $\Gamma^{(i)} = K^{(i)}(\Gamma)$.

Proof. Let x be a vertex where f has the maximum absolute value. Without loss of generality, assume $f(x) = 1$; so, $|f(y)| \leq 1$ for any vertex y. We partition the vertex set of Γ into the sets $\Gamma^{(0)}(x)$, $\Gamma^{(1)}(x)$, ..., $\Gamma^{(\mathrm{diam}(\Gamma))}(x)$. The sum of f over $\Gamma^{(i)}(x)$ is $K^{(i)}(\theta)$ because f is a $K^{(i)}(\theta)$-eigenfunction of $\Gamma^{(i)} = K^{(i)}(\Gamma)$. Therefore, f cannot have less than $\lceil |K^{(i)}(\theta)| \rceil$ nonzeros in $\Gamma^{(i)}(x)$. Summing over all i, we get the required bound. □

By Propositions 1.18 and 1.23, the following bound on the difference of two perfect colorings holds.

Corollary 1.13. *Let f and g be different perfect colorings of a distance-regular graph Γ with the same quotient $k \times k$ matrix S. Assume that there is a θ-eigenvector $(a_0, \ldots, a_{k-1})^{\mathrm{T}}$ of S whose entries a_i, $i = 0, \ldots, k-1$, are all distinct. Then f and g differ in at least* (1.20) *number of vertices.*

1.2.3.2 Mutual Distributions of Colorings

The two following theorems are elementary from an algebraic point of view; nevertheless, they are very significant for the perfect structures.

Proposition 1.24. *Let F and G be S- and R- perfect structures over A and A^{T} respectively (F, G, A, S, and R are $N \times \nu$, $N \times \mu$, $N \times N$, $\nu \times \nu$, and $\mu \times \mu$ matrices). Then $G^{\mathrm{T}}F$ is a perfect structure over R^{T} with parameters S. Briefly,*

$$(AF = FS) \,\&\, (A^{\mathrm{T}}G = GR) \Rightarrow (R^{\mathrm{T}} \cdot G^{\mathrm{T}}F = G^{\mathrm{T}}F \cdot S).$$

Proof. $R^{\mathrm{T}}G^{\mathrm{T}}F = (GR)^{\mathrm{T}}F = (A^{\mathrm{T}}G)^{\mathrm{T}}F = G^{\mathrm{T}}AF = G^{\mathrm{T}}FS.$ □

Note that if A is the adjacency matrix of some graph, then $A = A^{\mathrm{T}}$.

Proposition 1.25. *If the matrix* $B = \{B_{i,j}\}_{i,j=0}^{\rho}$ *satisfies* $B_{i,i+1} \neq 0$, $B_{i,j} = 0$ *for any* $i = 0, \dots, \rho - 1$, $j > i + 1$, *then any* S*-perfect structure* H *over* B *is uniquely defined by its first row* h_0. *Moreover, the rows* $h_0, \dots, h_\rho$ *of* H *satisfy the recursive relation*

$$h_{i+1} = (h_i S - \sum_{j=0}^{i} B_{i,j} h_j)/B_{i,i+1}, \qquad i = 0, \dots, \rho - 1, \tag{1.21}$$

and, by induction,

$$h_{i+1} = h_0 \Pi_B^{(i+1)}(S) \tag{1.22}$$

where $\Pi_B^{(i+1)}(x)$ *is a degree-*$(i + 1)$ *polynomial in* x.

Proof. From $Bh = hS$ we have $\sum_{j=0}^{n} B_{i,j} h_j = h_i S$. Applying the hypothesis on $B_{i,j}$, we get (1.21). □

Remark 1.10. *In the important subcase of a tridiagonal matrix* B, *the recursive relation* (1.21) *has the following form, where* $B_{0,-1}$ *and* h_{-1} *are assumed to be zero:*

$$h_{i+1} = (h_i S - B_{i,i} h_i - B_{i,i-1} h_{i-1})/B_{i,i+1}, \qquad i = 0, \dots, \rho - 1. \tag{1.23}$$

The role of Proposition 1.24 for equitable partition is unclear until we understand the role of the matrix $G^{\mathrm{T}} F$.

Definition 1.46 (mutual distribution of colorings). *For of two vertex colorings* f *and* g *of the same graph with the incident matrices* F *and* G, *respectively, we call the matrix* $G^{\mathrm{T}} F$ *the* distribution of f with respect to g.

Remark 1.11. *Obviously, the* (i, j)*th element of* $G^{\mathrm{T}} F$ *equals the number of vertices having color* i *in the first coloring and color* j *in the second.*

Now, the special case of Proposition 1.24 for perfect colorings sounds as follows:

Corollary 1.14. *For perfect colorings* f *and* g *of the same graph with quotient matrices* S *and* R, *respectively, the distribution of* f *with respect to* g *is an* S*-perfect structure over* R^{T}.

If g is the distance coloring with respect to an R-CR code, then $B = R^{\mathrm{T}}$ satisfies the hypothesis of Proposition 1.25 and Remark 1.25.

Corollary 1.15. *For a CR-*ρ *code* C *with quotient matrix* $R = (R_{i,j})$ *and an equitable* S*-partition* $(P_0, \dots, P_{k-1})$ *in the same graph, the weight distributions* $\overline{W}_t = (W_t^0, \dots, W_t^\rho)$ *(see Definition 1.35) of the cells* P_i *with respect to* C *depend only on the intersection cardinalities* $W_j^0 = |P_j \cap C|$, $j = 0, \dots, k - 1$, *and the quotient matrices* R *and* S. *Moreover,*

$$h_{i+1} = (h_i S - R_{i,i} h_i - R_{i-1,i} h_{i-1})/R_{i+1,i}, \qquad i = 0, \dots, \rho - 1,$$

where $h_i = (W_0^i, \dots, W_{k-1}^i)$, $i = 0, \dots, \rho$, *and, for convenience,* $h_{-1} = (0, \dots, 0)$ *and* $R_{-1,0} = 0$.

Remark 1.12. *Proposition 1.24 can be potentially used to prove the nonexistence of perfect colorings with some putative quotient matrix S. Indeed, if, for some* existing *perfect coloring with quotient matrix T, the matrix equation $T^{\mathrm{T}} \cdot H = H \cdot S$ has no integer nonnegative matrix solution F whose sum equals the graph order, then there are no equitable S-partitions in the same graph. Even the very special case of this principle considered in the next section can be applied to reject some putative quotient matrices.*

1.2.3.3 Orthogonality of Colorings

Definition 1.47 (orthogonality of colorings). *Two vertex colorings of the same graph (or the corresponding vertex partitions) are called* orthogonal *if for their incidence matrices F, G the matrix $G^{\mathrm{T}}F$ has rank* 1.

Remark 1.13. *Orthogonality of two colorings f and g means that they do not correlate with each other, i.e., the density of vertices with f-color j among the vertices with g-color i equals the density of vertices with f-color j among all vertices of the graph.*

Proposition 1.26. *If the quotient matrices S, R of two perfect colorings of the same connected d-regular graph have only one common eigenvalue, then the colorings are orthogonal.*

Proof. Since d is the eigenvalue of both S and R, it is their unique common eigenvalue. By Corollary 1.6, the eigenvalue d has multiplicity 1 in each of S, R. Also by Corollary 1.6, there is a basis $\varphi_0, \ldots, \varphi_{s-1}$ from eigenvectors of S. Let θ_i be the eigenvalue corresponding to φ_i and $\theta_0 = d$.

Denote $H = G^{\mathrm{T}}F$, where F and G are the incidence matrices of the considered colorings. By Proposition 1.24, we have $R^{\mathrm{T}}H = HS$. Since φ_i is a θ_i-eigenvector, we have $S\varphi_i = \varphi_i\theta_i$ and

$$R^{\mathrm{T}} \cdot H\varphi_i = HS\varphi_i = H\varphi_i\theta_i,$$

so $H\varphi_i$ is either all-zero or a θ_i-eigenvector of R. But θ_i is not an eigenvalue of R if $i \neq 0$. Therefore, the rows of H are orthogonal to all φ_i, $i \neq 0$. We conclude that the rows of H are collinear to φ_0 and H has rank 1. □

More generally, for any partition of the vertex set, we can define the eigenspectrum as the set of graph eigenvalues corresponding to nontrivial components in the decomposition of the cell characteristic functions into eigenfunctions of the graph. If the eigenspectra of two arbitrary (not necessarily equitable) partitions of the vertex set of a regular graph intersect in only one eigenvalue, then the partitions are orthogonal.

Remark 1.14. *Proposition 1.26 gives an alternative proof of Corollary 1.5. Indeed, any set S inducing a $H(n-t,q)$-subgraph of $H(n,q)$ is a CR code with eigenvalues $n(q-1)$, $n(q-1)-q$, ..., $n(q-1)-qt$ (moreover, it is a cell of a covering of* *the graph $H(t,q)+(q-1)(n-t)$ with these eigenvalues). If an equitable partition $(C_0,\ldots,C_{k-1})$ avoids all these eigenvalues except $n(q-1)$, then by Proposition 1.26 it is orthogonal to the distance partition with respect to S, the density of each cell C_i in S coincides with its density in $H(n,q)$, and it is an orthogonal array of strength t, by the definition.*

Remark 1.15. *Proposition 1.26 can be used to prove the nonexistence of some perfect colorings. Indeed, if two perfect colorings are orthogonal, then we can uniquely determine, from the quotient matrices, the number of vertices having ith color in the first coloring and jth color in the second coloring. If this number is not integer, then the existence of a coloring with the first quotient matrix implies the nonexistence for the second one. This principle does work to reject some quotient matrices, see e.g. [94]. Moreover, even its special case, Corollary 1.5, was used [51] to prove the nonexistence of some perfect codes (e.g., one-perfect codes in $H(19,6)$).*

1.2.3.4 Local Weight Distributions in the Cartesian Product of Graphs

The property of equitable partitions in the Cartesian product of two distance-regular graph declared in Proposition 1.27 is called the local distance invariance. It was studied in [54,128,129,131,133] for Hamming graphs and in [68] in general.

Definition 1.48 (bi-weight distribution, local weight distribution). *Let $\Gamma=\Gamma'\Box\Gamma''$ be the Cartesian product of graphs $\Gamma'=(V',E')$ and $\Gamma''=(V'',E'')$ with selected vertices o' and o'', respectively. The* bi-weight distribution *(with respect to (o',o'')) of a set C of vertices of Γ is the collection $(W_{i,j}(C))_{i=0,\ldots,\mathrm{diam}(\Gamma'),\ j=0,\ldots,\mathrm{diam}(\Gamma'')}$, where*

$$W_{i,j}(C)=|\{(x,y)\in C:\ d_{\Gamma'}(x,o')=i,\ d_{\Gamma''}(y,o'')=j\}|.$$

The subcollections $(W_{i,0}(C))_{i=0,\ldots,\mathrm{diam}(\Gamma')}$ and $(W_{0,j}(C))_{j=0,\ldots,\mathrm{diam}(\Gamma'')}$ are called local weight distributions *of C (Γ'-local and Γ''-local, respectively).*

Proposition 1.27. *The bi-weight distributions of the cells of an equitable S-partition $\bar{P}=(P_0,\ldots,P_{k-1})$ of the Cartesian product $\Gamma=\Gamma'\Box\Gamma''$ of two distance-regular graphs $\Gamma'=(V',E')$ and $\Gamma''=(V'',E'')$ (however, Γ is not required to be distance-regular itself) is completely determined by the Γ'-local weight distributions of all cells of $\bar{P}$ and the parameters n, q, m, and S. Moreover, it is sufficient to know only the components up to weight $\min\{\mathrm{diam}(\Gamma'),\mathrm{diam}(\Gamma'')\}$ of the Γ'-local weight distributions, i.e., $(W_{i,0}(P_s))_{i=0,\ldots,\min\{\mathrm{diam}(\Gamma'),\mathrm{diam}(\Gamma'')\}}$, $s=0,\ldots,k-1$.*

There are compact formulas connecting the Γ'- and Γ''-local weight distributions in $H(n,q)$, see [54,133]. General formulas given in [68] are not so nice, involve computations over large matrices, and use the fact that the bi-weight distribution (it can also be generalized to the distribution with respect to the direct product of two perfect colorings, of Γ' and of Γ'') can be represented as a perfect structure.

Using local weight distributions essentially reduced the amount of computation during the enumeration of optimal 1-error-correcting binary codes of lengths 12 and 13 [77] (such codes can be represented as equitable partitions, see Section 1.4.2.1).

1.2.3.5 Interweight Distributions and Strong Distance Invariance

In this section, we consider one of the strongest known invariants for equitable partitions. Unfortunately, it only exists in very special distance-regular graphs.

Definition 1.49. *Let $M = (V, d)$ be a finite metric space with set of distances $d(M)$. Let $\bar{C} = (C_0, \ldots, C_{s-1})$ be tuple of subsets of the vertex set of M, and let o be some vertex in M. The collection $(W^{d_0,d_1,d_2}_{a_1,a_2}(o))_{\substack{d_0,d_1,d_2 \in d(M) \\ a_1,a_2 \in \{0,\ldots,s-1\}}}$ of numbers*

$$W^{d_0,d_1,d_2}_{a_1,a_2}(o) = |\{(x_1, x_2) \in C_{a_1} \times C_{a_2} : d(o, x_1) = d_1,\ d(o, x_2) = d_2,\ d(x_1, x_2) = d_0\}|$$

is called the interweight distribution *of $\bar{C}$ with respect to o.*

Such $\bar{C}$ is called strongly distance invariant *if for every $i \in \{0, \ldots, s-1\}$, the interweight distribution of $\bar{C}$ with respect to $o \in C_i$ does not depend on the choice of o in C_i (but can depend on i); so, in this case, the elements of the distribution will be also denoted by $W^{d_0,d_1,d_2}_{i;a_1,a_2} = W^{d_0,d_1,d_2}_{a_1,a_2}(o)$, $o \in C_i$.*

A family of k-tuples of sets in M is called strongly distance invariant *if the interweight distribution is also independent of the choice of a concrete tuple of sets from this family.*

Proposition 1.28 ([127] for one-perfect codes, [129] in general). *For a given quotient matrix S, the family of all equitable S-partitions of $H(n, 2)$ is strongly distance invariant.*

Different formulas (recursive (see the proposition below), direct formulas, the interweight enumerator in the form of tree-variable polynomial) to calculate the interweight distributions of equitable partitions of $H(n, 2)$ where derived in [70]. Strong distance invariance with those formulas gives a rather strong test for non-existence, exploited in Chapter 7: if for some putative quotient matrix S the formulas give a negative value $W^{d_0,d_1,d_2}_{a_0;a_1,a_2}$ for some a_0, a_1, a_2, d_0, d_1, d_2, then S-perfect colorings do not exist in the hypercube.

Proposition 1.29 (recursive formulas for the interweight distribution in $H(n, 2)$ [70, Theorem 2]). *Let $\bar{C} = (C_0, \ldots, C_{k-1})$ be an equitable partition of $H(n, 2)$ with quotient matrix $S = (S_{i,j})_{i,j=0}^{k-1}$. Then*

$$\sum_{i=0}^{k-1} \mathcal{W}^{r_0,r_1,r_2}_{a_0;a_1,i} \cdot S_{i,a_2} = (r_0 + 1) \cdot \mathcal{W}^{r_0+1,r_1-1,r_2}_{a_0;a_1,a_2} + (r_1 + 1) \cdot \mathcal{W}^{r_0-1,r_1+1,r_2}_{a_0;a_1,a_2}$$
$$+ (n - r_0 - r_1 - r_2 + 1) \cdot \mathcal{W}^{r_0,r_1,r_2-1}_{a_0;a_1,a_2} + \underline{(r_2 + 1) \cdot \mathcal{W}^{r_0,r_1,r_2+1}_{a_0;a_1,a_2}},$$

$$\sum_{i=0}^{k-1} \mathcal{W}^{r_0,r_1,r_2}_{a_0;i,a_2} \cdot S_{i,a_1} = (r_0 + 1) \cdot \mathcal{W}^{r_0+1,r_1,r_2-1}_{a_0;a_1,a_2} + (r_2 + 1) \cdot \mathcal{W}^{r_0-1,r_1,r_2+1}_{a_0;a_1,a_2}$$
$$+ (n - r_0 - r_1 - r_2 + 1) \cdot \mathcal{W}^{r_0,r_1-1,r_2}_{a_0;a_1,a_2} + \underline{(r_1 + 1) \cdot \mathcal{W}^{r_0,r_1+1,r_2}_{a_0;a_1,a_2}},$$

$$\sum_{i=0}^{k-1} \mathcal{W}^{r_0,r_1,r_2}_{i;a_1,a_2} \cdot S_{a_0,i} = (r_1+1)\cdot \mathcal{W}^{r_0,r_1+1,r_2-1}_{a_0;a_1,a_2} + (r_2+1)\cdot \mathcal{W}^{r_0,r_1-1,r_2+1}_{a_0;a_1,a_2}$$
$$+ (n-r_0-r_1-r_2+1)\cdot \mathcal{W}^{r_0-1,r_1,r_2}_{a_0;a_1,a_2} + \underline{(r_0+1)\cdot \mathcal{W}^{r_0+1,r_1,r_2}_{a_0;a_1,a_2}},$$

where $\mathcal{W}^{r_0,r_1,r_2}_{a_0;a_1,a_2} = 0$ *if* $r_0 < 0$, $r_1 < 0$, $r_2 < 0$, *or* $r_0 + r_1 + r_2 > n$ *and* $\mathcal{W}^{r_0,r_1,r_2}_{a_0;a_1,a_2} = W^{r_1+r_2,r_0+r_1,r_0+r_2}_{a_0;a_1,a_2}$ *otherwise.*

We see that the first formula allows to increase r_2 not increasing $r_0 + r_1$; the second formula does the same with r_1; the third, with r_0.

Proposition 1.28 cannot be generalized from hypercubes to the class of all distance-regular graphs (there are counterexamples even in non-binary Hamming graphs, see [70]). However, we can prove this property for some class of equitable partitions of the Johnson graph $J(n,w)$ where w is close to $n/2$.

Theorem 1.4 (strong distance invariance of some combinatorial designs). (i) *If* S *is a quotient matrix with only two eigenvalues* $w(n-w)$ *and* $-w$ *(the maximum and minimum eigenvalues of* $J(n,w)$*),* $w = \lfloor n/2 \rfloor$*, then the family of equitable* S*-partitions of* $J(n,w)$ *is strongly distance invariant.*

(ii) *The family of* $\{w(n-w)-w; w\}$*-CR codes in* $J(n,w)$*,* $w = \lfloor n/2\rfloor - 1$*, is strongly distance invariant.*

The equitable partitions considered in Theorem 1.4 are essentially partitions of the vertex set of $J(n,w)$ into algebraic t-designs, where $t = w-1$. Algebraic t-designs in $J(n,w)$, $w \le n/2$, in their turn, are combinatorial t-designs [31, Theorem 4.7], t-(n,w,λ) designs, see, e.g., [23] for definitions and notations. So, Theorem 1.4 claims that

i. t-$(n,\lfloor n/2\rfloor,\lambda)$ designs, $t = \lfloor n/2\rfloor - 1$, are strongly distance invariant;

ii. t-$(n,\lfloor n/2\rfloor - 1, 1)$ designs, $t = \lfloor n/2\rfloor - 2$ (Steiner systems $S(\lfloor n/2\rfloor - 2, \lfloor n/2\rfloor - 1, n)$), are strongly distance invariant. Unfortunately, the last class of designs is very rare. The largest known case is $n = 10$.

As was established in [65,73], there are Steiner systems $S(k-1,k,n)$ with $k = 4,5,6$ that are not strongly distance invariant.

Proof (a sketch). Denote by $J(n,i,i+1)$ the subgraph of $H(n,2)$ induced by the binary n-words of weights i and $i+1$. Note that both $J(n,i)$ and $J(n,i+1)$ are halved graphs of $J(n,i,i+1)$. Moreover, for odd n and $i = \frac{n-1}{2}$, $J(n,i,i+1)$ is a bipartite antipodal distance-regular (and distance-transitive) graph, called the doubled Odd graph.

(i) Consider an odd n. For an equitable S-partition $\bar{C} = (C_0,\ldots,C_{s-1})$ of $J(n,w)$, denote by $\bar{C}' = (C'_0,\ldots,C'_{s-1})$ the self-antipodal equitable partition of $J(n,w,w+1)$ constructed from $\bar{C}$ as in Proposition 1.33. For $i \in \{0,\ldots,n\}$, we denote by B_i the set vertices of $H(n,2)$ of weight i.

We claim that

$$\bar{B} = (C'_0, \ldots, C'_{s-1}, B_{\frac{n-3}{2}} \cup B_{\frac{n+3}{2}}, B_{\frac{n-5}{2}} \cup B_{\frac{n+5}{2}}, \ldots, B_0 \cup B_n)$$

is an equitable partition of $H(n, 2)$. To show this, it remains to check that for $i = 0, \ldots, s-1$, the vertices from $B_{\frac{n-3}{2}}$ have a constant number of neighbors from C_i. But this is the defining property of an $\frac{n-3}{2}$-$(n, \frac{n-1}{2}, \lambda)$ design (another explanation is that the set of weight-w neighbors of a vertex x of weight $w - 1$ is a Delsarte clique in $J(n, w)$, see Section 1.4.2.2, which are orthogonal to $\bar{C}$ by Propositions 1.45 and 1.26).

Clearly, the quotient matrix of the equitable partition $\bar{B}$ is derived from S. Since the family equitable partitions of $H(n, 2)$ is strongly distance invariant, it remains to note that $W^{d_0,d_1,d_2}_{a_0;a_1,a_2}$ in $\bar{C}$ (with metric in $J(n, w)$) is equal to $W^{2d_0,2d_1,2d_2}_{a_0;a_1,a_2}$ in $\bar{B}$ (with metric in $H(n, 2)$).

For even n, the proof is even easier because we no not need $\bar{C}'$ and construct $\bar{B}$ from the cells of $\bar{C}$ and $B_i \cup B_{n-i}$, $i = \frac{n-2}{2}, \frac{n-4}{2}, \ldots, 0$.

(ii) Consider a $\{w(n-w)-w; w\}$-CR code C in $J(n, w)$, $w = \lfloor n/2 \rfloor - 1$, and the corresponding equitable 2-partition (C, C_1). Consider the partition of the vertex set of the bipartite graph $J(n, w, w+1)$ induced by (C, C_1) as in Definition 1.40; since C is an independent set, this partition has only 4 cells, C, C_1, and two new, say D and D_1. It is not difficult to see that C is a CR-3 code in $J(n, w, w + 1)$ with distance partition (C, D, C_1, D_1); moreover, (D, D_1) is an equitable 2-partition of $J(n, w + 1)$. Now, consider the self-antipodal partition

$$\bar{B} = (C', C'_1, D', D'_1, B_{\frac{n-5}{2}} \cup B_{\frac{n+5}{2}}, B_{\frac{n-7}{2}} \cup B_{\frac{n+7}{2}}, \ldots, B_0 \cup B_n)$$

of the vertex set of $H(n, 2)$ such that $C \subset C'$, $C_1 \subset C'_1$, $D \subset D'$, $D_1 \subset D'_1$. Similarly to the arguments in (i), we find that (D', D'_1) is an equitable partition in the doubled Odd graph $J(n, w+1, w+2)$, the partition $(C, C_1, B_{\frac{n-5}{2}})$ is equitable in $J(w-1, w)$, and finally $\bar{B}$ is equitable in $H(n, 2)$. Then, from the strong distance invariance of equitable partitions of $H(n, 2)$, we derive the strong distance invariance of equitable partitions of $J(n, w)$ with the considered parameters.

For even n, the proof is similar. □

From Proposition 1.28 and Theorem 1.4, we have the following corollaries for graphs related with $H(n, 2)$ and $J(2w + 1, w)$:

Corollary 1.16. *Equitable partitions of the folded hypercube* $\widetilde{H(n, 2)}$ *with given quotient matrix are strongly distance invariant.*

If n *is odd, then equitable partitions of the halved hypercube* $\frac{1}{2}H(n, 2)$ *with given quotient matrix are strongly distance invariant.*

Equitable partitions of the Odd graph O_{w+1} *(doubled Odd graph* DO_{w+1}*) with given quotient matrix having only two eigenvalues* $w+1$ *and* -1 *are strongly distance invariant.*

Proof. Indeed, O_{w+1} is the distance-k graph of $J(2w+1,w)$ and $J(2w+1,w)$ is the distance-2 graph of O_{w+1}; the relations correspondence of the eigenvalues $-w$ in $J(2w+1,w)$ and -1 in O_{w+1} is by straightforward calculations.

For DO_{w+1}, the statement is essentially about the partition $\bar{C}'$ considered in part (i) of the proof of Theorem 1.4.

Every equitable partition of $\widetilde{H(n,2)}$ corresponds to an antipodal equitable partition of its cover $H(n,2)$. Moreover, the preimages of two vertices x, y, $d(x,y)=i$ of $\widetilde{H(n,2)}$ under the covering are four vertices with distances i, i, $n-i$, $n-i$, n, n. This gives the correspondence between the interweight distributions of equitable partitions of $\widetilde{H(n,2)}$ and self-antipodal equitable partitions of $H(n,2)$.

If n is odd, then $\frac{1}{2}H(n,2)$ is the distance-2 graph of $H(n,2)$, while $\widetilde{H(n,2)}$ is the distance-$\lfloor n/2 \rfloor$ graph of $\frac{1}{2}H(n,2)$. □

The eigenvalue -1 plays a special role for CR-1 codes. Independent sets that are CR-1 with eigenvalue -1 are precisely one-perfect codes. Such codes in Odd O_{w+1} and doubled Odd DO_{w+1} graphs are equivalent to Steiner systems $S(w-1,w,2w+1)$, the existence of which is known only for $w=3$ and $w=5$ and open in an infinite number of cases.

1.2.3.6 Uniformly Packed Property of CR Codes

Definition 1.50. *A code C of covering radius ρ in a graph $\Gamma=(V,E)$ is called* uniformly packed in the wide sense *[8] if there are numbers α_0, ..., α_ρ such that for every $x \in V$ it holds*

$$\sum_{i=0}^{\rho} \alpha_i A_i = 1, \tag{1.24}$$

where A_i is the number of codewords at distance i from x.

CR codes in distance-regular graphs are uniformly packed. Moreover, we have the following formula for the packing coefficients α_i (essentially, it was shown in [8]).

Proposition 1.30. *In a distance-regular graph Γ, a CR-ρ code C with quotient matrix S is uniformly packed in the wide sense with coefficients α_i uniquely defined by the equations*

$$\sum_{i=0}^{\rho} \alpha_i S^{(i)}_{j,0} = 1, \quad j=0,\ldots,\rho, \tag{1.25}$$

where $(S^{(i)}_{j,l}) = S^{(i)} = K^{(i)}(S)$ for the degree-i polynomial $K^{(i)}$ such that $\Gamma^{(i)} = K^{(i)}(\Gamma)$.

Proof. By Proposition 1.21, for x at distance j from C the coefficients A_i in (1.24) equal $S^{(i)}_{j,0}$, $i=0,\ldots,\rho$. Hence (1.24) for all x is equivalent to (1.25) for all $j \in \{0,\ldots,\rho\}$. It remains to note that (1.25) is uniquely resolvable because $S^{(i)}_{j,0}=0$ for

$i < j$ and $S_{j,0}^{(j)} \neq 0$ (indeed $S_{j,0}^{(i)}$ is the number of codewords at distance i from a given vertex in $C^{(j)}$). □

The uniformly packed property can be useful in the study and classification of CR codes; however, one should be aware that not all wide-sense uniformly packed codes (even with the same coefficients α_i as for some CR codes) are CR.

1.2.4 Equitable Partitions of Imprimitive Distance-Regular Graphs

A graph Γ is called *imprimitive* if for some $i \in \{1, \ldots, \text{diam}(\Gamma)\}$ the graph $\Gamma^{(i)}$ is not connected. It is known (see, e.g., [13, Theorem 4.2.1]) that an imprimitive distance-regular graph of degree at least 3 is bipartite or antipodal (or both).

1.2.4.1 Antipodality of Equitable Partitions

If f is a perfect coloring of an antipodal distance-regular graph Γ, then by Proposition 1.21 the color $f(x)$ uniquely determines the f-spectrum $\bar{f}(X)$ of the antipodal class X containing x.

Corollary 1.17. *Assume that f is a perfect coloring of an antipodal distance-regular graph Γ. Then the f-spectra $\bar{f}(X)$, $\bar{f}(Y)$ of any two antipodal classes X, Y are either equal (as multisets) or disjoint. Moreover, the coloring $\bar{f}$ of $\widetilde{\Gamma}$ (folded Γ) into the f-spectra of the corresponding antipodal classes of Γ is perfect.*

An easy example is the characterization of perfect coloring of a regular k-partite graph $K_{q,\ldots,q}$, where every coloring is perfect if and only if the spectra of any two parts are either equal or disjoint.

The case of CR codes in 2-antipodal graphs is analyzed in [62, Lemma 2.8]. Using Proposition 1.22, we can generalize Corollary 1.17 as follows:

Proposition 1.31. *Assume that f is a perfect coloring of a graph Γ and for some polynomial Π, $\Pi(\Gamma)$ is the direct sum of complete graphs. Then the f-spectra $\bar{f}(X)$, $\bar{f}(Y)$ of any two connected components X, Y of $\Pi(\Gamma)$ are either equal or disjoint. Moreover, the coloring $\bar{f}$ of the graph on connected components where the multiplicity of an edge $\{X, Y\}$ equals the number of edges $\{x, y\}$ in Γ such that $x \in X$ and $y \in Y$ is perfect.*

1.2.4.2 Equitable Partitions of Bipartite Graphs and CR Codes in Halved Graphs

We first mention a well-known and obvious fact.

Proposition 1.32. *If $\bar{P} = (P_0, \ldots, P_{k-1})$ is an equitable k-partition of a connected bipartite graph with parts B' and B'', then either $\bar{P}$ is a bipartite partition (Definition 1.16) or $(P_0 \cap B', P_0 \cap B'', \ldots, P_{k-1} \cap B', P_{k-1} \cap B'')$ is a bipartite equitable $2k$-partition.*

It is easy to see that for an amply regular graph Γ with parameters (v, k, λ, μ), we have $\Gamma^{(2)} = (\Gamma^2 - \lambda\Gamma - k)/\mu$. If, additionally, Γ is bipartite, then by Proposition 1.22

the retract of f on each of the parts is a perfect coloring of $\frac{1}{2}\Gamma$ (it has less colors than f if f is a bipartite coloring). If the quotient matrix of the resulting coloring is tridiagonal, then we have a CR code. In particular:

Corollary 1.18. *Assume that C is a CR-ρ code in a bipartite amply regular graph Γ and, moreover, C lies in one of the bipartite parts of Γ. Then C and $C^{(\rho)}$ are CR-$\lfloor \rho \rfloor$ codes in the corresponding $\frac{1}{2}\Gamma$, and $C^{(1)}$ and $C^{(\rho-1)}$ are CR-$\lfloor \rho - 1 \rfloor$ codes.*

However, there are examples of CR codes C in $\frac{1}{2}\Gamma$, where Γ is a distance-regular graph, that cannot be represented as a cell of an equitable partition of the whole Γ.

1.2.4.3 Equitable Partitions of Antipodal Bipartite Distance-Regular Graphs

We consider antipodal bipartite distance-regular graphs of odd diameter, which are necessarily 2-antipodal because two vertices at maximum distance belong to different bipartite parts. In such graphs, perfect colorings of $\frac{1}{2}\Gamma$ are in one-to-one correspondence with self-antipodal perfect colorings of Γ and also with bipartite perfect colorings of Γ (the correspondence between the last two classes is obvious from Corollary 1.17 and Proposition 1.32).

Proposition 1.33. *Let Γ be an antipodal bipartite distance-regular graph of odd diameter with parts V', V'', and for a vertex x of Γ let $d(x)$ denote the vertex at distance* $\mathrm{diam}(\Gamma)$ *from x. A k-coloring f' of V' is perfect in $\frac{1}{2}\Gamma$ if and only if the k-coloring*

$$f(x) = \begin{cases} f'(x) & \text{if } x \in V' \\ f'(d(x)) & \text{if } x \in V'' \end{cases}$$

is perfect in Γ.

Proof. *If* f is a perfect coloring of Γ, then it is a perfect coloring of $\Gamma^{(2)}$, and f' is the retract of this coloring on $\frac{1}{2}\Gamma$, one of the connected components of $\Gamma^{(2)}$. Hence, f' is a perfect coloring of $\frac{1}{2}\Gamma$.

Only if. Assume f' is a perfect coloring of $\frac{1}{2}\Gamma$. Then f is a perfect coloring of the disconnected graph $\Gamma^{(2)}$. Since $\frac{1}{2}\Gamma$ is distance-regular, by Proposition 1.21 f' is a perfect coloring of $(\frac{1}{2}\Gamma)^{(\mathrm{diam}(\frac{1}{2}\Gamma))}$ and f is a perfect coloring of $(\Gamma^{(2)})^{(\mathrm{diam}(\frac{1}{2}\Gamma))}$, which is $\Gamma^{(\mathrm{diam}(\Gamma)-1)}$. It remains to note that $\Gamma^{(\mathrm{diam}(\Gamma)-1)}(x) = \Gamma(d(x))$, and from the self-antipodality of the coloring f we conclude that it is a perfect coloring of Γ. □

It is also easy to see from Proposition 1.21 that the quotient matrices S' and S of f' and f in $\frac{1}{2}\Gamma$ and Γ, respectively, can be represented as polynomials from each other.

1.2.5 Reconstruction from Spheres

Recall that a *sphere* of radius r centered in a vertex x consists of all vertices at distance r from x.

Proposition 1.34 (reconstruction inside a double-sphere). *Let Γ be a distance-regular graph with diameter d and $\Gamma^{(d)} = \Pi(\Gamma)$. Suppose that f_1 and f_2 are eigenfunctions of Γ with eigenvalue θ, $\Pi(\theta) \neq 0$. If, for some vertex o and $t < d/2+2$, we have $f_1(x) = f_2(x)$ for each vertex x at distance t or $t-1$ from o, then we also have $f_1(z) = f_2(z)$ for each vertex z at distance less than $t-1$ from o.*

Proof. Consider vertex z such that $d(z,o) < t-1$. By the triangle inequality, the hypothesis $t < d/2+2$ implies $d(o,\Gamma^{(d)}(z)) \geq t-1$. Consider the function

$$f_{12}(x) = \begin{cases} f_1(x), & \text{if } d(o,x) \leq t \quad \text{(in particular, if } x = z) \\ f_2(x), & \text{if } d(o,x) \geq t-1 \quad \text{(in particular, if } x \in \Gamma^{(d)}(z)). \end{cases}$$

It is easy to see that f_{12} is also a θ-eigenfunction of Γ (indeed, on each radius-1 ball, f_{12} identically coincides with at least one of f_1, f_2). From Proposition 1.22 and the definition of f_{12}, we find

$$\Pi(\theta)f_2(z) = \sum_{x\in\Gamma^{(d)}(z)} f_2(x) = \sum_{x\in\Gamma^{(d)}(z)} f_{12}(x) = \Pi(\theta)f_{12}(z) = \Pi(\theta)f_1(z).$$

Since $\Pi(\theta) \neq 0$, we get $f_1(z) = f_2(z)$. □

Hamming graph $H(n,q)$ has eigenvalues $\theta_k = n(q-1) - qk$, where $k = 0,\dots,n$. Usually, an eigenfunction with eigenvalue θ_k is completely defined by its values on any sphere with radius k. Some additional sufficient conditions are established in [132]. For the case $q = 2$, necessary and sufficient conditions for a θ_k-eigenfunction of $H(n,2)$, $\theta_k = n-2k$, to be reconstructed from its values on a sphere of radius k were found in [130, Theorem 3]; namely, for all $i = 0,\dots,k$, $K^{(i)}(\theta_i) \neq 0$, where $K^{(i)}$ is the (Krawtchouk) polynomial of degree i such that $K^{(i)}(H(n-2(k-i),2)) = H(n-2(k-i),2)^{(i)}$. Anyway, every eigenfunction of $H(n,q)$ with eigenvalue θ_k, $k < n/2$, can be reconstructed from its values on a ball of radius k (see [132, Lemma 5]). Thus, by Proposition 1.34 it is easy to prove that every eigenfunction with eigenvalue θ_k, $k < n/2$, is completely defined by its values on a pair of concentric spheres of radius k and $k-1$. By Proposition 1.18, we can restore a perfect coloring of Hamming graph $H(n,q)$ with known quotient matrix from its values on a sphere or on the pair of concentric spheres if the same sufficient conditions are provided. In the case of the binary Hamming graph $H(n,2)$ and odd n, every perfect coloring with known quotient matrix is restored by its values on a sphere of radius $(n-1)/2$; this corollary of Proposition 1.34 is a simple generalization of the theorem for one-perfect codes in [4].

1.3 BASIC CONSTRUCTIONS OF EQUITABLE PARTITIONS

1.3.1 Weisfeiler–Leman Algorithm

There is a natural algorithm that constructs the coarsest equitable refinement of a given partition of the vertex set of a graph. In the modern literature, this algorithm is often called the Weisfeiler–Leman (WL) algorithm (to be exact, the one-dimensional Weisfeiler–Leman algorithm), after the work [136], although it was independently discovered many times, including earlier works, see e.g. [96]. The algorithm is used in the software that recognizes the isomorphism of graphs, see [91].

We describe the algorithm in terms of colorings. Assume f_0 is a coloring of the vertex set of a connected graph $\Gamma = (V, E)$. The algorithm starts with $i = 1$ and does the following.

i. Stopping condition: If the coloring f_{i-1} is perfect, the algorithm stops and returns f_{i-1}.

ii. Step i: Define the vertex coloring $f_i : x \to (f_{i-1}(x), \bar{f}_{i-1}(\Gamma(x)))$ (recall that, according to Definition 1.16 $\bar{f}_{i-1}(\Gamma(x))$ is the f_{i-1}-spectrum of the multi-neighborhood $\Gamma(x)$ of x (if Γ is simple, the neighborhood of x)).

iii. Increase i by 1 and go to (i).

For simplicity, we can rename the colors of f_i after the ith step, making them consequent nonnegative integers. On the ith step of the algorithm, since f_{i-1} is not perfect, we have $f_{i-1}(x) = f_{i-1}(y)$ and $\bar{f}_{i-1}(\Gamma(x)) \neq \bar{f}_{i-1}(\Gamma(y))$ for some $x, y \in V$, and hence f_i has more colors than f_{i-1}. It follows that the algorithm stops after less than $|V|$ steps.

Proposition 1.35. *(a) The output coloring of the WL algorithm is perfect. Moreover, it is a refinement of the input coloring.*

(b) Every perfect coloring that is refinement of the input coloring is a refinement of the output coloring of the WL algorithm.

Proof. (a) The stopping condition (i) guarantees that the resulting coloring is perfect. Moreover, from (ii) we see that $f_{i-1}(x) \neq f_{i-1}(y)$ implies $f_i(x) \neq f_i(y)$, that is, f_i is a refinement of f_{i-1}. By induction, the resulting coloring is a refinement of f_0.

(b) Assume that g is a perfect coloring and a refinement of f_{i-1}. If for two vertices x and y we have

$$g(x) = g(y),$$

then, since g is a refinement of f_{i-1}, we also have

$$f_{i-1}(x) = f_{i-1}(y). \tag{1.26}$$

Moreover, by the definition of a perfect coloring,

$$\bar{g}(\Gamma(x)) = \bar{g}(\Gamma(y)),$$

which yields

$$\bar{f}_{i-1}(\Gamma(x)) = \bar{f}_{i-1}(\Gamma(y)) \tag{1.27}$$

because $\bar{g}$ is a refinement of $\bar{f}_{i-1}$. From (1.26), (1.27), and (ii), we conclude that

$$f_i(x) = f_i(y),$$

which means that g is a refinement of f_i as well. So, by induction, g is a refinement of the coloring produced at each step of the algorithm. □

Remark 1.16. *Essentially, the WL algorithm finds an A-invariant subspace closed with respect to the Hadamard product, where A is the adjacency matrix. By Proposition 1.19(c), such subspaces are equivalent to equitable partitions. Initially, we have the subspace spanned by the characteristic functions of the cells of the starting coloring. Each step acts on the subspace by A and then finds the Hadamard-product closure of the resulting subspace. The algorithm stops when we get a subspace closed both under the action of A and under the Hadamard product.*

1.3.2 Orbit Partitions

If $\mathcal{A}$ is a subgroup of the automorphism group of a graph $\Gamma = (V, E)$, then the vertex orbits under $\mathcal{A}$ form an equitable partition of Γ. In particular, for a code $C \subset V$, the orbit partition with respect to $\mathrm{Aut}(C)$ is an equitable partition. However, it is not necessarily the coarsest equitable refinement of the partition $(C, V \backslash C)$.

Example 1.3. *The structure of orbit partitions that contain the unique optimal distance-4 code of size* 40 *in $H(10, 2)$ (the Best code) has been studied in [19].*

1.3.3 Polynomials

Proposition 1.22 means that any S-perfect coloring of a graph Γ is also a $\Pi(S)$-perfect coloring of $\Pi(\Gamma)$ for any polynomial Π such that $\Pi(\Gamma)$ is a graph (if $\Pi(\Gamma)$ is not a graph, which means that we get negative multiplicities, then the statement can still be treated in terms of perfect structures over matrices). This fact is valuable even in the case of linear polynomials, because they can be effectively combined with coverings, see the next paragraph.

Another special case is when $\Pi(\Gamma)$ is the distance-i graph $\Gamma^{(i)}$ for some i. In Sections 1.2.4.2 and 1.2.4.3, for bipartite distance-regular Γ, we discuss perfect colorings of its halved graph, $\frac{1}{2}\Gamma$, which is a connected component of $\Gamma^{(2)}$.

1.3.4 Graph Covering and Concatenation of Colorings

Proposition 1.36. (i) *Let ϕ be a covering of a graph Δ by a graph Γ. A function f on the vertex set of Δ is an S-perfect coloring for some matrix S if and only if the composition $f \circ \phi$ is an S-perfect coloring of Γ.*

(ii) *More generally, if there is a T-perfect k-coloring ϕ of a graph Γ and f is an s-valued function on the colors of ϕ, then $f \circ \phi$ is an S-perfect coloring of Γ if and only if $TF = FS$, where F is the incidence matrix of f.*

Proof. Clearly, (i) is a special case of (ii). For the incidence matrix Φ of ϕ, and the adjacency matrix A of Γ, we have $A\Phi = \Phi T$.

If $TF = FS$, then $A\Phi F = \Phi TF = \Phi FS$ and ΦF is a perfect structure over Γ. Since both Φ and F are coloring matrices, ΦF is also a coloring matrix and the incidence matrix of some S-perfect coloring of Γ. On the other hand, ΦF is the incidence matrix of $f \circ \phi$.

Inversely, if $f \circ \phi$ is an S-perfect coloring, then $A\Phi F = \Phi F S$. Since $A\Phi = \Phi T$, we have $\Phi T F = \Phi F S$ and then $\Phi^{\mathrm{T}}\Phi T F = \Phi^{\mathrm{T}}\Phi F S$. Since Φ is a coloring matrix, the square matrix $\Phi^{\mathrm{T}}\Phi$ is full-rank, and we get $TF = FS$. □

Proposition 1.36(i) is a powerful way to construct perfect colorings and CR codes. Many graphs that are Cayley graphs of vector spaces over $\mathbb{F}_q$ are covered by $H(n, q)$ for an appropriate n. Actually, from Lemma 1.1, it is easy to see the following.

Lemma 1.9. *Let U be the vector space over $\mathbb{F}_q$ and S be a multiset of elements of U such that $S = \mathbb{F}_q^* \cdot S_0$, for some multiset S_0, where $\mathbb{F}_q^*$ is the set of all nonzero elements of $\mathbb{F}_q$. Then there is a covering from $H(|S_0|, q)$ to* $\mathrm{Cay}(U^+, S)$.

For example, for every prime power q and positive integer s, m, n, there is the following sequence of linear coverings:

$$H(n''', q) \to H(n'', q^s) \to H(n', q^{ms}) \to B_{q^s}(m, n),$$

$n''' = \frac{q^s-1}{q-1} \cdot n''$, $n'' = \frac{q^{sm}-1}{q^s-1} \cdot n'$, $n' = \frac{q^{sn}-1}{q^s-1}$, and even a singleton in the bilinear forms graph $B_{q^s}(m, n)$ results in nontrivial CR codes with the same intersection array in Hamming graphs. The covering $H(n''', q) \to H(n'', q^s)$ explains a known concatenation construction [140] of q-ary one-perfect codes from q^s-ary one-perfect codes and shows that this concatenation works for CR codes and equitable partitions as well.

Proposition 1.36 in combination with Section 1.3.3 has nontrivial consequences even for linear polynomials P. For example, if we have an S-perfect coloring of the Hamming graph $H(n, q)$, then it is a $(kS + t(q-1)I)$-perfect coloring of the graph $kH(n, q) + t(q-1)$, obtained from $H(n, q)$ by setting multiplicity k to every edge and adding $t(q-1)$ loops to every vertex. Since $kH(n, q) + t(q-1)$ is covered by $H(kn + t, q)$, we have a perfect $(kS + t(q-1)I)$-coloring of $H(kn + t, q)$.

1.3.5 Kramer–Mesner Method

The special case of use of Proposition 1.36(ii) when ϕ is an orbit partition is known as the Kramer–Mesner [64] method in the theory of combinatorial designs. Although we do not claim that all applications of this method can be treated in terms of equitable partitions, for many kinds of designs it is so. Assume that we want to find an equitable S-partition of a graph Γ with a prescribed automorphism group $\mathcal{A}$ (as a subgroup of the full automorphism group of the partition). The cells of such a partition should consist of vertex orbits under $\mathcal{A}$. So, we consider first the partition of the vertex set into orbits, which is equitable, and find the corresponding quotient matrix T. Then, by Proposition 1.36(ii), we need to find a coloring of orbits whose incidence matrix F satisfies the equation $TF = FS$. Clearly, the matrix T is smaller than the original adjacency matrix A of Γ, and sometimes a solution can be found manually, theoretically, or computationally.

1.3.6 Computational Approach

There are many methods of computational search and classification of combinatorial structures [55], many of which are special cases of equitable partitions (see, e.g., Section 1.4). Here, we want to mention only two basic principles, which can be effective in some cases.

First of all, the problem of finding a coloring matrix F satisfying the equation $AF = FS$ ($TF = FS$ if we combine with Proposition 1.36) is a special case of ILP (integer linear programming) problem, with $\{0,1\}$-valued variables, entries of F. It can be solved by ILP solvers for reasonably small F. Often, some additional knowledge or assumptions can reduce the number of variables by assigning constants to some entries of F, for example, if we know that the required perfect coloring is a refinement of some concrete coloring.

The second approach can be called "local reconstruction". Assume that we know or guess the retract of a required putative partition $(C_0, \dots, C_k)$ on the ball $B_r(o) = \sum_{l=0}^{r} \Gamma^{(r)}(o)$ of radius r centered in some vertex o. The retract of the ith cell of the required partition on the ball $B_{r+1}(o)$ of larger radius $r+1$ is a solution of an instance of the exact cover problem with multiplicities of covering. Indeed, for each vertex $x \in \Gamma^{(r)}(o)$, say $x \in C_j$, we know how many neighbors it has in $B_r(o) \cap C_i$ (namely $|\Gamma(x) \cap B_r(o) \cap C_i|$) and how many neighbors in $B_{r+1}(o) \cap C_i$ it needs (equals the corresponding entry $S_{j,i}$ of the quotient matrix). So additionally it needs to collect $s_x = S_{j,i} - |\Gamma(x) \cap B_r(o) \cap C_i|$ neighbors in C_i at distance $r+1$ from o. To satisfy this condition for all $x \in \Gamma^{(r)}(o)$, we construct an $|\Gamma^{(r+1)}(o)| \times |\Gamma^{(r)}(o)|$ *covering* $\{0,1\}$-matrix M indexed by vertices in $\Gamma^{(r+1)}(o)$ (rows) and $\Gamma^{(r)}(o)$ (columns) where $M_{y,x} = 1$ if and only if $\{x,y\}$ is an edge. The set $C_i \cap \Gamma^{(r+1)}(o)$ corresponds to a set of rows of this matrix with sum equal to the vector $(s_x)_{x \in \Gamma^{(r)}(o)}$. All such sets can be found using an exact covering [56] or integer programming solver. After finding all possible continuations of the retract partition onto the ball $B_{r+1}(o)$, we can keep only nonequivalent solutions and continue to the next radius. More details can be found in one of the latest works [74] on this topic.

1.4 OPTIMAL STRUCTURES INDUCE PERFECT COLORINGS

In this section, we consider classes of combinatorial objects that are equivalent to perfect colorings, i.e. they can be alternatively defined as a cell of an equitable partition of special parameters. Most of such relations are like this: *the numeric characteristic [A] of a set attains the bound [B] if and only if this set is the first cell of an equitable partition with quotient matrix [S].*

1.4.1 Recognizing Equitable Partitions by Distance Distribution

The following criterion, in terms of equality in the bound (1.28), helps to recognize an equitable partition if we are given some elements of the distance distribution of a partition, namely, the number of edges between each two cells and the number of pairs of vertices at mutual distance 2 in each cell. This criterion was used in the proof of some results discussed in Section 1.4.2.1.

Proposition 1.37 ([69, Lemma 3]). *Let $\Gamma = (V, E)$ be an amply regular graph with parameters (v, k, λ, μ) (Definition 1.23). Let $\overline{C} = (C_0, \ldots, C_{s-1})$ be a partition of V with distance distribution $((\overline{A}_l^{ij})_{i,j=0}^{s-1})_{l=0,1,\ldots}$, where $\overline{A}_l^{ij}$ denotes the average, over x in C_i, number of C_j-vertices at distance l from x. Then for each i in $\{0, \ldots, s-1\}$, the numbers $\overline{A}_1^{ji}$, $j = 0, \ldots, s-1$, and $\overline{A}_2^{ii}$ satisfy*

$$|C_i|(\lambda\overline{A}_1^{ii} + \mu\overline{A}_2^{ii}) \geq \sum_{j=0}^{s-1} |C_j| \cdot \overline{A}_1^{ji}(\overline{A}_1^{ji} - 1). \tag{1.28}$$

Moreover, the following three assertions are equivalent:

a. *the partition $\overline{C}$ is equitable.*

b. *for all i, (1.28) holds with equality;*

c. *there is at least one equitable partition of Γ with the same numbers $\overline{A}_1^{ij}$ and $\overline{A}_2^{ii}$, $i, j = 0, \ldots, s-1$, in the distance distribution.*

The proof is based on the Cauchy–Bunyakovsky–Schwarz inequality.

1.4.2 Error-Correcting Codes

We consider some infinite classes of unrestricted codes attaining some bound that can be equivalently defined as classes of perfect colorings with some special quotient matrix.

The most famous such class is the class of perfect codes. A code C is r-perfect if every radius-r ball intersects with C in exactly one element. In a graph with distance regularity of radius at least r (for $r = 1$, just a regular graph, for $r = 2$, amply regular graph, ...), including distance-regular graphs, r-perfect codes are necessarily CR.

In distance-regular graphs, perfect codes belong to a wider class of codes whose code parameters yield the complete regularity by the following test of Delsarte:

Proposition 1.38 (Theorem 5.13 in [31]). *Let C be a distance-d code in a distance-regular graph, and let R be the external distance of C (Definition 1.37). If*

$$d \in \{2R - 1, 2R, 2R + 1\}, \tag{1.29}$$

then C is a CR code.

The classes of codes whose code parameters yield (1.29) include:

- perfect, extended perfect, nearly perfect binary (Preparata-like, shortened one-perfect) codes; note that extended nearly perfect binary codes (extended Preparata-like codes, extended shortened one-perfect codes) are CR but do not satisfy (1.29) in general.

- Diameter-perfect distance-2 and distance-3 codes with respect to CR anti-codes:
  - in $H(n,q)$: distance-2 and distance-3 MDS codes;
  - in $B_q(m,n)$ (bilinear form graph, rank-metric space): distance-2 and distance-3 MRD codes;
  - in $J(n,k)$ (Johnson graph): Steiner $\mathrm{S}(k-1,k,n)$ and $\mathrm{S}(k-2,k,n)$ systems;
  - in $J_q(n,k)$ (q-ary Johnson, or Grassmann, graph): Steiner $\mathrm{S}_q(k-1,k,n)$ and $\mathrm{S}_q(k-2,k,n)$ subspace systems.
- in $H(q(q^m-1)/(q-1),q)$: shortened-one-perfect-like codes (with parameters of shortened one-perfect codes, but not necessarily shortened one-perfect [107]).

1.4.2.1 Linear-Programming Bound

In [10], the following bound was proved for a binary 1-error-correcting code C of length n:

$$|C| \le \frac{2^n}{n+1+i}, \qquad n \equiv 3-i \bmod 4, \quad i=0,1,2,3. \tag{1.30}$$

For $i=0$ and $i=1$, the bound is a special case of the Hamming and Johnson bounds, respectively, and the codes attaining the bound are necessarily CR (perfect or nearly perfect). The cases $i=2$ and $i=3$ are more interesting.

Proposition 1.39 (on doubly-shortened-one-perfect-like codes [67]). *A code C in $H(n,2)$ attains bound (1.30) with $i=2$ (i.e., C has cardinality $2^n/(n+3)$ and distance at least 3) if and only if C (its extension in $H(n',2)$, $n'=n+1$), is the first cell of an equitable S-partition (respectively, S'-partition), where*

$$S=\begin{pmatrix}0&1&n-1&0\\1&0&n-1&0\\1&1&n-4&2\\0&0&n-1&1\end{pmatrix},\quad S^+=\begin{pmatrix}0&1&n-1&0&0\\1&0&n-1&0&0\\1&1&n-4&1&1\\0&0&n-1&0&1\\0&0&n-1&1&0\end{pmatrix},\quad S'=\begin{pmatrix}0&0&n'&0&0\\0&0&n'&0&0\\1&1&0&n'-2&0\\0&0&n'-2&0&2\\0&0&0&n'&0\end{pmatrix}.$$

(Remark: C is a doubly-shortened one-perfect code if and only if the corresponding equitable S-partition can be refined to an equitable S^+-partition, which is not always possible [77]).

Proposition 1.40 (on extended triply-shortened-one-perfect-like codes [69]). *A code C in $H(n,2)$ attains bound (1.30) with $i=3$ if and only if its extension in $H(n',2)$, $n'=n+1$, is the first cell of an equitable S'-partition, where*

$$S' = \begin{pmatrix} 0 & 0 & 0 & 1 & n'-1 & 0 \\ 0 & 0 & 0 & 1 & n'-4 & 3 \\ 0 & 0 & 0 & 0 & n'-1 & 1 \\ 1 & n'-1 & 0 & 0 & 0 & 0 \\ 1 & n'-4 & 3 & 0 & 0 & 0 \\ 0 & n'-1 & 1 & 0 & 0 & 0 \end{pmatrix}.$$

Without the extension, we can guarantee that C (attaining (1.30) with $i = 3$) is a cell of an equitable partition only if it is antipodal; in general, however, we can still guarantee that C induces some generalized equitable partition.

Proposition 1.41 (on triply-shortened-one-perfect-like codes [69]). *A code C in $H(n, 2)$ attains bound (1.30) with $i = 3$ (i.e., C has cardinality $2^n/(n + 4)$ and distance at least 3) if and only if it is the first cell of a 3-partition (C, D, E) of the vertex set of $H(n, 2)$ such that $(C, D, E, \overline{C}, \overline{D}, \overline{E})$, where $\overline{Y} = Y + 11\ldots1$, is an equitable family (Definition 1.20) with quotient matrix S',*

$$S' = \begin{pmatrix} 0 & n & 0 & 0 & 0 & 0 \\ 1 & n-4 & 3 & 0 & 0 & 0 \\ 0 & n-2 & 2 & 0 & 2 & -2 \\ 0 & 0 & 0 & 0 & n & 0 \\ 0 & 0 & 0 & 1 & n-4 & 3 \\ 0 & 2 & -2 & 0 & n-2 & 2 \end{pmatrix}, \quad S = \begin{pmatrix} 0 & n & 0 \\ 1 & n-4 & 3 \\ 0 & n & 0 \end{pmatrix}.$$

For codes C in $H(n, 2)$ with additional antipodal property $C = C + 11..1$, attaining bound (1.30) with $i = 3$ is equivalent to C being CR with quotient matrix S.

Since antipodal codes in $H(n, 2)$ are equivalent to codes in the folded n-cube $\overline{H(n, 2)}$, the last claim of Proposition 1.41 implies the following.

Corollary 1.19. *A code C in the folded n-cube $\overline{H(n, 2)}$ is a code of cardinality $2^{n-1}/(n + 4)$ with distance at least 3 if and only if C is a $\left(\begin{smallmatrix} 0 & n & 0 \\ 1 & n-4 & 3 \\ 0 & n & 0 \end{smallmatrix}\right)$-CR code.*

We illustrate the technique of the proof of the results in this section by proving a new similar fact.

Proposition 1.42. *Let C be an even-weight code C in $H(n, 2)$, $n \geq 3$, such that every odd-weight vertex of $H(n, 2)$ is adjacent to at most two codewords. If $|C| = \frac{2^n}{n+1}$, then C is a cell of an equitable partition with quotient matrix $\left(\begin{smallmatrix} 0 & 0 & 1 & n-1 \\ 0 & 0 & 2 & n-2 \\ 1 & n-1 & 0 & 0 \\ 2 & n-2 & 0 & 0 \end{smallmatrix}\right)$.*

Proof (a sketch). Assume C, $|C| = \frac{2^n}{n+1}$, satisfies the hypothesis. Since $|C|$ is integer, $n + 1 = 2^k$ for some integer $k \geq 2$. We define the partition (C, C', D, D') of the vertices of $H(n, 2)$ as follows: $C \cup C'$ is the set of all even-weight vertices, $D = C + \overline{1}$, $D' = C' + \overline{1}$ (note that n is odd).

The code C attains the bound [79, Theorem 5(c)] on the cardinality of a two-fold packing of radius-1 balls in $H(n,2)$. By [79, Corollary 2], the weight distribution with respect to a codeword x in C is uniquely determined by n and does not depend on the concrete code C or codeword x. From the definitions of C', D, and D', we see that their weight distributions with respect to $x \in C$ are also uniquely determined. The same is true for $x \in D$, because D is the antipode of C.

For each l, we know the number of elements of C' at distance l from a codeword $x \in C$. Hence, we know the number of pairs (x,y), $x \in C$, $y \in C'$, at distance l from each other, which means that we know the average number of elements of C at distance l from an element y of C'. Similarly, for the considered 4-partition, we know all $\overline{A}_l^{ij}$ from Proposition 1.37. By p.(c) of that proposition, it remains to show the existence of an equitable partition with the same average weight distribution. Such partition can be constructed by taking the even-weight and odd-weight subcodes of an $\{n-1;2\}$-CR code in $H(n,2)$, $n = 2^k - 1$, for the roles of C and D, respectively. □

1.4.2.2 Cliques and Cocliques as Distance-2 Codes

Independent sets (cocliques) in graphs can be considered as distance-2 codes. The well-known Delsarte–Hoffman upper bound [53] (see also [13, Proposition 1.3.2(i)]) on the size of a maximum independent set in a regular graph is the special case $l = 0$ of the following bound.

Proposition 1.43 (generalized independent sets [102]). *Suppose that $\Gamma = (V,E)$ is an k-regular graph with minimum eigenvalue (of the adjacency matrix) $\theta_{\min}$. If every vertex in a subset C of V is adjacent to at most ℓ, $\ell < k$, vertices of C, then*

$$\frac{|C|}{|V|} \le \frac{\ell - \theta_{\min}}{k - \theta_{\min}}. \tag{1.31}$$

A vertex set C attains (1.31) *if and only if it is* $\begin{pmatrix} \ell & k-\ell \\ \ell - \theta_{\min} & k - \ell + \theta_{\min} \end{pmatrix}$*-CR.*

So, in regular graphs, CR-1 codes with eigenvalue $\theta_{\min}$ are exactly the sets with inner degree not exceeding l whose cardinality attains upper bound (1.31).

Cliques are diameter-1 sets in a graph, or, equivalently, independent sets (codes with distance at least 2) in the complement graph. If a graph is distance-regular or vertex- and edge-transitive, then every clique C satisfies the Delsarte bound.

Proposition 1.44 (see e.g. [40, Lemma 7.1]). *The size of every clique C in a distance-regular graph with degree k, diameter D, and minimum eigenvalue θ_D satisfies the bound*

$$|C| \le 1 - \frac{k}{\theta_D}. \tag{1.32}$$

Moreover, every clique attaining this bound is a CR code with covering radius $D-1$ and all eigenvalues of Γ except θ_D.

Proof. Let Γ be a distance-regular graph of order n, diameter D, eigenvalues, θ_0, $\ldots$, θ_D, in decreasing order, intersection matrix S, and distance-i matrices $A^{(i)}$, $i = 0, 1, \ldots, D$ ($A^{(i)}$ is the adjacency matrix of $\Gamma^{(i)}$).

Let $v = (v_0, \ldots, v_D)$ be a θ_D-eigenvector of the quotient matrix. Clearly,

$$0 \neq \theta_D v_0 = k v_1. \tag{1.33}$$

Denote $E = \sum_{i=0}^{D} v_i A^{(i)}$ (it is the Dth idempotent of the Bose–Mesner algebra). Straightforwardly, the xth row (similarly, column) $\bar{e}_x$ of E is an eigenvector of the adjacency matrix A, with value $v_{d(x,y)}$ in a position y. Now consider E^2. Since its rows (and columns) are linear combinations of rows of E, they are also eigenvectors of A. On the other hand, E^2 is also a linear combination of A^i, $i = 0, \ldots, D$. Therefore, it can only be cE for some constant c. For convenience, we make $c = 1$ by scaling the eigenvector v. So, $E^2 = E$.

Since $\langle \bar{e}_x, \bar{e}_y \rangle$ is the (x, y)th element of E^2, and $E^2 = E$, it follows that

$$\langle \bar{e}_x, \bar{e}_y \rangle = v_{d(x,y)}. \tag{1.34}$$

Let C be a clique in Γ. From (1.34), we have

$$0 \leq \Big\langle \sum_{x \in C} \bar{e}_x, \sum_{x \in C} \bar{e}_x \Big\rangle = \sum_{x,y \in C} \langle \bar{e}_x, \bar{e}_y \rangle = |C| \cdot (v_0 + (|C| - 1) \cdot v_1)$$

and hence $v_0 + (|C| - 1) v_1 \geq 0$. Taking into account (1.33), this means $k + (|C| - 1)$ $\theta_D \geq 0$, which is equivalent to (1.32) because θ_D is negative.

□

The cliques attaining bound (1.32) are called *Delsarte cliques.*

Proposition 1.45 (see e.g. [40, Lemma 7.2]). *A Delsarte clique C in a distance-regular graph Γ of diameter D is CR with covering radius $D-1$ and all eigenvalues of the graph except the minimum one.*

Proof. Assume that C meets (1.32) with equality. From the previous proof, we have $\big\langle \sum_{x \in C} \bar{e}_x, \sum_{x \in C} \bar{e}_x \big\rangle = 0$ and hence $\sum_{x \in C} \bar{e}_x = \bar{0}$. In particular, because the yth element of $\bar{e}_x$ is $v_{d(x,y)}$, we have

$$\sum_{x \in C} v_{d(x,y)} = 0. \tag{1.35}$$

for every vertex y. Denote by α_y the number of elements of C at distance $d(y, C)$ from y. The remaining $|C| - \alpha_y$ elements of the clique C are at distance $d(y, C) + 1$ from y, and (1.35) turns to

$$\alpha_y \cdot v_{d(y,C)} + (|C| - \alpha_y) \cdot v_{d(y,C)+1} = 0. \tag{1.36}$$

Obviously, there are vertices y at distance 0, 1, ..., $D-1$ from C. By induction on $d(y,C)$, we see from (1.36) that all v_i, $i=0,\dots,D$, are nonzero. If there is y at distance D from C, then (1.36) turns to the contradiction $|C|\cdot v_D=0$; hence, the covering radius of C is $D-1$. Finally, we see from (1.36) that α_y depends only on $d(y,C)$, which means that C is CR by Delsarte's definition (Corollary 1.4).

It remains to find the D eigenvalues of C. Take x in C. For each eigenvalue θ of Γ, there is an eigenfunction $\bar{e}'_x$ of Γ that equals k in vertex x and equals θ in any vertex adjacent to x (it is constructed similarly to $\bar{e}_x$ in the proof of Proposition 1.44). If it is orthogonal to the characteristic function χ_C of C, then $k+(|C|-1)\cdot\theta=0$. Substituting $|C|=1-\frac{k}{\theta_D}$, we find $\theta=\theta_D$. We see that χ_C is not orthogonal to some eigenfunction for each eigenvalue from θ_0 to θ_{D-1}. □

1.4.2.3 Diameter-Perfect Codes

The concept of diameter-perfect codes is related to the following upper bound on the size of a code, known as the *code–anticode bound.* If S is a finite metric space, then *sometimes* we can claim that the following inequality holds for every code C with minimum distance d and every set (anticode) A of diameter smaller than d:

$$|C|\cdot|A|\le|S|. \tag{1.37}$$

Here, "sometimes" means that the bound is not true for all metric spaces in general, and to validate it we need to require some additional regularity properties from the metric space S. One of known properties that guarantee (1.37) is based of symmetries of S: the code-anticode bound holds if the isometry group $\mathrm{Aut}(S)$ of the space acts transitively on points, that is, if for every two points x and y there is an isometry of S that sends x to y. The proof is based on double-counting the number of pairs (π,c) such that $\pi\in\mathrm{Aut}(S)$, $c\in C$, and $c\in\pi(A)$. Delsarte proved that (1.37) is also true for distance-regular graphs with natural (shortest-path) distance, even without any assumption on the automorphism group of the graph (isometry group of the corresponding metric space).

If the code–anticode bound holds in the given metric space, then every code attaining it (for some anticode of diameter smaller than the code distance) is called *diameter-perfect.* Sometimes, (at least, in the case of the transitive group of isometries of the space), we can say the following:

Proposition 1.46. *Assume* (1.37) *holds with equality for some finite metric space S whose isometry group acts transitively on points, for some code C with minimum distance d, and for some set A of diameter smaller than d. Then for any set A' of diameter smaller than d such that $|A'|=|A|$ it holds $|A'\cap C|=1$.*

Based on the proposition above, we can define diameter-perfect code as a code of distance d that intersects with every maximum set of diameter smaller than d (clearly, the intersection with each such set cannot be larger than one point).

This definition can be applied to some infinite metric spaces as well, and below we consider it in a slightly more general form.

Remark 1.17. *One should be aware of the wrong treatment of the alternative definition of diameter-perfect code above. In general, it is not true that every set that intersects with each maximum set (anticode) of diameter smaller than d in exactly one point is necessarily a distance-d code. The reason is that there can be two vertices at distance smaller than d that do not lie in a maximum anticode. This can even happen in a distance-regular graph. For example, consider the Cartesian product of the Shrikhande graph (defined on $\mathbb{Z}_4^2$, two vertices are adjacent if their difference is in $\{01, 10, 11, 03, 30, 33\}$) and K_4, the complete graph on 4 vertices. This product is the distance-regular graph $D(1, 1)$, a Doob graph. The maximum sets of diameter smaller than 2 are 4-cliques, the graph has exactly 16 such cliques, and each two of them are disjoint. Now, if we take all 16 vertices of some Shrikhande subgraph of $D(1, 1)$ as the codewords, we have exactly 1 codeword in each 4-clique. However, the code distance is 1, not 2.*

Definition 1.51. *Assume we have a metric space S such that every set of a finite diameter D has cardinality bounded by some finite constant (depending on D). We also assume that all distances are integer; in particular the phrase "diameter less than d" is equivalent to "diameter at most $d-1$" and so on. For some positive integer d, let $|A|$ be the maximum cardinality of a set (anticode) of diameter at most $d-1$, and let $\mathcal{A}$ be a nonempty collection of such anticodes of cardinality $|A|$ and such that every point belongs to exactly α anticodes from $\mathcal{A}$ (in this case, we will say that the collection $\mathcal{A}$ is α-*regular *or just* regular*). A distance-d code C in S is called* diameter-perfect with respect to $\mathcal{A}$ *if $|C \cap A| = 1$ for every A in $\mathcal{A}$. A distance-d code C in S is called* diameter-perfect *if it is diameter-perfect with respect to some regular collection of anticodes of diameter at most $(d-1)$.*

Remark 1.18. *We note that it is not required that the collection $\mathcal{A}_{all}$ of all maximum anticodes is regular. For the definition to have sense, it is sufficient if $\mathcal{A}_{all}$ has a regular subcollection $\mathcal{A}$. However, if $\mathcal{A}_{all}$ is regular itself (for example, it is true for every transitive metric space), it is convenient to consider it for the role of $\mathcal{A}$ by default.*

The definition above has some advantages comparing to the known definitions of diameter-perfect codes in infinite metric spaces (graphs). For example, the definition in [138] is based on the concept of tiling, which can only be defined if the underlying set of the metric space forms a group; as a result, that definition is not a generalization of finite diameter-perfect codes, which are defined without such a restriction. If the *density* $\varrho(C)$ of a code C can be properly defined if the given metric space (as the limit of the proportion of codewords in a finite fragment of growing diameter; we do not discuss here the formal definition and conditions of its correctness), then the code–anticode bound can be sometimes proven in the form $\varrho(C) \cdot |A| \le 1$. Diameter-perfect codes from Definition 1.51 will obviously attain this bound, although the inverse will not be true for unrestricted codes because for example deleting one vertex will not change the density but obviously breaks the diameter-perfect property of a code.

Theorem 1.5. *Assume that we have a metric space S with integer-valued metric and $\mathcal{A}$ is an α-regular collection of sets (anticodes) of the same finite cardinality $|A|$ and diameter at most $d-1$. For each two different points x, y in the space, define $\mu(x,y)$ as the number of anticodes from $\mathcal{A}$ that contain both x and y. Then, in the graph $\Gamma_{\mathcal{A}}$ on S where the multiplicity of each pair of vertices x, y in the edge set is $\mu(x,y)$, every diameter-perfect distance-d code C is completely regular with quotient matrix $Q = \begin{pmatrix} 0 & \alpha(|A|-1) \\ \alpha & \alpha(|A|-2) \end{pmatrix}$.*

Proof. We first check that the graph $\Gamma_{\mathcal{A}}$ is regular and determine its degree. The degree of each vertex x equals

$$\sum_{y\in S\setminus\{x\}} \mu(x,y) = \sum_{y\in S\setminus\{x\}} \sum_{A\in\mathcal{A}:x,y\in A} 1 = \sum_{A\in\mathcal{A}:x\in A} \sum_{y\in A\setminus\{x\}} 1 = \sum_{A\in\mathcal{A}:x\in A} (|A|-1) = \alpha(|A|-1).$$

Since the distance between codewords is larger than the diameter of an anticode, any codeword has no codeword neighbors in $\Gamma_{\mathcal{A}}$. So, the first row of the quotient matrix Q is confirmed. Next, consider a non-code vertex x and find the number of its code neighbors (respecting edge multiplicities):

$$\begin{aligned}\sum_{y\in C} \mu(x,y) &= \sum_{y\in C} \sum_{A\in\mathcal{A}:x,y\in A} 1 = \sum_{A\in\mathcal{A}:x\in A} \sum_{y\in C\cap A} 1 \\ &= \sum_{A\in\mathcal{A}:x\in A} |C\cap A| = \sum_{A\in\mathcal{A}:x\in A} 1 = |\{A\in\mathcal{A} : x\in A\}| = \alpha.\end{aligned}$$

This confirms the second row of Q. □

However, we cannot say in general that every Q-CR code in $\Gamma_{\mathcal{A}}$ is a distance-d code.

Proposition 1.47. *If, under hypothesis and notation of Theorem 1.5, every two vertices x, y at distance smaller than d belongs to at least one anticode from $\mathcal{A}$, then every CR code in $\Gamma_{\mathcal{A}}$ with quotient matrix Q is diameter-perfect with distance d.*

Example 1.4. *Consider the infinite square grid graph on $\mathbb{Z}^2$, where two vertices (x,y) and (x',y') are adjacent if and only if $|x-x'|+|y-y'|=1$. Define A to be the union of two radius-2 balls centered in $(0,0)$ and $(1,0)$ (see the left picture below). It is an anticode of diameter 5. If we form $\mathcal{A}$ from all anticodes equivalent to A (by translation and rotation), then it will satisfy the hypothesis of Proposition 1.47 with $\alpha = 36$. An example of distance-6 diameter-perfect code is $a\cdot(3,3)+b\cdot(4,-2)$, $a,b\in\mathbb{Z}$ (see the right picture below).*

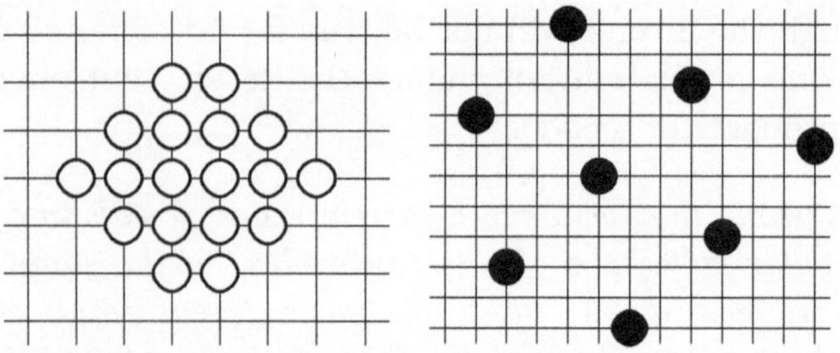

If we include in $\mathcal{A}$ only the translations of A, then the hypothesis of Proposition 1.47 is not satisfied: $(0,0)$ *and* $(0,5)$ *are at distance* 5 *from each other but do not belong to a common anticode. However, it is possible to prove that the conclusion of Proposition 1.47 is still true for this collection of anticodes. Of course, these examples of anticodes and diameter-perfect codes are well known, see, e.g., the recent work [138] on diameter-perfect Lee codes and references there.*

1.4.3 Orthogonal Arrays and Algebraic t-Designs

1.4.3.1 Generalized Bierbrauer–Friedman Bound

Here we prove some general bounds on t-designs in arbitrary regular graphs and show that sets attaining one of these bounds are CR.

Let $\Gamma = (V, E)$ be a graph (multiple edges and loops are allowed). For real-valued functions f, g on V, we denote

$$\langle f, g\rangle = \sum_{x\in V} f(x)g(x) \quad \text{and } |f| = \sum_{x\in V} f(x),$$

in particular, $|f^2| = \sum_{x\in V}(f(x))^2 = \langle f, f\rangle$.

Theorem 1.6. *Let f be a non-negative real-valued t-design in a k-regular graph $\Gamma = (V, E)$ of order N. And let θ be $(t+1)$th largest non-main eigenvalue of Γ. Then*

(a)

$$\frac{k-\theta}{|V|} \cdot |f| \geq -\theta \cdot \frac{|f^2|}{|f|} + \frac{\langle \Gamma(f), f\rangle}{|f|}, \tag{1.38}$$

with equality if and only if f is the sum of a constant and a θ-eigenfunction.

(b) *In particular (replacing the nonnegative value $\langle \Gamma(f), f\rangle$ by* 0*), we have*

$$\frac{|f|^2}{|f^2|} \geq -\frac{\theta}{k-\theta} \cdot |V|, \tag{1.39}$$

with equality if and only if the nonzeros of f are independent in Γ and f is the sum of a constant and a θ-eigenfunction.

(c) *Moreover, if f is the multiplicity function of some multiset C, then*

$$\frac{|C|}{|V|} \geq \frac{\kappa(C)-\theta}{k-\theta}, \tag{1.40}$$

where $\kappa(C)$ is the average number of C-neighbors of a C-vertex (the inner degree of C, respecting the multiplicities of edges and elements of C):

$$\kappa(C) = |C|^{-1} \sum_{c \in C} |\Gamma(c) \cap C| = |f|^{-1} \cdot \langle \Gamma(f), f \rangle;$$

(1.40) holds with equality if and only if the set C is simple (without multiplicities more than 1) and CR-1 with eigenvalue θ.

(d) *In particular,*

$$\frac{|C|}{|V|} \geq \frac{-\theta}{k-\theta}, \tag{1.41}$$

with equality if and only if the set C is simple (without multiplicities more than 1), independent, and CR-1 with eigenvalue θ.

(e) *Moreover, if C is a simple set, then*

$$\frac{|C|}{|V|} \leq \frac{\alpha(C)}{k-\theta}, \tag{1.42}$$

where $\alpha(C)$ is the average number of codeword neighbors of a non-codeword vertex:

$$\alpha(C) = |V \setminus C|^{-1} \sum_{x \in V \setminus C} |\Gamma(x) \cap C|;$$

(1.42) holds with equality if and only if C is a CR-1 set with eigenvalue θ.

Proof. We represent f as the sum

$$f = \phi_0 + \phi_1 + \ldots + \phi_l$$

of eigenfunctions $\phi_0, \phi_1, \ldots, \phi_l$ corresponding to eigenvalues $\lambda_0, \lambda_1, \ldots, \lambda_l$, respectively, where

$$k = \lambda_0 > \theta = \lambda_1 > \ldots > \lambda_l.$$

Recall that

$$\langle \phi_i, \phi_j \rangle = 0 \qquad \text{if } i \neq j. \tag{1.43}$$

Now, we evaluate

$$\begin{aligned}
\langle \Gamma(f), f \rangle &= \Big\langle \Gamma\Big(\sum_{i=0}^{l} \phi_i\Big), \sum_{j=0}^{l} \phi_j \Big\rangle = \Big\langle \sum_{i=0}^{l} \Gamma(\phi_i), \sum_{j=0}^{l} \phi_j \Big\rangle \\
&= \Big\langle \sum_{i=0}^{l} \lambda_i \phi_i, \sum_{j=0}^{l} \phi_j \Big\rangle \overset{(1.43)}{=} \sum_{i=0}^{l} \lambda_i \langle \phi_i, \phi_i \rangle \\
&= k\langle \phi_0, \phi_0 \rangle + \sum_{i=1}^{l} \lambda_i \langle \phi_i, \phi_i \rangle \leq k\langle \phi_0, \phi_0 \rangle + \theta \sum_{i=1}^{l} \langle \phi_i, \phi_i \rangle \\
&= (k-\theta)\langle \phi_0, \phi_0 \rangle + \theta \sum_{i=0}^{l} \langle \phi_i, \phi_i \rangle.
\end{aligned} \tag{1.44}$$

The first term of the last expression equals

$$(k-\theta)\cdot|V|\cdot\left(\frac{|f|}{|V|}\right)^2=(k-\theta)\cdot\frac{|f|^2}{|V|}$$

because ϕ_0 is the constant whose value is the average value of f over V. The second term equals $\theta|f^2|$ by Parseval's identity (Pythagorean theorem). So, we get

$$\langle\Gamma(f),f\rangle\le(k-\theta)\cdot\frac{|f|^2}{|V|}+\theta|f^2|,$$

which proves (1.38). Moreover, the only inequality in (1.44) turns to equality if and only if f is decomposed into the sum of only two eigenfunctions ϕ_0 and ϕ_1. This completes the proof of claim (a).

Next, we have $\langle\Gamma(f),f\rangle\ge 0$ because this scalar product is the sum of non-negative values. Trivially, $\langle\Gamma(f),f\rangle=0$ if and only if the set of non-zeros of f is independent. This proves (b).

Next, if f is the multiplicity function of a multiset C, then f has no values between 0 and 1. It follows that $(f(x))^2\ge f(x)$ for every vertex x, with equality if and only if $f(x)\in\{0,1\}$; integrally, $|f^2|\ge|f|$ with equality if and only if f is a $\{0,1\}$-function, i.e., C is a simple set. Replacing $|f^2|$ by $|f|$ in inequalities (1.38) and (1.39) and noting that trivially $\kappa(C)=\langle\Gamma(f),f\rangle/|f|$, we get (1.40) and (1.41). For claims (c) and (d), it remains to observe that a simple set is CR-1 if and only if its characteristic function is a constant plus some non-constant eigenfunction (Corollary 1.9).

Finally, we see that for a simple vertex set C the number of edges between C and its complement is $|C|\cdot(k-\kappa(C))$, on one hand, and $(|V|-|C|)\cdot\alpha(C)$ on the other. This gives

$$\kappa(C)=\frac{|C|\cdot k-(|V|-|C|)\cdot\alpha(C)}{|C|}=k-\frac{|V|-|C|}{|C|}\cdot\alpha(C)$$

Substituting this expression to (1.40), we get (1.42), after simple transformation. □

Algebraic t-designs of cardinality M in the Hamming graph $H(n,q)$ are known as orthogonal arrays OA(M,n,q,t) (see Proposition 1.9). In this case,

$$k=n(q-1),\quad \theta=k-(t+1)q,\quad |V|=q^n,$$

and (1.41) turns to the Bierbrauer–Friedman bound ([37, Theorem 2.1], $q=2$; [11], any q)

$$|C|\ge q^n\left(1-\frac{(q-1)n}{q(t+1)}\right), \tag{1.45}$$

on the parameters of OA$(|C|,n,q,t)$, while (1.42) turns to the following bound from [100]:

$$\frac{|C|}{q^n}\le\frac{\alpha(C)}{(t+1)q}.$$

It can be observed that (1.39) cannot be attained by a multiset with vertices of different multiplicities. Since the corresponding f has at least three different values and the nonzeros are independent, it is impossible for f to be an eigenfunction+constant. So, (1.39) can be improved for non-simple multisets.

Problem 1.1. *Improve* (1.39) *for multisets with a multiple vertex. Does the following non-simple* $OA(24, 6, 2, 3)$ *attain this hypothetical bound:*

$$\{\ 000000,\ 000000,\ 000111,\ 001011,\ 001101,\ 001110,\\ 010011,\ 010101,\ 010110,\ 011001,\ 011010,\ 011100,\\ 100011,\ 100101,\ 100110,\ 101001,\ 101010,\ 101100,\\ 110001,\ 110010,\ 110100,\ 111000,\ 111111,\ 111111\ \}\ ?$$

As shown in [9], q-ary orthogonal arrays attaining the Bierbrauer–Friedman bound can be constructed as additive codes over $\mathbb{F}_q$, $q = p^s$. In this case, the size of the array is a power of p. In the binary case, there is a construction [35] of orthogonal arrays (in terms of perfect colorings/equitable partitions) that gives orthogonal arrays attaining the Friedman bound of size that is not a power of 2. For example, OA$(1536, 13, 2, 7)$ (see also [72]). However, some putative parameters remain unsolved.

Problem 1.2. *In this direction, the first two open questions are about the existence of equitable* $\begin{pmatrix} 0 & 25 \\ 7 & 18 \end{pmatrix}$*- and* $\begin{pmatrix} 0 & 27 \\ 5 & 22 \end{pmatrix}$*-partitions of* $H(25, 2)$ *and* $H(27, 2)$*, respectively (equivalently, orthogonal arrays* $OA(7 \cdot 2^{20}, 25, 2, 15)$ *and* $OA(5 \cdot 2^{22}, 27, 2, 15)$*).*

Remark 1.19. *The possibility for* Γ *to have multiple edges essentially expands the scope of applicability of Theorem 1.6. For example, in [78], this theorem was applied to specially constructed generalizations of the Hamming graph to prove different variations of the Bierbrauer–Friedman bound, including its version for mixed-level orthogonal arrays.*

1.4.3.2 Improved Friedman Bound for Binary OA of Even Strength

As was noted by Levenshtein (cited in [12] as a private communication), since every OA$(N, n, 2, t)$ with even t is a shortened OA$(2N, n+1, 2, t+1)$, (1.45) implies

$$N \geq 2^n \left(1 - \frac{n+1}{2(t+2)}\right) \tag{1.46}$$

for a binary orthogonal array OA$(N, n, 2, t)$ of even strength t.

Proposition 1.48 ([72, Theorem 1]). *Assume that* C *is an orthogonal array* $OA(N, n, 2, t)$ *of even strength* t *attaining* (1.46)*. If* $C' = C + \bar{1} = \{\bar{c} + \bar{1} : \bar{c} \in C\}$ *and* $C'' = \{0,1\}^n \backslash (C \cup C')$*, then* (C, C', C'') *is an equitable* S*-partition, where*

$$S = \begin{pmatrix} 0 & 2t{-}n{+}2 & 2n{-}2t{-}2 \\ 2t{-}n{+}2 & 0 & 2n{-}2t{-}2 \\ 2t{-}n{+}3 & 2t{-}n{+}3 & 3n{-}4t{-}6 \end{pmatrix} = \begin{pmatrix} 0 & a & n{-}a \\ a & 0 & n{-}a \\ a{+}1 & a{+}1 & n{-}2a{-}2 \end{pmatrix}, \quad a = 2t{-}n{+}2. \tag{1.47}$$

Corollary 1.20. *Binary orthogonal arrays $OA(N, n, 2, t)$ of even strength t attaining bound* (1.46) *are in one-to-one correspondence with S'-CR codes in $H(n, 2)$, where*

$$S' = \begin{pmatrix} a & n{-}a & 0 \\ a{+}1 & n{-}2a{-}2 & a{+}1 \\ 0 & n{-}a & a \end{pmatrix}, \quad a = 2t - n + 2. \tag{1.48}$$

Proof. Each OA satisfying the hypothesis corresponds to an equitable S-partition (C, C', C'') as in Proposition 1.48. We now partition C (similarly, C') into two subsets, C_{ev} and C_{odd}, depending on the parity of the weight of codewords. It is easy to see that $(C_{\mathrm{ev}} \cup C'_{\mathrm{odd}}, C'', C_{\mathrm{odd}} \cup C'_{\mathrm{ev}})$ is an equitable S'-partition. Readily, this correspondence is bijective. □

1.4.3.3 Correlation-Immunity Bound

The next bound (also discussed in Section 1.2.1.2) was proved for correlation immune Boolean functions, essentially simple binary orthogonal arrays; it is very important in the theory of CR-1 codes (perfect 2-colorings) in $H(n, 2)$.

Proposition 1.49 (correlation-immunity bound [36]). *If S is a simple $OA(|S|, n, 2, t)$ and $|S| \notin \{0, 2^{t-1}, 2^t\}$, then $t \le \frac{2n}{3} - 1$. Moreover, a set S of vertices of $H(n, 2)$, $|S| \notin \{0, 2^{t-1}, 2^t\}$, is an $OA(|S|, n, 2, \frac{2n}{3} - 1)$ if and only if it is CR-1 with eigenvalue $-\frac{n}{3}$ (equivalently, $\{b; c\}$-CR for some b and c such that $b + c = \frac{4n}{3}$).*

Remark 1.20. *Khalyavin [58] generalized the bound above to a non-simple binary OA with $0 < |S| < 2^{n-1}$. However, in the unrestricted case, we cannot relate the equality in $t \le \frac{2n}{3} - 1$ with CR-1 codes, because there are non-simple orthogonal arrays (e.g., $OA(24, 6, 2, 3)$ and $OA(1536, 12, 2, 7)$ [74, Section 4.2]) attaining this bound.*

A known open problem is generalizing the correlation-immunity bound to the non-binary case (for perfect 2-colorings in particular, or (simple) orthogonal arrays in general).

Finally, we observe that there is a class of designs in distance regular graphs Γ that, almost immediately from the definition, are orthogonal to the Delsarte cliques of the corresponding graphs. They are algebraic $(\mathrm{diam}(\Gamma) - 1)$-designs and hence are CR with covering radius 1 and the minimum eigenvalue of the graph. Examples: t-$(v, t + 1, \lambda)$-designs in the Johnson graph $J(v, t + 1)$ and their subspace analogs in the Grassmann graph $J_q(v, t+1)$; orthogonal arrays of strength $n - 1$ (including distance-2 MDS codes), Latin and frequency squares, Latin and frequency hypercubes (in Hamming graphs $H(n,q)$); distance-2 MRD codes (in bilinear-forms graphs $B_q(m,n)$); transversals (and multifold transversals = k-plexes) of Latin squares and of systems of mutually orthogonal Latin squares MOLS (in the corresponding Latin-square graph or MOLS graphs); extended one-perfect binary codes (in the halved hypercube $\frac{1}{2}H(n,2)$), ...

1.4.4 Edge Cuts and Similar Questions

Let $\Gamma = (V,E)$ be a graph. For nonempty proper subsets $C_1, C_2 \subset V$, denote by $E(C_1,C_2)$ the multiset

$$\{\{c_1,c_2\} : \ c_1 \in C_1, \ c_2 \in C_2, \ \{c_1,c_2\} \in E\}$$

of edges of Γ that connect C_1 and C_2. In particular, if $c_1, c_2 \in C_1 \cap C_2$, then the multiplicity of $\{c_1,c_2\}$ in $E(C_1,C_2)$ is twice larger than in $E(\Gamma)$. The set $E(C,V\backslash C)$ is called an *edge cut*. Below, we consider a sequence of bounds (in the form of lemmas) on the value $E(C_1,C_2)$ and correspondences (in propositions) between the equality in the bounds and CR-1 codes. Note that in k-regular graphs, CR-1 codes with eigenvalue θ, $\theta \neq k$ are $\{b;c\}$-CR for some b, c such that $b+c=k+\theta$.

Lemma 1.10 ([47]). *If $\Gamma = (V,E)$ is a k-regular graph with the second largest eigenvalue θ_1 and the minimal eigenvalue $\theta_{\min}$, then*

$$\theta_{\min} \cdot |C| + \frac{(k-\theta_{\min}) \cdot |C|^2}{|V|} \ \leq \ |E(C,C)| \ \leq \ \theta_1 \cdot |C| + \frac{(k-\theta_1) \cdot |C|^2}{|V|}. \qquad (1.49)$$

Note that if C is an independent set in Γ, then $|E(C,C)| = 0$ and the left inequality in (1.49) turns to the Delsarte–Hoffman bound on the size of an independent set.

Proposition 1.50 ([101, Corollary 3]). *If, under the hypothesis of Lemma 1.10, the left (right) inequality in* (1.49) *holds with equality, then C is CR-*1 *code with eigenvalue $\theta_{\min}$ (respectively, θ_1).*

Lemma 1.11 ([2,115]). *If $\Gamma = (V,E)$ is a k-regular graph with the second largest eigenvalue θ_1 and the minimum eigenvalue $\theta_{\min}$, then*

$$\frac{(k-\theta_1) \cdot |C| \cdot |V\backslash C|}{|V|} \ \leq \ |E(C,V\backslash C)| \ \leq \ \frac{(k-\theta_{\min}) \cdot |C| \cdot |V\backslash C|}{|V|}. \qquad (1.50)$$

Proposition 1.51 ([43, Theorem 2.4] left, [101, Corollary 2] right). *If, under the hypothesis of Lemma 1.11, the left or right part of* (1.50) *holds with equality, then C is a CR-*1 *code and eigenvalue θ_1 or θ_{min} respectively.*

Lemma 1.12 (expander mixing lemma [32]). *If $\Gamma = (V, E)$ is a connected k-regular graph with the second largest, in absolute value, eigenvalue λ (if Γ is bipartite, then $\lambda = -k$), then for any $C_1, C_2 \subset V$ it holds*

$$\left| |E(C_1, C_2) - \frac{k \cdot |C_1| \cdot |C_2|}{|V|} \right| \leq |\lambda| \sqrt{|C_1| \cdot |C_2| \cdot \left(1 - \frac{|C_1|}{|V|}\right) \cdot \left(1 - \frac{|C_2|}{|V|}\right)}. \quad (1.51)$$

An earlier version of inequality (1.51), proved in [1], does not contain the two factors in the parenthesis and hence is not tight.

Proposition 1.52 ([101, Lemma 1]). *If, under the hypothesis of Lemma 1.12,* (1.51) *holds with equality and, additionally, $-\lambda$ is not an eigenvalue of Γ, then $C_2 \in \{C_1, V \setminus C_1\}$ and C_1 is a CR-1 code with eigenvalue λ.*

1.4.5 Bent Functions

A *bent function* is a function $b : \mathbb{F}_2^n \to \mathbb{F}_2$ such that for every non-zero a in $\mathbb{F}_2^n$ it holds

$$|\{x \in \mathbb{F}_2^n : \ b(x) = b(x + a)\}| = 2^{n-1}.$$

Such functions are known to exist if and only if n is even. They have $2^{n-1} + 2^{\frac{n}{2}-1}$ or $2^{n-1} - 2^{\frac{n}{2}-1}$ ones and are known to be at largest possible distance from all affine functions from $\mathbb{F}_2^n$ to $\mathbb{F}_2$. Bent functions and their generalizations are extensively studied because of their nice properties that can be applied in different applications see e.g. [20].

Proposition 1.53 ([102, Theorem 2]). *Let b be a $\mathbb{F}_2$-valued function on $\mathbb{F}_2^n$ such that $b(\bar{0}) = 1$. Define a function f on 2-dimensional subspaces S whose value is the number of ones of b in S: $f(S) = |\{x \in S : \ b(x) = 1\}|$. The function f is a perfect 4-coloring of $J_2(n, 2)$ with quotient matrix*

$$\begin{pmatrix} 3(2^{n-3} - 2^{\frac{n}{2}-2} - 1) & 3 \cdot 2^{n-2} - 3 & 3(2^{n-3} + 2^{\frac{n}{2}-2}) & 0 \\ 2^{n-2} - 2^{\frac{n}{2}-1} & 5 \cdot 2^{n-3} - 2^{\frac{n}{2}-2} - 5 & 2^{n-1} + 2^{\frac{n}{2}-1} & 2^{n-3} + 2^{\frac{n}{2}-2} - 1 \\ 2^{n-3} - 2^{\frac{n}{2}-2} & 2^{n-1} - 2^{\frac{n}{2}-1} - 1 & 5 \cdot 2^{n-3} + 2^{\frac{n}{2}-2} - 3 & 2^{n-2} + 2^{\frac{n}{2}-1} - 2 \\ 0 & 3(2^{n-3} - 2^{\frac{n}{2}-2}) & 3 \cdot 2^{n-2} & 3(2^{n-3} + 2^{\frac{n}{2}-2} - 2) \end{pmatrix}$$

if and only if b is a bent function with $2^{n-1} + 2^{\frac{n}{2}-1}$ ones.

1.4.6 Strongly Regular Graphs

Proposition 1.54 ([103]). *Let V be a finite set, E the set of all 2-subsets of V, and T the set of 3-subsets of E of form $\{\{a, b\}, \{b, c\}, \{a, c\}\}$, $a, b, c \in V$. Let f be a surjective function from E to $\{0, 1\}$. Denote $E_s = f^{-1}(s)$, $s = 0, 1$. The coloring f of the 3-uniform hypergraph (E, T) is perfect if and only if one of the following assertions takes place:*

i. (V, E_1) *is a strongly regular graph;*

ii. (V, E_1) *is the complete bipartite graph* $K_{1,|V|-1}$ *or its complement.*

Proof (a sketch). The "if" part of the statement is straightforward. Let us prove the "only if" part.

By the hypothesis, f is a perfect coloring of the hypergraph (E, T) and hence the induced coloring of $E \cup T$ is perfect in the incidence graph. Let $S = (S_{i,j})$ be the corresponding quotient matrix. Let a hyperedge in T containing exactly 0, 1, 2, or 3 ones of f have color $\bar{0}$, $\bar{1}$, $\bar{2}$, or $\bar{3}$ in the induced coloring, respectively. If for some i there are no hyperedges of color $\bar{i}$, we define $S_{0,\bar{i}} = S_{1,\bar{i}} = 0$, although neither $S_{0,\bar{i}}$ nor $S_{1,\bar{i}}$ denotes an entry of the quotient matrix S (because $\bar{i}$ is not really a color of the induced coloring).

1. If $\Gamma = (V, E_1)$ is not connected, then $S_{0,\bar{2}} = 0$. It is not difficult to conclude that in this case all connected components of Γ are complete graphs. Further, considering $S_{0,\bar{1}}$, we find that either the components are of the same order (and Γ is strongly regular) or there are only two components. Finally, for two components of different order, considering $S_{1,\bar{3}}$, we see that one of them has order 1 and we have (ii).

2. If $\Gamma = (V, E_1)$ is connected and bipartite, then $S_{0,\bar{2}} > 0$, from which we find that Γ is complete bipartite. Then, considering $S_{0,\bar{0}}$ we see that either the two bipartite parts have equal size (Γ is strongly regular in this case) or one of these sets is a singleton (and we have (ii)).

3. If $\Gamma = (V, E_1)$ is connected and not bipartite, then it must be regular. Indeed, for $\{a, b\} \in E_1$, the sum of degrees of a and b equals $2 + 2S_{1,\bar{3}} + S_{1,\bar{2}}$. It is not difficult to find that different degrees yield the bipartiteness of Γ, contradicting the hypothesis of p. 3. So, Γ is regular, and the definition of a strongly regular graph is automatically satisfied with $\lambda = S_{1,\bar{3}}$ and $\mu = S_{0,\bar{2}}$. □

Specifically, it is easy to relate the parameters of a strongly regular graph and the parameters of the corresponding perfect coloring as follows.

Proposition 1.55. *Let V be a finite set, E the set of all 2-subsets of V, and T the set of 3-subsets of E of form $\{\{a, b\}, \{b, c\}, \{a, c\}\}$, $a, b, c \in V$. For a proper subset E_1 of E, the graph (V, E_1) is an* srg(v, k, λ, μ) *if and only if E_1 is the first cell of an equitable S-partition of the incidence graph of (E, T), where S is obtained from the following matrix by removing the all-zero columns (if any) and the corresponding rows:*

$$\begin{pmatrix} 0 & 0 & 0 & v-2k+\lambda & 2(k-1-\lambda) & \lambda \\ 0 & 0 & v-2-2k+\mu & 2(k-\mu) & \mu & 0 \\ 0 & 3 & 0 & 0 & 0 & 0 \\ 1 & 2 & 0 & 0 & 0 & 0 \\ 2 & 1 & 0 & 0 & 0 & 0 \\ 3 & 0 & 0 & 0 & 0 & 0 \end{pmatrix}.$$

1.4.7 Width and Dual Width

The meaning of "width" in the paper [14] of Brouwer, Godsil, Koolen, and Martin coincides with the meaning of "diameter" in the current survey.

Proposition 1.56 ([14, Theorem 1]). *For a set C of vertices of a distance-regular graph Γ, it holds*

$$\mathrm{diam}(C) \geq \mathrm{diam}(\Gamma) - s^*(C), \tag{1.52}$$

where $s^*(C)$ is the external distance of C (Definition 1.37). Equality holds if and only if C is a CR code of covering radius $\mathrm{diam}(\Gamma) - \mathrm{diam}(C)$.

Note that by Proposition 1.21 the diameter of C is uniquely determined from the intersection arrays of Γ and C.

Bound (1.52) is straightforward from $\rho(C) \leq s^*(C)$ [31, Theorem 5.10] and the trivial $\rho(C) \geq \mathrm{diam}(\Gamma) - \mathrm{diam}(C)$, where $\rho(C)$ is the covering radius of C. If C is CR, then $\rho(C) = s^*(C)$ [108]. However, the inverse is not true, see, e.g., the example following [31, Theorem 6.10]. So, the second claim makes (1.52) a more interesting bound in the context of Section 1.4.

In [14], also a nice dual version of Proposition 1.56 is proved, including the bound

$$\mathrm{diam}^{\perp}(C) \geq \mathrm{diam}(\Gamma) - s(C)$$

for Q-polinomial distance-regular graphs Γ and a description of the sets C attaining it, in terms of cometric association schemes. Here, $s(C)$ is the number of different nonzero distances in C and $\mathrm{diam}^{\perp}(C)$ is the maximum i such that the projection of the characteristic function of C on the ith eigenspace of Γ is nonzero (the ordering of eigenspaces should be Q-polynomial). That description is out of the topic of our survey, but we want to show here one of other results of [14] involving both $\mathrm{diam}(C)$ and $\mathrm{diam}^{\perp}(C)$.

Proposition 1.57 ([14, Theorem 8]). *For a set C of vertices of the Hamming graph $H(D,q)$ or a Johnson graph $J(n,D)$, $2D \leq v$ (in both cases, D is the diameter of the graph) and $\mathrm{diam}^{\perp}(C)$ defined with respect to the natural descending ordering of eigenvalues, the equality $\mathrm{diam}(C) + \mathrm{diam}^{\perp}(C) = D$ holds if and only if C is a CR-$\mathrm{diam}^{\perp}(C)$ code inducing a subgraph isomorphic to $H(D-\delta,q)$ in $H(D,q)$ or $J(n-\delta, D-\delta)$ in $J(n,D)$, where $\delta = D - \mathrm{diam}(C)$.*

Tanaka [113,114] proved similar results for Grassmann graphs, bilinear forms graphs, dual polar graphs, and other families of distance-regular graphs of unbounded diameter and used them to characterize maximum sets of given diameter (Erdős–Ko–Rado theorems) in the first two families of graphs.

Remark 1.21. *The parameter $\mathrm{diam}^{\perp}(C)$ with respect to the natural descending ordering of the eigenvalues coincides with the degree of the characteristic function of C for the classes of graphs, including Hamming, Johnson, and Grassmann graphs, where the degree of functions was defined and studied, see e.g. [28] and references there.*

1.5 CR CODES IN DIFFERENT DISTANCE-REGULAR GRAPHS

In this section, without pretending to be complete or regular, we mention some results on CR codes in different classes of distance-regular graphs. A good survey can also be found in [125, Section 12]. CR codes in Hamming, Johnson, and Grassmann graphs are discussed in Chapters 2 and 3 of this book. In the literature, a lot of attention is paid to the study of CR codes of covering radius 1, named intriguing sets in [29], and also studied under the name perfect 2-colorings (equitable 2-partitions). For intriguing sets in strongly regular graphs, see recent works [16,26,45,112] and references there. Equitable 2-partitions of strongly regular graphs such that each cell induces a strongly regular subgraph (including the cases of complete and edge-free subgraphs) or a divisible-design subgraph are considered in [38,48,105,110].

CR-1 codes in halved hypercubes $\frac{1}{2}H(n,2)$ were studied in [66,76]. The main tools for constructing such codes are retracting CR-1 codes in $H(n,2)$; retracting the distance partition of an even-weight CR-ρ codes for $\rho \leq 4$ (Corollary 1.18); merging disjoint CR-1 codes (especially in [66]); equivalence between linear CR-1 codes in $\frac{1}{2}H(n,2)$ and special difference sets (forming the column set of a check matrix of the code); solving, for small n, the corresponding ILP problem by computer; covering $t^2 \cdot \frac{1}{2}H(n,2) + n\binom{t}{2}$ by $\frac{1}{2}H(tn,2)$; a special splitting construction, a variant of the main construction in [36]. The parameters of CR-1 codes in halved hypercubes $\frac{1}{2}H(n,2)$ are characterized for $n \leq 12$.

CR-1 codes in Latin-square graphs with eigenvalue -3 are essentially transversals of Latin squares and their multifold generalizations, called k-plexes, see the survey [135]. Equitable partitions of the Latin-square graphs of order n^2 without eigenvalue -3 (essentially, partitions into CR-1 codes with eigenvalue $n-3$) were characterized in [7]. A special case of a Latin-square graph is the bilinear forms graph $B_2(2,n)$.

In bilinear forms graphs, similarly to Hamming graphs, linear codes with distance at least 3 are CR-2 if and only if its dual code is two-weight [31] (the set of nonzero weights of codewords has size 2). While projective two-weight codes in $H(n,q)$ are classical objects [17] (a code is projective if its dual is at least distance-3), two-weight codes in bilinear forms graphs have only recently been studied [34,104]. However, nondegenerate $\mathbb{F}_{q^m}$-linear codes in $B_q(m,n)$ considered in [34,104] have dual codes with guaranteed distance at least 2. It needs an additional clarification which of those dual are of distance at least 3 and hence CR-2. We can certainly say that MRD codes with distance 3 in $B_q(m,n)$, $m \geq n$, are CR-2 because they are dual to two-weight MRD codes with minimum distance $n-1$ (another explanation is that the preimage of a distance-3 MRD code under the covering of $B_q(m,n)$ by $H(\frac{q^n-1}{q-1}, q^m)$ via Lemma 1.9 is a distance-3 MDS code, which is in turn CR-2 for the same reasons). Another simple observation about bilinear forms graphs is the following.

Proposition 1.58. *Deleting the last column of each vertex matrix of* $B_q(m, n+1)$ *is a covering of* $q \cdot B_q(m,n) + (q^m - 1)$ *by* $B_q(m, n+1)$. *Similarly, deleting the last row of a matrix is a covering of* $q \cdot B_q(m,n) + (q^n - 1)$ *by* $B_q(m+1, n)$.

Taking into account results in Sections 1.3.3 and 1.3.4, we have the following corollary.

Corollary 1.21. *If there is a* $\{b_0, \ldots, b_{\rho-1}; c_1, \ldots, c_\rho\}$*-CR code in* $B_q(m,n)$, *then there is a* $\{q^{m'+n'}b_0, \ldots, q^{m'+n'}b_{\rho-1}; q^{m'+n'}c_1, \ldots, q^{m'+n'}c_\rho\}$*-CR code in* $B_q(m+m', n+n')$ *for all* $m', n' = 0, 1, 2, \ldots$.

Below, for each of the 13 degree-3 (cubic) distance-regular graphs, we list all intersection arrays for which CR codes exist (for each intersection array B, we first show the cardinalities of the cells of the distance partition with respect to a B-CR code C, in the form $|C|/|C^{(1)}|/\ldots/|C^{(\rho)}|$, then the array B, then the spectrum of the corresponding quotient matrix).

K_4: 2/2, $\{2;2\}$, $[-1,3]$; 1/3, $\{3;1\}$, $[-1,3]$.

$K_{3,3}$: 2/4, $\{2;1\}$, $[0,3]$; 3/3, $\{3;3\}$ $[\pm 3]$; 1/3/2, $\{3,2;1,3\}$, $[0,\pm 3]$.

Cubical graph $H(3,2)$ of order 8:

4/4, $\{1;1\}$, $[1,3]$; 4/4, $\{3;3\}$, $[\pm 3]$;
4/4, $\{2;2\}$, $[-1,3]$; 2/4/2, $\{2,1;1,2\}$, $[\pm 1,3]$;
2/6, $\{3;1\}$, $[-1,3]$; 1/3/3/1, $\{3,2,1;1,2,3\}$, $[\pm 1,\pm 3]$.

Petersen graph of order 10:

5/5, $\{1;1\}$, $[1,3]$; 2/4/4, $\{2,2;1,2\}$, $[1,-2,3]$;
4/6, $\{3;2\}$, $[-2,3]$; 1/3/6, $\{3,2;1,1\}$, $[1,-2,3]$.

Heawood graph of order 14. CR codes in Heawood graph can be illustrated by the corresponding CR codes in the infinite hexagonal grid because the Heawood graph is covered by this grid (the covering is obtained by identifying vertices differing in period $2v + r_{30}(v)$, where v is a period of the grid drown on the plain in a regular way and r_{30} is the 30° rotation), see Figure 1.1. All CR codes in the infinite hexagonal grid are classified in [5].

7/7, $\{3;3\}$, $[\pm 3]$; 2/6/6, $\{3,1;1,1\}$, $[\pm\sqrt{2},3]$;
2/4/8, $\{2,2;1,1\}$, $[\pm\sqrt{2},3]$; 1/3/6/4, $\{3,2,2;1,1,3\}$, $[\pm\sqrt{2},\pm 3]$.

Pappus graph of order 18. CR codes in Pappus graph can be illustrated by the corresponding CR codes in the infinite hexagonal grid because the Pappus graph is covered by this grid (the covering is obtained by identifying vertices differing in period $3v$, where v is a period of the grid drown on the plain in a regular way), see Figure 1.2.

6/12, $\{2;1\}$, $[0,3]$;
9/9, $\{3;3\}$, $[\pm 3]$;
6/6/6, $\{2,1;2,1\}$, $[\pm\sqrt{3},3]$;
3/9/6, $\{3,2;1,3\}$, $[0,\pm 3]$;
2/4/8/4, $\{2,2,1;1,1,2\}$, $[0,\pm\sqrt{3},3]$;
1/3/6/6/2, $\{3,2,2,1;1,1,2,3\}$, $[0,\pm\sqrt{3},\pm 3]$.

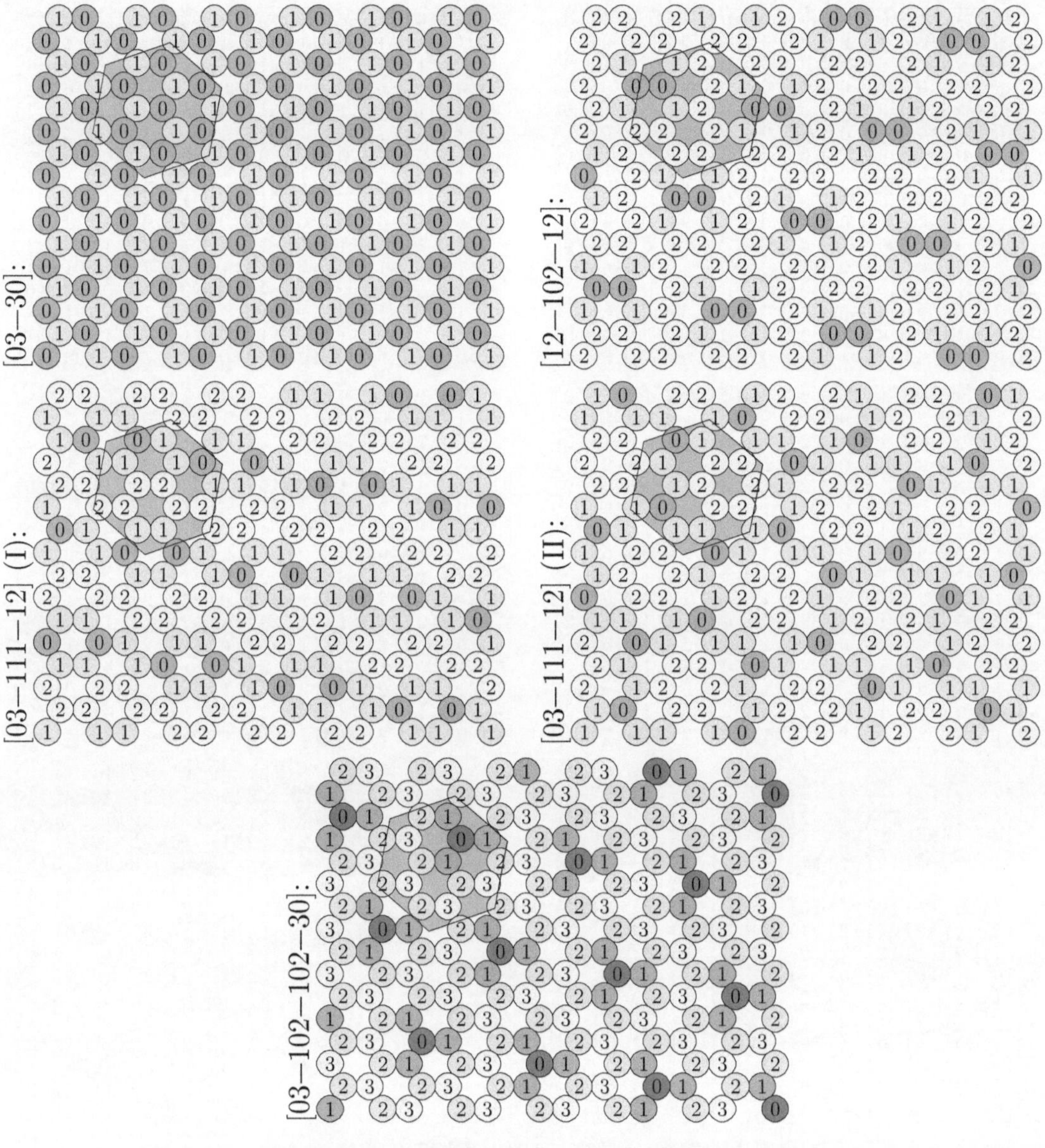

Figure 1.1 CR codes in the infinite hexagonal grid illustrating CR codes in the Heawood graph; note that colorings [03−111−12](I) and [03−111−12](II) of the grid are nonequivalent but the corresponding colorings of the Heawood graph are equivalent; the same story happens with [12−201−12](I,II) in Figure 1.2 (the colorings are named as in [5]).

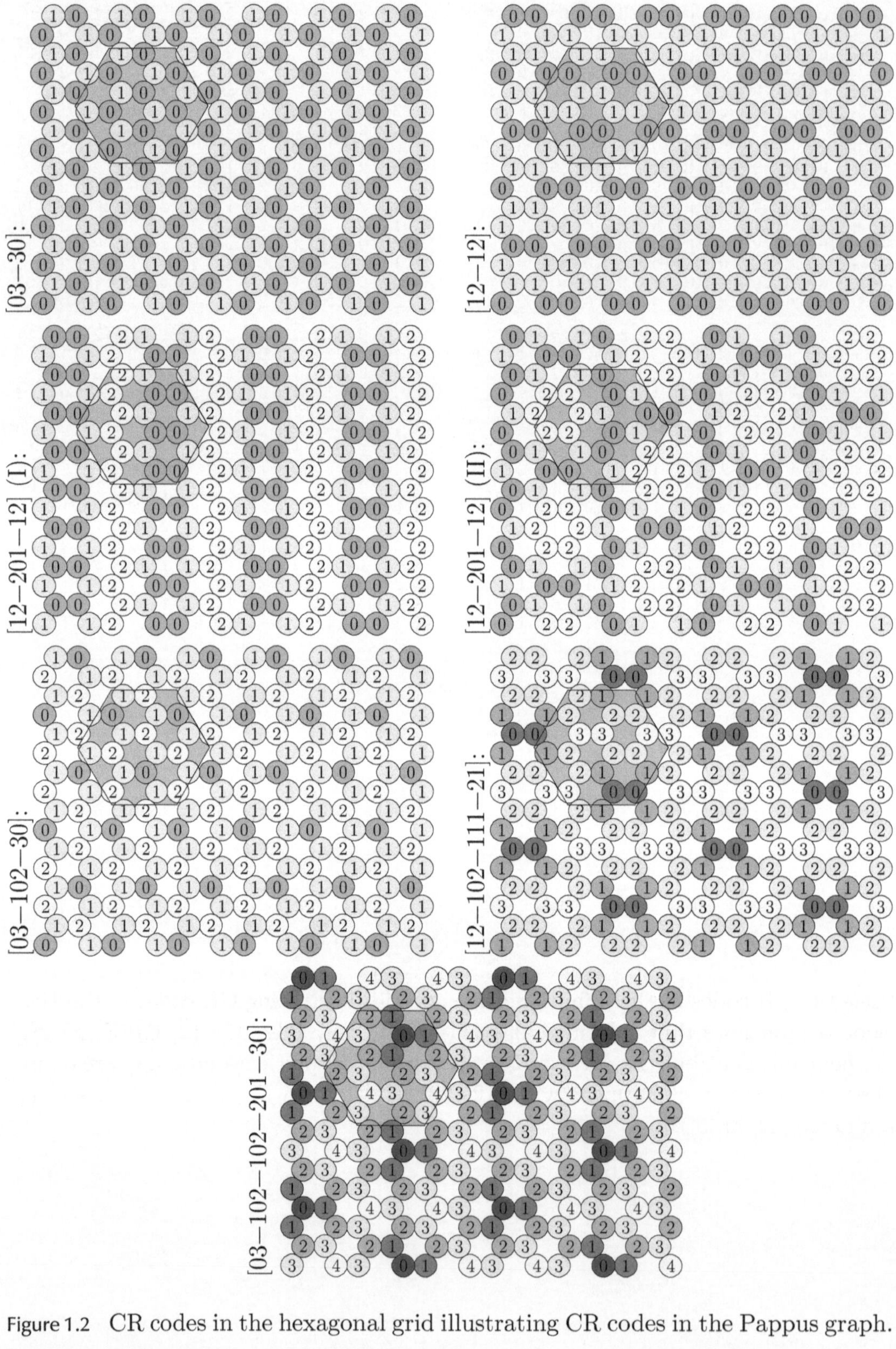

Figure 1.2 CR codes in the hexagonal grid illustrating CR codes in the Pappus graph.

Dodecahedral graph of order 20 (equitable partitions are classified in [57,60]), see Figure 1.3:
$10/10$, $\{1;1\}$, $[1,3]$;
$8/12$, $\{3;2\}$, $[-2,3]$;
$4/8/8$, $\{2,2;1,2\}$, $[1,-2,3]$;
$4/12/4$, $\{3,1;1,3\}$, $[0,-2,3]$;
$2:12$, $\{3,2;1,1\}$, $[1,-2,3]$;
$5/5/5/5$, $\{1,2,1;1,2,1\}$, $[1,\pm\sqrt{5},3]$;
$1/3/6/6/3/1$, $\{3,2,1,1,1;1,1,1,2,3\}$, $[0,1,-2,\pm\sqrt{5},3]$.

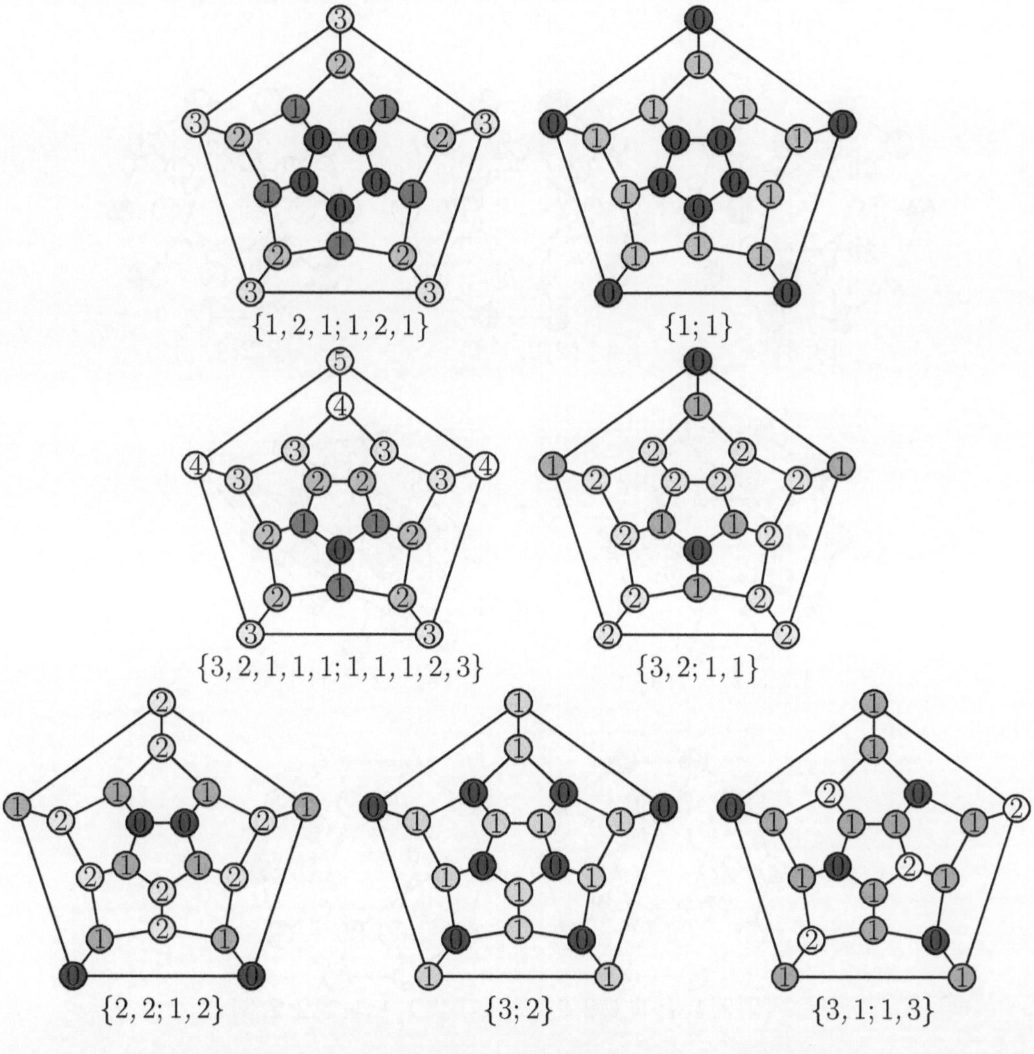

Figure 1.3 Completely regular codes in the Dodecahedral graph.

Desargues graph of order 20, see Figure 1.4:
10/10, $\{1;1\}$, $[1,3]$;
10/10, $\{2;2\}$, $[-1,3]$;
8/12, $\{3;2\}$, $[-2,3]$;
10/10, $\{3;3\}$, $[\pm 3]$;

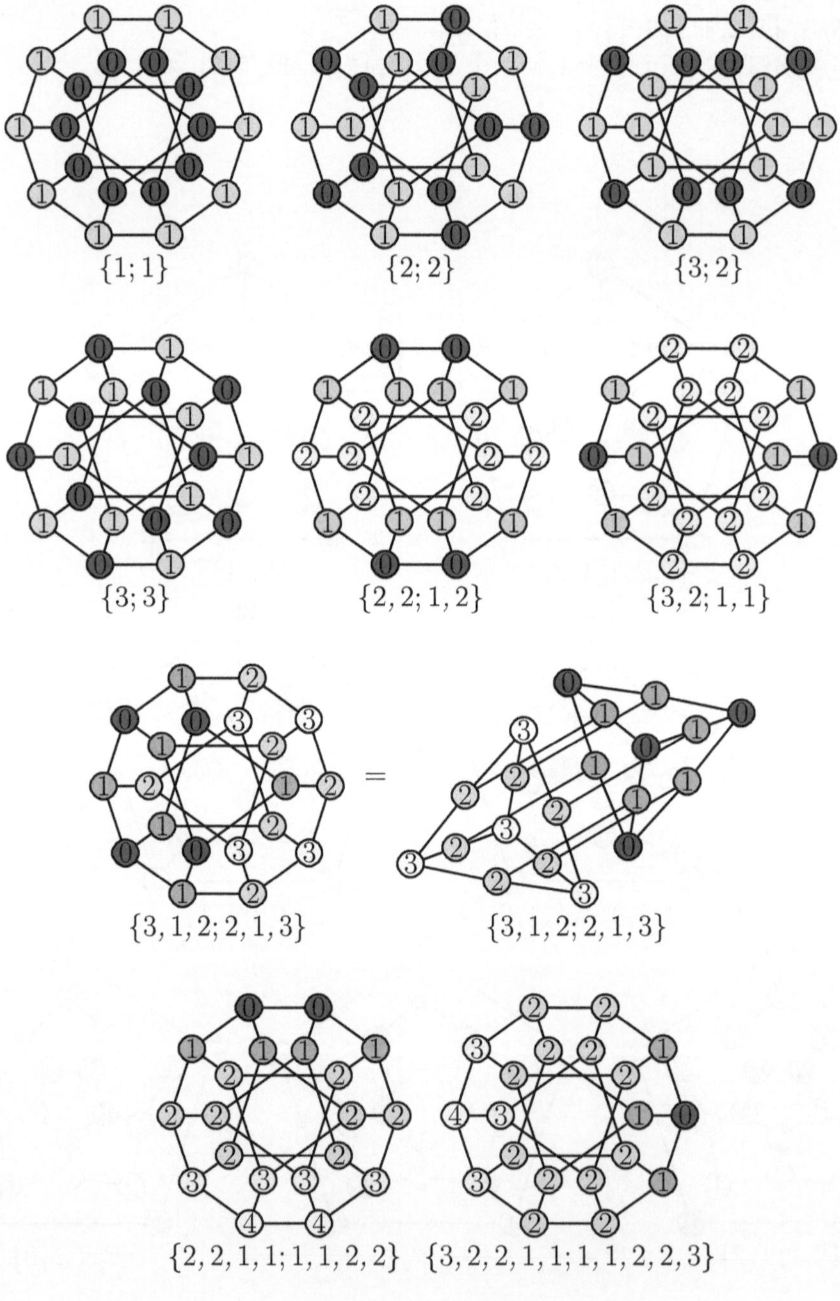

Figure 1.4 Completely regular codes in the Desargues graph.

4/8/8, $\{2,2;1,2\}$, $[1,-2,3]$;
2/6/12, $\{3,2;1,1\}$, $[1,-2,3]$;
4/6/6/4, $\{3,1,2;2,1,3\}$, $[\pm 2,\pm 3]$;
2/4/8/4/2, $\{2,2,1,1;1,1,2,2\}$, $[\pm 1,\pm 2,3]$;
1/3/6/6/3/1, $\{3,2,2,1,1;1,1,2,2,3\}$, $[\pm 1,\pm 2,\pm 3]$.
Coxeter graph of order 28, see Figure 1.5:
7/21, $\{3;1\}$, $[-1,3]$;
7/7/14, $\{1,2;1,1\}$, $[-1,2,3]$;
4/12/12, $\{3,2;1,2\}$, $[-1\pm\sqrt{2},3]$;

Figure 1.5 Completely regular codes in the Coxeter graph (colorings with the same intersection arrays are equivalent).

4/8/8/8, $\{2,2,1;1,2,1\}$, $[-1\pm\sqrt{2},2,3]$;
1/3/6/12/6, $\{3,2,2,1;1,1,1,2\}$, $[-1\pm\sqrt{2},-1,2,3]$.
Tutte–Coxeter graph of order 30, see Figure 1.6:
10/20, $\{2;1\}$, $[0,3]$;
12/18, $\{3;2\}$, $[-2,3]$;
15/15, $\{3;3\}$, $[\pm 3]$;
10/10/10, $\{1,1;1,1\}$, $[0,2,3]$;
6/18/6, $\{3,1;1,3\}$, $[0,-2,3]$;
5/15/10, $\{3,2;1,3\}$, $[0,\pm 3]$;
2/4/8/16, $\{2,2,2;1,1,1\}$, $[0,\pm 2,3]$;
6/9/9/6, $\{3,1,2;2,1,3\}$, $[\pm 2,\pm 3]$;

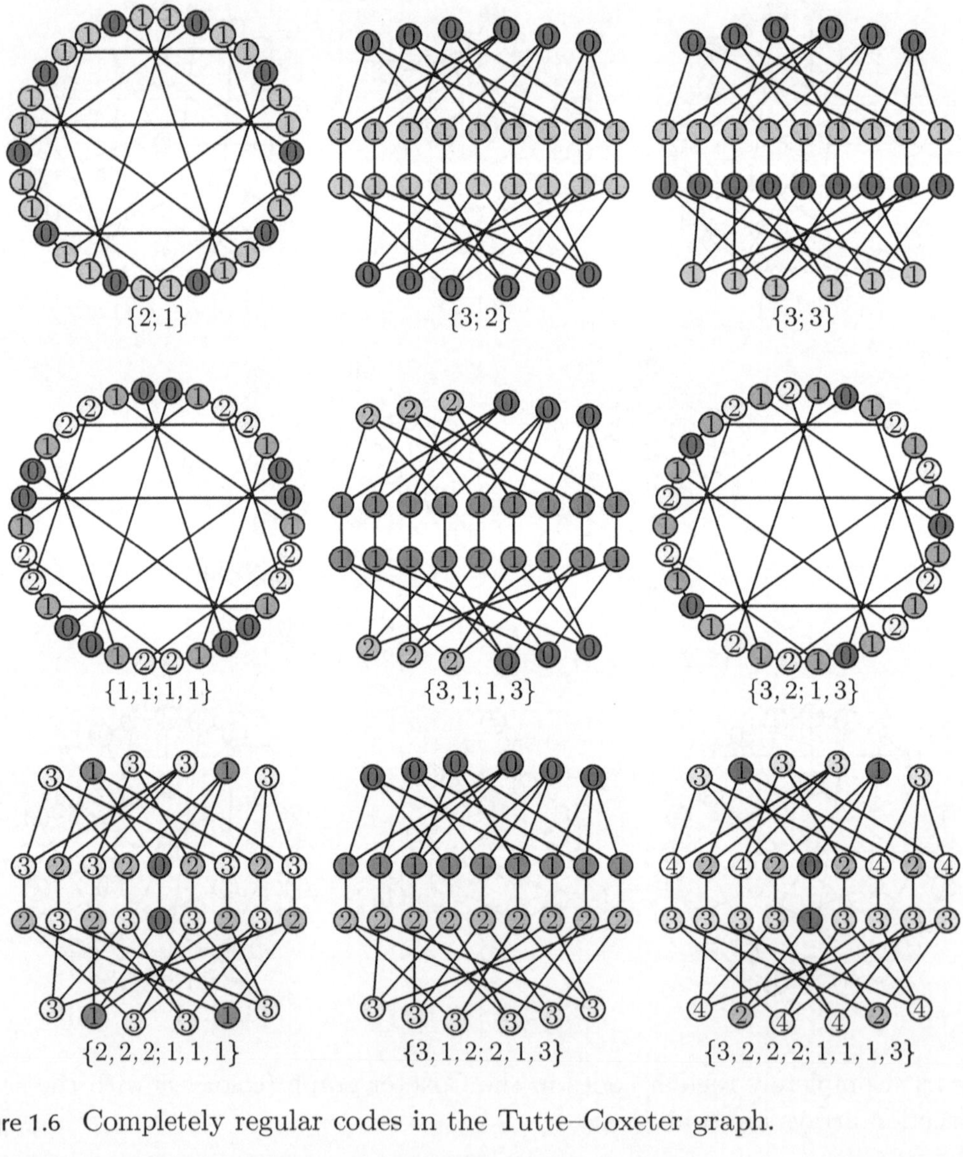

Figure 1.6 Completely regular codes in the Tutte–Coxeter graph.

1/3/6/12/8, $\{3,2,2,2;1,1,1,3\}$, $[0,\pm 2,\pm 3]$.

Foster graph of order 90:

30/60, $\{2;1\}$, $[0,3]$;
36/54, $\{3;2\}$, $[-2,3]$;
45/45, $\{3;3\}$, $[\pm 3]$;
30/30/30, $\{1,1;1,1\}$, $[0,2,3]$;
18/54/18, $\{3,1;1,3\}$, $[0,-2,3]$;
15/45/30, $\{3,2;1,3\}$, $[0,\pm 3]$;
6/12/24/48, $\{2,2,2;1,1,1\}$, $[0,\pm 2,3]$;
18/27/27/18, $\{3,1,2;2,1,3\}$, $[\pm 2,\pm 3]$;
10/20/20/20/20, $\{2,2,1,1;1,2,1,1\}$, $[0,1,\pm\sqrt{6},3]$;
10/15/15/30/20, $\{3,1,2,2;2,1,1,3\}$, $[0,\pm\sqrt{6},\pm 3]$;
3/9/18/36/24, $\{3,2,2,2;1,1,1,3\}$, $[0,\pm 2,\pm 3]$;
2/4/8/16/32/16/8/4, $\{2,2,2,2,1,1,1;1,1,1,1,2,2,2\}$, $[0,\pm 1,\pm 2,\pm\sqrt{6},3]$;
1/3/6/12/24/24/12/6/2, $\{3,2,2,2,2,1,1,1;1,1,1,1,2,2,2,3\}$, $[0,\pm 1,\pm 2,\pm\sqrt{6},\pm 3]$.

Biggs–Smith graph of order 102 (see [61] for the characterization of CR codes):

34/68, $\{2;1\}$, $[0,3]$;
34/34/34, $\{1,1;1,1\}$, $[0,3]$;
34/17/17/34, $\{1,1,2;2,1,1\}$, $[0,\frac{1\pm\sqrt{17}}{2},3]$;
1/3/6/12/24/24/24/8, $\{3,2,2,2,1,1,1;1,1,1,1,1,1,3\}$, $[0,2,3,\frac{1\pm\sqrt{17}}{2},\theta_1,\theta_2,\theta_3]$, $\theta_i^3+3\theta_i^2-3=0$.

Tutte 12-cage of order 126:

63/63, $\{3;3\}$, $[\pm 3]$;
21/63/42, $\{3,2;1,3\}$, $[0,\pm 3]$;
9/27/54/36, $\{3,2,2;1,1,3\}$, $[\pm\sqrt{2},\pm 3]$;
14/21/21/42/28, $\{3,1,2,2;2,1,1,3\}$, $[0,\pm\sqrt{6},\pm 3]$;
2/4/8/16/32/64, $\{2,2,2,2,2;1,1,1,1,1\}$, $[0,\pm\sqrt{2},\pm\sqrt{6},3]$;
1/3/6/12/24/48/32, $\{3,2,2,2,2,2;1,1,1,1,1,3\}$, $[0,\pm\sqrt{2},\pm\sqrt{6},\pm 3]$.

Additionally, we list all intersection arrays of CR codes in the strongly regular **Hoffman–Singleton graph** of order 50, found in [87]:

15/35, $\{7;3\}$, $[7,-3]$;
20/30, $\{6;4\}$, $[7,-3]$;
25/25, $\{5;5\}$, $[7,-3]$;
10/40, $\{4;1\}$, $[7,2]$;
20/30, $\{3;2\}$, $[7,2]$;
10/25/15, $\{5,3;2,5\}$, $[7,2,-3]$;
5/25/20, $\{5,4;1,5\}$, $[7,2,-3]$;
2/12/36, $\{6,6;1,2\}$, $[7,2,-3]$;
8/28/14, $\{7,2;2,4\}$, $[7,2,-3]$;
1/7/42, $\{7,6;1,1\}$, $[7,2,-3]$.

Finally, we mention some results concerning perfect codes in distance-regular graphs. In the Hamming graphs $H(n,q)$, the problem of complete characterization of parameters of perfect codes is solved only for the case when q is a prime power [120, 139]: there are no nontrivial perfect codes except the e-perfect repetition codes in $H(2e+1,2)$, the 3- and two-perfect Golay codes [42] in $H(23,2)$ and $H(11,3)$, respectively, and the one-perfect codes in $H((q^k-1)/(q-1),q)$. In the case of a non-prime-power q, no nontrivial perfect codes are known, and the parameters for which the nonexistence is not proven are restricted by 1- and two-perfect codes (the last case is solved for some values of q), see [50] for a survey of the known results in this area. All admissible parameters of perfect codes in Doob graphs are characterized in [71], see also the detailed survey in Chapter 6 of this book (one of the important observations is that one-perfect codes can exist in Doob graphs where there are no linear or additive one-perfect codes). The problem of the (non)existence of perfect codes in the Johnson graphs $J(n, w)$ is known as Delsarte's conjecture, see [33,44] and Chapter 3 of this book for the known nonexistence results; in general, the problem remains open. An interesting open problem is connected with the problem of existence of one-perfect codes in the Odd graph O_{w+1} (equivalently, in the doubled Odd graph DO_{w+1}): the existence of such codes is equivalent to the existence of Steiner systems $S(w-1, w, 2w+1)$ or, equivalently, of Steiner systems $S(w, w+1, 2w+2)$; in particular, the Steiner quadruple system $S(3, 4, 8)$ and the small Witt design $S(5, 6, 12)$ [21,137] correspond to nontrivial one-perfect codes in O_4 and O_6; the nonexistence of Steiner systems $S(4, 5, 15)$ [92] and $S(4, 5, 17)$ [98] implies the nonexistence of one-perfect codes in O_{10} and O_{12} (in general, the problem remains open, with the first open case in O_{16}). In the Grassmann graphs $J_q(n, w)$, dual polar graphs, the bilinear forms graphs $B_q(m, n)$, and the other forms graphs (with possibly some exceptions of concrete form) nontrivial perfect codes do not exist [22], see also [90] for a simpler "anticode bound" arguments for Grassmann and bilinear forms graphs. Some perfect codes in dual polar graphs are discussed in [111, p. 659], including the examples of one-perfect codes found in [119] in graphs of diameter 3. Studying diameter-3 antipodal distance-regular graphs with one-perfect codes (usually, with some assumptions on the graph automorphisms) is a separate topic, see [24,41,84–86,121–124] and references there.

Acknowledgments

The authors thank Evgeny Bespalov, Alexander Gavrilyuk, Evgeny Gorkunov, Jack Koolen, Bill Martin, Ivan Mogilnykh, Gábor Nagy, Santu Pal, Patrick Solé, Anna Taranenko, Yuriy Tarannikov, Anastasia Vasil'eva, and all participants of the Seminar "2024-ary Quasigroups and Related Questions" for fruitful discussions.

The new results in Sections 1.2.3.5, 1.4.2.1, and 1.4.3.1 were funded by the Russian Science Foundation (22-11-00266). The other work of the first author was carried out within the framework of the state contract of the Sobolev Institute of Mathematics (Project FWNF-2022-0017).

Note

1. A code from only one vertex has no well-defined distance. Usually, this does not lead to any problems, but sometimes it is convenient to think that it has any distance larger than diam(Γ). This assumption allows to include one-vertex codes to some infinite sequences of codes, for example, perfect $(2^m-1, 2^{2^m-m-1}, 3)_2$ codes, $m = 1, 2, 3, \ldots$.

REFERENCES

[1] N. Alon and F. R. K. Chung. Explicit construction of linear sized tolerant networks. Discrete Math., 72(1–3):15–19, 1988. https://doi.org/10.1016/0012-365X(88)90189-6.

[2] N. Alon and V. D. Milman. λ_1, isoperimetric inequalities for graphs, and superconcentrators. *J. Comb. Theory Ser. B*, 38(1):73–88, 1985. https://doi.org/10.1016/0095-8956(85)90092-9.

[3] D. Angluin and A. Gardiner. Finite common coverings of pairs of regular graphs. *J. Comb. Theory Ser. B*, 30(2):184–187, 1981. https://doi.org/10.1016/0095-8956(81)90062-9.

[4] S. V. Avgustinovich. On a property of perfect binary codes. In A. D. Korshunov, editor, *Operations Research and Discrete Analysis*, volume 391 of Applied Mathematics, pp. 13–15. Kluwer, Dordrecht, 1997.

[5] S. V. Avgustinovich, D. S. Krotov, and A. Yu. Vasil'eva. Completely regular codes in the infinite hexagonal grid. *Sib. Èlektron. Mat. Izv.*, 13:987–1016, 2016. https://doi.org/10.17377/semi.2016.13.079.

[6] S. V. Avgustinovich and I. Yu. Mogilnykh. Induced perfect colorings. *Sib. Èlektron. Mat. Izv.*, 8:310–316, 2011.

[7] R. A. Bailey, P. J. Cameron, A. L. Gavrilyuk, and S. V. Goryainov. Equitable partitions of Latin-Square graphs. *J. Comb. Des.*, 27(3):142–160, 2019. https://doi.org/10.1002/jcd.21634.

[8] L. A. Bassalygo, G. V. Zaitsev, and V. A. Zinoviev. On uniformly packed codes. *Probl. Inf. Transm.*, 10(1):6–9, 1974. Translated from *Probl. Peredachi Inf.*, 10(1):9–14, 1974.

[9] E. A. Bespalov, D. S. Krotov, A. A. Matiushev, A. A. Taranenko, and K. V. Vorob'ev. Perfect 2-colorings of Hamming graphs. *J. Comb. Des.*, 29(6):367–396, 2021. https://doi.org/10.1002/jcd.21771.

[10] M. R. Best and A. E. Brouwer. The triply shortened binary Hamming code is optimal. *Discrete Math.*, 17(3):235–245, 1977. https://doi.org/10.1016/0012-365X(77)90158-3.

[11] J. Bierbrauer. Bounds on orthogonal arrays and resilient functions. *J. Comb. Des.*, 3(3):179–183, 1995. https://doi.org/10.1002/jcd.3180030304.

[12] J. Bierbrauer, K. Gopalakrishnan, and D. R. Stinson. Orthogonal arrays, resilient functions, error-correcting codes, and linear programming bounds. *SIAM J. Discrete Math.*, 9(3):424–452, 1996. https://doi.org/10.1137/S0895480194270950.

[13] A. E. Brouwer, A. M. Cohen, and A. Neumaier. *Distance-Regular Graphs*. Springer-Verlag, Berlin, 1989. https://doi.org/10.1007/978-3-642-74341-2.

[14] A. E. Brouwer, C. D. Godsil, J. H. Koolen, and W. J. Martin. Width and dual width of subsets in polynomial association schemes. *J. Comb. Theory Ser. A*, 102(2):255–271, 2003. https://doi.org/10.1016/S0097-3165(03)00006-2.

[15] A. E. Brouwer and W. H. Haemers. *Spectra of Graphs*. Springer-Verlag, New York, 2012. https://doi.org/10.1007/978-1-4614-1939-6.

[16] A. E. Brouwer and H. Van Maldeghem. *Strongly Regular Graphs*, volume 182 of *Encycl. Math. Appl.* Cambridge: Cambridge University Press, 2022.

[17] A. R. Calderbank and W. M. Kantor. The geometry of two-weight codes. *Bull. Lond. Math. Soc.*, 18(2):97–122, 1986. https://doi.org/10.1112/blms/18.2.97.

[18] P. Camion, B. Courteau, and P. Delsarte. On r-partition designs in Hamming spaces. *Appl. Algebra Eng. Commun. Comput.*, 2(3):147–162, 1992. https://doi.org/10.1007/BF01294330.

[19] R. Canogar. An example related to the Best code of reconstructing the cells of a partition design from its adjacency matrix. *Discrete Math.*, 242(1-3):269–276, 2002. https://doi.org/10.1016/S0012-365X(01)00319-3.

[20] C. Carlet. Boolean functions for cryptography and error-correcting codes. In Y. Crama and P. L. Hammer, editors, *Boolean Models and Methods in Mathematics, Computer Science, and Engineering*, volume 134 of Encyclopedia of Mathematics and its Applications, Chapter 8, pp. 257–397. Cambridge University Press, Cambridge 2010. https://doi.org/10.1017/CBO9780511780448.012.

[21] R. D. Carmichael. Configurations of rank two. *Am. J. Math.*, 53(1):217–240, 1931. https://doi.org/10.2307/2370885.

[22] L. Chihara. On the zeros of the Askey–Wilson polynomials, with applications to coding theory. *SIAM J. Math. Anal.*, 18(1):191–207, 1987. https://doi.org/10.1137/0518015.

[23] C. J. Colbourn and J. H. Dinitz, editors. *Handbook of Combinatorial Designs (Discrete Mathematics and Its Applications)*, Second edition. Chapman & Hall/CRC Press, Boca Raton, FL, London, New York, 2006.

[24] G. Coutinho, C. Godsil, H. Shirazi, and H. Zhan. Equiangular lines and covers of the complete graph. *Linear Algebra Appl.*, 488:264–283, 2016. https://doi.org/10.1016/j.laa.2015.09.029.

[25] M. B. Crawford, D. J. Marchette, W. Maxwell, and S. S. Mendelson. Spectral properties of random graphs with fixed equitable partition. E-print 2311.07675, arXiv.org, 2023. https://doi.org/10.48550/arXiv.2311.07675.

[26] D. Crnković, F. Pavese, and A. Švob. Intriguing sets of strongly regular graphs and their related structures. *Contrib. Discrete Math.*, 18(1):66–89, 2023.

[27] D. M. Cvetković, M. Doob, and H. Sachs. *Spectra of Graphs: Theory and Application.* Academic Press, New York, San Francisco, CA, London, 1980.

[28] J. De Beule, J. D'haeseleer, F. Ihringer, and J. Mannaert. Degree 2 Boolean functions on Grassmann graphs. *Electr. J. Comb.*, 30(1):#P1.31(1–23), 2023. https://doi.org/10.37236/11040.

[29] B. De Bruyn and H. Suzuki. Intriguing sets of vertices of regular graphs. *Graphs Comb.*, 26(5):629–646, 2010. https://doi.org/10.1007/s00373-010-0924-y.

[30] S. De Winter and K. Metsch. Perfect 2-colorings of the Grassmann graph of planes. *Electr. J. Comb.*, 27(1):#1.21(1–19), 2020. https://doi.org/10.37236/8672.

[31] P. Delsarte. *An Algebraic Approach to Association Schemes of Coding Theory*, volume 10 of Philips Research Reports: Supplements. N.V. Philips' Gloeilampenfabrieken, Eindhoven, 1973.

[32] K. Devriendt and P. Van Mieghem. Tighter spectral bounds for the cut size, based on Laplacian eigenvectors. *Linear Algebra Appl.*, 572:68–91, 2019. https://doi.org/10.1016/j.laa.2019.02.025.

[33] T. Etzion. Configuration distribution and designs of codes in the Johnson scheme. *J. Comb. Des.*, 15(1):15–34, 2007. https://doi.org/10.1002/jcd.20102.

[34] P. Santonastaso, F. Zullo, O. Polverino, and J. Sheekey. Two-weight rank-metric codes. E-print 2405.02841, arXiv.org, 2024. https://doi.org/10.48550/arXiv.2308.05229.

[35] D. G. Fon-Der-Flaass. A bound on correlation immunity. *Sib. Èlektron. Mat. Izv.*, 4:133–135, 2007. Online: http://mi.mathnet.ru/eng/semr149.

[36] D. G. Fon-Der-Flaass. Perfect 2-colorings of a hypercube. *Sib. Math. J.*, 48(4):740–745, 2007. Translated from *Sib. Mat. Zh.*, 48(4):923-930, 2007. https://doi.org/10.1007/s11202-007-0075-4.

[37] J. Friedman. On the bit extraction problem. In *IEEE Annual Symposium on Foundations of Computer Science*, pp. 314–319, Los Alamitos, CA, 1992. IEEE Computer Society. https://doi.org/10.1109/SFCS.1992.267760.

[38] A. L. Gavrilyuk and V. V. Kabanov. Strongly regular graphs decomposable into a divisible design graph and a Hoffman coclique. *Des. Codes Cryptography*, 92(5):1379–1391, 2024. https://doi.org/10.1007/s10623-023-01348-9.

[39] C. Godsil and G. Royle. *Algebraic Graph Theory*. Springer-Verlag, New York, 2001. https://doi.org/10.1007/978-1-4613-0163-9.

[40] C. D. Godsil. *Algebraic Combinatorics*. Chapman & Hall, New York, 1993.

[41] C. D. Godsil, R. A. Liebler, and C. E. Praeger. Antipodal distance transitive covers of complete graphs. *Eur. J. Comb.*, 19(4):455–478, 1998. https://doi.org/10.1006/eujc.1997.0190.

[42] M. J. E. Golay. Notes on digital coding. *Proc. IRE*, 37(6):657, 1949. https://doi.org/10.1109/JRPROC.1949.233620.

[43] K. Golubev. Graphical designs and extremal combinatorics. *Linear Algebra Appl.*, 604:490–506, 2020. https://doi.org/10.1016/j.laa.2020.07.012.

[44] D. M. Gordon. Perfect single error-correcting codes in the Johnson scheme. *IEEE Trans. Inf. Theory*, 52(10):4670–4672, 2006. https://doi.org/10.1109/TIT.2006.881744.

[45] S. Goryainov and D. Krotov. Completely regular codes in graphs covered by a Hamming graph. arXiv eprint 2411.09698, DOI10.48550/arXiv.2411.09698, 2024.

[46] J. L. Gross and T. W. Tucker. Generating all graph coverings by permutation voltage assignments. *Discrete Math.*, 18(3):273–283, 1977. https://doi.org/10.1016/0012-365X(77)90131-5.

[47] W. H. Haemers. Eigenvalue techniques in design and graph theory. Dissertation, Technische Hogeschool Eindhoven, Amsterdam, 1979.

[48] W. H. Haemers and D. G. Higman. Strongly regular graphs with strongly regular decomposition. *Linear Algebra Appl.*, 114–115:379–398, 1989. https://doi.org/10.1016/0024-3795(89)90471-0.

[49] A. S. Hedayat, N. J. A. Sloane, and J. Stufken. *Orthogonal Arrays: Theory and Applications*, Springer Series in Statistics. Springer, New York, 1999. https://doi.org/10.1007/978-1-4612-1478-6.

[50] O. Heden. On perfect codes over non prime power alphabets. In A. A. Bruen and D. L. Wehlau, editors, *Error-Correcting Codes, Finite Geometries and Cryptography*, volume 523 of Contemporary Mathematics, pp. 173–184. AMS, Providence, RI, 2010.

[51] O. Heden and C. Roos. The non-existence of some perfect codes over non-prime power alphabets. *Discrete Math.*, 311(14):1344–1348, 2011. https://doi.org/10.1016/j.disc.2011.03.024.

[52] D. G. Higman. Coherent configurations. Part I: Ordinary representation theory. *Geometriae Dedicata*, 4(1):1–32, 1975. https://doi.org/10.1007/BF00147398.

[53] A. J. Hoffman. On eigenvalues and colourings of graphs. In B. Harris, editor, *Graph Theory and Its Applications*, pp. 79–91. Academic Press, New York, 1970.

[54] J. Y. Hyun. Local duality for equitable partitions of a Hamming space. *J. Comb. Theory Ser. A*, 119(2):476–482, 2012. https://doi.org/10.1016/j.jcta.2011.10.006.

[55] P. Kaski and P. R. J. Östergård. *Classification Algorithms for Codes and Designs*, volume 15 of Algorithms and Computation in Mathematics. Springer, Berlin, 2006. https://doi.org/10.1007/3-540-28991-7.

[56] P. Kaski and O. Pottonen. libexact user's guide, version 1.0. Technical Report 2008-1, Helsinki Institute for Information Technology HIIT, 2008.

[57] D. V. Kerkesner. Perfect colorings of a generalized Petersen graph and their properties. Master thesis, Novosibirsk State University, Novosibirsk, 2022. In Russian.

[58] A. V. Khalyavin. Estimates of the capacity of orthogonal arrays of large strength. *Mosc. Univ. Math. Bull.*, 65(3):130–131, 2010. https://doi.org/10.3103/S002713221 0030101.

[59] T. E. Kireeva. Perfect orientation colorings of cubic graphs. *Sib. Èlektron. Mat. Izv.*, 15:1353–1360, 2018. https://doi.org/DOI 10.17377/semi.2018.15.111 [Russian, English abstract].

[60] D. A. Kondakova. *Algorithms for Constructing Perfect Colorings and Their Parameters Matrices for the Generalized Petersen Graph.* Bachelor's thesis, Novosibirsk State University, Novosibirsk, 2022. In Russian.

[61] J. H. Koolen. On a conjecture of Martin on the parameters of completely regular codes and the classification of the completely regular codes in the Biggs–Smith graph. *Linear Multilinear Algebra*, 39(1–2):3–17, 1995. https://doi.org/10.1080/03081089508818376.

[62] J. H. Koolen, W. S. Lee, and W. J. Martin. Characterizing completely regular codes from an algebraic viewpoint. In R. A. Brualdi, S. Hedayat, H. Kharaghani, G. B. Khosrovshahi, and S. Shahriari, editors, *Combinatorics and Graphs: The Twentieth Anniversary Conference of IPM Combinatorics, May 15–21, 2009, Tehran, Iran*, volume 531 of Contemporary Mathematics, pp. 223–242. AMS, Providence, RI, 2010.

[63] J. H. Koolen, W. S. Lee, W. J. Martin, and H. Tanaka. Arithmetic completely regular codes. *Discrete Math. Theor. Comput. Sci.*, 17(3):59–76, 2009. https://doi.org/10.46298/dmtcs.2150.

[64] E. S. Kramer and D. M. Mesner. t-designs on hypergraphs. *Discrete Math.*, 15(3):263–296, 1976. https://doi.org/10.1016/0012-365X(76)90030-3.

[65] D. Krotov and A. Vasil'eva. On strong distance invariance of some Steiner quadruple systems. In *2023 XVIII International Symposium "Problems of Redundancy in Information and Control Systems" (REDUNDANCY)*, pp. 159–162. IEEE, 2023. https://doi.org/10.1109/Redundancy59964.2023.10330167.

[66] D. S. Krotov. On perfect colorings of the halved 24-cube. *Diskretn. Anal. Issled. Oper.*, 15(5):35–46, 2008. In Russian; translated at http://arxiv.org/abs/0803.0068.

[67] D. S. Krotov. On the binary codes with parameters of doubly-shortened 1-perfect codes. *Des. Codes Cryptography*, 57(2):181–194, 2010. https://doi.org/10.1007/s10623-009-9360-5.

[68] D. S. Krotov. On weight distributions of perfect colorings and completely regular codes. *Des. Codes Cryptography*, 61(3):315–329, 2011. https://doi.org/10.1007/s10623-010-9479-4.

[69] D. S. Krotov. On the binary codes with parameters of triply-shortened 1-perfect codes. *Des. Codes Cryptography*, 64(3):275–283, 2012. https://doi.org/10.1007/s10623-011-9574-1.

[70] D. S. Krotov. On calculation of the interweight distribution of an equitable partition. *J. Algebr. Comb.*, 40(2):373–386, 2014. https://doi.org/10.1007/s10801-013-0492-3.

[71] D. S. Krotov. The existence of perfect codes in Doob graphs. *IEEE Trans. Inf. Theory*, 66(3):1423–1427, 2020. https://doi.org/10.1109/TIT.2019.2946612.

[72] D. S. Krotov. On the OA(1536,13,2,7) and related orthogonal arrays. *Discrete Math.*, 343(2):111659/1–11, 2020. https://doi.org/10.1016/j.disc.2019.111659.

[73] D. S. Krotov. Strong distance invariance of some designs. In *International Conference "Mal'tsev Meeting"*. November 14–18, 2022. Collection of Abstracts, p. 42, Novosibirsk, 2022. Sobolev Institute of Mathematics, Novosibirsk State University. Online: http://old.math.nsc.ru/conference/malmeet/22/maltsev22.pdf.

[74] D. S. Krotov. Equitable [[2,10],[6,6]]-partitions of the 12-cube. *Cryptogr. Commun.*, 16, 2024. https://doi.org/10.1007/s12095-024-00716-z.

[75] D. S. Krotov, I. Yu. Mogilnykh, and V. N. Potapov. To the theory of q-ary Steiner and other-type trades. *Discrete Math.*, 339(3):1150–1157, 2016. https://doi.org/10.1016/j.disc.2015.11.002.

[76] D. S. Krotov, I. Yu. Mogilnykh, and A. Yu. Vasil'eva. On completely regular codes of covering radius 1 in the halved hypercubes. E-print 1812.03159, arXiv.org, 2018.

[77] D. S. Krotov, P. R. J. Östergård, and O. Pottonen. On optimal binary one-error-correcting codes of lengths $2^m - 4$ and $2^m - 3$. *IEEE Trans. Inf. Theory*, 57(10):6771–6779, 2011. https://doi.org/10.1109/TIT.2011.2147758.

[78] D. S. Krotov, F. Özbudak, and V. N. Potapov. Generalizing the Bierbrauer–Friedman bound for orthogonal arrays. E-print 2411.16559, arXiv.org, 2024. https://doi.org/10.48550/arXiv.2411.16559.

[79] D. S. Krotov and V. N. Potapov. On multifold packings of radius-1 balls in Hamming graphs. *IEEE Trans. Inf. Theory*, 67(6):3585–3598, 2021. https://doi.org/10.1109/TIT.2020.3046260.

[80] D. S. Krotov and K. V. Vorob'ev. On unbalanced Boolean functions with best correlation immunity. *Electr. J. Comb.*, 27(1):#P1.45(1–24), 2020. https://doi.org/10.37236/8557.

[81] F. T. Leighton. Finite common coverings of graphs. *J. Comb. Theory Ser. B*, 33(3):231–238, 1982. https://doi.org/10.1016/0095-8956(82)90042-9.

[82] M. A. Lisitsyna and S. V. Avgustinovich. Perfect colorings of the prism graph. *Sib. Èlektron. Mat. Izv.*, 13:1116–1128, 2016. https://doi.org/10.17377/semi.2016.13.088.

[83] S. P. Lloyd. Binary block coding. *Bell Syst. Tech. J.*, 36(2):517–535, 1957. https://doi.org/10.1002/j.1538-7305.1957.tb02410.x.

[84] A. A. Makhnev, D. V. Paduchikh, and L. Yu. Tsiovkina. Edge-symmetric distance-regular coverings of cliques: The affine case. *Sib. Math. J.*, 54(6):1076–1087, 2013. https://doi.org/10.1134/S0037446613060141.

[85] A. A. Makhnev, D. V. Paduchikh, and L. Yu. Tsiovkina. Arc-transitive distance-regular covers of cliques with $\lambda = \mu$. *Proc. Steklov Inst. Math.*, 284(Suppl.1):124–134, 2014. https://doi.org/10.1134/S0081543814020114.

[86] A. A. Makhnev, D. V. Paduchikh, and L. Yu. Tsiovkina. Edge-symmetric distance-regular coverings of complete graphs: The almost simple case. *Algebra Logic*, 57(2):141–152, 2018. https://doi.org/10.1007/s10469-018-9486-5.

[87] I. A. Mamchits. Distance regular colorings of the Hoffman–Singleton graph. In *International Conference "Mal'tsev Meeting"*, November 14–18, 2022, Collection of Abstracts, p. 42, Novosibirsk, 2022. Sobolev Institute of Mathematics, Novosibirsk State University. Online: http://old.math.nsc.ru/conference/malmeet/22/maltsev22.pdf (in Russian).

[88] W. J. Martin. Completely regular subsets. PhD thesis, University of Waterloo, Waterloo, Ontario, Canada, 1992. Online: http://users.wpi.edu/~martin/RESEARCH/THESIS/.

[89] W. J. Martin. Completely regular designs. *J. Comb. Des.*, 6(4):261–273, 1998. https://doi.org/10.1002/(SICI)1520-6610(1998)6:4<261::AID-JCD4>3.0.CO;2-D.

[90] W. J. Martin and X. J. Zhu. Anticodes for the Grassman and bilinear forms graphs. *Des. Codes Cryptography*, 6(1):73–79, 1995. https://doi.org/10.1007/BF01390772.

[91] B. D. McKay. Practical graph isomorphism. In *Proceedings of the 10th Manitoba Conference*, Winnipeg/Manitoba 1980, volume 30 of Congressus Numerantium, pp. 45–87. Combinatorial Press, Winnipeg, MB, 1981.

[92] N. S. Mendelsohn and S. H. Y. Hung. On the Steiner systems $S(3,4,14)$ and $S(4,5,15)$. *Util. Math.*, 1:5–95, 1972.

[93] I. Mogilnykh. Completely regular codes in Johnson and Grassmann graphs with small covering radii. *Electr. J. Comb.*, 29(2):#P2.57(1–15), 2022. https://doi.org/10.37236/10083.

[94] I. Yu. Mogilnykh. Codes in star and bubble-sort graphs. In *Proceedings of 2021 International Conference Automatics and Informatics* (*ICAI*, 30 September–2 October 2021, Varna, Bulgaria, online), December 2021. https://doi.org/10.1109/ICAI52893.2021.9639869.

[95] I. Yu. Mogilnykh and K. V. Vorob'ev. On completely regular codes with minimum eigenvalue in geometric graphs. *Discrete Math.*, 346(7):113357, 2023. https://doi.org/10.1016/j.disc.2023.113357.

[96] H. L. Morgan. The generation of a unique machine description for chemical structures: A technique developed at chemical abstracts service. *J. Chem. Doc.*, 5(2):107–113, 1965. https://doi.org/10.1021/c160017a018.

[97] A. Neumaier. Completely regular codes. *Discrete Math.*, 106-107:353–360, 1992. https://doi.org/10.1016/0012-365X(92)90565-W.

[98] P. R. J. Östergård and O. Pottonen. There exists no Steiner system $S(4,5,17)$. *J. Comb. Theory Ser. A*, 115(8):1570–1573, 2008. https://doi.org/10.1016/j.jcta.2008.04.005.

[99] M. Petersdorf and H. Sachs. über spektrum, automorphismengruppe und teiler eines graphen. *Wiss. Z. Tech. Hochsch. Ilmenau*, 15(4–5):123–128, 1969.

[100] V. N. Potapov. On perfect 2-colorings of the q-ary n-cube. *Discrete Math.*, 312(6):1269–1272, 2012. https://doi.org/10.1016/j.disc.2011.12.004.

[101] V. N. Potapov. On extremal properties of perfect 2-colorings. E-print 2204.03308, arXiv.org, 2022. Available at http://arxiv.org/abs/2204.03308.

[102] V. N. Potapov and S. V. Avgustinovich. Combinatorial designs, difference sets, and bent functions as perfect colorings of graphs and multigraphs. *Sib. Math. J.*, 61(5):867–877, 2020. Translated from *Sib. Mat. Zh.*, 61(5):1087–1100, 2020. https://doi.org/10.1134/S0037446620050109.

[103] V. N. Potapov and S. V. Avgustinovich. Combinatorial designs, difference sets, and bent functions as perfect colorings of graphs and multigraphs. E-print 2403.02904, arXiv.org, 2024. Available at http://arxiv.org/abs/2403.02904.

[104] R. Pratihar and T. H. Randrianarisoa. Antipodal two-weight rank metric codes. *Des. Codes Cryptography*, 92(3):753–769, 2024. https://doi.org/10.1007/s10623-023-01283-9.

[105] A. D. Sankey. On strongly regular designs admitting fusion to strongly regular decomposition. *J. Comb. Des.*, 29(10):703–722, 2021. https://doi.org/10.1002/jcd.21798.

[106] H. S. Shapiro and D. L. Slotnick. On the mathematical theory of error correcting codes. *IBM J. Res. Dev.*, 3(1):25–34, 1959.

[107] M. Shi, R. Wu, and D. S. Krotov. On q-ary shortened-1-perfect-like codes. *IEEE Trans. Inf. Theory*, 68(11):7100–7106, 2022. https://doi.org/10.1109/TIT.2022.3187004.

[108] P. Solé. Completely regular codes and completely transitive codes. *Discrete Math.*, 81(2):193–201, 1990. https://doi.org/10.1016/0012-365X(90)90152-8.

[109] E. Sotnikova and A. Valyuzhenich. Minimum supports of eigenfunctions of graphs: A survey. *Art Discrete Appl. Math.*, 4(2):P2.09(1–34), 2021. https://doi.org/10.26493/2590-9770.1404.61e.

[110] Z. Stanić. Star complementary strongly regular decompositions of strongly regular graphs. *Linear Multilinear Algebra*, 68(12):2448–2461, 2020. https://doi.org/10.1080/03081087.2019.1586821.

[111] D. Stanton. Some q-Krawtchouk polynomials on Chevalley groups. *Am. J. Math.*, 10(4):625–662, 1980. https://doi.org/10.2307/2374091.

[112] X. Sun and J. Lu. On intriguing sets in five classes of strongly regular graphs. *J. Comb. Des.*, 30(6):384–405, 2022. https://doi.org/10.1002/jcd.21832.

[113] H. Tanaka. Vertex subsets with minimal width and dual width in Q-polynomial distance-regular graphs. *Electr. J. Comb.*, 18(1):#P167(1–32), 2011. https://doi.org/10.37236/654.

[114] T. Tanaka. Classification of subsets with minimal width and dual width in Grassmann, bilinear forms and dual polar graphs. *J. Comb. Theory Ser. A*, 113(5):903–910, 2006. https://doi.org/10.1016/j.jcta.2005.08.006.

[115] R. M. Tanner. Explicit concentrators from generalized n-gons. *SIAM J. Algebraic Discrete Methods*, 5(3):287–293, 1984. https://doi.org/10.1137/0605030.

[116] A. A. Taranenko. On a metric property of perfect colorings. *Sib. Èlektron. Mat. Izv.*, 18(1):640–646, 2021. https://doi.org/10.33048/semi.2021.18.046.

[117] A. A. Taranenko. Perfect colorings of hypergraphs. E-print 2208.03447, arXiv.org, 2022. https://doi.org/10.48550/arXiv.2208.03447.

[118] A. A. Taranenko. Algebraic properties of perfect structures. *Linear Algebra Appl.*, 607:286–306, 2020. https://doi.org/10.1016/j.laa.2020.08.012.

[119] J. Thas. Two infinite classes of perfect codes in metrically regular graphs. *J. Comb. Theory Ser. B*, 23:236–238, 1977.

[120] A. Tietavainen. On the nonexistence of perfect codes over finite fields. *SIAM J. Appl. Math.*, 24(1):88–96, 1973. https://doi.org/10.1137/0124010.

[121] L. Yu. Tsiovkina. On G-vertex- transitive covers of complete graphs having at most two G-orbits on the arc set. *Ural Math. J.*, 10(1):147–158, 2024. https://doi.org/10.15826/umj.2024.1.013.

[122] L. Yu. Tsiovkina. On affine distance-regular covers of complete graphs. *Sib. Èlektron. Mat. Izv.*, 12:998–1005, 2015. https://doi.org/10.17377/semi.2015.12.086.

[123] L. Yu. Tsiovkina. Two new infinite families of arc-transitive antipodal distance-regular graphs of diameter three with $\lambda = \mu$ related to groups $Sz(q)$ and ${}^2G_2(q)$. *J. Algebr. Comb.*, 41(4):1079–1087, 2015. https://doi.org/10.1007/s10801-014-0566-x.

[124] L. Yu. Tsiovkina. Arc-transitive antipodal distance-regular covers of complete graphs related to $SU_3(q)$. *Discrete Math.*, 340(2):63–71, 2017. https://doi.org/10.1016/j.disc.2016.08.001.

[125] E. R. van Dam, J. H. Koolen, and H. Tanaka. Distance-regular graphs. *Electr. J. Comb.*, DS22:1–156, 2016. https://doi.org/10.37236/4925.

[126] A. Vasil'eva. Linear binary completely regular codes with distance 2. In *Proceedings of the 2008 IEEE Region 8 International Conference on Computational Technologies in Electrical and Electronics Engineering "SIBIRCON 2008"*, pp. 12–15, Novosibirsk, Russia, July 2008.

[127] A. Yu. Vasil'eva. Strong distance invariance of perfect binary codes. *Diskretn. Anal. Issled. Oper. Ser. 1*, 9(4):33–40, 2002. In Russian.

[128] A. Yu. Vasil'eva. Local spectra of perfect binary codes. *Discrete Appl. Math.*, 135(1-3):301–307, 2004. Translated from *Diskretn. Anal. Issled. Oper. Ser. 1*, 6(1):3–11, 1999. https://doi.org/10.1016/S0166-218X(02)00313-X.

[129] A. Yu. Vasil'eva. Local and interweight spectra of completely regular codes and of perfect colorings. *Probl. Inf. Transm.*, 45(2):151–157, 2009. Translated from *Probl. Peredachi Inf.*, 45(2):84–90, 2009. https://doi.org/10.1134/S0032946009020069.

[130] A. Yu. Vasil'eva. On reconstructive sets of vertices in the Boolean cube. *J. Appl. Ind. Math.*, 6(3):393–402, 2012. https://doi.org/10.1134/S1990478912030155.

[131] A. Yu. Vasil'eva. Local distributions and reconstruction of hypercube eigenfunctions. *Probl. Inf. Transm.*, 49(1):32–39, 2013. Translated from *Probl. Peredachi Inf.*, 49(1):37–45, 2009. https://doi.org/10.1134/S0032946013010031.

[132] A. Yu. Vasil'eva. Reconstruction of eigenfunctions of a q-ary n-dimensional hypercube. *Probl. Inf. Transm.*, 51(3):231–239, 2015. Translated from *Probl. Peredachi Inf.*, 51(3):31–40, 2015. https://doi.org/10.1134/S0032946015030035.

[133] A. Yu. Vasil'eva. Local distributions for eigenfunctions and perfect colorings of q-ary hamming graph. *Des. Codes Cryptography*, 87(2–3):509–516, 2019. https://doi.org/10.1007/s10623-018-0559-1.

[134] A. G. Vizing. Distributive coloring of graph vertices. *Diskretn. Anal. Issled. Oper.*, 2(4):3–12, 1995. In Russian. Online: http://mi.mathnet.ru/eng/da/v2/i4/p3.

[135] I. M. Wanless. Transversals in Latin squares: A survey. In R. Chapman, editor, *Surveys in Combinatorics 2011*, volume 392 of London Mathematical Society Lecture Note Series, pp. 403–437. Cambridge University Press, Cambridge, 2011.

[136] B. Yu. Weisfeiler and A. A. Leman. The reduction of a graph to canonical form and the algebra which appears therein. *Nauchno-Technicheskaya Informatsia*, 2(9):12–16, 1968. In Russian, translated in https://www.iti.zcu.cz/wl2018/pdf/wl_paper_translation.pdf.

[137] V. E. Witt. Uber Steinersche systeme. *Abh. Math. Semin. Univ. Hamb.*, 12(1):265–275, 1937. https://doi.org/10.1007/BF02948948.

[138] T. Zhang and G. Ge. On linear diameter perfect Lee codes with distance 6. *J. Comb. Theory Ser. A*, 201:105816(1–26), 2024. https://doi.org/10.1016/j.jcta.2023.105816.

[139] V. Zinoviev and V. Leontiev. The nonexistence of perfect codes over Galois fields. *Probl. Control Inf. Theory*, 2(2):123–132, 16–24[Engl. transl.], 1973.

[140] V. A. Zinoviev. Combinatorial methods for the construction and analysis of nonlinear error-correcting codes. Dr.sc. thesis, Moscow, 1988. In Russian.

CHAPTER 2

Completely Regular Codes over Finite Fields

Victor A. Zinoviev

2.1 INTRODUCTION

This chapter is mainly based on a recent survey [32], where completely regular codes have been carefully studied. The author of this chapter cardially thanks Josep Rifà and Joaquim Borges for their kind permission to use freely the all results of [32].

In 1959, Shapiro and Slotnick [139] proved that binary perfect codes are completely regular (without using this term). Later, in 1969, Semakov, Zinoviev and Zaitsev [133,134,136] proved that Preparata codes, the extended Preparata codes, binary perfect codes, and extended binary perfect codes are completely regular also without using this term. In 1973, *completely regular codes* in the Hamming metric have been introduced by Delsarte [57]. Such codes have quite good packing density and possess combinatorial properties similar to perfect codes. The class of completely regular codes includes not only perfect and extended perfect codes [133,134,136] but also uniformly packed codes [80,133,134,136,153], codes obtained by their extensions [133,134,136], and completely transitive codes [70,71,143]. Known completely regular codes are, for example, Hamming, Golay, Preparata, binary BCH codes (of length $2^{2m+1} - 1$ with $d = 5$) and two Hadamard codes (of lengths 11 and 12). The combinatorial properties of completely regular codes allow to establish different relations with other combinatorial structures such as distance-regular graphs, association schemes, and combinatorial designs. A comprehensive text about these relations is a monograph of Brouwer, Cohen, and Neumaier (Chapter 11 in [38]) complemented in a survey of van Dam, Koolen, and Tanaka [55]. A table of possible parameters of completely regular codes of finite lengths and their intersection arrays due to Koolen, Krotov and Martin can be found in [95] (see Chapter 7 of this book). More recently, Koolen, Lee, Martin and Tanaka [96] studied and classified the class of arithmetic completely regular codes.

It is known that completely regular codes exist for arbitrary large covering radius (see, e.g., the direct constructions of Solé [143] and Rifà and Zinoviev [126–128]).

DOI: 10.1201/9781003393931-2

However, there are no known nontrivial completely regular codes with large error-correcting capability. Concretely, there are no known completely regular codes with minimum distance $d > 8$ and more than two codewords. In 1973, it has been proven the nonexistence of unknown nontrivial perfect codes over finite fields independently by Tietäväinen [151] and by Zinoviev and Leontiev [162]. The same result was obtained in 1975 for quasi-perfect uniformly packed codes by Goethals and van Tilborg [80,153] and Lindström [107] (while infinite families of such codes have been classified earlier in [136]). For the particular case of completely regular codes, namely for binary linear completely transitive codes, Borges, Rifà and Zinoviev [24,25] also proved the nonexistence for $d > 8$ and more than two codewords in 2001. Therefore, a natural conjecture seems to be the nonexistence of nontrivial completely regular codes for $d > 8$. In 1992, Neumaier [118] conjectured that the only completely regular code containing more than two codewords with $d \geq 8$ is the binary extended perfect Golay code. However, Borges, Rifà and Zinoviev [26] found a counterexample of Neumaier's conjecture. More precisely, they proved that the even half of the binary perfect Golay $[23, 12, 7]$-code (composed of its codewords of even weight, i.e. the $[23, 11, 8]$-code) is also completely regular. Nevertheless, the existence of unknown nontrivial completely regular codes for $d \geq 8$ remains a major open question in this area. In this survey, we consider completely regular codes only for finite lengths and finite cardinalities (see, e.g., [4] and references therein) for completely regular codes in the n-dimensional grids, or in the infinite grids.

The aim of this chapter looks as follows. Section 2.2 is devoted to preliminary results on completely regular codes in Hamming schemes. We give the main definitions and basic properties of completely regular codes (shortly CR codes), including equivalences and connections with several classes of uniformly packed (UP) codes. There are three main classes of such codes, which we call UP in narrow sense (UPNS) codes, which are inside of UP codes and UP in wide sense (UPWS) codes, which, in turn, contain UP codes. From one side, UPWS codes contain all CR codes, and, from the other side, CR codes include all UP codes. In Section 2.3, we consider the necessary conditions for the existence of completely regular codes, in particular, two different Lloyd theorems, the sphere packing conditions, the weight polynomials for CR codes, and some combinatorial and geometric properties (where we mostly follow Neumaier [118]). Then in Section 2.4, we consider extension of binary codes, in particular, uniformly packed, completely regular codes, and completely transitive codes. The main problem here is *to find the necessary and sufficient conditions when the extension of a binary CR code with odd distance is a CR code.* For the binary UP in wide sense codes, it is solved in [10]; see Section 2.4.2. In Section 2.5, we study completely transitive codes (shortly CT codes), a particular case of completely regular codes. In the remaining sections, we collect all known different constructions of completely regular codes in Hamming schemes. In particular, many new CR and CT codes with interesting properties are obtained by applying simple well-known constructions to perfect Hamming codes (Sections 2.6 and 2.7). Infinite families of CR and CT codes with growing covering radius (with growing length) are considered in Sections 2.6.3, 2.7.1, 2.7.2, and 2.8.1.

We emphasize here only one interesting result on CR and CT codes obtained by the Kronecker product. For any prime power q there exists a growing (with increasing length of codes) number of CR and CT codes with different parameters, but an identical intersection array (Section 2.7.1). It means that the same distance regular and distance transitive graphs can be obtained as coset graphs in many different ways. In Sections 2.9–2.11, we consider CR codes with covering radius $\rho \leq 2$. Some of these CR codes are CT. In total, we present exact constructions of more than 50 infinite families of CR codes.

2.2 PRELIMINARY RESULTS ON COMPLETELY REGULAR CODES

2.2.1 Introduction to Codes

We will consider codes over the finite alphabet $E = E_q = \{0, 1, \ldots, q-1\}$, which in most cases is a finite field $E = \mathbb{F}_q$, where q being a prime power. We agree that in the case $E = \mathbb{F}_q$ the elements of $\mathbb{F}_q$ are also denoted by $0, 1, \ldots, q-1$, where 0 and 1 denote the elements of $\mathbb{F}_q$, but all the operations are executed in $\mathbb{F}_q$. For the case when the size q of the alphabet E_q is not a prime power, we assume that E_q is at least an abelian group, or the ring $\mathbb{Z}_q$ of residues modulo q, depending on what is said in each of these cases. The Hamming distance between two vectors $x, y \in \mathbb{F}_q^n$, denoted by $d(x, y)$, is the number of coordinates in which they differ. The whole space $\mathbb{F}_q^n$ with the partition $R = \{R_0, R_1, \ldots, R_n\}$ of the Cartesian square $(\mathbb{F}_q^n)^2$ given by the condition $(x, y) \in R_i$ if and only if $d(x, y) = i$, is called the *Hamming association scheme* (or simply the *Hamming scheme*) $H(q, n)$. A code C is a subset of $\mathbb{F}_q^n$ equipped with the Hamming metric. For codes over rings $\mathbb{Z}_q$, the Lee metric is often used. In the most interesting case $q = 4$, such codes can be viewed as binary codes under Gray map; hence, we can consider them as binary codes again with the Hamming distance. As usual, for a code $C \subset \mathbb{F}_q^n$, we denote by n, d, and e, the length, the minimum distance, and the packing radius

$$e = \lfloor (d-1)/2 \rfloor$$

(or *error-correcting capability* of this code). We shall use the standard notation $(n, M, d)_q$ to denote a q-ary code C of length n, *size* or *cardinality* (i.e., the number of codewords in this code) M and minimum distance d,

$$d = \min\{d(x, y) : x, y \in C, x \neq y, \}$$

where we always assume that d is reached if we did not say opposite. We say that C is a *trivial* code if $|C| \leq 2$ or $C = \mathbb{F}_q^n$. If $E = \mathbb{F}_q$ and the code C is a linear space, then the notation is $[n, k, d]_q$. Then k denotes its dimension (or the *number of information symbols*), so in this case $M = q^k$. A code with packing radius e is often called an e-code. For the binary case ($q = 2$), we usually omit the subscript q. Unless stated otherwise, we assume that C is a *distance invariant code* (i.e., the weight distribution of codewords does not depend on the choice of zero codeword), containing the zero vector. In this chapter, by the concept of *additive code*, we mean an Abelian subgroup in the Abelian group E_q^n with an additive coordinate-wise operation in E_q (so, of course, these codes include linear codes).

For a vector $x \in E^n$, $x = (x_1, \ldots, x_n)$ define its *support*, denoted by Supp(x), the set of indexes of its nonzero positions:

$$\mathrm{Supp}(x) = \{j : x_j \neq 0,\ j = 1, \ldots, n\}.$$

Hence, the (Hamming) weight of the vector x, denoted by wt(x), is the size of its support: $\mathrm{wt}(x) = |\,\mathrm{Supp}(x)|$. Similarly, for a set $X \subseteq E^n$ denote its support as a set

$$\mathrm{Supp}(X) = \bigcup_{x \in X} \mathrm{Supp}(x).$$

Given any vector $\mathbf{v} \in \mathbb{F}_q^n$, its *distance to the code* C is

$$d(\mathbf{v}, C) = \min_{\mathbf{x} \in C}\{d(\mathbf{v}, \mathbf{x})\},$$

where $d(\mathbf{v}, \mathbf{x})$ denotes the distance between $\mathbf{v}$ and $\mathbf{x}$, and the *covering radius* $\rho = \rho(C)$ of the code C is

$$\rho = \max_{\mathbf{v} \in \mathbb{F}_q^n}\{d(\mathbf{v}, C)\}.$$

If we also want to specify the covering radius of the code, then we write $(n, M, d; \rho)_q$ for a nonlinear code, or $[n, k, d; \rho]_q$ for a linear code. For a q-ary linear $[n, k, d]_q$ code C, its dual code $C^\perp$ is defined in the standard way:

$$C^\perp = \{x \in \mathbb{F}_q^n : \langle x \cdot y \rangle = 0, \text{ for } \quad \text{every } y \in C\},$$

where $\langle x \cdot y \rangle$ denotes the inner product of vectors x and y in $\mathbb{F}_q^n$.

Denote by $S^{(t)} = S^{(t)}(x)$ a *sphere* at center $x \in E_q^n$ of radius t, i.e. the set of points $y \in E_q^n$, which are at distance t from x, i.e.

$$S^{(t)}(x) = \{y \in E_q^n : d(x, y) = t\}.$$

Similarly, denote by $B^{(t)} = B^{(t)}(x)$ a *ball* at center $x \in E_q^n$ of radius t, i.e. the set of points $y \in E_q^n$, which are at distance d, where $d \leq t$, from x, i.e. the union of spheres $S^{(i)}$ at center $x \in E_q^n$ of radius i:

$$B^{(t)}(x) = \{y \in E_q^n : d(x, y) \leq t\} = \cup_{i=0}^{t} S^{(i)}(x).$$

For two sets $X \subseteq E_q^n$ and $Y \subseteq E_q^n$ under $X + Y$ we mean here the set

$$X + Y = \{x + y : x \in X, y \in Y\},$$

where the addition is a coordinate-wise addition in E_q.

For two binary vectors $x = (x_1, \ldots, x_n)$ and $y = (y_1, \ldots, y_n)$ we say that x *covers* y, denoted $y \preceq x$, if the condition $y_i \neq 0$ implies $x_i \neq 0$ for all $i \in \{1, \ldots, n\}$.

We need one more combinatorial concept. A q-ary matrix V of size $N \times n$ is called an *orthogonal array* denoted $OA(N, n, q, t)$ of *strength* t*, index* $\lambda = N/q^t$ *and* n *constraints (or columns)*, if every $N \times t$ submatrix of V contains as a row every q-ary vector of length t exactly λ times (see, for example, [18]).

For two binary vectors $x = (x_1, \ldots, x_n)$ and $y = (y_1, \ldots, y_n)$ denote by $z = x \circ y$ the vector $z = (z_1, \ldots, z_n)$, where $z_i = x_i y_i$. In these terms, a vector x covers a vector y if the following equality holds $x \circ y = y$. Define the *k-face* $\Gamma_{[x]}$, induced by the vector x, as the following set of vectors

$$\Gamma_{[x]} = \{y \in \mathbb{F}_2^n : x \circ y = x\},$$

where $k = n - |\operatorname{Supp}(x)|$.

A subset K of a binary code C of length n with distance d is called an *i-component* of C if the set

$$C' = \{C \setminus K\} \bigcup \{K + e_i\}$$

is also a code with distance d, where e_i is a binary vector of length n with only one nonzero ith coordinate position, $i \in \{1, \ldots, n\}$.

Recall several simple commonly known operations over codes which permit to obtain new codes from already known. For a binary code C of length n, with minimum distance $d = 2e + 1$ denote by C^* the code obtained from C by adding one more coordinate position, the *overall parity check position*, such that C^* has codewords only of even weight. The code C^* (the *extension* of C) has length $n+1$, the minimum distance $d^* = d + 1$, and the same cardinality $M^* = M$. For a q-ary code C, the extension means the code obtained by adding one more position to every codeword such that in the resulting code, for every codeword, the sum in E_q of its coordinates is equal to zero.

For a code C of length n, the *i-punctured code* is obtained by deleting the coordinate i in all codewords. When the coordinate i is not specified, it is assumed that the parameters of the resulting code do not depend on the choice of the ith deleted position. The *a-shortened code* of C is obtained by taking all codewords that have the value a in a fixed coordinate followed by deleting this coordinate. Similarly, we say that the code is obtained by shortening code C, if its parameters do not depend on the choice of a. In a more general case, for fixed vectors $x_1, \ldots, x_r$, such that

$$S = \bigcup_{i=1}^{r} \{\operatorname{Supp}(x_i)\}$$

where the set of coordinates S is of size $j = |S| < n$, we define the $\{x_1, \ldots, x_r\}$-shortened code of C (with minimum distance d) obtained by fixing $j = |\operatorname{Supp}(S)|$ nonzero coordinate positions of the set S, taking all codewords of C that have one of the vectors $x_1, \ldots, x_r$ in these j coordinates, and then deleting the j fixed coordinates of the set S. Thus, the resulting code has length $n-j$ and the minimum distance

$$d' \geq d - \max_{\substack{i,i' \in \{1,\ldots,r\} \\ i \neq i'}} d(x_i, x_{i'}).$$

Similarly, we write $\langle x_1, \ldots, x_r \rangle$-shortening of C when, instead of vectors $x_1, \ldots, x_r$ on the positions $\mathrm{Supp}(S)$, we choose all vectors from the linear span of these vectors $x_1, \ldots, x_r$, i.e. the code generated by these vectors. In this case, we have to take into account the maximal distance in the code generated by the vectors $x_1, \ldots, x_r$.

We will use later more complicated constructions, such as lifting of the alphabet, concatenation, the Kronecker product and modified concatenation, which we will define in the corresponding sections.

Let M be a monomial matrix, i.e. a matrix with exactly one nonzero entry in each row and column. If q is prime, then the *automorphism group* of C, $\mathrm{Aut}(C)$, consists of all monomial $(n \times n)$-matrices M over $\mathbb{F}_q$ such that $cM \in C$ for all $c \in C$. If q is a power of a prime number, then $\mathrm{Aut}(C)$ also contains any field automorphism of $\mathbb{F}_q$ which preserves C. The group $\mathrm{Aut}(C)$ acts on the set of cosets of C in the following way: for all $\pi \in \mathrm{Aut}(C)$ and for every vector $v \in \mathbb{F}_q^n$ we have $\pi(v+C) = \pi(v)+C$. For a binary code C, the automorphism group of C is the set of coordinate permutations that fixes C as a set.

2.2.2 Completely Regular Codes

Let $C \subset E_q^n$ be a code. For a given vector $x \in E_q^n$, we denote by $B_{x,i}$ the number of codewords at distance i (where $0 \leq i \leq n$) from x. The *outer distribution matrix* of C is a $q^n \times (n+1)$ matrix B, with entries

$$B_{x,i} = \mid \{v \in C : d(x,v) = i\} \mid .$$

Hence the row B_x, i.e. an integer vector

$$B_x = (B_{x,0}, \ldots, B_{x,n}),$$

is the weight distribution of the translate $C + x$. Denote by $b+1$ the number of distinct rows of B.

For a code C, define the sets

$$C(i) = \{x \in \mathbb{F}_q^n : d(x,C) = i\},$$

where $d(x,C)$ is the distance of x to the code C, defined above. The sets C, $C(1)$, ..., $C(\rho)$ are called *subconstituents* in [118] and *layers* by some other authors. Note that $C(0) = C$ and $C(t) \neq \emptyset$ if and only if $t \leq \rho$.

We say that a binary code C of length n is *antipodal*, if for every codeword $c \in C$, there exists the codeword $\bar{c}$, which is at distance n from c, that is $\bar{c} = c + \mathbf{1}$. Obviously, if C is a binary distance invariant code and contains the codeword of weight n, then C is antipodal. But if $\mathbf{1} \notin C$, then C is non-antipodal.

Definition 2.1 ([80]). *A code C with covering radius ρ is t-*regular*, where $0 \leq t \leq \rho$, if for all $i, i = 0, \ldots, \rho$, the value $B_{x,i}$ depends only on i and on the distance $d(x, C)$ of x to C, for all x such that $d(x, C) \leq t$.*

In other words, C is t-regular if B_x depends only on $d(x, C)$ for $d(x, C) \leq t$. By definition, if C is t-regular, then it is j-regular, for all $j = 0, \ldots, t$. Intuitively, C is t-regular if we "see" the same amount of codewords at the same distances from any vector at distance at most t from the code. For example, a 0-regular code is exactly a distance invariant code.

There are several equivalent definitions of *completely regular* code (CR code, for short).

Definition 2.2 ([57]). *A code C is CR if it is ρ-regular.*

Clearly, the following definitions are equivalent.

1. A code C is CR if for all $x \in E_q^n$, the value $B_{x,i}$ depends only on i and $d(x, C)$.
2. A code C is CR if for all $x \in E_q^n$, the weight distribution B_x depends only on $d(x, C)$.
3. A code C is CR if $b = \rho$.

However, it is not that straightforward to see the equivalence of the definitions above with the next definition. Two vectors x and y in E_q^n are said to be *neighbors* if they are at distance $d(x, y) = 1$ from each other.

Definition 2.3 ([118]). *A code C is* completely regular *if, for all $\ell \geq 0$, every vector $x \in C(\ell)$ has the same number c_ℓ of neighbors in $C(\ell - 1)$ and the same number b_ℓ of neighbors in $C(\ell + 1)$, where we set $c_0 = b_\rho = 0$.*

It is clear that the sets C, $C(1), \ldots, C(\rho)$ form a partition, called *distance partition* of the space $\mathbb{F}_q^n$. If the conditions of Definition 2.3 are satisfied, then the partition is said to be *equitable*. Hence, C is CR if and only if the distance partition $C, \ldots, C(\rho)$ is equitable.

For the equivalence between Definitions 2.2 and 2.3, see Theorem 4.1 in [118], which we formulate later as Theorem 2.15. For $\ell \geq 0$, let $a_\ell = n(q - 1) - b_\ell - c_\ell$. Thus, a_ℓ is the number of neighbors in $C(\ell)$ of any vector in $C(\ell)$. The parameters a_ℓ, b_ℓ and c_ℓ are called the *intersection numbers* and the sequence $\mathrm{IA} = \{b_0, \ldots, b_{\rho-1}; c_1, \ldots, c_\rho\}$ is referred to as the *intersection array* of C.

Let us give examples of CR codes. The first example of CR code clearly is the whole set E_q^n, or the Hamming scheme $H(q, n)$. We give here the intersection numbers of this scheme, which we will need later and which we denote for all this chapter as $a_i^{(h)}$, $b_i^{(h)}$ and $c_i^{(h)}$:

$$a_i^{(h)} = i(q - 2), \quad b_i^{(h)} = (n - i)(q - 1), \quad c_i^{(h)} = i, \quad i = 0, \ldots, n. \tag{2.1}$$

Recall that a code C is *perfect* if $\rho = e$. It means that for any $x \in E_q^n$ there exists the unique codeword $v \in C$, such that $d(x, v) \leq e$. In other words, the union of all balls $B^{(e)}(v)$ of radius e centered in the all codewords v of C covers the whole space E_q^n without intersections:

$$E_q^n = \bigcup_{v \in C} B^{(e)}(v).$$

A code C with minimum distance $d = 2e+1$ is *quasi-perfect* if $\rho = e+1$. In this case the balls of radius $e+1$ centered in the all codewords of C covers the whole space E_q^n with intersections, since there are codewords of C at the minimum distance $2e+1$.

There are the following trivial perfect codes:

1. The whole space $C = E_q^n$ with $d = 1$ and $e = 0$.
2. Single codeword codes $C = \{x\}$ where x is an arbitrary vector in E_q^n with $e = n$.
3. Binary repetition codes of odd length which consist of two antipodal vectors of length n with $d = n$ and $e = (n-1)/2$.

Nontrivial perfect codes exist only for $e \leq 3$ [151,162]. In $\mathbb{F}_q^n$, the following perfect codes are known:

4. The binary Golay code with $n = 23$, $k = 12$, $d = 7$, which we denote $\mathcal{G}_{23}$.
5. The ternary Golay code with $n = 11$, $k = 6$, $d = 5$, which we denote $\mathcal{G}_{11}$.
6. Perfect codes of size $M = q^{n-m}$ and length $n = (q^m-1)/(q-1)$ with minimum distance $d = 3$.

In the last case, for every prime power q and for every natural $m \geq 2$, there is a unique linear version which is the Hamming code of length n_m, where

$$n_m = \frac{q^m - 1}{q - 1}$$

which we denote for the whole chapter by $\mathcal{H}_m$.

2.2.3 Completely Regular and Uniformly Packed Codes

An interesting class of codes, closely related to CR codes, is the class of *uniformly packed* (UP) codes, which were introduced in 1969 by Semakov, Zinoviev and Zaitsev [133,136]. There are three mostly used different classes of UP codes. We will start with the first class of such codes introduced in the papers mentioned above.

Definition 2.4 ([133,136]). *A binary quasi-perfect code C with minimum distance $d = 2e+1$ is* uniformly packed in the narrow sense *if there exists a natural number μ, such that*

$$B_{x,e} + B_{x,e+1} = \mu$$

for any vector $x \in C(e) \cup C(e+1)$.

The parameter μ was called *packing density* in [133,136]. These UP codes are called by some authors also as *strongly uniformly packed.* We emphasize that all binary perfect codes fall into this category with value $\mu = (n+1)/(e+1)$, which is maximum possible value. Indeed, it can be seen that a UP code C of length n, with minimum distance $d = 2e+1$ and with $\mu = (n+1)/(e+1)$ is perfect [136]. For all other codes, μ is bounded above as:

$$\mu \leq \left\lfloor \frac{n+1}{e+1} \right\rfloor. \tag{2.2}$$

From the definition it follows immediately (by sphere packing arguments) that μ is uniquely defined by the parameters $n, d = 2e+1$ and M of the code C. Indeed, since C is quasi-perfect we have from the definition [136]

$$M \cdot \left(\sum_{i=0}^{e-1} \binom{n}{i} + \frac{1}{\mu} \cdot \left(\binom{n}{e} + \binom{n}{e+1} \right) \right) = 2^n,$$

implying

$$\mu = \frac{M \cdot \binom{n+1}{e+1}}{2^n - M \cdot \sum_{i=0}^{e-1} \binom{n}{i}}.$$

Uniformly packed in the narrow sense codes include:

1. As we already mentioned, binary perfect codes [136].
2. 1-shortened binary perfect codes (which have $\mu = (n-e)/(e+1)$) [136].
3. Preparata codes of length $n = 2^{2m}-1$ for which $\mu = n/(e+1) = n/3$, i.e. these codes are the most closest to perfect codes relative to the value of packing density, what was mentioned in [136].
4. The binary primitive BCH codes with designed distance 5 of length $n = 2^{2m+1} - 1$ with $\mu = (n-1)/(e+1) = (n-1)/6$ [12].
5. The Hadamard $(11, 24, 5)$-code with $\mu = 3$ [80].

In [136] it was proved, in terms of Definition 2.2, that UP codes in the narrow sense and codes obtained as their extensions (which are not uniformly packed in this sense but are UP in the wide sense; see Definition 2.6 below) are CR.

UP codes in the narrow sense are a subclass of UP codes, introduced in 1975 by Goethals and van Tilborg [80].

Definition 2.5 ([80]). *A quasi-perfect q-ary code C with minimum distance $d = 2e+1$ is called* uniformly packed *if there exist natural numbers λ and μ such that for any vector x the value $B_{x,e+1}$ takes two values:*

$$B_{x,e+1} = \begin{cases} \lambda & \text{if } d(x,C) = e \\ \mu & \text{if } d(x,C) = e+1. \end{cases}$$

For the case $\mu = \lambda + 1$ any such binary uniformly packed code is uniformly packed in the narrow sense. Note that binary extended perfect codes fall into this category with values $\lambda = 0$ and $\mu = (n+1)/(e+1)$ [136]. These codes include also the ternary perfect Golay code, its extension, and the code obtained by shortening. Nonexistence of unknown infinite familes of UPNS codes was proven in 1971 by Semakov, Zinoviev and Zaitsev in [136] and for all such codes in 1977 by Lindström in [107]. Furthermore, van Tilborg [153] in 1976 showed that no other nontrivial UP codes exist for $e > 3$. He also determined all UP codes with $e = 3$ and a subclass of those with $e = 1, 2$.

Calderbank [45] proved nonexistence of putative linear UP code C with parameters

$$n = 70, \;\; k = 58, \;\; d = 5, \;\; \rho = 3.$$

The parameters of this code were found by a computer search in [12] and C should have the following packing coefficients

$$\beta_0 = \beta_1 = 1, \;\; \beta_2 = -\frac{3}{5}, \;\; \beta_3 = \frac{1}{10}$$

and the roots of Lloyd polynomial (see Section 2.3.3)

$$\xi_1 = 24, \;\; \xi_2 = 32, \;\; \xi_3 = 40,$$

which we give here, since the existence of such nonlinear code is an open question. The proof is based on examining of partial geometries, associated with the dual 3-weight code $C^\perp$. This binary UP code with $d = 5$ (i.e. $e = 2$) was the last open case in the classification of all binary UP codes with $e = 2$ by Calderbank and Goethals in [47].

The following result is a generalization of the result in [136] to the case of UP codes.

Proposition 2.1 ([80]). *A uniformly packed code is completely regular.*

Note that the extension of any UP code, which is not perfect, may not be UP. This question was asked in 1989 in [38]. Before, in 1977 Bassalygo and Zinoviev [10] presented an example of such code. We will look at this example in more detail later (see the beginning of Section 2.4.2).

Now we are going to define a more general class of uniformly packed codes, which includes the whole class of completely regular codes and which was introduced in 1973 by Bassalygo, Zaitsev and Zinoviev [11,12].

Definition 2.6 ([11,12]). *A code $C \subseteq E_q^n$ with covering radius ρ is* uniformly packed in the wide sense *if there exist rational numbers $\beta_0, \ldots, \beta_\rho$ such that for any $x \in E_q^n$ the following equality takes place:*

$$\sum_{i=0}^{\rho} \beta_i B_{x,i} = 1. \tag{2.3}$$

The numbers $\beta_0, \ldots, \beta_\rho$ are called *packing coefficients*. For example, for perfect codes all these numbers are equal: $\beta_i = 1$ for all $i = 0, 1, \ldots, \rho = e$. The above concept is much more general than those in Definitions 2.4 and 2.5. In other words, a code C is UP in the wide sense if the (all-one) column $\mathbf{1}^t$ is a linear combination of the first ρ columns of the outer distribution matrix B. Later we will see that a CR code is uniformly packed in a wide sense. To simplify the code terminology, for all this chapter, we use the following abbreviated names for these three classes of uniformly packed codes, which we introduced:

UP codes ↔ uniformly packed codes;

UPNS codes ↔ uniformly packed in a narrow sense codes;

UPWS codes ↔ uniformly packed in a wide sense codes.

It follows from the definition that the parameters $\beta_0, \beta_1, \ldots, \beta_\rho$ of a UPWS code C with covering radius ρ satisfy the following system of linear equations for any fixed $\rho + 1$ elements $x_i \in C(i)$, $i = 0, \ldots, \rho$:

$$\left\{ \begin{array}{ccccccccc} & & & & & \beta_\rho B_{x_\rho,\rho} & = & 1, \\ & & \beta_\rho B_{x_{\rho-1},\rho} & & + & \beta_{\rho-1} B_{x_{\rho-1},\rho-1} & = & 1, \\ \ldots & & \ldots & & & \ldots & & \ldots \\ \beta_\rho B_{x_0,\rho} & + & \beta_{\rho-1} B_{x_0,\rho-1} & + \ldots & + & \beta_0 B_{x_0,0} & = & 1. \end{array} \right. \tag{2.4}$$

Since the matrix of the system (2.4) is nondegenerate, the parameters β_i are uniquely determined for a given UPWS code C. It also follows from (2.4) that the numbers β_i, $i = 0, 1, \ldots, \rho$, are rational and upper and lower bounds can be obtained for them from the corresponding bounds for the values $B_{x_i,j}$. Thus, for quasi-perfect uniformly packed codes (i.e., when $\rho = e + 1$), recognizing that

$$1 \leq B_{x_{e+1},e+1} < \frac{n(q-1)}{e+1}, \quad 0 \leq B_{x_e,e+1} \leq \frac{(n-t)(q-1)}{e+1}, \quad B_{x_e,e} = 1,$$

we obtain from (2.4) [12]

$$\beta_0 = \beta_1 = \ldots = \beta_{e-1} = 1, \quad \beta_e = 1 - \frac{k}{m}, \quad \beta_{e+1} = \frac{1}{m}, \tag{2.5}$$

where the positive integers k and m are bounded by the inequalities

$$0 < m < \frac{(q-1)n}{e+1}, \; 0 \leq k \leq \frac{(q-1)(n-e)}{e+1}. \tag{2.6}$$

Now we have to mention two more classes of UP codes which form two special cases of UP codes in the narrow and wide senses.

A binary quasi-perfect code of length n and distance $d = 2e + 1$ for which

$$\lambda + 1 = \mu = \left\lfloor \frac{n+1}{e+1} \right\rfloor,$$

(i.e. the inequality (2.2) becomes an equality) have been introduced in 1972 by Goethals and Snover in [79]. Such codes called *nearly perfect* include perfect codes and in turn belong to uniformly packed in the narrow sense codes. Clearly they include, except perfect codes, Preparata codes and codes, obtained by shortening of perfect.

Goethals and van Tilborg [80] in 1975 suggested to generalize UP codes to the codes with the following packing coefficients:

$$\begin{array}{lllll} \beta_0 & = \ldots = & \beta_{d-\rho-1} & = & 1, \\ \beta_{d-\rho} & = \ldots = & \beta_e & = & 1-\lambda/\mu, \\ \beta_{e+1} & = \ldots = & \beta_\rho & = & 1/\mu, \end{array}$$

where $\rho = \rho(C)$ and $e = \lfloor (d-1)/2 \rfloor$. These codes are special case of UPWS codes and form an intermediate class of codes between UP codes and UPWS codes. Remark that there are many examples of UPWS codes, which do not satisfy the definitions of Goethals and van Tilborg [80]. For example, the unique binary self-dual $[48, 24, 12]$-code C^* with $\rho^* = 8$ (which is a classical extended quadratic-residue code) as a UPWS code, has the following packing coefficients:

$$\begin{array}{lllllll} \beta_0 = \beta_1 = \beta_2 = & \beta_3 & = 1, & \beta_4 & = & 5/27, \\ \beta_5 = \beta_6 = \beta_7 = & 1/9, & & \beta_8 & = & 1/54. \end{array}$$

This code C^* is an extension of the binary UPWS $[47, 24, 11]$-code C with covering radius $\rho = 7$ by adding an overall parity check symbol. The fact that this $[47, 24, 11]$-code is an UPWS was mentioned by Goethals and van Tilborg in [80], where the following packing coefficients were also derived:

$$\beta_0 = \beta_1 = \beta_2 = \beta_3 = 1, \quad \beta_4 = \beta_5 = \beta_6 = \beta_7 = 1/9.$$

Surely, this UPWS code can be considered as a uniformly packed code in the sense of Goethals and Tilborg [80]. Generally speaking, binary uniformly packed codes do not save their property to be uniformly packed under extension. Later on, we will consider the extension of binary UPWS and CR codes more in detail (see Section 2.4.2); for UPWS codes this problem is completely solved in [10].

As it was mentioned in [80], if a binary (putative) self-dual $(72, 2^{36}, 16)$ code C^* with $\rho = 12$ would exist, then its punctured $(71, 2^{36}, 15)$ code C with $\rho = 11$ would be a UPWS code in the sense of Goethals and Tilborg with $\lambda = 78$ and $\mu = 91$. If the code C^* would exists, then it would be an UPWS code (not uniformly packed in the sense Goethals and Tilborg) whose the packing coefficients can be written down using Theorem 2.18 in Section 2.4.2.

Now, we give some easy examples of CR codes.

1. As we have already mentioned, any perfect code is a CR code [139].
2. The set of all even weight vectors is a linear CR code with $\rho = 1$ [136].
3. An extended binary perfect code is a quasi-perfect UP code, and so CR [136].

4. For any $q \geq 2$ a q-ary repetition $(3, q, 3; 2)_q$ code is CR with intersection array

$$\mathrm{IA} = \{3(q-1), 2(q-2); 1, 6\}.$$

5. For any CR code C, the subconstituent $C(\rho)$ is also CR, with reversed intersection array [118].

Before finishing this section, we will provide one more result for UP codes (and hence for CR codes). Denote by $L_{n,d}$ the number of different UPWS (n, M, d) codes with $\rho \leq d$, where n and d are odd. Tokareva [155] proved that

$$L_{n,d} \leq 2^{2^{n-\frac{d}{2}\log n + o(\log n)}},$$

where the logarithm is binary and $o(\log n)$ is going to zero when n is growing to infinity. So, it is similar to the upper bound for the number of binary perfect codes, where we have also the lower bound with twice less constant in the double exponent. Hence the following open question arises: *what is the best lower bound for the number $L_{n,d}$?*

2.2.4 Introduction to Designs and Distance Regular Graphs

Two related combinatorial structures connected with CR codes are of special interest: t-designs and distance-regular graphs.

Definition 2.7. *A $T(v, k, t, \lambda)$-design T (also called a $t-(v, k, \lambda)$-design or simply a t-design) is an incidence structure $(\mathcal{X}, \mathcal{B})$ where $\mathcal{X}$ is a v-set of elements (called* points*) and $\mathcal{B}$ is a collection of k-subsets of points (called* blocks*) such that every t-subset of points is contained in exactly $\lambda \geq 1$ blocks, where $1 \leq t \leq k \leq v$.*

In terms of a (binary) incidence matrix, a $T(v, k, t, \lambda)$-design is a binary code C of length $n = v$ with codewords of weight $w = k$ such that any binary vector of length n and weight t is covered by exactly λ codewords. A $T(v, k, t, \lambda)$-design is *simple* if it has no repeated blocks, and *nontrivial* if it is simple and does not contain all the k-subsets of points (in particular, this means that k belongs to the following interval : $2 \leq k \leq v-2$). A design with $\lambda = 1$ is called a *Steiner system* and is denoted also by $S(v, k, t)$. The following properties are well known and can be found, e.g., in monographs [18,22,88].

Proposition 2.2. *For a given $T(v, k, t, \lambda)$-design, every i-subset of points $(0 \leq i \leq t)$ is contained in exactly λ_i blocks, where*

$$\lambda_i = \lambda \frac{\binom{v-i}{t-i}}{\binom{k-i}{t-i}}.$$

Corollary 2.1. *Let T be a $T(v, k, t, \lambda)$-design then:*

1. *T is a $T(v, k, i, \lambda_i)$-design for any $0 \leq i \leq t$;*

2. $\lambda = \lambda_t$;

3. *The number of blocks of* T *is* $b = \lambda_0$;

4. *Each point of* X *is contained in the same number of blocks, namely,* $r = \lambda_1 = bk/v$ *(the number* r *is called the* replication (or repetition) number*).*

Designs with $t \leq 5$ are known from 1938 (starting from famous Witt designs [54], i.e., classical Steiner systems $S(24, 8, 5)$ and $S(12, 6, 5)$). The first nontrivial designs for $t = 6$ were found in 1983 by Magliveras and Leavitt [113]. In 1987, L. Teirlinck [146] proved that there are nontrivial simple t-designs for any natural number t.

There is a natural q-ary generalization of such designs (see [1,57,80,139,161–163]). Let $E = \{0, 1, \dots, q-1\}$. A collection $\mathcal{B}$ of b vectors $x_1, \dots, x_b$ of length v and weight k over E is called a q-ary $T(v, k, t, \lambda)_q$ design, if for every vector y over E of length v and weight t there are exactly λ vectors $x_{i_1}, \dots, x_{i_\lambda}$ from $\mathcal{B}$ such that $d(y, x_{i_j}) = k - t$ for all $j = 1, \dots, \lambda$ (in other words, two vectors y and x_{i_j} coincide at non-zero positions of vector y). If $\lambda = 1$, then we obtain a q-ary Steiner system, denoted $S(v, k, t)_q$.

For a code C denote by C_w the set of all codewords of C of weight w. Regularity of a code C implies that the sets C_w induce t-designs as we see in the next statement.

Theorem 2.1 ([80]). *Let* C *be a* t*-regular code with minimum distance* $d \geq 2t$*, then the supports of any nonempty set* C_w *form the blocks of a* t*-design.*

Directly from the definition of CR codes, we have the following common known results.

Theorem 2.2. *Let* C *be a* q*-ary CR code of length* n *with distance* d*.*

1. *If* $d = 2e + 1$ *then any nonempty set* C_w *is a* $T(n, w, e, \lambda_w)_q$*-design.*

2. *If* $d = 2e + 2$ *then any nonempty set* C_w *is a* $T(n, w, e+1, \lambda_w)_q$*-design.*

3. *If* C *is a* q*-ary perfect code then any nonempty set* C_w *is a* $T(n, w, e+1, \lambda_w)_q$*-design and* C_d *is a Steiner system* $S(n, 2e+1, e+1)_q$*.*

4. *If* C *is an extended binary perfect code then any nonempty set* C_w *is a* $T(n, w, e+2, \lambda_w)$*-design and* C_d *is a Steiner system* $S(n, 2e+2, e+2)$*.*

Let Γ be a finite connected simple graph (i.e., undirected, without loops and multiple edges). Let $d(\gamma, \delta)$ be the distance between two vertices γ and δ (i.e., the number of edges in the minimal path between γ and δ). The *diameter* D of Γ is its largest distance. Two vertices γ and δ from Γ are *neighbors* if $d(\gamma, \delta) = 1$. Denote

$$\Gamma_i(\gamma) \;=\; \{\delta \in \Gamma : \; d(\gamma, \delta) = i\}.$$

An *automorphism* of a graph Γ is a permutation π of the vertex set of Γ such that, for all $\gamma, \delta \in \Gamma$ we have $d(\gamma, \delta) = 1$ if and only if $d(\pi(\gamma), \pi(\delta)) = 1$.

Definition 2.8 ([38]). *A simple connected graph* Γ *is called* distance-regular *if it is regular of valency* k*, and if for any two vertices* $\gamma, \delta \in \Gamma$ *at distance* i *apart, there are precisely* c_i *neighbors of* δ *in* $\Gamma_{i-1}(\gamma)$ *and* b_i *neighbors of* δ *in* $\Gamma_{i+1}(\gamma)$*. Furthermore, this graph is called* distance-transitive, *if for any pair of vertices* γ, δ *at distance* $d(\gamma, \delta)$ *there is an automorphism* π *from* $Aut(\Gamma)$ *which moves this pair* (γ, δ) *to any other given pair* γ', δ' *of vertices at the same distance* $d(\gamma, \delta) = d(\gamma', \delta')$.

The sequence $\{b_0, b_1, \ldots, b_{D-1}; c_1, c_2, \ldots, c_D\}$, where D is the diameter of Γ, is called the *intersection array* of Γ. The numbers c_i, b_i, and a_i, where $a_i = k - b_i - c_i$, are called *intersection numbers.* Clearly $b_0 = k, \;\; b_D = c_0 = 0, \;\; c_1 = 1$.

Definition 2.9 ([38]). *Let* C *be an additive CR code with covering radius* ρ *and intersection array given by* $IA = \{b_0, \ldots, b_{\rho-1}; c_1, \ldots c_\rho\}$*. Let* $\{A\}$ *be the set of cosets of* C*. Define the graph* Γ_C*, which is called the* coset graph of C, *taking all different cosets* $A = C + x$ *as vertices, with two vertices* $\gamma = \gamma(A)$ *and* $\gamma' = \gamma(A')$ *adjacent if and only if the cosets* A *and* A' *contain neighbor vectors, i.e., there are* $v \in A$ *and* $u \in A'$ *such that* $d(v, u) = 1$.

2.3 NECESSARY CONDITIONS ON CR CODES

2.3.1 Characteristic Numbers and Characteristic Polynomial of a Code

We need some more tools for codes, namely, characteristic numbers, external distances, and characteristic polynomials. To define these concepts, first, recall characters of codes (see [10,58,80]).

For the case $E_q = \mathbb{F}_q$, where $q = p^m$, p prime and $m = 1, 2, 3, \ldots$, denote by $\chi_a(b)$ a nonprincipal additive character of $\mathbb{F}_q$, namely, for any elements $a, b \in \mathbb{F}_q$, we define

$$\chi_a(b) = \exp\left\{\frac{2\pi i \mathrm{Tr}(ab)}{p}\right\},$$

where $i^2 = -1$ and $\mathrm{Tr}(x)$ denote the trace function from $\mathbb{F}_q$ into $\mathbb{F}_p$, i.e.

$$\mathrm{Tr}(x) = x + x^p + x^{p^2} + \cdots + x^{p^{m-1}}.$$

We extend in a natural way this definition for any vectors $x = (x_1, \ldots, x_n)$ and $y = (y_1, \ldots, y_n)$ over $\mathbb{F}_q$:

$$\chi_x(y) = \exp\left\{\frac{2\pi i \mathrm{Tr}(\langle x \cdot y\rangle)}{p}\right\}, \;\; \forall \; x, y \in \mathbb{F}_q^n;$$

we recall that by $\langle x \cdot y \rangle$ we denote the inner product of vectors x and y over $\mathbb{F}_q$.

For the case when the size q of the alphabet $E_q = \{0, 1, \ldots, q-1\}$ is not a prime power, as we agreed before, we assume that E_q is the ring $\mathbb{Z}_q$ of residues modulo q. In this case for vectors $u = (u_1, \ldots, u_n)$ and $v = (v_1, \ldots, v_n)$ from $\mathbb{Z}_q^n$ we set

$$\chi_u(v) \; = \; \exp\left\{\frac{2\pi i}{q}\sum_{j=1}^{n} u_j\, v_j\right\}, \;\; \forall \; v, u \in \mathbb{Z}_q^n,$$

where $\chi_u(v)$ is a character of the additive group $\{0, 1, \ldots, q-1\}$ with addition modulo q.

Recall that we denote by $S^{(j)} = S^{(j)}(\mathbf{0})$ a sphere of the radius j at center $\mathbf{0}$, i.e. the set of all vectors x of E_q^n of Hamming weight j. The polynomial

$$P_k(n,x) = \sum_{j=0}^{k} (-1)^j (q-1)^{k-j} \binom{x}{j} \binom{n-x}{k-j}$$

is the Krawtchouk polynomial over x of degree k, where, for any real number a, we set

$$\binom{a}{i} = a\,(a-1)\ldots(a-i+1)\,\frac{1}{i!}.$$

Definition 2.10. *Now for any code $C \subseteq E^n$ of length n with minimum (Hamming) distance d, define the following concepts:*

1. *The* characteristic numbers $B_0, B_1, \ldots, B_n$ *of the code C [58]:*

$$B_j = \frac{1}{|C|^2} \sum_{u \in S^{(j)}} \left| \sum_{v \in C} \chi_u(v) \right|^2,$$

2. *The* external distance $s = s(C)$ *of the code C [58] which is a number of nonzero B_j with j satisfying $1 \leq j \leq n$. If C is a linear code, then $s(C)$ is a number of nonzero weights of the dual code $C^{\perp}$.*

3. *The* characteristic polynomial $F_C(x)$ *of the code C [58,80]*

$$F_C(x) = \frac{q^n}{|C|} \prod_{j \geq 1\,|\,B_j \neq 0} \left(1 - \frac{x}{j}\right) = \sum_{i=0}^{s(C)} \beta_i\, P_i\,(n,x),$$

where β_i are the coefficients of the Krawtchouk expansion of the characteristic polynomial $F_C(x)$.

Remark that the polynomial $F_C(x)$ was called *minimal polynomial* by Delsarte [58], while Goethals and van Tilborg [80] called it by *characteristic polynomial* and we follow this term.

The external distance s is a key parameter as we will see in several theorems in the next section. But now we give a nice property of the characteristic polynomial $F_C(x)$ for an arbitrary code C, which we will need later. Recall that by $B_{x,i}$ we denote the number of codewords of C which are at distance i from any vector $x \in E_q^n$.

Proposition 2.3 ([58,80]). *Let $\beta_0, \beta_1, \ldots, \beta_n$ be the coefficients of the Krawtchouk expansion of the characteristic polynomial $F_C(x)$. Then for any $x \in E_q^n$*

$$\sum_{i=0}^{n} \beta_i B_{x,i} = 1.$$

Therefore, we have the following necessary and sufficient conditions for a code C to be UPWS.

Proposition 2.4 ([10]). *The necessary and sufficient condition for a code C with covering radius ρ to be UPWS with packing coefficients $\beta_0, \beta_1, \ldots, \beta_\rho$ is the following equality*

$$F_C(x) = \sum_{i=0}^{\rho} \beta_i P_i(n, x),$$

where $F_C(x)$ is its characteristic polynomial.

2.3.2 Basic Properties of CR Codes

Recall that B is the outer distribution matrix of the code C and $b+1$ (introduced by Solé in [142,143]) denotes the number of different rows of B.

Theorem 2.3. *The following statements hold:*

1. *[58]* $\mathrm{Rank}(B) = s + 1$.
2. *[142,143]* $s \leq b$.
3. *[58]* $\rho \leq s$.

Hence, the following nice chain of inequalities is valid:

$$e \leq \rho \leq s \leq b.$$

Now, we have the following characterizations of perfect, CR, UP, and UPWS codes.

Theorem 2.4.

1. *[58] A code C is perfect if and only if $e = s$.*
2. *[80,136] A code C is UP if and only if $s = e + 1$.*
3. *[142,143] A code C is CR if and only if $\rho = b$.*
4. *[142,143] If a code C is CR then $\rho = s$.*
5. *[10] A code C is UPWS if and only if $\rho = s$.*

The converse of (4) is not true. Delsarte [58] gives the example of a $[48, 24, 12]$ extended quadratic residue code with $\rho = s = 8$ and $b = 14$. Therefore, by Theorem 2.4,(iii) this code is not CR, since $\rho \neq b$. However, by the same Theorem 2.4,(v) this code is UPWS, since $\rho = s$ (see the end of Section 2.2.3). Therefore, we obtain [38] the following well-known result.

Corollary 2.2. *If C is CR, then C is UPWS.*

Many UPWS codes which are not CR are constructed in [30,126,127]. In most cases, such codes belong to the three wide families of codes, obtained by the Kronecker product construction [126], by the lifting of the alphabet of perfect codes [127], and by the concatenation of Hamming codes [30] which we consider later in Sections 2.6.3, 2.7.1 and 2.7.3.

The following properties of CR codes are due to Delsarte [58].

Theorem 2.5 ([58]).

1. *If $t \geq d - s \geq 0$, then C is t-regular.*

2. *If C is t-regular, with $t \geq s - 1$, and $d \geq 2s - 1$, then C is CR.*

We can strengthen these conditions if weights of all codewords of C are even (such code is called *even*).

Theorem 2.6. *Let C be an even code. Then,*

1. *[10] If $t \geq d - s + 1 \geq 0$, then C is t-regular.*

2. *[38] If $d \geq 2s - 2$, then C is CR.*

Non-antipodal CR codes (i.e. when codes do not contain the codeword **1** of all ones) have the following interesting properties:

Theorem 2.7. *Let C be a non-antipodal CR code with zero codeword $\mathbf{0} = (0, \ldots, 0)$ and with covering radius ρ.*

1. *[26] The set $C(\rho)$ is a translate of C by the vector from all ones $\mathbf{1} = (1, \ldots, 1)$, i.e. $C(\rho) = C + \mathbf{1}$.*

2. *[128] The code $A = C \cup C(\rho)$ is a CR code with covering radius $\rho_a = \lfloor \rho/2 \rfloor$.*

Except all perfect binary codes which satisfy this property [26], currently it is known only one family of CR codes with $d = 3$ also having it [128] (they are so-called *binomial codes*, which we will consider later in Section 2.8.1. Hence the next open question arises: *which binary CR codes satisfy this property?*

Gillespie [72] observed that non-antipodal codes cannot have a large minimum distance, proving the following result.

Theorem 2.8 ([72]). *Let C be a binary CR code with $|C| > 1$ and minimum distance $d > max\{2, n/2\}$. Let C contain the zero codeword $\mathbf{0}$ and do not contain the all one vector $\mathbf{1}$, i.e. C is non-antipodal. Then either $d = n$ and C is equivalent to the binary repetition $(n, 2, n)$ code, or $(n, d) = (7, 4)$ and C ie equivalent to the even half of the Hamming $[7, 3, 4]$ code $\mathcal{H}_3$.*

Therefore we see that the existence of the codeword from all ones, i.e. the word $\mathbf{1} = (1, 1 \ldots, 1)$ in a CR code is quite important. For perfect codes, this question was derived in [130] and this problem was solved for UPNS codes with $d \geq 5$ and for their extensions (see Theorem 9 in [136]). Similar arguments, which were used in [136], show that for $d \geq 5$ only CR codes with large covering radius ρ can be non-antipodal. In particular, Gillespie [72] proved the following

Proposition 2.5. *Let C be a non-antipodal binary CR code with parameters (n, M, d). Let C contain the zero word $\mathbf{0}$ and let $d \geq 3$. Then C has the covering radius $\rho \geq d - 1$.*

We can slightly improve this result.

Proposition 2.6. *Let C be a non-antipodal binary CR code with parameters $(n, M, d = 2e + 1)$ and covering radius ρ, where $e \geq 1$, and let w be the largest weight of codewords of C. Then we are in one of two following cases: (i) C has a covering radius $\rho \geq 2e + 1$; (ii) The set $C_w + \mathbf{1}$ is a Steiner system $S(n, \rho, e)$, implying that $\rho \geq 2e$ and $n - e + 1$ is multiple to $e + 1$.*

Proof: Assume in contrary that $\rho \leq 2e - 1$. Since C does not contain the word $\mathbf{1}$, consider the set C_w of codewords with the largest weight w. Recall that by Theorem 2.7 we have that $C(\rho) = C + \mathbf{1}$. Hence by the definition of covering radius we obtain that $w \geq n - \rho$. But the set C_w is a e-design $T = T(n, w, e, \lambda_w)$ (Theorem 2.2), which we present in terms of its incidence $(v \times b)$-matrix A_T (i.e. the rows of A_T correspond to coordinate positions of C and the columns correspond to blocks of the T, or codewords of the C of the weight w). It means that the set of all columns of A_T forms a constant-weight (n, w, d, b) code with minimum distance $d(T) \geq 2e + 2$. Consider the complementary design $T' = T(n, \rho, e, \lambda)$ formed by complementary blocks of T, which we also present in terms of its incidence matrix $A_{T'}$. Since T' is formed by the complementary words of C, any two (binary) vectors from $A_{T'}$ are at distance, at least, $2e + 2$, i.e. $d(A_{T'}) \geq 2e + 2$.

Now we have to consider two different cases. First, assume that $\lambda > 1$. Clearly the minimum (Hamming) distance $d(T)$ between corresponding vectors (i.e. between columns of its incidence matrix $A_{T'}$) in any e-design $T = T(n, \rho, e, \lambda)$ with $\lambda > 1$ satisfies the inequality $d(T) \leq 2(\rho - e)$ (indeed, any vector of weight e is covered by λ different vectors from T of weight ρ. For the e-design $T' = T(n, w', e, \lambda)$, we have, from one side, that $d(A_{T'}) \geq 2e+2$. From the other side, $d(A_{T'}) \leq 2(\rho - e)$, implying that $\rho \geq 2e + 1$.

The case $\lambda = 1$ implies that $T' = T(n, \rho, e, \lambda)$ is a Steiner system $S(n, \rho, e)$. In this case, by definition of the Steiner system, the distance $d(T')$ is equal $2(\rho - e + 1)$, but should be equal to $2e + 2$, which gives that $\rho \geq 2e$ and the value $(n - e + 1)$, evidently, should be divided by the number $\rho - e + 1$, that finishes the proof. △

Some more interesting results on non-antipodal CR codes, we will see later in Section 2.5.

2.3.3 Necessary Conditions on CR Codes

For a given CR code C with covering radius ρ and intersection numbers a_i, b_i, c_i, a tridiagonal matrix $L = [L_{i,j}]$ (indexed by $0, 1, \ldots, \rho$) which is called the *intersection matrix* of C is defined as follows:

$$L = \begin{bmatrix} a_0 & b_0 & 0 & \dots & 0 & 0 \\ c_1 & a_1 & b_1 & \dots & 0 & 0 \\ 0 & c_2 & a_2 & \dots & 0 & 0 \\ & \dots & \dots & \dots & \dots & \\ 0 & 0 & 0 & \dots & a_{\rho-1} & b_{\rho-1} \\ 0 & 0 & 0 & \dots & c_\rho & a_\rho \end{bmatrix}. \tag{2.7}$$

The *eigenvalues* of a completely regular code C are the eigenvalues of its intersection matrix L; which are distinct real numbers, since L is tridiagonal and nonnegative. Recall that the eigenvalues of the Hamming scheme $H(q,n)$ are equal to $(q-1)n-qj$, where $j = 0, 1, \dots, n$.

The following statement due to Neumaier [118] is called Lloyd's theorem for CR codes.

Theorem 2.9 ([118]). *Let C be a CR code of length n with covering radius ρ and with intersection matrix L. Then L has $\rho+1$ distinct integer eigenvalues, which are eigenvalues of $H(q,n)$, i.e. belong to the interval $-n, -(n-q), \dots, 0, 1, \dots, (q-1)n$.*

But for UPWS codes there exists another variant of such kind of theorem and which is a natural generalization of classical Lloyd's theorem [111] for perfect codes. That theorem was the main tool for the proofs of nonexistence theorems for perfect codes (see [151,162] and references there). Since any CR code is UPWS, it should satisfy both theorems which in some cases might be more useful.

The following result [11,12] is a variant of Lloyd's theorem for UPWS codes.

Theorem 2.10 ([12]). *Let C be a UPWS code of length n with covering radius ρ and with packing coefficients $\beta_0, \beta_1, \dots \beta_\rho$, defined by Proposition 2.3. Then generalized Lloyd's polynomial in ξ of degree ρ*

$$L_\rho(n,\xi) = \sum_{r=0}^{\rho} \beta_r P_r(n,\xi), \tag{2.8}$$

where $P_r(n,\xi)$ is the Krawtchouk polynomial, has ρ distinct roots, which are natural numbers between 0 and n.

Since any CR code is UPWS (Corollary 2.2), we obtain that for any CR code C we have two different Lloyd's theorems. Emphasize that in order to use Lloyd's Theorem 2.9 for CR code C we have to use the packing coefficients $\beta_0, \beta_1, \dots \beta_\rho$ of this code, which satisfies the equation (2.3) for any point x in E^n. Besides, these coefficients β_i are exactly those numbers in Proposition 2.3.

We have to emphasize once more that Theorem 2.9 was the main tool for proving nonexistence results for perfect codes in $H(q,n)$ (see [9,108,109,150–152,161,162]) and UPNS codes (see [107,153]). In particular, the essential property of Lloyd's polynomial $L_\rho(n,\xi)$ is that we can write two very useful equalities for its roots.

First, note that this polynomial has very simple expressions for the first and last nontrivial coefficients. If we set

$$L_\rho(n,\xi) = \sum_{j=0}^{\rho} L_j \xi^j$$

and denote the roots of $L_\rho(n,\xi)$ by $\xi_1,\ldots,\xi_\rho$, then, for the case of perfect codes (when $\beta_0 = \cdots = \beta_\rho = 1$, $d = 2e+1$ and $\rho = e$) from the coefficients L_0 and $L_{\rho-1}$ of the Lloyd polynomial $L_\rho(n,\xi)$ we immediately obtain the following two expressions:

$$L_0 = \xi_1\,\xi_2 \ldots \xi_e \;=\; \frac{e!}{q^e}\sum_{i=0}^{e} \beta_i (q-1)^i \binom{n}{i} \;=\; e\,!\,q^{n-k-e}, \tag{2.9}$$

and

$$L_{\rho-1} = \xi_1 + \xi_2 + \cdots + \xi_e \;=\; e \cdot \frac{q-1}{q}\left(n+1-\frac{(e+1)(q-2)}{2\,(q-1)}\right). \tag{2.10}$$

which implies the upper bound for the length n of a perfect code which generally less than the lower bound (see [151,162]). Currently, only the case of perfect codes over an arbitrary alphabet $q \neq p^m$ (where p prime) with distances 3 and 5 remains open. In particular, for the case $d = 5$, only one general result is known about the non-existence of perfect codes for $q = 2^a\,3^b$, where a and b are arbitrary natural numbers.

Theorem 2.11 ([9])**.** *There are no unknown nontrivial q-ary perfect codes with* $d = 5$ *for any* $q = 2^a 3^b$*, where a and b are arbitrary natural numbers.*

Consider the smallest nontrivial example of CR code.

Example 2.1. *Let C be the even half of perfect (Hamming) code* $\mathcal{H}_3$ *of length* 7*, i.e. the following* $[7,3,4]$ *code, which is CT (hence UPWS and CR) with* $\rho = 3$*:*

$$C = \left\{ \begin{array}{cc} (0000000), & (0101110), \\ (0010111), & (1001011), \\ (1100101), & (1110010), \\ (0111001), & (1011100). \end{array} \right\}$$

The code has the intersection array $IA = \{7,6,1;1,6,7\}$*. As an UPWS code, C has the following packing coefficients* β_i *(see (2.3)):*

$$\beta_0 = 1,\;\; \beta_1 = \frac{3}{7},\;\; \beta_2 = \frac{1}{3},\;\; \beta_3 = \frac{1}{7}.$$

Lloyd;s theorem (in the sense of Neumaier [118]) states that the intersection matrix L

$$L = \begin{bmatrix} 0 & 7 & 0 & 0 \\ 1 & 0 & 6 & 0 \\ 0 & 6 & 0 & 1 \\ 0 & 0 & 7 & 0 \end{bmatrix}. \tag{2.11}$$

has the $\rho+1$ integer eigenvalues θ_i in the interval $[-7,7]$. It is easy to check that the matrix L above has the eigenvalues $\theta_{1,2} = \pm 1$ and $\theta_{3,4} = \pm 7$.

Lloyd's theorem (in the sense of UPWS codes) states that generalized Lloyd's polynomial

$$\begin{aligned} L_3(7,\xi) &= \sum_{r=0}^{3} \beta_r P_r(7,\xi) \\ &= 1 + \frac{3}{7}(2\xi - 7) + \frac{1}{3}(2\xi^2 - 14\xi + 21) \\ &+ \frac{1}{7}(4\xi^3 - 38\xi^2 + 128\xi - 105) \end{aligned}$$

has 3 *integer roots in the interval* $[1,7]$. *This is the case, as it is easy to check that the roots of the* $L_3(7,\xi)$ *are* $3,4,7$.

In connection with two Lloyd's theorems for CR codes, the following two open questions arise: *(i) Is there a connection between these two Lloyd's theorems? (ii) Is there a direct connection between eigenvalues θ_i of the intersection matrix L and the roots ξ_j of the Lloyd polynomial of the CR code C"?*

2.3.4 CR Codes and Equitable Partitions

A very interesting fact that we cannot miss is that Lloyd's Theorem in the sense of Neumaier is valid for a more general combinatorial object, which is closely related to CR codes and which is actively being investigated at the present time. Now we want to introduce this classical object, namely, *equitable partition* for Hamming schemes $H(q,n)$ and, in general, for regular graphs (see [49,75,76] and references there).

Definition 2.11. *An ordered r-partition $P = P_1, \dots, P_r$ of the Hamming scheme $H(q,n)$,*

$$H(q,n) = \bigcup_{i=1}^{r} P_i$$

where P_i are non-intersections subsets of $H(q,n)$, is called equitable r-partition, if for any $i,j \in \{1,\dots,r\}$ there is an non-negative integer $S_{i,j}$, such that any vector from P_i has exactly $S_{i,j}$ neighbors in P_j.

In the literature, the concept of equitable partition of $H(q,n)$ is also referred to as "*perfect r-colorings*", "*regular partition*", or "*r-partition designs*". The matrix $S = [S_{i,j}]$ is called *quotient matrix*. It is clear that a q-ary CR code of length n and with the covering radius ρ induces a $(\rho+1)$-partition of the space $H(q,n)$. Indeed, its distance partition into subconstituents $C(i), i = 0,\dots,\rho$, as one can easily see, is an equitable partition. Evidently, the inverse statement, in the general case, is not valid, i.e. an arbitrary equitable partition does not necessarily induce a CR code.

Lloyd's Theorem 2.9 for CR codes as mentioned by Neumaier [118] (see also [76]) is also valid for equitable partitions. In this way, we have a very nice picture: *all CR codes and equitable partitions satisfy Lloyd's Theorem 2.8 and all CR and UPWS codes satisfy another Lloyd's theorem, namely, Theorem 2.9.*

The concept of regular partitions introduced in Definition 2.11 can be both strengthened and weakened. We will give both of these concepts, since they make sense and there are interesting examples in both cases. Let's start with a stronger concept given in [38]. A partition $P = (P_0, \ldots, P_{r-1}$ of $H(q,n)$ is *completely regular* if it is regular and all $P_i \in P$ are completely regular with the same outer distribution numbers. A weaker concept looks as follows [98]: a collection $P = (P_0, \ldots, P_{r-1}$ of subsets of $H(q,n)$ is called an *equitable family*, if there is a quotient matrix $S = [S_{i,j}]$ such that any x has exactly

$$\sum_{i \in I(x)} S_{i,j}$$

neighbors from P_j for every $j = 0, \ldots, r-1$, where $I(x)$ consists of indices of subsets P_i containing this vector x, i.e. $I(x) = \{i : x \in P_i\}$. If all P_i are disjoint and cover all space $H(q,n)$ then such equitable family is an equitable partition.

2.3.5 Sphere Packing Conditions and Weight Polynomials

For a CR $(n, N, d)_q$-code C with covering radius ρ and with intersection numbers a_i, b_i, c_i, denote by k_i the cardinality of $C(i)$. Define the real numbers $\kappa_1, \ldots, \kappa_\rho$ as

$$\kappa_i = \frac{k_i}{|C|} = \frac{|C(i)|}{|C|}.$$

In terms of κ_i the distance partition, induced by the code C, implies as in [118] the relation

$$|C| = \frac{q^n}{1 + \kappa_1 + \cdots + \kappa_\rho}. \tag{2.12}$$

Proposition 2.7 ([118]). *Let C be a CR $(n, N, d)_q$-code with covering radius ρ and with intersection numbers a_i, b_i, c_i. Then*

1. *If $C(i)$ and $C(i+1)$ are two subconstituents of C for $0 \le i < \rho$, then*

$$b_i |C(i)| = c_{i+1} |C(i+1)|. \tag{2.13}$$

2. *The following expression for κ_i is valid:*

$$\kappa_i = \frac{b_0 \cdots b_{i-1}}{c_1 \cdots c_i}, \quad i = 1, \ldots, \rho, \tag{2.14}$$

The next theorem generalizes the classical sphere packing condition for perfect codes for UPWS codes (hence, for all CR codes).

Theorem 2.12 ([12]). *Let C be a q-ary UPWS code of length n with packing coefficients $\beta_0, \beta_1, \ldots \beta_\rho$, defined by Proposition 2.3, i.e. for any $x \in E^n$ the equality (2.3) holds. Then*

$$|C| = \frac{q^n}{\sum_{i=0}^{\rho} \beta_i (q-1)^i \binom{n}{i}}. \tag{2.15}$$

Thus, CR codes satisfy two Lloyd's theorems and two sphere packing conditions. In both cases, these statements are based on two different tuples of packing coefficients, namely, $\beta_0, \beta_1, \ldots \beta_\rho$, defined by Proposition 2.3, i.e. the coefficients of the Krawtchouk expansion of the characteristic polynomial $F_C(x)$, and $\kappa_1, \ldots, \kappa_\rho$, defined by the intersection numbers a_i, b_i, c_i (which are absent for UPWS codes).

One of the important tools for studying codes is the weight function of codes to the consideration of which we proceed for CR and UPWS codes. Define the weighting function of a set $A \subseteq E^n$ as the following *weight polynomial* $W_A(x)$,

$$W_A(x) = \sum_{\mathbf{a} \in A} x^{\mathrm{wt}(\mathbf{a})} = \sum_{i=0}^{n} \eta_i x^i,$$

where $\mathrm{wt}(\mathbf{a})$ is the weight of the vector $\mathbf{a}$. The coefficient η_i of x^i in the weight polynomial $A(x)$ is equal to the number of vectors of weight i in the set A.

For an arbitrary UPWS code C with covering radius ρ denote by ξ_i, where $i = 1, \ldots, \rho$, the roots of Lloyd's polynomial $L_\rho(n, \xi)$. The knowledge of the roots ξ_i for the code C allows to write out the exact expression for the weight polynomial of any translate of C. Denote by $C^{[r]}$ the translate of C, i.e. $C^{[r]} = C + y$, such that the minimum weight of codewords of $C^{[r]}$ equals to r, where the integer r evidently belongs to the interval: $0 \leq r \leq \rho$.

Theorem 2.13 ([12]). *The weight polynomial $W_{C^{[r]}}(x)$ of a UPWS code $C^{[r]}$ is given by the expression*

$$W_{C^{[r]}}(x) = \frac{(1+(q-1)x)^n}{\sum_{i=0}^{\rho} \beta_i (q-1)^i \binom{n}{i}} + \sum_{i=1}^{\rho} A_{r,i} (1+(q-1)x^{n-\xi_i})(1-x)^{\xi_i}, \tag{2.16}$$

where $A_{r,i}$ are constants depending on the initial conditions of the translate $C^{[r]}$ (i.e., on the known coefficients of $W_{C^{[r]}}(x)$) and therefore determined by solving the corresponding system of linear equations. If C is CR, then $W_{C^{[r]}}(x)$ is uniquely defined by the r.

For example, for the code $C^{[r]}$ with minimum distance $d = 2e+1$ and covering radius ρ in the interval $e+1 < \rho < d$, we will have for any $r \leq e$:

$$A_{r,0} = \cdots = A_{r,r-1} = A_{r,r+1} = \cdots = A_{r,e} = 0, \;\; A_{r,r} = 1.$$

Consequently, for a UPWS code $C^{[r]} = C + y$ the weight polynomial $W_{C^{[r]}}(x)$ is independent on the choice of that vector y (but only depends on its weight $\mathrm{wt}(y)$) and completely determined by the fundamental parameters of the code. The number of different weight polynomials equals the number of different rows of the matrix B, i.e. $b+1$. For a CR code $C^{[r]}$ the weight polynomial $W_{C^{[r]}}(x)$ is uniquely determined by the value of the first nonzero coefficient, implying that a CR code C with covering radius ρ has exactly $\rho+1$ different weight polynomials (one polynomial for every value of the minimum weight $r \in \{0, 1, \ldots, \rho\}$ of translates of the code C.

2.3.6 Combinatorial Properties

CR codes (as a natural generalization of perfect codes) are interesting objects in Hamming schemes $H(q,n)$ which satisfy many combinatorial and geometric properties. In particular, there are more integer parameters denoted $p^{\ell}_{i,j}$ [118], whose integrality conditions are also necessary conditions for the existence of corresponding CR codes. Recall that $a^{(h)}_i$, $b^{(h)}_i$, and $c^{(h)}_i$ are the intersection numbers of $H(q,n)$ given in (2.1) and $S^{(i)}(x)$ denotes the sphere in $H(q,n)$ of radius i, and centered in x.

Theorem 2.14 ([118]). *Let C be a CR code in $H(q,n)$ with intersection matrix $L=[L_{i,j}]$. Then there are integer constants $p^{\ell}_{i,j}$ such that*

$$|S^{(i)}(x) \cap C(j)| = p^{\ell}_{i,j} \text{ for all } x \in C(\ell), \tag{2.17}$$

which are computable from the initial expressions $p^{\ell}_{0,j} = \delta_{j,\ell}$ and $p^{\ell}_{1,j} = a_{\ell,j}$ and the following recurrence relation:

$$p^{\ell-1}_{i,j} c_\ell + p^{\ell}_{i,j} a_\ell + p^{\ell+1}_{i,j} b_\ell = p^{\ell}_{i+1,j} c^h_{i+1} + p^{\ell}_{i,j} a^h_i + p^{\ell-1}_{i,j} b^h_{i-1} \tag{2.18}$$

Using the introduced constants $p^{\ell}_{i,j}$, an arbitrary row of the matrix B, corresponding to a vector $x \in C(\ell)$ takes the form:

$$(0, \ldots, 0, p^{\ell}_{\ell,0}, p^{\ell}_{\ell+1,0}, \ldots, p^{\ell}_{\rho,0})$$

It is proper to see two initial expressions $p^{\ell}_{0,j} = \delta_{j,\ell}$ and $p^{\ell}_{1,j} = L_{\ell,j}$ above, what is very simple. First, consider the parameter $p^{\ell}_{0,j}$. Chose any point x in $C(\ell)$. Then $S^{(0)}(x)$ is this point x if $\ell = j$, implying $p^{\ell}_{0,\ell} = 1$, or $S^{(0)}(x)$ is an empty set if $\ell \neq j$, implying $p^{\ell}_{0,j} = 0$. Now consider $p^{\ell}_{1,j}$. Again choose any point x in $C(\ell)$ and consider all points y which are at distance 1 from x. Clearly, they are a_ℓ points in $C(j)$ if $\ell = j$, or c_ℓ points in $C(j-1)$ if $\ell = j-1$, or b_ℓ points in $C(j+1)$ if $\ell = j+1$. So, in any case we can write $p^{\ell}_{1,j} = a_{j,\ell}$.

One of the problems for CR codes is to check complete regularity of some code. As Neumaier observed [118], in many cases, it is easier to check complete regularity of a code in terms of the following numbers:

$$\lambda_i = p^{i-1}_{i,0}, \text{ and } \mu_i = p^{i}_{i,0}. \tag{2.19}$$

In other words, μ_i (respectively, λ_i) is equal to the number of codewords at distance i from $x \in C(i)$ (respectively, from $x \in C(i-1)$. The next result gives the equivalence between Definitions 2.2 and 2.3 (more exactly, the nontrivial transient from Definition 2.2 to 2.3).

Theorem 2.15 ([118]). *A code C in $H(q,n)$ with covering radius ρ is CR if for all i, where $0 \leq i \leq \rho$, the following condition $(P[i])$ holds:*
$(P[i])$: If $d(x,C) = i$ then C contains exactly μ_i codewords at distance i from x and λ_{i+1} codewords at distance $i+1$ from x.
In this case, the intersection numbers of C are given by the following expressions:

$$b_i = b_i^{(h)} + c_i^{(h)} - \frac{\lambda_{i+1} c_{i+1}^{(h)}}{\mu_i} + \frac{(\lambda_i - \mu_i) c_i^{(h)}}{\mu_{i-1}}, \quad i < \rho, \tag{2.20}$$

and

$$c_i = \frac{\mu_i c_i^{(h)}}{\mu_{i-1}}, \quad i \leq \rho. \tag{2.21}$$

As an example, the *repetition code* $C = \{\mathbf{0}, \mathbf{1}\}$ of length n is a CR code with covering radius $\rho = \lfloor n/2 \rfloor$ which has $\mu_i = 1$ and $\lambda_i = 0$ for all $i \leq \rho - 2$ and the following parameters μ_i and λ_i for $i \in \{\rho-1, \rho\}$ in depending on the parity of the code length n:
n even, $\rho = n/2$:

$$\begin{array}{rcl} i &=& \rho - 1: \quad \mu_i = 1, \quad \lambda_i = 0, \\ i &=& \rho: \quad \mu_i = 2, \quad \lambda_i = 0, \end{array}$$

n odd, $\rho = (n-1)/2$ (in this case the code C is perfect):

$$\begin{array}{rcl} i &=& \rho - 1: \quad \mu_i = 1, \quad \lambda_i = 1, \\ i &=& \rho: \quad \mu_i = 1, \quad \lambda_i = 0. \end{array}$$

These introduced parameters λ_i and μ_i imply the following properties of code parameters [118].

Proposition 2.8 ([118]). *Let a (n, M, d)-code C satisfies the condition $(P[i])$ of Theorem 2.13 for all i. Then*

i. $d \leq 2\rho + 1$,

ii. If $2i \leq d$, *then* $\lambda_i = 0$,

iii. If $2i + 1 \leq d$, *then* $\mu_i = 1$,

iv. $\mu_i \geq \mu_{i-1}$

The case of equality in (i) of Proposition 2.8 is the most interesting. We give here the results from [118] and some more results using also some results from [12,136], and some commonly known results without references.

Theorem 2.16. *Let C be a $(n, M, d = 2e+1)$-code in $H(q,n)$. Then the following properties are equivalent:*

i. [118] $d = 2\rho + 1$,

ii. *[118] C is CR with* $\lambda_i = 0, \mu_i = 1$ *for all* $i = 0, 1, \ldots, \rho$,

iii. *[136] C is UPNS with packing density* $\mu = \mu_{mm} = (n+1)/(e+1)$,

iv. *[12] C is UPWS with packing numbers* $\beta_0 = \cdots = \beta_\rho = 1$,

v. *The code C is perfect, i.e. the spheres of radius e around all codewords partition* $H(2, n)$,

vi. *The code C has cardinality*

$$M = \frac{q^n}{\sum_{i=0}^{e}(q-1)^i\binom{n}{i}}.$$

Having a CR code C with the intersection array $IA = \{b_0, \ldots, b_{\rho-1}; c_1, \ldots, c_\rho\}$, it is quite natural to find conditions under which two arbitrary sequences of natural integers are intersection numbers of some CR codes. In other words, to find the necessary conditions that the intersection numbers of an arbitrary CR code must satisfy.

There is some progress in this direction. In particular. having two sequences $b_0, \ldots, b_{\rho-1}$ and $c_1, \ldots, c_\rho$ of natural numbers, it is natural to assume, that the following monotonicity conditions are valid:

$$b_0 \geq b_1 \geq b_{\rho-1} \text{ and } c_1 \leq c_2 \leq \ldots \leq c_\rho.$$

The counter-examples (a CR code with codewords, formed by the vertex set of any pentagon in the dodecahedron, has the intersection array $\{1, 2, 1; 1, 2, 1\}$) found by Koolen [94] (see also [114]) show that this property does not necessarily hold for completely regular codes in arbitrary distance-regular (i.e. not in $H(q, n)$) graphs. Moreover, Koolen [94] proved that the monotonicity conditions do hold for completely regular codes in Hamming or Johnson schemes. For Hamming schemes, a stronger monotonicity condition is given in [97].

Some more interesting properties of CR codes (which are also necessary conditions) can be found in [38,118].

Now we want to consider the quite natural question of how the intersection numbers and the packing parameters of the CR code (which, as we know, is UPWS) are related. Let we have a CR $(n, M, d)_q$ code with the covering radius ρ and the intersection array IA $= (b_0, \ldots, b_{\rho-1}; c_1, \ldots, c_\rho)$. By Theorem 2.12 we can find from this IA all the parameters μ_i, λ_i and $p^\ell_{i,j}$.

Proposition 2.9 ([12]). *For a given CR* $(n, M, d)_q$ *code C with covering radius* ρ *and with its intersection array* $IA = (b_0, \ldots, b_{\rho-1}; c_1, \ldots, c_\rho)$, *the packing coefficients* $\beta_0, \ldots, \beta_\rho$ *of C, as an UPWS code, are uniquely determined as the solution of the following system of linear equations:*

$$\left\{\begin{array}{ccccccccc} & & & & & \beta_\rho B_{x_\rho,\rho} & = & 1, \\ & & \beta_\rho B_{x_{\rho-1},\rho} & + & & \beta_{\rho-1} B_{x_{\rho-1},\rho-1} & = & 1, \\ & & \ldots\ldots & & & \ldots\ldots & & \ldots \\ \beta_\rho B_{x_0,\rho} & + & \beta_{\rho-1} B_{x_0,\rho-1} & + & \ldots & + \ \beta_0 B_{x_0,0} & = & 1. \end{array}\right. \tag{2.22}$$

where $x_i \in C(i)$, $i = 0, \ldots, \rho$ are any fixed $\rho + 1$ vectors.

Proof: By the definition of CR code C, the corresponding matrix $B = [B_{x,i}]$ contains exactly ρ distinct nonzero rows and for any $x \in C(i)$, this row can be written, in terms of parameters $p^i_{s,j}$ from (2.16), as follows:

$$p^i_{i,0} = \mu_i,\ p^i_{i+1,0} = \lambda_i,\ p^i_{i+2,0}, \ldots, p^i_{\rho,0}\,.$$

Let x_i be an arbitrary vectors from $C(i)$, i.e. $d(x_i, C) = i$ for $i = 0, \ldots, \rho$. Since C is CR, it is UPWS (Corollary 2.2), and, therefore, by the Definition 2.6, for any such vector x_i the equality (2.3) holds. In particular, for x_ρ this equality takes the form:

$$\mu_\rho \beta_\rho = 1.$$

Since $\mu_\rho \neq 0$, we conclude that

$$\beta_\rho = \frac{1}{\mu_\rho}.$$

For $x_{\rho-1}$ the equality (2.3) looks as

$$\mu_{\rho-1}\beta_{\rho-1} + \lambda_\rho \beta_\rho = 1.$$

and we deduce by (2.3), that

$$\beta_{\rho-1} = \frac{1 - \lambda_\rho \beta_\rho}{\mu_{\rho-1}} = \frac{\mu_\rho - \lambda_\rho}{\mu_\rho \mu_{\rho-1}}.$$

(indeed, by Definition 2.6 all parameters β_i are non-zero; this also follows from the non-negativity of all parameters b_i for $i = 0, 1, \ldots, \rho - 1$). For the next step we consider $x_{\rho-2}$ for which the equality (2.3) becomes

$$\mu_{\rho-2}\beta_{\rho-2} + \lambda_{\rho-1}\beta_{\rho-1} + p^{\rho-2}_{\rho,0}\beta_\rho = 1,$$

implying

$$\beta_{\rho-2} = \frac{1 - \lambda_{\rho-1}\beta_{\rho-1} - p^{\rho-2}_{\rho,0}\beta_\rho}{\mu_{\rho-2}}\,.$$

Since all constants $p^\ell_{i,j}$ can be uniquely computed, in this way we uniquely obtain all packing coefficients β_j of the CR code C as an UPWS code. △

2.4 EXTENSION OF CR CODES

2.4.1 When the Extended Code Is CR

For a given $(n, M, d)_q$-code C with odd distance $d = 2e + 1$, we say that $(n + 1, M, d^*)_q$-code C^* with distance $d^* \in \{d, d + 1\}$, is obtained by extension of code C, if we add to the all codewords of C the parity checking position, i.e.

$$C^* = \{(c_1, \ldots, c_n, c_{n+1}) : (c_1, \ldots, c_n) \in C\}, \text{ where } c_{n+1} = \sum_{i=1}^{n} c_i.$$

For the case $d = 3$, we know the exact answer for q-ary Hamming codes, which we denoted by $\mathcal{H}_m$. For a prime power q and a natural m denote by n_m the integer $n_m = (q^m - 1)/(q - 1)$. So, a (linear) code $\mathcal{H}_m$ is a $[n_m, n_m - m, 3]_q$-code. The following result is well known and can be found, for example, in [112].

Lemma 2.1. *Let $\mathcal{H}_m$ be a $[n_m, k, 3]_q$ Hamming code. Then, the extended code $\mathcal{H}^*_m$ has minimum distance 4 if and only if*

i. $q = 2$ *and* $m \geq 2$, *or*

ii. $q = 2^r \geq 4$ *and* $m = 2$, *i.e.* $n_m + 1 = q + 2$ *and* $k = q - 1$.

Recall that for a binary code C with odd distance $d = 2e + 1$, the extended code C^* (with even distance $d^* = 2e + 2$) is obtained by adding a parity (or antiparity) check bit to each codeword. Clearly, the extension of all space $\mathbb{F}_2^n$ gives the binary CR code with distance $d = 2$. But for a binary CR code C with $d = 2$ we can say nothing. Some results for binary CR codes with $d = 2$ were obtained by Vasil'eva [157], which we shortly give here. Denote by $C(n, k)$ the code, which is dual to the k-multiple repetition of the space $\mathbb{F}_2^n$. It can be seen [157] that $C(n, k)$ is a CR $[nk, n(k - 1), 2]$ code with $\rho = n$ and intersection array

$$IA = \{nk, (n - 1)k, \ldots, k; k, 2k, \ldots, nk\}.$$

Denote by Γ_C the graph of minimal distances of a linear CR code C.

Theorem 2.17 ([157]). *If C is a binary linear code of length nk with distance 2 and $c_1 = k$ then every connected component of the graph Γ_C is isomorphic to the code $C(n, k)$.*

One interesting open question for arbitrary q-ary CR codes is the following one: *given a CR $(n, M, d)_q$-code C with odd distance $d = 2e + 1$ under which conditions its extension, i.e., the code C^*, is again a CR code?*. Here we give some results for binary codes, although it seems that many results can be extended to the non-binary case.

2.4.2 Extension of UPWS Codes

In [39] it is proven (answering a question posed in [12]) that puncturing an even CR code gives also a CR code. Therefore, if C^* is CR, so is C. However, the converse is not true, in general. Bassalygo and Zinoviev [10] gave an example of a CR code C (with $d = 2e + 1$) such that C^* (with distance $d^* = 2e + 2$) is not CR. Namely, the $[21, 12, 5]$-code C obtained by double puncturing of the binary Golay code $\mathcal{G}_{23}$ is CR but its extension C^* is not. Moreover, in this case, the extended code C^* is not even UPWS. Therefore, the extension of a UPWS code could be not a UPWS code. A necessary and sufficient condition, for UPWS code, stay UPWS code under extension, gives the following

Theorem 2.18 ([10]). *A binary UPWS code C of length n with covering radius ρ and with packing coefficients $\beta_0, \ldots, \beta_\rho$, i.e. satisfying equations (2.3), remains to be UPWS under extension, if and only if the following system of equations holds:*

$$\beta_{\rho-2i} = \beta_{\rho-2i-1}, \ \forall \ i, \ 0 \leq i \leq \left\lfloor \frac{\rho-1}{2} \right\rfloor. \tag{2.23}$$

Furthermore, the packing coefficients, denoted $\gamma_0, \ldots, \gamma_\rho, \gamma_{\rho+1}$ of the extended code C^* are defined by the following formulas:

$$\gamma_{\rho-2i} = \beta_{\rho-2i}, \ \forall \, i = 0, 1, \ldots, \left\lfloor \frac{\rho}{2} \right\rfloor, \tag{2.24}$$

$$\gamma_{\rho-2i+1} = \frac{1}{n+1}\big((\rho+1-2i)\beta_{\rho-2i} + (n-\rho+2i)\beta_{\rho-2i+2}\big), \ \forall \, i = 0, \ldots, \left\lfloor \frac{\rho+1}{2} \right\rfloor. \tag{2.25}$$

The condition in the previous theorem is obviously necessary for a code to be CR. Hence we have the following interesting open question: *whether or not the condition in Theorem 2.18 is sufficient for CR codes?*

We give also a useful necessary condition for CR codes. As we know (Theorem 2.2), the set C_d of a binary UP code C of length n and distance $d = 2e+1$ induces a $T(n, d, e, \lambda)$-design.

Proposition 2.10 ([10]). *Let C^* be the extension of a binary UP code of length n and minimum distance $d = 2e+1$. If C^* is UPWS, then the set C^*_{d+1} induces a $T(n+1, d+1, e+1, \lambda)$-design.*

In the case when C and C^* are UP and the covering radius of C is $\rho = e+1$, both codes are also CR, what we have from the following

Proposition 2.11 ([136]). *Let C be a quasi-perfect UP code (so, CR). If C^* is UP, then C^* is CR.*

Another necessary condition is the following.

Proposition 2.12 ([32]). *If C is a CR code of even length and C is antipodal, then C^* is not UPWS (and hence not CR).*

We give a proof, since it is very clear and short. Assume that C^* is UPWS. Then the external distance is $s^* = s+1$ implying that for each weight w in the dual distribution, the number $n+1-w$ is also a weight in the dual distribution. But w must be even (since C contains the vector $\mathbf{1}$), hence $n+1-w$ is odd, getting a contradiction.

2.4.3 Extension of Some Known CR Codes

The following property is a strengthening of a result in [10].

Proposition 2.13 ([29]). *Let C be a binary linear CR code of length $n = 2^m - 1$, minimum distance $d = 3$, covering radius $\rho = 3$ and intersection array $\{n, b_1, 1; 1, c_2, n\}$. Let the dual code $C^\perp$ have nonzero weights w_1, w_2 and w_3. Then the extended code C^* is CR with covering radius $\rho^* = 4$ if and only if $w_1 + w_3 = 2w_2 = n + 1$. In such case, the intersection array of C^* is $\{n+1, n, b_1, 1; 1, c_2, n, n+1\}$.*

Hence, the weights of the dual code (when C is linear) play an important role. Another important fact is that puncturing the extended code C^* at any coordinate i should give almost the same code C independently of the chosen position i.

Proposition 2.14 ([143]). *Let C be a binary CR code and C^* be its extension. Assume that puncturing C^* at any coordinate gives the same code C. Let s^* be the external distance of C^*. If $s^* \leq s + 1$, then C^* is CR.*

Corollary 2.3 ([143]). *If C is a binary linear CR code, the weights of $C^\perp$ are even and symmetrical with respect to $(n+1)/2$, and if* $\mathrm{Aut}(C^*)$ *is transitive, then C^* is CR.*

Let C be a double-error-correcting BCH code with parameters $[2^m - 1, 2^m - 1 - 2m, 5]$, $m \geq 3$ odd (see Section 2.12.2). The weights of the dual code satisfy the hypothesis of Corollary 2.3 and C^* is left invariant under the action of the affine group. Hence C^* is CR with $\rho^* = s^* = 4$ and packing radius $e^* = 2$. This result can be also deduced from [12], where it was shown that these BCH codes are UPNS with packing coefficients

$$\beta_0 = \beta_1 = 1, \quad \beta_2 = \beta_3 = \frac{6}{n-1}.$$

Now, according to Theorem 2.18, extended BCH codes are UPWS and, therefore, by Proposition 2.11 these codes are CR.

Consider the codes studied by Calderbank and Goethals in [46,47], namely, the three weight cyclic subcodes of the shortened second-order Reed-Muller code and denote by C the dual of one of these codes. The code C is CR and C^* is left invariant by the affine group. Again, the three weights of $C^\perp$ satisfy the condition of Proposition 2.13. Hence C^* is CR with $\rho^* = s^* = 4$ and $e^* = 1$. We will return to these interesting codes in Section 2.6.2, as members of nested families of CR codes with $\rho = 3$.

Corollary 2.3 can be generalized to the nonlinear case.

Corollary 2.4 ([143]). *Let C be a binary code whose dual distances are even and symmetrical with respect to $(n+1)/2$. If C is CR and Aut(C^*) is transitive, then C^* is CR.*

Dumer [63] in 1976 generalized the original Preparata codes [121] by constructing a growing number of Preparata-like codes for every length $n = 2^{2m}$ (more exactly, the number of different codes of length n equals the number of integers of the form $2^i + 1$, which are mutually prime to $2^{2m} - 1$). In [6] this result was reopened in a very nice presentation (see Section 2.11), from which we deduce that the automorphism group of an extended generalized Preparata code is transitive. The punctured code, i.e. the Preparata code (of length $2^{2m} - 1$) is CR [136] and its dual distances, i.e. the weights of codewords of the Kerdock code [137], satisfy the condition of Corollary 2.4. The conclusion is that the extended Preparata code is CR [136] with $\rho^* = s^* = 4$ and $e^* = 2$.

Let C^* be a Hadamard $(12, 24, 6)$ code. It is known that C is UP with dual distances $4, 6, 8$ [49,80]. The group $\mathrm{Aut}(C^*)$ is isomorphic to the Mathieu group M_{12} (see [73]). Hence C^* is CR with $\rho^* = s^* = 4$ and $e^* = 2$. From [73] we have the following

Proposition 2.15 ([73]). *Let C be a binary CR code with parameters (n, d).*

i. If $(n, d) = (12, 6)$ then C is equivalent to the Hadamard code.

ii. If $(n, d) = (11, 5)$ then C is equivalent to the (punctured) Hadamard code.

The results of Solé [143] were slightly improved by Brouwer [40].

Proposition 2.16 ([40]). *Let C be a binary CR code. If the outer distribution matrices of all codes obtained from C^* by deleting one coordinate position have the same set of rows and, in particular, if C^* admits a group which is transitive on the set of coordinate positions, then C^* is CR.*

Corollary 2.5 ([40]). *A binary code C^* is CR if and only if all codes obtained from it by deleting one coordinate position are CR with the same outer distribution.*

2.4.4 Limitations for the Dual Distances

It is interesting, of course, to find necessary and sufficient conditions on C for C^* to be CR. From the last corollary one can derive some necessary conditions on the punctured codes of C^*, in particular, on C.

For any binary vector $v = (v_1, \ldots, v_n)$ and for each $i = 1, \ldots, n$, we define a vector

$$\tau_i(v) = (v_1, \ldots, v_{i-1}, p(v), v_{i+1}, \ldots, v_n),$$

where $p(v)$ denotes the parity of v, i.e.,

$$p(v) = \sum_{i=1}^{n} v_i \pmod{2}.$$

For every position $i = 1, \ldots, n$, define the code $C_{[i]} = \{\tau_i(x) \mid x \in C\}$.

Lemma 2.2 ([32]). *C^* is a CR code if and only if C and $C_{[i]}$ are CR, for all $i \in \{1, \ldots, n\}$.*

The proof of this statement is very short and we give it now.

Proof: For any code D, denote by $D^{(i)}$ the punctured code obtained by deleting the ith coordinate in D. Also, denote by $\sigma_{i,j}$ the transposition of the coordinates i and j. Assume that the parity check coordinate is at position $n+1$. Hence, it is clear that

$$C_{[i]} = (\sigma_{i,n+1}(C^*))^{(i)} \text{ and } C = (C^*)^{(n+1)}.$$

The result then follows by Corollary 2.5. □

Proposition 2.17 ([32]). *If C^* is a CR code, then for all $i = 1, \ldots, n$*

i. The weight distributions of C and $C_{[i]}$ coincide.

ii. The minimum distances of C and $C_{[i]}$ coincide and are odd.

iii. The external distances of C and $C_{[i]}$ coincide.

iv. The covering radius of C and $C_{[i]}$ coincide.

Now, we have the following necessary condition on C (or $C_{[i]}$).

Corollary 2.6 ([32]). *If C^* is CR code of length $n+1$ with minimum distance $d^* = 2e+2 \geq 4$ then, for all odd w*

$$(n-w)A_w = (w+1)A_{w+1}, \tag{2.26}$$

where A_w is the number of codewords of weight $w \geq 2e+1$ in C (or $C_{[i]}$).

In particular, any perfect code must satisfy (2.26). For the case of binary perfect codes with $d=3$, this recursion is well known (see, for example, [112]):

$$(n-i+1)A_{i-1} + A_i + (i+1)A_{i+1} = \binom{n}{i}. \tag{2.27}$$

By combining (2.26) and (2.27), we obtain

Corollary 2.7 ([32]). *For any binary perfect code with $d=3$ containing the zero codeword, the number of codewords of weight i is:*

$$A_i = \begin{cases} \binom{n}{i}\frac{1}{n-i+1} - A_{i-1} & \text{if } i \text{ is odd;} \\ \frac{n-i+1}{i}A_{i-1} & \text{if } i \text{ is even.} \end{cases} \tag{2.28}$$

By [26], we know that the even half of a binary Golay code $\mathcal{G}_{23}$, say C^*, is CR. Puncturing at any coordinate gives a code C with weight distribution:

A_0	A_7	A_8	A_{11}	A_{12}	A_{15}	A_{16}
1	176	330	672	616	176	77

Of course, C is CR [26] and hence it verifies (2.26) as can be readily seen.

2.4.5 On Extension of Perfect q-Ary Codes

Here we shortly consider the recent results from [15], where the non-existence of extended perfect codes in $H(q,n)$ is proven for $q = 3,4$ and $n > q+2$ or for the case when both n and q are odd. These results are based on the following theorem which we formulate in terms of CR codes.

Theorem 2.19 ([15]). *i. A code C in $H(q,n)$ is an extended one-perfect if and only if C is CR with intersection matrix L*

$$L = \begin{bmatrix} 0 & n(q-1) & 0 \\ 1 & q-2 & (n-1)(q-1) \\ 0 & n & n(q-2) \end{bmatrix}$$

ii. The matrix L has the following eigenvalues:

$$q-2, \ -n, \ n(q-1).$$

From these results, one can deduce the following statements.

Corollary 2.8 ([15]).

i. Let C be an extended one-perfect code in $H(q,n)$. Then n is even.

ii. If q is odd, then there are no MDS codes with distance 4 in $H(q,q+2)$.

iii. Let q be an odd prime power, and let C be an extended one-perfect code in $H(q,n)$. Then

$$n = \frac{q^\ell + q - 2}{q-1}$$

for some odd ℓ.

Proposition 2.18 ([15]).

i. Let C be an extended one-perfect code in $H(3,n)$, where $n = (3^\ell+1)/2$. Then $\ell \leq 2$.

ii. Let C be an extended one-perfect code in $H(4,n)$, where $n = (4^\ell+2)/3$. Then $\ell \leq 3$.

iii. There are no extended one-perfect codes in $H(4,22)$.

In this way, we arrive to the following theorem:

Theorem 2.20 ([15]).

i. An extended one-perfect code in $H(3,n)$ exists if and only if n =2.

ii. An extended one-perfect code in $H(4,n)$ exists if and only if $n = 2$ or $n = 6$.

iii. For any q, there are no extended one-perfect codes in $H(q,n)$ if n is odd.

2.5 COMPLETELY TRANSITIVE CODES

Completely transitive (shortly, CT) codes were first introduced in 1987 by Solé [142,143] as a subclass of binary linear CR codes. If C is a binary linear code, then consider the natural action of $\mathrm{Aut}(C)$ on the quotient space $\mathbb{F}_2^n/C$: for any coset $C+x$ and any $\sigma \in \mathrm{Aut}(C)$, set

$$\sigma(C+x) = \sigma(C) + \sigma(x) = C + \sigma(x).$$

Definition 2.12 ([142,143]). *A binary linear code C with covering radius ρ is CT if the group $\mathrm{Aut}(C)$ induces $\rho+1$ orbits under action on $\mathbb{F}_2^n/C$.*

Since two cosets in the same orbit have identical weight distributions, we obtain

Proposition 2.19 ([142,143]). *If C is a binary linear CT code, then C is CR.*

The following fact is a strengthening of Theorem 2.7.

Theorem 2.21 ([128]). *If C is a non-antipodal binary linear CT code, then $C \cup C(\rho)$ is CT too.*

There is a strong relation between CR codes and distance-regular graphs. In particular, the following results hold.

Theorem 2.22. *Let C be a binary linear CR code with covering radius ρ and intersection array $IA = \{b_0, \ldots, b_{\rho-1}; c_1, \ldots c_\rho\}$ and let Γ_C be the coset graph of C. Then*

i. [38] The coset graph Γ_C is distance-regular of diameter $D = \rho$ with the same IA.

ii. [123] If C is CT, then Γ_C is distance-transitive.

Therefore, by Theorem 2.22 any CR or CT code induces a distance regular or distance transitive coset graph, respectively. Throughout this section, we will not deal with the associated distance regular graphs or distance transitive graphs to CR or CT codes, respectively. Those graphs, with the same intersection array than the respective codes can be seen in Chapter 6, in corresponding references or in [38,95].

For a given permutation group G of degree n (acting on an n-set), we say that G is t-transitive (respectively t-homogeneous) if it sends any t-tuple (respectively t-set) to any t-tuple (respectively t-set). G is transitive if it is 1-transitive. A result of Livingston and Wagner [110] states that if G is i-homogeneous, with $i \leq n/2$, then G is also j-homogeneous, for $j \leq i$. This fact implies

Proposition 2.20 ([143]). *Let C be a binary linear code of length n and covering radius $\rho \leq n/2$. If $Aut(C)$ is ρ-homogeneous then C is CT.*

Using this property we obtain that the following codes are CT:

i. All perfect binary linear codes (repetition codes, Hamming codes, and the binary Golay code).

ii. All extended binary linear perfect codes.

Proposition 2.20 gives us a sufficient (but not necessary) condition for complete transitivity. It can be seen from the binary $[9, 5, 3]$ code C which is dual to the code obtained by the Kronecker product of two $[3, 2, 2]$ parity check codes. The code C is UPNS [136,143] with $\rho = s = 2$ and $e = 1$. Moreover, C is CT, however $\mathrm{Aut}(C)$ is transitive but not 2-homogeneous.

A necessary condition is the following one.

Proposition 2.21 ([143]). *If C is a binary linear CT code with minimum distance $d = 2e + 1$ then $\mathrm{Aut}(C)$ is e-homogeneous.*

As an example of CR code which is not CT, consider a binary primitive BCH $[2^m - 1, 2^m - 2m - 1, 5; 3]$ code, i.e., a cyclic code C with $m > 4$ and m odd. Such a code is CR [10,12]. The intersection array is

$$\mathrm{IA} = \{n, n-1, \frac{n+3}{2}; 1, 2, \frac{n-1}{2}\},$$

where $n = 2^m - 1$ (these are the codes for $\ell = 3$ in the family of Section 2.12; see also Section 2.4.3). As can be seen in [13], $\mathrm{Aut}(C)$ is the semilinear group of $\mathbb{F}_{2^m}$ over $\mathbb{F}_{2^m}$ (recall that the semilinear group, denoted by $\Gamma L(t, q)$, consists of all invertible semilinear transformations of $\mathbb{F}_q^t$ over $\mathbb{F}_q$). If $q = p^r$, then it is well known that $|\Gamma L(t, q)| = r|\mathrm{GL}(t, q)|$, where $\mathrm{GL}(t, q)$ denotes the general linear group (see, for example, pages 161-163 in [18]); in our case $t = 1$. Hence, the order of $\mathrm{Aut}(C)$ is

$$|\Gamma L(1, 2^m)| = m|\mathrm{GL}(1, 2^m)| = m(2^m - 1).$$

Since C has packing radius $e = 2$, we know that C has exactly $(2^m - 1)(2^{m-1} - 1)$ cosets of minimum weight 2. Therefore, as the number of such cosets is greater than $|\mathrm{Aut}(C)|$, it is not possible that they are in the same orbit by $\mathrm{Aut}(C)$. We conclude that C is not CT.

Similarly to perfect and quasi-perfect UP codes, the nonexistence of CT codes for $e > 3$ was also established. In 2000, Borges and Rifà [24] proved:

Theorem 2.23 ([24]). *If C is a nontrivial binary linear CT code with minimum distance $d = 2e + 1$, then $e \leq 4$.*

The proof was based on the following fundamental results on nonexistence of highly transitive groups (see, for example, [51]).

Theorem 2.24. *Let G be a finite t-transitive group of degree n.*

i. If $t > 5$, then G is the symmetric or the alternating group of degree n.

ii. If $t = 5$ and G is not the symmetric or the alternating group, then G is one of the Mathieu groups M_{12} (of degree 12) or M_{24} (of degree 24).

Using the Griesmer bound for linear codes and the nonexistence of certain t-designs, the result was improved in 2001 by Borges, Rifà and Zinoviev [25] giving the following

Theorem 2.25 ([25]). *If C is a nontrivial binary linear CT code with minimum distance $d = 2e + 1$, then $e \leq 3$.*

We bring several statements which were used for the proof of this theorem. First, we have to formulate the two following classification results on 4-homogeneous and 4-transitive groups.

Theorem 2.26. *Let G be a finite t-transitive group of degree n.*

i. [51] Let G be a finite 4-transitive group of degree n, which is not the symmetric or the alternating group. Then G is one of the Mathieu groups M_{11} (of degree 11) or M_{23} (of degree 23).

ii. [91] Let G be a finite 4-homogeneous group of degree $n \geq 8$ which is not 4-transitive. Then G is similar to $PSL(2;8)$, $P\Gamma L(2;8)$ or $P\Gamma L(2;32)$, in their usual permutation representations.

The next observation has an interest of its own [32].

Lemma 2.3. *Let $C \subset \mathbb{F}_2^n$ be a binary code with minimum distance $d \geq 3$. If $Aut(C)$ is the symmetric group S_n or the alternating group A_n, then $|C| \leq 2$.*

Clearly, Definition 2.12 can be extended to nonbinary linear codes. Giudici and Praeger [70,71] studied this more general case. They called *coset-completely transitive* these codes, including the binary CT codes introduced by Solé [142,143]. In the previous preprint, Godsil and Praeger generalized the concept of complete transitivity to the nonlinear case. A newer version of this preprint is [77].

Definition 2.13 ([77]). *A code $C \subset \mathbb{F}_q^n$ is G-*completely transitive*, or simply,* completely transitive *if there exists a subgroup G of* $\mathrm{Aut}(\mathbb{F}_q^n)$ *such that each subconstituent $C(i)$ of the distance partition is an orbit under action of G.*

Proposition 2.22 ([71]). *If $C \subset \mathbb{F}_q^n$ is CT in the sense of Definition 2.13 then it is CR.*

Note that, in general, a coset-completely transitive code may be not $\mathrm{Aut}(C)$-completely transitive in the sense of Definition 2.13. This is because $\mathrm{Aut}(C)$ is often not even transitive on C, for example, when C has codewords of different weights.

For a linear code $C \subset \mathbb{F}_q^n$, define F_C as the set of all translations of $\mathbb{F}_q^n$ by vectors in C, i.e., $F_C = \{\tau_x \mid x \in C\}$, where $\tau_x(v) = v + x$, for every $v \in \mathbb{F}_q^n$. Clearly, F_C is a subgroup, since C is linear. Define now the semidirect product $G = F_C \rtimes \mathrm{Aut}(C)$. The group G fixes C set-wise and the F_C-orbits are the cosets of C, in particular, C is a G-orbit [71]. From all these observations, the following result is obtained:

Theorem 2.27 ([71]). *Let $C \subset \mathbb{F}_q^n$ be a linear code. Then C is coset-completely transitive if and only if C is $(F_C \rtimes Aut(C))$-completely transitive.*

However, for $q \leq 3$, the concepts are equivalent:

Theorem 2.28 ([71]). *Let $C \subset \mathbb{F}_q^n$ be a linear code, where $q \leq 3$. Then C is coset-completely transitive if and only if C is completely transitive.*

Let $q \geq 7$ be a prime power, $q \neq 8$, and let C be the repetition code in $\mathbb{F}_q^3$. Taking $G = S_q \rtimes S_3$, it one can directly verify that C is G-completely transitive, however C is not coset-completely transitive [71].

Completely transitivity is a quite special property, which has no relation to the optimality of codes. For example, Preparata codes (which have the maximal possible packing number $\mu = n/(e+1)$ after perfect binary codes (for which $\mu = (n+1)/(e+1)$ [136]) are not completely transitive (except when they coincide with Kerdock codes, i.e., when the code is the Nordstrom–Robinson code $\mathcal{NR}$ [119]) (see also [131]). This interesting fact was proven in [74].

Theorem 2.29 ([74]). *Let C and C^* be the Preparata code of length n and its extension of length $n+1$, respectively. Then these codes are CT if and only if $n = 15$.*

We have to add here that the Preparata code of length 15 (as well as its extension) is unique, what is not the case for larger lengths, what could be the explanation of this exceptional case. We consider these remarkable codes in Section 2.11 and the code $\mathcal{NR}$ in Section 2.13.2.

2.6 CR AND CT CODES FROM PERFECT CODES

In the recent survey [32], we presented many infinite families of known to us CR codes. Here we follow mostly to this survey [32] with some additions. In particular, Sections 2.7, 2.9, 2.10, and 2.12 are extended, and Sections 2.11 and 2.13 are new. In the survey [32] we gave infinite families of the CR codes using consecutive numbering (F.i) for infinite families and (S.i) for sporadic such codes. Here we use double numbering (F.i.j) and (S.i.j), where i denotes the section number, and j is the number of the family (or code) in this section. Recall that q is always a prime power.

2.6.1 CR Codes from Perfect Codes by Extension, Shortening and Puncturing

(F.6.1) Any q-ary perfect $(n, q^{n-m}, 3; 1)_q$ code of length $n = (q^m - 1)/(q - 1)$ is a CR code. It is CT if it is linear. Its intersection array is

$$\text{IA} = \{(q-1)n; 1\}, \text{ where } q \geq 2, n \geq 3.$$

(F.6.2) Any q-ary extended perfect $(n + 1, q^{n-m}, 4; 2)_q$ code is a CR code (and CT in the binary linear case) with

$$\text{IA} = \{(q-1)(n+1), (q-1)n; 1, 4\}, \text{ where } q \geq 2,\ n \geq 4.$$

The code is also CT when $q = 4$ and $m = 2$.

(F.6.3) Any q-ary $(n - 1, q^{n-m}, 2; 1)_q$ code obtained by puncturing any perfect $(n, q^{n-m}, 3; 1)_q$ code is a CR and a CT (if it is linear) code with

$$\text{IA} = \{(q-1)(n-1); q)\}, \text{ where } q, n \geq 3.$$

Let B be any q-ary perfect $(n, M, 3)_q$-code with n odd and let C be any $(n, M/q, 4)_q$ subcode of B with the following property: for any choice of the zero codeword in B, the set C_4 is a q-ary $T(n, 4, 2, \lambda)_q$-design where $\lambda = (n-3)/2$. Then, from [29,163], we have:

(F.6.4) For $n \geq 4$ the code C is a CR code with

$$\text{IA} = \{n(q-1), (n-1)(q-1), 1; 1, (n-1), n(q-1)\}.$$

(F.6.5) In particular, for any prime power $q = 2^s$ where $s \in \{2, 3\ldots\}$, there exists a CR $[q+1, q-2, 4; 3]_q$ code, a subcode of a perfect $[q+1, q-1, 3]_q$-code, with

$$\text{IA} = \{q^2 - 1, q(q-1), 1; 1, q, q^2 - 1\}.$$

(F.6.6) Any q-ary $(n, N, 3; 2)_q$ code, obtained by shortening a perfect $(n + 1, qN, 3; 1)_q$ code C, is a CR and CT (in the linear case) code with

$$\text{IA} = \{(q-1)n, q-1; 1, (q-1)n\}.$$

As it was indicated in [26], there are several different halves of a binary perfect code giving CR codes with different intersection arrays.

Let C be a binary perfect $(n, N, 3)$ code. From [26], we have that the even or odd half of C is a CR code (included in the family (F.6.4)). Also from [26], we have that the punctured code of the even half of C is a CR code (included in family (F.6.6)).

Other halves of binary Hamming codes can be obtained with the following procedure. Let H_m denote the parity check matrix of a binary Hamming code of length $n = 2^m - 1$ and let h_i denote the ith column of H_m. For a given even $m \geq 4$ and any $i_1, i_2 \in \{0, 1, 2, 3\}$, where $i_1 \neq i_2$, denote by v_{i_1,i_2} the binary vector whose i-th position v_i is a function of the value of the weight of the column h_i:

$$v_i = \begin{cases} 1 & \text{if} \quad \mathrm{wt}(h_i) \equiv i_1, \text{ or } i_2 \pmod 4, \\ 0 & \text{otherwise.} \end{cases}$$

Theorem 2.30 ([28,32]). *Let C_{i_1,i_2} be the binary linear $[n = 2^m - 1, k = n - m - 1, 3]$ code with m even and parity check matrix $H_m(v_{i_1,i_2})$, obtained from H_m by adding one more row v_{i_1,i_2}. Then, for the resulting code C_{i_1,i_2} we have [28]:*

(F.6.7) If $\{i_1, i_2\} = \{1, 2\}$ or $\{i_1, i_2\} = \{2, 3\}$, then C_{i_1,i_2} is an antipodal half of a Hamming code, which is a CR and CT $[n, n - m - 1, 3; 3]$-code with

$$IA = \{n, (n + 1)/2, 1; 1, (n + 1)/2, n\}.$$

(F.6.8) If $\{i_1, i_2\} = \{0, 1\}$ or $\{i_1, i_2\} = \{0, 3\}$, then C is a non-antipodal half of a Hamming code, which is a CR and CT $[n, n - m - 1, 3; 3]$-code with

$$IA = \{n, (n - 3)/2, 1; 1, (n - 3)/2, n\}.$$

As can be seen in [28], if $\{i_1, i_2\} = \{0, 2\}$, then C_{i_1,i_2} is the even half of a Hamming code, hence included in family (F.6.4). If $\{i_1, i_2\} = \{1, 3\}$, then C_{i_1,i_2} is the Hamming code $[n, n - m, 3]$.

Extensions of the codes (F.6.7) are also CR [28]:

(F.6.9) The extension of the code (F.6.7) is an antipodal CR and CT $[n + 1, n - m, 4; 4]$ code with

$$\mathrm{IA} = \{n + 1, n, (n + 1)/2, 1; 1, (n + 1)/2, n, n + 1\}.$$

From [32,124,163], we have several families obtained by shortening binary perfect or extended perfect codes. Let C^* be any binary extended perfect $(n^*, N^*, 4)$ code, $n^* = 2^m \geq 8$.

(F.6.10) Let C be the $(n = n^* - 2, N = N^*/2, 2; 3)$ code obtained from C^* by $\{(00), (11)\}$-shortening. Then C is a CR code and also CT with

$$\mathrm{IA} = \{n, n - 2, 2; 2, n - 2, n\}, \text{ where } n = 2^m - 2 \geq 6.$$

(F.6.11) Let C be the $(n = n^* - 3, N^*/4, 1; 2)$ code obtained from C^* by $\{(000), (111)\}$-shortening. Then C is a CR code and also CT with

$$\mathrm{IA} = \{n - 1, 3; 1, n - 1\}, \quad \text{where } n = 2^m - 3 \geq 5.$$

(F.6.12) Let D be any binary perfect $(n, N, 3)$ code of length $n = 2^m - 1 \geq 7$ and let C be the $(n-2, N/2, 1; 2)$ code obtained by $\{(00), (11)\}$-shortening D. Then C is a CR code and also CT with

$$\mathrm{IA} = \{n-3, 2; 2, n-3\}.$$

Also from [32,124,163], we have a family obtained by shortening q-ary extended perfect codes:

(F.6.13) Let C^* be any q-ary extended perfect $[n^*, k^*, 4; 2]_q$ code where $q = 2^m \geq 4$, $n^* = q + 2$ and $k^* = q - 1$. Let C be the $[n = q, k = q - 2, 2; 2]_q$ code obtained by S-shortening C^*, where $S = \{(\alpha, \alpha) : \alpha \in \mathbb{F}_q\}$. Then C is a CR code with

$$\mathrm{IA} = \{q(q-1), (q-1)(q-2); 2, q\} \text{ where } q = 2^m \geq 4.$$

Now we give sporadic CR codes, which come from Golay codes [26,32,122,124]. The complete regularity and completely transitivity of the codes (S.6.5) and (S.6.6) were stated in [26]. Recall that $\mathcal{G}_{23}$ denote the perfect binary $[23, 12, 7]$ Golay code and $\mathcal{G}_{24}$ denote the extended perfect binary $[24, 12, 8]$ Golay code.

(S.6.1) [139] *The code* $\mathcal{G}_{23}$. This perfect $[23, 12, 7; 3]$ code is CR and CT with

$$\mathrm{IA} = \{23, 22, 21; 1, 2, 3\}.$$

(S.6.2) [136] *The punctured code* $\mathcal{G}_{23}$. This $[22, 12, 6; 3]$ code is CR and CT with

$$\mathrm{IA} = \{22, 21, 20; 1, 2, 6\}.$$

(S.6.3) [136] *The code* $\mathcal{G}_{24}$. This $[24, 12, 8; 4]$ code is CR and CT with

$$\mathrm{IA} = \{24, 23, 22, 21; 1, 2, 3, 24\}.$$

(S.6.4) [124] *The double punctured code* $\mathcal{G}_{23}$. This $[21, 12, 5; 3]$ code is CR and CT with

$$\mathrm{IA} = \{21, 20, 16; 1, 2, 12\}.$$

(S.6.5) [26] *The half of the code* $\mathcal{G}_{23}$. This $[23, 11, 8; 7]$ code is CR and CT with

$$\mathrm{IA} = \{23, 22, 21, 20, 3, 2, 1; 1, 2, 3, 20, 21, 22, 23\}.$$

(S.6.6) [26] *The punctured of the half of the code* $\mathcal{G}_{23}$. This $[22, 11, 7; 6]$ code is CR and CT with

$$\mathrm{IA} = \{22, 21, 20, 3, 2, 1; 1, 2, 3, 20, 21, 22\}.$$

(S.6.7) [32,124] *The* $\{(00,11)\}$*-shortened code code* $\mathcal{G}_{24}$. This $[22,11,6;7]$ code is CR and CT with

$$\text{IA} = \{22,21,20,16,6,2,1;1,2,6,16,20,21,22\}.$$

(S.6.8) [32,124] *The* $\{(000,111)\}$*-shortened code code* $\mathcal{G}_{24}$. This is a CR and CT $[21,10,5;6]$ code with

$$\text{IA} = \{21,20,16,9,2,1;1,2,3,16,20,21\}.$$

(S.6.9) [32,124] *The* $\{(00,11)\}$*-shortened code* $\mathcal{G}_{23}$. This $[21,11,5;6]$ code is CR and CT with

$$\text{IA} = \{21,20,16,6,2,1;1,2,6,16,20,21\}.$$

(S.6.10) [122] It is a (linear) $[15,10,4;3]$ code over $\mathbb{F}_2$ which is CR and CT with intersection array

$$\text{IA} = \{15,14,1;1,14,15\},$$

and with generator matrix:

$$\begin{bmatrix} 100000000001110 \\ 010000000001101 \\ 001000000011001 \\ 000100000010011 \\ 000010000000111 \\ 000001000010110 \\ 000000100011111 \\ 000000010011010 \\ 000000001011100 \\ 000000000110101 \end{bmatrix},$$

obtained by $\langle S\rangle$-shortening of the code $\mathcal{G}_{24}$, where

$$S = \left\{ \begin{array}{llll} (100001000), & (010001000), & (001001000), & (000100001), \\ (000010001), & (000000101), & (000000011) & \end{array} \right\}.$$

(S.6.11) [122] The code is a linear $[15,9,3;3]$ code with generator matrix

$$\begin{bmatrix} 100000000100101 \\ 010000000010011 \\ 001000000111001 \\ 000100000110011 \\ 000010000100111 \\ 000001000111101 \\ 000000100011111 \\ 000000010001111 \\ 000000001001001 \end{bmatrix}.$$

It is CR and CT code with intersection array

$$\mathrm{IA} = [15, 12, 1; 1, 4, 15],$$

obtained by $\langle S\rangle$-shortening of the code $\mathcal{G}_{24}$, where

$$S = \left\{ \begin{array}{ccc} (100000011), & (010000011), & (001000010), \\ (000100011), & (000010001), & (000001111), \end{array} \right\}.$$

(S.6.12) [122] It is a (linear, binary) [15, 10, 3] code, which is is CR and CT with intersection array

$$[15, 6, 1; 1, 6, 15]$$

and with generator matrix:

$$\left[\begin{array}{c} 100000000001110 \\ 010000000000110 \\ 001000000001100 \\ 000100000010011 \\ 000010000010010 \\ 000001000011101 \\ 000000100010100 \\ 000000010010001 \\ 000000001011100 \\ 000000000110101 \end{array} \right],$$

obtained by $\langle S\rangle$-shortening of the code $\mathcal{G}_{24}$, where

$$S = \left\{ \begin{array}{cccc} (100000011), & (010000010), & (001000010), & (000100001), \\ (000010011), & (000001010), & (000000110) & \end{array} \right\}.$$

There are many other CR, CT and UPWS codes (in particular, with $\rho \leq 2$ and $d \leq 2$) which can be obtained by some shortening and puncturing of binary Golay codes $\mathcal{G}_{23}$ and $\mathcal{G}_{24}$. We consider only one more important case, which was mentioned in 1969 by Semakov and Zinoviev [132].

Proposition 2.23 ([132]). *Let the set S look as*

$$S = \left\{ \begin{array}{cccc} (00000000), & (11000000), & (10100000), & (10010000), \\ (10001000), & (10000100), & (10000010), & (10000001) \end{array} \right\}.$$

Then the S-shortening of extended binary Golay code $\mathcal{G}_{24}$, where $\mathrm{Supp}(S) = \mathrm{Supp}(c)$, *such that $c \in \mathcal{G}_{24}$, is a (binary) nonlinear $(16, 256, 6)$-code, i.e. Nordstrom–Robinson code $\mathcal{NR}$ [119] (see also [132]), which is CR and CT with $IA = \{16, 15, 14, 1; 1, 2, 15, 16\}$.*

Goethals [78] also found in 1971 this construction of $\mathcal{NR}$-code. As we know this code is CR [133,136] and also CT [74] (Theorem 2.29). We will return to this code and also to the binary Golay code $\mathcal{G}_{24}$ in Section 2.13, where we describe new modified concatenation constructions of both these remarkable codes.

Now we turn to the consideration of ternary CR and CT codes that can be obtained from the ternary Golay code. Here we also follow [32,124] and [122]. Let $\mathcal{G}_{11}$ (respectively, $\mathcal{G}_{12}$) denote the ternary perfect Golay $[11,6,5]_3$-code (respectively, the ternary extended perfect Golay $[12,6,6]_3$-code and denote by $\mathcal{G}_{11}^{(0)}$ the subcode of $\mathcal{G}_{11}$, formed by all codewords of $\mathcal{G}_{11}$ with overall parity checking 0. It is easy to see that $\mathcal{G}_{11}^{(0)}$ is the $[11,5,6]_3$ code, formed by all codewords of weights 6 and 9. Call this code *the third part of the ternary Golay code* $\mathcal{G}_{11}$.

(S.6.13) [32,124] *The code* $\mathcal{G}_{11}$. This perfect $[11,6,5;2]_3$ code is CR and CT with

$$\text{IA} = \{22,20;1,2\}.$$

(S.6.14) [32,124] *The punctured code* $\mathcal{G}_{11}$. This $[10,6,4;2]_3$ code is CR and CT with

$$\text{IA} = \{20,18;1,6\}.$$

(S.6.15) [32,124] *The code* $\mathcal{G}_{12}$. This $[12,6,6;3]_3$ code is CR and CT with

$$\text{IA} = \{24,22,20;1,2,12\}.$$

(S.6.16) [32,124] *The code* $\mathcal{G}_{11}^{(0)}$. The $[11,5,6;5]_3$-code $G_{11}^{(0)}$ is CR and CT with

$$\text{IA} = \{22,20,18,2,1;1,2,9,20,22\}.$$

(S.6.17) [32,124] *The punctured code* $\mathcal{G}_{11}^{(0)}$. The $[10,5,5;4]_3$ code is CR and CT with

$$\text{IA} = \{20,18,4,1;1,2,18,20\}.$$

(S.6.18) [122] *The* $\langle(1011),(0112)\rangle$*-shortened code* $\mathcal{G}_{12}$. This is a CR $[8,4,3;2]_3$ code with

$$\text{IA} = \{16,8;1,2\}.$$

and with generator matrix:

$$\begin{bmatrix} 10000220 \\ 01020002 \\ 00120001 \\ 00001120 \end{bmatrix}.$$

(S.6.19) [122] *The* $\langle(10001),(01002),(00101),(00011)\rangle$*-shortened code* $\mathcal{G}_{12}$. This is a CR $[7,5,1;2]_3$ code with

$$\mathrm{IA} = \{12,4;2,12\},$$

and with generator matrix:

$$\begin{bmatrix} 1000210 \\ 0100210 \\ 0010020 \\ 0001200 \\ 0000001 \end{bmatrix}$$

(S.6.20) [122] *The* $\langle(10100),(01100),(00012)\rangle$*-shortened code* $\mathcal{G}_{12}$. This is a CR $[6,4,2;2]_3$ code with

$$\mathrm{IA} = \{12,4;2,12\},$$

and with generator matrix:

$$\begin{bmatrix} 100011 \\ 010011 \\ 001002 \\ 000110 \end{bmatrix}.$$

There are also many UP and UPWS codes (and also CR with $\rho = 1$ or (and) $d = 1$), obtained by shortening of the ternary Golay codes $\mathcal{G}_{11}$ and $\mathcal{G}_{12}$. We do not give the parameters of all such codes.

2.6.2 Nested Families of CR Codes

Recall that by $\mathcal{H}_m$ we denote a binary Hamming code of length $n = 2^m - 1$. Let $m = 2u$, $q = 2^u$, $r = 2^u + 1$ and $\bar{r} = 2^u - 1$. We can think of the parity check matrix H_m of $\mathcal{H}_m$ as the binary representation of $[\alpha^0, \alpha^1, \ldots, \alpha^{n-1}]$, where $\alpha \in \mathbb{F}_{2^m}$ is a primitive element. Now present the elements of $\mathbb{F}_{2^m}$ as elements in a quadratic extension of $\mathbb{F}_{2^u}$. Let $\beta = \alpha^r$ be a primitive element of $\mathbb{F}_{2^u}$ and let $\mathbb{F}_{2^m} = \mathbb{F}_{2^u}[\alpha]$. Let E_m be the binary representation of the matrix $[\alpha^0, \alpha^r, \ldots, \alpha^{(n-1)r}]$. Take the matrix P_m as the vertical join of H_m and E_m:

$$P_m = \begin{bmatrix} H_m \\ E_m \end{bmatrix}.$$

It is well known [46] that the code $C^{(u)}$ with parity check matrix P_m is a cyclic binary CR code with covering radius $\rho = 3$, minimum distance $d = 3$ and dimension $n - (m + u)$.

It can be seen in [29] that the number of cosets $C^{(u)} + v$, of weight three, is $\bar{r}$. Indeed, their syndromes $S(v)$ are the nonzero elements of $\mathbb{F}_{2^u}$. For every $i \in \{0, \ldots, u\}$, taking $u - i$ cosets $C^{(u)} + v^{(1)}, \ldots, C^{(u)} + v^{(u-i)}$ with independent syndromes $S(v^{(1)}), \ldots, S(v^{(u-i)})$ (independent, means that they are linearly independent binary vectors in $\mathbb{F}_2^u$), we can generate a linear binary code $C^{(i)}$ as the following linear span:

$$C^{(i)} = \langle C^{(u)}, v^{(1)}, \ldots, v^{(u-i)} \rangle.$$

Let A_{u-i} be the linear subspace of $\mathbb{F}_2^u$ generated by the syndromes $S(v^{(1)})$, ..., $S(v^{(u-i)})$. The dimension of code $C^{(i)}$ is $\dim(C^{(i)}) = u - i + \dim(C^{(u)})$, where $\dim(C^{(u)}) = n - m - u$. Note that the maximum number of independent syndromes we can take is u, so the biggest code we can obtain is of dimension $u + \dim(C^{(u)}) = n - m$, which is the Hamming code $C^{(0)} = \mathcal{H}_m$. All the constructed codes $C^{(i)}$ contain $C^{(u)}$ and, at the same time, they are contained in the Hamming code $C^{(0)}$.

The number of different codes $C^{(u-i)}$ equals the number of subspaces of dimension i we can take in $\mathbb{F}_2^u$, so the Gaussian binomial coefficient

$$|\{C^{(u-i)}\}| = \binom{u}{i}_2 = \frac{(2^u - 1)(2^u - 2) \cdots (2^u - 2^{i-1})}{(2^i - 1)(2^i - 2) \cdots (2^i - 2^{i-1})}.$$

The number of different nested families of codes between $C^{(u)}$ and $C^{(0)} = \mathcal{H}_m$, we can construct, equals

$$\prod_{i=0}^{u-1} (2^{u-i} - 1).$$

The following property was stated in [46] for the code $C^{(u)}$, but it can be extended to all codes $C^{(i)}$, for $i \in \{1, \ldots, u\}$.

Proposition 2.24 ([29]). *For every $i \in \{1, \ldots, u\}$ the cosets of weight three of $C^{(i)}$ are at distance three from each other and $C^{(i)} \cup C^{(i)}(3)$ is the Hamming code.*

Recall that $C^{(i)*}$ is obtained from $C^{(i)}$ by extension and that the property CT implies the property CR.

Theorem 2.31 ([29]).

i. $C^{(1)}$ is a CT code with covering radius 3.

ii. $C^{(u)}$ is a CT code with covering radius 3.

iii. For every $i \in \{0, \ldots, u\}$, the code $C^{(i)*}$ is CT when $C^{(i)}$ is CT.

iv. For every $i \in \{1, \ldots, u\}$, the code $C^{(i)}$ is a subcode of $C^{(i-1)}$, and $C^{(i)*}$ is a subcode of $C^{(i-1)*}$.

(F.6.14) For every $i \in \{0, \ldots, u\}$, the code $C^{(i)}$ is CR with

$$IA = \{2^m - 1, 2^m - 2^{m-i}, 1; 1, 2^{m-i}, 2^m - 1\}.$$

(F.6.15) For every $i \in \{0, \ldots, u\}$, the extended code $C^{(i)*}$ is CR with

$$IA = \{2^m, 2^m - 1, 2^m - 2^{m-i}, 1; 1, 2^{m-i}, 2^m - 1, 2^m\}.$$

Note that, for $i \in \{0, 1, u\}$ the codes $C^{(i)}$ and $C^{(i)*}$ are CT. Also, for $m = 6$, all codes $C^{(i)}$ and $C^{(i)*}$ are CT. In [90] for the graphs (to be distance transitive) coming from such codes (to be CT) were obtained the following divisibility conditions:

$$2^m \text{ is a power of } 2^i, \text{ or } 2^m = 2^i, \text{ or } 2^i - 1 \text{ divides } 2m.$$

Therefore, we conjecture that when one of such divisibility conditions is satisfied, then $C^{(i)*}$ is a CT code. Moreover, in such cases, we conjecture that $C^{(i)}$ is also CT.

Therefore, we have the following open question: *for which i, such that $i \neq 0, 1, u$, the codes $C^{(i)}$ and $C^{(i)*}$ are CR and CT?*

In Section 2.7.4, we will give another representation of some of these codes (namely, as concatenated codes), which makes it possible to generalize the known Kasami codes to the non-binary case, what was proposed in the work [141]

2.6.3 Lifting Alphabets of Hamming Codes

A wide class of q-ary CR codes with growing covering radius can be obtained by lifting the alphabet of Hamming codes [127]. Denote by $H_m = H_m^q$ the parity check matrix of the Hamming code $\mathcal{H}_m = \mathcal{H}_m^q$ of length $n = (q^m - 1)/(q - 1)$ over $\mathbb{F}_q$ (we omit q from designation of the term, when it is fixed). For any natural number $r \geq 2$, define a new linear code, denoted by $C_{(m,r)}$ of the same length n over $\mathbb{F}_{q^r}$ with this parity check matrix H_m^q. We will start with the following two simple statements [127].

Proposition 2.25 ([127]). *Let C be any completely regular code, C^* be its extension and $C_{(m,r)}$ and $C^*_{m,r}$ be the corresponding lifted codes of C and C^*, respectively, over the field $\mathbb{F}_{q^r}$ where $r \geq 2$ be any natural number. Let α be a primitive element of $\mathbb{F}_{q^r}$. Then*

i. The code $C_{(m,r)}$ can be presented as follows:

$$C_{(m,r)} = C + \alpha C + \alpha^2 C + \cdots + \alpha^{r-1} C.$$

ii. The following equality is valid

$$(C^*)_{m,r} = (C_{m,r})^*.$$

If $q = 2$ the code $C_{(r,m)}$ is a completely transitive and, therefore, completely regular $[n, n - r, 4]_{2^m}$ code with intersection numbers

$$b_i = (2^r - 2^i)(2^m - 2^i), \;\; c_i = 2^{i-1}(2^i - 1),$$

where $i = 0, \ldots, \rho$ and where $\rho = \min\{r, m\}$. Furthermore, in the case $m \neq r$, the codes $C_{(r,m)}$ and $C_{(m,r)}$ are not equivalent, but they have the same intersections arrays. More exactly, they have the same intersections numbers b_i, c_i, but different numbers a_i (indeed, the codes $C_{(r,m)}$ and $C_{(m,r)}$ have different lengths).

Theorem 2.32 ([127]).

(F.6.16) The code $C_{(m,r)}$ is a CT (and, hence, CR) $[n, n-m, 3; \rho]_{q^r}$ code with $\rho = min\{r, m\}$ and intersection numbers:

$$b_i = \frac{(q^r - q^i)(q^m - q^i)}{(q-1)}, \ i = 0, \ldots, \rho - 1; \ c_i = q^{i-1}\frac{q^i - 1}{q-1}, \ i = 1, \ldots, \rho.$$

When $r \neq m$, codes $C_{(m,r)}$ and $C_{(r,m)}$ are not equivalent, but they have the same intersection array.

Note that Hamming codes are the only codes whose lifting give CR codes.

Theorem 2.33 ([127]). *Let C be a nontrivial code of length n over the field $\mathbb{F}_q$ with minimum distance $d \geq 3$, with covering radius $\rho \geq 1$ and let C_r be its lifting over $\mathbb{F}_{q^r}$. Then C_r is CR, if and only if the initial code C is a Hamming code $\mathcal{H}_m^q$ of length n.*

Lifting of the extended Hamming codes can be also exactly formulated. In particular, for the extended codes the following results were obtained in [127].

Theorem 2.34 ([127]). *Let H_m be the parity check matrix of the Hamming code $\mathcal{H}_m$ of length $n = (q^m - 1)/(q-1)$ and H_m^* be the parity check matrix of the extended Hamming code $\mathcal{H}_m^*$ of length $n = 1 + (q^m - 1)/(q-1)$ with some distance $d^* \in \{3, 4\}$. Let $C_{(m,r)}$ and $C^*_{(m,r)}$) be the codes, obtained by lifting of the codes $\mathcal{H}_m$ and $\mathcal{H}_m^*$, respectively, over the field $\mathbb{F}_{q^r}$, where $r \geq 2$ be any natural number, i.e. these q^r-ary codes have parity check matrices H_m and H_m^*, respectively. Then*

i. The code $\mathcal{H}_m^*$ is UP if and only if $d^* = 4$.

ii. The code $C^*_{(m,r)}$ with covering radius $\rho^* = 1 + min\{r, m\}$ is the extension of the code $C_{(m,r)}$ whose covering radius equals to $\rho = min\{r, m\}$.

iii. The code $C^*_{(m,r)}$ is not CR.

As a corollary, taking into account that we know that $\mathcal{H}_m^*$ is UP only in two cases ((i) $q = 2$ and $m \geq 2$ and (ii) $q = 2^u \geq 4$ and $n^* = q + 2$) we obtain the following result, which generalizes results of [10,12] to the nonbinary case.

Proposition 2.26 ([127]). *Let C be the q-ary Hamming $[n, n-m, 3]$ code, $n = (q^m - 1)/(q-1)$ and let C^* be its extended code. The code C^* is UP if and only if it has minimum distance 4. In other words, the code C^* is UP if and only if $q = 2$ with $m \geq 2$ or $q = 2^u \geq 4$ with $m = 2$.*

Theorem 2.35 ([127]). *Let C be the Hamming $[n, n-m, 3]_q$-code of length $n = (q^m - 1)/(q-1)$ and let C^* be the extended code. The lifted code $C^*_{(m,r)}$ is a UPWS code if and only if C^* is UP. Hence, the lifted code $C^*_{(m,r)}$ is a UPWS code if and only if $q = 2$ with $m \geq 2$ or $q = 2^u \geq 4$ with $m = 2$.*

2.7 DIRECT SUMS, PRODUCT AND CONCATENATION OF CODES

These are very powerful methods of construction of CR and CT codes, which give many interesting families of such codes. In particular, the most interesting result is the growing (with increasing length) number of CR and CT codes for any prime power q, which have different code parameters q, n, M and d, but the same intersection arrays [126,129].

2.7.1 CR Codes by Kronecker Product

In [126] a Kronecker construction of CR codes has been investigated and, later, in [129] the construction has been extended taking different alphabets in the initial component codes. It turned out that this approach is also connected with lifting constructions of CR codes. A very interesting thing is that as a result we obtain a growing number of CR and CT codes with different parameters, but identical intersection array. In terms of graphs, all resulting distance-regular graphs defined on cosets of these CR codes are isomorphic, but they are constructed from different CR codes with different parameters q, n and M.

Recall that the Kronecker product of two matrices $A = [a_{r,s}]$ over $\mathbb{F}_{q^u}$ and B over $\mathbb{F}_q$ is a new matrix $H = A \otimes B$, obtained by replacing every element $a_{r,s}$ in A by the matrix $a_{r,s}B$.

In this section, it is more convenient for us to use the following notation for codes and their parity check matrices. Denote by H^q_m the parity check matrix of a perfect Hamming $[n, k, 3]_q$-code $C = C(H^q_m)$ over $\mathbb{F}_q$, where $n = (q^m - 1)/(q-1)$ and $k = n - m$. By $C(H)$ we denote the code C over $\mathbb{F}_q$ induced by H as a parity check matrix. In the same notation, we denote by $C_r(H^q_m)$ the code over $\mathbb{F}_{q^r}$ with parity check matrix H^q_m over $\mathbb{F}_q$, i.e. obtained by lifting the field $\mathbb{F}_q$ to the field $\mathbb{F}_{q^r}$.

Theorem 2.36 ([126,129]). *Let $C(H^{q^u}_{m_a})$ and $C(H^q_{m_b})$ be two Hamming codes with parameters $[n_a, n_a - m_a, 3]_{q^u}$ and $[n_b, n_b - m_b, 3]_q$, respectively, where*

$$n_a = \frac{q^{u\,m_a} - 1}{q^u - 1}, \quad n_b = \frac{q^{m_b} - 1}{q - 1},$$

where q is a prime power, $m_a, m_b \geq 2$, and $u \geq 1$.

(F.7.1) The code $C = C(H)$ with parity check matrix $H = H^{q^u}_{m_a} \otimes H^q_{m_b}$, the Kronecker product of $H^{q^u}_{m_a}$ and $H^q_{m_b}$, is a CT, and hence CR code with parameters $[n, k, d; \rho]_{q^u}$, where

$$n = n_a\, n_b, \;\; k = n - m_a\, m_b, \;\; d = 3, \;\; \rho = min\{u\, m_a,\; m_b\}. \tag{2.29}$$

and with intersection numbers:

$$b_\ell = \frac{(q^{u\,m_a} - q^\ell)(q^{m_b} - q^\ell)}{(q-1)}, \quad \ell = 0, 1, \ldots, \rho - 1,$$

and

$$c_\ell = q^{\ell-1}\frac{q^\ell - 1}{q-1}, \quad \ell = 1, 2, \ldots, \rho.$$

The lifted code $C_{m_b}(H^q_{um_a})$ is CR with the same intersection array as C.

Remark that in the above Theorem 2.36 we can not take the code $C_{m_b}(H^{q^u}_{m_a})$ (instead of $C_{m_b}(H^q_{um_a})$), which seems to be natural. We emphasize that the codes $C_{m_b}(H^q_{um_a})$ and $C_{m_b}(H^{q^u}_{m_a})$ are not only different CR codes, but they induce different distance-regular graphs with different intersection arrays. So, the code $C_{m_b}(H^q_{um_a})$ suits to the codes in the family (F.7.1) in the sense that it has the same intersection array. For example, the code $C_2(H^{2^2}_3)$ induces a distance-regular graph with intersection array $\{315, 240; 1, 20\}$ and the code $C_2(H^2_6)$ gives a distance-regular graph with intersection array $\{189, 124; 1, 6\}$. To obtain these results in both cases we use the same Theorem 2.32.

The above Theorem 2.36 can not be extended to the more general case when the alphabets $\mathbb{F}_{q^a}$ and $\mathbb{F}_{q^b}$ of component codes C_A and C_B, respectively, neither $\mathbb{F}_{q^a}$ is a subfield of $\mathbb{F}_{q^b}$ or vice versa $\mathbb{F}_{q^b}$ is a subfield of $\mathbb{F}_{q^a}$. We illustrate it by considering the smallest nontrivial example. Take two Hamming codes, the $[5, 3, 3]$ code C_A over $\mathbb{F}_{2^2}$ with parity check matrix $H^{2^2}_2$, and the $[9, 7, 3]$ code C_B over $\mathbb{F}_{2^3}$ with parity check matrix $H^{2^3}_2$. Then the resulting $[45, 41, 3]$ code $C = C(H^{2^2}_2 \otimes H^{2^3}_2)$ over $\mathbb{F}_{2^6}$ is not even UPWS, since it has covering radius $\rho = 3$ and external distance $s = 7$, which can be checked by considering the parity check matrix of C.

Theorem 2.37 ([126,129]). *Let q be any prime number and let u, a, b be any natural numbers. Then there exist the following CR $[n, k, d; \rho]_{q'}$ codes with different parameters n, k, q', where q' is a power of q, the minimum distance is $d = 3$, and the covering radius is $\rho = min\{ua, b\}$:*

(F.7.2) $C_{ua}(H^q_b)$ over $\mathbb{F}_{q^{ua}}$ with $n = \frac{q^b-1}{q-1}$, $k = n - b$;

(F.7.3) $C_b(H^q_{ua})$ over $\mathbb{F}_{q^b}$ with $n = \frac{q^{ua}-1}{q-1}$, $k = n - ua$;

(F.7.4) $C(H^q_b \otimes H^q_{ua})$ over $\mathbb{F}_q$ with $n = \frac{q^b-1}{q-1} \cdot \frac{q^{ua}-1}{q-1}$, $k = n - bua$;

(F.7.5) $C(H^q_b \otimes H^{q^a}_u)$ over $\mathbb{F}_{q^a}$ with $n = \frac{q^b-1}{q-1} \cdot \frac{q^{ua}-1}{q^a-1}$, $k = n - bu$;

(F.7.6) $C(H^q_b \otimes H^{q^u}_a)$ over $\mathbb{F}_{q^u}$ with $n = \frac{q^b-1}{q-1} \cdot \frac{q^{ua}-1}{q^u-1}$, $k = n - ba$;

All the above codes have the same intersection numbers

$$b_\ell = \frac{(q^b - q^\ell)(q^{ua} - q^\ell)}{(q-1)},\ \ell = 0,\dots,\rho-1,\ \ c_\ell = q^{\ell-1}\frac{q^\ell - 1}{q-1},\ \ell = 1,\dots,\rho.$$

All codes above coming from Kronecker constructions (i.e., codes from families (F.7.4) – (F.7.6)) are CT.

Denote by $\tau(n)$ the number of different natural divisors of a natural number n, and let $r^* = n/r$ for any divisor r of n.

Corollary 2.9 ([129]). *Given a prime power q choose any two natural numbers a, $b > 1$. Then, we can build $\tau(a) + \tau(b)$ following different CR codes with identical intersection array and covering radius $\rho = min\{a, b\}$:*

i. CT codes $C(H_r^{q^{r^*}} \otimes H_b^q)$ over $\mathbb{F}_{q^{r^*}}$, for any proper divisor r of a, with $rr^* = a$.

ii. CT codes $C(H_a^q \otimes H_r^{q^{r^*}})$ over $\mathbb{F}_{q^{r^*}}$, for any proper divisor r of b, with $rr^* = b$.

iii. The CR code $C_a(H_b^q)$ over $\mathbb{F}_{q^a}$.

iv. The CR code $C_b(H_a^q)$ over $\mathbb{F}_{q^b}$.

This construction gives also UPWS codes which are not CR. Let $R_t = [I_{t-1}\,|\,\mathbf{1}^t]$ be a matrix of the size $(t-1) \times t$, where I_{t-1} is a binary diagonal matrix of order $t-1$, and $\mathbf{1}^t$ is the column vector of all ones.

Theorem 2.38 ([129]). *For any natural $m \geq 2$ and $u \geq 1$ and any prime power q, let $C(H_m^{q^u})$ be the q^u-ary Hamming $[n, k, 3]_{q^u}$-code of length $n_a = (q^{um}-1)/(q^u-1)$ and $C(R_{n_b})$ be the repetition $[n_b, 1, n_b]_q$-code, where $4 \leq n_b \leq (q^u - 1)n_a + 1$.*

i. The code $C = C(H_m^{q^u} \otimes R_{n_b})$ is a q^u-ary UPWS $[n, k, d]_{q^u}$-code with covering radius $\rho = n_b - 1$ and parameters

$$n = n_a\, n_b,\quad k = n - m\,(n_b - 1),\quad d = 3. \tag{2.30}$$

ii. The code C is not CR.

2.7.2 CR Codes by Concatenation of Hamming Codes

We start this section with the results from [30,31] dealing with CR codes constructed using concatenations of Hamming codes.

For any vector $x = (x_1, \dots, x_n) \in \mathbb{F}_q^n$, denote by $\sigma(x)$ the right cyclic shift of x, i.e. $\sigma(x) = (x_n, x_1, \dots, x_{n-1})$. Define recursively $\sigma^i(x) = \sigma(\sigma^{i-1}(x))$, for $i = 2, 3, \dots$ and $\sigma^1(x) = \sigma(x)$. For $j < 0$, we define $\sigma^j(x) = \sigma^\ell(x)$, where $\ell = j \mod n$.

The next constructions are described in [30], although the dual codes of the resulting family of q-ary completely regular codes are well known as the family SU2 in [48].

Let H be the parity check matrix of a q-ary cyclic Hamming code $\mathcal{H}_m$ of length $n = (q^m - 1)/(q - 1)$ (recall that a cyclic version of $\mathcal{H}_m$ exists exactly when n and $q - 1$ are coprime numbers). Clearly the simplex code generated by H is also a cyclic code. For any $r \in \{1, 2, \ldots, n\}$, consider the code $B^{(r)}$ of length nr with the following parity check matrix $H_b(r)$ (we call it Construction I in [30]):

$$H_b(r) = \begin{bmatrix} H & H & \cdots & H \\ H_1 & H_2 & \cdots & H_r \end{bmatrix}, \quad r = 1, \ldots, n; \tag{2.31}$$

where H_i, for $1 \leq i \leq n$, is the matrix $H\alpha^i$ (here $\alpha \in \mathbb{F}_{q^m}$ is a primitive nth root of unity), hence H_i is obtained from H by cyclically shifting i times its columns to the right (in these terms $H = H_n$). In [30] we also presented Construction II. In this case, the corresponding code of length $n(r + 3)$, which we denote $A^{(r)}$, has parity check matrix $H_a(r)$ of the form:

$$H_a(r) = \begin{bmatrix} H & O & H & H & H & \cdots & H \\ O & H & H & H_1 & H_2 & \cdots & H_r \end{bmatrix}. \tag{2.32}$$

For the codes $A^{(r)}$ we fix the following interval for r: $r \in \{-2, -1, 0, 1, \ldots, n - 1\}$ (for $r = n$ we would have pairs of linear dependent columns), where $A^{(0)}$, $A^{(-1)}$ and $A^{(-2)}$ are defined by the parity check matrices, respectively

$$H_a(0) = \begin{bmatrix} H & O & H \\ O & H & H \end{bmatrix}, \quad H_a(-1) = \begin{bmatrix} H & O \\ O & H \end{bmatrix}, \quad H_a(-2) = \begin{bmatrix} H \\ O \end{bmatrix}. \tag{2.33}$$

Note that the codes $B^{(1)}$ and $A^{(-2)}$ are the Hamming code of length $(q^m - 1)/(q - 1)$.

Theorem 2.39 ([30,31]). *Let $m \geq 2$ be any natural number and q be any prime power.*

(F.7.8) The code $B^{(r)}$ with parity check matrix given in (2.31), where $r \in \{1, \ldots, n\}$, is a CR code with parameters $[nr, nr - 2m, 3; 2]_q$ and intersection array

$$\mathrm{IA} = \{(q - 1)nr, ((q - 1)n - r + 2)(r - 1); 1, r(r - 1)\}.$$

1. *The code $B^{(r)}$ in the binary case (i.e. when $q = 2$) and for any value of m (so $n = 2^m - 1$), is CT for the case $r \in \{2, 3, n - 1, n\}$.*
2. *In general, in the q-ary case, for any value of m (so $n = q^m - 1$), the code $B^{(r)}$ is CT only when $r = 2$.*

So almost all codes described in family (F.7.8) are not CT codes. By extending the codes given in family (F.7.8) we do not obtain CR codes, except for the binary case when r equals $2^{m-1} + 1$. In this case, the resulting extended $[n(2^{m-1} + 1) + 1, n(2^{m-1} + 1) - 2m, 4; 3]$ codes coincide with the codes from the family (F.7.10) below.

In [30] it was proved that all codes $A^{(r)}$ are also CR.

Theorem 2.40 ([30,31]). *Let $m \geq 2$ be any natural number and q be any prime power.*

(F.7.9) *The code $A^{(r)}$ with parity check matrix given in (2.32) and (2.33), where $r \in \{-1, 0, 1, \ldots, n-1\}$, is a CR code with parameters $[(r+3)n, (r+3)n - 2m, 3; 2]_q$ and with intersection array*

$$IA = \{(r+3)(q-1)n, (r+2)((q-1)n-1-r); 1, (r+2)(r+3)\}.$$

1. *The code $A^{(r)}$ in the binary case (i.e. when $q = 2$) and for any value of m (so $n = 2^m - 1$), is CT for the case $r \in \{2^m - 5, 2^m - 4, 2^m - 3\}$.*
2. *In general, in the q-ary case, for any value of m (so $n = q^m - 1$), the code $A^{(r)}$ is CT when $r = 2$.*

In the binary case, when $r = n - 1$ the code $A^{(n-1)}$ coincides with the $[2^{2m} - 1, 2^{2m} - 1 - 2m, 3; 1]$ Hamming code.

If we consider the extension of the codes $A^{(r)}$ from the family (F.7.9) we obtain new CR codes only in the binary case for two values of c, namely, $c \in \{2^{m-1} - 2, 2^m - 2\}$.

(F.7.10) Let $A^{(r)}$ be the code with parity check matrix $H_a(r)$ and let $A^{(r)*}$ be its extended code. For $r = 2^{m-1} - 2$, the code $A^{(r)*}$ is a $[(r+3)n+1, (r+3)n - 2m, 4; 3]$ CR code with

$$\text{IA} = \{(r+3)n+1, (r+3)n, 2^{2m-2}; 1, (r+2)(r+3), (r+3)n+1\}.$$

For $r = 2^m - 2$, the code $A^{(r)*}$ is a $[(r+3)n+1, (r+3)n - 2m, 4; 2]$ CR code which coincides with the extended $[2^{2m}, 2^{2m} - 2m, 4; 2]$ Hamming code.

For $r \notin \{2^{m-1} - 2, 2^m - 2\}$ the code $A^{(r)*}$ is not CR.

Let m and r be natural numbers, such that $1 \leq r \leq n-1 = 2^m - 2$ (so $n = 2^m - 1$). Let $C = A^{(n-1)}$ be the code with parity check matrix H_c. Code C is a Hamming code of length $2^{2m} - 1$. Let $B^{(n-1-r)}$ denote the code with parity check matrix H_c after removing the columns in $H_a(r)$. Let $H_b(n-1-r)$ denote the parity check matrix of the code $B^{(n-1-r)}$, which coincides (up to permutation equivalence) with the matrix (2.31). The above three codes $A^{(r)}$, $B^{(n-1-r)}$ and C have lengths $n_a = (r+3)(2^m - 1)$, $n_b = (n-1-r)(2^m - 1)$, and $n_c = n_a + n_b = 2^{2m} - 1$, respectively. Therefore, the matrix H_c can be presented as follows:

$$H_c = \left[\; H_a(r) \;\middle|\; H_b(n-1-r) \;\right].$$

The code $B^{(n-1-r)}$ is equivalent to the code with parity check matrix as in (2.31); we replace r with $n-1-r$. Therefore, the codes $A^{(r)}$ and $B^{(n-1-r)}$ are CR and belong to families (F.7.8) and (F.7.9), respectively. Moreover, both codes are CT or not simultaneously.

Recall that for $n-1-r=0$ the code $A^{(r)}$ and for $n-1-r=1$ the code $B^{(n-1-r)}$ are Hamming codes.

For the binary case, let $\mathrm{Aut}(B)$ be the automorphism group of the code $B^{(n-1-r)}$ and $\mathrm{Aut}(A)$ be the automorphism group of the code $A^{(r)}$ for every $1 \le r \le n-1$. Then the groups $\mathrm{Aut}(A)$ and $\mathrm{Aut}(B)$ are isomorphic [31].

Let $\mathcal{C}_t$ be the cyclic group of order t, $\mathcal{S}_t$ be the symmetric group of order t and $\mathrm{GL}(m,2)$ mean the binary general linear group. In [31] it is derived that:

For $n-1-r=2$ *we have* $\mathrm{Aut}(A) = \mathrm{Aut}(B) = \mathrm{GL}(m,2) \times \mathrm{GL}(m,2) \times \mathcal{C}_2$.

For $n-1-r=3$ *we have* $\mathrm{Aut}(A) = \mathrm{Aut}(B) = \mathrm{GL}(m,2) \times \mathcal{S}_3$.

The codes $A^{(r)}$ *and* $B^{(n-1-r)}$ *are CT codes if and only if* $n-1-r \in \{1,2,3\}$.

To finish this section, we give a few sporadic examples of CR codes constructed by using concatenation methods.

(S.7.1) Let

$$H = \begin{bmatrix} 1 & 0 & 1 \\ 0 & 1 & 1 \end{bmatrix}$$

be the parity check matrix of the binary Hamming code of length 3 and denote by H_1 (respectively, H_2) the matrices obtained by one cyclic shift of the columns of $H = H_0$ (respectively, by two cyclic shifts). The binary $[15,9,3;3]$-code C with parity check matrix

$$\begin{bmatrix} H & 0 & 0 & H & H \\ 0 & H & 0 & H & H_1 \\ 0 & 0 & H & H & H_2 \end{bmatrix}$$

is CR with

$$\mathrm{IA} = \{15, 12, 1; 1, 4, 15\}.$$

So, this code has the same parameters as the code (S.6.11) and these codes are equivalent [122].

(S.7.2) The binary $[16,9,4;4]$-code, obtained by extension of the above mentioned $[15,9,3]$ code, is also CR with

$$\mathrm{IA} = \{16, 15, 12, 1; 1, 4, 15, 16\}.$$

Denote by $D(u,q)$ a difference matrix (see [18]), i.e., a square matrix of the order qu over an additive group of the order q, such that the component-wise difference of any two rows contains any element of the group exactly u times. Take the difference matrix $D = D(2,3)$:

$$D = \begin{bmatrix} 0 & 0 & 0 & 0 & 0 & 0 \\ 0 & 0 & 1 & 2 & 2 & 1 \\ 0 & 1 & 0 & 1 & 2 & 2 \\ 0 & 2 & 1 & 0 & 1 & 2 \\ 0 & 2 & 2 & 1 & 0 & 1 \\ 0 & 1 & 2 & 2 & 1 & 0 \end{bmatrix}.$$

(S.7.3) Let H be a binary (12×18) matrix obtained from D by changing any element i by the matrix H_i (which is defined under describing the code (S.7.1). Then the $[18, 12, 3; 2]$ code with parity check matrix H, is a CR code with $\text{IA} = \{18, 15; 1, 6\}$.

Do the same construction as in description of the code (S.7.3) for the matrix D', which is the difference matrix $D(2,3)$ without the trivial column. The resulting $[15, 9, 3; 3]$ code is CR with

$$\text{IA} = \{15, 12, 1; 1, 4, 15\}.$$

This code coincides by parameters and intersection arrays with the code (S.7.1) above and also with the code (S.6.11). All these codes are mutually equivalent to each other what was checked by computer [122].

2.7.3 CR Codes by Concatenation of Simplex Codes

For two codes A and $B = \{\mathbf{y}_j : j = 0, 1, \ldots, N_b - 1\}$ with parameters $(n_a, N_a, d_a)_{q_a}$ and $(n_b, N_b, d_b)_{q_b}$, such that $Q_{q_b} \subseteq Q_{q_a}$ and $N_b = q_a$, define a new code C over Q_{q_b} (which is called a *concatenated code*, or a *concatenation of A and B*), such that

$$C = \{(\mathbf{y}_{x_1}, \mathbf{y}_{x_2}, \cdots, \mathbf{y}_{x_{n_a}}) : \mathbf{x} = (x_1, x_2, \cdots, x_{n_a}) \in A\}, t$$

where the every symbol $i \in Q_{q_a}$ of codewords of A we change by codewords $\mathbf{y}_i$ of B (with number i). The code C has parameters $[n, N, d]_q$, where

$$n = n_a\, n_b, \;\; d \geq d_a\, d_b, \;\; N = N_a, \;\; q = q_b \,. \tag{2.34}$$

Clearly as an inner code, we can use a set of different codes (or matrices) putting in a one-to-one correspondence between the symbols of the external code and the corresponding internal code (or matrix). The construction of CR codes in the previous section can be considered as concatenation of Hamming codes. Indeed, for the given $(2 \times r)$-matrix H over the field $\mathbb{F}_q$, we build a new matrix replacing each nonzero element of the field $\mathbb{F}_q$ with the corresponding matrix of the inner code obtained by cyclic shifts of the original parity check matrix H_m of the Hamming code $\mathcal{H}_m$. The resulting matrix of the size $2q \times r(q^m - 1)$ was considered as a parity check matrix of a new CR code.

In this section, we consider cascading of simplex codes, which gives a new description of the well-known Kasami codes, which was considered in a recent paper. In [141] the authors used a similar idea for the construction and generalization of classical Kasami CR codes. But as inner codes they considered simplex codes, i.e., codes whose dual are Hamming codes $\mathcal{H}_m$ over $\mathbb{F}_p$. We follow the notation that was used in the paper [141]. Let p be a prime power and let $q = p^m$. Denote by $\mathcal{S}_q$ a p-ary Simplex $[n, k, d]_p$ code with parameters

$$n = \frac{q-1}{p-1}, \;\; k = m, \;\; d = \frac{q}{p},$$

which is cyclic for the case, when m and $p-1$ are mutually prime. Denote by $M_q^{\perp}$ the cyclic code of length $q+1$ over $\mathbb{F}_q$ with parity check polynomial

$$h(x) = (x-1)(x-\zeta)(x-\zeta^q),$$

where ζ is a root of order $q+1$ over $\mathbb{F}_{q^2}$. As it is shown in [141], the code $M_q^{\perp}$ is a MDS $[q+1, 3, q-1]_q$-code.

Theorem 2.41 ([141]). *The concatenation of $M_q^{\perp}$ with S_q is a p-ary code $K_q^{p\perp}$ with parameters $[\frac{q^2-1}{p-1}, 3m]$ with nonzeros α and $\alpha^{(1+q)(p-1)}$, where α is a primitive root of $\mathbb{F}_{q^2}$.*

i. If m and $p-1$ are mutually prime then this code K_q^p is cyclic.

ii. The coset graphs of K_q^p and M_q are isomorphic.

iii. If $p = 2$ then K_q^p is a completely regulat code with covering radius 3 and intersection array

$$\{q^2-1,\, q^2-q,\, 1;1,\, q,\, q^2-1\}.$$

Note that for even p the result (iii) was obtained in [29,30,32].

2.7.4 CR Codes by Direct Sum Construction

Let C_1 and C_2 be two codes, not necessarily linear, of the same length n. The direct sum of C_1 and C_2 is the code defined by:

$$C_1 \oplus C_2 = \{(c_1, c_2) : c_1 \in C_1,\, c_2 \in C_2\}.$$

In [12] this construction was used for $u = 2$ to obtain the class of (quasi-perfect) UPNS codes (hence CR) taking as C_1 and C_2 two arbitrary perfect codes with the same parameters. Then Solé [142,143] used this construction taking u times an arbitrary (linear or not) perfect code with $d = 3$ containing the zero codeword. The next statement from [32] generalizes these results for arbitrary CR codes with covering radius $\rho = 1$ and the same intersection numbers and gives the sufficient conditions in order to the resulting code was also CT.

Theorem 2.42 ([12,32,143]). *Let u be any positive integer and let C_i, $i = 1, 2, ..., u$ be q-ary CR $(n, M, d; 1)_q$ codes with the same intersection array $(b_0; c_1)$.*

(F.7.7) Then, for any $u \geq 1$, their direct sum

$$C = C_1 \oplus \cdots \oplus C_u$$

is a CR $(nu, M^u, d; u)_q$ code with intersection numbers for $i = 0, 1, ..., u$

$$a(u)_i = n(q-1) - (u-i)b_0 - ic_1,\;\; b(u)_i = (u-i)b_0,\;\; c(u)_i = ic_1.$$

Moreover, if all codes C_i are CT and permutation equivalent to each other, then C is also CT.

We also give here somewhat more general results from [96] about the direct sum of CR codes. Clearly that a direct sum can be defined for schemes $H(q_1, n_1)$ and $H(q_2, n_2)$ with different sizes of alphabets and lengths. In this case, the neighborhood property corresponds to the direct sum of these schemes, i.e., in the direct sum $H(q_1, n_1) \oplus H(q_2, n_2)$. Two vectors x_1 and x_2, $x_i = (y_i, z_i), i = 1, 2$ where $y_i \in H(q_1, n_1)$ and $z_i \in H(q_2, n_2)$ are neighbors in $H(q_1, n_1) \oplus H(q_2, n_2)$, if either y_1 and y_2 are neighbors in $H(q_1, n_1)$, or z_1 and z_2 are neighbors in $H(q_2, n_2)$.

Proposition 2.27 ([96]). *Let C_1 and C_2 be non-trivial codes in the Hamming schemes $H(q_1, n_1)$ and $H(q_2, n_2)$, respectively. If the code $C = C_1 \oplus C_2)$ is CR in $H(q_1, n_1) \oplus H(q_2, n_2)$, then both C_1 and C_2 themselves must be CR in their respective schemes.*

Moreover, the distance partition $\Pi = \{C(0), C(1), \ldots, C(\rho)$, where $\rho = \rho_1 + \rho_2$, induced by the code C is also a direct sum of the partitions $\Pi_i, i = 1, 2$, induced by C_1 and C_2 [96]. In particular, the intersection matrix L of C looks as follows

Lemma 2.4 ([96]). *Let C_i, $i = 1, 2$, be a CR code in $H(q_i, n_i)$ with distance partition $\Pi_i = \{C_i, C_i(1), \ldots, C_i(\rho_i)\}$, for $i = 1$ and $i = 2$. Then*

i. The partition $\Pi = \Pi_1 \oplus \Pi_2$ is an equitable partition in the product scheme $H(q_1, n_1) \oplus H(q_2, n_2)$.

ii. A codeword $c \in (C^{(1)}(i) \oplus C^{(2)}(j)$ has all its neighbors in

$$\{\{C_1(i-1) \cup C_1(i) \cup C_1(i+1)\} \oplus C_2(j)\}$$
$$\bigcup \{C_1(i) \oplus \{C_2(j-1) \cup C_2(j) \cup C_2(j+1)\}\}$$

iii. If C_i has intersection numbers $a_j^{(s)}, b_j^{(s)}, c_j^{(s)}$, $s = 1, 2, j = 0, \ldots, \rho_s$, then in the product scheme $H(q_1, n_1) \oplus H(q_2, n_2)$, a codeword $c_{i,j} \in C_i^{(1)} \oplus C_j^{(2)}$ has:
$c_i^{(1)}$ neighbors in $C_1(i-1) \oplus C_2(j)$,
$b_i^{(1)}$ neighbors in $C_1(i+1) \oplus C_2(j)$,
$c_j^{(2)}$ neighbors in $C_1(i) \oplus C_2(j-1)$,
$b_j^{(2)}$ neighbors in $C_1(i) \oplus C_2(j+1)$,
$a_i^{(1)} + a_j^{(2)}$ neighbors in $C_1(i) \oplus C_2(j)$,

2.7.5 Arithmetic CR Codes

The next lemma is a useful result for tridiagonal matrices (see [96,147]).

Lemma 2.5. *Let k, b, c and ρ be natural numbers. The tridiagonal matrix*

$$L = \begin{bmatrix} a_0 & \rho b & & & & \\ c & a_1 & (\rho-1)b & & & \\ & 2c & a_2 & (\rho-2)b & & \\ & \cdot & \cdot & \cdot & \cdot & \cdot \\ & \cdot & \cdot & \cdot & \cdot & \cdot \\ & & & (\rho-1)c & a_{\rho-1} & b \\ & & & & \rho c & a_\rho \end{bmatrix}$$

where $a_i = k - ic - (\rho - i)b$ for $i = 0, \ldots \rho$, has eigenvalues $\mathrm{Spec}(L) = \{k - ti : 0 \leq i \leq \rho\}$, *where $t = b + c$.*

According to [96] a CR code is called *arithmetic* if its eigenvalues form an arithmetic progression.

Proposition 2.28 ([96]). *Let C_1 and C_2 be non-trivial CR codes in the Hamming schemes $H(q_1, n_1)$ and $H(q_2, n_2)$, respectively, with the intersection numbers $a_j^{(s)}, b_j^{(s)}, c_j^{(s)}$, $s = 1, 2, j = 0, \ldots, \rho_s$, where $s = 1, 2$. Then*

i. The code $C = C_1 \oplus C_2$ is a CR code in $H(q_1, n_1) \oplus H(q_2, n_2)$, if and only if there exist integers c and b satisfying (a) and (b):
(a) $c_i^{(s)} = ic$, for $i = 0, \ldots, \rho_s$ where $s = 1, 2$ and
(b) $b_{\rho_s - i}^{(s)} = ib$, for $i = 0, \ldots, \rho_s$ where $s = 1, 2$;

ii. In this case, $C_1 \oplus C_2$ has covering radius $\rho = \rho_1 + \rho_2$ and intersection numbers

$$c_i = ic, \quad b_i = (\rho - i)b, \quad \text{for } i = 0, \ldots, \rho,$$

iii. All three codes C_1, C_2, and $C_1 \oplus C_2$ are arithmetic CR codes.

Unfortunately, we cannot give more detailed results on these questions here (in particular, the main classification theorem in [96]) (which classify *the possible quotients $H(q, n)/P$, where P is a completely regular partition of the Hamming scheme $H(q, n)$ into arithmetic completely regular codes*), since their formulation would require the definitions of several new terms from graph theory, which is beyond the scope of this chapter (see [96,147] and references there).

2.8 DIRECT METHODS FOR CONSTRUCTION OF CR CODES

2.8.1 Binomial CR Codes

Denote by $H^{(m,\ell)}$ the binary matrix of size $m \times \binom{m}{\ell}$, whose columns are all different vectors of length m and weight ℓ. Define the binary linear code $C^{(m,\ell)}$, which we call *binomial*, with parity check matrix $H^{(m,\ell)}$.

Theorem 2.43 ([125]). *Let m and ℓ be two natural numbers such that $2 \leq \ell \leq m - 2$. The code $C^{(m,\ell)}$ is CT (and, therefore, CR) exactly in the following four cases:*

(F.8.1) For any $m \geq 4$, the code $C^{(m,2)}$ is a $[n,k,d;\rho]$-code with parameters:

$$n = \binom{m}{2}, \quad k = n - m + 1, \quad d = 3, \quad \rho = \lfloor m/2 \rfloor .$$

and with intersection numbers, for $i = 0, \ldots, \rho$:

$$a_i = 2i(m-2i), \; b_i = \binom{m-2i}{2}, \; c_i = \binom{2i}{2}.$$

(S.8.1) The code $C^{(5,3)}$ is the $[10,5,4;3]$-code with

$$IA = \{10,9,4;1,6,10\}.$$

(S.8.2) The code $C^{(6,4)}$ is the $[15,10,3;3]$-code with

$$IA = \{15,8,1;1,8,15\}.$$

(S.8.3) The code $C^{(7,4)}$ is the $[35,29,3;2]$-code with

$$IA = \{35,16;1,20\}.$$

In fact, from the codes with $\ell = 2$ we have some more new CR codes. First, we divide these codes into two families [125]. Denote $C^{(m)} = C^{(m,2)}$.

Theorem 2.44 ([125,128]). *Let m be a natural number, $m \geq 3$. The code $C^{(m)}$ is antipodal if m is odd and non-antipodal if m is even.*

Since for even m the code $C^{(m)}$ is non-antipodal, its covering set $C^{(m)}(\rho)$ is a translate of $C^{(m)}$ (Theorem 2.7). Consider the new (linear) code $C^{[m]} = C^{(m)} \cup C^{(m)}(\rho)$. The generating matrix $G^{[m]}$ of this code has a very symmetric structure:

$$G^{[m]} = \left[\begin{array}{c|c} I_{k-1} & H^t_{m-1} \\ \hline 0\ldots 0 & 1\ldots 1 \end{array}\right],$$

where H^t_{m-1} is the transpose matrix $H^{(m-1,2)}$. Using that $C^{(m)}(\rho) = C^{(m)} + (1,1,\ldots,1)$ (Theorem 2.7), we obtain

Theorem 2.45 ([128]). *Let m be even, $m \geq 6$ and let $C^{[m]} = C^{(m)} \cup C^{(m)}(\rho)$.*

(F.8.2) The code $C^{[m]}$ is CT (and hence, CR) $[n,k,d;\rho]$ code with parameters

$$n = m(m-1)/2, \quad k = n - m + 2, \quad d = 3, \quad \rho = \lfloor m/4 \rfloor.$$

The intersection numbers of $C^{[m]}$ for $m \equiv 0 \pmod 4$ and $\rho = m/4$ are

$$b_i = \binom{m-2i}{2}, \quad c_i = \binom{2i}{2}, \quad i = 0,1,\ldots,\rho-1, \quad c_\rho = 2\binom{2\rho}{2},$$

and for $m \equiv 2 \pmod 4$ and $\rho = (m-2)/4$ are

$$b_i = \binom{m-2i}{2}, \quad c_i = \binom{2i}{2}, \quad i = 0,1,\ldots,\rho.$$

2.8.2 CR Codes from Combinatorial Configurations

All codes, which we give here, were indicated in paper [32].

(F.8.3) *One Latin square codes.* For any $q \geq 2$ a q-ary MDS $(3, q^2, 2; 1)_q$ code is CR with

$$\mathrm{IA} = \{3(q-1); 3\}.$$

In this case the set $C(\rho)$ is the rest of $\mathbb{F}^3$: $C(\rho) = \mathbb{F}^3 \setminus C$. Thus, this is nothing but a perfect 2-coloring, which we will consider in Section 2.9.

(F.8.4) *Two Latin square codes.* For any $q \geq 3$ and $q \neq 6$ a q-ary MDS $(4, q^2, 3; 2)_q$ code is CR with

$$\mathrm{IA} = \{4(q-1), 3(q-3); 1, 12\}.$$

(S.8.4) *Three Latin square codes.* Three mutually orthogonal Latin squares of order 4 form the equidistant $[5, 2, 4; 3]_4$ code (i.e., any two different codewords are at distance d from each other). This code is CR with

$$\mathrm{IA} = \{15, 12, 3; 1, 4, 15\}.$$

The punctured $[4, 2, 3]_4$ code, obtained from the code above by deleting any one position is also a CR code and it belongs to the family (F.8.4).

(S.8.5) *Four Latin squares code.* Four mutually orthogonal Latin squares of order 5 form the equidistant $[6, 2, 5; 3]_5$ code. This code is CR with

$$\mathrm{IA} = \{24, 20, 13; 1, 2, 6\}.$$

The $[5, 2, 4]_5$ code $C^{(i)}$, obtained by puncturing the above code, is not CR. The subset C_4 is not a 2-design. But the $[4, 2, 3; 2]_5$ code, obtained by double puncturing, is CR, and belongs to the family (F.8.4).

(S.8.6) *The Hadamard code.* The unique (up to equivalence) Hadamard matrix of order 12 induces the CR binary $(11, 24, 5; 3)$ code H (see [80]) with

$$\mathrm{IA} = \{11, 10, 3; 1, 2, 9\}.$$

In [73] it was proved that H is a CT code. Recall that it was also shown that H is unique as a binary CR code with $(n, d) = (11, 5)$ (see Proposition 2.15).

(S.8.7) *The extended Hadamard code.* The extension H^* of the code H considered above is an $(12, 24, 6; 4)$ code. According to [33], H^* is UP and therefore a CR code with

$$\mathrm{IA} = \{12, 11, 10, 3; 1, 2, 9, 12\}.$$

In [35] it was derived that H^* is also a CT code and also is unique among all binary CR codes with $(n, d) = (12, 6)$ (see Proposition 2.15).

(F.8.5) *Johnson scheme in Hamming scheme.* For any natural w, $w \geq 1$, the trivial constant weight $(n, M, d; \rho)$ code (formed by all binary vectors of fixed length n and weight w), (or the Johnson scheme $J(n, w)$) with parameters

$$n = 2w, \; M = \binom{2w}{w}, \; d = 2, \; \rho = w$$

is CR with

$$\mathrm{IA} = \{2w, 2w-1, 2w-2, \ldots, 1; w+1, w+2, \ldots, 2w\}.$$

Clearly, this code is CT, since its automorphism group is the full symmetric group S_n. In Chapter 3, we discuss CR codes in Johnson schemes. This example shows that, in this special case, the Johnson scheme $J(2w, w)$ is CR and CT in the Hamming scheme $H(2, 2w)$.

2.9 CR CODES WITH $\rho = 1$

The case $\rho = 1$ is completely solved, but only for the case of linear codes, which is where we want to start.

2.9.1 Linear CR Codes with $\rho = 1$

Linear q-ary CR codes with $\rho = 1$ are fully classified in [27] by using the two simple following constructions.

Construction $I(u)$. Let C be a $[n, k, d]_q$ code with a parity check matrix H. Define a new code C^{+u} with parameters $[n+u, k+u, 1]_q$ as the code with parity check matrix H^{+u}, obtained by adding $u > 0$ zero columns to H.

Theorem 2.46 ([27]). *The codes C and C^{+u} have the same covering radius and, moreover, C^{+u} is CR if and only if C is CR. In this case, both codes have the same intersection numbers $b_i = b'_i$ and $c_i = c'_i$, i.e.,*

$$a'_i \;=\; a_i + (q-1)u, \;\; b'_i \;=\; b_i, \;\; c'_i \;=\; c_i, \;\; i = 0, 1, \ldots, \rho$$

(here a_i, b_i, c_i are intersection numbers of C).

Construction $II(\ell)$. Let C be a $[n, k, d]_q$ code with parity check matrix H. Let $C^{\times \ell}$ be the code with parameters $[n\,\ell, k + (\ell-1)n, 2]_q$, whose parity check matrix, denoted $H^{\times \ell}$, is ℓ times the repetition of H (or monomially equivalent matrices of H), i.e.,

$$H^{\times \ell} \;=\; [H^{(1)} \,|\, H^{(2)} \,|\, \cdots \,|\, H^{(\ell)}],$$

where, for all $i = 1, \ldots, \ell$, $H^{(i)}$ is the parity check matrix of a code equivalent to the code C.

Theorem 2.47 ([27]). *An $[n, k, d]_q$ code $C^{\times \ell}$ is CR with covering radius $\rho = 1$ if and only if C is CR with covering radius $\rho' = 1$.*

Now for linear CR codes with $\rho = 1$ we have the following result.

Theorem 2.48 ([27]). *Let $C = C(H)$ be a nontrivial $[n, k, d]_q$ code with covering radius $\rho = 1$ and with parity check matrix H.*

(F.9.1) The code C is a $[n, n-m, d; 1]_q$ CT (and CR) code, where $n = n_m \ell + u$ and $n_m = (q^m - 1)/(q-1)$, if and only if the matrix H is of the form (up to monomial equivalence)

$$H = \left((H^q_m)^{\times \ell}\right)^{+u}$$

where H^q_m is a parity check matrix of a Hamming code of length n_m over $\mathbb{F}_q$.

Furthermore,

$d = 3$, if and only if $u = 0$, $\ell = 1$, $n = n_m$ and C is a Hamming code, in which case these codes are included in the family (FP.1).

(F.9.2) $d = 2$, if and only if $u = 0$, $\ell \geq 2$, $n = n_m \ell$.

(F.9.3) $d = 1$, if and only if $u > 0$, $\ell \geq 1$.

In all cases the code C has

$$IA = \{(q-1)\ell\, n_{n-k}; \ell\}.$$

Another approach to description of such codes can be found in [96] in terms of arithmetic CR codes.

2.9.2 Nonlinear CR Codes with $\rho = 1$

A coset of a linear CR code with $\rho = 1$ obviously gives a nonlinear such code with the same $\rho = 1$. Apart from these trivial codes, there are not too many results on this topic. First of all, we should refer to the results on binary such codes obtained due to Fon-Der-Flaas [65–67] and which we consider here. As we mentioned in Section 2.3.4 the equivalent definition for the construction of CR codes with given ρ is the perfect $(\rho + 1)$-coloring of a hypercube. Especially simple perfect colorings are defined for the case $\rho = 1$, i.e., for 2-colorings. Recall that by $H(q, n)$ we denote a q-ary hypercube of dimension n (or q-ary Hamming scheme). Its vertices are all q-ary vectors of length n, and two vertices are neighbors, if the corresponding vectors are at distance 1 from each other. Coloring of its vertices into (usually) white and black colors is called a perfect 2-coloring with intersection matrix

$$\begin{pmatrix} a & b \\ c & d \end{pmatrix}, \tag{2.35}$$

if every black vertex has a black and b white neighbors, and every white vertex has c black and d white neighbors. Clearly, $a+b = c+d = n(q-1)$. In our terminology, $a_0 = a$, $b_0 = b$, $c_1 = c$, and $a_1 = d$ and, hence, it is a nonlinear CR code with $\mathrm{IA} = \{b; c\}$.

The following result gives the lower bound for the value $a = n - b$ for the binary case $q = 2$ (this is the best known bound for so-called *correlation immunity*, which any Boolean function has and what we will consider in the next section (see [65,103,106,120,145] and references there).

Theorem 2.49 ([65]). *Let C be a binary CR code of length n and $\rho = 1$ with $IA = \{b; c\}$. If $b \neq c$, then*

$$c - a \leq \frac{n}{3}. \tag{2.36}$$

In [65] it is established that in the case of the equality (2.36), the code C induced a perfect 2-coloring. The other necessary condition is the following result from [66].

Theorem 2.50 ([66]). *Let C be a binary CR code of length n with $\rho = 1$ and intersection array $\{b; c\}$. Then $b \neq 0 \neq c$, and*

$$\textit{the number } \frac{b+c}{\gcd(b,c)} \textit{ is a power of } 2, \tag{2.37}$$

where $\gcd(b, c)$ *is the greatest common divisor.*

We will prove this last result, since it is very clear and short. Denote by T_w (respectively, T_b) the set of white (respectively, black) vertices in $H(2, n)$. Consider the bipartite graph, induced by these two sets [8]. It is biregular and vertices from T_w have degree b and from T_b have degree c. We conclude that $|T_w| = 2^n c/(b + c)$ should be an integer, implying the result.

Some more necessary conditions for existence of perfect 2-coloring were obtained in [16].

Proposition 2.29 ([16]).

i. Each color of a (b, c)-coloring of $H(q, n)$ forms an orthogonal array of strength $(b + c)/q - 1$.

ii. Assume that there exists a perfect (b, c)-coloring of $H(q, n)$. Then $b'+c'$ divides the cardinality q^k of a k-face, where

$$k = n - \frac{b+c}{q} + 1, \quad b' = \frac{b}{\gcd(b,c)}, \quad c' = \frac{c}{\gcd(b,c)}.$$

iii. Assume that there exists a perfect $(b, 1)$-coloring of $H(q, n)$. Then b is divisible by $q - 1$. Moreover, if $q = p^s$ for some prime p, then $b + 1 = q^r$ for some natural r.

We will discuss the condition (i) in the next section.

Both constructions for linear codes with $\rho = 1$ considered in the previous section work for nonlinear codes also. The following statement from [32] generalizes the corresponding results from [27] for nonlinear case and from [65,66] for nonbinary case (note that Theorem 4.7 in [16] is statement (iii) below).

Proposition 2.30 ([32]).

i. *For every* $n = (q^m - 1)/(q - 1)$, $m \geq 2$, *and any* k, $1 \leq k \leq (q - 1)n$, *there exists a* q*-ary CR code* C *with* $\rho = 1$ *and* $\mathrm{IA} = \{(q-1)n - k + 1; k\}$, *formed by* k *arbitrary translates of* q*-ary perfect code of length* n *and minimum distance* $d = 3$.

ii. *The existence of* q*-ary CR code* $C \subset \mathbb{F}_q^n$ *of length* n *with* $\mathrm{IA} = \{b; c\}$ *implies the existence of CR code* $C^{+k} \subset \mathbb{F}_q^{n+k}$ *with* $\mathrm{IA} = \{b; c\}$ *for any* $k \geq 1$, *formed by changing every codeword* $c \in C$ *by* q^k *codewords of the form:* $(c \,|\, x)$ *where* x *runs over* $\mathbb{F}_q^k$.

iii. *The existence of a CR code* C *of length* n *with* $\mathrm{IA} = \{b; c\}$ *implies the existence, for any* $k \geq 1$, *of a CR code* $C^{\times k}$ *with* $\mathrm{IA} = \{(kb; kc\}$. *Instead of every codeword* $v = (v_1, \ldots, v_n) \in C$, *we consider all vectors of this form*

$$(v_{1,1}, v_{1,2}, \ldots, v_{1,k}, \ldots, v_{1,1}, v_{1,2}, \ldots, v_{1,k}) \in \mathbb{F}_q^{k\,n}$$

such that

$$\sum_{j=1}^{k} v_{i,j} = v_i.$$

The codes which meet the bound given in (2.36) are the most interesting. Such codes of length $n = 3u$ should have $\mathrm{IA} = \{3u - a; a + u\}$ where $a \geq 0$. Two infinite families of codes with such intersection arrays are known (see [65] and references there): $\{3k; k\}$ and $\{5k; 3k\}$. These families come from codes with intersection arrays $\{3; 1\}$ and $\{5; 3\}$ applying Proposition 2.30. The first code is the trivial binary perfect code of length 3, and the second of length 6 (constructed by Tarannikov [145]) can be constructed from the first one using the two following lemmas due to Fon-Der-Flaass [66].

Before, we have to remind one more concept for Hamming scheme $H(2, n)$, namely, so-called k*-faces*. For two such vectors x and y, where x covers y, denote by $\Gamma^y_{[x]}$ the face induced by x and y, i.e. the following set of binary vectors from $H(2, n)$:

$$\Gamma^y_{[x]} = \{z + y + x : \ z \in H(2, n),\ x \preceq z\}.$$

In other words, the set $\Gamma^y_{[x]}$ consists of all $2^{n - \mathrm{wt}(x)}$ (hence $k = n - \mathrm{wt}(x)$) vectors from $H(2, n)$, that cover y and coincide with y on the support of x.

Lemma 2.6 ([66]). *Let* C *be a CR code with* $\rho = 1$ *and intersection matrix (2.35) Let* C *can be partitioned into non-intersecting* k*-faces, where* $0 \leq k \leq a$. *Then there exists a CR code* C' *with* $\rho = 2$ *and intersection matrix*

$$\begin{pmatrix} a - k & a + k & 2b \\ a + k & a - k & 2b \\ c & c & 2d \end{pmatrix}.$$

Lemma 2.7 ([66]). *Let C be a CR code with $\rho = 2$ and intersection matrix*

$$\begin{pmatrix} a-k & a+k & 2b \\ a+k & a-k & 2b \\ c & c & 2d \end{pmatrix},$$

where $c \geq a+k$. Then there exists a CR code C' with $\rho = 1$ and intersection matrix

$$\begin{pmatrix} a-k & 2b+c \\ c & 2d+2c-a-k \end{pmatrix}.$$

In connection with condition (2.37) a natural question arises [66]: *to find the value $a^*(b,c)$ for all pairs b, c, satisfying the condition (2.37).* Here under $a^*(b,c)$ we mean the minimum integer a, such that there exists a CR code with IA $= \{b; c\}$, if and only if $a \geq a^*(b,c)$. There are lower and upper bounds for the value $a^*(b,c)$ (see [66] and references there). The next statements give the best known such bounds.

Theorem 2.51 ([66]). *The following lower bounds hold.*

i. If $c < b < 2c$, then

$$a^*(b,c) \geq \frac{1}{2} + \sqrt{c(b-c) + \frac{1}{4}} - (b-c).$$

ii. If $2c < b < 2c + \sqrt{3c-2}$, then $a^*(b,c) \geq 1$.

For given natural integers x, y, such that $x + y = 2^k - 1$, and one of them odd with ℓ consecutive ones in its binary expansion (where $1 \leq \ell < k$), define a function $z(x,y) = k - 1$.

Theorem 2.52 ([66]). *The following upper bounds hold.*

i. $a^*(2b+c, c) \leq \max(0, a^*(b,c) - 1)$.

ii. Let b and c satisfy the condition (2.37). Then

$$a^*(c,b) = a^*(b,c) + b - c.$$

iii. Let b and c satisfy the condition (2.37) and $b \geq c$, let $m = \gcd(b,c), b = m(2b'+1)$, and $c = m(2c'+1)$. If $c' = 0$, then $a^*(b,c) = 0$. Otherwise

$$a^*(b,c) \leq \max(0, c - m(z(b',c') + 1)),$$

Following [16], we say that the parameters b, c and q are admissible if there is an $n_0 = n_0(b,c,q)$ such that (b,c)-colorings of $H(q,n)$ exist if and only if $n \geq n_0$. The results of the calculations in [16] supported the following conjecture: *parameters q, and $b \geq c \neq 1$ are admissible if and only if $(b+c)/\gcd(b,c)$ divides q^k for some natural k.* In particular, the following result holds:

Theorem 2.53 ([16]). *Let $q = p^m$ be a power of prime p. If, for given b and c, we have*

$$b' + c' = q^k \text{ where } b' = \frac{b}{gcd(b,c)}, \quad c' = \frac{c}{gcd(b,c)},$$

for some natural k, then the parameters b and c are admissible. Moreover, the threshold $n_0 = n_0(b, c; q)$ satisfies the inequalities

$$max\left\{\frac{b}{q-1}, \frac{b+c}{q} + k - 1\right\} \leq n_0 \leq \frac{b + c - \gcd(b,c)}{q-1}.$$

Theorem 2.53 implies that the conjecture above holds. It shows the following results.

Theorem 2.54 ([16]). *Let q be a prime.*

1. *The parameters b, c and q are admissible if and only if $b + c = q^k$ for some natural k.*

2. *If b and c satisfy*

$$\frac{b+c}{gcd(b,c)} = q,$$

   then $n_0 = gcd(b, c)$.

3. *If b and c are relatively prime such that $b + c = q^2$, then $n_0 = q + 1$.*

Now turn to the binary 2-partitions of small lengths. An optimal binary CR code of length $n = 12$ with intersection matrix $[[3k, 9k], [7k, 5k]]$ and with IA $= \{9; 7\}$ meeting the bound (2.36) was constructed in [67]. In the same paper, it was proven also that the putative CR code of length $n = 12$ with IA $= \{11; 5\}$ does not exist. As we already mentioned, these results are important from the study of parameters of equitable 2-partitions of the $H(2, n)$. They closed the smallest open cases remaining after the general paper [66]. After that, all quotient matrices of equitable 2-partitions of the $H(2, n)$ were characterized for any $n \leq 23$. A significant step in this problem was made in [106], where the following new necessary condition for the existence of a perfect 2-coloring was obtained.

Theorem 2.55 ([106]). *If there exists an equitable partition of $H(2, n)$ with quotient matrix (2.35) (i.e. $S = [[a; b]; [c; d]]$, attaining the correlation-immunity bound, where $b \neq c$, then either*

$$\frac{b}{gcd(b;c)} \quad or \quad \frac{c}{gcd(b;c)}$$

is divisible by 3.

From this statement we obtain [106] the non-existence of the following two putative perfect 2-colorings of $H(2, 24)$ with intersection matrices: $[[2; 22]; [10; 14]]$ and $[[5; 19]; [13; 11]]$.

After the paper [106], for $n = 24$, the remaining questionable intersection matrices are

$$[[a,b],[c,d]] = [[1;23];[9;15]],\ [[3;21];[11;13]],\ [[7;17];[15;9]],$$

and it is important that all these parameters correspond to unbalanced Boolean functions with maximal correlation immunity $15 = 2n/3 - 1$.

2.9.3 Perfect 2-Coloring, Orthogonal Arrays and Boolean Functions

In the binary case, when $q = 2$, which we will focus on now, a perfect 2-coloring is associated with two important concepts that we need to remind. First we notice, that 2-coloring means that we label all vertices of a binary cube $H(2,n)$ with two colors. But this is nothing but a Boolean function, that is, such a coloring is described by a Boolean function. For any Boolean function, its so-called *immune resistance* can be determined. A Boolean function $f : \{0,1\}^n \to \{0,1\}$ is called *unbalanced* if the number of its ones is not equal to $0, 2^{n-1}$ and 2^n. It is called *t-th order correlation immune* if the number of ones (equivalently, zeros) $(x_1,\ldots,x_n)$: $f(x_1,\ldots,x_n) = 1$ is statistically independent on the values of any t arguments. More exactly, a function $f : \{0,1\}^n \to \{0,1\}^m$ is t-resilient if for every x from $\{0,1\}^m$ the function

$$f_x(y) = \begin{cases} 1 & \text{if} \quad f(y) = x \\ 0 & \text{if} \quad f(y) \neq x \end{cases}$$

is correlation immune of order t with 2^{n-m} ones. The resilient functions (which are important for cryptography) for $m = 2$ gives the inequality

$$t \leq \frac{2n}{3} - 1,$$

If $m = 2$ and $t = 2n/3 - 1$, then the function f_x corresponds to the equitable partition of $H(2,n)$ with the quotient matrix $S = [[0,3u],[u,2u]]$, where $u = n/3$. That is we obtain the 2-coloring mentioned in [65] as most interesting [2,3].

Fon-Der-Flaass [65] proved that the correlation-immunity order of an unbalanced Boolean function in n variables cannot exceed $2n/3 - 1$; moreover, any unbalanced Boolean function f with such order of correlation-immunity corresponds to a perfect 2-coloring of the $H(2,n)$ with intersection matrix (2.35), i.e. $[[a;b];[c;d]]$, where $a + b = c + d = n$ and $a - c = -n/3$.

Consider a little more this important concept and try to explain its connection with orthogonal arrays and perfect 2-colorings. In particular, a connection of Boolean functions having the optimal immune resistance with orthogonal arrays (which are also meet the corresponding bounds), and finally how it is connected with perfect 2-colorings. This interesting subject we will give here following mostly to the papers [100,106].

This concept is easy to define, remembering orthogonal arrays, which we defined in the introduction. Such an array can be induced by a perfect 2-coloring, as we can see from the following statement (see [43,65,100])

Proposition 2.31. *Each cell P of an equitable partition of $H(q,n)$ with quotient matrix $S = [S_{i,j}]$ is a simple $OA(|P|, n, q, t)$, where*

$$t = \frac{n(q-1) - \theta}{q} - 1,$$

and θ is the largest second eigenvalue of quotient matrix S. In particular,

$$t = \frac{b+c}{q} - 1,$$

if S looks as $S = [[a, b], [c, d]]$.

The bound obtained by Bierbrauer in [20] (and by Friedman in [69] for $q = 2$) for q-ary orthogonal arrays states that the size N of an $OA(N, n, t, q)$ satisfies the inequality:

$$N \geq q^n \left(1 - \frac{(q-1)n}{q(t+1)}\right). \tag{2.38}$$

This bound gives us a criterion for the quality of colorings in terms of the corresponding orthogonal arrays, which can be obtained from these colorings. Taking into account, that any 2-coloring induces a Boolean function, thereby we get an estimate for its immune resistance.

The orthogonal arrays $\mathrm{OA}(N, n, q, t) = \mathrm{OA}(2^{n-2}, n, 2, 2n/3-1)$ $\mathrm{OA}(2n \ldots 2; n; 2; 2n = 3 \ldots 1)$ lie on the bound (2.38). Hence, they are simple and, as follows from the results of [65], correspond to the equitable partitions with quotient matrix $[[0; 3T]; [T; 2T]]$, $T = n = 3$. In particular, the results of classification [106] imply that there are exactly 16 inequivalent $\mathrm{OA}(1024; 12; 2; 7)$ and exactly 37 inequivalent $\mathrm{OA}(512; 11; 2; 6)$ (such arrays are obtained from $\mathrm{OA}(1024; 12; 2; 7)$ by shortening). Some more results, connected with equitable partitions (in particular, with the quotient matrix $[[0; 3T]; [T; 2T]]$), orthogonal arrays, and Boolean functions can be found in [99,104].

The equitable partitions considered in the current section are related with several classes of combinatorial configurations. The next statement summarizes several known results about such relations.

Lemma 2.8 ([106]). *The following objects are in one-to-one correspondence:*

I. the equitable partitions of the $H(2,n)$ with quotient matrix (2.35) with $[a,b] = [0,n]$ and $[c,d] = [c, n-c]$, where $c < n$;

II. the orthogonal arrays $OA(N, n, 2, t)$, where

$$t = \frac{n+c}{2} - 1 \text{ and } N = 2^n \left(1 - \frac{n}{2(t+1)}\right),$$

so, the parameters attain the bound (2.38);

III. the orthogonal arrays $OA(N/2; n-1; 2; t-1)$;

IV., V. the equitable partitions of the $H(2, n-1)$ *with quotient matrices*

$$\begin{bmatrix} 0 & c-1 & n-c \\ c-1 & 0 & n-c \\ c & c & n-2c-1 \end{bmatrix} \text{ and } \begin{bmatrix} 0 & c-1 & n-c \\ c-1 & 0 & n-c \\ c & c & n-2c-1 \end{bmatrix};$$

VI. the CR codes in $H(2, n)$ *with the intersection array* $IA = (n-c, c; c, n-c)$.

Taking into account the previous lemma, it makes sense to give some final results, obtained in [100] on the existence and classification of orthogonal arrays, related to perfect 2-colorings of binary cube $H(2, n)$. With a computer-aided approach based on connection with equitable partitions, the following results were established in [100]:

i. the uniqueness of the orthogonal array $\mathrm{OA}(1536, 13, 2, 7)$, constructed in Fon-Der-Flaass [66] as an equitable partition of the $H(2, 13)$ with quotient matrix $[[0, 13], [3, 10]]$.;

ii. by shortening the $\mathrm{OA}(1536, 13, 2, 7)$, it was obtained 3 inequivalent orthogonal arrays $\mathrm{OA}(768, 12, 2, 6)$, which is a complete classification for these parameters too;

iii. as a result of computer search, the first parameters of unclassified binary orthogonal arrays $\mathrm{OA}(N, n, 2, t)$ attending the Friedman bound (2.38), i.e. $\mathrm{OA}(2048, 14, 2, 7)$, which can be obtained by puncturing any binary one-perfect code of length 15;

iv. the orthogonal arrays $\mathrm{OA}(2048, 14, 2, 7)$ and OA with similar parameters, which can not be obtained by puncturing of one-perfect codes, were constructed;

v. it is proven that any $\mathrm{OA}(N, n, 2, t)$ with even t meeting the bound

$$N \geq 2^n - 2^{n-2} \frac{n+1}{\lceil (t+1)/2 \rceil}$$

induces an equitable 3-partition of the $H(2, n)$.

According to [100], the first unclassified case was $\mathrm{OA}(2048, 14, 2, 7)$, recently solved in [102].

2.9.4 New Constructions of Perfect 2-Coloring

After a detailed survey [32] in 2019, several papers with new constructions of perfect 2-colorings have been published. In particular, many of these works can be found in a recent survey paper [16] on perfect 2-colorings. We will give several results from this paper (without detailed descriptions of their construction methods) that seem to us very promising directions for constructing CR codes with the covering radius $\rho \geq 2$.

First, we consider the construction connected with additive codes [16]. This construction is based on the following statement.

Theorem 2.56 ([16]). *Let C be an additive code in $(E^n$ where $E = \mathbb{Z}_p^s$ (hence C is a linear code over $\mathbb{F}_p$), with the $m \times sn$ parity check matrix H, whose sn columns denoted by*

$$h_{1,1}, \ldots, h_{1,s}, \; h_{2,1}, \ldots, h_{2,s}, \; \ldots\ldots, h_{n,1}, \ldots, h_{n,s}.$$

For i from the set $\{1, ..., n\}$, denote by V_i the vector subspace of $\mathbb{F}_p^m$, spanned by $h_{i,1}, \ldots, h_{i,s}$. The following assertions are equivalent.

i. All the spaces V_i, $i = 1, ..., n$, are of dimension s, and every nonzero element of $\mathbb{F}_p^m$ belongs to exactly c of them, where

$$c = \frac{(p^s - 1)n}{p^m - 1}.$$

ii. The characteristic function of C in $(\mathbb{F}_p^s)^n$ is a perfect 2-coloring of $H(p^s, n)$ with quotient matrix $[[0, b], [c, b - c]]$, where

$$b = (p^s - 1)n, \quad c = \frac{b}{p^m - 1}.$$

iii. C is an orthogonal array $OA(N, n, q, t)$ with parameters

$$N = p^{sn-m}, \quad q = p^s, \quad t = \frac{p^{m-s}(p^s - 1)n}{p^{m-1} - 1} \pm 1.$$

The next two theorems from [16] are main constructions of the type *Splitting constructions I.*

Theorem 2.57 ([16]). *Assume that there is a one-perfect code in $H(q, q+1)$. If f is a (b, c)-coloring in $H(q, n)$ with the second eigenvalue $\theta \leq 0$, then for all $\tau_1, \tau_2 = 0, \ldots, q$, $\tau_1 + \tau_2 \neq 0, 2q$, there exists a $(q(b+c) - (c\tau_1 + b\tau_2), c\tau_1 + b\tau_2)$-coloring F of $H(q, qn - \tau)$. Moreover, F has the same second eigenvalue θ.*

Theorem 2.58 ([16]). *Assume that there is a one-perfect code in $H(q, q+1)$. Let f be a (b, c)-coloring in $H(q, n)$ with the second eigenvalue θ. Suppose that the vertex set of the first color of f can be partitioned into k-faces. Then the following holds:*

i. If $\theta + k \leq 0$ then for all $\tau = 1, \ldots, q$ there is a $(q(b+c) - c\tau, c\tau)$-coloring F of $H(q, qn - \theta - k)$. The second eigenvalue of F is $\theta - k(q-1)$.

ii. If $\theta \leq k(q-1)$ then for all $\tau = 1, \ldots, q$ there is a $(q(b+c) - c\tau, c\tau)$-coloring F of $H(q, qn - \theta + k(q-1))$. The second eigenvalue of F is $\theta + k(q-1)^2$.

Here is another theorem that applies to the type *Splitting constructions II.*

Theorem 2.59 ([16]). *Let f be a $(q^2-1,1)$-coloring of $H(q,q+1)$ corresponding to a one-perfect code such that the vertex set of the second (non-code) color can be partitioned into lines. Then for every natural p and for each τ, $0 \leq \tau \leq p-1$ there exists a $((q^2-1)(p-\tau),(q^2-1)\tau+p)$-coloring of $H(pq,q+1)$ (with the second eigenvalue $q(p-1)-1$).*

We should also mention the works related to the construction and enumeration of perfect 2-colorings with restrictions on eigenvalues. Denote by $\lambda_i(q,n)$ the ith eigenvalue of the Hamming scheme $H(q,n)$, i.e. $\lambda_i(q,n)=(q-1)n-qi$. Meyerovitz [115] classified all perfect 2-colorings, which have the eigenvalues $\lambda_1(q,n)$ (indeed, by Theorem 2.8 the eigenvalue of perfect 2-coloring is an eigenvalue of $H(q,n)$. In the recent paper [117] the authors enumerated all equitable 2-partitions of $H(q,n)$ with the second eigenvalue $\lambda_2(q,n)$.

The weight spectra of such equitable partitions and, in particular, their local weight spectra are being investigated quite actively, which generalizes Vasil'eva's early papers on local spectra for perfect codes (see [156] and references there). Unfortunately, we cannot dwell on these problems here (which are outside of our subject), and we only refer to several papers where these questions are studied (see [16,89,101,158] and references there).

2.10 CR CODES WITH $\rho = 2$

In the recent paper [34], we could solve the classification problem of CR codes C with covering radius $\rho = 2$ in a very special case, when their dual codes $C^{\perp}$ are antipodal. These results come from the paper [27] and from the recent paper [36]. We will consider these results in the next section.

2.10.1 Linear CR Codes with $\rho = 2$ and Antipodal Dual

We will start by defining the *difference matrices*, which we need here and later in order to describe one special class of linear CR codes with covering radius $\rho = 2$. Recall that a code C over E is *additive* if it is an abelian group over component-wise addition.

Definition 2.14. *Let G be an abelian group of order q written additively. A square matrix D of order $q\mu$ with elements from G is called a difference matrix and denoted by $D(q,\mu)$, if the component-wise difference of any two different rows of D contains every element of G exactly μ times.*

Clearly the property of being a difference matrix is invariant under addition of the constant vector $(a,a,\ldots,a)$, where $a \in G$, to any row or to any column of D. By doing such operations we can obtain the *normalized* difference matrix which has the zero first row and the zero first column.

From [41] (see also [135]) we have the following result.

Lemma 2.9. *For any prime power number q and any natural numbers ℓ and h there exists a difference matrix $D(q^\ell, q^h)$. If ℓ divides h then rows of $D(q^\ell, q^h)$ form a linear equidistant $[q^{\ell+h}, 1 + h/\ell, q^h(q^\ell - 1)_{q^\ell}$ code over $\mathbb{F}_{q^\ell}$.*

We describe the construction of all such difference matrices $D(q^\ell, q^h)$ from [135]. For any natural $m \geq 1$ fix one-to-one correspondence between the elements of $\mathbb{F}_{q^m}$ and the vector space $\mathbb{F}_q^m$. For any natural numbers ℓ and h, denote $u = \ell + h$. For the Galois field $\mathbb{F}_{q^u}$ with elements $\{f_0 = 0, f_1 = 1, f_2, \ldots, f_{q^u-1}\}$ denote by $F = [f_{i,j}]$ the matrix of size $q^u \times q^u$, whose rows and columns are indexed by the elements of $\mathbb{F}_{q^u}$, where $f_{i,j} = f_i f_j$, i.e. F is a multiplicative table of the elements of $\mathbb{F}_{q^u}$. Define the operator $\Phi = \Phi_{u \to \ell}$, which map the elements $x = (x_1, \ldots, x_u)$ of $\mathbb{F}_q^u$ into the elements $x^{(\ell)} = (x_1, \ldots, x_\ell)$ of $\mathbb{F}_q^\ell$ by cutting the last right $u - \ell$ positions of vectors from $\mathbb{F}_q^u$:

$$\Phi(x_1, \ldots, x_\ell, \ldots, x_u) = (x_1, \ldots, x_\ell).$$

Denote by $F^{[\ell]}$ the matrix obtained from F by applying the operator Φ to all elements of F,

$$F^{[\ell]} = [f_{i,j}^{[\ell]}] : \; f_{i,j}^{[\ell]} = \Phi_{u \to \ell}(f_{i,j}).$$

Now we obtain ([41] (see also [135])) the following

Lemma 2.10. *For any prime power number q and any natural numbers ℓ and h the matrix $F^{[\ell]}$ is an additive difference matrix $D = D(q^\ell, q^h)$. If ℓ divides h, i.e. $N = q^{h/\ell+1}$ then $F^{[\ell]}$ is a vector space, implying that the difference matrix D is a linear code.*

According to Delsarte [57] (see Theorem 2.3), the dual code of any linear two-weight code should be a CR code with $\rho = 2$. There are many different families of such codes, and their classification is not still finished (see [48] for a survey of such codes).

The classification of linear CR codes with covering radius $\rho = 1$, enabled the description of structures of linear CR codes with covering radius $\rho = 2$, whose dual codes are antipodal. The next statement explains how it works.

Theorem 2.60 ([27]). *Let C be a nontrivial $[n, k, d]_q$ code with parity check matrix H. Then, C is CR with covering radius $\rho = 2$ and the dual code $C^\perp$ is antipodal if and only if its parity check matrix H looks, up to equivalence, as follows:*

$$H = \begin{bmatrix} 1 & \cdots & 1 \\ & M & \end{bmatrix},$$

where M generates an equidistant code E with minimum distance d', which satisfies the following property: for any nonzero codeword $v \in E$, every symbol $\alpha \in \mathbb{F}_q$, which occurs in a coordinate position of v, occurs in this codeword exactly $n - d'$ times. Moreover, up to equivalence, C is the extension of a CR code C' with covering radius $\rho' = 1$.

Using this Theorem 2.60, in [36] we could enumerate all additive two-weight codes with distances d and n, which allowed to classified all linear CR codes with covering radius $\rho = 2$, whose dual codes are antipodal [34]. We list all these CR codes, but before we have to recall some well-known facts.

Let C_1 and C_2 be linear q-ary codes of the same cardinality, with generator matrices G_1 and G_2, where q is any prime power. Say that C_1 and C_2 are *complementary codes* if we can choose the matrix G_2 in such a form, that the matrix $G = [G_1 \mid G_2]$ generates a *simplex code*, i.e. the commonly known equidistant code with the following parameters (we mentioned these codes in Section 2.7.3):

$$n = (q^m - 1)/(q-1), \;\; k = m, \;\; d = q^{m-1}$$

Recall that by H_m we denote the parity check matrix of Hamming $[n = (q^m - 1)/(q-1), \;\; k = n - m, \;\; d = 3]$ code $\mathcal{H}_m$. In this terminology, the well-known fact is that a difference matrix $D(q, q^{m-1})$ induces the Hamming code $\mathcal{H}_m$ as a complementary to its dual code.

Proposition 2.32 ([27,48]). *The following codes are CR with covering radius $\rho = 2$ whose dual codes are antipodal.*

(F.10.1) *The binary extended perfect $[n, k, 4; 2]_2$ code $\mathcal{H}_m^*$ of length $n = 2^m$, where $k = n - m - 1$ and $m \geq 2$ with*

$$IA = \{n, n-1; 1, n\}.$$

The dual code is a linear Hadamard $[n, m+1, n/2]$-code.

(F.10.2) *The extended q-ary perfect $[n, k, 4; 2]_q$ code $\mathcal{H}_m^*$ of length $n = q+2$ and $k = q - 1$ with*

$$IA = \{(q+2)(q-1), q^2 - 1; 1, q+2\},$$

where $q = 2^r \geq 4$, and $m = 2$ [41,56]. The dual codes belong to the family $TF1$ in [48] and Bose-Bush codes in [34]

(F.10.3) *The $[n, k, 3; 2]_q$ code with*

$$IA = \{n(q-1), n-1; 1, n(q-1)\},$$

whose dual is the code generated by a difference matrix $D(q, q^m)$. This code has length $n = q^m$ and dimension $k = n - (m+1)$, where $m \geq 1$, and $q \geq 3$ is any prime power (all codes of such type generated by the difference matrices have been constructed in [41] (see also [135]). The complementary code to this code is the Hamming code $\mathcal{H}_m$.

(F.10.4) *The $[n, n-2, 3; 2]_q$-codes, whose dual are well-known Latin squares $[n, 2, n-1; 2]_q$-codes, where $3 \leq n \leq q$ and $q \geq 3$ is any prime power [56]. The intersection array is*

$$IA = \{n(q-1), (q-n+1)(n-1); 1, n(n-1)\}.$$

(F.10.5) *A* $[n = q(q-1)/2, k = n-3, 4; 2]_q$ *code for* $q = 2^r \geq 4$ *[56] (Delsarte code; see also [34]).*

(F.10.6) *A code with parameters* $[n = 1 + (q+1)(h-1), k = n-3, 4]_q$*, where* $1 < h < q$ *and* h *divides* q*, for* $q = 2^r \geq 4$ *with*

$$IA = \{(q-1)n, (q+1)(h-1)(q-h+1); 1, (h-1)n\}.$$

The dual code belongs to the family $TF2$ *in [48] and it is called by Denniston–Thas code in [34], since these codes are based on maximal arcs in projective planes* $PG(2, q)$*. The corresponding constructions were given by Denniston in [59] and Thas in [148,149]. The nonexistence of maximal arcs for odd* q *was proven in [7], which implies the nonexistence of similar codes for odd* q*.*

As we already mentioned from [34] we have

Theorem 2.61 ([34]). *Let* C *be a linear nontrivial completely regular* $[n, k, d]_q$*-code* C *with covering radius* $\rho = 2$*, where* $k \geq 2$ *and whose dual code* $C^\perp$ *is antipodal, i.e. a* $[n, n-k, \{d^\perp, n\}]_q$*-code* $C^\perp$*. Then this code* C *belongs to one of the families (F.10.1)–(F.10.6) above.*

Remark that Bose-Bush and Delsarte codes are special cases of Denniston–Thas codes, who constructed such codes in terms of maximal arcs in projective planes (see [34,36] and references there). Clearly all codes, which are complementary to Denniston–Thas codes (F.10.6), also have covering radius 2, but their dual are not antipodal. Hence we have

Proposition 2.33 ([27,48]). *The following codes which are complementary to Denniston–Thas codes (F.10.6) are CR with covering radius* $\rho = 2$*:*

(F.10.7) *A code with parameters* $[n = q(q-h+1)/h, k = n-3, 4]_q$*, where* $1 < h < q$ *and* h *divides* q*, for* $q = 2^r \geq 4$ *with*

$$IA = \{(q-1)n, (q+1)(q-h)(q(h-1)+h)/h^2; 1, q(q-h)(q-h+1)/h^2\}.$$

These complementary codes belong to the family $TF2^d$ in [48] (this is why we mention these codes here) and it is a Denniston–Thas code in [34]). Indeed, as easy to see, h in the code (F.10.6) and h' in the code (F.10.7) are connected, namely, $hh' = q$. More results on complementary codes with covering radius $\rho = 2$ we will give in Section 2.10.3.

2.10.2 Self-Dual CR Codes with $\rho = 2$

Some of RC codes, considered in the previous section, with $\rho = 2$ are self-dual, and we give here several families of such codes from [27] (see also [32]). Recall that by $C_{(m,r)}$ we denote the linear code over $\mathbb{F}_{q^r}$ with parity check matrix H_q^m over $\mathbb{F}_q$ of the Hamming code $\mathcal{H}_m$ of length $n = (q^m - 1)/(q-1)$.

Theorem 2.62 ([27]). *Let $C_{(m,r)} \subset \mathbb{F}_{q^r}^n$ be the lifted code from a perfect $[n, n-m, 3]_q$ Hamming code $\mathcal{H}_m$. Then,*

(F.10.8) for any r, the code $C_{(m,r)}$ is CR. Moreover, $C_{(m,r)}$ is self-dual if and only if $\mathcal{H}_m$ is a ternary $[4, 2, 3]_3$ Hamming code.

Theorem 2.63 ([27]). *Let $\{0, 1, \xi_2, \ldots, \xi_{q-1}\}$ be the set of elements of the field $\mathbb{F}_q$, where $q \geq 4$ is any prime power. Let the matrix H,*

$$H = \begin{bmatrix} 1 & 1 & 1 & 1 \\ 0 & 1 & \xi_i & \xi_j \end{bmatrix},$$

be a parity check matrix for the code C and a generator matrix for the code $C^\perp$, where $\xi_i, \xi_j \in \mathbb{F}_q^*$ are two different elements such that $\xi_i + \xi_j + 1 = 0$. Then

(F.10.9) The code C, as well as $C^\perp$, is a linear antipodal CR $[4, 2, 3; 2]_q$ code with

$$IA = \{4\,(q-1), 3\,(q-3); 1, 12\}.$$

(F.10.10) For the case $q = 2^r \geq 4$, these two equivalent codes coincide: $C = C^\perp$, i.e., C is self-dual.

Note that the family (F.10.9) is a linear subfamily of two Latin squares codes (F.8.4) for the case when q is a prime power. Similarly, the family (F.10.10) is a subfamily of codes (F.10.9) for the case of even $q \geq 4$.

Now consider the smallest example of such codes, when $q = 4$.

(S.10.1) Present the code C as a union of four following subcodes C^i:

$$C = C^1 \cup C^2 \cup C^3 \cup C^4$$

where

$$C^1 = \begin{bmatrix} (0\,0\,0\,0) \\ (0\,1\,2\,3) \\ (0\,2\,3\,1) \\ (0\,3\,1\,2) \end{bmatrix}, \quad C^2 = \begin{bmatrix} (1\,0\,3\,2) \\ (1\,1\,1\,1) \\ (1\,2\,0\,3) \\ (1\,3\,2\,0) \end{bmatrix}, \quad C^3 = \begin{bmatrix} (2\,0\,1\,3) \\ (2\,1\,3\,0) \\ (2\,2\,2\,2) \\ (2\,3\,0\,1) \end{bmatrix},$$

$$C^4 = \begin{bmatrix} (3\,0\,2\,1) \\ (3\,1\,0\,2) \\ (3\,2\,1\,0) \\ (3\,3\,3\,3) \end{bmatrix}.$$

The code C, as well as $C^\perp$, is a linear antipodal CR $[4, 2, 3; 2]_4$ code with

$$\text{IA} = \{12, 3; 1, 12\}.$$

Since it belongs to two Latin squares codes (F.8.4), it is interesting to see the corresponding Latin squares. So, write out these two Latin squares L_1 and L_2 of order 4: the ith column of the matrix C_j is the jth row of the Latin square L_{i-2}, where $i = 3, 4$. It gives the following squares L_1 and L_2:

$$L_1 = \begin{bmatrix} 0231 \\ 3102 \\ 1320 \\ 2013 \end{bmatrix}, \quad L_2 = \begin{bmatrix} 0312 \\ 2130 \\ 3021 \\ 1203 \end{bmatrix}$$

Comparing these two Latin squares with codewords of C we can see the following interesting facts: (i) *the Latin squares L_1 and L_2 are not only mutually orthogonal Latin squares in the classical sense, but they have mutually orthogonal rows. (ii) Furthermore, any two rows from L_i and L_j, where $i, j \in \{1, 2\}$, including the case $i = j$, are orthogonal over $\mathbb{F}_4$.*

I wonder, if there exist such Latin squares for any q, not necessarily the power of a prime.

2.10.3 Linear Complementary CR and CT Codes

Given a parity check matrix H_m of a q-ary Hamming code, which we denote $\mathcal{H}_m$, we consider here a partition of the columns of H_m into two subsets. Then, we consider the two codes that have these submatrices as parity check matrices. We say that anyone of these two codes is the *complementary* code of the other one. We obtain that if one of these codes is a Hamming code, then the complementary code is completely regular and completely transitive. If one of the codes is completely regular with covering radius $\rho = 2$, then the complementary code is also completely regular with covering radius at most 2. Moreover, in this case, either both codes are completely transitive, or both are not. With this technique, we obtain infinite families of completely regular and completely transitive codes which are quasi-perfect uniformly packed.

In this section, we denote as n_m, where $m > 1$, the length of the Hamming code $\mathcal{H}_m$, and we recall that

$$n_m = \frac{q^m - 1}{q - 1}. \tag{2.39}$$

Take a non-empty subset of $n_A < n_m$ columns of H_m as the parity check matrix of a code A. Call B the complementary code that has as parity check matrix the remaining $n_B = n_m - n_A$ columns of H_m. In this section, we consider the following question: *under which conditions the fact that A or B is a completely regular (or completely transitive) code with covering radius $\rho(A) \leq 2$, implies the same fact for the complementary code.*

Let M be a monomial matrix, i.e. a matrix with exactly one nonzero entry in each row and column. Such a matrix can be written as $M = DP$, where D is a monomial diagonal matrix and P is a permutation matrix. If q is prime, then the automorphism group of C, $\mathrm{Aut}(C)$, consists of all monomial $(n \times n)$-matrices M

over $\mathbb{F}_q$ such that $xM \in C$ for all $x \in C$. If q is a power of a prime number, then the monomial automorphism group of C is denoted by $\mathrm{MAut}(C)$, however, $\mathrm{Aut}(C)$ also contains any field automorphism of $\mathbb{F}_q$ which preserves C. The following result can be found in [87].

Lemma 2.11 (Theorem 1.7.9 in [87]). *If DP is the corresponding matrix automorphism α of a code C (where D is a monomial diagonal matrix and P is a permutation matrix), then $D^{-1}P$ corresponds to an automorphism α' of the dual code $C^{\perp}$.*

First consider the case $\rho(A) = 1$. Since there are no two linearly dependent columns in H_m, we have that, for $n_A \geq 3$, the minimum distance of A (and of B, for $n_A \leq n-3$) is at least three and thus, the packing radius $e = \lfloor (d-1)/2 \rfloor$ is at least 1. If $\rho(A) = 1$ (hence $n_A \neq 2$), then $e = \rho(A) = 1$ for $n_A \geq 3$, by Theorem 2.3. Therefore, A is a perfect Hamming code for $n_A > 1$.

For $u \in \{1, \ldots, m-1\}$, the matrix H_m can be written as:

$$H_m = \left[\ H'_u \mid H_{u,m}\ \right], \tag{2.40}$$

where the first u rows of H'_u are as the parity check matrix of $\mathcal{H}_u$ and the remaining $m-u$ rows are all-zero vectors. For the case $u = 1$, the matrix H'_u is simply the column vector $(1, 0, \ldots, 0)^T$. We denote by $B_{u,m} = B$ the code that has parity check matrix $H_{u,m}$. Note that for $n_A > 1$, we have $A = \mathcal{H}_u$. Clearly we have [33] that the dual code of $B_{u,m}$, i.e. the code $B^{\perp}_{u,m}$ generated by $H_{u,m}$, has exactly two nonzero weights, namely, $w_1 = q^{m-1}$ and $w_2 = q^{m-1} - q^{u-1}$. It implies that the code $B_{u,m}$ has the following parameters [33]:

$$n_B = \frac{q^m - q^u}{q-1}, \quad k = \frac{q^m - q^u}{q-1} - m, \quad \rho = 2,$$

and

$$d = \begin{cases} 4 & \text{if } u = m-1, \text{ and } q = 2; \\ 3 & \text{otherwise.} \end{cases}$$

Taking into account the sizes of the subconstituents $B_{u,m}(1)$ and $B_{u,m}(2)$ of $B_{u,m}$ [33]:

$$|B_{u,m}(1)| = q^{n_B - m}(q^m - q^u) \quad \text{and} \quad |B_{u,m}(2)| = q^{n_B - m}(q^u - 1),$$

(where $n_B = (q^m - q^u)/(q-1)$ is the length of $B_{u,m}$), we arrive to the following statement:

Proposition 2.34 ([33]). *The code $B_{u,m}$ is quasi-perfect UP (hence CR) with intersection array:*

$$IA = \{q^m - q^u, q^u - 1; 1, q^m - q^u\}.$$

The matrix H_m given by (2.40) can be written as:

$$H_m = \left[\begin{array}{c|c|c|c|c} H_u & H_u & \cdots & H_u & \mathbf{0}_{u,n_{m-u}} \\ \hline \mathbf{0}_{m-u,n_u} & G_1 & \cdots & G_{q^{m-u}-1} & H_{m-u} \end{array}\right], \tag{2.41}$$

where $\mathbf{0}_{i,j}$ stands for the all-zero matrix of size $i \times j$ and $G_1, \ldots, G_{q^{m-u}-1}$ are $(m-u) \times n_u$ matrices, each one with identical nonzero columns and such that no two columns of two distinct matrices G_i and G_j are equal. To see that the matrix (2.41) is equivalent to the matrix (2.40), note that no two columns of the matrix (2.41) are linearly dependent. Therefore, the matrix (2.41) is a parity check matrix for $\mathcal{H}_m$. Indeed the total number of columns is $q^{m-u}n_u + n_{m_u} = n_m$. The structure of the matrix (2.41) implies the following connection between automorphisms of both codes.

Lemma 2.12 ([33]). *If $\alpha \in Aut(\mathcal{H}_u)$ (acting on the coordinates $\{1, \ldots, n_u\}$), then there exists $\beta \in \mathrm{Aut}(B_{u,m})$ (acting on the coordinates $\{n_u + 1, \ldots, n_m\}$) such that $\gamma = (\alpha \mid \beta) \in Aut(\mathcal{H}_m)$.*

This statement above permits [33] to prove that

Proposition 2.35. *The automorphism group $Aut(B_{u,m})$ is transitive (on the set of vectors of weight one with coordinates in $\{n_u + 1, \ldots, n_m\}$).*

In fact, Proposition 2.35 shows that the action of $\mathrm{MAut}(B_{u,m})$ on the set of vectors of weight one is transitive. As a consequence (see [33]) the full automorphism group $\mathrm{Aut}(B_{u,m})$ is also transitive. In this way we arrive to the following result for the case $\rho(A) = 1$.

Theorem 2.64 ([33]). *Let m, u be any natural numbers, such that $m \geq 2$ and $1 \leq u \leq m-1$. Let $B_{u,m}$ be a code with parity check matrix $H_{u,m}$ defined in (2.40) where q is any prime power. Then*

(F.10.11) *the q-ary code $B_{u,m}$ is CT, hence CR with covering radius $\rho = 2$, with parameters*

$$n_B = \frac{q^m - q^u}{q-1}, \quad k = \frac{q^m - q^u}{q-1} - m, \quad \rho = 2,$$

where

$$d = \begin{cases} 4 & \textit{if } u = m-1, \textit{ and } q = 2; \\ 3 & \textit{otherwise.} \end{cases}$$

and with intersection array:

$$IA = \{q^m - q^u, q^u - 1; 1, q^m - q^u\}.$$

For the case $\rho(A) = 2$, we have the following results.

Theorem 2.65. *If the code A has dimension $n_A - m$ and is completely regular with $\rho(A) = 2$, then the complementary code B, of length n_B, is completely regular with $\rho(B) \leq 2$.*

If the length of A verifies $n_A > n_{m-1}$, then the zero vector cannot be a row of the parity check matrix of A, otherwise H_m would have two linearly dependent columns. Hence, the zero vector could not be generated by the rows of the parity check matrix of A and, as a consequence, the dimension of $A^\perp$ would be m. Therefore, the condition $n_A > n_{m-1}$ implies that the dimension of A is $n_A - m$. Note that the converse statement is not true (see the next example).

Example 2.2. *Let A be the ternary Golay $[11, 6, 5; 2]_3$ code $\mathcal{G}_{11}$. Consider the ternary matrix H_5, which is the parity check matrix of a ternary Hamming $[121, 116, 3; 1]_3$ code. Let B be the complementary code which has length $n_B = 110$. Since A is perfect (so completely regular) with covering radius $\rho(A) = 2$, we have that B is a completely regular code. Clearly, B is not perfect, thus $\rho(B) = 2$. Therefore, the parameters of B are $[110, 105, 3; 2]_3$. Moreover, we have computationally verified that*

(S.10.2) *The code B with parameters $[110, 105, 3; 2]_3$ and covering radius 2 is CT and hence CR with intersection array*

$$IA = \{220, 20; 1, 200\}.$$

Note that the hypothesis about the dimension of A in Theorem 2.65 cannot be relaxed, as the next example shows.

Example 2.3. *Let A be the punctured ternary Golay $[10, 6, 4; 2]_3$ code. As in Example 2.2, consider H_5, the parity check matrix of a ternary Hamming $[121, 116, 3; 1]_3$ code. Now, let B be the complementary code with length $n_B = 111$. In this case, the dimension of A is $6 \neq n_A - m = 5$.*

The code A is completely regular and completely transitive with intersection array

$$IA = \{20, 18; 1, 6\}.$$

The code B has parameters $[111, 106, 3; 2]_3$ and it is not completely regular (even not UPWS) since its external distance is $s(B) = 4$.

The construction described in Theorem 2.65 does not work for covering radius $\rho(A) = 3$. For example, let A be the extended ternary Golay code and consider the ternary matrix H_6, which is the parity check matrix of a ternary Hamming $[364, 258, 3; 1]_3$ code. Let B be the complementary code. The code A is completely transitive with $\rho(A) = 3$. The code B has parameters $[352, 346, 3; 2]_3$ and it is not completely regular since its external distance is $s(B) = 3$.

We also give the expressions of the intersection numbers of A and B in terms of the lengths n_A and n_B and the parameter b_1.

Proposition 2.36. *Assume that the code A is completely regular with dimension $n_A - m$, covering radius $\rho(A) = 2$, and the complementary (completely regular) code B has also covering radius $\rho(B) = 2$.*

i. The code B is completely regular with dimension $n_B - m$.

ii. The code A has intersection array

$$IA(A) = \{b_0,\ b_1; c_1,\ c_2\} = \{(q-1)n_A,\ b_1; 1, \frac{n_A}{n_B},\ b_1\}.$$

iii. The code B has intersection array

$$IA(B) = \{b'_0,\ b'_1; c'_1,\ c'_2\} = \{(q-1)n_B,\ (q-1)n_A - \frac{n_A}{n_B} b_1; 1,\ (q-1)n_B - b_1\}.$$

Finally, we will also show that the codes A and B, under the hypothesis of Proposition 2.36, are both completely transitive or both are not. But before we give a connection between automorphism groups of these codes.

Proposition 2.37 ([33]). *Assume that the code A has dimension $n_A - m$ and the complementary code B has dimension $n_B - m$. Also assume that the covering radius are $\rho(A) = \rho(B) = 2$. Then $Aut(A)$ and $Aut(B)$ are isomorphic as abstract groups.*

We provide a very clear proof of this result (which has an independent meaning) following [33]. Let $\phi \in \mathrm{Aut}(A)$. Let H_A (respectively H_B) be the parity check matrix of A (respectively B). Note that since the dimension of A is $n_A - m$, the minimum distance of A is not less than three. Let $\mathbf{h}_1, \ldots, \mathbf{h}_s$ be a set of s columns in H_A such that

$$\sum_{i=1}^{s} \alpha_i \mathbf{h}_i = 0, \quad \text{where } \alpha_i \in \mathbb{F}_q.$$

Hence, we are assuming that the columns $\mathbf{h}_1, \ldots, \mathbf{h}_s$ are the support of a codeword in A. Since ϕ is an automorphism of A we should have

$$\sum_{i=1}^{s} \alpha_i \phi(\mathbf{h}_i) = 0$$

and so the action of ϕ is linear over the columns in H_A. Since the dimension of A is $n_A - m$ we can extend, by linearity, the action of ϕ over all columns in H_m obtaining $\phi^{(e)} \in \mathrm{Aut}(\mathcal{H}_m)$. The projection of $\phi^{(e)}$ over the columns of H_B gives $\phi_B^{(e)} \in \mathrm{Aut}(B)$. The map $\phi \longrightarrow \phi_B^{(e)}$ is a bijective homomorphism as we can see repeating the above argumentation switching the roles of A and B. In fact, both $\mathrm{Aut}(A)$ and $\mathrm{Aut}(B)$ are isomorphic to the subgroup of $\mathrm{Aut}(\mathcal{H}_m)$ which stabilizes both A and B.

This result permits to prove the following [33]:

Theorem 2.66. *If the code A is completely transitive with dimension $n_A - m$ and covering radius $\rho(A) = 2$, then the complementary code B is also CT.*

It is worth mentioning that, in some cases, the construction used here can be equivalent to the construction described in [30]. But this is not always the case as the following examples show.

Example 2.4. *Consider the binary matrix H_6, which is the parity check matrix of the binary Hamming code of length 63. Take 35 columns of H_6 following the procedure described in [30], as the parity check matrix of the code A. The code A has parameters $[35, 29, 3; 2]_2$ and the complementary code B has parameters $[28, 22, 3; 2]_2$. Computationally, we have verified that both codes are CR but not CT.*

Example 2.5. *Now, consider the binomial code $A' = C^{(7,4)}$, whose parity check matrix $H^{(7,4)}$ contains as columns all the binary vectors of length 7 and weight 4 (see the code (S.8.3))). Note that adding all the rows of $H^{(7,4)}$ gives the zero vector, hence the dimension of $(A')^{\perp}$ is 6. The code A' has parameters $[35, 29, 3, 2]_2$ and the complementary code B' has parameters $[28, 22, 3; 2]_2$. In this case, we have computationally verified that both codes are CR and also CT.*

The codes A and A' (respectively B and B') in Examples 2.4 and 2.5 are completely regular codes with the same parameters and IA. However, A and A' (resp. B and B') are not equivalent since A' (respectively, B') is completely transitive, but A (respectively, B) is not.

Now we give some infinite families of CR and CT codes with $\rho = 2$, whose complementary codes also give CR and CT codes with $\rho = 2$. One family of linear such codes with $\rho = 2$ comes from Theorem 2.36, i.e. from the Kronecker product of parity check matrices of two Hamming codes. To have for the resulting code $\rho = 2$ (see Theorem 2.36) we have to consider the case $m_a = m_b = 2$ and $u = 1$. It gives

Proposition 2.38 ([32,126]).

(F.10.12) The code $C = C(H)$ with parity check matrix $H = H_2^q \otimes H_2^q$, the Kronecker product of H_2^q and H_2^q, is CT, and hence CR code with parameters $[n, k, d; \rho]_{q^u}$, where

$$n = (q+1)^2, \;\; k = n - 4, \;\; d = 3, \;\; \rho = 2. \tag{2.42}$$

and with intersection numbers:

$$b_\ell \;=\; \frac{(q^2 - q^\ell)^2}{(q-1)}, \quad \ell = 0, 1,$$

and

$$c_\ell \;=\; \frac{q^\ell - 1}{q - 1}\, q^{\ell - 1}, \quad \ell = 1, 2,$$

One more family of linear CR codes with $\rho = 2$ comes from Theorem 2.32, i.e. from lifting of Hamming codes (see Theorem 2.32). To have $\rho = 2$ we have to consider the case $m = r = 2$, i.e. the linear code $C_{m,r}$ over $\mathbb{F}_{q^r}$ with parity check matrix H_2^q. It gives CR and CT code with the same intersections numbers as the code from Proposition 2.38, but with different parameters.

Proposition 2.39 ([32,126]).

(F.10.13) The codes $C_{2,2}$ over $\mathbb{F}_{q^2}$ with parity check matrix H_2^q, i.e. the lifted Hamming codes $\mathcal{H}_2^q$, are $[q+1, q-1, 3]_{q^2}$ codes which are CR and CT, with the same intersection numbers as the codes in Proposition 2.38.

We finish this subsection giving one more family of codes, whose dual are the well-known codes described as family SU1 in [48]. Let H_m be the parity check matrix of q-ary Hamming code $\mathcal{H}_m$ of length $n = (q^m - 1)/(q-1)$. For a given prime power q and natural numbers u and m, $u < m$ define a code $C_{u,m}$, whose parity check matrix is obtained from H_m by deleting all $(q^u - 1)/(q-1)$ columns (up to scalar multiple) of the parity check matrix H_u. The q-ary code $C_{u,m}$ has parameters

$$n = \frac{q^m - q^u}{q-1}, \quad k = \frac{q^m - q^u}{q-1} - m, \quad \rho = 2,$$

where

$$d = \begin{cases} 4 & \text{if } u = m-1, \text{ and } q = 2; \\ 3 & \text{otherwise.} \end{cases}$$

Proposition 2.40 ([32,126]).

(F.10.14) For any prime power q and natural $m > 3$ and $u < m$ the code $C_{u,m}$ is CT and, hence, CR with $\rho = 2$ and with

$$IA = \{q^m - q^u, q^u - 1; 1, q^m - q^u\}.$$

2.10.4 CR Codes with $d \geq 4$

Calderbank [44] determined all uniformly packed (linear) $[n, k, 4]$ codes over $\mathbb{F}_2$ and derived non-trivial necessary condition for the existence of uniformly packed $[n, k, 4]$ codes over $\mathbb{F}_q$, where $q \neq 2$ is a prime power. This condition allows to classify uniformly packed $[n, k, 4]$ codes over $\mathbb{F}_4$. The main result of the paper [44] is the following theorem:

Theorem 2.67. *Let C be an $[n, k]$ code over $\mathbb{F}_q$ with exactly two non-zero weights w_1 and w_2 and with the property that the minimum weight in the dual code $C^\perp$ is at least 4.*

If $k = 2$, then $n = 2$, $w_1 = 1$ and $w_2 = 2$.
If $k \geq 3$ and $q = 2$, then either

1. $n = 2^{k-1}, w_1 = 2^{k-2}$ *and* $w_2 = 2^{k-1}$, *or*

2. $k = 4,\ n = 5,\ w_1 = 2$ *and* $w_2 = 4$.

If $k \geq 3$ *and* $q \neq 2$, *then either*

3. $k = 3, q = 2^m,\ n = 2^m + 2, w_1 = 2^m$ *and* $w_2 = 2^m + 2$,

4. $k = 4, n = q^2 + 1, w_1 = (q-1)q$, *and* $w_2 = q^2$,

5. $n = u \cdot \frac{q^{k/2}+1}{q-1},\ w_1 = u\,q^{(k-2)/2}$, *and* $w_2 = (u+1)\,q^{(k-2)/2}$, *where* u *is a natural number and*
$$(2u+3)^2 = 4q^{k/2} + 4q + 1,$$
or

6. $2(q-1)n = (2u+1)q^{(k-1)/2} + (q-2) - \frac{u(u-1)}{q},\quad w_1 = u\,q^{(k-3)/2}$ *and* $w_2 = (u+1)\,q^{(k-3)/2}$, *where* u *is a natural number and*
$$(2u + (2q+1))^2 = 4q^{(k+1)/2} + 4q + 1.$$

If (5) or (6) in Theorem 2.67 holds then there is an integer solution (y, a) of the

$$y^2 = 4q^{a/2} + 4q + 1 \qquad (2.43)$$

Since $(y, a) = (2q+1, 4)$ is a solution, case (5) of Theorem 2.67 includes case (4).

Corollary 2.10 ([44]). *Let* C *be an* $[n, k]$ *code over* $\mathbb{F}_4$ *with exactly two non-zero weights* w_1 *and* w_2 *and with the property that the minimum weight in the dual code is at least* 4. *Then either*

1. $k = 2,\quad n = 2,\ w_1 = 1$ *and* $w_2 = 2$

2. $k = 3,\quad n = 6,\ w_1 = 4$ *and* $w_2 = 6$

3. $k = 4,\quad n = 17,\ w_1 = 12$ *and* $w_2 = 16$

4. $k = 5,\quad n = 78,\ w_1 = 56$ *and* $w_2 = 64$

5. $k = 6,\quad n = 430,\ w_1 = 320$ *and* $w_2 = 352$

We give the proof of these nice results following to Calderbank [44]. If $q = 4$, then (2.43) becomes

$$y^2 - 17 = 2^b. \qquad (2.44)$$

Beukers [19] has shown that the only integer solutions (y, b) of (2.44) are

$$(y, b) = (5, 3),\ \ (7, 5),\ \ (9, 6),\ \ (23, 9)$$

If C arises in part (5) of Theorem 2.67 then

$$(2u+3, k+2) = (5,3),\ (7,5),\ (9,6),\ \text{or}\ (23,9)$$

and so $(u,k) = (2,3),\ (3,4)$ or $(10,7)$. If C arises in part (6) of Theorem 2.67 then

$$(2u+9, k+3) = (5,3),\ (7,5),\ (9,6),\ \text{or}\ (23,9)$$

and so $(u,k) = (7,6)$. The result now follows from Theorem 2.67.

There exist codes corresponding to the set of parameters given as (1), (2), (3) and (4) of Corollary 2.10. The existence of the $[430,7]$ code was an open problem in [44]. The $[78,6]$ code was constructed by R. Hill (see the reference in [44]). Calderbank in [44] conjectured that, for the case $q \neq 2,3,4$, the only integer solution of the equation (2.43) is $y = 2q+1,\ a = 4$.

Now consider the case $q = 3$, where UP (hence CR) codes were also classified in [37]. The next theorem yields a characterization of UP $[n,k,4]$ codes over $\mathbb{F}_3$.

Theorem 2.68 ([37]). *Let C be an $[n,k]$ code over $\mathbb{F}_3$ with two nonzero weights w_1 and w_2 and with the property that the minimum weight in the dual code $C^\perp$ is at least 4. If $k > 1$, then either*

1. $k=2,\ n=2,\ w_1=1$ *and* $w_2=2$, *or*
2. $k=4,\ n=10,\ w_1=6$ *and* $w_2=9$, *or*
3. $k=6,\ n=56,\ w_1=36$ *and* $w_1=45$, *or*
4. $k=5,\ n=11,\ w_1=6$, *and* $w_2=11$.

Furthermore, we have to note that codes with each of the above sets of parameters there exist (see [37] and references there).

2.11 PREPARATA CODES

In this section, we consider one of the most important and interesting class of binary nonlinear codes, without a doubt, with the most interesting completely regular codes, which has been the object of intensive research since 1969 to the present time.

2.11.1 Preparata-Like Codes

Original Preparata $(n = 2^{2m}-1, d = 5, M = 2^{n+1-4m})$-codes were constructed by Preparata [121] in 1968 for any $m = 2,3,\ldots$. In 1969 these codes were studied in detail and included in the class of uniformly packed in narrow sense (UPNS) codes in [133,134,136], where many of their interesting properties were established and considered (in particular, that they are completely regular codes and this property is preserved when they are extended). It was found that any code with such parameters has such interesting properties.

We will start with statements that are valid for any such code and which explain why any such code $(2^m-1, M = 2^{n+1-2m}, 5)$ should be a Preparata-like code.

Theorem 2.69. *Let we have a binary code C with parameters $n = 2^m - 1$, $M = 2^{n+1-2m}$, $d = 5$ and let C^* be its extension, i.e. the $(n+1, 2^{n+1-2m}, 6)$ code. Then*

1. *[133,136] The code C (as well as C^*) is optimal. More precisely, the number of its codewords meets the integer Johnson upper bound.*
2. *[136] The number m is even.*
3. *[160] The codes C and C^* have the covering radii 3 and 4, respectively.*
4. *[6,62,79] The codes C and C^* are non-linear.*

We say several words concerning the proof, which explains why the Preparata codes exist for lengths $n = 2^m - 1$, where m is an even number. Denote by $A(n,d)$ the maximum possible cardinality M of the (n,M,d) code C, and by $A(n,w,d)$ the maximum possible cardinality M of the constant weight (n,M,w,d) code C, whose all codewords are of weight w. The known Johnson upper bound [112] states that

$$A(n,5)\left(1+\binom{n}{1}+\binom{n}{2}+\frac{\binom{n}{3}-\binom{5}{2}A(n,5,6)}{\lfloor\frac{n}{3}\rfloor}\right)\leq 2^n.$$

From the Johnson bound (see [112]) for $A(n,w,d)$, we have that

$$A(n,5,6)\leq\left\lfloor\frac{n}{5}\left\lfloor\frac{n-1}{4}\left\lfloor\frac{n-2}{3}\right\rfloor\right\rfloor\right\rfloor=\frac{n(n-1)}{5\cdot 4}\cdot\frac{n-3}{3},$$

It gives

$$A(n,5,6)\leq\begin{cases}\frac{n(n-1)}{5\cdot 4}\cdot\frac{n-3}{3} & \text{for even} \quad m,\\ \frac{n(n-1)}{5\cdot 4}\cdot\frac{n-4}{3} & \text{for odd} \quad m,\end{cases}$$

implying the following expression for $A(n,5)$:

$$A(n,5)\leq\begin{cases}\frac{2^{n+1}}{(n+1)^2} & \text{for even} \quad m,\\ \frac{2^{n+1}}{(n+1)(n+2)} & \text{for odd} \quad m.\end{cases}$$

Thus, for even m, the cardinality M of C strictly achieves the integer Johnson upper bound (what gives (1)), but, for odd m, the cardinality $M = 2^{n+1}/(n+1)^2$ exceeds the upper bound $A(n,5) \leq 2^{n+1}/(n+1)(n+2)$, what is impossible (implying (2)). This is an explanation of the fact, that Preparata codes have length $n = 2^m - 1$, where m is even.

To see (3) consider the following average arrangement of vectors from $C(3)$ around the zero word. Denote by $c_1, \ldots, c_u$ the codewords from C_5, which have the two fixed first nonzero positions. Since $d(c_i, c_j) = 6$ we deduce that $u \leq \lfloor (n-2)/3 \rfloor \leq (n-3)/3$. Indeed, n is a multiple of 3. So we see that there is only one free position for the vector $x \in C(3)$, i.e. there is no place for the vector $y \in C(4)$, neighbor to x, which should be for the code with covering radius $\rho_C \geq 4$. We deduce that $\rho_C = 3$.

The statement (4) directly follows from the antipodality of such codes C [136], but was first stated in [79] and then was reopened independently in [6] and then in [62].

The fundamental properties of such codes, given in Theorem 2.69, are the reasons for the following definition.

Definition 2.15 ([136,138]). *We call a binary code with the following parameters:*

$$n = 2^m - 1, \;\; M = 2^{n+1-2m}, \;\; d = 5, \;\; m \text{ even}$$

a Preparata-like code, which we denote for all this section by P. Respectively, a binary code with parameters

$$n^* = 2^m, \;\; M = 2^{n^*-2m}, \;\; d = 6$$

will be called an extended Preparata-like code denoted by P^*.

Clearly, according to Theorem 2.69, the puncturing of a Preparata-like $(n+1, M, 6)$ code P^* in any position is a Preparata-like $(n, M, 5)$ code P. Now from [133,134,136] and [77] we have

Theorem 2.70. *We have the following CR codes.*

(F.11.1) *[133,134,136] A Preparata-like $(n = 2^{2m} - 1,\, M = 2^{n+1-4m},\, 5; 3)$ code P is an antipodal CR code with intersection array*

$$IA = \{n, n-1, 1; 1, 2, n\}$$

(F.11.2) *[133,134,136] An extended Preparata-like $(n+1 = 2^{2m},\, M = 2^{n+1-4m},\, 6; 4)$ code P^* is an antipodal CR code with*

$$IA = \{n+1, n, n-1, 1; 1, 2, n, n+1\}$$

(S.11.1) *[77] For the case of $n = 15$, i.e. when these Preparata-like codes become the Nordstrom–Robinson code $\mathcal{NR}$ and its extension, these codes are CT (but they are not CT for $n > 15$).*

We can recall more results for arbitrary Preparata-like codes. The rank of an arbitrary binary code C is the dimension of the space over a binary field generated by its codewords and which we denote by $\mathrm{Rank}(C) = \langle C \rangle$. One of the most interesting geometric properties of any Preparata-like code P (respectively, P^*) is the following fact [136,138]: *this code is a subcode of some perfect code of the same length.* In this section, we denote by H_P (respectively, H_{P^*}) such a perfect code. Clearly, $H_{P^*} = H_P^*$.

Theorem 2.71. *Let P (respectively, P^*) be a Preperata-like code (respectively, an extended Preperata-like code) and let $P, P(1), P(2), P(3)$ be its subconstituents (respectively, $P^*, P^*(1), P^*(2), P^*(3), P^*(4)$).*

1. *[138] The code P is a subcode of the binary perfect code H_P of the same length n. Respectively, P^* is a subcode of the corresponding extended code H_P^*.*

2. *[138] The union of code P and the subconstituent $P(3)$ is a binary perfect code H_P, i.e. $H_P = P \cup P(3)$. Respectively, $H_P^* = P^* \cup P^*(4)$.*

3. *[23,62,138] The codes P and H_P (respectively, P^* and H_P^*) have the same ranks over $\mathbb{F}_2$.*

4. *[136] The code P is a CR code not only in $H(2, n)$ but also in the code H_P which contains this code P.*

5. *[132,134,136] Let P contain the zero codeword. Then for any natural $w, 5 \leq w \leq n-5$, the set P_w (i.e. all codewords of weight w) is a 2-design $T(n, w, 2, \lambda_w)$, where*
$$\lambda_5 = \frac{n-3}{3}.$$
Respectively, for a given P^* of length $n^* = n+1$, which contains the zero codeword, the set P_w^* is a 3-design $T(n^*, w, 3, \lambda_w^*)$, for any even w, $6 \leq w \leq n-6$, where
$$\lambda_6^* = \frac{n^*-4}{3}.$$

6. *[138] Let $P^{*[4]}$ be a translate of P^*, such that any its codeword c is of weight* $\mathrm{wt}(c) \geq 4$. *Then for any even natural $w, 4 \leq w \leq n-6$, the set $P_w^{*[4]}$ is a 2-design $T(n^*, w, 2, \lambda_w)$, where $\lambda_4 = 1$, i.e. the set $P_4^{*[4]}$ is a Steiner system $S(n^*, 4, 2)$.*

7. *[154] Let T be an i-component of P. Then this component is embedded in the unique way in a component of the code H_P.*

We will give some more results from [12,138]. The extended Preparata-like UPWS [136] codes P^* with parameters ($n^* = 2^{2m}$, $M = 2^{n^*-4m}$, $d = 6$) for $m = 2, 3, \ldots$, have the following intersection matrix

$$A = \begin{bmatrix} 0 & n^* & 0 & 0 & 0 \\ 1 & 0 & n^*-1 & 0 & 0 \\ 0 & 2 & 0 & n^*-2 & 0 \\ 0 & 0 & n^*-1 & 0 & 1 \\ 0 & 0 & 0 & n^* & 0 \end{bmatrix},$$

with eigenvalues $\pm\sqrt{n^*}$ and $\pm n^*$. Its packing coefficients $\beta_i, i = 0, \ldots, 4$, are

$$\beta_0 = \beta_1 = 1, \;\; \beta_2 = \frac{5(n^*-1)-3}{n^*(n^*-1)}, \;\; \beta_3 = \frac{3}{n^*-1}, \;\; \beta_4 = \frac{12}{n^*(n^*-1)}. \qquad (2.45)$$

Corresponding Lloyd's polynomial $L_\rho(n,\xi)$,

$$L_4(n^*,\xi) = \sum_{i=0}^{4} \beta_i P_i(n^*,\xi),$$

has four (distinct) roots ξ_j, namely,

$$\xi_1 = n^*/2 - \sqrt{n^*}/2, \;\; \xi_2 = n^*/2, \;\; \xi_3 = n^*/2 + \sqrt{n^*}/2, \;\; \xi_4 = n.$$

Using these roots we can write out the weight polynomial $W_P^{*[r]}(x)$ for the code $P^{*[r]}$ for any $r \in \{0,1,2,3,4\}$, i.e. a translate of P^* with minimum weight of codewords equal to r (see (2.16)).

Consider such translate of the code P^* for $r = 4$, i.e. $P^{*[4]}$. For this translate the zero codeword $\mathbf{0}$ belongs to $P^{*[4]}(4)$, i.e. $d(\mathbf{0}, P^{*[4]}) = 4$. Hence for this vector $x = \mathbf{0}$ we have from the (2.3) that

$$\sum_{i=0}^{4} \beta_i B_{\mathbf{0},i} = 1 = \beta_4 B_{\mathbf{0},4}$$

(indeed, $B_{\mathbf{0},i} = 0$ for $i = 0,1,2,3$). We conclude that

$$B_{\mathbf{0},4} = \frac{1}{\beta_4} = \frac{n^*(n^*-1)}{12},$$

So, we obtain that the number of codewords of weight 4 in the code $P^{*[4]}$ equals to $n^*(n^*-1)/12$, i.e. the cardinality of a Steiner system $S(n^*,4,2)$ (this is what we claimed in (7) of Theorem 2.71). Indeed, this is the case, since all these codewords form a constant weight code with parameters $(n, M = n(n-1)/12, w = 4, d = 6)$ and, therefore, should be the incidence matrix of a Steiner system $S(n,4,2)$ [138]. It is what we claimed already in (6) of Theorem 2.71.

2.11.2 Generalized Preparata Codes

Recall the basic steps in the construction of Preparata-like codes. Dumer [63] in 1976 generalized the original Preparata codes. In the known construction of the Preparata codes [121], where the binary cyclic primitive BCH codes, with generating polynomial $g(x) = m_1(x)m_3(x)$, are used as subcodes of length $(n+1)/2)$ (where n is the length of the Preparata code P), he used codes with a generating polynomial $m_1(x)m_{2^t+1}(x)$ under the condition that $2^t \pm 1$ and $2^m - 1$ are mutually prime to each other. Here m is an odd natural number, $m_i(x)$ is the minimal function of α^i, and α is a primitive element of $\mathbb{F}_{2^m}$. Then Baker, van Lint and Wilson [6] in 1983 rediscovered the result of Dumer in a very nice form, which we are going to consider it here.

For the admissible values of t and m, we are going to construct the code P^* of the length $n = 2^{m+1}$. The codewords of P^* consist of two equal halves

$c = (c_\ell \mid c_r)$, where both halves are characteristic vectors of two subsets of elements of $\mathbb{F} = \mathbb{F}_{2^m}$. Set as locators of codewords of P^* the two copies of $\mathbb{F}$ and denote by c_ℓ the characteristic vector $c_\ell = \text{Char}(X)$ of the subset X of the left copy of $\mathbb{F}$ and by c_r the characteristic vector $c_r = \text{Char}(Y)$ of the subset Y of the right copy. So, a codeword $c \in P^*$ can be written as $c = (\text{Char}(X) \mid \text{Char}(Y))$. For any subset X of $\mathbb{F}$ we denote by $|X|$ its cardinality, i.e. the size of X.

Theorem 2.72 ([6]). *For any natural odd number $m > 1$, let natural t be such that the numbers $2^t \pm 1$ and $2^m - 1$ be mutually prime to each other. Then the extended Preparata code P^*, which we denote by $P^*_{m,t}$ of length $n = 2^{m+1}$ consists of the codewords $c = (Char(X) \mid Char(Y))$, described by all pairs (X, Y) satisfying*

1.
$$|X| \equiv |Y| \equiv 0 \pmod 2,$$

2.
$$\sum_{x \in X} x = \sum_{y \in Y} y,$$

3.
$$\sum_{x \in X} x^\sigma + \left(\sum_{x \in X} x\right)^\sigma = \sum_{y \in Y} y^\sigma,$$

where we denote $\sigma = 2^t + 1$.

The original (extended) Preparata code [121] comes from this construction for the case $t = 1$. This description permits to write out the permutations from the automorphism group $\text{Aut}(P^*_{m,t})$ of the code $P^*_{m,t}$.

Theorem 2.73 ([6]). *The automorphism group $Aut(P^*_{m,t})$ contains the following permutations:*

1. $(X \mid Y) \to (X + a \mid Y + a),\ \forall a \in \mathbb{F}$,
2. $(X \mid Y) \to (Y \mid X)$,
3. $(X \mid Y) \to (aX \mid aY),\ \forall a \in \mathbb{F}^*$,
4. $(X \mid Y) \to (X^g \mid Y^g),\ \forall g \in Aut(\mathbb{F}^*)$.

Using these results, Kantor [92] found the automorphism groups for the codes $P_{m,t}$ and their extension $P^*_{m,t}$.

Theorem 2.74 ([92]).

1. *If $m > 3$, then $Aut(P_{m,t})$ is of the order $(2^m - 1)\,m$ induced by permutations (iii) and (iv) from Theorem 2.73.*

2. *If $m > 3$, then $Aut(P^*_{m,t})$ is of the order $2^{m+1}(2^m - 1)\, m$ induced by permutations (i) – (iv) from Theorem 2.73.*

3. *The group $Aut(P^*_{m,t})$ is transitive on $(\mathbb{F} \mid \mathbb{F})$ and has the normal subgroup induced by permutations (iii) from Theorem 2.73.*

4. *The all punctured codes of $P^*_{m,t}$ are equivalent to $P_{m,t}$.*

The next fundamental result of Kantor [92] completely solves the equivalence problem between the codes $P_{m,t}$ and $P_{m,r}$ and their extensions.

Theorem 2.75 ([92]). *For any natural odd number $m > 3$ let natural t and r be such that there exist the Preparata codes $P_{m,t}$ and $P_{m,r}$. Let $2^t = \sigma$ and $2^r = \chi$. Then*

1. *$P_{m,t}$ and $P_{m,r}$ are equivalent to each other if and only if $\sigma = \chi^{\pm 1}$.*

2. *$P^*_{m,t}$ and $P^*_{m,r}$ are equivalent to each other if and only if $\sigma = \chi^{\pm 1}$.*

All these generalized Preparata codes $P^*_{m,t}$ are inside of the extended Hamming code (which we denote in this chapter) $\mathcal{H}^*_{m+1}$ of length $n^* = 2^{m+1}$, i.e. in this case the code H^*_P coincides with $\mathcal{H}^*_{m+1}$. of length $n = 2^{m+1}$ [6,63,138].

Theorem 2.76 ([6,63,138]). *For any natural odd number $m > 3$ let natural t be such that there exists the Preparata code $P^* = P^*_{m,t}$. Then*

1.
$$P^* \cup P^*(4) = \mathcal{H}^*_{m+1}. \tag{2.46}$$

2. *The code $\mathcal{H}^*_{m+1}$ is partitioned into translates of P^**
$$\mathcal{H}^*_{m+1} = \bigcup_{i=1}^{2^m} \{P^* + h_i\}, \tag{2.47}$$
where $h_i \in P^*(4)$.

2.11.3 $\mathbb{Z}_4$-Linear Preparata Codes

Now we want to consider a fundamental result obtained in 1994 by Hammons Jr., Kumar, Calderbank, Sloane, and Solé [84]. They proved that a wide class of Preparata-like codes comes from linear codes over $\mathbb{Z}_4$ with Lee distance 6 [84] (see Chapter IV). A similar result was obtained also for nonlinear Goethals codes with distance 8. Goethals [81] constructed a remarkable class of binary nonlinear UPWS $(n = 2^{2m}, N = 2^{n-6m+1}, 8)$ codes with covering radius $\rho = 6$, which has 4 times more codewords in comparison with binary BCH codes with the same length and distance.

To distinguish between the codes over $\mathbb{Z}_4$ and over $\mathbb{Z}_2$, we will use the following notation: a code C over $\mathbb{Z}_4$ with the minimum distance Lee we denote by $C^{(\ell)}$, and a code C over $\mathbb{Z}_2$ obtained from $C^{(L)}$ with the minimum Hamming distance we denote by $C^{(h)}$. It means that a code $C^{(h)}$ always has a $\mathbb{Z}_4$-description.

To consider in more detail such Preparata-like codes having a $\mathbb{Z}_4$-description, we will start with the definition. Recall the usual Gray map from $\mathbb{Z}_4$ onto $\mathbb{Z}_2^2$:

$$\varphi(0) = (0,0), \ \ \varphi(1) = (0,1), \ \ \varphi(2) = (1,1), \ \ \varphi(3) = (1,0),$$

which we extend to vectors $x = (x_1, \ldots, x_n) \in \mathbb{Z}_4^n$ in a natural way:

$$\varphi(x) = (\varphi(x_1), \ldots, \varphi(x_n)).$$

Let $d_L(x,y)$ denote the Lee distance between vectors $x = (x_1, \ldots, x_n)$ and $y = (y_1, \ldots, y_n)$ over $\mathbb{Z}_4$, i.e.

$$d_L(x,y) = \sum_{i=1}^{n} d_L(x_i, y_i), \ \text{ where } d_L(x_i, y_i) = \min\{|x_i - y_i|, 4 - |x_i - y_i|\}.$$

An important property of the Gray map is the following well-known fact: *for any two vectors x, y from $\mathbb{Z}_4^n$ with the distance Lee d_L their binary images $\varphi(x)$ and $\varphi(y)$ (under the Gray map) are at the same Hamming distance*, i.e.

$$d_L(x,y) = d(\varphi(x), \varphi(y)). \tag{2.48}$$

Let $C^{(h)} = \varphi(C^{(\ell)})$ where the codewords of $C^{(\ell)}$ look as $c = (c^{(2)} \mid c^{(4)})$ where $c^{(2)} \in \mathbb{Z}_2^{n_1}$ and $c^{(4)} \in \mathbb{Z}_4^{n_2}$. The codewords of code $C^{(h)}$ are derived from the words of code $C^{(\ell)}$ by Gray map of the right parts of words:

$$\varphi(c) = (c^{(2)} \mid \varphi(c^{(4)})).$$

Definition 2.16 ([23]). *We call the (binary) code $C^{(h)}$ by Preparata-like $\mathbb{Z}_2|\mathbb{Z}_4$-linear code if $C^{(h)}$ has the proper parameters, i.e. $n = n_1 + 2n_2 = 2^m$, (m even) $M = 2^{n+1-2m}$, $d(C^{(h)}) = 6$, and if its preimage, i.e. the code $\varphi^{-1}(C^{(h)}) = C^{(\ell)}$ is an abelian group under coordinate-wise addition in $\mathbb{Z}_2$ on the first n_1 positions and in $\mathbb{Z}_4$ on the right n_2 positions. If $n_1 = 0$, then the binary code $C^{(h)}$ is called $\mathbb{Z}_4$-linear.*

Theorem 2.77 ([23]). *Let P^* be a Preparata-like code of length $n = 2^m$. Then*

1. *If P^* is a $\mathbb{Z}_2|\mathbb{Z}_4$-linear, then P^* is a $\mathbb{Z}_4$-linear.*

2. *If P^* is a $\mathbb{Z}_4$-linear, i.e. $P^* = P^{*(h)}$ then $H^* = P^* \cup P^*(4)$ is a $\mathbb{Z}_4$-linear too, i.e. $H^* = H^{*(h)}$.*

3. *The code $H^* = P^* \cup P^*(4)$ is an extended binary perfect code of the same length n, which can be presented as a code $H^{*(\ell)}$ of length $n/2 = 2^{m-1}$ over $\mathbb{Z}_4$ with Lee distance $d_L = 4$.*

4. *The code H^* is partitioned into translates of P^**

$$H^* = \bigcup_{i=1}^{2^{m-1}} \{P^* + h_i\}, \;\; h_i \in P^*(4). \tag{2.49}$$

Since any Preparata-like code $P^{*(\ell)}$ over $\mathbb{Z}_4$ is an abelian group under coordinate-wise addition we can define its dual code over $\mathbb{Z}_4$, what implies the existence of the $\mathbb{Z}_4$-linear Kerdock-like code [84], which we denote here by $K^{*(\ell)}$. The Gray image of the $K^{*(L)}$ is a Kerdock code $K^{*(h)}$ with the following parameters n, M, d dual (in the sense of MacWilliams transform [112] to the parameters of a Preparata code:

$$n = 2^m, \;\; M = n^2, \;\; d = \frac{n}{2} - \frac{\sqrt{n}}{2}. \tag{2.50}$$

Theorem 2.78 ([23]). *Let P^* be a Preparata-like code of length $n = 2^{2m}$ and let $H^* = P^* \cup P^*(4)$ be an extended (binary) perfect code, containing P^*. Then*

1. *[84] If P^* is a $\mathbb{Z}_4$-linear Preparata-like code, then its dual is a $\mathbb{Z}_4$-linear Kerdock-like code K^*.*

2. *[23] If P^* is a $\mathbb{Z}_4$-linear Preparata-like code. Then*

$$Rank(P^*) = Rank(H^*) = \begin{cases} 2^{2m} - 2m & for \quad m > 2, \\ 2^{2m} - 2m - 1 & for \quad m = 2, \end{cases}$$

3. *[23] If K^* is a $\mathbb{Z}_4$-linear Kerdock-like code. Then*

$$Rank(K^*)) = 2m^2 + m + 1.$$

As we many times mentioned, any Preparata-like code P of length n is contained in a binary perfect code H of the same length n. The important result was obtained in [105], where binary perfect codes were explicitly constructed that do not contain Preparata-like codes. These are Vasil'ev codes obtained by switching one nth component from a Hamming code.

According to Theorem 2.77,(2), any $\mathbb{Z}_4$-linear Preparata-like code P^* is a subset of a $\mathbb{Z}_4$-linear binary extended perfect code H^*. Since a $\mathbb{Z}_2|\mathbb{Z}_4$-linear Preparata-like code is a $\mathbb{Z}_4$-linear, we deduce immediately the following corollary.

Corollary 2.11. *A $\mathbb{Z}_2|\mathbb{Z}_4$-linear binary extended perfect code H^* could not contain any $\mathbb{Z}_4$-linear Preparata-like code.*

2.11.4 Preparata-Like Codes and Resolvable Steiner Quadruple Systems

First define a classical concept for Steiner systems (which exists, of course for all t-designs).

Definition 2.17. *A system $S = S(v,k,t)$ is called* ℓ-resolvable, $0 \leq \ell \leq t$, *if the set of blocks B can be partitioned into different non-intersecting subsets B_i, i.e.*

$$B = \bigcup_{i=1}^{r} B_i, \; r = \frac{|B|}{|B_i|} = \frac{\binom{v-\ell}{t-\ell}}{\binom{k-\ell}{t-\ell}}. \tag{2.51}$$

where for every i the set B_i is a Steiner system $S(v,k,\ell)$.

A problem r-resolvability of arbitrary Steiner system $S(v,k,t)$ is an actual open problem (see the survey [85] and references there). Recall two known facts concerning codes which we mentioned above: (i) *if the minimum weight of codewords of the Preparata-like code P^* of length $n = 2^m$ equals 4, then the set P^*_4 is a Steiner system $S(n,4,2)$*; (ii) *for the binary extended perfect code H^*, the set H^*_4 is a Steiner system $S(n,4,3)$*. From the point of view of these facts, the statements we have given in the Theorems 2.76 and 2.77 on the partition of codes H^* into translates of codes P^* mean the following fact: the system $S(n,4,3)$ is partitioned into $(n-2)/2$ (indeed, the code P^* has no words of weight 4) nonoverlapping systems $S(n,4,2)$:

$$S(n,4,3) = \bigcup_{i=1}^{(n-2)/2} S_i, \tag{2.52}$$

where, for every i the set S_i is a Steiner system $S(n,4,2)$ and where $n = 2^m$.

Hence any such partition induces a 2-resolvable system $S(n,4,3)$. The first such family of 2-resolvable systems $S(n,4,3)$ was constructed in [138], based on the partitioning of the extended binary Hamming codes into translates of the original Preparata codes. Since the set H^*_4 of all words of weight 4 of the code H^* coincides with the set of all lines of the affine geometry $AG(2,n)$, such a partition, in other words, means a partitioning of all planes of the $AG(2,n)$ into disjoint spreads [136,138] (see also [5] where these results were independently derived).

We have to say that all currently known different families of Preparata-like codes P^* induce partitions of binary extended perfect codes H^*, which contain them, into translates of these codes P^* (see [6,23,63,84,138,164]). In connection with this, the following open question arises: *do we have such a phenomenon for an arbitrary Preparata-like code?*

In fact constructed by us systems $S(n,4,3)$ are double resolvable. Let 2-resolvable Steiner system $S(n,4,3)$ be given, i.e. the partition (2.52) is valid. If now every system $S_i(n,4,2)$ is 1-resolvable (in standard terminology, resolvable), i.e. for any $i = 1,\ldots,(n-2)/2$ this system can be presented in the form

$$S_i(n,4,2) = \bigcup_{j=1}^{n/4} S_{i,j}(n,4,1), \tag{2.53}$$

then $S(n,4,3)$ is called *double resolvable* [85]. In such case the system $S(n,4,3)$ can be presented as

$$S(n,4,3) = \bigcup_{i=1}^{\frac{n-2}{2}} \bigcup_{j=1}^{\frac{n}{4}} S_{i,j}(n,4,1). \tag{2.54}$$

For the original Preparata codes, this fact that all systems $S_i(n,4,2)$ are resolvable, and, hence, the system $S(n,4,3)$ is double resolvable was mentioned in [5]. Actually all constructed above systems $S(n,4,3)$ are double resolvable.

Concerning these remarkable Preparata-like codes we have to mention one more result on construction of the generalized Preparata-like codes with noncommutative group operation [164]. It gives the growing number of mutually nonequivalent such Preparata-like codes. These codes induce new partitions of corresponding perfect codes into cosets of such Preparata-like codes. Constructed partitions induce partitions (2.52), or new 2-resolvable Steiner quadruple systems $S(n,4,3)$, i.e. systems $S(n,4,3)$, which are partitioned into non-intersecting Steiner systems $S(n,4,2)$. It is proved in [164] that all constructed there 2-resolvable Steiner quadruple systems $S(n,4,3)$ are double resolvable, and they are not equivalent to the known such double resolvable systems. We formulate only one exact result from [164] concerning these Steiner systems. The construction of noncommutative Preparata-like codes derived in [164] permits to choose $(n-4)^2/144$ different parallel classes of the every system $S_i(n,4,2)$ in (2.53) and so to construct

$$\left(\frac{n-4}{12}\right)^{n-2}, \quad n = 2^m,$$

different partitions of the type (2.54) for one the same double resolvable system $S(n,4,3)$.

2.12 LINEAR CR CODES WITH $\rho \geq 3$.

2.12.1 CR Codes from Bent and Semi-Bent Functions

Let F be any function from $\mathbb{F}_2^m$ to $\mathbb{F}_2^{m'}$. For any $(a,b) \in \mathbb{F}_2^m \times \mathbb{F}_2^{m'}$, define the Fourier transform of F as

$$\mu_F(a,b) \;=\; \sum_{\mathbf{x}\in\mathbb{F}_2^m} (-1)^{\langle b\cdot F(x)\rangle + \langle a\cdot x\rangle}\,, \tag{2.55}$$

recall that by "$\langle\cdot\rangle$" we denote the inner product.

For even m, a function F from $\mathbb{F}_2^m$ to $\mathbb{F}_2^{m'}$ is a *bent function* if its Fourier transform is $\mu_F(a,b) = \pm 2^{m/2}$, for all $(a,b) \in \mathbb{F}_2^m \times \mathbb{F}_2^{m'}$ where $b \neq 0$. Bent functions are known to exist only if $m' \leq m/2$ [118]. For odd m, a function F from $\mathbb{F}_2^m$ to $\mathbb{F}_2^{m'}$ is a *semi-bent function* (the old term *almost bent*, shortly AB) if its Fourier transform is $\mu_F(a,b) \in \{0, \pm 2^{(m+1)/2}\}$, for all $(a,b) \in \mathbb{F}_2^m \times \mathbb{F}_2^{m'}$ where $b \neq 0$.

Let F be any function from $\mathbb{F}_{2^m}$ to $\mathbb{F}_{2^{m'}}$ such that $F(0) = 0$. Define

$$\Omega_m = \begin{cases} \{2^{m-1},\ 2^{m-1} \pm 2^{m/2-1}\} & \text{if } m \quad \text{even}, \\ \{2^{m-1},\ 2^{m-1} \pm 2^{(m-1)/2}\} & \text{if } m \quad \text{odd}. \end{cases}$$

For a linear binary code C define the set of integers W_C as the set of all weights of its nonzero codewords:

$$W_C = \{\mathrm{wt}(c) : \; c \in C, \; c \neq \mathbf{0}\}.$$

For any function F define the matrix H_F:

$$H_F = \begin{bmatrix} 1 & \alpha & \alpha^2 & \cdots & \alpha^{n-1} \\ F(1) & F(\alpha) & F(\alpha^2) & \cdots & F(\alpha^{n-1}) \end{bmatrix}. \tag{2.56}$$

The statements of the next theorem can be found in [53].

Theorem 2.79 ([53]). *Let F be a function from $\mathbb{F}_{2^m}$ to $\mathbb{F}_{2^{m'}}$ such that $F(0) = 0$. Let $C = C_F$ be the $[n = 2^m - 1, k, d]$ code defined by the parity check matrix H_F of the form (2.56).*

1. *If m is odd, the function F is semi-bent if and only $W_{C^\perp} = \Omega_m$.*
2. *If m is even, the function F is bent if and only if $W_{C^\perp} = \Omega_m$.*
3. *The code C_F is UPWS if and only if $|W_{C^\perp}| = 3$.*

From bent and semi-bent functions F, we obtain the following CR codes C_F:

(F.12.1) If m is even, F is bent from $\mathbb{F}_2^m$ to $\mathbb{F}_2^{m'}$ such that $F(0) = 0$, then the code C_F is CR with

$$\mathrm{IA} = \{n, \; (n+1)(2^{m'} - 1)/2^{m'}, \; 1; \; 1, \; (n+1)/2^{m'}, \; n\}.$$

(F.12.1') If m is even, F is bent from $\mathbb{F}_2^m$ to $\mathbb{F}_2$ such that $F(0) = 1$, then the code C_F is CR with

$$\mathrm{IA} = \{n, \; (n-3)/2, \; 1; \; 1, \; (n-3)/2, \; n\}.$$

(F.12.1") If C_F is a CR code from family (F.12.1), then its extension C_F^* is CR with

$$\mathrm{IA} = \{n+1, \; n, \; (n+1)(2^{m'} - 1)/2^{m'}, \; 1; \; 1, \; (n+1)/2^{m'}, \; n, \; n+1\}.$$

(F.12.2) If m is even, F is semi-bent from $\mathbb{F}_2^m$ to $\mathbb{F}_2^m$ such that $F(0) = 0$, then the code C_F is CR with

$$\mathrm{IA} = \{n, \; n-1, \; (n+3)/2; \; 1, \; 2, \; (n-1)/2\}.$$

Now using the corresponding results from [28,29] we obtain the following result.

Proposition 2.41. *Let $C = C_F$ of length $n = 2^m - 1$, where m is odd, be defined by the parity check matrix (2.56) and let $C^\perp$ be its dual code with the set of weights $W_{C^\perp}$. Then*

1. *If C is UP and $W_{C^\perp} = \Omega_m$, then the code C^* is UP.*

2. *If C is CR with minimum distance $d \in \{3,5\}$ and C^* is UP, then C^* is CR.*

The first statement comes directly from [10]. For the second statement, the case $d = 3$ is known [29]. Consider the case $d = 5$. Since C is CR, its covering radius is $\rho = 3$. Hence C is quasi-perfect. Now the result follows from Proposition 2.11.

Taking a power function F in (2.56) for m odd, we obtain binary primitive cyclic $[n = 2^m - 1, n-2m, d; \rho]$ codes with generator polynomials of type $g(x) = m_1(x)m_\ell$.

In the case $d = 5$ they are CR [53] with $\rho = 3$ and with

$$\mathrm{IA} = \{n, n-1, (n+3)/2; 1, 2, (n-1)/2\}. \tag{2.57}$$

We found in the literature the following cases.

1. BCH codes with $d = 5$ where $m = 2t + 1$ and $\ell = 3$ [12].

2. Gold codes [82], $\ell = 2^t + 1$, $\gcd(t, m) = 1$.

3. Kasami codes [93], $\ell = 2^{2t} - 2^t + 1$, $\gcd(t, m) = 1$.

4. Welch codes [166] (see also [50,86], $\ell = 2^{\frac{m-1}{2}} + 3$.

5. Niho codes [60]:

$$\ell = \begin{cases} 2^t + 2^{t/2} - 1 & \text{if } t \text{ even}, \\ 2^t + 2^{(3t+1)/2} - 1 & \text{if } t \text{ odd}. \end{cases}$$

The existence of the aforementioned codes comes from known bent and semi-bent functions. Some new such functions, which are not power functions and which provide new CR codes can be found, for example, in [14,17,42,52,61]. The question on complete transitivity of all these codes needs additional attention (for small n we give it later).

(F.12.3) The extended codes of all binary codes above are CR with $\rho = 4$ and with

$$\mathrm{IA} = \{n+1, n, n-1, (n+3)/2; 1, 2, (n-1)/2, n+1\}. \tag{2.58}$$

2.12.2 BCH Codes with $d = 5$ and $d = 7$

We say several words about BCH codes. The binary (primitive, in narrow sense) BCH codes are cyclic codes of length $n = 2^m - 1$ with generator polynomial $g(x)$,

$$g(x) = m_1(x)m_3(x)\cdots m_{2t-1}(x)$$

have $k \geq n - mt$ information symbols and (constructive) minimum distance $d \geq 2t + 1$, where $m_i(x)$ is a minimal polynomial of α^i and α is a primitive element of the Galois field $\mathbb{F}_{2^m}$. Two classes of such codes are very interesting for RC and UPWS codes. In [12] it was shown that the BCH codes of length $n = 2^{2m+1} - 1$ and distance $d = 5$, for $m = 2, 3, \ldots$, are UPNS and, since they have $\rho = 3$ (see reference in [112]), hence they are RC codes. These codes (as UPNS) have the following packing coefficients β_i [12]:

$$\beta_0 = \beta_1 = 1,\ \beta_2 = \beta_3 = \frac{6}{(n-1)},$$

implying that their extension is also UPNS (Theorem 2.18), and hence CR. They have the following roots ξ_i of Lloyd's polynomial [12]:

$$\xi_1 = \frac{n+1}{2} - \sqrt{\frac{n+1}{2}},\ \xi_2 = \frac{n+1}{2},\ \xi_3 = \frac{n+1}{2} + \sqrt{\frac{n+1}{2}}.$$

Goethals and van Tilborg [80] proved that the BCH codes of length $n = 2^{2m+1} - 1$ and distance $d = 7$, for $m = 2, 3, \ldots$, are UPWS (but not CR) with $\rho = 5$ with the following parameters λ and μ:

$$\lambda = \mu + 1,\ \ \mu = (2^{2m} - 1)(2^{2m-2} - 1) \cdot \frac{2}{15},$$

More exactly, they proved that these codes are uniformly packed in the sense, which we mentioned in Section 2.2.3, i.e. they have the following packing coefficients β_i [12]:

$$\beta_0 = \beta_1 = 1,\ \beta_2 = \beta_3 = 1 - \frac{\lambda}{\mu},\ \beta_4 = \beta_5 = \frac{1}{\mu}.$$

We again deduce that their extensions are also UPWS (but not CR) with $\rho = 6$ by the same Theorem 2.18.

2.12.3 Binary Cyclic CR Codes with $\rho \geq 3$

Now we want to give some results concerning binary cyclic codes with generating polynomials of the form $m_i(x)m_j(x)$, where $m_i(x)$ denotes the minimal function of the element α^i, and α is a primitive element of the field $\mathbb{F}_{2^m}$ and where i and j are arbitrary natural numbers in the region $0 \leq i, j \leq 2^m - 2$. Denote by $C_{i,j}$ the cyclic code with the generator polynomial $m_i m_j$. Since a code $C_{i,j}$ has the distance $d(C_{i,j}) \leq 5$, any such code with $d(C_{i,j}) = 5$ and the covering radius $\rho = 3$ is quasi-perfect and, therefore, all such UPNS codes are CR codes.

Note that for the field $\mathbb{F} = \mathbb{F}_{2^m}$ any elements $a \in \mathbb{F}$ and a^2 have the same minimal polynomial. If $\alpha \in \mathbb{F}$ is a primitive element, then for any natural r, which is mutually prime to $2^m - 1$, the element α^r is also a primitive element of $\mathbb{F}$. These evident facts imply the following

Proposition 2.42. *Let $n = 2^m - 1$ and α be a primitive element of $\mathbb{F}_{2^m}$.*

i. The codes $C_{i,j}$ and $C_{i,2j}$ coincide.

ii. *For any natural* r*, which is mutually prime to* n*, the codes* $C_{i,j}$ *and* $C_{ri,rj}$ *are equivalent.*

iii. *Let two codes* $C_{i,j}$ *and* $C_{s,\ell}$ *be equivalent to each other and let* r *be a natural number, such that* r *and* $n = 2^m - 1$ *are mutually prime. Then the codes* $C_{ri,rj}$ *and* $C_{rs,r\ell}$ *are equivalent.*

iv. *A code* $C_{i,j}$ *is equivalent to the code* $C_{-i,-j}$*.*

For given m and any natural odd numbers i and j, where $i \neq j$ and $1 \leq i, j \leq 2^m - 3$, define the orbit $\mathcal{O}_{i,j}$, consisting of codes $C_{ri,rj}$ and $C_{-ri,-rj}$, where r, $1 \leq r \leq 2^m - 3$ and r is coprime with $2^m - 1$. From Proposition 2.42 we deduce that all codes $C_{i',j'}$ from the orbit $\mathcal{O}_{i,j}$ are mutually equivalent to each other.

Now we want to give some results obtained with the help of a computer search [122] for the first several values of m. We give the code parameters, the covering radius, the type of code (i.e. CR, CT, or UPWS) its intersection array, as well as the corresponding parameters of the extended code, except the intersection index, which is obtained from the original code C only by adding two new parameters, namely, $b_0^* = c_{\rho^*}^* = n + 1$, where n is the length of the code C and $\rho^* = \rho + 1$ is the covering radius of C^*.

For $m = 3$ we have only common known (such as perfect, even halves of perfect, codes with parity checking, etc.) CR (which are all CT) codes.

For the case $m = 4$ we have $x^{15} - 1 = m_0(x)m_1(x)m_3(x)m_5(x)m_7(x)$: where

$$m_0 = x + 1, \;\; m_1 = x^4 + x + 1, \;\; m_3 = x^4 + x^3 + x^2 + x + 1, \;\; m_5 = x^2 + x + 1,$$
$$m_7 = x^4 + x^3 + 1.$$

For this case, we have the following CT (and, hence, CR) codes $C_{i,j}$ [122], which are not mentioned before.

(S.12.1) [122] The CT $[15, 10, 2]$ code $C_{0,3}$ with $\rho = 5$ and with

$$\text{IA} = \{15, 12, 9, 6, 3; 3, 6, 9, 12, 15\}.$$

The extended code is not UPWS.

(S.12.2) [122] The CT $[15, 12, 2]$ code $C_{0,5}$ with $\rho = 3$ and with

$$\text{IA} = \{15, 10, 5; 5, 10, 15\}.$$

The extended code is not UPWS.

(S.12.3) [122] The CT $[15, 9, 3]$ codes $C_{5,i}$, $i = 1, 7$ with $\rho = 3$ and with

$$\text{IA} = \{15, 12, 1; 1, 4, 15\}.$$

The codes $C_{5,1}$ and $C_{5,7}$ are equivalent to each other and these codes are equivalent to the code (S.6.11).

(S.12.4) [122] The extended [16, 9, 4] codes $C^*_{5,i}, i = 1, 7$ are also CT with $\rho = 4$ and

$$\mathrm{IA} = \{16, 15, 12, 1; 1, 4, 15, 16\}.$$

For $m = 5$ and, hence, at length $n = 31$, all codes $C_{i,j}$ fit into the following cases: not UPWS, CR, and CT. With the field $\mathbb{F}_{2^5}$ generated by $m_1 = x^5 + x^2 + 1$ we have (except common known codes) only CR [31, 21, 5] codes $C_{i,j}$ with $j \neq i, -i$ from the family (F.12.2) with intersection array (2.57) and extended such CR ([32, 21, 6] codes from the family (F.12.3) with intersection array (2.58). All such codes form two orbits of equivalence, namely, $\mathcal{O}_{1,3}$ and $\mathcal{O}_{1,5}$ [122].

Similar situation we have for $m = 7$, since $127 = 2^7 - 1$ is prime and, therefore, the all minimal functions are of degree 7. There are only CR [127, 113, 5] codes $C_{i,j}$, where $j \neq i, -i$ from the family (F.12.2) with intersection array (2.57) and extended such CR [128, 113, 6) codes from the family (F.12.3) with intersection array (2.58). For the field $\mathbb{F}_{2^7}$ with $m_1(x) = x^7 + x + 1$, we have five orbits of equivalence [122], namely, with generators

$$C_{1,3},\ C_{1,5},\ C_{1,9},\ C_{1,11},\ C_{1,23}.$$

Remark, that the all $C_{i,j}$ codes above with $j \neq i, -i$ of length 31 and 127 are CR, but not CT.

The case $m = 6$ is very interesting. Let $\mathbb{F}_{2^6}$ is generated by $m_1(x) = x^6 + x^4 + x^3 + x + 1$. Then we have except m_0 the following functions of degree less than 6:

$$m_9(x) = x^3 + x + 1 \quad m_{-9} = m_{27} = x^3 + x^2 + 1, \quad m_{21} = x^2 + x + 1.$$

This case is quite rich, since the length of the codes is not a prime number. We have here the following equivalence classes of CR and (or) CT codes $C_{i,j}$ and their extensions [122].

(S.12.5) The CT cyclic [63, 60, 2] code $C_{0,21}$ with $\rho = 3$ and with

$$\mathrm{IA} = \{63, 42, 21; 21, 42, 63\}.$$

(S.12.6) The CT cyclic [63, 59, 2] codes $C_{0,i}$ for $i = 9, 27$ with $\rho = 3$ and with

$$\mathrm{IA} = \{63, 54, 9; 9, 54, 63\}.$$

(S.12.7) The CT cyclic [63, 56, 4] codes $C_{0,i}$ for $i = 1, 5, 11, 13, 23, 31$ with $\rho = 3$ and with

$$\mathrm{IA} = \{63, 62, 1; 1, 62, 63\}.$$

(S.12.8) The CT cyclic [63, 55, 2] codes $C_{7,21}$ with $\rho = 4$ and with

$$\mathrm{IA} = \{63, 56, 49, 42; 7, 14, 21, 28\}.$$

(S.12.9) The CT cyclic $[63, 57, 2]$ codes $C_{9,27}$ with $\rho = 3$ and with

$$\mathrm{IA} = \{63, 54, 45; 9, 18, 27\}.$$

(S.12.10) The CT cyclic $[63, 54, 2]$ codes $C_{9,15}$ and $C_{3,27}$ with $\rho = 3$ and with

$$\mathrm{IA} = \{63, 42, 21; 3, 6, 9\}.$$

(S.12.11) The CT cyclic $[63, 54, 3]$ codes $C_{3,9}$ and $C_{15,27}$ with $\rho = 3$ and with

$$\mathrm{IA} = \{63, 60, 48; 3, 6, 36\}.$$

(S.12.12) The CT extended cyclic $[63, 54, 3]$ codes $C^*_{9,i}$ for $i = 1, 11, 23$ and $C^*_{27,j}$ for $j = 5, 13, 31$ with $\rho = 3$ and with

$$\mathrm{IA} = \{63, 56, 1; 1, 8, 63\}.$$

(S.12.13) The CT extended cyclic $[64, 54, 4]$ codes $C^*_{9,i}$ for $i = 1, 11, 23$ and $C^*_{27,j}$ for $j = 5, 13, 31$ with $\rho = 4$ and with

$$\mathrm{IA} = \{64, 63, 56, 1; 1, 8, 63, 64\}.$$

(S.12.14) The CR (not CT) cyclic $[63, 51, 5]$ codes $C_{i,j}$ for

$$(i, j) \in \{(1, 5), (5, 11), (1, 13), (23, 31), (13, 23), (11, 31)\}$$

with $\rho = 3$ and with

$$\mathrm{IA} = \{63, 60, 49; 1, 4, 15\}.$$

(S.12.15) The CR (not CT) extended cyclic $[64, 51, 6]$ codes $C^*_{i,j}$ for

$$(i, j) \in \{(1, 5), (5, 11), (1, 13), (23, 31), (13, 23), (11, 31)\}$$

with $\rho = 4$ and with

$$\mathrm{IA} = \{64, 63, 60, 49; 1, 4, 15, 64\}.$$

Open questions: *find the necessary and sufficient conditions for the code $C_{i,j}$ to be UPWS (but not CR), to be CR (but not CT), to be CT.*

2.13 CR CODES BY SPECIAL CONSTRUCTIONS

2.13.1 New UP, UPWS and CR Codes from Code of Length 12

In this section, we will describe several interesting new codes, constructed in recent paper [140] and which partially solve the old open problem of the existence of UP (hence CR) codes. In 1974 the following hypothetical parameters

$$q = 4, \;\; n = \frac{2^{2m+1}+1}{3}, \;\; k = n - 2m - 1, \;\; d = 5, \tag{2.59}$$

of UP codes with covering radius $\rho = 3$, the packing coefficients

$$\beta_0 = \beta_1 = 1, \;\; \beta_2 = \beta_3 = \frac{3}{2^{2m}-1},$$

and the roots of Lloyd's polynomial

$$\xi_1 = 2^{2m-1} - 2^{m-1}, \;\; \xi_2 = 2^{2m-1}, \;\; \xi_3 = 2^{2m-1} + 2^{m-1},$$

where $m = 2, 3, 4, \ldots$, were found in the paper [12] using a computer search. For $m = 2$ the corresponding code from this family has the parameters

$$q = 4, \;\; n = 11, \;\; k = 6, \;\; d = 5, \;\; \mu = \lambda + 1 = 5, \tag{2.60}$$

the packing coefficients

$$\beta_0 = \beta_1 = 1, \;\; \beta_2 = \beta_3 = \frac{1}{5}$$

and the roots of Lloyd's polynomial

$$\xi_1 = 6, \;\; \xi_2 = 8, \;\; \xi_3 = 10.$$

Then in [64], such linear 4-ary codes with parameters (2.59) were constructed, but as it turned out, they were not UP codes. Still we give a construction of the first code (2.60) from this family. Let α be a root of the primitive over $\mathbb{F}_4$ polynomial $f(x) = x^5 + x^4 + x^2 + x + \zeta$, where $\mathbb{F}_4 = \{0, 1, \zeta, \zeta^2\}$ and $\zeta^2 = \zeta + 1$. Hence α is a primitive element of $\mathbb{F}_{4^5}$. Then the polynomial

$$g(x) = (x+\gamma)(x+\gamma^4)(x+\gamma^{16})(x+\gamma^{-2})(x+\gamma^{-8}) = x^5 + \zeta^2 x^4 + x^3 + \zeta x + 1,$$

where $\gamma = \alpha^{31}$ has the order 33 and generates a cyclic linear 4-ary $[11, 6, 5]_4$ code, which we denote by $\mathcal{C}_{11}$ and which after being extended by adding an overall parity check position, becomes a $[12, 6, 6]_4$ code $\mathcal{C}^*_{11}$. It is valid for all family of codes (2.59), i.e. these codes can be extended to the codes of length $n^* = n+1$ with the minimum distance $d^* = 6$. Furthermore, the mapping of the extended $[12, 6, 6]_4$ code (2.60) over $\mathbb{F}_4$ to the binary code of length 24 (by mapping $\mathbb{F}_4$ into $(\mathbb{F}_2)^2$) results in the extended binary perfect Golay $[24, 12, 8]$ code $\mathcal{G}_{24}$ (see [64]).

Now turn to the paper [140]. Take the so-called the *dodecacode*, i.e. a nonlinear additive $(12, 4^6, 6)_4$ code over $\mathbb{F}_4$ (see reference in [140]). So this code has the same parameters as the code (2.60), but it is not equivalent to the linear code (2.60). By puncturing this code at any coordinate, we obtain an additive $(11, 4^6, 5)_4$ code, denoted by D^- in [140]. This code is UP, since it has the outer distance $s = 3$ (i.e. the dual code has non-zero codewords only of weights 6, 8 and 10) and the covering radius $\rho_{D^-} = 3$, implying that $s = \rho$ and, hence, it is UP (Theorem 2.4,(5)) and, therefore CR, since it is quasi-perfect (Proposition 2.1) (equivalently since $\rho = b$ again by Theorem 2.4,(3). Taking into account that D^- contains 198 codewords of weight 5 [140] and these codewords form a $T(11, 5, 2, 4)_4$ design (Theorem 2.2), one can deduce that the intersection numbers of D^- look as follows [140]:

$$b_0 = 33,\ b_1 = 30,\ b_2 = 15; c_1 = 1,\ c_2 = 2,\ c_3 = 15 \tag{2.61}$$

Theorem 2.80 ([140]). 1. *(S.13.1) The additive $(11, 4^6, 5)_4$-code D^- is CR with covering radius $\rho_{D^-} = 3$ and with intersection numbers (2.61).*

2. *The generator matrix G_{D^-} of this code is of the form:*

$$G_{D^-} = \begin{bmatrix} 0 & 0 & 0 & 0 & 0 & w^2 & w^2 & 0 & w & 1 & w \\ 0 & 0 & 0 & 0 & 0 & w & 0 & w & w & w & 1 \\ 1 & 0 & 0 & 0 & 0 & 1 & 0 & 1 & w^2 & w^2 & 1 \\ w & 0 & 0 & 0 & 0 & 0 & w & 1 & w & w & w \\ 0 & 1 & 0 & 0 & 0 & 0 & 1 & 1 & 1 & w^2 & w^2 \\ 0 & w & 0 & 0 & 0 & w^2 & 1 & w^2 & w & w & 0 \\ 0 & 0 & 1 & 0 & 0 & w & w & 1 & 1 & 0 & 1 \\ 0 & 0 & w & 0 & 0 & w & 1 & w & w^2 & w^2 & 0 \\ 0 & 0 & 0 & 1 & 0 & w & w & w & 0 & 1 & w \\ 0 & 0 & 0 & w & 0 & w^2 & 1 & 1 & w^2 & 0 & w \\ 0 & 0 & 0 & 0 & 1 & w^2 & w^2 & w^2 & 1 & 0 & w^2 \\ 0 & 0 & 0 & 0 & w & w & 1 & 0 & 1 & w & w \end{bmatrix}$$

3. *The full automorphism group $Aut(D^-)$ is the product of the monomial automorphism group with the group of translations and has the order $54 \cdot 2^{12}$. The monomial automorphism group is generated by the following two monomial matrices:*

$$\begin{bmatrix} 0 & 0 & 0 & 0 & 0 & w & 0 & 0 & 0 & 0 & 0 \\ 1 & 0 & 0 & 0 & 0 & 0 & 0 & 0 & 0 & 0 & 0 \\ 0 & 0 & w^2 & 0 & 0 & 0 & 0 & 0 & 0 & 0 & 0 \\ 0 & 0 & 0 & 0 & 0 & w & 0 & w^2 & 0 & 0 & 0 \\ 0 & 0 & 0 & 0 & 0 & 0 & 0 & 0 & 0 & 1 & 0 \\ 0 & 0 & 0 & 0 & w & 0 & 0 & 0 & 0 & 0 & 0 \\ 0 & 0 & 0 & 0 & 0 & 0 & 0 & 0 & 0 & 0 & w \\ 0 & 0 & 0 & 1 & 0 & 0 & 0 & 0 & 0 & 0 & 0 \\ 0 & w^2 & 0 & 0 & 0 & 0 & 0 & 0 & 0 & 0 & 0 \\ 0 & 0 & 0 & 0 & 0 & 0 & 0 & 0 & w^2 & 0 & 0 \\ 0 & 0 & 0 & 0 & 0 & 0 & 1 & 0 & 0 & 0 & 0 \end{bmatrix},$$

$$\begin{bmatrix} 1 & 0 & 0 & 0 & 0 & 0 & 0 & 0 & 0 & 0 & 0 \\ 1 & 0 & 0 & 0 & 0 & 0 & 0 & 0 & 0 & 0 & w^2 \\ 0 & 0 & 0 & 0 & 0 & w^2 & 0 & 0 & 0 & 0 & 0 \\ 0 & 0 & 0 & 0 & 0 & 0 & 0 & w^2 & 0 & 0 & 0 \\ 0 & 0 & 0 & 0 & 0 & 0 & 0 & 0 & w & 0 & 0 \\ 0 & 0 & w & 0 & w & 0 & 0 & 0 & 0 & 0 & 0 \\ 0 & 0 & 0 & 0 & 0 & 0 & 0 & 0 & 0 & 1 & 0 \\ 0 & 0 & 0 & w & 0 & 0 & 0 & 0 & 0 & 0 & 0 \\ 0 & 0 & 0 & 0 & w^2 & 0 & 0 & 0 & 0 & 0 & 0 \\ 0 & 0 & 0 & 0 & 0 & 0 & 1 & 0 & 0 & 0 & 0 \\ 0 & w & 0 & 0 & 0 & 0 & 0 & 0 & 0 & 0 & 0 \end{bmatrix},$$

More results concerning the code D^-,in particular, the weight distributions of this code and its dual, can be found in [140].

Now we will explain how from the generator matrix G_{D^-} of the additive quaternary code above, to obtain the binary matrix G, which will be the generator matrix of a new binary CR code. Let α be a primitive element in $\mathbb{F}_4$. For every row in G_{D^-} substitute any coordinate in the following way:

$$0 \longrightarrow 000 \text{ or } 111; \; 1 \longrightarrow 100 \text{ or } 011; \alpha \longrightarrow 001 \text{ or } 110; \; \alpha^2 \longrightarrow 010 \text{ or } 101.$$

This means that any row in G gives rise to 2^{11} binary rows in H. In this way, the matrix H generates

(S.13.2) a linear binary $[33, 23, 3; 3]$ code which is CR (but not CT) with intersection array

$$\mathrm{IA} = \{33, 30, 15; 1, 2, 15\}.$$

Open questions: *do we have the same results for all the codes of family* (2.59)*?*

2.13.2 Two Best CT Codes by Modified Concatenation

In the recent paper [165] the two best known codes (namely, the nonlinear binary Nordstrom–Robinson code, which we denote $\mathcal{NR}$ code, and the binary extended perfect Golay code $\mathcal{G}_{24}$ were constructed by a new modified concatenation construction. Since it gives the best known CT codes there is a sense to give both these constructions (since they are slightly different from each other).

The nonlinear binary $\mathcal{NR}$ code with parameters

$$n = 16, \qquad N = 2^8, \qquad d = 6$$

was constructed in 1968 by Nordstrom and Robinson [119] and independently in [131]. The binary extended perfect Golay code with parameters

$$n = 24, \qquad N = 2^{12}, \qquad d = 8$$

was constructed in 1949 by Golay [83]. For both codes there are numerous publications, the significant part of which can be found in [112]. In particular, Snover

[144] proved that the $\mathcal{NR}$ code and the Golay $\mathcal{G}_{24}$ code are unique up to equivalence. This essentially simplifies our task (namely, it is enough to construct OK codes with parameters of these codes). We will only refer to the more recent papers, mostly dedicated to the $\mathcal{NR}$ code, in connection with the new interesting results (see [21,68,74,116] and references there). We mention only two of these results. In [84] the all nonlinear binary Preparata codes (including the $\mathcal{NR}$ code) were presented as linear codes over $\mathbb{Z}_4$. In [64] the concatenated presentation for the Golay code $\mathcal{G}_{24}$ was given. The code was obtained by the mapping of the $[12, 6, 6]_4$-code from [64] to the binary code. construct

The generalized concatenation construction [159] is based on two families of codes, which are called *inner codes* and *outer codes*. So first, we have to describe these codes.

We start with inner codes for $\mathcal{NR}$ code. Let B denote the trivial $(4, 16, 1)$ code (i.e., all binary vectors of length 4). The code B is partitioned into two trivial subcodes: $(4, 8, 2)$ code B_1 with the zero overall parity checking and $(4, 8, 2)$ code B_2 with the nonzero overall parity checking. In turn, both of these codes are further partitioned into the trivial $(4, 2, 4)$ codes $B_{i,j}$. We present these codes along with the numbering of its codewords (the vector $\mathbf{b(i,j,k)}$, where $k = 1, 2$, belongs to the code $B_{i,j}$):

$$\begin{aligned}
B_{1,1} &= \{\mathbf{b(1,1,1)} = (\mathbf{0\,0\,0\,0}),\ \mathbf{b(1,1,2)} = (\mathbf{1\,1\,1\,1})\},\\
B_{1,2} &= \{\mathbf{b(1,2,1)} = (\mathbf{1\,1\,0\,0}),\ \mathbf{b(1,2,2)} = (\mathbf{0\,0\,1\,1})\},\\
B_{1,3} &= \{\mathbf{b(1,3,1)} = (\mathbf{1\,0\,1\,0}),\ \mathbf{b(1,3,2)} = (\mathbf{0\,1\,0\,1})\},\\
B_{1,4} &= \{\mathbf{b(1,4,1)} = (\mathbf{1\,0\,0\,1}),\ \mathbf{b(1,4,2)} = (\mathbf{0\,1\,1\,0})\},\\
B_{2,1} &= \{\mathbf{b(2,1,1)} = (\mathbf{1\,0\,0\,0}),\ \mathbf{b(2,1,2)} = (\mathbf{0\,1\,1\,1})\},\\
B_{2,2} &= \{\mathbf{b(2,2,1)} = (\mathbf{0\,1\,0\,0}),\ \mathbf{b(2,2,2)} = (\mathbf{1\,0\,1\,1})\},\\
B_{2,3} &= \{\mathbf{b(2,3,1)} = (\mathbf{0\,0\,1\,0}),\ \mathbf{b(2,3,2)} = (\mathbf{1\,1\,0\,1})\},\\
B_{2,4} &= \{\mathbf{b(2,4,1)} = (\mathbf{0\,0\,0\,1}),\ \mathbf{b(2,4,2)} = (\mathbf{1\,1\,1\,0})\}.
\end{aligned}$$

For the outer codes, choose two MDS codes A_1 and V_1 with parameters $(4, 16, 3)_4$, which are partitioned into the following subcodes $A_{1,i}$, $i = 1, 2, 3, 4$, and $V_{1,j}$, $j = 1, 2, 3, 4$, with distances 4:

$$A_{1,1} = \begin{bmatrix} (0000) \\ (1111) \\ (2222) \\ (3333) \end{bmatrix}, \quad A_{1,2} = \begin{bmatrix} (0321) \\ (3012) \\ (2103) \\ (1230) \end{bmatrix}, \quad A_{1,3} = \begin{bmatrix} (0213) \\ (2031) \\ (1302) \\ (3120) \end{bmatrix}, \quad A_{1,4} = \begin{bmatrix} (0132) \\ (1023) \\ (3201) \\ (2310) \end{bmatrix} \tag{2.62}$$

and

$$V_{1,1} = \begin{bmatrix} (0\,0\,0\,0) \\ (1\,1\,1\,1) \\ (2\,2\,2\,2) \\ (3\,3\,3\,3) \end{bmatrix}, \quad V_{1,2} = \begin{bmatrix} (0\,1\,2\,3) \\ (1\,0\,3\,2) \\ (2\,3\,0\,1) \\ (3\,2\,1\,0) \end{bmatrix}, \quad V_{1,3} = \begin{bmatrix} (0\,2\,3\,1) \\ (2\,0\,1\,3) \\ (3\,1\,0\,2) \\ (1\,3\,2\,0) \end{bmatrix}, \quad V_{1,4} = \begin{bmatrix} (0\,3\,1\,2) \\ (3\,0\,2\,1) \\ (1\,2\,0\,3) \\ (2\,1\,3\,0) \end{bmatrix}. \tag{2.63}$$

The intersection of codes A_1 and V_1 are the subcodes $A_{1,1}$ and $V_{1,1}$, and all the other subcodes $A_{1,i}$, $i = 2,3,4$, and $V_{1,j}$, $j = 2,3,4$, are at a distances 2 or 3, i.e.,

$$d(A_{1,i}, V_{1,j}) = \begin{cases} 0 & \text{if } i = j = 1, \\ 2 & \text{if } i, j \in \{2,3,4\}, \\ 3 & \text{if } 1 \in \{i,j\},\ i \neq j. \end{cases} \tag{2.64}$$

For the outer codes A_2 and V_2, choose the two binary $(4,8,2)$ codes (i.e., the inner codes B_1 and B_2): A_2 (with the zero overall parity checking) and V_2 (with the nonzero overall parity checking), which are partitioned into the subcodes $A_{2,i}$, $i = 1,2$, and $V_{2,j}$, $j = 1,2$, with distance 2:

$$A_{2,1} = \begin{bmatrix} (0\,0\,0\,0) \\ (1\,1\,0\,0) \\ (1\,0\,1\,0) \\ (1\,0\,0\,1) \end{bmatrix}, \quad A_{2,2} = \begin{bmatrix} (1\,1\,1\,1) \\ (0\,0\,1\,1) \\ (0\,1\,0\,1) \\ (0\,1\,1\,0) \end{bmatrix} \tag{2.65}$$

and

$$V_{2,1} = \begin{bmatrix} (1\,0\,0\,0) \\ (0\,1\,0\,0) \\ (0\,0\,1\,0) \\ (0\,0\,0\,1) \end{bmatrix}, \quad V_{2,2} = \begin{bmatrix} (0\,1\,1\,1) \\ (1\,0\,1\,1) \\ (1\,1\,0\,1) \\ (1\,1\,1\,0) \end{bmatrix}. \tag{2.66}$$

Construct the following three OC codes [159]:

- the code C based on the outer codes $A = \{(0\,0\,0\,0),\ (1\,1\,1\,1)\}$, $A_1 \cup V_1$ (interpreting this union as a set with repetitions), the codes $A_2 \cup V_2$ and the inner code $B = B_1 \cup B_2$;
- The code C_1 based on the outer codes A_1 and A_2 and the inner code B_1;
- The code C_2 based on the outer codes V_1 and V_2 and the inner code B_2.

Let us explain shortly the construction of C_1. Choose two codewords: $\mathbf{a} = (\mathbf{a_1}, \mathbf{a_2}, \mathbf{a_3}, \mathbf{a_4})$ from the code A_1 and $\mathbf{x} = (\mathbf{x_1}, \mathbf{x_2}, \mathbf{x_3}, \mathbf{x_4})$ from the code A_2. These codewords induce the codeword $\mathbf{c} = (\mathbf{c_1} \,|\, \mathbf{c_2} \,|\, \mathbf{c_3} \,|\, \mathbf{c_4})$ of code C_1. As an i-th block $\mathbf{c_i}$, $i = 1,2,3,4$, choose the codeword $\mathbf{b}$ of code B_1 with the number $\mathbf{c_i} = \mathbf{b}(\mathbf{1}, \mathbf{a_i} + \mathbf{1}, \mathbf{x_i} + \mathbf{1})$ (the addition is performed over the field of real numbers).

When $\mathbf{a}$ and $\mathbf{x}$ run over all codewords of A_1 and A_2, respectively, the resulting codeword $\mathbf{c}$ runs over the whole new code C_1. The construction implies that C_1 has the following parameters

$$n = 4 \cdot 4 = 16, \qquad N = 16 \cdot 8 = 128, \qquad d = \min\{2 \cdot 3, 4 \cdot 2\} = 6.$$

The code C_2 is constructed similarly from the outer codes V_1 and V_2 and the inner code B_2 and has the same parameters. Let's explain the construction of C. When the codeword $(0\,0\,0\,0)$ is selected, the codes A_1 and A_2 are utilized, and when the codeword $(1\,1\,1\,1)$ is selected, the codes V_1 and V_2 are used.

Our next goal is to show, that C_1 and C_2 are apart from each other at a distance 6. The fact that they are at a distance 4, follows from the fact that C which has the minimum distance 4 (by the concatenated construction) is the union of codes C_1 and C_2.

Recall, that all codewords $\mathbf{c}$ of code C have the block structure: $\mathbf{c} = (\mathbf{c_1}\,|\,\mathbf{c_2}\,|\,\mathbf{c_3}\,|\,\mathbf{c_4})$, where the blocks $\mathbf{c_i}$ are the codewords of code B. Since the codewords B_1 and B_2 are different, it follows that the codewords C_1 and C_2, differ in the every block. In particular, it follows that for any two codewords $\mathbf{c} = (\mathbf{c_1}\,|\,\mathbf{c_2}\,|\,\mathbf{c_3}\,|\,\mathbf{c_4})$ from the code C_1 and $\mathbf{c}' = (\mathbf{c_1'}\,|\,\mathbf{c_2'}\,|\,\mathbf{c_3'}\,|\,\mathbf{c_4'})$ from the code C_2 the following inequality is valid:

$$d(\mathbf{c_i}, \mathbf{c_i'}) \geq \mathbf{1}, \quad \text{for each } \mathbf{i} \in \{\mathbf{1}, \mathbf{2}, \mathbf{3}, \mathbf{4}\}. \tag{2.67}$$

The partitions (2.62), (2.63), and (2.65) of codes A_1 and V_1 into the subcodes $A_{1,i}$, $i = 1, 2, 3, 4$, and $V_{1,j}$, $j = 1, 2, 3, 4$, respectively induce the partitions of codes C_1 and C_2 into the subcodes $C_{1,i}$ and $C_{2,j}$. The properties of codes C_1 and C_2, that we need, are given in the following

Lemma 2.13 ([165]). *Let* $\mathbf{x} = (\mathbf{x_1}\,|\,\mathbf{x_2}\,|\,\mathbf{x_3}\,|\,\mathbf{x_4})$ *be an arbitrary codeword of* $C_1 \setminus C_{1,1}$, *and* $\mathbf{y} = (\mathbf{y_1}\,|\,\mathbf{y_2}\,|\,\mathbf{y_3}\,|\,\mathbf{y_4})$ *be an arbitrary codeword of* $C_2 \setminus C_{2,1}$. *Then the block vectors* $\mathbf{x_i}$ *and* $\mathbf{y_j}$ *satisfy the following properties:*

1. *If* $\mathrm{wt}(\mathbf{x}) = \mathbf{6}$, *then there exists an index* $i_0 \in \{1,2,3,4\}$, *such that* $i_0 \in \mathrm{Supp}(\mathbf{x_i})$ *for all nonzero* $\mathbf{x_i}$. *If an index* s *differs from* i_0, *then it is covered by the support of* $\mathbf{x_i}$ *only once, i.e., only for one* i;

2. *If* $\mathrm{wt}(\mathbf{x}) = \mathbf{10}$, *then there exists an index* i_0, *such that* $i_0 \in \mathrm{Supp}(\mathbf{x_i})$ *only once, i.e., for one* i. *Any index* i, *which differs from* i_0, *is covered by* $\mathbf{x_i}$ *exactly three times;*

3. *If* $\mathrm{wt}(\mathbf{y}) = \mathbf{6}$, *then there exists an index* $j_0 \in \{1,2,3,4\}$, *which is not covered by any of the supports of* $\mathbf{y_j}$. *All other indexes* $j \neq j_0$ *are covered exactly two times by the supports of* $\mathbf{y_j}$;

4. *If* $\mathrm{wt}(\mathbf{y}) = \mathbf{10}$, *then there exists an index* j_0, *which is covered by the supports of all vectors* $\mathbf{y_j}$. *All other indexes* $j \neq j_0$ *are covered exactly two times by the supports of* $\mathbf{y_j}$.

Define the following three permutations π_1, π_2 and π_3, which act on the set of vectors of length 4 as follows: for any vector $\mathbf{x} = (\mathbf{x_1}, \mathbf{x_2}, \mathbf{x_3}, \mathbf{x_4})$ set

$$\pi_1(\mathbf{x}) = (\mathbf{x_2}, \mathbf{x_1}, \mathbf{x_4}, \mathbf{x_3}), \qquad \pi_2(\mathbf{x}) = (\mathbf{x_4}, \mathbf{x_3}, \mathbf{x_2}, \mathbf{x_1}), \qquad \pi_3(\mathbf{x}) = (\mathbf{x_1}, \mathbf{x_4}, \mathbf{x_2}, \mathbf{x_3}).$$

Denote by $\mathcal{G}$ the group, induced by the permutations π_1, π_2 and π_3.

Let $\mathbf{c} = (\mathbf{c_1} \,|\, \mathbf{c_2} \,|\, \mathbf{c_3} \,|\, \mathbf{c_4})$ be an arbitrary codeword of C_1, which is constructed from the codewords $\mathbf{a} \in \mathbf{A_1}$ and $\mathbf{x} \in \mathbf{A_2}$, such that one can write $\mathbf{c} = \mathbf{c}(\mathbf{a}, \mathbf{x})$. Similarly we will write $\mathbf{c} = \mathbf{c}(\mathbf{v}, \mathbf{y})$ for the codeword $\mathbf{c} \in \mathbf{C_2}$, constructed from the codewords $\mathbf{v} \in \mathbf{V_1}$ and $\mathbf{y} \in \mathbf{V_2}$.

Define the action of the group $\mathcal{G}$ on the codeword c as follows: for any $g \in \mathcal{G}$ set

$$g(\mathbf{c}) = \begin{cases} \mathbf{c}(\mathbf{g}(\mathbf{a}), \mathbf{g}(\mathbf{x})) & \text{if } \mathbf{c} \in \mathbf{C_1}, \\ \mathbf{c}(\mathbf{g}(\mathbf{v}), \mathbf{g}(\mathbf{y})) & \text{if } \mathbf{c} \in \mathbf{C_2}. \end{cases}$$

Lemma 2.14 ([165]).

1. *The group $\mathcal{G}$ stabilizes the codes A_i and V_i for $i = 1, 2$ and acts transitively on the subcodes $A_1 \setminus A_{1,1}$ and $V_1 \setminus V_{1,1}$;*
2. *The group $\mathcal{G}$ stabilizes the codes C_1 and C_2 and acts transitively on the subcodes $C_1 \setminus C_{1,1}$ and $C_2 \setminus C_{2,1}$;*
3. *For any $c \in C_1$ and $c' \in C_2$ and for any $g \in \mathcal{G}$ the following equality takes place :*
$$d(g(c), g(c')) = d(c, c'); \tag{2.68}$$
4. *The codes C_1 and C_2 are invariant under the addition of the vector of all ones, i.e.,*
$$C_i + (1\,1\, \ldots\, 1) = C_i, \quad i = 1, 2. \tag{2.69}$$

Theorem 2.81 ([165]). *The union of codes C_1 and C_2 , i.e., the code*

$$C = C_1 \cup C_2$$

is the $(16, 256, 6)$-code, i.e., the code $\mathcal{NR}$ is an OC code of the order three.

Now consider the construction of the Golay code $\mathcal{G}_{24}$. Building the binary extended perfect Golay code $[24, 12, 8]$ is slightly different from the previous construction of the N-R code. As an inner code we choose the same binary $[4, 4, 1]$ code B with the same partition into two subcodes $[4, 3, 2]$: the $[4, 3, 2]$ code B_0 (with the zero overall parity checking) and the $(4, 8, 2)$ code B_1 (with the nonzero overall parity checking). However in order to linearize the mapping from $\mathbb{F}_4 \times \mathbb{F}_2$ into the codewords of code B_0 (which what we need for the proof of the linearity of the resulting code, but what we couldn't do to build the code $\mathcal{NR}$) we change word numbering as follows:

$$\begin{aligned}
B_{0,0} &= \{\mathbf{b}(0,0,0) = (\mathbf{0\,0\,0\,0}),\ \mathbf{b}(0,0,1) = (\mathbf{1\,1\,1\,1})\},\\
B_{0,1} &= \{\mathbf{b}(0,1,0) = (\mathbf{1\,1\,0\,0}),\ \mathbf{b}(0,1,1) = (\mathbf{0\,0\,1\,1})\},\\
B_{0,2} &= \{\mathbf{b}(0,2,0) = (\mathbf{1\,0\,1\,0}),\ \mathbf{b}(0,2,1) = (\mathbf{0\,1\,0\,1})\},\\
B_{0,3} &= \{\mathbf{b}(0,3,0) = (\mathbf{0\,1\,1\,0}),\ \mathbf{b}(0,3,1) = (\mathbf{1\,0\,0\,1})\}.
\end{aligned}$$

For the code B_1 the numbering is as follows:

$$\begin{aligned}
B_{1,0} &= \{\mathbf{b}(1,0,0) = (\mathbf{1\,0\,0\,0}),\ \mathbf{b}(1,0,1) = (\mathbf{0\,1\,1\,1})\},\\
B_{1,1} &= \{\mathbf{b}(1,1,0) = (\mathbf{0\,1\,0\,0}),\ \mathbf{b}(1,1,1) = (\mathbf{1\,0\,1\,1})\},\\
B_{1,2} &= \{\mathbf{b}(1,2,0) = (\mathbf{0\,0\,1\,0}),\ \mathbf{b}(1,2,1) = (\mathbf{1\,1\,0\,1})\},\\
B_{1,3} &= \{\mathbf{b}(1,3,0) = (\mathbf{0\,0\,0\,1}),\ \mathbf{b}(1,3,1) = (\mathbf{1\,1\,1\,0})\}.
\end{aligned}$$

Define the outer codes: let $A = \{(0\,0\,0\,0\,0\,0),\ (1\,1\,1\,1\,1\,1)\}$ to be the trivial code and let $[6,3,4]_4$ be the MDS code A_1 over $\mathbb{F}_4$, and the same two binary codes: the $[6,5,2]$ code A_2 with the zero overall parity checking and the $(6,32,2)$ code V_2 with the nonzero overall parity checking.

We shall explain the construction of the $[6,3,4]_4$ MDS code A_1 over $\mathbb{F}_4$. Let

$$\mathbb{F}_4 = \{0,1,\xi,\xi^2\}, \quad \text{where} \quad \xi^2 + \xi + 1 = 0,$$

and let $\mathbb{F}_{4^2}$ be obtained from $\mathbb{F}_4$ by means of the primitive polynomial

$$f(x) = x^2 + x + \xi.$$

Let α be the root of this polynomial, i.e., a primitive element of the field $\mathbb{F}_{4^2}$. Here and throughout the rest of the section instead of the elements $0, 1, \xi, \xi^2$ of the field $\mathbb{F}_4$ we use the numbers $0, 1, 2, 3$, respectively, which we use for the codeword numbering of the inner codes. Denote by $m_i(x)$ the minimal function of the element α^i. Since

$$x^5 + 1 = (x^2 + 3x + 1)(x^2 + 2x + 1)(x + 1), \tag{2.70}$$

we can define the following cyclic MDS $[5,3,3]_4$ code C, which has the generating polynomial

$$g_a(x) = m_3(x) = x^2 + 3x + 1.$$

Denote by A_1, the code obtained from C by extension, i.e., by adding to every codeword $\mathbf{a}' = (\mathbf{a_1}, \mathbf{a_2}, \mathbf{a_3}, \mathbf{a_4}, \mathbf{a_5})$ of code C one more position a_6, i.e. an overall parity checking:

$$a_6 = \sum_{i=1}^{5} a_i.$$

In such way we obtain the $[6,3,4]_4$ code A_1, formed by codewords $\mathbf{a} = (\mathbf{a_1}, \mathbf{a_2}, \mathbf{a_3}, \mathbf{a_4}, \mathbf{a_5}, \mathbf{a_6})$. it is convenient to bring this code using the following five codewords (generators):

$$(1\,1\,1\,1\,1\,|\,1), \quad (2\,1\,2\,0\,0\,|\,1), \quad (3\,3\,0\,1\,0\,|\,1), \quad (1\,2\,2\,1\,0\,|\,0), \quad (1\,2\,3\,2\,1\,|\,3). \tag{2.71}$$

The first generator of (2.71) clearly generates three codewords. Each of the other four generators induces 15 codewords by multiplication by scalar (i.e., ξ and ξ^2) and by cyclic shifts of the first five positions. If $\mathbf{a} = (\mathbf{a_1}, \mathbf{a_2}, \mathbf{a_3}, \mathbf{a_4}, \mathbf{a_5} \,|\, \mathbf{a_6})$ is one of such generators, then 15 corresponding codewords are induced by the cyclic shifts (of the first five positions, but the sixth position a_6 is fixed) of the following three vectors $\mathbf{a}$, $\xi\mathbf{a}$ and $\xi^2\mathbf{a}$.

For the convenience of the reader, we give a list of all the codewords of the code A_1:

(000000)	(111111)	(222222)	(333333),
(212001)	(330101)	(122100)	(123213),
(021201)	(033011)	(012210)	(112323),
(002121)	(103301)	(101220)	(211233),
(200211)	(010331)	(210120)	(321123),
(120021)	(301031)	(221010)	(232113),
(323002)	(110202)	(233200)	(231321),
(032302)	(011022)	(023320)	(223131),
(003232)	(201102)	(202330)	(322311),
(300322)	(020112)	(320230)	(132231),
(230032)	(102012)	(332020)	(313221),
(131003)	(220303)	(311300)	(312132),
(013103)	(022033)	(031130)	(331212),
(001313)	(302203)	(303110)	(133122),
(100133)	(030223)	(130310)	(213312),
(310013)	(203023)	(113030)	(121332).

Construct three following OC codes:

- a code G (of order 3), based on the inner code B and the outer codes A, A_1 and $A_2 \cup V_2$ (when the word $(0\,0\,0\,0\,0\,0)$ is selected, we use code A_2, and when the word $(1\,1\,1\,1\,1\,1)$ is selected we use V_2);
- a code G_1 (of order 2), based on the inner code B_0 and two outer codes A_1 and A_2;
- a code G_2 (of order 2), based on the inner code B_1 and two outer codes A_1 and V_2.

Denote by φ_0 the mapping from set $\mathbb{F}_4 \times \mathbb{F}_2$ into the words of code B_0 and by φ_1 the mapping into words of code B_1:

$$\varphi_i(j,k) = \mathbf{b}(i,j,k), \quad i = 0,1.$$

Define these mappings in a natural way for the vectors $\mathbf{x} = (\mathbf{x_1}, \ldots, \mathbf{x_n})$ over $\mathbb{F}_4 \times \mathbb{F}_2$:

$$\varphi_i(\mathbf{x}) = (\varphi_\mathbf{i}(\mathbf{x_1}), \varphi_\mathbf{i}(\mathbf{x_2}), \ldots, \varphi_\mathbf{i}(\mathbf{x_n})), \quad \mathbf{i} = \mathbf{0}, \mathbf{1}.$$

Lemma 2.15 ([165]).

1. *The mapping φ_0 from $\mathbb{F}_4 \times \mathbb{F}_2$ to the codewords of code B_0 is linear for both indexes, i.e., if $\mathbf{b}'$ and $\mathbf{b}''$ are codewords of code B_0 with the numbers $(0, j', k')$ and $(0, j'', k'')$ respectively, then their sum $\mathbf{b} = \mathbf{b}' + \mathbf{b}''$ belongs to the code B_0 and the codeword $\mathbf{b}$ has number*

$$(i, j, k) = (0, j' + j'', k' + k''),$$

where the indexes j' and j'' are summed in the field $\mathbb{F}_4$ (taking into account the notation for the elements of the field $\mathbb{F}_4$), and the indexes k' and k'' are added in the field $\mathbb{F}_2$.

2. *The mapping φ_1 from $\mathbb{F}_4 \times \mathbb{F}_2$ to the codewords of code B_1, is linear for the first index and is linear for the second index up to the correction coefficient $k_3 \in \{0, 1\}$ (depending upon the number of appearances of element 3 in the set $\{j', j''\}$), i.e., if $\mathbf{b}'$ and $\mathbf{b}''$ are codewords of code B_1 with numbers $(1, j', k')$ and $(1, j'', k'')$, respectively, then their sum $\mathbf{b} = \mathbf{b}' + \mathbf{b}''$ belongs to code B_0 and the codeword $\mathbf{b}$ has number*

$$(i, j, k) = (0, j' + j'', k' + k'' + k_3),$$

where k_3 is the number of appearances of element 3 in the set $\{j', j''\}$, taken modulo 2 and interpreted as an element of the field $\mathbb{F}_2$.

Lemma 2.16. [165]

1. The codes G_1 and G_2 have the minimum distances 8, besides $\text{wt}(\mathbf{g}) \geq \mathbf{8}$ for every codeword $\mathbf{g}$ of code G_2;

2. the code G_1 is a linear code, i.e., it is the $[24, 11, 8]$-code ;

3. the code G_2 is invariant under the shifts by the word of code G_1.

Lemma 2.17. [165] The codes G_1 and G_2 are at distance 8 from each other.

In fact we proved already, that the union of G_1 and G_2 is the binary extended Golay code $\mathcal{G}_{24}$, since it is known [144], that any binary code with parameters $n = 24$, $N = 2^{12}$, and $d = 8$ is the Golay code $\mathcal{G}_{24}$. However, we will give the second proof of this fact using the linearity of this code.

Lemma 2.18. [165] The code G_2 is a coset of the code G_1.

Thus, we have proved the following

Theorem 2.82. [165] The union of codes G_1 and G_2 , i.e., the code

$$G = G_1 \cup G_2$$

is the $(24, 2^{12}, 8)$ Golay code, i.e., a OCC code of the third order G, is the binary extended perfect $[24, 12, 8]$ Golay code $\mathcal{G}_{24}$.

Open question; *are there other CT codes by similar constructions?*

REFERENCES

[1] Assmus F., Goethals J. M., Mattson H. F., Generalized t-designs and majority decoding of linear codes, *Information and Control*, 1976, vol. 32, 43–60.

[2] Avgustinovich S. V., Mogilnykh I. Yu., Perfect colorings of Johnson graphs $J(8, 3)$ and $J(8, 4)$ in two colours, *Discrete Analysis and Operation Investigation*, 2010, vol. 17, no. 2, 3–19.

[3] Avgustinovich S. V., Mogilnykh I. Yu., Induced perfect colorings, *Siberian Electronic Mathematical Reports*, 2011, vol. 8, 310–316.

[4] Avgustinovich S. V., Krotov D. S., Vasil'eva A. Yu., Completely regular codes in the infinite hexagonal grid, *Siberian Electronic Mathematical Reports*, 2015, vol. 12, 144–170 (http://semr.math.nsc.ru).

[5] Baker R. D., Partitioning the planes $\mathrm{AG}_{2m}(2)$ into 2-designs, *Discrete Mathematics*, 1976, vol. 15, no. 3, 205–211.

[6] Baker R. D., van Lint J. H., Wilson R. M., On the Preparata and Goethals codes, *IEEE Transactions on Information Theory*, 1983, vol. 29, no. 1, 342–348.

[7] Ball S., Blokhuis A., Mazzocca F., Maximal arcs in Desarguesian planes of odd order do not exist, *Combinatorica*, 1997, vol. 17, no. 1, 31–41.

[8] Bannai E., Codes in bipartite distance-regular graphs, *Journal of the London Mathematical Society*, 1977, vol. 2, 197–202.

[9] Bassalygo L. A., Leontiev V. K., Zinoviev V. A., Feldman N.I., Nonexistence of perfect codes over some composite alphabets, *Problems of Information Transmission*, 1975, vol. 11, no. 3, 3–13.

[10] Bassalygo L. A., Zinoviev V. A., Remark on uniformly packed codes, *Problems of Information Transmission*, 1977, vol. 13, no. 1, 178–180.

[11] Bassalygo L. A., Zaitsev G. V., Zinoviev V. A., On uniformly packed codes, in: *Proceedings of Third International Symposium on Information Theory*, Moscow-Tallinn, part 2, 1973, pp. 29–32.

[12] Bassalygo L. A., Zaitsev G. V., Zinoviev V. A., Uniformly packed codes, *Problems of Information Transmission*, 1974, vol. 10, no. 1, 9–14.

[13] Berger Th., The automorphism group of double-error-correcting BCH codes, *IEEE Transactions on Information Theory*, 1994, vol. 40, no. 2, 538–542.

[14] Berger Th., Canteaut A., Charpin P., Laigle-Chapuy Y., On almost perfect nonlinear functions over GF(2^n), *IEEE Transactions on Information Theory*, 2006, vol. 52, no. 9, 4160–4170.

[15] Bespalov Evgeny A., On the non-existence of extended 1-perfect codes and MDS codes, *Journal of Combinatorial Theory, Series A*, 2022, vol. 189, 105607.

[16] Bespalov Evgeny A., Krotov Denis S., Matiushev Aleksandr A., Taranenko Anna A., Vorob'ev Konstantin V., Perfect 2-colorings of Hamming graphs, *Journal of Combinatorial Designs*, 2021, vol. 29, no. 6, 367–396.

[17] Beth T., Ding C., On almost perfect nonlinear permutations, in: Helleseth T. (ed.), *Advances in Cryptology-EUROCRYPT'93*, Lecture Notes in Computer Science, vol. 765, pp. 65–76. Springer-Verlag: New York, 1993.

[18] Beth T., Jungnickel D., Lenz H., *Design Theory*. Cambridge University Press: Cambridge, 1993.

[19] Beukers F., The generalized Ramanujan-Nagel equation, Thesis, The University of Leiden, 1979.

[20] Bierbrauer J., Bounds on Orthogonal arrays and resilient functions, *Journal of Combinatorial Designs*, 1995, vol. 3, no. 3, 179–183.

[21] Bierbrauer J., Nordstrom–Robinson code and A_7-geometry, *Finite Fields and Their Applications*, 2007, vol. 13, no. 1, 158–170.

[22] Blake Ian F., Mullin Ronald C., *The Mathematical Theory of Coding*. Academic Press: Cambridge, MA, 2014.

[23] Borges J., Phelps K. T., Rifà J., Zinoviev V. A., On $\mathbb{Z}_4$-linear preparata-like and Kerdock-like codes, *IEEE Transactions on Information Theory*, 2003, vol. 49, no. 11, 2834–2843.

[24] Borges J., Rifà J., On the nonexistence of completely transitive codes, *IEEE Transactions on Information Theory*, 2000, vol. 46, no. 1, 279–280.

[25] Borges J., Rifà J., Zinoviev V. A., Nonexistence of completely transitive codes with error-correcting capability $e > 3$, *IEEE Transactions on Information Theory*, 2001, vol. 47, no. 4, 1619–1621.

[26] Borges J., Rifà J., Zinoviev V. A., On non-antipodal binary completely regular codes, *Discrete Mathematics*, 2008, vol. 308, no. 16, 3508–3525.

[27] Borges J., Rifà J., Zinoviev V. A., On q-ary linear completely regular codes with $\rho = 2$ and antipodal dual, *Advances in Mathematics of Communications*, 2010, vol. 4, no. 4, 567–578.

[28] Borges J., Rifà J., Zinoviev V. A., New families of completely regular codes and their corresponding distance regular coset graphs, *Designs, Codes and Cryptography*, 2014, vol. 70, 139–148.

[29] Borges J., Rifà J., Zinoviev V. A., Families of nested completely regular codes and distance-regular graphs, *Advances in Mathematics of Communications*, 2015, vol. 9, 233–246.

[30] Borges J., Rifà J., Zinoviev V. A., Completely regular codes by concatenating Hamming codes, *Advances in Mathematics of Communications*, 2018, vol. 12, no. 2, 337–349.

[31] Borges J., Rifà J., Zinoviev V. A., On new infinite families of completely regular and completely transitive binary codes, *Proceedings of the 16th International Workshop on Algebraic and Combinatorial Coding Theory*, Svetlogorsk (Kaliningrad region, Russia), 2018.

[32] Borges J., Rifà J., Zinoviev V. A., On completely regular codes, *Problems of Information Transmission*, 2019, vol. 55, no. 4, 1–45, DOI: 10.1134/S0032946019010010.

[33] Borges J., Rifà J., Zinoviev V. A., On completely regular and completely transitive complementary codes, *Discrete Mathematics*, 2020, vol. 343, no. 3, 1–8.

[34] Borges J., Zinoviev D., Zinoviev V. A., On the classification of completely regular codes with covering radius two and antipodal dual, *Problems of Information Transmission*, 2023, submitted.

[35] Boyvalenkov P., Delchev K., Zinoviev D. V., Zinoviev V. A., On q-ary codes with two distances d and $d+1$, *Problems of Information Transmission*, 2020, vol. 56, 33–44.

[36] Boyvalenkov P., Delchev K., Zinoviev D. V., Zinoviev V. A., On codes with distances d and n, *Problems of Information Transmission*, 2022, vol. 58, no 4, 352–371.

[37] Bremner A., Calderbank R.,Hanlon P., Mortom P., Wolfskill J., Two-weight ternary codes and the equation $y^2 = 4 \times 3^a + 13$, *Journal of Number Theory*, 1983, vol. 16, 212–234.

[38] Brouwer Andries E., Cohen Arjeh M, Neumaier A., *Distance-Regular Graphs*. Springer: Berlin, Heidelberg, 1989.

[39] Brouwer Andries E., A note on completely regular codes, *Discrete Mathematics*, 1990, vol. 83, no. 1, 115–117.

[40] Brouwer Andries E., On complete regularity of extended codes, *Discrete Mathematics*, 1993, vol. 117, no. 1–3, 271–273.

[41] Bose R. C., Bush K. A., Orthogonal arrays of strenth two and three, *Annals of Mathematical Statistics*, 1952, vol. 23, no. 4, 508–524.

[42] Budaghyan L., Carlet C., Pott A., New classes of almost bent and almost perfect nonlinear polynomials, *IEEE Transactions on Information Theory*, 2006, vol. 52, no. 3, 1141–1152.

[43] Bush Kenneth A., Orthogonal arrays of index unity, *The Annals of Mathematical Statistics*, 1952, vol. 23, no. 3, 426–434.

[44] Calderbank A. R., On uniformly packed $[n, n-k, 4]$ codes over $GF(q)$ and a class of caps in $PG(k-1, q)$, *Journal of the London Mathematical Society*, 1982, vol. 26, no. 2, 365–384.

[45] Calderbank A. R., Nonexistence of a uniformly packed $[70, 58, 5]$ code, *IEEE Transactions on Information Theory*, 1986, vol. 32, no. 6, 828–833.

[46] Calderbank A. R., Goethals J. -M., Three-weight codes and association schemes, *Philips Journal of Research*, 1984, vol. 39, no. 4–5, 143–152.

[47] Calderbank A. R., Goethals J.-M., On a pair of dual subschemes of the Hamming scheme $H_n(q)$, *European Journal of Combinatorics*, 1985, vol. 6, no. 2, 133–147.

[48] Calderbank R., Kantor W. M., The geometry of two-weight codes, *Bulletin of the London Mathematical Society*, 1986, vol. 18, no. 2, 97–122.

[49] Camion P., Courteau B., Delsarte P., On r-partition designs in Hamming spaces, *Applicable Algebra in Engineering, Communication and Computing*, 1992, vol. 2, 147–162.

[50] Canteaut A., Charpin P., Dobbertin H., Binary m-sequences with three-valued cross-correlation: A proof of Welch's conjecture, *IEEE Transactions on Information Theory*, 2000, vol. 46, no. 1, 4–8.

[51] Camerun P., Finite permutation groups and finite simple groups, *Bulletin of the London Mathematical Society*, 1981, vol. 13, 1–22.

[52] Carlet C., Boolean functions for cryptography and error-correcting codes, in: Crama Y., Hammer P. L. (eds.), *Boolean Models and Methods in Mathematics, Computer Science, and Engineering*, Encyclopedia of Mathematics and its Applications, vol. 134, Chapter 8, pp. 257–397. Cambridge University Press: Cambridge, 2010.

[53] Carlet C., Charpin P., Zinoviev V. A., Codes, bent functions and permutations suitable for DES-like cryptosystems, *Designs, Codes and Cryptography*, 1998, vol. 15, no. 2, 125–156.

[54] Colbourn C. J., Dinitz J. H., *Handbook of Combinatorial Designs*, 2nd edition. Chapman and Hall/CRC: Boca Raton, FL, 2007.

[55] van Dam E. R., Koolen J. H., Tanaka H., Distance-regular graphs, *The Electronic Journal of Combinatorics*, DS22, 1–156.

[56] Delsarte P., Two-weight linear codes and strongly regular graphs, MBLE Laboratoire de Recherches, 1971.

[57] Delsarte P., An algebraic approach to the association schemes of coding theory, PhD thesis, Philips Research Laboratories, 1973.

[58] Delsarte P., Four fundamental parameters of a code and their combinatorial significance, *Information and Control*, 1973, vol. 23, no. 5, 407–438.

[59] Denniston R. H. F., Some maximal arcs in finite projective planes, *Journal of Combinatorial Theory*, 1969, vol. 6, 317–319.

[60] Dobbertin H., Almost perfect nonlinear power functions over $GF(2^n)$: The Niho case, *Information & Computation*, 1999, vol. 151, 57–72.

[61] Dobbertin H., Almost perfect nonlinear power functions on $GF(2^n)$: A new case for n divisible by 5, *Finite Fields and Their Applications*, 2001, vol. 7, 113–121.

[62] Dodunekov S. M., Zinoviev V. A., A note on preparata codes, in: *Proceedings of 6th International Symposium on Information Theory*, Moscow–Tashkent, (September 18–22), 1984, Part 2, pp. 78–80.

[63] Dumer I. I., Some new uniformly packed codes, *Proceedings of Moscow Institute of Physics and Technology*, Moscow, 1976, pp. 72–78, (in Russian).

[64] Dumer I. I., Zinoviev V. A., Some new maximal codes over Galya field $GF(4)$, *Problems of Information Transmission*, 1978, vol. 14, no. 3, 24–34.

[65] Fon-Der-Flaass D. G., A bound on correlation immunity, *Siberian Electronic Mathematical Reports*, 2007, vol. 4, 133–135. Online: http://mi.mathnet.ru/eng/semr149.

[66] Fon-Der-Flaass D. G., Perfect 2-colorings of a hypercube, *Siberian Mathematical Journal*, 2007, vol. 48, no. 4, 740–745, DOI: 10.1007/s11202-007-0075-4.

[67] Fon-Der-Flaass D. G., Perfect colorings of the 12-cube that attain the bound on correlation immunity, *Siberian Electronic Mathematical Reports*, 2007, vol. 4, 292–295. English translation: https://arxiv.org/abs/1403.8091.

[68] Forney G. D., Jr., Sloane N. J. A., Trott M. D., The Nordstrom–Robinson code is the binary image of the octacode, in: Calderbank R., Forney G.D., Jr., Moayeri N. (eds.), *Coding and Quantization* (*Proceedings of DIMACS/IEEE Workshop*, Princeton University, NJ, October 19–21, 1992), Providence, RI: American Mathematical Society, 1993, pp. 19–26.

[69] Friedman J., On the bit extraction problem, in: *Foundations on Computer Science, IEEE Annual Symposium on IEEE Computer Society*, Les Alamites, CA, 1992, pp. 314–319, https://doi.org/10.1109/SFCS.1992.267760.

[70] Giudici M., Completely transitive codes in Hamming graphs, PhD thesis, The University of Western Australia, Perth, Australia, 1998.

[71] Giudici M., Praeger C. E., Completely transitive codes in Hamming graphs, *European Journal of Combinatorics*, 1999, vol. 20, no. 7, 647–662.

[72] Gillespie N. I., A note on binary completely regular codes with large minimum distance, *Discrete Mathematics*, 2013, vol. 313, 5132–5134.

[73] Gillespie N. I., Praeger C. E., Uniqueness of certain completely regular Hadamard codes, *Journal of Combinatorial Theory, Series A*, 2013, vol. 120, no. 7, 1394–1400.

[74] Gillespie N. I., Praeger C. E., New characterisations of the Nordstrom–Robinson codes, *Bulletin of the London Mathematical Society*, 2016, vol. 49, no. 2, 320–330.

[75] Godsil C., Equitable partitions, *Combinatorics, Paul Erdos is Eighty*, 1983, vol. 1, 173–192.

[76] Godsil C. D., McKay B. D., Feasibility conditions for the existence of walk-regular graphs, *Linear Algebra and Its Applications*, 1980, vol. 30, 51–61.

[77] Godsil C. D., Praeger C. E., Completely transitive designs, ArXiv e-print 1405.2176, May 2014.

[78] Goethals J.-M., On the golay binary perfect code, *Journal of Combinatorial Theory*, 1971, vol. 11, 178–186.

[79] Goethals J.-M., Snover S. L., Nearly perfect binary codes, *Discrete Mathematics*, 1972, vol. 3, 65–88.

[80] Goethals J. M., van Tilborg H. C. A., Uniformly packed codes, *Philips Research Reports*, 1975, vol. 30, no. 1, 9–36.

[81] Goethals J. M., Nonlinear codes defined by quadratic forms over $GF(2)$, *Information and Control*, 1976, vol. 31, no. 1, 43–74.

[82] Gold R., Maximal recursive sequences with 3-valued recursive cross-correlation functions, *IEEE Transactions on Information Theory*, 1968, vol. 14, no. 1, 154–156.

[83] Golay M. J. E., Notes on digital coding, *Proceedings IRE*, 1949, vol. 37, 657.

[84] Hammons A. R., Jr., Kumar P. V., Calderbank A. R., Sloane N. J. A., Solé P., The Z_4-linearity of Kerdock, Preparata, Goethals, and related codes, *IEEE Transactions on Information Theory*, 1994, vol. 40, no. 2, 301–319.

[85] Hartman A., Phelps K. T., Steiner quadruple systems, in: Dinitz J. H., Stinson D. R. (eds.), *Contemporary Design Theory: A Collection of Surveys*, Chapter 6, pp. 205–240. Wiley: New York, 1992.

[86] Hollmann H. D. L., Xiang Q., A proof of the Welch and Niho conjectures on cross-correlations of binary m-sequences, *Finite Fields and Their Applications*, 2001, vol. 7, no. 2, 253–286.

[87] Huffman W. C., Pless V., *Fundamentals of Error-Correcting Codes.* Cambridge University Press: Cambridge, 2003.

[88] Hughes D. R., Piper F. C., *Design Theory.* Cambridge University Press: Cambridge, 1985.

[89] Hyun J. Y., Local duality for equitable partitions of a Hamming space, *Journal of Combinatorial Theory, Series A*, 2012, vol. 119, no. 2, 476–482, DOI: 10.1016/j.jcta.2011.10.006.

[90] Ivanov A. A., Liebler R. A., Penttila T., Praeger C. E., Antipodal distance-transitive covers of complete bipartite graphs, *European Journal of Combinatorics*, 1997, vol. 18, no. 1, 11–33.

[91] Kantor W. M., k-homogeneous groups, *Mathematische Zeitschrift*, 1972, vol. 124, 261–265.

[92] Kantor W. M., On the inequivalence of generalized Preparata codes, *IEEE Transactions on Information Theory*, 1983, vol. 29, no. 3, 345–348.

[93] Kasami T., The weight enumerators for several classes of subcodes of the 2nd order binary Reed-Muller codes, *Information and Control*, 1971, vol. 18, no. 14, 369–394.

[94] Koolen J. H., On a conjecture of Martin on the parameters of completely regular codes and the classification of the completely regular codes in the Biggs-Smith graph, *Linear and Multilinear Algebra*, 1995, vol. 39, 3–17.

[95] Koolen J., Krotov D., Martin B., Completely regular codes, 2016, https://sites.google.com/site/completelyregularcodes.

[96] Koolen J. H., Lee W. S., Martin W. J., Tanaka H., Arithmetic completely regular codes, *Discrete Mathematics and Theoretical Computer Science (DMTCS)*, 2016, vol. 17, no. 3, 59–76.

[97] Koolen J. H., Lee W. S., Martin W. J., Characterizing completely regular codes from an algebraic viewpoint, *Contemporary Mathematics*, 2010, vol. 531, 223–242.

[98] Krotov D. S., On the binary codes with parameters of triply-shortened 1-perfect codes, *Designs, Codes and Cryptography*, 2012, vol. 64, 275–283, DOI: 10.1007/s10623-011-9574-1.

[99] Krotov D. S., On $(2n/3-1)$-resilient $(n/2)$-functions, in: *IEEE International Symposium on Information Theory*, Paris, France, July 7–12, 2019, pp. 2957–2961, IEEE, DOI: 10.1109/ISIT.2019.8849584.

[100] Krotov D. S., On the $OA(1536, 13, 2, 7)$ and related orthogonal arrays, *Discrete Mathematics*, 2020, vol. 343, 111659.

[101] Krotov D. S., On weight distributions of perfect colorings and completely regular codes, *Designs, Codes and Cryptography*, 2011, vol. 61, no. 4, 315–329.

[102] Krotov D. S., The classification of orthogonal arrays $OA(2048, 14, 2, 7)$ and some completely regular codes, *Discrete Mathematics*, 2024, vol. 347, 113923.

[103] Krotov D. S., Equitable $[[2, 10]; [6, 6]]$-partitions of 12-cube, *Cryptography and Communications*, 2024, vol. 16, 975–996.

[104] Krotov D. S., Valyuzhenich A. A. On degree-3 and $(n-4)$-correlation-immune perfect colorings of n-cubes, *Discrete Mathematics*, 2024, vol. 347, 114138.

[105] Krotov D. S., Vasil'eva A. Yu., On perfect codes that do not contain preparata-like codes, *Problems of Information Transmission*, 2016, vol. 52, no. 3, 284–288.

[106] Krotov D. S., Vorob'ev K. V., On unbalanced Boolean functions with best correlation immunity, *The Electronic Journal of Combinatorics*, 2020, vol. 27, no. 1, 1–24, DOI: 10.37236/8557.

[107] Lindström K., All nearly perfect codes are known, *Information and Control*, 1977, vol. 35, no. 1, 40–47.

[108] Van Lint J. H., In the nonexistence of perfect 2- and 3-Hamming-error-correcting codes, *Information and Control*, 1970, vol. 16, no. 4, 396-401.

[109] Van Lint J. H., On the nonexistence of perfect 5-, 6- and 7-Hamming-error-correcting codes over CG(q). Report 70-WSK-06, Technological University Eindhoven, 1970.

[110] Livingstone D., Wagner A., Transitivity of finite permutation groups on unordered sets, *Mathematische Zeitschrift*, 1965, vol. 90, no. 5, 393–403.

[111] Lloyd S. P., Binary block coding, *Bell Labs Technical Journal*, 1957, vol. 36, no. 2, 517–535.

[112] MacWilliams F. J., Sloane N. J. A., *The Theory of Error-Correcting Codes.* Elsevier, Amsterdam, Netherlands, 1977.

[113] Magliveras S. S, Leavitt D. W., Simple six designs exist, *Congressus Numerantium*, 1983, vol. 40, 195–205.

[114] Martin W. J., Completely regular codes: A viewpoint and some problems, in: *Proceedings of 2004 Com2MaC Workshop on Distance-Regular Graphs and Finite Geometry*, Pusan, Korea, 2004, pp. 43–56.

[115] Meyerowitz A. D., Cycle-balanced partitions in distance-regular graphs, *Discrete Mathemathics*, 2003, vol. 264, no. 2, 149–165.

[116] Mogilnykh I. Yu., On extending propelinear structures of the Nordstrom–Robinson code to the Hamming code, *Problems of Information Transmission*, 2016, vol. 52, no. 3, 289–298.

[117] Mogilnykh I. Yu., Valyuzhenich A. A., Equitable 2-partitions of the Hamming graphs with the second eigenvalue, *Discrete Mathematics*, 2020, vol. 343, no. 11, 112039(1–9), DOI: 10.1016/j.disc.2020.112039.

[118] Neumaier A., Completely regular codes, *Discrete Mathematics*, 1992, vol. 106, 353–360.

[119] Nordstrom A. W., Robinson I. P., An optimum nonlinear code, *Information and Control*, 1968, vol. 11, no. 5–6, 613–616.

[120] Potapov V. N., On perfect 2-colorings of the q-ary n-cube, *Discrete Mathematics*, 2012, vol. 312, 1269–1272, http://dx.doi.org/10.1016/j.disc.2011.12.014.

[121] Preparata F. P., A class of optimum nonlinear double-error-correcting codes, *Information and Control*, 1968, vol. 13, no. 4, 378–400.

[122] Rifà J., Personal Communication, March 2023.

[123] Rifà J., Pujol J., Completely transitive codes and distance transitive graphs, *International Symposium on Applied Algebra, Algebraic Algorithms, and Error-Correcting Codes*, 1991. Springer-Verlag, Berlin, Heidelberg, pp. 360–367. https://doi.org/10.1007/3-540-54522-0_124.

[124] Rifà J., Zinoviev V. A., On completely regular codes from perfect codes, In *Proceedings of the 10th International Workshop on Algebraic and Combinatorial Coding Theory*, Zvenigorod, Russia, 2006, pp. 225–229.

[125] Rifà J., Zinoviev V. A., On a class of binary linear completely transitive codes with arbitrary covering radius, *Discrete Mathematics*, 2009, vol. 309, no. 16, 5011–5016.

[126] Rifà J., Zinoviev V. A., New completely regular q-ary codes based on Kronecker products, *IEEE Transactions on Information Theory*, 2010, vol. 56, no. 1, 266–272.

[127] Rifà J., Zinoviev V. A., On lifting perfect codes, *IEEE Transactions on Information Theory*, 2011, vol. 57, no. 9, 5918–5925.

[128] Rifà J., Zinoviev V. A., On a family of binary completely transitive codes with growing covering radius, *Discrete Mathematics*, 2014, vol. 318, 48–52.

[129] Rifà J., Zinoviev V. A., Completely regular codes with different parameters and the same distance-regular coset graphs, *Discrete Mathematics*, 2017, vol. 340, 1649–1656.

[130] Roos J. E., An algebraic study of group and nongroup error correcting codes, *Information and Control*, 1965, vol. 8, no. 2, 195–214.

[131] Semakov N. V., Zinoviev V. A., Perfect and quasi-perfect constant weight codes, *Problems of Information Transmission*, 1969, vol. 5, no. 2, 11–13].

[132] Semakov N. V., Zinoviev V. A., Constant weight codes and tactical configurations, *Problems of Information Transmission*, 2016, vol. 5, no. 3, 28–36.

[133] Semakov N. V., Zinoviev V. A., Zaitsev G. V., Uniformly packed codes and tactical configurations, in: *Proceedings of the Forth All-Union Conference on Information and Coding Theory*, Moscow-Tashkent, 1969, part 1, pp. 215–220.

[134] Semakov N. V., Zinoviev V. A., Zaitsev G. V., Weight spectrums of uniformly packed binary codes, in: *Proceedings of the Forth All-Union Conference on Information and Coding Theory*, Moscow-Tashkent, 1969, part 1, pp. 221–226.

[135] Semakov N. V., Zinoviev V. A., Zaitsev G. V., Class of maximal equidistant codes, *Problems of Information Transmission*, 1969, vol. 5, no. 2, 65–68.

[136] Semakov N. V., Zinoviev V. A., Zaitsev G. V., Uniformly packed codes, *Problems of Information Transmission*, 1971, vol. 7, no. 1, 30–39.

[137] Semakov N. V., Zinoviev V. A., Zaitsev G. V., On duality of Preparata and Kerdock codes, in: *Proceedings of Fifth All-Union Conference on Coding Theory*, Moscow - Gorkyi, 1972, Part 2, pp. 55–58.

[138] Semakov N. V., Zinoviev V. A., Zaitsev G. V., Interrelation of Preparata and Hamming codes and extension of Hamming codes to new double-error-correcting codes, in: *Proceedings of 2nd International Symposium on Information Theory* Tsahkadsor, Armenia, USSR, September 2–8, 1971, Acad. Kiado, Budapest, Hungary, 1973, pp. 257–263.

[139] Shapiro H. S., Slotnick D. L., On the mathematical theory of error correcting codes, *IBM Journal of Research and Development*, 1959, vol. 3, no. 1, 25–34.

[140] Shi M., Krotov D. S., Solé P., A new distance-regular graph of diameter 3 on 1024 vertices, arXiv:1806.07069v3 [math.CO]. November 2018.

[141] Shi M., Krotov D. S., Solé P., A new approach to the Kasami codes of types 2, *IEEE Transactions on Information Theory*, 2020, vol. 66, no. 4, 2456–2465.

[142] Solé P., Completely regular codes and completely transitive codes, *INRIA, Institut National de Recherche en Informatique et en Automatique, Rapports de Recherche*, 1987, vol. 727, no. 2, 1–9.

[143] Solé P., Completely regular codes and completely transitive codes, *Discrete Mathematics*, 1990, vol. 81, no. 2, 193–201.

[144] Snover S. L., The uniqueness of the Nordstrom–Robinson and the golay binary codes, PhD Thesis, Department of Mathematics, Michigan State University, 1973.

[145] Tarannikov Y. V., On resilient boolean functions with maximal possible nonlinearity, in: Roy, Bimal (ed.) et al., *Proceedings of Indocrypt 2000*. Lecture Notes in Computer Science, vol. 1977, pp. 19–30. Springer-Verlag, 2000.

[146] Teirlinck L., Non-trivial t-designs without repeated blocks exist for all t, *Discrete Mathematics*, 1987, vol. 65, no. 3, 301–311.

[147] Terwilliger P., The subconstituent algebra of an association scheme I, *Journal of Algebraic Combinatorics*, 1992, vol. 1, no. 4, 363–388.

[148] Thas J. A., Construction of maximal arcs and partial geometry, *Geometriae Dedicata*, 1974, vol. 3, no. 1, 61–64.

[149] Thas J. A., Construction of maximal arcs and dual ovals in translation planes, *European Journal of Combinatorics*, 1980, vol. 1, no. 2, 189–192.

[150] Tietäväinen A., On the nonexistence of perfect 4-error-correcting codes. *Annales Academiae Scientiarum Fennicae. Series A I. Mathematica*, 1970, 485, pp. 1–6. https://doi.org/10.5186/aasfm.1971.000.

[151] Tietäväinen A., On the nonexistence of perfect codes over finite fields, *SIAM Journal on Applied Mathematics*, 1973, vol. 24, no. 1, 88–96.

[152] Tietäväinen A., Perko A., There are no unknown perfect binary codes. *Annales Universitatis Turkuensis, Series A*, 1971, vol. 148, 1–10.

[153] van Tilborg H. C. A., Uniformly packed codes, PhD thesis, Technische Hogeschool Eindhoven, 1976.

[154] Tokareva N. N., On components of Preparata codes, *Problems of Information Transmission*, 2004, vol. 40, no. 2, 159–164.

[155] Tokareva N. N., On upper bound for the number of binary uniformly packed codes, *Discrete Analysis and Operation Research*, 2014, vol. 7, no. 3, 70–79.

[156] Vasil'eva A. Y., Local spectra of perfect binary codes, *Discrete Applied Mathematics*, 2004, vol. 135, 301–307.

[157] Vasil'eva A., Linear binary completely regular codes with distance 2, *2008 IEEE Region 8 International Conference on Computational Technologies in Electrical and Electronics Engineering*, Novosibirsk, Russia, 2008, pp. 12–15, DOI: 10.1109/SIBIRCON.2008.4602615.

[158] Vasil'eva A., Local and interweight spectra of completely regular codes and of perfect colorings, *Problems of Information transmission*, 2009, vol. 45, no. 2, 151–157.

[159] Zinoviev V. A., Generalized concatenated codes, *Problems of Information Transmission*, 1976, vol. 12, no. 1, 5–15.

[160] Zinoviev V. A., Combinatorial methods of construction of nonlinear codes, Doctor of Sciences Dissertation, Institute for Problems of Information Transmission of the Russian Academie of Sciences, Moscow 1985.

[161] Zinoviev V. A., Leontiev V. K., On perfect codes, *Problems of Information Transmission*, 1972, vol. 8, no. 1, 26–35.

[162] Zinoviev V. A., Leontiev V. K., The nonexistence of perfect codes over Galois fields, *Problems of Control and Information Theory* , 1873, vol. 2, no. 2, 123–132.

[163] Zinoviev V. A., Rifà J., On new completely regular q-ary codes, *Problems of Information Transmission*, 2007, vol. 43, no. 2, 97–112.

[164] Zinoviev V. A., Zinoviev D. V., On generalized Preparata codes and 2-resolvable Steiner systems, *Problems Information Transmission*, 2016, vol. 52, no. 1, 15–36.

[165] Zinoviev V. A., Zinoviev D. V., On generalized concatenated construction of Nordstrom-Robinson code and binary Golay code, *Problems Information Transmission*, 2021, vol. 57, no. 4, 34–44.

[166] Welch L. R., Trace mappings in finite fields and shift register cross-correlation properties, University of Southern California, Electrical Engineering Department, Report, 1969.

CHAPTER 3

Completely Regular Codes in the Johnson Graphs

Alexander L. Gavrilyuk and Victor A. Zinoviev

IN THIS CHAPTER, we are interested in completely regular codes in the Johnson graphs.

3.1 PRELIMINARIES

Let us first recall some basic definitions, notation, and results on designs and distance-regular graphs.

3.1.1 Designs and Codes

We assume that the reader is familiar with some basic theory of combinatorial designs; the purpose of this section is merely to fix the notation. For a set X, let $\binom{X}{i}$ denote the set of all i-element subsets of X.

Definition 3.1. *A* t-(n, w, λ)-design *(or simply a* t-design*) is an incidence structure* $(X, \mathcal{B})$ *where* X *is an* n*-element set of* points *and* $\mathcal{B}$ *is a collection of* w*-subsets of points (called* blocks*) such that every* t*-subset of points is contained in exactly* $\lambda > 0$ *blocks,* $0 \leq t \leq w \leq n$.

A design is said to be *simple* if it has no repeated blocks. For simplicity, we will often identify a simple design $(X, \mathcal{B})$ (which is formally an incidence structure) with the collection $\mathcal{B} \subseteq \binom{X}{w}$ of its blocks. A design is said to be *nontrivial* if it is simple and $\mathcal{B} \neq \binom{X}{w}$.

A design with $\lambda = 1$ is called a *Steiner system*, sometimes denoted by $S(n, w, t)$. Among Steiner systems, the most celebrated ones have parameters $S(11, 5, 4)$, $S(12, 6, 5)$, $S(22, 6, 3)$, $S(23, 7, 4)$, and $S(24, 8, 5)$, which are often referred to as the *Mathieu designs* or the *Witt systems*, see [103] or [52, Section 10.11].

DOI: 10.1201/9781003393931-3

The following types of designs will be of our particular interest:

- a projective plane of order q, which is an $S(q^2+q+1, q+1, 2)$,
- a *symmetric* design, i.e., a one with $|X| = |\mathcal{B}|$ (in particular, a projective plane is a symmetric design),
- a *quasi-symmetric* design, i.e., a one in which every two distinct blocks intersect in only two possible numbers of points, i.e., $|\{|b \cap b'|: b \neq b' \in \mathcal{B}\}| = 2$ holds. The class of quasi-symmetric designs [25, IV.37] includes (non-symmetric) Steiner systems, strongly resolvable designs, residual designs of biplanes, and the two Witt designs on 22 and 23 points (and their complements) among others.

The following properties of designs are well known [12,25,59].

Proposition 3.1. *Given a t-(n, w, λ)-design T, every i-subset of points $(0 \leq i \leq t)$ is contained in exactly λ_i blocks, where*

$$\lambda_i = \lambda \frac{\binom{v-i}{t-i}}{\binom{k-i}{t-i}}. \tag{3.1}$$

In particular:

i. T is an i-(n, w, λ_i)-design for every $0 \leq i \leq t$;

ii. $\lambda = \lambda_t$;

iii. The number of blocks of T equals $b = \lambda_0$;

iv. Each point of X is contained in the same number of blocks, namely, $r = \lambda_1 = bk/v$ (the number r is called the replication number*).*

The first nontrivial designs with $t = 6$ were found in 1983 by Magliveras and Leavitt [70]. In 1987, Teirlinck [98] proved that there exist nontrivial simple t-designs for any natural number t. The existence of nontrivial Steiner system with $t \geq 6$ and the existence of infinitely many of those with $t = 4$ or 5 were long-standing problems in design theory; both seem to be resolved by Keevash [63] (although the proof is non-constructive and currently no actual Steiner systems are known for large values of t).

A design can also be viewed as a set of binary codewords, each of length n and weight w, such that the support of every binary vector of length n and weight t is contained in the supports of exactly λ codewords. This gives a rough idea of a connection between error-correcting codes and designs. In this regard, recall (see [69]), for example, that the "big" Witt designs with parameters

$$5-(24,8,1), \quad 4-(23,7,1), \quad 3-(22,6,1)$$

are related to the low weight codewords of the Golay code of length 24, while the "small" Witt designs with parameters

$$5-(12,6,1), \quad 4-(11,5,1)$$

are related to the low weight codewords of the ternary Golay code of length 12. More results on an interplay between designs and completely regular codes in the Hamming spaces are given in Chapter 2.

3.1.2 Distance-Regular Graphs and the Johnson Graphs

3.1.2.1 Basic Theory

We consider only finite *simple* (i.e., undirected, without loops or multiple edges) graphs. Let Γ be a connected graph. The distance $\partial(x,y) := \partial_\Gamma(x,y)$ between any two vertices x, y of Γ is the length of a shortest path connecting x and y in Γ. For a subset X of the vertex set of Γ, we will also write X for the subgraph of Γ induced by X. For a vertex $x \in \Gamma$, define $\Gamma_i(x)$ to be the set of vertices that are at distance precisely i from x $(0 \leq i \leq D)$, where $D := \max\{\partial(x,y) \mid x, y \in \Gamma\}$ is the *diameter* of Γ. (Note that $\Gamma_{-1}(x) = \Gamma_{D+1}(x) = \emptyset$.) The subgraph of Γ induced by $\Gamma_1(x)$ is called the *neighborhood* or the *local graph* of a vertex x. We often write $\Gamma(x)$ instead of $\Gamma_1(x)$ for short. A graph Γ is *regular* with *valency* k if the local graph $\Gamma(x)$ contains precisely k vertices for all $x \in \Gamma$.

The *eigenvalues* of a graph Γ are the eigenvalues of its $(0,1)$-adjacency matrix $A := A(\Gamma)$. In what follows, we will sometimes treat an eigenvector, say $\mathbf{u}$, of A for an eigenvalue θ as an θ-*eigenfunction*, i.e., a real–valued function $\mathbf{u}\colon \Gamma \to \mathbb{R}$ whose $\mathbf{u}(x)$-value is simply the x-coordinate of $\mathbf{u}$, for every $x \in \Gamma$. If, for an eigenvalue θ of Γ, the θ-eigenspace contains a vector orthogonal to the all-ones vector, we say that the eigenvalue θ is *non-principal*. If Γ is regular with valency k, then all its eigenvalues are non-principal unless the graph is connected and then the only eigenvalue that is principal is its valency k. The *spectrum* of a graph Γ is the multiset of its eigenvalues, which is written as $[\theta_0]^{m_0}$, $[\theta_1]^{m_1}$, ..., $[\theta_D]^{m_D}$, where $\theta_0 > \theta_1 > \ldots > \theta_D$ are all distinct eigenvalues of Γ, and $m_0, m_1, \ldots, m_D$ are their respective multiplicities. For a comprehensive treatment of graph spectra theory, the reader is referred to [18].

Definition 3.2. *A simple connected graph Γ of diameter D is called* distance-regular *if there exist integers b_i and c_i, $0 \leq i \leq D$, such that, for any pair of vertices $x, y \in \Gamma$ with $\partial(x,y) = i$, there are precisely c_i neighbors of y in $\Gamma_{i-1}(x)$ and b_i neighbors of y in $\Gamma_{i+1}(x)$. The array*

$$\{b_0, b_1, \ldots, b_{D-1}; c_1, c_2, \ldots, c_D\}$$

is called the intersection array *of the distance-regular graph Γ.*

Recall the following monotonicity property of these numbers for any distance-regular graph:

$$b_0 > b_1 \geq b_2 \geq \ldots \geq b_{D-1} \quad \text{and} \quad c_1 \leq c_2 \leq \ldots \leq c_D. \tag{3.2}$$

One can see that a distance-regular graph is regular with valency $k := b_0 = c_i + a_i + b_i$, where $a_i := k - b_i - c_i$, $1 \leq i \leq D$. Note that $b_D = c_0 = 0$, $c_1 = 1$ and $a_i = |\Gamma(y) \cap \Gamma_i(x)|$ holds for any pair of vertices x, y with $\partial(x, y) = i$, i.e., a_i is the valency of the subgraph $\Gamma_i(x)$. We also define $k_i := \frac{b_0 \cdots b_{i-1}}{c_1 \cdots c_i}$, $1 \leq i \leq D$, and note that $k_i = |\Gamma_i(x)|$ for all $x \in \Gamma$ (so that $k = k_1$).

Alternatively, a distance-regular graph Γ can be defined as follows: for all integers h, i, j with $0 \leq h, i, j \leq D$ and all pairs of vertices $x, y \in \Gamma$ with $\partial(x, y) = h$, the number

$$p^h_{ij} := |\{z \in \Gamma \mid \partial(x, z) = i,\ \partial(y, z) = j\}| = |\Gamma_i(x) \cap \Gamma_j(y)|$$

does not depend on the particular choice of x, y. The numbers p^h_{ij} are called the *intersection numbers* of Γ. Note that $k_i = p^0_{ii}$, $c_i = p^i_{1i-1}$, $a_i = p^i_{1i}$, and $b_{i-1} = p^{i-1}_{1i}$, $1 \leq i \leq D$, and all intersection numbers p^h_{ij} can be calculated from the intersection array of Γ, see [19, Lemma 4.1.7].

An *automorphism* of a graph Γ is a permutation π of the vertex set of Γ, which preserves the adjacency relation, i.e., for all $x, y \in \Gamma$, x and y are adjacent if and only if $\pi(x)$ and $\pi(y)$ are adjacent. Clearly, the set of all automorphisms of Γ forms a group denoted by $\mathrm{Aut}(\Gamma)$. A graph Γ of diameter D is called *distance-transitive*, if for every distance $0 \leq i \leq D$, the group $\mathrm{Aut}(\Gamma)$ acts transitively on the set of ordered pairs of vertices at distance i. A distance-transitive graph is distance-regular (but the converse is not true in general).

Let Γ be a distance-regular graph of diameter D. By $\mathbb{R}^\Gamma$, we mean the vector space over $\mathbb{R}$ whose standard basis is indexed by the vertex set of Γ. For every integer i with $0 \leq i \leq D$, define the i-th *distance matrix* $A_i \in \mathbb{R}^{\Gamma \times \Gamma}$ of Γ whose entries are given by:

$$(A_i)_{x,y} = \begin{cases} 1 \text{ if } \partial(x, y) = i, \\ 0 \text{ if } \partial(x, y) \neq i. \end{cases}$$

Then $A_0 = I$, which is the identity matrix, and $A := A_1$ is the *adjacency matrix* of Γ; furthermore, the matrices A_i $(0 \leq i \leq D)$ sum to the all-ones matrix J. The distance-regularity of Γ can be then expressed by the following equation:

$$A_i A_j = \sum_{h=0}^{D} p^h_{ij} A_h \quad (0 \leq i, j \leq D). \tag{3.3}$$

In particular, $AA_i = b_{i-1}A_{i-1} + a_iA_i + c_{i+1}A_{i+1}$ $(1 \leq i \leq D-1)$, and $AA_D = b_{D-1}A_{D-1} + a_DA_D$, and this three-term recurrence implies that

$$A_i = p_i(A) \tag{3.4}$$

for some polynomial p_i of degree i $(0 \leq i \leq D)$ (for their exact formulas, see, e.g., [19, Chapter 4]).

Further, equation (3.3) says that the real vector space $\mathcal{M}$ spanned by $\{A_0, A_1, \ldots, A_D\}$ is closed under matrix multiplication and, in fact, under entry-wise (also known as Schur or Hadamard) multiplication as well. The corresponding matrix subalgebra of the full matrix algebra $\mathbb{R}^{\Gamma\times\Gamma}$ is called the *Bose-Mesner* algebra of Γ. It follows that $\mathcal{M}$ has dimension $D+1$, and $\{A_0, A_1, \ldots, A_D\}$ forms a basis of $\mathcal{M}$. Since the algebra $\mathcal{M}$ is semi-simple and commutative, it also has a basis of pairwise orthogonal idempotents $E_0 := \frac{1}{|\Gamma|}J, E_1, \ldots, E_D \in \mathbb{R}^{\Gamma\times\Gamma}$ (the so-called *primitive idempotents* of $\mathcal{M}$) satisfying:

$$E_iE_j = \delta_{ij}E_i \quad (0 \leq i, j \leq D), \quad E_i = E_i^\top \quad (0 \leq i \leq D),$$

$$E_0 + E_1 + \cdots + E_D = I.$$

Further, Γ has precisely $D+1$ distinct eigenvalues, which are, by convention, usually ordered in decreasing order: $\theta_0 > \theta_1 > \ldots > \theta_D$. The eigenvalues and their multiplicities can be calculated from the intersection array of Γ, see [19, Section 4.1.B]. We may naturally order the primitive idempotents of Γ in such a way that E_j $(0 \leq j \leq D)$ is the matrix of rank $m_j := \mathsf{Trace}(E_j)$ representing orthogonal projection onto the eigenspace of A corresponding to eigenvalue θ_j, with multiplicity m_j of Γ. In other words, one can uniquely write the spectral decomposition

$$A = \sum_{j=0}^{D} \theta_j E_j,$$

and hence A_i $(0 \leq i \leq D)$ has eigenvalues $P_{ji} := p_i(\theta_j)$ $(0 \leq j \leq D)$, i.e.,

$$A_i = \sum_{j=0}^{D} P_{ji}E_j. \tag{3.5}$$

Conversely, one has

$$E_j = \frac{1}{|\Gamma|}\sum_{i=0}^{D} Q_{ij}A_i, \tag{3.6}$$

for some scalars Q_{ij} (which are sometimes called the "dual" eigenvalues). In particular, (3.4) and (3.6) imply that every orthogonal projection E_j is a polynomial in the adjacency matrix A_1.

The change-of-basis matrices $P = (P_{ji})_{j,i=0}^{D}$ and $Q = (Q_{ij})_{i,j=0}^{D}$ are called the *first* and *second* eigenmatrices of Γ, respectively.[1] Composing (3.5) and (3.6), one obtains

$$\frac{1}{|\Gamma|}PQ = I,$$

which is called the orthogonality relation.

Distance-regular graphs give rise to a special type of association scheme: the so-called *metric* or *P-polynomial* ones. In fact, much of the study of distance-regular graphs generalizes to the theory of association schemes. The interested reader is referred to [9,10,19] for a comprehensive treatment of the latter one.

3.1.2.2 The Johnson Graphs

Let $[n]$ denote the set $\{1, 2, \ldots, n\}$. For an integer w, $0 \leq w \leq n$, the *Johnson graph* $J(n, w)$ is defined on the set $V := \binom{[n]}{w}$ of all w-element subsets of $[n]$, with two such subsets being adjacent if and only if they share exactly $w-1$ elements. More generally, $x, y \in V$ are at distance i in the graph $J(n, w)$ if and only if they share exactly $w-i$ elements. The graphs $J(n, 2)$ are also known as the *triangular graphs*, often denoted by $T(n)$, and the graphs $J(n, 3)$ are sometimes called the *tetrahedral graphs*. Note that the Johnson graph $J(n, w)$ can also be seen as a graph on the set of all binary vectors of length n and weight w, with two vertices being adjacent if they are at Hamming distance 2 from each other (thus, the *Johnson graph distance* ∂_J equals twice the Hamming graph distance).

It is easily seen that the graphs $J(n, w)$ and $J(n, n-w)$ are isomorphic, hence in what follows we consider only the case $w \leq n/2$. Then the Johnson graph $J(n, w)$ has diameter w and it is distance-regular and even distance-transitive: $\mathrm{Aut}(J(n, w))$ contains a distance-transitive subgroup isomorphic to $\mathrm{Sym}(n)$. The intersection numbers of $J(n, w)$ are given by

$$b_{i-1} = (w-i+1)(n-w-i+1), \quad c_i = i^2, \quad (i = 1, \ldots, w) \tag{3.7}$$

and its adjacency matrix A has precisely $w+1$ distinct eigenvalues

$$\theta_i = \theta_i(n, w) := (w-i)(n-w-i) - i, \quad (i = 0, \ldots, w). \tag{3.8}$$

All of this information is derived by Delsarte [34, Section 4.2] (see also [19, Chapter 9]), who was, in fact, the first to consider the Johnson graphs as an algebraic framework for the combinatorial design theory and to analyze them as association schemes. Understanding the eigenspaces of $J(n, w)$ plays a fundamental role in this approach. Indeed, a collection C of w-element subsets of $[n]$, i.e., a vertex subset of $J(n, w)$, can be viewed as a t-design for some appropriate value of t (possibly 0). We define the *strength* of C to be the largest integer t such that C is a t-design. Decomposing the characteristic vector of a design over the eigenspaces of $J(n, w)$, Delsarte [34] obtained an "algebraic" characterization of its strength, see Theorem 3.5.

Let $I_{a,b}$ denote the *inclusion matrix* (also known as a *Wilson matrix*) of a-subsets versus b-subsets of $[n]$, $a \geq b$. Delsarte proved the following relationship between the column spaces $\mathsf{Col}(\cdot)$ of inclusion matrices $I_{w,j}$ and those of orthogonal projections of the Johnson graph $J(n, w)$:

$$\mathsf{Col}(I_{w,j}) = \mathsf{Col}(E_0) \oplus \mathsf{Col}(E_1) \oplus \ldots \oplus \mathsf{Col}(E_j), \tag{3.9}$$

whence the j-th eigenspace $\mathsf{Col}(E_j)$ equals $\mathsf{Col}(I_{w,j}) \cap \mathsf{Col}(I_{w,j-1})^{\perp}$.

One can see that the left multiplication by $I_{\ell,w}$ is a linear mapping from $\mathsf{Col}(I_{w,j})$ to $\mathsf{Col}(I_{\ell,j})$. Since $I_{\ell,w}$ has full rank by a result of Kantor [62], it follows that this mapping is in fact an isomorphism between these spaces and hence it induces an isomorphism between the corresponding j-th eigenspaces of $J(n, w)$ and $J(n, \ell)$, called the *inducing map*.

Theorem 3.1. *Let* $\mathbf{u}$ *be an eigenvector of the Johnson graph* $J(n,w)$ *with eigenvalue* $\theta_i(n,w)$*. Then for any* $\ell > w$ *the vector* $I_{\ell,w}\mathbf{u}$ *is an eigenvector of the Johnson graph* $J(n,\ell)$ *with eigenvalue* $\theta_i(n,\ell)$*.*

The concept of inducing maps has been treated by Avgustinovich and Mogilnykh [3] in a more general context of the so-called distance-2 biregular graphs. Here we just give another interpretation of the above theorem by using their approach. Namely, consider a bipartite graph Γ with the vertex set $\binom{[n]}{w} \cup \binom{[n]}{w+1}$ and two vertices being adjacent if one is a proper subset of the other one (i.e., the adjacency relation is the symmetrized inclusion relation of w-subsets vs. $(w+1)$-subsets). Note that the graph on the same vertex set, in which two vertices are adjacent if and only if they are at distance two in Γ, is the disjoint union of $\Gamma' = J(n,w)$ and $\Gamma'' = J(n,w+1)$. Let $\mathbf{u}'$ be an eigenfunction of Γ' and define

$$\mathbf{u}''(x) = \sum_{y \in \Gamma(x)} \mathbf{u}(y) \quad (\forall x \in \Gamma''), \tag{3.10}$$

called the function *induced in* Γ'' *by* $\mathbf{u}'$ in [3]. Avgustinovich and Mogilnykh [3, Theorem 1] showed that the induced function $\mathbf{u}''$ is then an eigenfunction of Γ''.

Going in a somewhat opposite direction, given a real-valued function $\mathbf{v}$ defined on vertices of $J(n,w)$ and distinct elements $i,j \in [n]$, define the *partial difference* $\mathbf{v}_{ij}$ as follows:

$$\mathbf{v}_{ij}(y) = \mathbf{v}(y \cup \{i\}) - \mathbf{v}(y \cup \{j\}), \quad \forall\, y \in \binom{[n] \setminus \{i,j\}}{w-1}. \tag{3.11}$$

Clearly, $\mathbf{v}_{ij}$ can be treated as a function defined on the vertices of $J(n-2,w-1)$. In the case when $\mathbf{v}$ is an eigenfunction of $J(n,w)$, Vorobiev, Mogilnykh, and Valyuzhenich [102] showed, among other things, the following nice characterization of the partial differences of $\mathbf{v}$.

Lemma 3.1. *Let* $\mathbf{v}$ *be a* $\theta_k(n,w)$*-eigenfunction of* $J(n,w)$*, and* $i,\, j \in [n]$*. Then* $\mathbf{v}_{ij}$ *is a* $\theta_{k-1}(n-2,w-1)$*-eigenfunction of* $J(n-2,w-1)$ *or the all-zeros function.*

We finish this section with the following properties of the local structure of the Johnson graph $J(n,w)$, which will be needed later in Section 3.3.3. They can be easily understood from the definition of the graph. Recall that a *c-clique* C of a graph is a complete subgraph (i.e., every two distinct vertices of C are adjacent) with exactly c vertices. We say that C is a *clique* if it is a c-clique for some c. By the $\ell \times m$-grid, we mean the Cartesian product of two cliques on ℓ and m vertices. A clique is said to be *maximal* if it is maximal with respect to inclusion. The Johnson graph $J(n,w)$ has two types of maximal cliques: the cliques of the first type are the collections of the w-subsets contained in some $(w+1)$-set, while the cliques of the other type are the collections of the w-subsets containing some $(w-1)$-set. The latter ones are the maximum cliques of $J(n,w)$, also called the *Delsarte cliques*, as their cardinality $n-w+1$ attains the Delsarte bound (see [19, Proposition 4.4]).

Lemma 3.2. *The neighborhood of every vertex in the Johnson graph* $J(n,w)$ *is the* $w \times (n-w)$*-grid.*

3.1.3 Completely Regular Codes

3.1.3.1 Codes in Graphs: Radius and Dual Degree

By a *code* in a graph Γ, we simply mean an arbitrary subset C of the vertex set of Γ. For a code C with $|C| \geq 2$, the *minimum distance* δ of C is

$$\delta = \delta_C := \min\{\partial(x, y) \mid x, y \in C,\ x \neq y\}.$$

and the *covering radius* of C is

$$\rho = \rho_C := \max\{\partial(x, C) \mid x \in \Gamma\},$$

where $\partial(x, C) = \min\{\partial(x, y) \mid y \in C\}$ is the distance between C and a vertex x. One can see that $\delta \leq 2\rho + 1$ (consider two balls of radius ρ and recall the triangle inequality.[2]) The following definition generalizes the concept of perfect codes in Hamming graphs to that in an arbitrary graph Γ.

Definition 3.3. *A code C is said to be* perfect *in Γ if $\delta = 2\rho + 1$.*

A code C determines an ordered partition of the vertex set of Γ, called the *distance* partition (with respect to C), as follows:

$$\Pi_C = (\Gamma_0(C), \Gamma_1(C), \ldots, \Gamma_\rho(C))$$

with cells

$$\Gamma_i(C) = \{x \in \Gamma \mid \partial(x, C) = i\} \quad (0 \leq i \leq \rho).$$

We also put $\Gamma_{-1}(C) = \Gamma_{\rho+1}(C) = \varnothing$.

From now on, we assume that Γ is a distance-regular graph of diameter D. Given a code C of Γ, let T denote the *outer distribution* matrix with respect to C, whose rows are indexed by the vertex set of Γ and whose columns are labeled by all possible distances from 0 to D, with $(T)_{x,i}$ being equal to $|\Gamma_i(x) \cap C|$, i.e., the number of "codewords" at distance i from x. Further, let $\chi_C \in \mathbb{R}^\Gamma$ denote the *characteristic* vector of C, i.e., the x-entry of χ_C equals 1 if $x \in C$ and 0 otherwise, and let $\mathcal{M}$ denote the Bose-Mesner algebra of Γ. Then one can see that the *outer distribution module* of C, which is a subspace of $\mathbb{R}^\Gamma$ defined by

$$\mathcal{M}\chi_C = \{M\chi_C \mid M \in \mathcal{M}\},$$

is the column space of the outer distribution matrix T. Indeed, the i-th column of D equals $A_i\chi_C$:

$$T = [A_0\chi_C\ A_1\chi_C\ \cdots\ A_D\chi_C]. \tag{3.12}$$

Since $\{A_0, A_1, \ldots, A_D\}$ forms a basis of $\mathcal{M}$, we have that $\mathcal{M}\chi_C$ is spanned by the columns of T. Now, the integer

$$s^* = s^*(C) := \mathrm{rk}(D) - 1$$

is called the *dual degree* (also known as *external distance*) of C. Since the graph Γ is connected by our assumption, it contains a geodetic path P of length ρ, which starts at a vertex in C and ends at a vertex at distance ρ from C. One can see that the $(\rho+1)\times(\rho+1)$-submatrix of the outer distribution matrix T restricted to the rows indexed by the vertices in P, and columns indexed by $0, 1, \ldots, \rho$, is upper triangular and has rank $\rho+1$; thus,

$$\rho \leq s^*, \tag{3.13}$$

which holds for an arbitrary code C.

3.1.3.2 Equitable Partitions and Complete Regularity

Let Γ be a connected graph and $\Pi = (C_0, \ldots, C_\rho)$ be a partition of its vertex set. We also denote by Π the *characteristic* matrix of this partition:

$$\Pi = [\chi_{C_0}\ \chi_{C_1}\ \cdots\ \chi_{C_\rho}],$$

where $\chi_{C_i} \in \mathbb{R}^{\Gamma}$ are the characteristic vectors of the classes of Π.

Definition 3.4. *Partition Π is said to be* equitable[3] *if there exist constants $a_{i,j}$ such that every vertex from C_i has precisely $a_{i,j}$ neighbors in C_j for all $0 \leq i, j \leq \rho$.*

Given an ordering of the elements of an equitable partition Π, we define the *quotient matrix* of Π as follows:

$$A/\Pi = (a_{i,j})_{i,j=0}^{\rho}.$$

One can interpret the numbers $a_{i,j}$ as those obtained by averaging the row-sums in each block of the adjacency matrix A of Γ which is block-partitioned according to Π. Moreover, the following matrix equality holds, and it is the defining property of equitable partitions (see, e.g., [18, Chapter 2.3]):

$$A \cdot \Pi = \Pi \cdot (A/\Pi), \tag{3.14}$$

which is attributed to Sachs by Martin [71, Theorem 2.1.2]. Using (3.14), it is not hard to get the following result, which is well-known (see, e.g., [18, Lemma 2.3.1]) and useful.

Lemma 3.3. *For an equitable partition Π of a graph with adjacency matrix A:*

i. if $\mathbf{v}$ is an eigenvector of A/Π for an eigenvalue θ, then $\Pi\mathbf{v}$ is an eigenvector of A for the same eigenvalue θ;

ii. if there exists an eigenvector of A for an eigenvalue θ that is constant on the cells of Π, then θ is an eigenvalue of A/Π.

In what follows, an eigenvector of A, satisfying Part (*ii*) of the above lemma will be referred to as an eigenvector *associated with* an equitable partition Π.

Lemma 3.3 says that the spectrum of A consists of the spectrum of the quotient matrix A/Π with eigenvectors in the column space of Π, i.e., constant on the cells of the partition, together with the eigenvalues belonging to eigenvectors orthogonal to the columns of Π (i.e., summing to zero on each cell of the partition). We will also see a special version of this lemma in Theorem 3.4.

Recall two natural examples of equitable partitions:

- Let G be an automorphism group of Γ. Then the partition of the vertex set into the G-orbits is equitable. An equitable partition of this type is called an *orbit partition.*
- Suppose that Γ is a distance-regular graph of diameter D. Then the *distance* partition $(\Gamma_0(x), \Gamma_1(x), \ldots, \Gamma_D(x))$ with respect to any vertex x is equitable. (Note that such a distance partition is an orbit partition if and only if Γ is distance-transitive.)

The following definition can be seen as a generalization of perfect codes (more on this later in Section 3.2) and uniformly packed codes (in the setting of arbitrary distance-regular graphs), and it is central in the present book.

Definition 3.5. *A code C in a distance-regular graph Γ is said to be* completely regular *if the distance partition of Γ with respect to C is equitable.*

A somewhat trivial example can be seen immediately from the above discussion: any vertex (viewed as a singleton) in a distance-regular graph is a completely regular code. Further, as the distance partition with respect to the code

$$C^{\mathrm{opp}} := \Gamma_\rho(C),$$

called the *opposite* of C, can be obtained by "reversing" the partition Π_C, it immediately follows by definition that completely regular codes come in pairs.

Proposition 3.2. *A code C is completely regular if and only if so is C^{opp}.*

Note that by the triangle inequality, every vertex in $\Gamma_i(C)$ has all its neighbors in $\Gamma_{i-1}(C) \cup \Gamma_i(C) \cup \Gamma_{i+1}(C)$. Therefore, when a code C is completely regular, the quotient matrix of Π_C is tridiagonal.

Recall that the i-th distance matrix A_i of Γ is the polynomial p_i in the adjacency matrix A. Combining (3.4) and (3.14), one obtains

$$A_i \cdot \Pi_C = \Pi_C \cdot (A_i/\Pi_C), \tag{3.15}$$

for some matrix A_i/Π_C, which is a polynomial (in fact, the same, p_i) in A/Π_C. This means that the distance partition Π_C induces an equitable partition of the graph Γ_i defined by the adjacency matrix A_i. In other words, the entries of the quotient matrix A_i/Π_C have the following combinatorial interpretation: $(A_i/\Pi_C)_{k,\ell}$ is the number of vertices in $\Gamma_\ell(C)$ at distance i from an arbitrary vertex in $\Gamma_k(C)$.

The case of equality in (3.13) with one additional assumption provides another characterization of completely regular codes, as the following theorem (attributed by Martin [71] to Neumaier) shows.

Theorem 3.2 ([71, Theorem 2.1.10]). *Let Γ be a distance-regular graph, C be a non-empty code of Γ with covering radius ρ, and Π_C be the distance partition of Γ with respect to C. In the above notation, the following are equivalent:*

i. C is a completely regular code, i.e., the distance partition Π_C is equitable;

ii. the outer distribution module $\mathcal{M}\chi_C$ is equal to the column space of the characteristic matrix Π_C (in particular, $\rho = s^*$);

iii. the outer distribution module $\mathcal{M}\chi_C$ is closed under entrywise multiplication and satisfies $\rho = s^*$;

iv. the outer distribution matrix with respect to C has precisely $\rho + 1$ distinct rows.

Theorem 3.2 says that a code C is completely regular if and only if the outer distribution numbers $(T)_{x,i}$ only depend on i and $j := \partial(x, C)$ but not on the particular choice of x at distance j from C. Note that Delsarte introduced the concept of a completely regular code in terms of the outer distribution matrix with respect to code. The definition in terms of equitable partitions was considered by Neumaier. The interested reader can find an in-depth discussion of Theorem 3.2 and some strengthening of it in [71, Section 2.1.4].

Further, for an arbitrary code C in Γ, it follows from $\delta_C \leq 2\rho_C + 1$ (see the discussion before Definition 3.3) and (3.13) that $\delta_C \leq 2s^* + 1$. The following result provides us with a useful sufficient condition for complete regularity.

Theorem 3.3 ([34, Theorem 5.13]). *In the above notation, if $\delta_C \geq 2s^* - 1$, then a code C is completely regular.*

In particular, Theorem 3.3 shows that perfect codes in distance-regular graphs are completely regular. One can see that the quotient matrix of the distance partition of a perfect code with radius ρ in a distance-regular graph Γ is

$$\begin{bmatrix} 0 & b_0 & 0 & \dots & 0 \\ c_1 & a_1 & b_1 & & \\ 0 & c_2 & a_2 & b_2 & \\ \vdots & \ddots & \ddots & \ddots & \\ 0 & \dots & 0 & c_\rho & k - c_\rho \end{bmatrix}. \tag{3.16}$$

3.1.3.3 Intersection Array, Spectrum and Strength

We now take a look at the quotient matrix A/Π_C of the distance partition of a completely regular code C in a distance-regular graph Γ. In general, the matrix A/Π_C has the following tridiagonal form:

$$A/\Pi_C = \begin{bmatrix} \alpha_0 & \beta_0 & 0 & \dots & 0 \\ \gamma_1 & \alpha_1 & \beta_1 & & \\ 0 & \gamma_2 & \alpha_2 & \beta_2 & \\ \vdots & \ddots & \ddots & \ddots & \\ 0 & \dots & 0 & \gamma_\rho & \alpha_\rho \end{bmatrix} \tag{3.17}$$

where $\alpha_i, \beta_i, \gamma_i$, $(i = 0, \dots, \rho)$ are the *intersection numbers* of C. That is, each vertex from $\Gamma_i(C)$ has precisely α_i neighbors in $\Gamma_i(C)$, β_i neighbors in $\Gamma_{i+1}(C)$, and γ_i neighbors in $\Gamma_{i-1}(C)$. We continue this analogy with distance-regular graphs and call the ordered sequence

$$\{\beta_0, \beta_1, \dots, \beta_{\rho-1}; \gamma_1, \gamma_2, \dots, \gamma_\rho\}$$

the *intersection array* of C. One might then expect an analogue of (3.2) to hold:

$$\beta_0 \geq \beta_1 \geq \dots \geq \beta_{\rho-1} \quad \text{and} \quad \gamma_1 \leq \gamma_2 \leq \dots \leq \gamma_\rho. \tag{3.18}$$

which was, in fact, conjectured by Martin [71, Section 2.2.2]. Koolen [64] gave a very simple counterexample showing that it is not true in arbitrary distance-regular graphs; however, in the same paper, he proved that, if a graph Γ admits a non-identity automorphism σ satisfying $\partial(x, x^\sigma) \leq 1$ for all vertices x, then (3.18) holds for any completely regular code in Γ. In particular, this shows that the monotonicity condition (3.18) holds for all completely regular codes in the Johnson graphs (and in many other families of classical distance-regular graphs, see Chapter 1).

In general, not much else is known about feasibility conditions for an intersection array $\{\beta_0, \beta_1, \dots, \beta_{\rho-1}; \gamma_1, \gamma_2, \dots, \gamma_\rho\}$. The following are some obvious conditions obtained by counting.

Proposition 3.3 ([71, Proposition 2.2.4]). *The intersection numbers of a completely regular code C satisfy:*

i. $\gamma_i + \alpha_i + \beta_i = k$ *for all* $0 \leq i \leq \rho$;

ii. $\beta_i|\Gamma_i(C)| = \gamma_{i+1}|\Gamma_{i+1}(C)|$ *for all* $0 \leq i \leq \rho - 1$;

iii. $\gamma_i \geq c_i$ *for all* $0 \leq i \leq \rho$;

iv. $\beta_i \leq b_i$ *for all* $0 \leq i \leq \rho - 1$.

Clearly, the intersection numbers and $|\Gamma_i(C)|$ must be non-negative integers. This observation also applies to the entries of the quotient matrices A_i/Π_C defined by (3.15) and might be useful in ruling out putative completely regular codes (see Section 3.2.2).

Recall a classical result in Coding Theory by Lloyd [67] that gives a strong necessary condition on the parameters of perfect binary codes. It then received numerous generalizations, in particular, the following one to completely regular codes by Neumaier [89] (cf. Lemma 3.3).

Theorem 3.4 ([89]). *For a completely regular code C with covering radius ρ in a distance-regular graph Γ, the tridiagonal matrix A/Π_C (see (3.17)) has precisely $\rho+1$ distinct eigenvalues and each of these is an eigenvalue of Γ.*

As usually, let $\theta_0 > \theta_1 > \ldots > \theta_D$ denote the $D+1$ distinct eigenvalues of Γ. Then one can see that θ_0 is always an eigenvalue of A/Π_C. The $\rho+1$ eigenvalues $\theta_0, \theta_{j_1}, \theta_{j_2}, \ldots, \theta_{j_\rho}$ of A/Π_C form the *spectrum* of the code, and the set of non-zero indices $\{j_1, \ldots, j_\rho\}$ of those eigenvalues of Γ appearing in the spectrum of C is called the *dual degree set* of C.

Using the definition of the first eigenmatrix P, one can verify that the outer distribution matrix T with respect to C (see (3.12)) satisfies $T = F \cdot P$ where

$$F = [E_0\chi_C \; E_1\chi_C \; \ldots \; E_D\chi_C].$$

Recall that the primitive idempotents E_j are pairwise orthogonal. Since the matrix T (and hence F) has rank s^*+1, it follows that F has s^*+1 non-zero columns.

Lemma 3.4 ([76, Lemma 1]). *For $j > 0$, $E_j\chi_C \neq 0$ holds if and only if θ_j is an eigenvalue of A/Π_C.*

Recall also the following algebraic characterization of combinatorial designs of strength t, given by Delsarte [34, Theorem 4.2].

Theorem 3.5. *Let $E_0, E_1, \ldots, E_w$ denote the primitive idempotents (in the natural order) of the Johnson graph $J(n,w)$. Then C is the set of block of a t-design if and only if*

$$E_1\chi_C = E_2\chi_C = \cdots = E_t\chi_C = 0.$$

It now follows from Lemma 3.4 and Theorem 3.5 that a completely regular code C in $J(n,w)$ is a t-design. The strength of this design is the smallest nonnegative t such that θ_{t+1} is in the spectrum of C [71, Corollary 4.1.2]. For this reason, completely regular codes in $J(n,w)$ are also referred to as *completely regular designs* [72] (in this context, in what follows we will freely switch between the code/design notions). In view of Proposition 3.1, we then have the following statement.

Corollary 3.1. *The vertex set of a completely regular code C in $J(n,w)$ is the set of blocks of a t-(n,w,λ) design where t is the smallest integer such that $\theta_{t+1}(n,w)$ is in the spectrum of C and $\lambda = |C| \cdot \binom{w}{t}/\binom{n}{t}$.*

The converse claim (i.e., whether the set of blocks of a t-design, viewed as a vertex subset of $J(n,w)$, is a completely regular code) is not true in general: we will discuss this question, among others, in Section 3.3.

For completely regular codes of radius one in $J(n,w)$, we have the following convenient corollary from the above results and Proposition 3.1.

Proposition 3.4 ([82, Corollary 5]). *Let C be a completely regular code in $J(n,w)$ with $\rho_C = 1$ and intersection array $\{\beta_0; \gamma_1\}$. Then $|C| = \frac{\gamma_1}{\gamma_1+\beta_0}\binom{n}{w}$, the spectrum of C consists of $\theta_0(n,w)$ and $\theta_{t+1}(n,w) = \theta_0 - \beta_0 - \gamma_1$ for some t. Moreover, the numbers $\frac{\gamma_1}{\gamma_1+\beta_0}\binom{n-i}{w-i}$ are integer for all $i = 0, \ldots, t$.*

Finally, it follows from Theorem 3.5 and the definition of the dual degree s^* that

$$t + s^* \leq w, \tag{3.19}$$

whence one has the following useful corollary of Theorem 3.3.

Corollary 3.2. *A simple t-(n, w, λ) design having minimum distance of at least $2(w - t) - 1$ is completely regular in $J(n, w)$.*

Actually, this only gives that $(w-1)$-(n, w, λ) and $(w-2)$-$(n, w, 1)$ designs are completely regular because the minimum distance is at most $w - t + 1$ for $\lambda = 1$ and at most $w - 1$ for $\lambda \geq 2$.

3.2 PERFECT CODES IN JOHNSON GRAPHS

Perfect codes are perhaps the most intriguing type of completely regular codes in the Johnson graphs. According to Definition 3.3, a code[4] C in the Johnson graph $J(n, w)$ with minimum distance $\delta_C = 2e + 1$ is a *perfect e-code* if, for any vertex $x \in J(n, w)$, there exists exactly one vertex $c \in C$ such that $\partial_J(x, c) \leq e$. In other words, all balls of (Johnson) radius e around vertices of C do not intersect each other and exhaust the set of all vertices of $J(n, w)$.

The following trivial examples (which are similar to those in the Hamming graphs) of perfect e-codes in $J(n, w)$ are known:

i. The Johnson graph $J(n, w)$ itself as the 0-code.

ii. Any vertex of $J(n, w)$ as a perfect w-code.

iii. For the case $n = 4e+2$ and $w = 2e+1$, any pair of vertices x and y of $J(n, w)$ at the distance $\partial_J(x, y) = w$ is a perfect e-code.

The existence problem of perfect nontrivial codes in Johnson graphs was formulated in 1973 by Delsarte [34, p. 55]:

> "*After having recalled that there are "very few" perfect codes in the Hamming schemes, one must say that, for* $1 < e < n$, *there is not a single one known in the Johnson schemes. It is tempting to risk the conjecture that such codes do not exist. Certain results contained in the present work could be useful to attack this problem; especially the generalized Lloyd theorem of Sec.* 5.2.2 (in [34]) *and Theorem* 4.7 *about t-designs.*"

Until now this problem (conjecture) is still open. In this section, we overview some results that support the conjecture. For further details, we refer the interested reader to a recent monograph by T. Etzion [40].

3.2.1 Overview of the Results Supporting the Delsarte Conjecture

Let us formulate a simple necessary existence condition. Denote by $\Phi_e(n,w)$ the cardinality of a ball (around any vertex) of radius e in $J(n,w)$. Clearly,

$$\Phi_e(n,w) = \sum_{i=0}^{e} \binom{w}{i}\binom{n-w}{i}.$$

Then, by the sphere packing condition, for a perfect e-code C in $J(n,w)$, one has

$$|C| = \frac{\binom{n}{w}}{\Phi_e(n,w)},$$

which means that the ratio on the right-hand side should be an integer: many results rely on this number-theoretical condition and/or combinatorial argument.

Recall that Lloyd's theorem [67] was the main tool in proving the classification of nontrivial perfect codes in the Hamming graphs (see [99,104] and references therein). However, a direct application of its generalization (Theorem 3.4) to perfect codes in the Johnson graphs did not lead to such a strong result (see also Section 3.2.2).

Exploiting a connection between perfect codes in a Johnson graph and uniformly packed codes in the corresponding double Odd graph ([19, Section 9.1D]), Bannai [8] proved the nonexistence of perfect e-codes in the Johnson graphs $J(2w+1,w)$ for all $e \geq 2$. Using an alternative, purely combinatorial argument, Hammond [58] improved this result as follows.

Theorem 3.6 ([8,58]). *For all $e \geq 1$ and all positive integers w, nontrivial perfect e-codes do not exist in the Johnson graph $J(n,w)$ with $n \in \{2w+1, 2w+2\}$.*

It should not come as a surprise that two nice combinatorial objects in the Johnson graphs $J(n,w)$, namely: Steiner systems $S(n,w,t)$ and perfect e-codes, may be related to each other. There are several results known which describe connections between them (see [38,87]).

Proposition 3.5 ([38]). *If a perfect e-code exists in $J(n,w)$, then Steiner systems $S(w, 2e+1, e+1)$ and $S(n-w, 2e+1, e+1)$ exist.*

Considering the blocks of a Steiner system as the vertices of a perfect code, one has the following result.

Proposition 3.6 ([38]). *Except for the (trivial) Steiner systems $S(2w,w,1)$ and $S(n,w,w)$, no other Steiner systems are perfect codes in the Johnson graph $J(n,w)$.*

Many nonexistence results for perfect codes in $J(n,w)$ were obtained by Etzion and Schwartz [38,39,41]. Using the above-mentioned relation between Steiner systems and perfect codes in $J(n,w)$ and some divisibility conditions resulting from it, Etzion [38] proved the following.

Theorem 3.7 ([38]). *Given $e \geq 1$ and w, nontrivial perfect e-codes do not exist in the Johnson graph $J(n, w)$ for all of the following values of n:*

i. $n = 2w + e + 1$;

ii. $n = 2w + p$, *where p is prime;*

iii. $n = 2w + 2p$, *where p is prime and $p \neq 3$;*

iv. $n = 2w + 3p$, *where p is prime, $p \notin \{2, 3, 5\}$.*

Shimabukuro [97] further added to this list the graphs $J(2w+9, w)$, $J(2w+5p, w)$ ($p \neq 3$), and $J(2w + p^2, w)$, for p prime, which do not admit the existence of nontrivial perfect e-codes.

Using the property (called the *k-regularity* in [41]) that a perfect code in $J(n, w)$ is a completely regular design (Corollary 3.1), Etzion and Schwartz [41] obtained the following statement.

Theorem 3.8 ([41]). *There are no perfect 1-codes in $J(n, w)$ for $n \leq 50,000$ and perfect 2-codes in $J(n, w)$ for $n \leq 40,000$.*

Their proof uses the divisibility conditions from Proposition 3.1 (cf. Proposition 3.4). In particular, for perfect 1-codes, $\Phi_1(n, w) = 1 + w(n - w)$ should divide $\binom{n-i}{n-w}$ for some range of consecutive values of i. Gordon [55] further improved the lower bound on the size of a 1-perfect code by considering possible prime divisors of $\Phi_1(n, w)$. He used the following two number-theoretical results, which we recall now. The first one is known as Kummer's lemma (1852).

Lemma 3.5 ([65]). *Let p be a prime. The number of times that p appears in the factorization of $\binom{a}{b}$ equals the number of carries when adding b to $a - b$ modulo p.*

Using Lemma 3.5, it was shown that $\Phi_1(n, w)$ is square-free and, furthermore, certain powers of every prime factor of $\Phi_1(n, w)$ should belong to the "short" interval $[w, w + \sqrt{w}]$. The second lemma is due to Loxton [68] (a gap in its proof was fixed by Bernstein [11]):

Lemma 3.6 ([11,68]). *The number of perfect powers (i.e., powers of the type a^m where $a, m > 1$) in the interval $[w, w + \sqrt{w}]$ is at most*

$$exp(40\sqrt{\log \log w \log \log \log w}).$$

Using Lemma 3.6 and a computer search for perfect powers in short intervals, Gordon [55] derived the following result.

Theorem 3.9 ([55]). *There are no perfect 1-codes in $J(n, w)$ for $n \leq 2^{250}$.*

Given radius e and weight w, the length n of perfect constant weight-w e-codes is bounded from above. This was shown by Movsisyan [86] in 1982 and Roos [94] in 1983 (see also [41] for a simpler proof of a slightly stronger result, which shows that their original inequality cannot be tight).

Theorem 3.10 ([41,86,94]). *If a perfect e-code in the Johnson graph $J(n,w)$ exists, then its length n is bounded from above by the following inequality:*

$$n < \frac{2e+1}{e}(w-1).$$

This upper bound immediately implies the following result.

Corollary 3.3 ([86]). *Given e and w, the number of perfect e-codes in $J(n,w)$ is finite.*

Let C be a perfect e-code in $J(n,w)$. For any vector y from $J(n,w+1)$, denote by $R(y)$ the set of vectors x from C, such that $\partial_J(x,y) = 2e+1$.

Theorem 3.11 ([86]). *If there exists a perfect e-code in $J(n,w)$, then, for every vector $y \in J(n,w+1)$, the set $R(y)$ has cardinality $(w+1)/(e+1)$, which implies that $e+1$ divides $w+1$.*

From Theorems 3.10 and 3.11, one can deduce immediately that the weight w of vectors in a perfect e-code is bounded from below by the inequality $w \geq 2e+1$. This bound was significantly improved in [41] and [86] as follows:

Proposition 3.7. *If there exists a nontrivial perfect e-code in $J(n,w)$, then*

$$w \geq \begin{cases} e^2+3e+1 & \text{if} \quad w > n/2 & & [85] \\ \frac{1}{2}e(e+1)(e+2)+2(e+1) & \text{if} \quad w < n/2 & \text{and } n \text{ odd} & [41] \\ e(e+1)(e+2)+2(e+1) & \text{if} \quad w < n/2 & \text{and } n \text{ even} & [41] \\ 2e^2+4e+1 & \text{if} \quad w = n/2 & & [41] \end{cases}$$

Finally, let us briefly discuss another interesting direction in the study of perfect constant weight codes. The existence of perfect codes in the Johnson graphs turns out to be deeply related to the problem of so-called D-representation of codes [66,88], which we will explain now. Let A be any subset of the binary vector space $\mathbb{F}_2^n$ and φ be an arbitrary function defined on the set of natural numbers. For a vector x from A, define the *Dirichlet region* $D_x(\varphi)$ as follows:

$$D_x(\varphi) = \{y \in \mathbb{F}_2^n : \varphi(\partial_J(x,y)) \leq \varphi(\partial_J(z,y)), \forall z \in A\}$$

and write just D_x when φ is the identity.

Definition 3.6. *A set A of vectors from $\mathbb{F}_2^n$ is called a $D(\varphi)$-*representable code *(D-representable, respectively), if and only if the Dirichlet regions $D_x(\varphi)$ (D_x, respectively) of all vectors of A are pairwise disjoint in $\mathbb{F}_2^n$.*

Proposition 3.8 ([66,88]). *A vertex set C is a perfect e-code in $J(n,w)$ if and only if C is $D(\varphi)$-representable for any strictly monotone function φ.*

3.2.2 Completely Regular Codes from Perfect Codes

Martin [71, Chapter 5] considered a linear-algebraic approach to the Delsarte conjecture via studying related completely regular codes. Among other things, he proved that the derived design of a perfect code is always a completely regular code (see Theorem 3.12), which may give more leverage to a number-theoretic attack on the Delsarte conjecture.

To warm up, consider first perfect 1-codes in $J(n, w)$. It follows from (3.16) that such a code C is completely regular with quotient matrix

$$\begin{bmatrix} 0 & w(n-w) \\ 1 & w(n-w)-1 \end{bmatrix}$$

which has eigenvalues $b_0 = w(n-w)$ and -1. It follows from (Lloyd's) Theorem 3.4 that -1 must be an eigenvalue of $J(n, w)$, say, θ_{t+1} (see (3.8)). One can then show (equating $\theta_{t+1} = -1$ and writing $t+1 = w-s$) that there exist integers r and s such that $w = rs+1$ and $n = 2rs+r-s+1$. By Theorem 3.5, C is a block design of strength $t = (r-1)s$. Using (3.1) in Proposition 3.1, Martin [71, Chapter 5] further examined the existence problem of perfect 1-codes and found that the numbers r, s must obey some non-trivial arithmetic conditions.

Next, consider another example given in [71, p. 40]. Suppose that $J(20, 9)$ contains a perfect 2-code C. The corresponding quotient matrix (see (3.7) and (3.16)) is

$$A/\Pi_C = \begin{bmatrix} 0 & 99 & 0 \\ 1 & 18 & 80 \\ 0 & 4 & 95 \end{bmatrix}$$

with eigenvalues $99, 19$, and -5, which are the eigenvalues θ_0, θ_5 and θ_8 of $J(20,9)$, respectively, so Lloyd's Theorem is satisfied. Computing $A_4/\Pi_C = p_4(A/\Pi_C)$, we obtain:

$$A_4/\Pi_C = \frac{1}{4}\begin{bmatrix} 0 & 13860 & 152460 \\ 140 & 8680 & 157500 \\ 77 & 7875 & 158368 \end{bmatrix}$$

from which it follows that $J(20, 9)$ contains no such perfect codes. (Their nonexistence also follows from the earlier result by Hammond, see Theorem 3.6).

Perfect codes are not only completely regular: Martin observed that a perfect code in $J(n, w)$ would also give rise to completely regular codes in two "neighboring" Johnson graphs. Consider the partition of $J(n+1, w+1)$ into the $(w+1)$-sets containing a fixed symbol, say "1", and the $(w+1)$-sets not containing it. This partition is equitable (i.e., each cell is a completely regular code of radius 1, see Section 3.3.2), and the subgraph induced by each cell is isomorphic to a Johnson graph (namely, $J(n, w)$ by the former one, and $J(n, w+1)$ by the latter one). Suppose $J(n+1, w+1)$ contains a perfect e-code C. Then the above-mentioned partition correspondingly splits C into two parts:

$$C' = \{x : 1 \in x \in C\},$$
$$C'' = \{x : 1 \notin x \in C\},$$

called the *derived* and *residual* designs of C, respectively, and which can be viewed as codes in $J(n, w)$ and in $J(n, w+1)$, respectively. Carefully inspecting the distance partitions with respect to these codes, Martin shows the following.

Theorem 3.12 ([71, Theorem 5.2.1]). *If C is a non-trivial perfect code in $J(n+1, w+1)$, then its derived and residual designs are completely regular codes in the respective Johnson graphs.*

Corollary 3.4 ([71, p. 123]). *If C is a non-trivial perfect code in $J(n, w)$, then $e+1$ divides $n-w+1$.*

Martin gives the intersection numbers for the derived completely regular design C' of a perfect e-code in $J(n, w)$. For perfect 1-codes, the consideration of the derived design does not give any further restrictions. For perfect 2-code in $J(n, w)$ (note that $J(n, 2)$ and $J(10, 5)$ possess trivial perfect 2-codes), applying Lloyd's theorem (Theorem 3.4) to the quotient matrix of C', Martin [71, Section 5.3.3] derives some arithmetic conditions on n, w, which are necessary for such codes to exist. As Martin observes, this rules out many, but not all, feasible parameter sets for perfect 2-codes in $J(n, w)$.

We finish this section with another result due to Avgustinovich and Mogilnykh [3], which produces completely regular codes from perfect ones; although we don't know if this can be used to eliminate any putative perfect codes. The proof is based on an inducing map between eigenspaces of $J(n, w)$ and $J(n, w+1)$ (see the discussion after Theorem 3.1 and also Section 3.3.4).

Theorem 3.13. *If there exists a perfect 1-code in $J(n, w)$, then there exists a completely regular code with covering radius 1 in $J(n, w+1)$ with the quotient matrix:*

$$\begin{bmatrix} (w+1)(n-w-2) & w+1 \\ w(n-w-1) & n-w-1 \end{bmatrix}$$

In the next section, we consider completely regular codes in Johnson graphs in the general context.

3.3 CONSTRUCTIONS AND CLASSIFICATIONS

In this section, we discuss the general case of completely regular codes in the Johnson graphs. As was observed in Section 3.1.3, the vertices of such a code represent the set of blocks of a combinatorial design of some strength. In Section 3.3.1, we first take a look at which "celebrated" designs may be completely regular. Further, much attention has been paid to completely regular designs of small strength. Those of strength 0 are fully classified, see Section 3.3.2. Completely regular designs of strength 1 have been classified when the minimum distance is at least 2 (see Section 3.3.3.1) or the covering radius ρ equals 1 (Section 3.3.3.2) or $w-1$ (Section 3.3.3.3), which are the minimum and maximum possible values of ρ, respectively. Finally, in Section 3.3.4, we describe various constructions and partial classification results known for completely regular designs in the Johnson graphs $J(n, w)$ with $w \leq 6$.

3.3.1 Which "Classical" Designs Are Completely Regular

Martin [71, Chapter 4] examined the completely regularity property in several well-known classes of designs, such as Steiner, symmetric and quasi-symmetric ones. Let C be a simple t-design on n points and with block size w, viewed as a subset of vertices of $J(n, w)$. Observe that, by definition, the minimum distance δ_C and maximum block intersection cardinality, say m, are related by $\delta_C = w - m$. Corollary 3.2 then immediately implies the following two results.

Theorem 3.14 ([71, Corollary 3.7]). *A simple t-design in $J(n, t+1)$ is completely regular.*

In other words, the completely regular codes in $J(n, w)$ of strength $w - 1$ coincides with the simple $(w-1)$-(n, w, λ) designs.

Theorem 3.15 ([71, Corollary 3.8]). *A Steiner t-design in $J(n, t+2)$ is completely regular.*

For example, Theorem 3.14 tells us that all triple systems (i.e., 2-$(n, 3, \lambda)$ designs) are completely regular in $J(n, 3)$ (see Section 3.3.3.2 for more results on completely regular codes in $J(n, 3)$), while Theorem 3.15 tells us that all Steiner quadruple systems (i.e., 2-$(n, 4, 1)$ designs) are completely regular in $J(n, 4)$ (see Section 3.3.4 for more results on completely regular codes in $J(n, 4)$).

Further, on the one hand, every t-subset of $[n]$ is contained in some vertex, say x, of $J(n, w)$, while on the other hand, every t-subset of $[n]$ belongs to some block y of C; thus, one has $\partial(x, C) \leq \partial(x, y) \leq w - t$ and the following result.

Proposition 3.9 ([71, Proposition 4.3.1]). *The covering radius of a t-(n, w, λ) design is at most $w - t$.*

A detailed consideration of two blocks intersecting in $m > 0$ points gives an upper bound on w.

Theorem 3.16 ([71, Theorem 4.3.2]). *In the above notation, if C is completely regular and $m > 0$, then $w \leq 3m + 1$ or, equivalently, $\delta_C \leq \frac{2w+1}{3}$.*

Corollary 3.5 ([73]). *If a Steiner system $S(n, w, t)$ is completely regular, then $w \leq 3t - 2$.*

Theorem 3.16 also implies that $w \leq 3\lambda + 1$ holds for a completely regular symmetric 2-(n, w, λ) design. In particular, it means that a projective plane of order q, which is a symmetric Steiner design $S(q^2+q+1, q+1, 2)$, is not completely regular whenever $q \geq 4$. Martin [71, Proposition 4.4.3] further shows that a symmetric design formed by the points and hyperplanes of the projective geometry PG(n, q) is not completely regular whenever $n \geq 3$. In fact, the following nice result by Martin [73, Theorem 4.4] classifies *all* completely regular symmetric designs.

Theorem 3.17 ([73]). *A completely regular symmetric design is a projective plane of order two or three (i.e., it is one of the two Steiner systems $S(7, 3, 2)$ or $S(13, 4, 2)$).*

This theorem follows from a somewhat technical result (also due to Martin), which is stated below. Let us first recall the concept of a split design introduced in [73]. For $0 \leq t_1 \leq w_1 \leq n_1$ and $0 \leq t_2 \leq w_2 \leq n_2$, a *split design* with parameters (t_1, t_2)-(n_1, w_1, n_2, w_2) is a collection $\mathcal{M}$ of ordered pairs (B_1, B_2) where $B_i \subseteq \binom{[n_i]}{w_i}$ $(i = 1, 2)$ having the property that for any ordered pair (S_1, S_2) where $S_i \subseteq \binom{[n_i]}{t_i}$ $(i = 1, 2)$, the number of pairs (B_1, B_2) satisfying $S_i \subseteq B_i$ $(i = 1, 2)$ does not depend on the particular choice of S_1 and S_2. Split designs can be seen as a special case of *mixed block designs* introduced in [74], which are designs in direct (also known as tensor) products of Johnson schemes. Delsarte's theory of designs in association schemes was extended to this case in [75]; in particular, with a little extra work, it follows from [75] that a split design satisfies

$$|\mathcal{M}| \geq \binom{n_1}{\lfloor t_1/2 \rfloor}\binom{n_2}{\lfloor t_2/2 \rfloor}, \tag{3.20}$$

which can be seen as a generalization of the Ray-Chaudhuri and Wilson bound (see, e.g., [25, IV.1.23]) for ordinary block designs.

Theorem 3.18 ([73]). *Let C be a completely regular design in $J(n, w)$ with minimum distance $\delta \geq 2h$. Choose $B \in C$ and define*

$$\mathcal{M} = \{(B \setminus B', B' \setminus B) \mid B' \in C\colon |B \cap B'| = w - \delta\}$$

Then $\mathcal{M}$ is a split design with parameters (h, h)-$(w, \delta, n - w, \delta)$. Consequently,

$$|C| \geq |\mathcal{M}| \geq \binom{w}{\lfloor h/2 \rfloor}\binom{n-w}{\lfloor h/2 \rfloor}.$$

Moreover, if C itself is an h-design with λ_h blocks on every h-set, then $\lambda_h \geq \binom{n-w}{\lfloor h/2 \rfloor}$.

Corollary 3.6 ([73]). *A completely regular 2-design in $J(n, w)$ having minimum distance $\delta \geq 4$ (i.e., no two blocks intersect in more than $w - 4$ points), satisfies $\lambda \geq n - w$.*

Now, in order to derive Theorem 3.17, observe that if a completely regular symmetric design has minimum distance at least $\delta \geq 4$, then, by Corollary 3.6 from Theorem 3.18, one has $\lambda = \lambda_2 \geq n - w$. This is impossible, since $n \geq 2w$ and $w = \lambda_1 > \lambda_2$. Thus, in order for a symmetric design to be completely regular, it must have minimum distance $\delta = w - \lambda \leq 3$. This implies $w \leq 5$ and leaves only the following parameter sets: 2-$(7, 3, 1)$ (the projective plane PG$(2, 2)$ of order 2), 2-$(13, 4, 1)$ (the projective plane PG$(2, 3)$ of order 3), and 2-$(11, 5, 2)$. The Fano plane PG$(2, 2)$ is completely regular in $J(7, 3)$ by Theorem 3.14. The design of PG$(2, 3)$ is completely regular in $J(13, 4)$ by Theorem 3.15. The last possibility, the so-called 2-$(11, 5, 2)$ biplane[5] is not completely regular in $J(11, 5)$, since its covering radius $\rho = 2$ is less than its dual degree $s^* = 3$ (see Theorem 3.2). The latter equality follows from the completely regularity of this biplane in the Odd graph O_6, where it has covering radius 3 (and hence $s^* = 3$) and the fact that $J(11, 5)$ and O_6 have the same Bose-Mesner algebra (but different orderings of the bases).

One may still construct completely regular designs from projective planes taking their so-called "4-shadow". Namely, given a Steiner t-design $S(n, w, t)$, say C, define the *h-shadow* of C by choosing as blocks all h-element subsets of points which are contained in blocks of C; this new design already lives in $J(n, h)$. In fact, this is a special case of a more general previously known idea (see, e.g., [95,96]) allowing one to construct more designs from a given design. Martin [71, Proposition 4.2.5] showed that the 4-shadow of a projective plane of order $q \geq 3$ is completely regular in $J(q^2 + q + 1, 4)$, and Godsil [73, Proposition 3.9] generalized this as follows.

Theorem 3.19. *The $(t+2)$-shadow of a Steiner system $S(n, w, t)$, $w \geq t+2$, is a completely regular design in $J(n, t+2)$ with covering radius $\rho = 2$.*

Further, combining Theorem 3.18 with the Ray-Chaudhuri/Wilson bound, Martin [73] obtained the following corollary for quasi-symmetric 2-designs.

Corollary 3.7 ([73]). *A completely regular quasi-symmetric design has minimum distance at most 7.*

Consider some classes of quasi-symmetric designs. Corollary 3.6 implies $w \leq 4$ for Steiner 2-designs, and we already know that Steiner triple and quadruple systems are completely regular in $J(n, 3)$ and $J(n, 4)$ by Theorems 3.14 and 3.15, respectively. As for residual designs of biplanes, such a completely regular design must have $w \leq 5$. In fact, it is either the residual of the unique 2-$(11, 5, 2)$ design, which is a triple system, or the residuals of some two of the three 2-$(16, 6, 2)$ biplanes (the residual of one of the three biplanes is not completely regular) [73]. Further, the quasi-symmetric Witt designs are those with parameters 3-$(22, 6, 1)$ and 4-$(23, 7, 1)$. The former one is not completely regular as it violates the bounds in Theorem 3.18. The latter one is completely regular by [73, Proposition 3.10].

Concerning Witt designs in general, it was already observed by Delsarte [34] that the 5-$(24, 8, 1)$ design is completely regular, as it has $s^* = 2$ and $\delta = 4$ (see Theorem 3.3). Finally, the 4-$(11, 5, 1)$ and 5-$(12, 6, 1)$ are both completely regular by Theorem 3.14. An interesting subclass of completely regular designs, the so-called *completely transitive* designs were considered by Godsil and Praeger [54], see Chapter 5 for further details. In particular, the Witt designs on 23 and 24 points are completely transitive [54].

We finish this section with one more technical yet strong result by Martin [73] which imposes necessary conditions on the number of blocks at minimum distance from a given block in an arbitrary completely regular design.

Theorem 3.20. *Let C be a completely regular design in $\Gamma = J(n, w)$ with minimum distance δ and put $h = \lfloor \delta/2 \rfloor$.*

i. The number $R_{0\delta}$ of blocks of the design at minimum distance from any block of C satisfies

$$R_{0\delta} \geq \frac{\binom{n-w}{h}\binom{w}{h}}{\binom{\delta}{h}^2}.$$

ii. *If δ is even, then the number R_{hh} of blocks of the design at distance h from any vertex in $\Gamma_h(C)$ satisfies*

$$R_{hh} \leq 1 + R_{0\delta} \frac{\binom{\delta}{h}^2}{\binom{n-w}{h}\binom{w}{h}}.$$

iii. *If δ is odd, then the number $R_{h,h+1}$ of blocks of the design at distance $h+1$ from any vertex in $\Gamma_h(C)$ satisfies*

$$R_{h,h+1} \leq R_{0\delta} \frac{\binom{\delta}{h}^2}{\binom{n-w}{h}\binom{w}{h}}.$$

Next, we consider completely regular designs of small strength.

3.3.2 Completely Regular Codes of Strength 0

We begin with a simple construction of completely regular designs of strength 0 in $J(n, w)$. For a nonempty subset $M \subseteq [n]$ having cardinality m, define the *shadow* of M to be the following set of vertices of $J(n, w)$:

$$Z_M = \left\{ x \in \binom{[n]}{w} \mid x \subseteq M \text{ or } M \subseteq x \right\}.$$

Note that the subgraph of $J(n, w)$ induced by Z_M is isomorphic to either $J(m, w)$ when $m > w$ or $J(n-m, w-m)$ when $m \leq w$. Thus, in either case, it is a connected graph (we will need this simple observation below in the proof of Theorem 3.21).

A vertex $x \in J(n, w)$ is at distance j from Z_M if and only if $|x \cap M| = w - j$ when $m > w$ or $|x \cap M| = m - j$ when $m \leq w$, i.e., $\partial(x, Z_M) = j = \min(m, w) - |x \cap M|$. Thus, an easy way to see that Z_M is a completely regular code is to observe that the stabilizer of M in the automorphisms group of $J(n, w)$ acts transitively on the set of vertices at distance j from M, for all j. In other words, the distance partition of $J(n, w)$ with respect to Z_M is an orbit partition and hence is equitable. Alternatively, one may carefully count the neighbors: x has precisely $|M \setminus x| \cdot |x \setminus M| = (m - |x \cap M|)(w - |x \cap M|)$ neighbors at distance $j - 1$ from Z_M, and $|[n] \setminus (x \cup M)| \cdot |x \cap M| = (n - (m + w - |x \cap M|))|x \cap M|$ neighbors at distance $j + 1$ from Z_M. Since the graph $J(n, w)$ is regular, the remaining neighbors of x are at distance j from Z_M.

Proposition 3.10. *In the above notation, Z_M is a completely regular design of strength 0 in $J(n, w)$ with covering radius $\rho = \min(n - m, w)$ if $m > w$, or $\rho = m$ if $m \leq w$.*

To prove that Z_M has strength 0, it suffices to show that the eigenvalue θ_1 of $J(n, w)$ is in the spectrum of the code Z_M (see Lemma 3.4 and Theorem 3.5). To this end, consider the following vector:

$$\mathbf{u}_M = (n - m) \sum_{i \in M} \mathbf{w}_i - m \sum_{i \notin M} \mathbf{w}_i, \tag{3.21}$$

where $\mathbf{w}_i$ is the characteristic vector of the shadow of $\{i\}$, $i \in [n]$. In other words, $\mathbf{w}_1, \mathbf{w}_2, \ldots, \mathbf{w}_n$ are the columns of the inclusion matrix of w-subsets versus 1-subsets of $[n]$. One can verify (see (3.9)) that $\mathbf{u}_M$ is an eigenvector of $J(n, w)$ with eigenvalue θ_1. For a vertex x at distance $j = \min(m, w) - |x \cap M|$ from Z_M, the coordinate $\mathbf{u}_M(x)$ equals

$$\mathbf{u}_M(x) = (n - m)|x \cap M| - m(m - |x \cap M|), \tag{3.22}$$

whence $\mathbf{u}_M$ is constant on the cells of distance partition with respect to this shadow. Lemma 3.3 then implies that θ_1 is in the spectrum of the code Z_M.

Meyerowitz [81] proves that no other completely regular designs can have strength 0.

Theorem 3.21. *Every completely regular design of strength* 0 *in* $J(n, w)$ *is the shadow of some subset of* $[n]$.

Let us outline the proof of this theorem. Let C be a completely regular design of strength 0 in $J(n, w)$, χ denote the characteristic vector of C, and Π_C the distance partition of $J(n, w)$ with respect to C and the corresponding characteristic matrix. Then $\mathbf{u} = E_1\chi$ is an eigenvector of $J(n, w)$ with eigenvalue θ_1 (note that $\mathbf{u} \neq \mathbf{0}$ by Lemma 3.4, as C has strength 0). Recall (see (3.6)) that the orthogonal projection E_1 is a polynomial in the adjacency matrix A, say $q(A)$. It follows from (3.14) that

$$E_1 \cdot \Pi_C = \Pi_C \cdot q(A/\Pi_C),$$

whence one can see that the vector $\mathbf{u}$ belongs to the column space of Π_C, i.e., the following lemma holds.

Lemma 3.7. *In the above notation, the vector* $\mathbf{u}$ *is constant on the cells of* Π_C.

Next, Meyerowitz [81] observes (see also (3.9)) that every vector $\mathbf{w}$ in the column space of E_1 can be written in the form

$$\mathbf{w} = \sum_{i=1}^{n} \alpha_i \mathbf{w}_i \quad \text{such that} \quad \sum_{i=1}^{n} \alpha_i = 0, \tag{3.23}$$

with $\mathbf{w}_i$ defined as above.

Using the representation of the vector $\mathbf{u}$ in the form of (3.23), one can then see that the coefficients α_i take at most two distinct values. Indeed, without loss of generality, consider $\alpha_1, \alpha_2, \alpha_3$ and choose a $(w-1)$-subset y of $[n] \setminus \{1, 2, 3\}$. Then the three vertices $x_i = y \cup \{i\}$, $i = 1, 2, 3$, are pairwise adjacent in $J(n, w)$; therefore, at least two of them are contained in the same cell of the distance partition Π_C. By Lemma 3.7, the coordinates $\mathbf{u}(x_i)$ can take at most two distinct values. Now,

$$\mathbf{u}(x_i) = \alpha_i + \sum_{j \in y} \alpha_j,$$

and hence the coefficients α_1, α_2 and α_3 take at most two distinct values.

Further, since $\sum_{i=1}^{n} \alpha_i = 0$ and $\mathbf{u} \neq \mathbf{0}$, the coefficients α_i take precisely two distinct values. Therefore, up to scalar multiple, $\mathbf{u}$ can be written in the form $\mathbf{u} = \mathbf{u}_M$ (see (3.21)) for some subset $M \subseteq [n]$. Thus, the vector $\mathbf{u}$ is constant of the cells of the distance partition Π_M of $J(n,w)$ with respect to the shadow of Z_M of M. Moreover, it follows from (3.22) that it takes distinct values on distinct cells of Π_M. Hence the partition Π_C is a refinement of Π_M. However, since the subgraph of $J(n,w)$ induced by Z_M is connected, it is not hard to see that either C or C^{opp} must, in fact, coincide with the first cell of Π_M, which is Z_M.

3.3.3 Completely Regular Codes of Strength 1

We first describe all known to us constructions of completely regular codes of strength 1 in the Johnson graphs. As we will see in this section, all such codes are classified in the following cases: $\delta \geq 2$, or $\rho = 1$, or $\rho = w - 1$, and the rest of this section is devoted to these results obtained by many authors in a series of papers.

Let $\{Y_1, \ldots, Y_q\}$ be a partition of $[n]$ into q parts, of size p each, i.e., $n = pq$.

- When $w \leq p$, let $\mathsf{M}(n,p,q,w)$ denote the following code:

$$\mathsf{M}(n,p,q,w) := \binom{Y_1}{w} \cup \binom{Y_2}{w} \cup \ldots \cup \binom{Y_q}{w},$$

  i.e., it is the union of the shadows of Y_i, for $i = 1, \ldots, q$.

- When $w \leq q$, let $\mathsf{GP}(n,p,q,w)$ denote the following code:

$$\mathsf{GP}(n,p,q,w) := \left\{ b \in \binom{[n]}{w} \mid |b \cap Y_i| \leq 1, \ \forall i \in [q] \right\},$$

  and when $p = 2$ and $w \geq p$, let $\mathsf{GP}_2(n,w) := \mathsf{GP}(n,2,n/2,w)$.

  Note that $(\mathsf{GP}(n,p,q,w))^{\text{opp}} = \mathsf{M}(n,p,q,w)$ if $p, q \geq w$, and if $p = 2$ then a vertex b belongs to $(\mathsf{GP}_2(n,w))^{\text{opp}}$ if and only if it has the following type:

$$b = \begin{cases} \text{a union of some } \frac{w}{2} \text{ sets } Y_i \text{ if } w \text{ is even, or} \\ \text{a union of some } \frac{w-1}{2} \text{ sets } Y_i \text{ and a point} \\ \quad \text{which is disjoint from them if } w \text{ is odd.} \end{cases} \tag{3.24}$$

- In the case when $w = 3$ and n is even, let us take $C_1 := \mathsf{M}(n, n/2, 2, 3)$ with Y_1 (Y_2, respectively) being the set of all odd (even, respectively) numbers from $[n]$, and $C_2 := \mathsf{GP}_2(n,3)$ with Y_i being $\{2i-1, 2i\}$, $i = 1, 2, \ldots, \frac{n}{2}$. Then define:

$$\mathsf{AM}(n) := \binom{[n]}{3} \setminus (C_1 \cup (C_2)^{\text{opp}}).$$

For the following four constructions, we assume $n = 2w$.

- Let $K = \{\{1\}, \{1,2\}, \{1,3\}, \{4,5\}\}$, and define

$$\mathsf{V}_1 := \left\{ b \in \binom{[n]}{w} \mid ([5] \cap b) \in K \text{ or } ([5] \setminus b) \in K \right\}.$$

- Let $K' = \{\{\varnothing\}, \{3\}, \{4\}, \{5\}, \{1,3\}, \{2,4\}, \{3,5\}, \{4,5\}\}$, and define

$$\mathsf{V}_2 := \left\{ b \in \binom{[n]}{w} \mid ([5] \cap b) \in K' \text{ or } ([5] \setminus b) \in K' \right\}.$$

- Let

$$\mathsf{V}_3 := \left\{ b \in \binom{[n]}{w} \mid |b \cap \{1,2\}| \in \{0,2\} \right\}.$$

- Let

$$\mathsf{V}_4 := \left\{ b \in \binom{[n]}{w} \mid |b \cap \{1,2,3\}| \in \{0,3\} \right\}.$$

The data in Table 3.1 exhaust all possibilities for p, q and w for which these codes in the Johnson graph $J(n, w)$ are completely regular ones of strength 1.

3.3.3.1 Minimum Distance at Least 2

Martin [72] expanded on the idea of Meyerowitz (that of describing spanning sets of eigenspaces via $(0,1)$-characteristic functions of subsets) in order to classify completely regular designs having strength one and minimum distance at least two. To formulate his result, let us define the following class of designs. Suppose that $n = qp$, $w = up$, where $q \geq 2u$, $p \geq 2$, and consider a partition of $[n]$ into q groups $Y_1, Y_2, \dots, Y_q$, each of size p. Define C to be the set of all vertices of $J(n, w)$ of the form $\cup_{i \in I} Y_i$, where I runs over all u-element subsets of $\{1, 2, \dots, q\}$:

$$C = \left\{ \cup_{i \in I} X_i \mid \forall I \in \binom{[q]}{u} \right\},$$

so that $|C| = \binom{q}{u}$. Then C is a design of strength 1 and minimum distance $p \geq 2$, called a *groupwise complete design* in [72]. Note that the covering radius of a groupwise complete design is at least 2 (as $q \geq 2u$) and the minimum distance equals p. Martin showed the following characterization.

Theorem 3.22 ([72, Theorem 3.1]). *In the above notation, a completely regular design in $J(n, w)$ having strength 1 and minimum distance at least 2 is a groupwise complete design satisfying one of the three cases below:*

i. $\mathrm{p} = w$ *and* $n = 2w$, *i.e., the design is a pair of antipodal vertices;*

ii. $\mathrm{p} = 2$;

iii. $\mathrm{p} = 3$ *and* $u = 1$.

TABLE 3.1 Some Completely Regular Codes of Strength 1 in the Johnson Graph $J(n, w)$

Type	p	q	w	ρ	Comment	Theorem
M	$\geq w$	2	any	$\lfloor \frac{w}{2} \rfloor$	[73, Theorem 3.2(i)] for $p = w$; [73, Proposition 3.4(i)]; [54, Example 2]; $GP.3, GP.3', GP.3''$ in [84]	3.22
M	≥ 2	≥ 3	2	1	$= (\mathsf{GP}(n, p, q, 2))^{\mathrm{opp}}$; [73, Proposition 3.4(ii)]; the opposite to [54, Example 5] or to [54, Example 3] if $p = 2$; $GP.4, GP.5$ in [84]	
M	≥ 3	≥ 3	3	2	$= (\mathsf{GP}(n, p, q, 3))^{\mathrm{opp}}$; [73, Theorem 3.2(ii)] for $p = 3$; [73, Proposition 3.4(ii)]; the opposite to [54, Example 4]; $GP.2$ in [84]	
GP_2	2	≥ 3	≥ 3	$\lfloor \frac{w}{2} \rfloor$	[54, Example 3]; the opposite to [73, Theorem 3.2(iii)] if w is even; the opposite to $GP.1, GP.1'$ (w odd) or to $M.1$ (w even) in [84]	
AM			3	1	$M.2$ in [84], Construction 3 in [4]	
V_i			≥ 3	1	[101, Section 5] $\mathsf{V}_3, \mathsf{V}_4$ are Construction 4, 5 in [4], respectively	

Conversely, a groupwise complete design is completely regular only if it satisfies one of the above cases.

Note that Theorem 3.22(i) corresponds to $\mathsf{M}(2w, w, 2, w)$, (ii) corresponds to $(\mathsf{GP}_2(2q, 2u))^{\mathrm{opp}}$, and (iii) corresponds to $\mathsf{M}(3q, 3, q, 3) = (\mathsf{GP}(3q, 3, q, 3))^{\mathrm{opp}}$.

Martin [72] also mentions the following completely regular codes in $J(n, w)$ (not satisfying the condition $\delta \geq 2$ in general): $(\mathsf{GP}_2(2q, 2u + 1))^{\mathrm{opp}}$ (see (3.24)) $\mathsf{M}(2p, p, 2, w)$ (with $p > w$), and $\mathsf{M}(pq, p, q, 3)$ (with $p, q > 3$), which he attributes to Godsil and Praeger [54].

3.3.3.2 Covering Radius 1

An ad hoc analysis of the classification problem of completely regular codes of covering radius 1 in the Johnson graphs $J(n, w)$ for some small values of n was conducted by Mogilnykh [83] and Avgustinovich and Mogilnykh [2,4]. In their study [4], they discovered three families of completely regular codes of radius 1 in $J(2m, 3)$.

Namely, they described an orbit partition of $J(2m,3)$, for all $m \geq 3$, into three parts X_1, X_2, X_3 with the following quotient matrix:

$$\begin{bmatrix} 3m-9 & 6 & 3m-6 \\ m-2 & 2m-1 & 3m-6 \\ m-2 & 6 & 5m-13 \end{bmatrix}. \tag{3.25}$$

We refer to [4,42] for a detailed presentation of this partition. Note that this quotient matrix has equal non-diagonal entries for each column. Thus, gluing any two parts of this equitable partition produces an equitable partition into two parts, i.e., a completely regular code with covering radius 1. Their quotient matrices also follow easily from this argument.

Proposition 3.11. *Let (X_1, X_2, X_3) be an equitable partition of $J(2m,3)$ with quotient matrix defined in (3.25). Then the partitions $\Pi_1 = (X_2 \cup X_3, X_1)$, $\Pi_2 = (X_1 \cup X_3, X_2)$ and $\Pi_3 = (X_3, X_1 \cup X_2)$ are equitable with the corresponding quotient matrices*

$$\begin{bmatrix} 5m-7 & m-2 \\ 3m & 3m-9 \end{bmatrix}, \quad \begin{bmatrix} 6m-15 & 6 \\ 4m-8 & 2m-1 \end{bmatrix}, \quad \begin{bmatrix} 5m-13 & m+4 \\ 3m-6 & 3m-3 \end{bmatrix},$$

respectively.

Note that the quotient matrices of Π_1, Π_2, and Π_3 all have eigenvalue $2m-7 = \theta_2(2m,3)$ (see (3.8)); thus, X_1, X_2, and X_3 are completely regular codes of strength one and radius one in $J(2m,3)$.

Note also that the code $(\mathsf{M}(2m,m,2,3))^{\text{opp}}$ has the quotient matrix of Π_1, the code $\mathsf{GP}_2(2m,3)$ that of Π_2, and $\mathsf{AM}(n)$ that of Π_3. Surprisingly, these codes exhaust all, up to equivalence, completely regular codes of strength 1 and covering radius 1 in $J(n,w)$ with $n > 2w$, $n \geq 11$, and all $w \geq 3$. This was shown in a series of papers as follows. First, Gavrilyuk and Goryainov [46] proved that the Johnson graphs $J(n,3)$ have no completely regular codes of strength 1 and covering radius 1 when n is odd. In their study, they used a combinatorial approach which enables one to analyze the local structure of an equitable partition into two parts. This will be sketched below in this section.

Later on, the same authors together with Evans and Vorobiev [42] elaborated on this approach to complete the classification in the case when n is even and $w = 3$. Furthermore, another proof was given in the same paper, which is based on the so-called partial differences, defined in Section 3.1.2.2. The partial differences approach was earlier developed by Vorobiev in a separate paper [101], where he first showed that $J(n,w)$ with $n > 2w$ and $w \geq 4$ has no completely regular codes of strength 1 and covering radius 1, and then characterized all of those in $J(2w,w)$ with $w \geq 7$. In doing so, he also discovered two new constructions $\mathsf{V}_1, \mathsf{V}_2$ (in addition to another two $\mathsf{V}_3, \mathsf{V}_4$, previously known by [4]), which we already described in the above. For now, note that the codes $\mathsf{V}_1, \mathsf{V}_2, \mathsf{V}_3, \mathsf{V}_4$ have the following respective quotient matrices:

$$Q_1 = \begin{bmatrix} w^2-3w+2 & 3w-2 \\ w & w^2-w \end{bmatrix}, \quad Q_2 = \begin{bmatrix} w^2-2w & 2w \\ 2w-2 & w^2-2w+2 \end{bmatrix}$$

$$Q_3 = \begin{bmatrix} w^2-2w & 2w \\ 2w-2 & w^2-2w+2 \end{bmatrix}, \quad Q_4 = \begin{bmatrix} w^2-3w & 3w \\ w-2 & w^2-w+2 \end{bmatrix}.$$

Theorem 3.23. *Suppose $n \geq 11$ and the Johnson graph $J(n,3)$ contains a completely regular code C of strength 1 and covering radius 1. Then n is even and C or C^{opp} is equivalent to* $\mathsf{M}(2m,m,2,3)$, $\mathsf{GP}_2(2m,3)$, *or* $\mathsf{AM}(n)$.

Theorem 3.24. *Suppose $n > 2w$, $n \geq 11$ and $w \geq 4$. Then the Johnson graph $J(n,w)$ has no completely regular codes of strength 1 and covering radius 1.*

Theorem 3.25. *Suppose $n = 2w$ and $w \geq 4$. Then the quotient matrix of a completely regular code C in the Johnson graph $J(n,w)$ of strength 1 and covering radius 1 or that of C^{opp} equals Q_1, Q_2, Q_3, or Q_4. Moreover, if $w \geq 7$, then C or C^{opp} is equivalent to* $\mathsf{V}_1, \mathsf{V}_2, \mathsf{V}_3$, *or* V_4.

To the rest of this section, let C be a completely regular code of strength one and radius one in $J(n,3)$, and (P_1,P_2) denote the equitable partition of the graph into the code $C = P_1$ and its opposite P_2, with quotient matrix

$$\begin{bmatrix} p_{11} & p_{12} \\ p_{21} & p_{22} \end{bmatrix},$$

where $\theta = p_{11} - p_{12}$ is a non-principal eigenvalue of $J(n,3)$ in the spectrum of C. Our assumption on the strength of C implies that $\theta = \theta_2(n,3) = n-7$ and

$$3(n-3) + \theta = 4(n-4) = p_{11} + p_{22}, \tag{3.26}$$

where, without loss of generality, we may assume that $p_{11} \geq 2n-8$.

Let us first describe a combinatorial approach [42,46] to study equitable partitions of $J(n,3)$. It was first used by Gavrilyuk and Goryainov [46] to prove Theorem 3.23 when n is odd or n is even and the quotient matrix is symmetric. In [42] this approach was extended to analyze the remaining cases of Theorem 3.23.

Let us denote $\Gamma = J(n,3)$ and, for a vertex $u \in \Gamma$, put

$$\overline{u} := \begin{cases} 1 & \text{if } u \in P_1, \\ 0 & \text{if } u \in P_2, \end{cases}$$

i.e., $\overline{u}$ is the value of the characteristic function of P_1 on u, and for a set U of vertices we put

$$\overline{U} := \sum_{u \in U} \overline{u}.$$

Slightly abusing notation, we let $ab\ldots$ denote the set $\{a,b,\ldots\}$, for any distinct elements $a,b,\ldots \in [n]$. By Lemma 3.2, the neighborhood $\Gamma(x)$ of any vertex x in $J(n,3)$ is isomorphic to the $3 \times (n-3)$-grid. In particular, there are three maximal cliques of size $n-3$ in the neighborhood of abc, given by $ab*$, $ac*$ and $bc*$, and $n-3$ maximal cliques of size 3, given by the triples $\{abi, aci, bci\}$, where $i \in [n] \setminus \{a, b, c\}$. The *ab-row* of $\Gamma(abc)$ is the set $ab* \setminus \{abc\}$. For an element $i \in [n] \setminus \{a, b, c\}$, the *i-column* of $\Gamma(abc)$ is the set $\{abi, aci, bci\}$, and i is the *index* of this column. One may refer to the following diagram, which is a $3 \times (n-3)$-array corresponding to the neighborhood $\Gamma(abc)$, and at each vertex $u \in \Gamma(abc)$ the corresponding entry of the array is equal to $\overline{u}$. The rows are given by the comments on the right. If needed, the indices of columns are given below the braces.

$$\Gamma(abc): \begin{array}{ccccccccccccl} \overline{abd} & \overline{abe} & \overline{abf} & \cdots & \cdots & \cdots & \cdots & \cdots & \cdots & \cdots & \cdots & \cdots & \leftarrow ab\text{-row} \\ \overline{acd} & \overline{ace} & \overline{acf} & \cdots & \cdots & \cdots & \cdots & \cdots & \cdots & \cdots & \cdots & \cdots & \leftarrow ac\text{-row} \\ \underbrace{\overline{bcd}}_{d} & \multicolumn{2}{c}{\underbrace{\overline{bce} \quad \overline{bcf}}_{\{e,f\}}} & \cdots & \cdots & \cdots & \cdots & \cdots & \cdots & \cdots & \cdots & \cdots & \leftarrow bc\text{-row} \end{array}$$

We then have the following identities, relating the characteristic function of P_1 on a vertex abc of Γ and the number of vertices in each row of $\Gamma(abc)$ that are in P_1. The following can be shown by using a simple counting argument (for the full details, see [46, Equation (2)]):

$$\overline{abc} \cdot (\theta + 3) + p_{21} = \overline{ab*} + \overline{ac*} + \overline{bc*}. \tag{3.27}$$

By (3.27), the following equalities hold:

$$\begin{aligned} \overline{abc} \cdot (\theta + 3) + p_{21} &= \overline{ab*} + \overline{ac*} + \overline{bc*}, \\ \overline{abd} \cdot (\theta + 3) + p_{21} &= \overline{ab*} + \overline{ad*} + \overline{bd*}, \\ -\overline{acd} \cdot (\theta + 3) - p_{21} &= -\overline{ac*} - \overline{ad*} - \overline{cd*}, \\ -\overline{bcd} \cdot (\theta + 3) - p_{21} &= -\overline{bc*} - \overline{bd*} - \overline{cd*}. \end{aligned}$$

Summing up these four equalities and dividing by 2, we obtain the next result (cf. [46, Lemma 3.5]).

Lemma 3.8. *For any four distinct elements $a, b, c, d \in [n]$, the following condition holds:*

$$\overline{ab*} - \overline{cd*} = \frac{\theta + 3}{2}(\overline{abc} + \overline{abd} - \overline{acd} - \overline{bcd}).$$

Let us now illustrate how the above lemma can be used in the proof of Theorem 3.23: we will show that n must be even.

Proposition 3.12. *Suppose the Johnson graph $J(n, 3)$, $n \geq 7$, contains a completely regular code C of strength 1 and covering radius 1. Then n is even.*

Proof: On the contrary, suppose n is odd. Without loss of generality, fix a vertex abc such that $abc \in P_1$, i.e., $\overline{abc} = 1$. Since $\frac{\theta+3}{2} = \frac{n-4}{2}$, it follows from Lemma 3.8 that $(\overline{abc} + \overline{abd} - \overline{acd} - \overline{bcd})$ must be even, i.e., $\overline{abd} + \overline{acd} + \overline{bcd}$ is odd. Denote by ℓ_i, $i = 0,1,2,3$, the number of columns $\{abd, acd, bcd\}$ of $\Gamma(abc)$ such that $\overline{abd} + \overline{acd} + \overline{bcd} = i$. Then $\ell_0 = \ell_2 = 0$ and

$$\ell_1 + \ell_3 = n - 3,$$
$$\ell_1 + 3\ell_3 = p_{11},$$

whence $\ell_3 = \frac{p_{11}-n+3}{2}$, $\ell_1 = \frac{3(n-3)-p_{11}}{2}$. If $\ell_1 = 0$, then $p_{11} = 3(n-3)$ and $P_2 = \varnothing$, a contradiction.

Next, suppose that $\Gamma(abc)$ does not have an all-ones row. Then one can choose a vertex $x \in P_2 \cap \Gamma(abc)$ such that the row of $\Gamma(x)$ containing abc has at least $\ell_3 + \ell_1/3 + 1$ vertices from P_1. By the above argument using Lemma 3.8, every column of $\Gamma(x)$ has 0 or 2 vertices from P_1. Hence we see that $\Gamma(x)$ contains at least $2(\ell_3 + \ell_1/3 + 1)$ vertices from P_1 with equality if and only if $\Gamma(x)$ contains an all-zeros row. This yields $2(\ell_3 + \ell_1/3 + 1) \leq p_{21}$. With the above equalities for ℓ_1, ℓ_3 and $p_{11} - p_{21} = \theta$, the latter inequality is equivalent to $p_{11} \geq 3n - 15$.

In the same manner, considering a vertex $y \in P_2$ and assuming that $\Gamma(y)$ does not contain an all-zeros row, one can show that $p_{22} \geq 3n - 15$. Then by (3.26), we get $4n - 16 = p_{11} + p_{22} \geq 6n - 30$, whence $n \leq 7$. The classification of completely regular codes of radius one in $J(7,3)$ was done in [2], from which we conclude that no such codes of strength one exist.

It remains to consider the case when some vertex neighborhoods contain all-zeros or all-ones rows. Suppose first that, for every vertex $y \in P_2$, $\Gamma(y)$ contains an all-zeros row, while $\Gamma(abc)$ does not contain an all-ones row. Then it follows from the above that $p_{11} = 3n - 15$ and thus the quotient matrix is

$$\begin{bmatrix} 3n-15 & 6 \\ 2n-8 & n-1 \end{bmatrix},$$

and then $|P_1| = \frac{p_{21}}{p_{12}+p_{21}}\binom{n}{3} = \frac{n(n-2)(n-4)}{6}$, which is not integer (recall that n is odd by our assumption), a contradiction.

Finally, suppose that, for every vertex from P_2 (from P_1, respectively), its neighborhood contains an all-zeros (all-ones, respectively) row. Counting vertices from P_2 and P_1 in a maximal $(n-3)$-clique corresponding to any other row, we then obtain $p_{21}/2 = n-3-p_{12}/2+1$. Hence $p_{21}+p_{12} = 2n-4$. Since $p_{11}+p_{12}+p_{21}+p_{22} = 6(n-3)$, we get $p_{11}+p_{22} = 4n-14$, a contradiction (see (3.26)), which completes the proof that n must be even. ■

The remainder of the proof of Theorem 3.23 in [42,46] is quite similar to the above arguments, but involves a more detailed case analysis. Again, by Lemma 3.8, the following equalities hold:

$$\overline{ab*} - \overline{cd*} = \frac{\theta+3}{2}(\overline{abc} + \overline{abd} - \overline{acd} - \overline{bcd}),$$
$$\overline{cd*} - \overline{be*} = \frac{\theta+3}{2}(\overline{bcd} + \overline{cde} - \overline{bce} - \overline{bde}),$$
$$\overline{be*} - \overline{ac*} = \frac{\theta+3}{2}(\overline{abe} + \overline{bce} - \overline{abc} - \overline{ace}).$$

Summing up these three equalities, we get the next one:

$$\overline{ab*} - \overline{ac*} = \frac{\theta+3}{2}(\overline{abd} + \overline{abe} + \overline{cde} - \overline{acd} - \overline{ace} - \overline{bde}),$$

whence, using the fact that $0 \leq \overline{ij*} \leq n-2$, we see that the difference $\overline{ab*} - \overline{ac*}$ is a multiple of $\frac{\theta+3}{2} = \frac{n-4}{2}$, say $h(n-4)/2$, where $h \in \{-2,-1,0,1,2\}$. Without loss of generality, we may assume that $\overline{ab*} \geq \overline{ac*} \geq \overline{bc*}$. We then have the following possibilities for the tuple $(\overline{ab*} - \overline{ac*}, \overline{ac*} - \overline{bc*})$:

(I) $(n-4, 0)$ (II) $(\frac{n-4}{2}, \frac{n-4}{2})$ (III) $(0, n-4)$

(IV) $(\frac{n-4}{2}, 0)$ (V) $(0, \frac{n-4}{2})$ (VI) $(0,0)$.

Some of these cases can be easily discharged for large enough values of n and some lead to the known completely regular codes in $J(n,3)$, described in Theorem 3.23. For example, if Case (I) occurs then $p_{11} = \overline{\Gamma(abc)}$ is at most $(n-3)+1+1 = n-1$. Therefore $n-1 \geq 2n-8$ by assumption, and so $n \leq 7$.

We now briefly discuss the second approach, based on partial differences of eigenfunctions, which was used in [42,101] to prove Theorems 3.24 and 3.25. Specifically, we are interested in the following function naturally associated with C:

$$\mathbf{v} = \frac{p_{12}}{p_{12}+p_{21}}\mathbf{1} - \chi_C, \quad \text{i.e.,} \quad \mathbf{v}(x) = \begin{cases} \frac{-p_{21}}{p_{12}+p_{21}}, x \in C \\ \frac{p_{12}}{p_{12}+p_{21}}, x \notin C. \end{cases} \tag{3.28}$$

One can easily verify (see, e.g., Lemma 3.3) that $\mathbf{v}$ is a θ-eigenfunction of $J(n,w)$.

Observe that, on the one hand, this eigenfunction takes two distinct values, hence every its partial difference (recall (3.11)) is a function with values from $\{-1,0,1\}$. On the other hand, by Lemma 3.1, such a partial difference (if not identically zero) is a $\theta_1(n-2,w-1)$-eigenfunction. Thus, a classification of eigenfunctions of the Johnson graphs with values in $\{-1,0,1\}$ plays a key role. For the general case $w \geq 2$, the corresponding classification can be found in [101, Theorem 1]; for the particular case $w=3$, we have the following result on such eigenfunctions of the triangular graphs.

Lemma 3.9. *Let* $\mathbf{u}$ *be a* $\theta_1(\ell,2)$*-eigenfunction of* $J(\ell,2)$ *with values in the set* $\{-1,0,1\}$, $\mathbf{u} \not\equiv 0$, $\ell \geq 5$. *Then* $\mathbf{u}$ *is equal to* $\mathbf{u}_1$ *or* $\mathbf{u}_2$, *defined as follows:*

1. *for some* $a, b \in [\ell]$, *let* $\mathbf{u}_1(x) = \begin{cases} 1, & a \in x,\ b \notin x, \\ -1, & a \notin x,\ b \in x, \\ 0, & \textit{otherwise}; \end{cases}$

2. *for* $M_1 \cup M_2 = [\ell]$, $|M_1| = |M_2| = \frac{\ell}{2}$, *let* $\mathbf{u}_2(x) = \begin{cases} 1, & x \subset M_1, \\ -1, & x \subset M_2, \\ 0, & \textit{otherwise}. \end{cases}$

The proof of this lemma (see [42]) uses the inducing map between eigenspaces of different Johnson graphs (see Section 3.1.2.2), which, in this particular case, gives us the following representation of $\mathbf{u}$:

$$\mathbf{u}(x) = \mathbf{z}(i) + \mathbf{z}(j), \quad (\text{for } x = \{i, j\})$$

where $\mathbf{z}$ is a $\theta_1(\ell, 1)$-eigenfunction of the Johnson graph $J(\ell, 1)$, which is a complete graph on ℓ vertices. (For simplicity, we can then consider $\mathbf{z}$ as a function on $[\ell]$.) As $\mathbf{z}$ is a $\theta_1(\ell, 1)$-eigenfunction, it must be orthogonal to a constant function, i.e.:

$$\sum_{i \in [\ell]} \mathbf{z}(i) = 0,$$

and, by the assumption of Lemma 3.9, the set $\{\mathbf{z}(i) + \mathbf{z}(j) | \, i, j \in [\ell], \, i \neq j\}$ does not contain any elements other than -1, 0 and 1. Then some simple case analysis shows the result.

Consider now some not identically zero partial difference $\mathbf{v}_{ij}$ of $\mathbf{v}$. Clearly, it takes a non-zero value on a vertex $y \in \binom{[n]}{2}$ whenever (necessarily adjacent) vertices $y \cup \{i\}$ and $y \cup \{j\}$ of $J(n, 3)$ belong to different parts of the equitable partition of $J(n, 3)$ into the code C and its opposite. Suppose, e.g., that $\mathbf{v}_{12}$ satisfies Lemma 3.9(2) with $M_1 = \{3, 4, 5, \ldots, \frac{n}{2} + 1\}$, $M_2 = \{\frac{n}{2} + 2, \frac{n}{2} + 3, \ldots, n\}$. Then, by the definition of $\mathbf{v}_{12}$, one can see that

$$1ab, 2cd \in P_2 \quad \text{and} \quad 2ab, 1cd \in P_1$$

for all distinct $a, b \in M_1$ and $c, d \in M_2$. This way, the structure of non-zero partial differences sheds light on the structure of the code C, and the proofs of Theorems 3.24 and 3.25 are based on a detailed case analysis of such eigenfunctions in conjunction with the next observation (Lemma 3.10). On the one hand, one knows how many non-zero values $\mathbf{v}_{ij}$ may take (depending on which part of Lemma 3.9 the function $\mathbf{v}_{ij}$ satisfies); on the other hand, one may count for how many partial differences $\mathbf{v}_{ij}$ given edge $\{y \cup \{i\}, y \cup \{j\}\}$ of $J(n, 3)$ determines the value of $\mathbf{v}_{ij}(y)$. Then double counting shows the following lemma.

Lemma 3.10. *With the above notation, suppose that the set*

$$\{\mathbf{v}_{ij} | \, 1 \leq i < j \leq n\}$$

contains d_1 and d_2 partial differences satisfying Lemma 3.9 (1) *and* (2)*, respectively, and $\binom{n}{2} - d_1 - d_2$ all-zeros partial differences. Then one has:*

$$p_{12}p_{21}n(n-2) = 24d_1(n-4) + 3d_2(n-2)(n-4).$$

3.3.3.3 Covering Radius $w - 1$

In this section, we discuss a recent nice result by Vorobiev and Mogilnykh [84], which completes the classification of completely regular codes of strength one in $J(n, w)$ in another extreme case, namely, when covering radius equals $\rho = s^* = w - 1$, which is maximum possible by (3.19).

Theorem 3.26. *The following holds.*

i. For all n, a completely regular code in $J(n,3)$ with covering radius 2 and strength 1 is equivalent to $\mathsf{M}(n,p,q,3)$, $p,q \geq 3$.

ii. For all $w \geq 4$, there are no completely regular codes in $J(n,w)$ with covering radius $w-1$ and strength 1.

A start-off for the proof in [84] is the following idea. First, observe that a Delsarte clique, say K, in $J(n,w)$ (see Section 3.1.2.2) is a completely regular code of strength 0 and covering radius $w-1$ by Proposition 3.10, as it is the shadow of a $(w-1)$-subset of $[n]$. By Theorem 3.4, the spectrum of K contains all but one eigenvalue of $J(n,w)$, namely, $\theta_w(n,w)$. In particular, the characteristic vector χ_K is orthogonal to the $\theta_w(n,w)$-eigenspace of $J(n,w)$.

Next, again by Theorem 3.4, the spectrum of a completely regular code C of strength 1 and covering radius $w-1$ contains w of the $w+1$ eigenvalues of $J(n,w)$, while $E_1\chi_C = 0$ by the assumption $t=1$ and Theorem 3.5. In particular, the eigenvalue $\theta_w(n,w)$ is in the spectrum of C and hence $E_w\chi_C \neq 0$ by Lemma 3.4.

Using the orthogonality of the vectors χ_K and $E_w\chi_C$, one may then control possible intersections of Delsarte cliques with the parts of the distance partition of the graph with respect to C. In fact, as shown in [84], this rather general observation works best for the class of distance-regular graphs, called *geometric* [53]. For our purpose here, we only note that the Johnson graph $\Gamma = J(n,w)$ is geometric, which means that there exists a collection $\mathcal{K}$ of Delsarte cliques such that every edge is contained in a unique clique from $\mathcal{K}$ (this is obvious by Lemma 3.2 if $n > 2w$ and not hard to see if $n = 2w$). The *clique graph* Δ of Γ is a graph defined on the set $\mathcal{K}$ with two cliques from $\mathcal{K}$ being adjacent if their intersection is non-empty.[6] It is a simple exercise to verify that the clique graph of $J(n,w)$ is isomorphic to the Johnson graph $J(n,w-1)$. Given a completely regular code C in $J(n,w)$, consider the following collection of cliques:

$$C' = \{K \in \mathcal{K} \mid K \cap C \neq \varnothing\},$$

viewed as a code in Δ. Identifying the clique graph Δ with $J(n,w-1)$, one can equivalently view C' as:

$$C' = \left\{y \in \binom{[n]}{w-1} \mid y \subset x \text{ for some } x \in C\right\}.$$

Furthermore, building on an earlier rather general result by Avgustinovich and Mogilnykh [3] on the so-called induced eigenvectors in distance-2 biregular graphs (see also the discussion after Theorem 3.1), Vorobiev and Mogilnykh [84, Corollary 1, Theorem 4] show that a linear transformation $I\colon \mathbb{R}^{\Gamma} \to \mathbb{R}^{\Delta}$ defined by:

$$I(\mathbf{u}) = I_{w-1,w} \cdot \mathbf{u},$$

sends χ_C to a non-zero scalar multiple of $\chi_{C'}$ (here $I_{w-1,w}$ is the inclusion matrix of the Delsarte cliques from $\mathcal{K}$ versus the vertices of Γ). Using the decomposition of χ_C over the eigenspaces of Γ, they determine the decomposition of $\chi_{C'}$ over the eigenspaces of Δ and then prove the following ([84, Corollary 2]).

Theorem 3.27. *Suppose that C is a completely regular code in the Johnson graph $J(n, w)$ with covering radius ρ, such that $\theta_w(n, w)$ is in the spectrum of C. Then, with the above notation, the code C' is completely regular in $J(n, w-1)$ with covering radius $\rho - 1$. The spectrum of C' consists of $\theta_i(n, w-1)$ for all $i \neq w$ such that $\theta_i(n, w)$ is in the spectrum of C. In particular, the strengths of C and C' coincide.*

Furthermore, as every vertex x of $\Gamma = J(n, w)$ is contained in w Delsarte cliques from $\mathcal{K}$, these cliques are pairwise adjacent, i.e., form a w-clique, say L_x, in $\Delta = J(n, w-1)$. Now, $x \in \Gamma_i(C)$ holds if and only if $L_x \subseteq \Delta_{i-1}(C') \cup \Delta_i(C')$; moreover, it is shown in [84, Lemma 2] that the numbers $|L_x \cap \Delta_{i-1}(C')|$ and $|L_x \cap \Delta_i(C')| = w - |L_x \cap \Delta_{i-1}(C')|$ do not depend on the particular choice of $x \in \Gamma_i(C)$. With this property in hand, one can then show the code C' in the triangular graph $J(n, 2)$, induced by a completely regular code C of strength 0 and radius 2 in $J(n, 3)$, can only be the one equivalent to $\mathsf{M}(n, p, q, 2)$, $p \geq 3$. This can be proved by a careful case analysis (see [84, Theorem 5(i)]), which is possible due to a relatively simple structure of the graph $J(n, 2)$. Conversely, starting from such a construction of C', one can show that C necessarily comes from $\mathsf{M}(n, p, q, 3)$, $p, q \geq 3$.

Vorobiev and Mogilnykh then complete the proof of Theorem 3.26, first showing that a completely regular code of strength 0 and radius 2 in $J(n, 3)$, that is equivalent to $\mathsf{M}(n, p, q, 3)$, $p, q \geq 3$, cannot give rise to a completely regular code of strength 0 and radius 3 in $J(n, 4)$, and then arguing by induction on n.

Combining this theorem by Vorobiev and Mogilnykh with Theorem 3.21 by Meyerowitz, one can now characterize all completely regular codes of covering radius $w - 1$ in $J(n, w)$, as such a code has strength 0 or 1 by (3.19).

3.3.4 Completely Regular Codes of Small Weight

In this section, we discuss completely regular codes of strength t and covering radius ρ in the Johnson graph $J(n, w)$ for some small values of w. Recall that, by Theorem 3.2, $\rho = s^*$ holds for a completely regular code, and clearly $1 \leq \rho \leq w$, while $t + s^* \leq w$ by (3.19). The classification of completely regular codes of strength $t = 0$ in $J(n, w)$ for all w is given in Theorem 3.21, so we may assume $t > 0$.

Suppose first that $w = 2$. Then $t = 1$ and thus $\rho = 1$. Since in this case $t = w - 1$, it follows from Theorem 3.14 that every simple 1-$(n, 2, \lambda)$ design is completely regular in $J(n, 2)$. Such designs exist if and only if n or λ are even. Indeed, using the fact that the triangular graph $J(n, 2)$ is the line graph of the complete graph $J(n, 1)$, Mogilnykh [83] observed that completely regular codes of covering radius 1 in $J(n, 2)$ are in a one-to-one correspondence with regular subgraphs of the complete graph. In particular, for every $\lambda \leq n - 2$ such that $n\lambda$ is even, there exists a completely regular code of $J(n,2)$ of covering radius one and strength one, and its quotient matrix is

$$\begin{bmatrix} 2(n-\lambda-2) & 2\lambda \\ 2(n-\lambda-1) & 2(\lambda-1) \end{bmatrix}.$$

Suppose now $w = 3$. Then $t + \rho \leq 3$. The case $t = 1$ and $\rho = 1$ is considered in Section 3.3.3.2, while the case $t = 1$ and $\rho = 2$ is handled in Section 3.3.3.3. If $t = 2 = w - 1$, then it follows from Theorem 3.14 that every simple 2-$(n, 3, \lambda)$ design is completely regular in $J(n, 3)$. It is a classical result by M. Dehon [33] that the following congruences:

$$\lambda(n-1) \equiv 0 \mod 2, \quad \lambda n(n-1) \equiv 0 \mod 6$$

are necessary and sufficient conditions for such a design to exist.

However, in Section 3.3.3.2, we did not discuss completely regular codes of covering radius one in $J(n, 3)$ for small values of n, so we do it now. By the above, it suffices to consider the case of strength one, hence, by Proposition 3.11, we may assume that $n \in \{6, 8, 10\}$. By [2, Section 6], a completely regular code with $t = \rho = 1$ in $J(6, 3)$ is a union of i, $1 \leq i \leq 5$ disjoint antipodal pairs of vertices (i.e., perfect codes in $J(6, 3)$). By [4, Construction 2], besides those listed in Theorem 3.23, the graph $J(8, 3)$ possesses one more completely regular code with $t = \rho = 1$, having quotient matrix

$$\begin{bmatrix} 5 & 10 \\ 4 & 11 \end{bmatrix}, \tag{3.29}$$

and such a code is likely unique, however, it is not stated explicitly in their classification result [4, Theorem 6]. Finally, in an unpublished work by Gavrilyuk, Goryainov and Mogilnykh [47], it was shown that $J(10, 3)$ possesses a completely regular code having quotient matrix

$$\begin{bmatrix} 12 & 9 \\ 9 & 12 \end{bmatrix}, \tag{3.30}$$

but not equivalent to $\mathsf{AM}(10)$ (see Proposition 3.11). Here is its description. We first construct an orbit partition of $J(10, 3)$ by using an auxiliary graph. Namely, consider the disjoint union of two 5-cycles whose vertices are labeled by the elements from [10], and let G denote the group of automorphisms of this graph. Then G acts transitively on the following sets of triples:

A : triples inducing 2-paths in either cycle;

B : an edge in one cycle union a vertex from the other one;

C : an edge in one cycle union a non-incident vertex from the same cycle;

D : a non-edge in one cycle union a vertex from the other cycle.

Clearly, $A \cup B \cup C \cup D = \binom{[10]}{3}$, so identifying these 3-subsets with the vertices of $J(10,3)$, we have the orbit partition of $J(10,3)$ under the action of G. One can verify that $A \cup B$ (or $C \cup D$) is then a completely regular code as required. It was shown by Tricot [100] that this construction and $\mathsf{AM}(10)$ exhaust, up to isomorphism, all completely regular codes with quotient matrix (3.30) in $J(10,3)$.

Concerning $w = 4$, Vorobiev and Mogilnykh [84, Theorem 8] obtained the following partial classification of codes of strength 1 (note that $\rho \leq 2$ by Theorem 3.26).

Theorem 3.28. *Let C be a completely regular code of strength* 1 *in $J(n,4)$. Then C or C^{opp} satisfies one of the following cases:*

i. $\rho = 1$, $n = 8$ and the code has the same intersection array as some of the codes constructed in [4,101]*;*

ii. $\rho = 2$, the spectrum of the code consists of $\theta_0(n,4)$, $\theta_2(n,4)$, and $\theta_4(n,4)$, and the code is equivalent to $\mathsf{M}(n, n/2, 2, 4)$ or $\mathsf{GP}_2(n,4)$;

iii. $\rho = 2$, the spectrum of the code consists of $\theta_0(n,4)$, $\theta_2(n,4)$, and $\theta_3(n,4)$, and the code has quotient matrix

$$\begin{bmatrix} 4n-16-\beta_0 & \beta_0 & 0 \\ -\frac{(\beta_0-2n+2)(\beta_0-3n+6)}{\beta_0-\gamma_2} & \beta_0+\gamma_2-n-8 & \frac{(\gamma_2-2n+2)(\gamma_2-3n+6)}{\beta_0-\gamma_2} \\ 0 & \gamma_2 & 4n-16-\gamma_2 \end{bmatrix}$$

for some distinct positive integers β_0 and γ_2.

Note that all admissible intersection arrays of completely regular codes of covering radius one in $J(8,4)$ are found in [4] (see also [101]); however, their classification up to equivalence is not stated.

Applying Theorem 3.27 in the condition of Part (*ii*) of the above theorem, one reduces the problem to recovering C from a completely regular code (of strength one) in $J(n,3)$; all of them are already known to us, as described above. The proof of Part (*iii*) essentially follows from Lloyd's Theorem 3.4; however, ruling out the case $\beta_0 = \gamma_2$ requires a special and quite involved argument. The existence of examples satisfying Part (*iii*) of Theorem 3.28 remains open. The following feasible quotient matrices for completely regular codes with covering radius 2 in $J(n,4)$ for small n are given in [84]:

$$\begin{bmatrix} 14 & 30 & 0 \\ 3 & 31 & 10 \\ 0 & 24 & 20 \end{bmatrix} (n=15), \quad \begin{bmatrix} 16 & 36 & 0 \\ 3 & 35 & 14 \\ 0 & 24 & 28 \end{bmatrix} (n=17), \quad \begin{bmatrix} 20 & 36 & 0 \\ 6 & 42 & 8 \\ 0 & 32 & 24 \end{bmatrix} (n=18).$$

Note that by using a combinatorial analysis of the subgraphs induced by a completely regular code of covering radius one in the vertex neighborhoods, Mogilnykh [83, Corollary 1, Theorem 4] ruled out the existence of codes with certain parameters in $J(n,w)$ with $n \leq 12$, $w \leq 6$. These cases are already covered by the above results.

Finally, let us discuss some infinite series of completely regular codes in the Johnson graphs $J(n,w)$ with $w \leq 6$. Recall (see the discussion after Theorem 3.1) the concept of induced eigenfunctions. Given an eigenfunction $\mathbf{u}'$ of $J(n,w)$, by (3.10), one obtains an eigenfunction $\mathbf{u}''$ of $J(n,w+1)$. Furthermore, under some additional regularity assumption, Avgustinovich and Mogilnykh [3, Theorem 2] proved that if $\mathbf{u}'$ corresponds to a completely regular code of covering radius one in Γ' (cf. (3.28)), then so does $\mathbf{u}''$ in Γ'', respectively. Now, recall that by Theorem 3.14, a simple t-design is completely regular in $J(n,t+1)$ with $\rho = 1$. Among simple t-designs, Steiner systems satisfy the additional condition needed in [3, Theorem 2], and hence, by the above, such a design induces a complete regular code of covering radius one in $J(n,t+2)$ (note that it is different from that obtained by Theorem 3.15).

Theorem 3.29. ([4,82]) *Let S be a Steiner system $S(n,w,w-1)$. Then*

$$\left\{b\colon b \in \binom{[n]}{w+1} \mid \binom{b}{w} \cap S = \varnothing\right\}$$

is a completely regular code of covering radius one in $J(n,w+1)$ with quotient matrix

$$\begin{bmatrix} (w+1)(n-2w-1) & (w+1)w \\ w(n-2w) & w^2-2w-1+n \end{bmatrix}.$$

Since there exist infinite families of Steiner triple 2-$(n,3,1)$ and quadruple 3-$(n,4,1)$ systems, Theorem 3.29 yields two infinite series of completely regular codes in the Johnson graphs $J(n,4)$ and $J(n,5)$, respectively.

Corollary 3.8. ([4]) *The following holds.*

i. For every n, $n \equiv 1,3 \pmod 6$, there exists a completely regular code in $J(n,4)$ with quotient matrix

$$\begin{bmatrix} 4(n-7) & 12 \\ 3(n-8) & n+2 \end{bmatrix}.$$

ii. For every n, $n \equiv 2,4 \pmod 6$ there exists a completely regular code in $J(n,5)$ with quotient matrix

$$\begin{bmatrix} 5(n-9) & 20 \\ 4(n-8) & n+7 \end{bmatrix}.$$

Further, using the codes from Corollary 3.8(i) induced by certain Steiner quadruple 3-$(n,4,1)$ systems, Mogilnykh [82] constructs completely regular codes in the Johnson graphs $J(n,6)$.

Theorem 3.30. ([82, Theorem 11]) *Let Q be a Steiner quadruple 3-$(n,4,1)$ system, satisfying $(b \setminus b') \cup (b' \setminus b) \in Q$ for all blocks $b, b' \in Q$ with $|b \cap b'| = 2$. Then*

$$M = \left\{x\colon x \in \binom{[n]}{6} \mid \binom{x}{4} \cap Q = \varnothing\right\}$$

is a completely regular code in $J(n,6)$ with covering radius 2.

From the condition on Q, it follows that every 6-subset of $[n]$ contains $0, 1$ or 3 blocks of Q. Thus, define:

$$C_0 = \left\{ x \colon x \in \binom{[n]}{6} \mid |\binom{x}{4} \cap Q| = 0 \right\},$$

$$C_1 = \left\{ x \colon x \in \binom{[n]}{6} \mid |\binom{x}{4} \cap Q| = 1 \right\},$$

$$C_2 = \left\{ x \colon x \in \binom{[n]}{6} \mid |\binom{x}{4} \cap Q| = 3 \right\}.$$

The proof goes by first showing that (C_0, C_1, C_2) is the distance partition of $J(n, 6)$ with respect to M, satisfying some regularity condition. Applying then the inducing map (see Theorem 3.1) to an eigenvector associated with the code obtained from Q by Theorem 3.29 yields an eigenvector associated with $\{C_0, C_1, C_2\}$ (in particular, the strength of both Q and M equals 3), which completes the proof by [82, Theorem 7].

An example of a Steiner quadruple 3-$(n, 4, 1)$ system satisfying the condition of Theorem 3.30 is given by the support of the codewords of weight 4 of the extended Hamming code of length n, see [82, Corollary 13]. Interestingly, Mogilnykh [82, Corollary 14] also shows that the codewords of weight 6 in the extended Hamming code of length 16 form a completely regular code of covering radius two in $J(16, 6)$.

3.4 THE q-ANALOGUE OF JOHNSON GRAPHS

Recall that, in general, a q-analogue is a mathematical concept (identity, theorem, etc.) that depends on a variable q and that generalises a known concept, to which it reduces in the limit $q \to 1$. In combinatorics it often arises by replacing a set and its subsets by a vector space over a finite field and its subspaces.

In this section, we briefly discuss completely regular codes in the q-analogue of Johnson graphs. As we will see, the situation here is more complicated compared to the Johnson graphs: despite great efforts, even a classification of completely regular codes of strength 0 is not yet available.

Let q be a prime power, $\mathbb{F}_q$ the finite field with q elements. Given positive integers n and k with $n \geq 2k$, let $J_q(n, k)$ denote a graph defined on the set of all k-dimensional subspaces of the n-dimensional vector space $\mathbb{F}_q^n$ over $\mathbb{F}_q$, with two vertices U_1, U_2 being adjacent if $\dim(U_1 \cap U_2) = k - 1$; the diameter of $J_q(n, k)$ is therefore k. Thus, these graphs can be seen as a q-analogue of the Johnson graphs; they are better known as the *Grassmann* graphs. The Grassmann graphs are distance-transitive; more details can be found in [19, Chapter 9].

Let $\mathrm{PG}(d, q)$ denote the d-dimensional[7] projective geometry over $\mathbb{F}_q$, whose *points*, *lines*, *planes*, etc., *hyperplanes* correspond to the 1-dimensional, 2-dimensional, 3-dimensional, etc., d-dimensional vector subspaces of $\mathbb{F}_q^{d+1}$, respectively. In particular, the Grassmann graph $J_q(4, 2)$ is the line concurrence graph of $\mathrm{PG}(3, q)$. Recall that a *spread* of $\mathrm{PG}(d, q)$ is a set of lines that partitions the set of points; it exists if and only if d is odd.

Many finite geometry problems can naturally be formulated as problems about the structure of Grassmann graphs. In particular, completely regular codes in the Grassmann graphs first received attention under the name of *Cameron-Liebler line classes* and in the special (and first non-trivial) case $n = 4$, $k = 2$. Note that the perfect codes in the Grassmann graphs are trivial [24].

3.4.1 Completely Regular Codes of Strength 0

Cameron and Liebler [23] studied irreducible collineation groups of $\mathrm{PG}(d, q)$ having equally many orbits on points and on lines (it is well known that a collineation group of a finite projective space always has at least as many orbits on lines as on points). For $d = 2$, any collineation group of a projective plane has this property; however, for $d > 2$, the situation is different: they conjectured that such a group is line-transitive, or fixes a hyperplane and acts line-transitively on it, or (dually) fixes a point and acts line-transitively on the quotient space. The conjecture was confirmed in [7]. Cameron and Liebler observed that such a group in question would give rise to a tactical decomposition of $\mathrm{PG}(d, q)$ with equally many point and line classes and, furthermore, such a decomposition induces a decomposition with the same property of any 3-dimensional subspace, i.e., $\mathrm{PG}(3, q)$. Every line class in such a decomposition of $\mathrm{PG}(3, q)$ enjoys many equivalent properties (see Theorem 3.31 below), among which the most interesting from the point of view of finite geometry is probably that it contains a fixed number (say, x) of lines of any spread. Taking this condition as definition, Cameron and Liebler introduced "special" classes of lines in $\mathrm{PG}(3, q)$: every class in a symmetric tactical decomposition of $\mathrm{PG}(3, q)$ is special, but not conversely. Subsequently, special classes of lines were named *Cameron-Liebler line classes* [90]; the number x is then called the *parameter* of a Cameron-Liebler line class.

Cameron and Liebler [23], and later Penttila [90] proved a series of algebraic and geometric characterizations of these sets. In the following theorem, Drudge [36] generalized these characterizations as well as the concept of a Cameron-Liebler line class in $\mathrm{PG}(3, q)$ to those in $\mathrm{PG}(d, q)$. Let P and L respectively denote the sets of point and lines of $\mathrm{PG}(d, q)$, and $I \in \mathbb{Q}^{\mathsf{P}\times\mathsf{L}}$ be the point-line incidence matrix. Let θ_i denote $q^i + \ldots + q + 1$.

Theorem 3.31. ([23,36,90]) *Let $\mathcal{L}$ be a set of lines in PG(d, q) with characteristic vector $\chi_{\mathcal{L}} \in \mathbb{Q}^{\mathsf{L}}$. Then the following are equivalent:*

1. *$\chi_{\mathcal{L}}$ is in the row space of I.*

2. *There exists x such that for any flag (P, X) with P a point, X an i-dimensional subspace of PG(d, q), we have*

$$|\mathsf{Star}(P) \cap \mathcal{L}| + \frac{\theta_{d-2}}{\theta_{i-1}\theta_{i-2}}|\mathsf{Line}(X) \cap \mathcal{L}| = x + \frac{\theta_{d-2}}{\theta_{i-2}}|\mathsf{Star}(P) \cap \mathsf{Line}(X) \cap \mathcal{L}|,$$

where $\mathsf{Star}(P)$, $\mathsf{Line}(X)$ are respectively the sets of lines on P or in X.

3. *There exists x such that, given a line $\ell \in \mathsf{L}$,*

$$|\{\text{all lines } m \in \mathcal{L} \text{ meeting } \ell,\ \ell \neq m\}| = x(q+1)+(q^{d-1}+q^{d-2}+\ldots+q^2-1)\chi_{\mathcal{L}}(\ell).$$

In addition, if d is odd, then the following conditions are also equivalent:

4. *There exists x such that $|\mathcal{S} \cap \mathcal{L}| = x$ for any line-spread $\mathcal{S}$ of* $\mathrm{PG}(d,q)$.

And if $d = 3$, *the above conditions are also equivalent to:*

5. *There exists x such that, for any skew lines ℓ, m,*

$$|\{\text{all lines } n \in \mathcal{L} \text{ meeting both } \ell \text{ and } m\}| = x + q(\chi_{\mathcal{L}}(\ell) + \chi_{\mathcal{L}}(m)).$$

In fact, it follows from the proof of the theorem [36] that the number x which occurs in the various conditions is the same for each of the conditions; this number is called the *parameter* of the line class $\mathcal{L}$.

Definition 3.7. *A set of lines in* $\mathrm{PG}(d,q)$ *satisfying any of the conditions in Theorem 3.31 is called a Cameron-Liebler line class (with parameter x).*

If d is odd, Condition (4) implies that the parameter x is an integer. For the case d is even, [36, Corollary 3.2] shows that $x(q+1)$ is an integer. Furthermore, x satisfies

$$0 \leq x \leq \frac{\theta_d}{\theta_1}$$

with equality implying that $\mathcal{L}$ consists either of all lines of $\mathrm{PG}(d,q)$ or the empty set. Also note that the complement of a Cameron-Liebler line class of parameter x is a line class of parameter $\frac{\theta_d}{\theta_1} - x$; thus, it suffices to consider $x \leq \frac{\theta_d}{2\theta_1}$.

From Condition (3), one can easily see that the partition $\{\mathcal{L}, \mathsf{L} \setminus \mathcal{L}\}$ of the line set of $\mathrm{PG}(d,q)$ induces an equitable partition of the Grassmann graph $J_q(d+1,2)$, i.e., by Definition 3.5, $\mathcal{L}$ is a completely regular code of radius 1. Furthermore, its strength equals 0. Other characterizations of Cameron-Liebler line classes consider them as those sets of lines of $\mathrm{PG}(d,q)$ which have the greatest possible number of intersecting line pairs [37] or Boolean degree one functions [45].

We now define the *trivial* examples of Cameron-Liebler line classes:

(CL0) the empty set, with parameter $x = 0$,

(CL1) all lines through a point P, with parameter $x = 1$,

(CL1′) all lines in a fixed hyperplane Π, also with parameter $x = 1$, and

(CL2) the union of (CL1) and (CL1′) when $P \notin \Pi$, with parameter $x = 2$,

(CL3) the complement of any of the above examples.

It was conjectured by Cameron and Liebler in [23] that no other examples exist in $\mathrm{PG}(3,q)$. Even though Drudge generalized Cameron-Liebler line classes in $\mathrm{PG}(3,q)$ to those in $\mathrm{PG}(d,q)$, it seems he did not explicitly state an analogous conjecture (nevertheless, no non-trivial examples are known if $d > 3$). Furthermore, in [35], he disproved the original conjecture by constructing a counterexample in $\mathrm{PG}(3,3)$ with parameter $x = 5$; it was generalized by Bruend and Drudge in [21] to an infinite (rather nice) family in $\mathrm{PG}(3,q)$ with parameter $x = \frac{q^2+1}{2}$ (q odd), which is the maximum possible value of x.

Many more non-trivial examples in $\mathrm{PG}(3,q)$ have been discovered later on: by Govaerts and Penttila [56]; by Rodgers in [92,93], some of them have been generalized to infinite families, see [31,43,44]; by Cossidente and Pavese [26–28]; by Gavrilyuk and Metsch [50], Gavrilyuk, Matkin and Penttila [49]. Nevertheless, non-trivial examples seem to be rare; the central problem for Cameron-Liebler line classes remains to determine for which values of x a (non-trivial) class may exist. Thus, a great deal of effort has been put in proving non-existence results. Probably one of the most powerful of those is given in the following theorem, known as the *modular equality* on x.

Theorem 3.32. *Suppose $\mathcal{L}$ is a Cameron-Liebler line class with parameter x in $\mathrm{PG}(3,q)$. Then there exists an integer n such that*

$$\binom{x}{2} + n(n-x) \equiv 0 \mod q+1.$$

Perhaps surprisingly, this condition excludes roughly half of the possible values of x; it was first observed in [51] and formally proved in [50].

There have also been a number of results in the form of lower bounds for x [22,29,57,78], the most recent and strongest one due to Metsch [79].

Theorem 3.33. *If $\mathcal{L}$ is a Cameron-Liebler line class with parameter x of $\mathrm{PG}(3,q)$, then $x \in \{0,1,2\}$ or $x > q\sqrt[3]{q/2} - \frac{2}{3}q$.*

The proof of the two above results relies on a counting technique developed in [51]. Let $\mathcal{L}$ be a Cameron–Liebler line class with parameter x in $\mathrm{PG}(3,q)$, and let ℓ be a line. Number the points on ℓ as $p_1, \dots, p_{q+1}$, and the planes containing ℓ as $\pi_1, \dots, \pi_{q+1}$. We define the *pattern* of ℓ with respect to $\mathcal{L}$ to be the $(q+1)\times(q+1)$ matrix $T(\ell)$, whose (i,j)-entry is given by $t_{i,j} = |\mathsf{Star}(p_i) \cap \mathsf{Line}(\pi_j) \cap \mathcal{L}|$. Roughly speaking, the pattern of ℓ with respect to $\mathcal{L}$ shows how the lines of $\mathcal{L}$ interface with the set of plane pencils containing ℓ.

Theorem 3.34. *For every line $\ell \in \mathrm{PG}(3,q)$, the pattern $T(\ell)$ has the following properties:*

1. $t_{i,j} \in \{0,\dots,q\}$ *for all* i,j;
2. $\sum_{i,j} t_{i,j} = x(q+1) + \chi_{\mathcal{L}}(\ell)(q^2-1)$;

3. $\sum_j t_{k,j} + \sum_i t_{i,l} = x + (q+1)t_{k,l} + (q-1)\chi_{\mathcal{L}}(\ell)$ *for all* k, l;

4. $\sum_{i,j} t_{i,j}^2 = (x - \chi_{\mathcal{L}}(\ell))^2 + q(x - \chi_{\mathcal{L}}(\ell)) + \chi_{\mathcal{L}}(\ell)q^2(q+1)$.

The set of all possible patterns may give some insight about the structure of a putative line class $\mathcal{L}$, sometimes enough to show its existence or uniqueness, or its non-existence, see [31,49–51]. Unfortunately, the number of admissible patterns quickly grows as q increases, which makes their analysis complicated in general. Nevertheless, this approach was used in [48] to determine all Cameron-Liebler line classes in PG$(3, q)$ with $q \in \{2, 3, 4, 5\}$ (see also [90] for $q = 2$, [35] for $q = 3$, and [51] for $q = 4$). Furthermore, once the classification of all Cameron-Liebler line classes in PG$(3, q)$ is available, one can attempt to classify those in PG(d, q) for $d > 3$, by using an inductive argument developed by Drudge [36]. This was done for $q \in \{3, 4, 5\}$ in [36,51,77], respectively (for $q = 2$ the result follows from the original paper by Cameron and Liebler). Besides this, a generalization (to some extent) of Theorem 3.32 for Cameron-Liebler line classes in PG(d, q), $d > 3$, was shown in [30].

Rodgers, Storme and Vansweevelt [91] generalized the concept of Cameron-Liebler classes of lines to *Cameron-Liebler sets* of k-spaces in PG(d, q), where the case $k = 1$ corresponds to line classes. The trivial examples (CL0)-(CL3) generalize straightforwardly. Focusing on the case $d = 2k + 1$, they also showed the respective analogue of Theorem 3.31. Blokhuis, De Boeck and D'haeseleer [15] proved the respective analogue of Theorem 3.31 for the general case $d \geq 2k+1$, from which it is clear that a Cameron-Liebler set of k-spaces in PG(d, q) induces a completely regular code of strength 0 and radius 1 in $J_q(d + 1, k + 1)$. Various bounds on the parameter x and classification results (no non-trivial examples are known) were shown in [13,15,16,60,80]. De Beule, Mannaert, and Storme [14] further improved some of these bounds and proved an analogue of the modular equality on x.

Filmus and Ihringer [45] developed an argument that gives a classification of completely regular codes of strength 0 and covering radius 1 in $J_q(d+1, k+1)$ once the classification of those in $J_q(d + 1, 2)$ is available. Using the above-mentioned results for Cameron-Liebler line classes in PG(d, q), $q \in \{2, 3, 4, 5\}$, they showed that such codes in $J_q(n, k + 1)$ are only the trivial ones. Furthermore, they conjectured that there exists an $n_0(k, q)$ such that, for any $n \geq n_0(k, q)$, Cameron-Liebler sets of k-spaces in $J_q(n, k+1)$ are trivial. In a recent breakthrough [60], Ihringer confirmed this conjecture with $n_0(k, q)$ much larger than 4, as the very nice proof is based on the Ramsey theory for geometric lattices. Thus, we want to believe there is still some room to explore for unexpected examples.

Finally, we would like to emphasize the diversity of mathematical ideas used in the study of Cameron-Liebler sets: from counting and the representation theory of association schemes to Erdos-Ko-Rado type theorems and Ramsey theory.

3.4.2 Completely Regular Codes of Strength ≥ 1

Starting again with the first non-trivial case $J_q(4,2)$ of the Grassmann graphs, their completely regular codes of strength 1 include spreads in $\mathrm{PG}(3,q)$. These are a classical object of interest in finite geometry, in particular, due to their connection with translation projective planes [1]. More generally, a completely regular code C of strength 1 in $J_q(d+1,2)$ (hence, the radius of C equals 1) is characterized by the property that every point of $\mathrm{PG}(d,q)$ is on the same number (say, t) of lines from C, see [36,37]. Such sets of lines are called *t-fold covers*, see [36] for further details.

De Winter and Metsch [32] studied completely regular codes with covering radius 1 and strength 1 in the Grassmann graphs $J_q(n,3)$. For even $n \geq 6$, they constructed two series of such codes. Mogilnykh [85] generalized some of these examples and constructed a few more, in particular, by using an integer programming in the smallest non-trivial case $J_2(6,3)$; although even in this case a full classification is not yet available, see [85, Section 7] for further details.

Note also that a generalization of Theorem 3.14 holds for Grassmann graphs (with a q-analogue of designs instead of combinatorial designs), e.g., see [85, Corollary 6]. Mogilnykh [85, Theorem 9] also generalized Theorem 3.29; thus, the first known example of a q-ary Steiner triple system constructed in [17] gives a completely regular code in $J_2(13,4)$.

3.5 CONCLUDING REMARKS

The following research problems could be accessible to students: classify (up to equivalence) completely regular codes of radius one in $J(8,3)$ with quotient matrix equation (3.29), or those in $J(8,4)$ (see Theorem 3.28(i)) (and more generally, in $J(2w,w)$ with $4 \leq w \leq 6$, see Theorem 3.25).

Apart from the above small cases, the completely regular designs in $J(n,w)$ were fully classified in the following cases: strength 0, strength 1 and minimum distance at least 2, strength 1 and covering radius 1, strength 1 and covering radius $w-1$. The case of strength 1 and minimum distance 1 remains unresolved in general. Perhaps, one could next try to eliminate (or construct) examples satisfying Theorem 3.28(iii).

In [76], Martin remarked that, for the Hamming graphs, any completely regular code with strength t has minimum distance at most $2t+1$ (with the exception of binary repetition codes) and argued that something similar should hold for the Johnson graphs. Another interesting question due to Martin is the existence of completely regular partitions of the Johnson graphs. These are equitable partitions into pairwise isometric cells, each of which is a completely regular code. Such a partition naturally gives rise to a quotient graph which is distance-regular. No examples are known; some partial results can be found in [71, Section 4.6].

It is interesting to see whether two known "sporadic" examples of Cameron-Liebler line classes (with parameters $x = 7$ in PG(3, 4) [56] or $x = 10$ in PG(3, 5) [50]) can be generalized to infinite series.

Completely regular codes have been intensely studied in other graphs related with finite geometries and more general incidence structures. In particular, for the collinearity graphs of polar spaces, their completely regular codes of strength 0 and 1 are known as i-tight sets and m-ovoids, respectively; the literature is vast, see [6,20] and references therein. Some results on codes in the bilinear forms graphs are found in [5,45].

The interested reader is further referred to a recent survey [61] by Ihringer, where completely regular codes (in various graphs) are treated from the position of Boolean functions analysis and related questions from theoretical computer science.

Notes

1. One can see from this definition that the first column in both P and Q is the all-ones vector.
2. In the sense of the usual graph distance metric.
3. At this point the reader should be warned that such partitions appear in Graph Theory under various names, including *perfect colourings*, *graph disivisors*, *intriguing sets*, *regular partitions*, *partitio designs*, etc. Some history of this terminology is given in Chapter 1.
4. It can also be viewed as a constant weight-w code in the Hamming graph $H(n, 2)$.
5. A *biplane* is a block design with $\lambda = 2$.
6. Note that $|K \cap K'| \leq 1$ for all distinct $K, K' \in \mathcal{K}$.
7. Recall that the projective dimension is one less than the vector space dimension.

REFERENCES

[1] J. Andre. Über nicht-Desarguessche Ebenen mit transitiver Translationsgruppe. *Math. Z.*, 60:156–186, 1954. URL: https://doi.org/10.1007/BF01187370.

[2] S. V. Avgustinovich and I. Yu. Mogilnykh. Perfect 2-colorings of Johnson graphs $J(6, 3)$ and $J(7, 3)$. *Lect. Notes Comp Sci*, 5228:11–19, 2008.

[3] S. V. Avgustinovich and I. Yu. Mogilnykh. Induced perfect colorings. *Sib. Electron. Math. Rep.*, 8:310–316, 2011.

[4] S. V. Avgustinovich and I. Yu. Mogilnykh. Perfect 2-colorings of Johnson graphs $J(8, 3)$ and $J(8, 4)$. *J. Appl. Industr. Math.*, 5(1):19–30, 2011.

[5] R. A. Bailey, P. J. Cameron, A. L. Gavrilyuk, and S. V. Goryainov. Equitable partitions of Latin-square graphs. *J. Comb. Des.*, 27(3):142–160, 2019. URL: https://doi.org/10.1002/jcd.21634.

[6] J. Bamberg, S. Kelly, M. Law, and T. Penttila. Tight sets and m-ovoids of finite polar spaces. *J. Comb. Theory Ser. A*, 114(7):1293–1314, 2007. URL: https://doi.org/10.1016/j.jcta.2007.01.009.

[7] J. Bamberg and T. Penttila. Overgroups of cyclic Sylow subgroups of linear groups. *Comm. Algebra*, 36(7):2503–2543, 2008.

[8] E. Bannai. Codes in bi-partite distance-regular graphs. *J. London Math. Soc.*, 2: 197–202, 1977.

[9] E. Bannai, E. Bannai, T. Ito, and R. Tanaka. *Algebraic Combinatorics.* Berlin, Boston: De Gruyter, 2021.

[10] E. Bannai and T. Ito. *Algebraic Combinatorics I: Association Schemes.* The Benjamin/Cummings Publishing Co., Inc., Menlo Park, CA, 1984.

[11] D. J. Bernstein. Detecting perfect powers in essentially linear time. *Math. Comp.*, 67:1253–1283, 1998.

[12] T. Beth, D. Jungnickel, and H. Lenz. *Design Theory.* Cambridge University Press, Cambridge, 1993.

[13] J. De Beule, J. Mannaert, and L. Storme. Cameron-liebler k-sets in subspaces and non-existence conditions. *Des. Codes Cryptogr.*, 90(3):633–651, 2022.

[14] J. De Beule, J. Mannaert, and L. Storme. On two non-existence results for Cameron-Liebler k-sets in $\mathrm{PG}(n, q)$, 2024. URL: https://arxiv.org/abs/arXiv:2403.00519.

[15] A. Blokhuis, M. De Boeck, and J. D'haeseleer. Cameron-Liebler sets of k-spaces in $\mathrm{PG}(n, q)$. *Des. Codes Cryptogr.*, 87:1839–1856, 2019.

[16] A. Blokhuis, M. De Boeck, and J. D'haeseleer. Correction to: Cameron-Liebler sets of k-spaces in $\mathrm{PG}(n, q)$. *Des. Codes Cryptogr.*, 90:477–487, 2022.

[17] M. Braun, T. Etzion, and R. J. Patric. Östergard, Alexander Vardy, and Alfred Wassermann. Existence of q-analogs of Steiner systems. *Forum Math. Pi*, 4:e7, 2016.

[18] A. E. Brouwer and W. H. Haemers. *Spectra of Graphs.* Springer, New York, 2012.

[19] A. E. Brouwer, A. M. Cohen, and A. Neumaier. *Distance-Regular Graphs.* Springer Berlin, Heidelberg, 1989.

[20] A. E. Brouwer and H. Van Maldeghem. *Strongly Regular Graphs*, volume 182 of Encyclopedia of Mathematics and Its Applications. Cambridge University Press, Cambridge, 2022. URL: https://doi.org/10.1017/9781009057226.

[21] A. A. Bruen and K. Drudge. The construction of Cameron-Liebler line classes in $\mathrm{PG}(3, q)$. *Finite Fields Appl.*, 5(1):35–45, 1999.

[22] A. A. Bruen and K. Drudge. On the non-existence of certain Cameron-Liebler line classes in $\mathrm{PG}(3, q)$. *Des. Codes Cryptogr.*, 14(2):127–132, 1998.

[23] P. J. Cameron and R. A. Liebler. Tactical decompositions and orbits of projective groups. *Linear Algebra Appl.*, 46:81–102, 1982.

[24] L. Chihara. On the zeros of the Askey-Wilson polynomials, with applications to coding theory. *SIAM J. Math. Anal.*, 18(1):191–207, 1987.

[25] C. J. Colbourn and J. H. Dinitz. *Handbook of Combinatorial Designs.* CRC Press, Boca Raton, FL, 1996.

[26] A. Cossidente and F. Pavese. New Cameron-Liebler line classes with parameter $\frac{q^2+1}{2}$. *J. Algebraic Comb.*, 49(2):193–208, 2019.

[27] A. Cossidente and F. Pavese. Intriguing sets of quadrics in $\mathrm{PG}(5,q)$. *Adv. Geom.*, 17(3):339–345, 2017.

[28] A. Cossidente and F. Pavese. Cameron-Liebler line classes of $\mathrm{PG}(3,q)$ admitting $\mathrm{PGL}(2,q)$. *J. Comb. Theory Ser. A*, 167:104–120, 2019.

[29] J. De Beule, A. Hallez, and L. Storme. A non-existence result on Cameron-Liebler line classes. *J. Comb. Des.*, 16(4):342–349, 2008.

[30] J. De Beule and J. Mannaert. A modular equality for Cameron-Liebler line classes in projective and affine spaces of odd dimension. *Finite Fields Appl.*, 82:102047, 2022.

[31] J. De Beule, J. Demeyer, K. Metsch, and M. Rodgers. A new family of tight sets in $\mathcal{Q}^+(5,q)$. *Des. Codes Cryptogr.*, 78(3):655–678, 2016.

[32] S. De Winter and K. Metsch. Perfect 2-colorings of the Grassmann graph of planes. *Electron. J. Comb.*, 27(1):Paper No. 1.21, 19, 2020. URL: https://doi.org/10.37236/8672.

[33] M. Dehon. On the existence of 2-designs $S_\lambda(2,3,v)$ without repeated blocks. *Discrete Math.*, 43(2–3):155–171, 1983.

[34] P. Delsarte. An algebraic approach to the association schemes of coding theory. *Philips Res. Reports Suppl.*, 10, 55 pp., 1973.

[35] K. Drudge. On a conjecture of Cameron and Liebler. *Eur. J. Comb.*, 20(4):263–269, 1999.

[36] K. W. Drudge. *Extremal Sets in Projective and Polar Spaces.* ProQuest LLC, Ann Arbor, MI, 1998.

[37] J. Eisfeld. On the common nature of spreads and pencils in $\mathrm{PG}(d,q)$. *Discrete Math.*, 189(1):95–104, 1998. URL: https://www.sciencedirect.com/science/article/pii/S0012365X98000041.

[38] T. Etzion. On the nonexistence of perfect codes in the Johnson scheme. *SIAM J. Discr. Math.*, 9(2):201–209, 1996.

[39] T. Etzion. Configuration distribution and designs of codes in the Johnson scheme. *J. Comb. Des.*, 15(1):15–34, 2004.

[40] T. Etzion. *Perfect Codes and Related Structures.* World Scientific, Singapore, 2022.

[41] T. Etzion and M. Schwarz. Perfect constant-weight codes. *IEEE Trans. Inform. Theory*, 50(9):2156–2165, 2004.

[42] R. J. Evans, A. L. Gavrilyuk, S. Goryainov, and K. Vorobiev. Equitable 2-partitions of the Johnson graphs $J(n,3)$. URL: arxiv.org/abs/2206.15341.

[43] T. Feng, K. Momihara, M. Rodgers, Q. Xiang, and H. Zou. Cameron-Liebler line classes with parameter $x = \frac{(q+1)^2}{3}$. *Adv. Math.*, 385:Paper No. 107780, 31, 2021.

[44] T. Feng, K. Momihara, and Q. Xiang. Cameron-Liebler line classes with parameter $x = \frac{q^2-1}{2}$. *J. Comb. Theory Ser. A*, 133:307–338, 2015.

[45] Y. Filmus and F. Ihringer. Boolean degree 1 functions on some classical association schemes. *J. Comb. Theory Ser. A*, 162:241–270, 2019. URL: https://www.sciencedirect.com/science/article/pii/S009731651830147X.

[46] A. L. Gavrilyuk and S. V. Goryainov. On perfect 2-colorings of Johnson graphs $J(v,3)$. *J. Comb. Des.*, 21(6):232–252, 2013.

[47] A. L. Gavrilyuk, S. V. Goryainov, and I. Yu. Mogilnykh. On non-isomorphic perfect 2-colorings of the Johnson graph $J(10,3)$. In *Contemporary Problems of Mathematics: Proceedings of the International (43rd All-Russian) School-Conference*, Yekaterinburg (Russia), pp. 17–19, 2021.

[48] A. L. Gavrilyuk and I. Matkin. Cameron-Liebler line classes in PG(3, 5). *J. Comb. Des.*, 26(12):563–580, 2018.

[49] A. L. Gavrilyuk, I. Matkin, and T. Penttila. Derivation of Cameron-Liebler line classes. *Des. Codes Cryptogr.*, 86(1):231–236, 2018.

[50] A. L. Gavrilyuk and K. Metsch. A modular equality for Cameron-Liebler line classes. *J. Comb. Theory Ser. A*, 127:224–242, 2014.

[51] A. L. Gavrilyuk and I. Yu. Mogilnykh. Cameron-Liebler line classes in $PG(n,4)$. *Des. Codes Cryptogr.*, 73(3):969–982, 2014.

[52] C. Godsil and G. Royle. *Algebraic Graph Theory.* Springer-Verlag, New York, 2001.

[53] C. D. Godsil. Geometric distance-regular covers. *New Zealand J. Math.*, 22(2):31–38, 1993.

[54] C. D. Godsil and C. E. Praeger. Completely transitive designs, 2014. URL: https://arxiv.org/abs/arXiv:1405.5583.

[55] M. D. Gordon. Perfect single error-correcting codes in the Johnson scheme. *IEEE Trans. Inf. Theory*, 52(10):4670–4672, 2006.

[56] P. Govaerts and T. Penttila. Cameron-Liebler line classes in PG(3, 4). *Bull. Belg. Math. Soc. Simon Stevin*, 12(5):793–804, 2005.

[57] P. Govaerts and L. Storme. On Cameron-Liebler line classes. *Adv. Geom.*, 4(3):279–286, 2004.

[58] P. Hammond. On the non-existence of perfect and nearly perfect codes. *Dis. Math.*, 39(1):105–109, 1982.

[59] D. R. Hughes and F. C. Piper. *Design Theory.* Cambridge University Press, Cambridge, 1985.

[60] F. Ihringer. The classification of Boolean degree 1 functions in high-dimensional finite vector spaces, 2024. *Proc. Amer. Math. Soc.*, 152(12):5355–5365, 2024.

[61] F. Ihringer. A Survey of Cameron-Liebler Sets and Low Degree Boolean Functions in Grassmann Graphs, 2024. URL: https://arxiv.org/abs/000.

[62] W. M. Kantor. On incidence matrices of finite projective and affine space. *Math. Z.*, 24:315–318, 1972.

[63] P. Keevash. The existence of designs, 2014. URL: https://arxiv.org/abs/1401.3665v3.

[64] J. H. Koolen. On a conjecture of Martin on the parameters of completely regular codes and the classification of the completely regular codes in the Biggs-Smith graph. *Linear Multilinear Algebra*, 39(1–2):3–17, 1995.

[65] E. Kummer. Uber die erganzungssatze zu den allgemeinen reciprocitatsgesetzen. *Journal fur die reine und angewandte Mathematik*, 44:93–146, 1852.

[66] V. K. Leontiev, G. L. Movsisyan, and Zh. G. Margaryan. Constant weight perfect and D-representablecodes. *Proceedings of the Yerevan State University: Physical and Mathematical Sciences*, vol. 1, pp. 16–19, 2012.

[67] S. P. Lloyd. Block coding. *Bell Syst. Tech. J.*, 36:517–535, 1957.

[68] J. H. Loxton. Some problems involving powers of integers. *Acta Arithmetica*, 46(1):113–123, 1986.

[69] F. J. MacWilliams and N. J. A. Sloane. *The Theory of Error-Correcting Codes.* North-Holland, Amsterdam, 1977.

[70] S. S. Magliveras and D. W. Leavitt. Simple six designs exist. *Congressus Numerantium*, 40:195–205, 1983.

[71] W. J. Martin. Completely regular subsets. PhD thesis, University of Waterloo, 1992.

[72] W. J. Martin. Completely regular designs of strength one. *J. Alg. Comb.*, 3:177–185, 1994.

[73] W. J. Martin. Completely regular designs. *J. Comb. Des.*, 6(4):261–273, 1998.

[74] W. J. Martin. Mixed block designs. *J. Comb. Des.*, 6(2):151–163, 1998.

[75] W. J. Martin. Designs in product association schemes. *Des. Codes Crypt.*, 16(3):271–289, 1999.

[76] W. J. Martin. Completely regular codes: A viewpoint and some problems. In *Proceedings of 2004 Com2MaC Workshop on Distance-Regular Graphs and Finite Geometry*, Pusan (South Korea), pp. 43–56, 2004.

[77] I. Matkin. Cameron-Liebler line classes in PG$(n, 5)$. *Trudy Inst. Mat. i Mekh. UrO RAN*, 24(2):158–172, 2018.

[78] K. Metsch. The non-existence of Cameron-Liebler line classes with parameter $2 < x \leqslant q$. *Bull. Lond. Math. Soc.*, 42(6):991–996, 2010.

[79] K. Metsch. An improved bound on the existence of Cameron-Liebler line classes. *J. Comb. Theory Ser. A*, 121:89–93, 2014.

[80] K. Metsch. A gap result for cameron–liebler k-classes. *Dis. Math.*, 340(6):1311–1318, 2017.

[81] A. D. Meyerowitz. Cycle-balanced partitions in distance-regular graphs. *Dis. Math.*, 264(2):149–165, 2003.

[82] I. Mogilnykh. Completely regular codes in Johnson and Grassmann graphs with small covering radii. *Electron. J. Comb.*, 29(2):P2.57, 2022.

[83] I. Yu. Mogilnykh. On nonexistence of certain perfect 2-colouring of Johnson graphs. *Dis. Analyz Oper. Invest.*, 16(5):52–68, 2009.

[84] I. Yu. Mogilnykh and K. V. Vorobiev. On completely regular codes with minimum eigenvalue in geometric graphs. *Dis. Math.*, 346(7):113357, 2023.

[85] I. Mogilnykh. Completely regular codes in Johnson and Grassmann graphs with small covering radii. *Electron. J. Comb.*, 29(2):Paper No. 2.57, 15, 2022. URL: https://doi.org/10.37236/10083.

[86] G. L. Movsisyan. Perfect codes in Johnson schemes. *Bull. Moscow State Univ. (MSU) Comput. Math. Cybern.*, 15(1):64–69, 1982.

[87] G. L. Movsisyan. Perfect constant weight codes and Steiner systems. *Problemy Peredachi Informatzii*, 19(2):109–112, 1983.

[88] G. L. Movsisyan and Z. G. Margaryan. D-representing code problem solution. In *Fundamentals of Computation Theory; International Conference (FCT-87)*, Kazan (USSR), pp. 328–331, 1987.

[89] A. Neumaier. Completely regular codes. *Dis. Math.*, 106/107:353–360, 1992.

[90] T. Penttila. Cameron-Liebler line classes in $\mathrm{PG}(3,q)$. *Geom. Dedicata*, 37(3):245–252, 1991.

[91] M. Rodgers, L. Storme, and A. Vansweevelt. Cameron-Liebler k-classes in $\mathrm{PG}(2k+1,q)$. *Combinatorica*, 38:739–757, 2018.

[92] M. Rodgers. Cameron-Liebler line classes. *Des. Codes Cryptogr.*, 68(1–3):33–37, 2013.

[93] M. J. Rodgers. On some new examples of Cameron-Liebler line classes. Thesis (Ph.D.)–University of Colorado at Denver. ProQuest LLC, Ann Arbor, MI, 2012.

[94] C. Roos. A note on the existence of perfect constant weight codes. *Dis. Math.*, 47(1):121–123, 1983.

[95] N. V. Semakov and V. A. Zinoviev. Constant weight codes and tactical configurations. *Probl. Inform. Transm.*, 5(3):28–36, 1969.

[96] N. V. Semakov, V. A. Zinoviev, and G. V. Zaitsev. Uniformly packed codes. *Probl. Inform. Transm.*, 7(1):30–39, 1971.

[97] O. Shimabukuro. On the nonexistence of perfect codes in $J(2w+p^2,w)$. *Ars Comb.*, 75:129–134, 2005.

[98] L. Teirlinck. Non-trivial t-designs without repeated blocks exist for all t. *Dis. Math.*, 65(3):301–311, 1987.

[99] A. Tietäväinen. On the nonexistence of perfect codes over finite fields. *SIAM J. Appl. Math.*, 24(1):88–96, 1973.

[100] P. Tricot. Symmetric perfect 2-colorings on $J(10,3)$, *Journal of Computational Algebra*, Volume 12, 2024, 100028. https://doi.org/10.1016/j.jaca.2024.100028.

[101] K. Vorobiev. Equitable 2-partitions of Johnson graphs with the second eigenvalue. URL: arxiv.org/abs/2003.10956.

[102] K. Vorobiev, I. Yu. Mogilnykh, and A. Valyuzhenich. Minimum supports of eigenfunctions of Johnson graphs. *Dis. Math.*, 341(8):2151–2158, 2018.

[103] E. Witt. Die 5-fach transitiven gruppen von mathieu. *Abh. Math. Sem. Univ. Hamburg*, 12:256–264, 1938.

[104] V. A. Zinoviev and V. K. Leontiev. The nonexistence of perfect codes over Galois fields. *Probl. Control Inform. Theory*, 2(2):123–132, 1973.

CHAPTER 4

Codes over Rings and Modules

Minjia Shi

IN THIS CHAPTER, we study equitable partitions in the Hamming graph obtained by codes over rings ($\mathbb{Z}_4$-codes) or codes over modules (additive $\mathbb{F}_4$-codes). Some related topics like two-weight codes (over $\mathbb{Z}_{2^k}$) and strongly walk regular graphs (as coset graphs of certain three-weight codes) are also considered. We follow closely [21] for $\mathbb{Z}_4$-codes, [80] for $\mathbb{Z}_{2^k}$-codes, [78] for additive $\mathbb{F}_4$-codes, and [75] for strongly walk regular graphs. In the first section, notations and definitions are reviewed. The second section studies the four topics mentioned in four subsections.

4.1 NOTATION AND DEFINITIONS

4.1.1 Codes over $\mathbb{Z}_4$

By a *quaternary code* $\mathcal{C}$ of length n, we shall mean a linear block code over $\mathbb{Z}_4$, i.e., an additive subgroup of $\mathbb{Z}_4^n$. Such codes have been studied recently both in connection with the construction of sequences with low correlation ([10,11,86,92]) and in a variety of other contexts (see [28] and the references contained therein).

We define an inner product on $\mathbb{Z}_4^n$ by $a \cdot b = a_1b_1 + \cdots + a_nb_n \pmod 4$, and then the notions of *dual* code ($\mathcal{C}^\perp$), *self-orthogonal* code ($\mathcal{C} \subseteq \mathcal{C}^\perp$) and *self-dual* code ($\mathcal{C} = \mathcal{C}^\perp$) are defined in the standard way (cf. [55,68]). For many applications, there is no need to distinguish between $+1$ components of codewords and -1 components, and so we say that two codes are *equivalent* if one can be obtained from the other by permuting the coordinates and (if necessary) changing the signs of certain coordinates. Codes differing by only a permutation of coordinates are called *permutation-equivalent*. The automorphism group $\mathrm{Aut}(\mathcal{C})$ of $\mathcal{C}$ consists of all permutations and sign-changes of the coordinates that preserve the set of codewords.

Any code is permutation-equivalent to a code $\mathcal{C}$ with generator matrix of the form

$$G = \begin{bmatrix} I_{k_1} & A & B \\ 0 & 2I_{k_2} & 2C \end{bmatrix}, \tag{4.1}$$

DOI: 10.1201/9781003393931-4

where A and C are $\mathbb{Z}_2$-matrices and B is a $\mathbb{Z}_4$-matrix. The code is then an elementary abelian group of type $4^{k_1}2^{k_2}$, containing $2^{2k_1+k_2}$ codewords. We shall indicate this by saying that $\mathcal{C}$ has type $4^{k_1}2^{k_2}$, or simply that $|\mathcal{C}| = 4^{k_1}2^{k_2}$.

Equation (4.1) illustrates a difference in point of view between ring theory and coding theory. Quaternary codes are $\mathbb{Z}_4$-modules, which is also denoted by $\mathbb{Z}_4$-code. A ring theorist would point out, correctly, that a quaternary code is not in general a free module [50], and so need not have a basis. Although this is true, equation (4.1) is a perfectly good generator matrix. Encoding is carried out by writing the information symbols in the form $u = u_1 \cdots u_{k_1} u_{k_1+1} \cdots u_{k_1+k_2}$, where $u_i \in \mathbb{Z}_4$ if $1 \le i \le k_1$, $u_i \in \mathbb{Z}_2$ if $k_1 + 1 \le i \le k_1 + k_2$, and mapping u to the codeword uG. The code $\mathcal{C}$ is a free $\mathbb{Z}_4$-module if and only if $k_2 = 0$.

If $\mathcal{C}$ has generator matrix equation (4.1), the dual code $\mathcal{C}^\perp$ has generator matrix

$$\begin{bmatrix} -B^{tr} - C^{tr}A^{tr} & C^{tr} & I_{n-k_1-k_2} \\ 2A^{tr} & 2I_{k_2} & 0 \end{bmatrix} \tag{4.2}$$

and type $4^{n-k_1-k_2}2^{k_2}$.

Formally, we define three maps from $\mathbb{Z}_4$ to $\mathbb{Z}_2$ by

c	$\alpha(c)$	$\beta(c)$	$\gamma(c)$
0	0	0	0
1	1	0	1
2	0	1	1
3	1	1	0

and extend them in the obvious way to maps from $\mathbb{Z}_4^n$ to $\mathbb{Z}_2^n$. The two-adic expansion of $c \in \mathbb{Z}_4$ is

$$c = \alpha(c) + 2\beta(c) \ . \tag{4.3}$$

Note that $\alpha(c) + \beta(c) + \gamma(c) = 0$ for all $c \in \mathbb{Z}_4$. We construct binary codes from quaternary codes using the *Gray map* $\phi : \mathbb{Z}_4^n \to \mathbb{Z}_2^{2n}$ given by

$$\phi(c) = (\beta(c), \gamma(c)), \quad c \in \mathbb{Z}_4^n \ . \tag{4.4}$$

When we speak of the binary image of a quaternary code $\mathcal{C}$, we will always mean its image $C = \phi(\mathcal{C})$ under the Gray map. We use script letters for quaternary codes, with the corresponding Latin letters for their binary images.

C is in general a nonlinear binary code of length $2n$. If C *is* linear, and $\mathcal{C}$ is defined by (4.1), then C has generator matrix

$$\begin{bmatrix} I_{k_1} & A & \alpha(B) & I_{k_1} & A & \alpha(B) \\ 0 & I_{k_2} & C & 0 & I_{k_2} & C \\ 0 & 0 & \beta(B) & I_{k_1} & A & \gamma(B) \end{bmatrix} . \tag{4.5}$$

We say that a binary code C is $\mathbb{Z}_4$-*linear* if its coordinates can be arranged so that it is the image under the Gray map ϕ of a quaternary code $\mathcal{C}$.

The crucial property of the Gray map is that it preserves distances.

Theorem 4.1. *ϕ is a distance-preserving map or* isometry *from*

$$(\mathbb{Z}_4^n, \text{ Lee distance}) \text{ to } (\mathbb{Z}_2^{2n}, \text{ Hamming distance}) .$$

Proof. It is easy to see from the definitions that

$$wt(\phi(a)) = wt_L(a), \quad a \in \mathbb{Z}_4^n , \tag{4.6}$$
$$d(\phi(a), \phi(b)) = d_L(a, b), \quad a, b \in \mathbb{Z}_4^n , \tag{4.7}$$

where $wt(\ )$ and $d(\ ,\ )$ are the usual Hamming weight and distance functions for binary vectors. ■

From (4.7), the Hamming distance between the binary images $\phi(a)$ and $\phi(b)$ is proportional to the squared Euclidean distance between the complex sequences i^a and i^b.

Two other binary codes $C^{(1)}$, $C^{(2)}$ are canonically associated with a quaternary code $\mathcal{C}$. These are the linear codes defined by

$$C^{(1)} = \{\alpha(c) : c \in \mathcal{C}\} , \tag{4.8}$$
$$C^{(2)} = \{\beta(c) : \ c \in \mathcal{C}, \ \alpha(c) = 0\} . \tag{4.9}$$

If $\mathcal{C}$ has generator matrix equation (4.1), then $C^{(1)}$ is an $[n, k_1]$ code with generator matrix

$$[I_{k_1} \ \ A \ \ \alpha(B)] , \tag{4.10}$$

while $C^{(2)} \supseteq C^{(1)}$ is an $[n, k_1 + k_2]$ code with generator matrix

$$\begin{bmatrix} I_{k_1} & A & \alpha(B) \\ 0 & I_{k_2} & C \end{bmatrix} \tag{4.11}$$

— compare (4.5). It is shown in [28] that given any binary codes C', C'' of length n with $C'' \supseteq C'$, there is a quaternary code $\mathcal{C}$ with $C^{(1)} = C'$, $C^{(2)} = C''$.

Since in general $C = \phi(\mathcal{C})$ is not linear, it need not have a dual. We define its $\mathbb{Z}_4$-*dual* to be $C_\perp = \phi(\mathcal{C}^\perp)$. The principal results to be derived here are the following:

1. C and $C_\perp$ are distance invariant.

2. The weight distributions of C and $C_\perp$ are MacWilliams transforms of one another.

A binary code C is said to be *distance invariant* [68, p. 40] if the Hamming weight distributions of its translates $u + C$ are the same for all $u \in C$.

Theorem 4.2. *If $\mathcal{C}$ is a (linear) quaternary code, then its binary Gray representation $C = \phi(\mathcal{C})$ is distance invariant.*

Proof. C is distance invariant (with respect to Lee distance) because it is linear, and the result then follows from Theorem 4.1. ■

Theorem 4.3. *The binary image $\phi(\mathcal{C})$ of a quaternary linear code $\mathcal{C}$ is linear if and only if*

$$a, b \in \mathcal{C} \Rightarrow 2\alpha(a) * \alpha(b) \in \mathcal{C} \tag{4.12}$$

Proof. This is an immediate consequence of the identity

$$\phi(a) + \phi(b) + \phi(a + b) = \phi(2\alpha(a) * \alpha(b)) \tag{4.13}$$

for all $a, b \in \mathbb{Z}_4^n$. ■

Several weight enumerators are associated with a quaternary code $\mathcal{C}$. The *complete weight enumerator* (or c.w.e.) of $\mathcal{C}$ is

$$\mathrm{cwe}_{\mathcal{C}}(W, X, Y, Z) = \sum_{a \in \mathcal{C}} W^{n_0(a)} X^{n_1(a)} Y^{n_2(a)} Z^{n_3(a)} , \tag{4.14}$$

where $n_j(a)$ is the number of components of a that are congruent to $j \pmod 4$ (cf. [55], [68, p. 141]). Permutation-equivalent codes have the same c.w.e., but equivalent codes may have distinct c.w.e.'s. The appropriate weight enumerator for an equivalence class of codes is the *symmetrized weight enumerator* (or s.w.e.), obtained by identifying X and Z in (4.14):

$$\mathrm{swe}_{\mathcal{C}}(W, X, Y) = \mathrm{cwe}_{\mathcal{C}}(W, X, Y, X) . \tag{4.15}$$

For a distance invariant code C of length n, the (Hamming) weight enumerator

$$\mathrm{Ham}_C(W, Z) = \sum_{c' \in C} W^{n-d(c',c)} X^{d(c',c)}$$

is independent of $c \in C$. If $C = \phi(\mathcal{C})$, it follows from Theorem 4.1 that

$$\mathrm{Ham}_C(W, X) = \mathrm{Lee}_{\mathcal{C}}(W, X) = \mathrm{swe}_{\mathcal{C}}(W^2, WX, X^2) . \tag{4.16}$$

We then have the following analogues of the MacWilliams identity, giving the weight enumerators for the dual code $\mathcal{C}^\perp$ ([28,55,56]):

$$\mathrm{cwe}_{\mathcal{C}^\perp}(W, X, Y, Z) = \frac{1}{|\mathcal{C}|}\mathrm{cwe}_{\mathcal{C}}(W + X + Y + Z, W + iX - Y - iZ,$$

$$W - X + Y - Z, W - iX - Y + iZ) , \tag{4.17}$$

$$\mathrm{swe}_{\mathcal{C}^\perp}(W, X, Y) = \frac{1}{|\mathcal{C}|}\mathrm{swe}_{\mathcal{C}}(W + 2X + Y, W - Y, W - 2X + Y) , \tag{4.18}$$

$$\mathrm{Lee}_{\mathcal{C}^\perp}(W, X) = \frac{1}{|\mathcal{C}|}\mathrm{Lee}_{\mathcal{C}}(W + X, W - X) , \tag{4.19}$$

$$\mathrm{Ham}_{\mathcal{C}^\perp}(W,X) = \frac{1}{|\mathcal{C}|}\mathrm{Ham}_{\mathcal{C}}(W+3X, W-X) \ . \tag{4.20}$$

There are also several analogues of Gleason's theorem, giving bases for the weight enumerators of self-dual codes — see [28,55].

By A_i we denote the number of codewords of C of homogeneous weight i and by B_i we denote the number of codewords of homogeneous weight i of its dual code $C^\perp$. Thus, we can easily rewrite the classical *MacWilliams identities* [68] or the *Pless power moments* [71] for the Hamming weight over $\mathbb{F}_2$. Using $A_0 = B_0 = 1$ and $B_1 = 0$ the first four *(Pless) power moments* for the $wt_L(\cdot)$ are given by:

$$\sum_{i>0} A_i = 2^{2k_1+k_2} - 1, \tag{4.21}$$

$$\sum_{i\geq 0} iA_i = 2^{2k_1+k_2}n, \tag{4.22}$$

$$\sum_{i\geq 0} i^2A_i = 2^{2k_1+k_2-1}(B_2 + n(2n+1)), \tag{4.23}$$

$$\sum_{i\geq 0} i^3A_i = 2^{2k_1+k_2-2}(3(B_2n - B_3) + 2n^2(2n+3)). \tag{4.24}$$

We remark that the first three equations (including B_1) can also be found in [85, Theorem 3.1].

Galois rings have been studied by MacDonald [67], Liebler and Mena [65], Shankar [74], Solé [86], Yamada [96], Boztaş, Hammons and Kumar [11], among others, and of course the general machinery of commutative algebra, as described for example in Zariski and Samuel [98], is applicable to these rings. We list here some of the basic facts we shall need; proofs may be found in the above references.

Let $h_2(X) \in \mathbb{Z}_2[X]$ be a primitive irreducible polynomial of degree m. There is a unique monic polynomial $h(X) \in \mathbb{Z}_4[X]$ of degree m such that $h(X) \equiv h_2(X)$ (mod 2) and $h(X)$ divides $X^n - 1$ (mod 4), where $n = 2^m - 1$ (see for example Yamada [96]). The polynomial $h(X)$ is a *primitive basic irreducible* polynomial, and may be found as follows.

Let $h_2(X) = e(X) - d(X)$, where $e(X)$ contains only even powers and $d(X)$ only odd powers. Then $h(X)$ is given by $h(X^2) = \pm(e^2(X) - d^2(X))$. This is Graeffe's method [86,91] for finding a polynomial whose roots are the squares of the roots of $h_2(X)$. For example, when $m = 3$, $n = 7$, we may take $h_2 = X^3 + X + 1$. Then $e = 1$, $d = -X^3 - X$, $e^2 - d^2 = -X^6 - 2X^4 - X^2 + 1$, so

$$h(X) = X^3 + 2X^2 + X - 1 \ . \tag{4.25}$$

Table I in [11] gives all primitive basic irreducible polynomials of degree $m \leq 10$.

Let ξ be a root of $h(X)$, so that $\xi^n = 1$. Then the *Galois ring* $GR(4^m)$ is defined to be $R = \mathbb{Z}_4[\xi]$. There are two canonical ways to represent the 4^m elements of R (just as there are two canonical ways, multiplicative and additive, to represent elements of $GF(q^m)$).

In the first representation, every element $c \in R$ has a unique 'multiplicative' or two-adic representation

$$c = a + 2b \ , \tag{4.26}$$

where a and b belong to the set

$$\mathcal{T} = \{0, 1, \xi, \xi^2, \ldots, \xi^{n-1}\} \ . \tag{4.27}$$

The map $\tau : c \mapsto a$ is given by

$$\tau(c) = c^{2^m}, \quad c \in R \ , \tag{4.28}$$

and satisfies

$$\begin{aligned} \tau(cd) &= \tau(c)\tau(d) \ , & (4.29) \\ \tau(c+d) &= \tau(c) + \tau(d) + 2(cd)^{2^{m-1}} & (4.30) \end{aligned}$$

(see [96]). Given c, one determines a from (4.28) and then b from (4.26).

In the second representation, each element $c \in R$ has a unique 'additive' representation

$$c = \sum_{r=0}^{m-1} b_r \xi^r, \quad b_r \in \mathbb{Z}_4 \ . \tag{4.31}$$

For example, if $m = 3$ and h is given by (4.25), the additive representations for the elements of $\mathcal{T}$ and $2\mathcal{T}$ are

element	b_0	b_1	b_2	$2b_0$	$2b_1$	$2b_2$
0	0	0	0	0	0	0
1	1	0	0	2	0	0
ξ	0	1	0	0	2	0
ξ^2	0	0	1	0	0	2
ξ^3	1	3	2	2	2	0
ξ^4	2	3	3	0	2	2
ξ^5	3	3	1	2	2	2
ξ^6	1	2	1	2	0	2

(4.32)

This table may be produced (just as for Galois fields) by a (modulo 4) shift register whose feedback polynomial is $h(X)$. By using (4.26), the table gives the additive representation of every element of R.

One essential difference between $R = GR(4^m)$ and a Galois field is that R contains zero divisors: these are the elements of the radical $2R$, the unique maximal ideal in R (R is a local ring). Let μ denote the map $R \to R/2R$. Then $\theta = \mu(\xi)$ is a root of $h_2(X)$, and we can identify $R/2R$ with $GF(2^m)$, taking the elements of $GF(2^m)$ to be

$$\mu(\mathcal{T}) = \{0, 1, \theta, \theta^2, \ldots, \theta^{n-1}\} \ . \tag{4.33}$$

We denote the set of *regular* or *invertible* elements of R by $R^* = R \setminus 2R$. Every element of R^* has a unique representation in the form $\xi^r(1+2t)$, $0 \leq r \leq n-1$, $t \in \mathcal{T}$. R^* is a multiplicative group of order $(2^m - 1)2^m$ which is a direct product $H \times \mathcal{E}$, where H is the cyclic group of order $2^m - 1$ generated by ξ, and $\mathcal{E}$ is the group of *principal units* of R, that is, elements of the form $1 + 2t$, $t \in \mathcal{T}$. $\mathcal{E}$ has the structure of an elementary abelian group of order 2^m and is isomorphic to the additive group of $GF(2^m)$.

The *Frobenius map* f from R to R is the ring automorphism that takes any element $c = a + 2b \in R$ to

$$c^f = a^2 + 2b^2 . \tag{4.34}$$

f generates the Galois group of R over $\mathbb{Z}_4$, and f^m is the identity map. The *relative trace* from R to $\mathbb{Z}_4$ is defined by

$$T(c) = c + c^f + c^{f^2} + \cdots + c^{f^{m-1}}, \quad c \in R . \tag{4.35}$$

For comparison, the usual trace from $GF(2^m)$ to $\mathbb{Z}_2$ is given by

$$tr(c) = c + c^2 + c^{2^2} + \cdots + c^{2^{m-1}}, \quad c \in GF(2^m) , \tag{4.36}$$

and the Frobenius map is simply the squaring map

$$f_2(c) = c^2, \quad c \in GF(2^m) . \tag{4.37}$$

The following commutativity relationships between these maps are easily verified:

$$\mu \circ f = f_2 \circ \mu , \tag{4.38}$$

$$\mu \circ T = tr \circ \mu . \tag{4.39}$$

In particular, since tr is not identically zero, it follows that the Galois ring trace is nontrivial. In fact, T is an onto mapping from R to $\mathbb{Z}_4$. The set of elements of R invariant under f is identical with $\mathbb{Z}_4$.

In this chapter, we are interested in structural results for l-weight codes where l is small. Mostly we will restrict our attention to projective codes only. In some cases, we will require additional constraints on the sum $\sum_{i=1}^{l} w_i$ of weights (which are motivated by strongly walk regular graphs, see e.g. [35] and see also [83], which studies the field case).

4.1.2 Codes over $\mathbb{Z}_{p^m}$

For simplicity, let R be a finite commutative ring. A *linear code* C over R of length n is an R-submodule of the standard R-module R^n. The *dual code* of C is defined as $C^\perp = \{x \in R^n \; x \cdot C = 0\}$, where $x \cdot y = x_1y_1 + \cdots + x_ny_n$ denotes the standard inner product on R^n.

Definition 4.1. *Let R be a finite ring. A function $w\colon R \to \mathbb{R}$ is called a* homogeneous weight, *if $w(0) = 0$ and* *i. if* $Rx = Ry$ *then* $w(x) = w(y)$ *for all* $x, y \in R$*;*

ii. there exists a real number $\gamma \neq 0$ *such that* $\sum_{y \in Rx} w(y) = \gamma|Rx|$ *for all* $x \in R \setminus \{0\}$.

It is known from the work in [20,42,52] that a homogeneous weight on R exists and is uniquely determined up to the normalization factor γ. If R is a Frobenius ring, it satisfies $\sum_{y \in I} w(y) = \gamma|I|$ for all nonzero ideals I in R. (This is trivially true for $R = \mathbb{Z}_{p^m}$ and other principal ideal rings.) In the sequel, we will denote the homogeneous weight by $\mathrm{w}_{\mathrm{hom}}$, assuming that γ has been fixed.

As usual we extend $\mathrm{w}_{\mathrm{hom}}$ to a weight function on R^n by

$$\mathrm{w}_{\mathrm{hom}}(x_1, \ldots, x_n) = \sum_{i=1}^{n} \mathrm{w}_{\mathrm{hom}}(x_i).$$

If R doesn't contain $\mathbb{Z}_2 \times \mathbb{Z}_2$ or $\mathbb{Z}_2 \times \mathbb{Z}_3 = \mathbb{Z}_6$ as a direct factor, the set R^n forms a metric space under the *homogeneous distance* $\mathrm{d}_{\mathrm{hom}}$ defined by $\mathrm{d}_{\mathrm{hom}}(x, y) = \mathrm{w}_{\mathrm{hom}}(x - y)$; cf. [52, Th. 3]. A code $C \subseteq \mathbb{R}^n$ is called a *(homogeneous) two-weight code* with weights $w_1 < w_2$ if every nonzero codeword has homogeneous weight w_1 or w_2 and both weights actually occur. If C is linear, then on C the homogeneous distance takes only the values 0, w_1, and w_2.

A matrix $G \in R^{\ell \times n}$ is called a *generator matrix* of the linear code C if the rows of G generate C as an R-module.

Definition 4.2. *Let* C *have* $\ell \times n$ *generator matrix* $G = [g_1| \ldots |g_n]$*. The code* C *is called:*

i. proper *if* $\mathrm{w}_{\mathrm{hom}}(c) = 0$ *implies* $c = 0$ *for all* $c \in C$*;*

ii. fat *if* $\{x \cdot g_i; x \in R^\ell\} = R$ *for* $i = 1, \ldots, n$*;*

iii. projective *if* $Rg_i \neq Rg_j$ *for any pair of distinct coordinates* $i, j \in \{1, \ldots, n\}$.

The term "fat linear code" was coined in [45] and alludes to the fact that the codewords of such a code, viewed as functions $\{1, \ldots, n\} \to R$, cover the whole cartesian product $\{1, \ldots, n\} \times R$. In the case $R = \mathbb{F}_q$ such codes are those of effective length n (i.e., without all-zero coordinates). Subsequent papers, e.g. [13], have used the less specific term "regular linear code" instead.

In the sequel, we shall be interested exclusively in the case $R = \mathbb{Z}_{p^m}$, the ring of integers modulo a prime power p^m. In this case, the homogeneous weight is given by

$$\mathrm{w}_{\mathrm{hom}}(x) = \begin{cases} 0 & \text{if } x = 0, \\ p^{m-1} & \text{if } 0 \neq x \in p^{m-1}\mathbb{Z}_{p^m}, \\ (p-1)p^{m-2} & \text{otherwise.} \end{cases}$$

The chosen normalization constant is $\gamma = (p-1)p^{m-2}$, which is the most convenient for our purposes. Since $\mathrm{w}_{\mathrm{hom}}(x) > 0$ for all $x \in \mathbb{Z}_{p^m} \setminus \{0\}$, all linear codes over $\mathbb{Z}_{p^m}$ are proper, and fatness reduces to the condition that every column of a generator matrix contains at least one entry not divisible by p.

Example 4.1. *The homogeneous weight on $\mathbb{Z}_4$ is precisely the Lee weight. The matrix*

$$\begin{bmatrix} 1 & 1 & 1 & 1 \\ 0 & 0 & 2 & 2 \\ 0 & 2 & 0 & 2 \end{bmatrix}$$

generates a fat, projective two-weight code over $\mathbb{Z}_4$ with 2^4 codewords and nonzero Lee weights $w_1 = 4$, $w_2 = 8$. The corresponding weight frequencies are $A_1 = 14$, $A_2 = 1$. The binary image of this code under the Gray map $\mathbb{Z}_4 \to \mathbb{Z}_2^2$ is the $[8,4,4]$ extended Hamming code.

Example 4.2. *The homogeneous weight on $\mathbb{Z}_9$ is given by $\mathrm{w}_{\mathrm{hom}}(0) = 0$, $\mathrm{w}_{\mathrm{hom}}(3) = \mathrm{w}_{\mathrm{hom}}(6) = 3$, and $\mathrm{w}_{\mathrm{hom}}(x) = 2$ for the remaining elements in $\mathbb{Z}_9$ which are the units of $\mathbb{Z}_9$. The matrix*

$$\begin{bmatrix} 1 & 1 & 1 & 1 & 1 & 1 & 1 & 1 & 1 \\ 0 & 0 & 0 & 3 & 3 & 3 & 6 & 6 & 6 \\ 0 & 3 & 6 & 0 & 3 & 6 & 0 & 3 & 6 \end{bmatrix}$$

generates a fat, projective two-weight code over $\mathbb{Z}_9$ with 3^4 codewords and nonzero homogeneous weights $w_1 = 18$, $w_2 = 27$. The corresponding frequencies are $A_1 = 78$, $A_2 = 2$. The ternary image of this code under the generalized Gray map $\mathbb{Z}_9 \to \mathbb{Z}_3^3$, $a = a_0 + 3a_1 \mapsto (a_0, a_1)\left(\begin{smallmatrix} 0 & 1 & 2 \\ 1 & 1 & 1 \end{smallmatrix}\right)$ is a nonlinear code with the same parameters as the $[27,4,18]$ first-order generalized Reed-Muller code $\mathrm{RM}_3(1,3)$.

These examples are special cases of the linear code over $\mathbb{Z}_{p^m}$ associated to a point neighbor class in the projective Hjelmslev plane over $\mathbb{Z}_{p^m}$

A key property of the homogeneous weight, which will be used later, is an expression in terms of a character sum. Recall that a *character* of the finite abelian group $(R,+)$ is a map $\chi\colon R \to \mathbb{C}^\times$ satisfying $\chi(x+y) = \chi(x)\chi(y)$ for $x, y \in R$. A character χ of $(R,+)$ is called a *generating character* of R if every character of $(R,+)$ is of the form $x \mapsto \chi(ax)$ for some $a \in R$. It is known from [95] that the finite rings possessing a generating character are precisely the finite Frobenius rings. A finite commutative ring is a Frobenius ring iff the local rings R_i in its block decomposition $R = \prod_{i=1}^{r} R_i$ have unique minimal ideals. In particular, this holds for residue class rings $\mathbb{Z}_n$ of integers, in which case we can take the generating character as $\chi(x) = \omega^x$ with $\omega = e^{2\pi i/n}$. The next lemma describes the homogeneous weight on $\mathbb{Z}_{p^m}$ in terms of this "canonical" generating character. See [43, Eq. (4), p. 412] for a proof.

Lemma 4.1. *For any $x \in \mathbb{Z}_{p^m}$, we have*

$$\begin{aligned} & \mathrm{w_{hom}}(x) \\ = \; & (p-1)p^{m-2} - \frac{1}{p} \sum_{u \in \mathbb{Z}_{p^m}^{\times}} \chi(ux) \\ = \; & (p-1)p^{m-2} - \frac{1}{p} \sum_{\substack{1 \le u \le p^m \\ \gcd(u,p)=1}} \epsilon^{2\pi i/u\, xu/p^m}. \end{aligned}$$

The next Lemma is a generalization from fields to rings of a classical result of Delsarte. See [3, Cor. 16] for a proof.

Lemma 4.2. *If C is a fat projective two-weight code over $\mathbb{Z}_{p^m}$ with nonzero weights $w_1 < w_2$, then there are integers $u \geq 1$ and $t \geq 0$ such that*

$$\begin{aligned} w_1 &= up^t \\ w_2 &= (u+1)p^t. \end{aligned}$$

We need some notions from graph theory.

Definition 4.3. *A simple graph on v vertices is called a* strongly regular graph *with parameters (v, η, λ, μ) if*

i. each vertex is adjacent to η vertices;

ii. for each pair of adjacent vertices, there are λ vertices adjacent to both;

iii. for each pair of non-adjacent vertices, there are μ vertices adjacent to both.

It is easily seen that the complementary graph of a strongly regular graph is again strongly regular and has parameters $(v, v-1-\eta, v-2\eta+\mu-2, v-2\eta+\lambda)$. Strongly regular graphs with $\mu \in \{0, \eta\}$ (unions of complete graphs of the same size and their complements, respectively) are said to be *trivial.* The parameters of a strongly regular graph are related by $\eta(\eta-\lambda-1) = \mu(v-\eta-1)$, which shows in particular that μ is determined by the other three parameters.

An *eigenvalue* of a graph Γ (i.e., an eigenvalue of its adjacency matrix) is called a *restricted eigenvalue* if there is a corresponding eigenvector which is not a multiple of the all-one vector $\mathbf{1}$. Note that for an η-regular connected graph, the restricted eigenvalues are simply the eigenvalues different from η.

It is known that a simple graph, not complete or edgeless, is strongly regular iff it has precisely two distinct restricted eigenvalues.

Lemma 4.3. *Let Γ be a strongly regular graph with parameters (v, η, λ, μ) and restricted eigenvalues $r > s$. Then:*

$$\begin{aligned} rs &= \mu - \eta, \\ r + s &= \lambda - \mu, \\ (r-s)^2 &= (\lambda-\mu)^2 + 4(\eta - \mu). \end{aligned}$$

These relations are well-known. They can be derived from the fact that the adjacency matrix A of Γ satisfies the equation $A^2 = \eta I_v + \lambda A + \mu(J_v - I_v - A)$, where I_v, J_v denote the $v \times v$ identity matrix and the $v \times v$ all-one matrix, respectively. See [2, Th. 9.1.3] or [32, Ch. 2] for details.

Definition 4.4. *Let T be a finite abelian group and $S \subseteq T$ a subset satisfying $S = -S$ and $0_T \notin S$. The corresponding* Cayley graph $\mathrm{C}(T, S)$ *has vertex set equal to T; two vertices $g, h \in T$ are adjacent in $\mathrm{C}(T, S)$ iff $g - h \in S$.*

The conditions imposed on S ensure that $\mathrm{C}(T, S)$ is a simple graph. The graph $\mathrm{C}(T, S)$ is regular of degree $|S|$, and it is connected iff S generates the group T.

Finally we provide an introduction to the correspondence between fat linear codes over $\mathbb{Z}_{p^m}$ and multisets of points in the projective Hjelmslev geometries $\mathrm{PHG}(k-1, \mathbb{Z}_{p^m})$; see [45]–[47] for further information and any terms not explained below.

The $(k-1)$-dimensional projective Hjelmslev geometry over $\mathbb{Z}_{p^m}$ is defined as the incidence structure whose points (lines) are the free rank-one (free rank-two, respectively) submodules of $(\mathbb{Z}_{p^m})^k$ and whose incidence relation is given by set inclusion. In contrast with classical geometry, the number of lines incident with two distinct points P, P' is not always one but depends on the isomorphism type ("shape") of the module $P + P'$. If $|P \cap P'| \geq p^i$ or, equivalently, $P + P' \cong \mathbb{Z}_{p^m} \oplus p^j\mathbb{Z}_{p^m}$ for some $j \in \{i, i+1, \ldots, m\}$, the points P, P' are said to be *i-neighbors* in $\mathrm{PHG}(k-1, \mathbb{Z}_{p^m})$; 1-neighbors are just called *neighbors*, and non-neighbors are said to be *distant*. Two points P, P' are incident with a unique line iff P, P' are distant, and incident with $t > 1$ lines if they are neighbors.

Free submodules of $(\mathbb{Z}_{p^m})^k$ are called *Hjelmslev subspaces* of $\mathrm{PHG}(k-1, \mathbb{Z}_{p^m})$ and are analogous to the flats in classical projective geometry. As in the classical case, free submodules of rank 3, 4, $k-1$ are called *planes*, *solids*, respectively, *hyperplanes* of $\mathrm{PHG}(k-1, \mathbb{Z}_{p^m})$. The neigborhood relations, which are equivalence relations, can be generalized to arbitrary Hjelmslev subspaces by calling S, T i-neighbors if $S \subseteq T + p^i(\mathbb{Z}_{p^m})^k$ or $T \subseteq S + p^i(\mathbb{Z}_{p^m})^k$.

Example 4.3. *The smallest example is* $\mathrm{PHG}(2, \mathbb{Z}_4)$, *the projective Hjelmslev plane over $\mathbb{Z}_4$. It has 28 points (lines), falling into 7 neighbor classes of size 4. The points are represented by the columns of the following 3×28 matrix, with neighbor classes separated by vertical bars.*

$$\left[\begin{array}{cccc|cccc|cccc|cccc|cccc|cccc|cccc}
\mathbf{1}&1&1&1&\mathbf{0}&0&2&2&\mathbf{0}&0&2&2&\mathbf{1}&1&1&1&0&0&2&\mathbf{2}&1&1&1&\mathbf{1}&1&1&1&\mathbf{1}\\
\mathbf{0}&0&2&2&\mathbf{1}&1&1&1&\mathbf{0}&2&0&2&\mathbf{1}&1&3&3&1&1&1&\mathbf{1}&0&0&2&\mathbf{2}&1&1&3&\mathbf{3}\\
\mathbf{0}&2&0&2&\mathbf{0}&2&0&2&\mathbf{1}&1&1&1&\mathbf{1}&3&1&3&1&3&1&\mathbf{3}&1&3&1&\mathbf{3}&0&2&0&\mathbf{2}
\end{array}\right].$$

Lines have equations $a_1x_1 + a_2x_2 + a_3x_3 = 0$ with at least one of a_1, a_2, a_3 a unit in $\mathbb{Z}_4$, and hence are represented by the columns as well.

Each point (line) is incident with 6 lines (points). The 7 points in bold type form a hyperoval $\mathfrak{H}$, *i.e., no three points of $\mathfrak{H}$ are collinear and hence every line through a point of $\mathfrak{H}$ is a secant. The point set $\mathfrak{H}$ gives rise to the Heptacode considered in the next example.*

With any n-multiset $\mathfrak{K}\colon \mathcal{P} \to \mathbb{N}_0$ of points in $\mathrm{PHG}(k-1, \mathbb{Z}_{p^m})$ we can associate a fat linear code C of length n over $\mathbb{Z}_{p^m}$ by choosing module generators $g_1, \ldots, g_n$ of the points and define C as the linear code generated by the matrix $G = [g_1|\ldots|g_n]$. The code C is projective iff the corresponding multiset of points is a set. The so-defined correspondence between multisets of points in $\mathrm{PHG}(k-1, \mathbb{Z}_{p^m})$ and linear codes over $\mathbb{Z}_{p^m}$ is many-to-many in general, but it becomes a 1-1 correspondence if we identify monomially isomorphic linear codes and linearly isomorphic multisets of points. The parameters of C can be computed from geometric data associated with $\mathfrak{K}$ as follows: Since the row and column spaces of any matrix over $\mathbb{Z}_{p^m}$ are isomorphic (see, e.g., [44, Cor. 1]), the isomorphism type of C (the row space of G) as a $\mathbb{Z}_{p^m}$-module is the same as that of $\langle\mathfrak{K}\rangle = \sum_{P;\mathfrak{K}(P)>0} P$ (the column space of G), and we may assume w.l.o.g. that both modules have rank k.

The $\mathfrak{K}$-type of a hyperplane H of $\mathrm{PHG}(k-1, \mathbb{Z}_{p^m})$ is the $(m+1)$-tuple $(a_0, a_1, \ldots, a_m)$, where a_i denotes the number of points in $\mathfrak{K}$ (counted with their multiplicities) that are exact i-neighbors (i.e., i-neighbors but not $(i+1)$-neighbors) of H. Then for each $(m+1)$-tuple $(a_0, a_1, \ldots, a_m)$ that occurs as the $\mathfrak{K}$-type of some hyperplane of $\mathrm{PHG}(k-1, \mathbb{Z}_{p^m})$, the code C contains codewords with weights

$$
\begin{aligned}
&(a_0 + \cdots + a_{m-2})(p^{m-1} - p^{m-2}) + a_{m-1}p^{m-1},\\
&(a_0 + \cdots + a_{m-3})(p^{m-1} - p^{m-2}) + a_{m-2}p^{m-1},\\
&\vdots\\
&a_0(p^{m-1} - p^{m-2}) + a_1 p^{m-1},\\
&a_0 p^{m-1},
\end{aligned}
$$

and all codeword weights except possibly zero arise in this way.

We consider two examples. The first example rather belongs to the realm of finite field geometry. It has been included here because it forms the smallest nontrivial instance of the hyperoval construction (Construction TF1) of classical two-weight codes [15] and illustrates an important difference between finite field and Hjelmslev geometry—interesting geometric objects, such as hyperovals, in the latter often give rise to three-weight codes rather than two-weight codes.

Example 4.4. *The* $[6,3,4]$ *Hexacode over* $\mathbb{F}_4 = \{0,1,a,b\}$ *is generated by*

$$
\mathbf{G} = \begin{bmatrix} 1 & 0 & 0 & 1 & 1 & 1\\ 0 & 1 & 0 & 1 & a & b\\ 0 & 0 & 1 & 1 & b & a \end{bmatrix}
$$

The columns of $\mathbf{G}$ *represent a hyperoval in the projective plane* $\mathrm{PG}(2, \mathbb{F}_4)$ *of order* 4. *Since such a hyperoval has* 15 *secants and* 6 *passants, the Hexacode is a two-weight code with nonzero weights* $w_1 = 4$, $w_2 = 6$ *and frequencies* $A_1 = 45$, $A_2 = 18$.

Example 4.5. *The hyperoval* $\mathfrak{H}$ *from Ex. 4.3 is associated with the* $(7, 4^3)$ *code* C *over* $\mathbb{Z}_4$ *generated by*

$$\mathbf{G} = \begin{bmatrix} 1 & 0 & 0 & 1 & 2 & 3 & 1 \\ 0 & 1 & 0 & 1 & 1 & 2 & 3 \\ 0 & 0 & 1 & 1 & 3 & 1 & 2 \end{bmatrix}$$

and called Heptacode. *(For reasons of symmetry, the other coordinate vector representing the 6th point of* $\mathfrak{H}$ *has been chosen as column* g_6.*) Since* $\mathfrak{H}$ *contains exactly one point from each neighbor class of* $\mathrm{PHG}(2, \mathbb{Z}_4)$, *the torsion code* $2C$ *is the* $[7, 4]$ *simplex code and contributes* 7 *codewords of Lee weight* 8. *Lines have* $\mathfrak{H}$*-type* $(4, 1, 2)$ *(the* $\binom{7}{2} = 21$ *secants of* $\mathfrak{H}$*) or* $(4, 3, 0)$ *(the* $28 - 21 = 7$ *passants of* $\mathfrak{H}$*) and contribute* $2 \cdot 21 = 42$ *codewords of Lee weight* 6, *respectively,* $2 \cdot 7 = 14$ *codewords of Lee weight* 10. *Thus* C *is a three-weight code with nonzero weights* $w_1 = 6$, $w_2 = 8$, $w_3 = 10$, *and corresponding frequencies* $A_1 = 42$, $A_2 = 7$, $A_3 = 14$.

The binary Gray image of C *is a nonlinear* $(14, 2^6, 6)$ *code and has higher minimum distance than any binary linear* $[14, 6]$ *code.*

From C *one can also obtain the famous Octacode* O *by prepending a zero column to* $\mathbf{G}$ *and then adding as first row the word* $(1, 0, 0, 0, 2, 1, 1, 1)$. *The point set* $\mathfrak{O}$ *in* $\mathrm{PHG}(3, \mathbb{Z}_4)$ *corresponding to* O *forms an* 8*-arc (i.e.,* 8 *points no four of which are coplanar) and can be constructed from the cone with vertex* $\mathbb{Z}_4(1, 0, 0, 0)$ *and base* $\mathfrak{H}$ *(embedded into* $\mathrm{PHG}(3, \mathbb{Z}_4)$ *by adding zero as first coordinate); for details see [46].*

We have restricted the discussion to the ring $R = \mathbb{Z}_{p^m}$, but projective Hjelmslev geometries $\mathrm{PHG}(k-1, R)$ can be defined for arbitrary finite chain rings R as the analogous incidence structures formed by the free submodules of R^k_R (or ${}_RR^k$) of ranks 1 and 2.

4.1.3 Additive Codes over $\mathbb{F}_4$

We consider codes of length n for the Hamming distance over $\mathbb{F}_4$, and denote the Hamming weight of x by $\mathrm{wt}(x)$. An additive code of length n over $\mathbb{F}_4$ is an additive subgroup of $\mathbb{F}_4^n$. It is a free $\mathbb{F}_2$ module with 4^k elements for some $k \leq n$ (here $2k$ is an integer, but k may be half-integral). There are two ways to specify such a code C. Firstly, using a *generator matrix* G, the code C can be cast as the $\mathbb{F}_2$-span of its rows. Secondly, a *parity check matrix* H can be defined by using the *trace inner product* given in dimension 1 by $x * y = \mathrm{Tr}(xy^2)$, where $\mathrm{Tr}(z) = z + z^2$, and extended coordinatewise to vectors of length n. The *dual* of a code C denoted by $C^\perp$ is understood with respect to that inner product. Define the $\star$-product of a $m \times n$ matrix M with rows M_i by a vector of length n as

$$M \star x = (M_1 * x, \cdots, M_m * x)^T,$$

the ^{T} denoting transposition. An additive code with 4^k codewords can then be specified by an $2n - 2k$ by n matrix H, the *parity check matrix* of C, as

$$C = \{x \in \mathbb{F}_4^n \mid H \star x = 0\}.$$

Note that H is a generator matrix for $C^\perp$. We call *syndrome* $s(x)$ of $x \in \mathbb{F}_4^n$ the column vector of size $2n - 2k$ defined by $s(x) = H \star x$. A code C^- is obtained by *puncturing* from a code C at the coordinate i if it is the projection of C on the remaining $n-1$ other coordinates. A *coset* of an additive code C is any translate of C by a constant vector. The definitions of coset leader and coset weight are the same as in the case of linear codes. The *coset graph* Γ_C of a code C is then the graph defined on the 4^{n-k} syndromes, two of them being connected if they differ by a syndrome $s(x)$ with $\mathrm{wt}(x) = 1$. By the obvious one-to-one correspondence between syndromes and cosets [49], this graph is also the graph on the cosets, two cosets being connected if they differ by a coset of weight one. We give without proof the following extension of [1, 11.1.11] from linear to additive codes.

Theorem 4.4. *If C is an additive quaternary code of minimum distance at least three, with dual weight distribution $[\langle i, A_i\rangle]$, then the spectrum of Γ_C is $\{(3n-4i)^{A_i}\}$. Thus A_i is the frequency of weight i in $C^\perp$ and the multiplicity of the eigenvalue $3n-4i$.*

The *monomial automorphism group* of a quaternary code C is the set of all monomial transformations that leave the code wholly invariant, where a *monomial transformation* [49] is the composition of a coordinate permutation and the multiplication of the value in every position by a nonzero scalar. The *complete automorphism group* $\mathrm{Aut}(C)$ of a code C is its stabilizer in the group of isometries of the Hamming space. The complete automorphism group includes the monomial automorphism group and the group of translations of the code, but can be larger than their product. An additive quaternary code is called *completely transitive* in the sense of [89] if the induced action of $\mathrm{Aut}(C)$ on the cosets of given weight of C is transitive. Such codes are completely regular, with a distance-transitive coset graph. For an alternative definition, see [39]. For a partial classification, see [5].

4.2 CONSTRUCTIONS

4.2.1 Preparata Codes over $\mathbb{Z}_4$

Let $h(X)$ be a primitive basic irreducible polynomial of degree m, as above, and let $g(X)$ be the reciprocal polynomial to $(X^n-1)/((X-1)h(X))$, where $n = 2^m - 1$.

Theorem 4.5. *let $\mathcal{K}^-$ be the cyclic code of length n over $\mathbb{Z}_4$ with generator polynomial $g(X)$, and let $\mathcal{K}$ be obtained from $\mathcal{K}^-$ by adjoining a zero-sum check symbol. Then for odd $m \geq 3$ the binary image $K = \phi(\mathcal{K})$ of $\mathcal{K}$ under the Gray map (4.4) is a nonlinear code of length 2^{m+1}, with 4^{m+1} words and minimal distance $2^m - 2^{(m-1)/2}$ that is equivalent to the Kerdock code. This code is distance invariant.*

Note that $\mathcal{K}^-$ has parity check polynomial $(X-1)h(X)$. There are two equivalent generator matrices for $\mathcal{K}$. The first is

$$\begin{bmatrix} 1 & 1 & 1 & 1 & \cdots & 1 \\ 0 & 1 & \xi & \xi^2 & \cdots & \xi^{n-1} \end{bmatrix}, \tag{4.40}$$

where the entries in the second row are to be replaced by the corresponding m-tuples $(b_0 b_1 \cdots b_{m-1})'$ (the prime indicating transposition) obtained from (4.31).

Alternatively, let $g(X) = \sum_{j=0}^{\delta} g_j X^j$, $\delta = 2^m - m - 2$, $g_j \in \mathbb{Z}_4$, and let $g_\infty = -\sum_{j=0}^{\delta} g_j$. Then the second form for the generator matrix for $\mathcal{K}$ is

$$\begin{bmatrix} g_\infty & g_0 & g_1 & \cdots & g_\delta & 0 & \cdots & 0 \\ g_\infty & 0 & g_0 & \cdots & g_{\delta-1} & g_\delta & \cdots & 0 \\ \cdot & \cdot & \cdot & & \cdot & \cdot & & \\ g_\infty & 0 & 0 & \cdots & g_0 & g_1 & \cdots & g_\delta \end{bmatrix}. \tag{4.41}$$

$\mathcal{K}$ is a code of type 4^{m+1}. The binary code $K^{(1)}$ associated with $\mathcal{K}$ (see (4.8)) is $RM(1, m)$.

In this section, we show that the binary image of the dual code $\mathcal{P} = \mathcal{K}^\perp$ is a Preparata-like code with essentially the same properties as Preparata's original code (yet is much simpler to construct).

4.2.1.1 The 'Preparata' Code is an Extended Cyclic Code over $\mathbb{Z}_4$.

Theorem 4.6. *Let $\mathcal{P}^-$ be the cyclic code of length $n = 2^m - 1$ with generator polynomial $h(X)$, and let $\mathcal{P}$ be obtained from $\mathcal{P}^-$ by adjoining a zero-sum check symbol, so that $\mathcal{P} = \mathcal{K}^\perp$. Then for odd $m \geq 3$ the binary image $P = \phi(\mathcal{P})$ of $\mathcal{P}$ under the Gray map (4.4) is a nonlinear code of length $l = 2^{m+1}$, with 2^{l-2m-2} codewords and minimal distance 6. This code is distance invariant and its weight distribution is the MacWilliams transform of the weight distribution of the Kerdock code of the same length.*

Note that $\mathcal{P}^-$ has parity check polynomial $g(X)$, and that (4.40) and (4.41) are equivalent parity check matrices for $\mathcal{P}$. Also $\mathcal{P}$ is a code of type 4^{2^m-m-1}. The code $P = \phi(\mathcal{P})$ is the $\mathbb{Z}_4$-dual of K, and we refer to it as a 'Preparata' code, using the quotes to distinguish it from Preparata's original code. It is known that the Preparata code (and P) contains more codewords than any linear code with the same minimal distance [12]. The binary code $P^{(1)}$ associated with $\mathcal{P}$ (see (4.8)) is $RM(m-2, m)$.

Proof of Theorem 4.6. It follows from Theorems 4.2 that P is distance invariant and its weight distribution is the MacWilliams transform of that of K. By Theorem 24 of [68], Chapter 15, P has the same weight distribution as the original Preparata code. ■

Semakov, Zinoviev and Zaitsev [73] had already shown in 1971 that any code with the same parameters as the Preparata code must be distance invariant.

The decoding algorithm given below provides an alternative proof that $\mathcal{P}$ has minimal Lee weight 6, for odd $m \geq 3$. For even $m \geq 2$, $\mathcal{P}$ contains words of Lee weight 4. For ξ satisfies $\xi^{3t} = 1$, where $t = (2^m - 1)/3$, and since $\xi^t - 1 \in R^*$, by (P1), $\xi^{2t} + \xi^t + 1 = 0$, yielding a word of Lee weight 3 in $\mathcal{P}^-$.

There is one essential difference between P and the original Preparata code. It is known that the latter is contained in the extended Hamming code spanned by its codewords.

Theorem 4.7. *For odd $m \geq 5$, P is contained in a nonlinear code with the same weight distribution as the extended Hamming code of the same length, and the linear code spanned by the codewords of P has minimal weight 2.*

Proof. The first assertion follows by considering the binary images of the following sequence of codes:

$$ZRM(1,m) \subseteq \mathcal{K} \subseteq ZRM(2,m) \subseteq \cdots \subseteq ZRM(2,m)^{\perp} \subseteq \mathcal{P} \subseteq ZRM(1,m)^{\perp} . \tag{4.42}$$

For the second assertion, we use the fact that $\mathcal{P}$ is an extended cyclic code with generator polynomial $h(X) = \sum_{j=0}^{m} h_j X^j$ (say). Let $h_\infty = -h(1) = \pm 1$, since $h_2(1) = 1$. Then $\mathcal{P}$ has a generator matrix of the form

$$\begin{array}{c} a \\ \\ \\ b \\ \\ \end{array} \begin{bmatrix} h_\infty & h_0 & h_1 & \cdots & h_m & & & \\ h_\infty & 0 & h_0 & \cdots & & h_m & & \\ \cdot & \cdot & \cdot & \cdots & & \cdot & & \\ h_\infty & 0 & 0 & \cdots & 0 & h_0 & \cdots & h_m \\ \cdot & \cdot & \cdot & \cdots & & & & \end{bmatrix} . \tag{4.43}$$

It follows from (4.13) that the linear span of $P = \phi(\mathcal{P})$ contains all words of the form $\phi(2\alpha(a) * \alpha(b))$, for $a, b \in \mathcal{P}$. Taking a and b to be as indicated in (4.43) produces a word of weight 2 in the linear span. ■

Again there is a result of Zaitsev, Zinoviev and Semakov that is relevant: they showed in [97] that any code with the same parameters as the Preparata code is a subcode of a possibly nonlinear code with the same parameter as an extended Hamming code. Theorem 4.7 answers a question raised in that paper, by providing an example where the Hamming-type code is indeed nonlinear.

As a quaternary linear code, $ZRM(1,m)^{\perp}$ (see (4.42)) is the union of $2^m - 1$ translates of $\mathcal{P}$, each nonzero translate having minimal Lee weight 4. The codewords of weight 4 in the binary image of $ZRM(1,m)^{\perp}$ (a nonlinear code with the same parameters as an extended Hamming code) form a Steiner system $S(3,4,2^{m+1})$. It is not difficult to show that this $S(3,4,2^{m+1})$ is identical to the Steiner system formed by the codewords of weight 4 in the classical extended Hamming code of length 2^{m+1}. The blocks of this design are divided equally among the binary images of the $2^m - 1$ nonzero cosets of $\mathcal{P}$. The blocks falling in the binary image of a fixed coset form a Steiner system $S(2,4,2^{m+1})$.

4.2.1.2 Transform-Domain Characterization of 'Preparata' Codes.

In spite of the previous theorem, in this section, we shall show that the 'Preparata' code $P = \phi(\mathcal{P})$ and Preparata's original code have similar characterizations by finite field transforms.

We define the *Galois ring transform* $\widehat{c} = (\widehat{c}(\lambda))$, $\lambda = 0, 1, \ldots, n-1, n = 2^m - 1$, of a quaternary sequence $c = (c_t)$, $t = 0, 1, \ldots, n-1$, by

$$\widehat{c}(\lambda) = \sum_{t=0}^{2^m-2} c_t \xi^{\lambda t} .$$

The inversion formula

$$c_t = -\sum_{\lambda=0}^{n-1} \widehat{c}(\lambda) \xi^{-\lambda t}$$

follows in the usual way from the fact that

$$\sum_{\lambda=0}^{n-1} \xi^{\lambda} = \frac{1-\xi^n}{1-\xi} = 0 .$$

We define the *finite field transform* $\tilde{a} = (\tilde{a}(\lambda))$, $\lambda = 0, 1, \ldots, n-1$, of a *binary* sequence $a = (a_t)$, $t = 0, 1, \ldots, n-1$, by

$$\tilde{a}(\lambda) = \sum_{t=0}^{n-1} a_t \theta^{\lambda t} ,$$

where $\theta \in GF(2^m)$ is the image of $\xi \in R$ after reduction modulo 2 (as in Section 3.1). We define the *half-convolution* $\mathcal{H}(\tilde{a}, \lambda) \in GF(2^m)$ of the sequence $\tilde{a}$ at lag λ by

$$\mathcal{H}(\tilde{a}, \lambda) = \sum_{\substack{\lambda_1 \leq \lambda_2 \\ \lambda_1 + \lambda_2 = \lambda}} \tilde{a}(\lambda_1) \tilde{a}(\lambda_2) ,$$

where λ_1, $\lambda_2 = 0, 1, \ldots, n-1$. The summation is a half rather than full convolution because we exclude the cases $\lambda_1 > \lambda_2$.

Theorem 4.8. *The quaternary 'Preparata' code $\mathcal{P}$ consists of all vectors $c = (c_t) \in \mathbb{Z}_4^n$, $t \in \{\infty, 0, 1, \ldots, n-1\}$ satisfying the Galois ring transform constraints*

$$\begin{aligned} c_\infty + \widehat{c}(0) &= 0 , \\ \widehat{c}(1) &= 0 . \end{aligned} \tag{4.44}$$

Proof. This follows from the definition of the Galois ring transform and the parity check matrix for $\mathcal{P}$ given in (4.40). ■

Theorem 4.9. *The binary 'Preparata' code P consists of all vectors $(b, a+b)$ for which $a, b \in \mathbb{Z}_2^n$ satisfy*

$$\begin{aligned} \tilde{a}(0) + a_\infty &= 0 , \\ \tilde{a}(1) &= 0 , \\ \tilde{b}(0) + b_\infty &= \mathcal{H}(\tilde{a}, 0) + a_\infty , \\ \tilde{b}(1) &= \mathcal{H}(\tilde{a}, 1) . \end{aligned} \tag{4.45}$$

Note that (4.44) are over R, whereas (4.45) are over $GF(2^m)$.

Proof. Consider a codeword $c = (c_t) \in \mathcal{P}$, where $c_t = a_t + 2b_t$, $t \in \{\infty, 0, \ldots, n-1\}$. It follows from the previous theorem that

$$\begin{aligned} a_\infty + 2b_\infty + \widehat{a}(0) + 2\widehat{b}(0) &= 0 , \\ \widehat{a}(1) + 2\widehat{b}(1) &= 0 . \end{aligned} \tag{4.46}$$

The next step is to identify the constraints that (4.46) places on $\tilde{a}(\lambda)$, $\tilde{b}(\lambda)$. Given $\lambda \in \{0, 1, \ldots, n-1\}$, let

$$\widehat{a}(\lambda) = e_\lambda + 2f_\lambda, \quad \text{where} \quad e_\lambda,\ f_\lambda \in \mathcal{T} . \tag{4.47}$$

We find f_λ indirectly, starting from the inversion formula

$$a_t = -\sum_{\lambda=0}^{n-1} \widehat{a}(\lambda)\xi^{-\lambda t} .$$

After squaring and also applying the Frobenius map we obtain

$$a_t = a_t^2 = \sum_{\lambda=0}^{n-1} e_\lambda^2 \xi^{-2\lambda t} + 2 \sum_{\lambda_1 < \lambda_2} e_{\lambda_1} e_{\lambda_2} \xi^{-(\lambda_1+\lambda_2)t}$$

and

$$\begin{aligned} a_t &= -\sum_{\lambda=0}^{n-1} (e_\lambda^2 + 2f_\lambda^2)\xi^{-2\lambda t} \\ &= \sum_{\lambda=0}^{n-1} e_\lambda^2 \xi^{-2\lambda t} + 2\sum_{\lambda=0}^{n-1} (e_\lambda^2 + f_\lambda^2)\xi^{-2\lambda t} \end{aligned}$$

respectively. Comparing these two expressions, and using the uniqueness of the Galois ring transform coefficients, we find

$$2(e_\lambda^2 + f_\lambda^2) = 2 \sum_{\substack{0 \le \lambda_1 < \lambda_2 \le n-1 \\ \lambda_1+\lambda_2 = 2\lambda}} e_{\lambda_1} e_{\lambda_2} . \tag{4.48}$$

Now $\mu(e_\lambda) = \tilde{a}(\lambda)$, $\mu(e_{2\lambda}) = \tilde{a}(2\lambda) = \tilde{a}(\lambda)^2 = \mu(e_\lambda)^2$, so (4.48) implies

$$\mu(f_\lambda^2) = \sum_{\substack{0 \le \lambda_1 \le \lambda_2 \le n-1 \\ \lambda_1+\lambda_2 = 2\lambda}} \tilde{a}(\lambda_1)\tilde{a}(\lambda_2) ,$$

an equation in $GF(2^m)$. Taking the square root of both sides we obtain

$$\mu(f_\lambda) = \mathcal{H}(\tilde{a}, \lambda) . \tag{4.49}$$

From (4.46), (4.47), (4.49) we see that

$$a_\infty + 2b_\infty + e_0 + 2f_0 + 2\widehat{b}(0) = 0 ,$$

which implies

$$a_\infty + \mu(e_0) = a_\infty + \tilde{a}(0) = 0 \ ,$$

$$\mu\left(\sqrt{a_\infty e_0}\right) + b_\infty + \mathcal{H}(\tilde{a}, 0) + \tilde{b}(0) = 0$$

and the first and third equations of (4.45) now follow. The second and fourth equations follow easily from the second equation of (4.46). ■

For comparison with (4.45), a transform characterization of Preparata's original code (of the same length 2^{m+1}) can be readily derived from the description given by Baker, van Lint and Wilson [8]: a vector $(b, a+b)$ is in this code if and only if

$$\begin{aligned} \tilde{a}(0) + a_\infty &= 0 \\ \tilde{a}(1) &= 0 \\ \tilde{b}(0) + b_\infty &= 0 \\ \tilde{b}(1)^3 &= \tilde{a}(3) \ . \end{aligned} \tag{4.50}$$

The similarity between (4.45) and (4.50) is evident. At length 16 (the case $m = 3$) the two descriptions must coincide, since the Nordstrom-Robinson code is unique (see Section 4.5). This may be verified directly as follows.

Theorem 4.10. *When $m = 3$ the 'Preparata' code P coincides with Preparata's original code.*

Proof. It is enough to show that $\mathcal{H}(\tilde{a}, 0) = a_\infty \tilde{a}(0)$ and $\mathcal{H}(\tilde{a}, 1) = \tilde{a}(3)^{1/3} = \tilde{a}(3)^5$. The cyclotomic cosets mod 7 are $\{0\}$, $\{1, 2, 4\}$ and $\{3, 5, 6\}$. Hence

$$\tilde{a}(2) = \tilde{a}(1)^2, \ \tilde{a}(4) = \tilde{a}(1)^4, \ \tilde{a}(6) = \tilde{a}(3)^2, \ \tilde{a}(5) = \tilde{a}(3)^4 \ .$$

Since $\tilde{a}(1) = 0$ and $\tilde{a}(0) = a_\infty$ are given, we have

$$\begin{aligned} \mathcal{H}(\tilde{a}, 0) &= \tilde{a}(0)^2 + \tilde{a}(1)\tilde{a}(6) + \tilde{a}(2)\tilde{a}(5) + \tilde{a}(3)\tilde{a}(4) \\ &= \tilde{a}(0)^2 \ = \ \tilde{a}(0)a_\infty \ , \\ \mathcal{H}(\tilde{a}, 1) &= \tilde{a}(0)\tilde{a}(1) + \tilde{a}(2)\tilde{a}(6) + \tilde{a}(3)\tilde{a}(5) + \tilde{a}(4)^2 \\ &= \tilde{a}(3)\tilde{a}(5) \ = \ \tilde{a}(3)^5 \ , \end{aligned}$$

as required. ■

As we have already seen before, the appropriate quaternary code in the case $m = 3$ is the self-dual octacode.

4.2.1.3 *Decoding the Quaternary 'Preparata' Code in the $\mathbb{Z}_4$ Domain.*

There is a very simple decoding algorithm for the 'Preparata' code $\mathcal{P}$, obtained by working in the $\mathbb{Z}_4$ domain. This is an optimal syndrome decoder: it corrects all error patterns of Lee weight at most 2, detects all errors of Lee weight 3, and

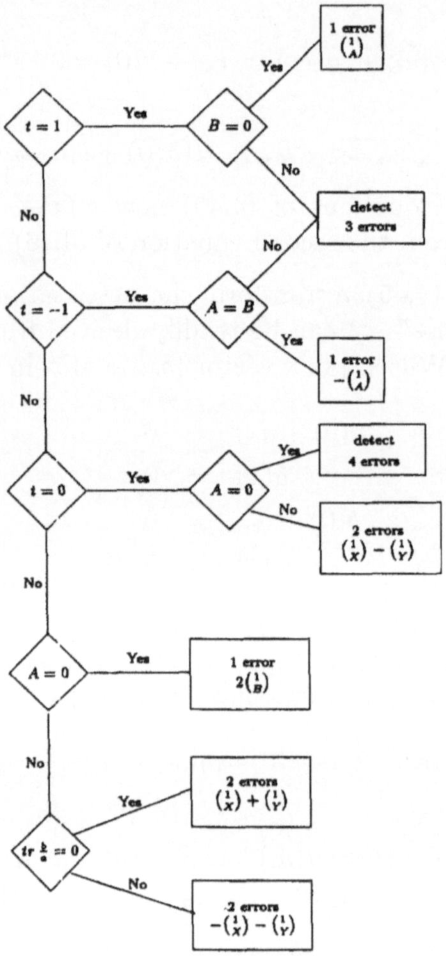

Figure 4.1 Decoding algorithm for 'Preparata' code

detects some errors of Lee weight 4. A decision tree for the algorithm is shown in Figure 4.1. We use the parity check matrix H given in (4.40), and assume m is odd and ≥ 3.

Let $v = (v_\infty, v_0, \ldots, v_{n-1}) \in \mathbb{Z}_4^{n+1}$ be the received vector. The syndrome Hv' has two components, which we write as

$$t = \sum_{j=0}^{n-1} v_j + v_\infty ,$$

$$A + 2B = \sum_{j=0}^{n-1} v_j \xi^j ,$$

where $A, B \in \mathcal{T}$.

In Theorem 13 of [48], we saw that exactly four nonzero weights occur in the Lee weight distribution of the quaternary Kerdock code $\mathcal{K} = \mathcal{P}^\perp$, and hence also in the Hamming weight distribution of K. It follows that the covering radius of P is at most 4 ([34]; [68], Theorem 21 of Chap. 6), i.e. the Lee distance $d_L(v, \mathcal{P})$ from a vector $v \in \mathbb{Z}_4^{n+1}$ to $\mathcal{P}$ satisfies $d_L(v, \mathcal{P}) \leq 4$. Note that $t = \pm 1$ if and only if $d_L(v, \mathcal{P}) = 1$ or 3.

Single errors of Lee weight 1 or 2. If $t = 1$ and $B = 0$, or if $t = -1$ and $A = B$, we decide that there is a single error of Lee weight 1 in column $(1, A)'$. If $t = 1$ and $B \neq 0$, or if $t = -1$ and $A \neq B$, then $d_L(v, \mathcal{P}) = 3$. If $t = 2$ and $A = 0$, we decide that there is a single error of Lee weight 2 in column $(1, B)'$.

Double errors of Lee weight 2. We begin by supposing that $t = 0$ and

$$A + 2B = X - Y \ ,$$

where $X, Y \in \mathcal{T}$ and $X \neq Y$. Note that $A \neq 0$ since by (P1) $X - Y$ is invertible. We have

$$\begin{aligned} A &= X + Y + 2X^{2^{m-1}} Y^{2^{m-1}} \ , \\ B &\equiv Y + X^{2^{m-1}} Y^{2^{m-1}} \pmod{2} \ . \end{aligned}$$

Let x, y, a, b respectively be the images of X, Y, A, B in $GF(2^m)$ after reduction mod 2 using the map μ. Then

$$a = x + y \ , \quad b = y + x^{2^{m-1}} y^{2^{m-1}} \ ,$$

which we rewrite as

$$a = x + y \ , \quad (b + y)^2 = xy \ .$$

The unique solution to these equations is $y = b^2/a$, $x = a + b^2/a$. Note that when $b = 0$ or $b = a$, the double error involves the first column of H. Next we suppose that $t = 2$ and that

$$A + 2B = X + Y \ ,$$

where $X, Y \in \mathcal{T}$, $X \neq Y$, $A \neq 0$. Proceeding as above we find

$$a = x + y \ , \quad b^2 = xy \ ,$$

and so x and y are distinct roots of the equation

$$u^2 + au + b^2 = 0 \ .$$

A necessary and sufficient condition for this equation to have distinct roots is that

$$tr(b^2/a^2) = tr(b/a) = 0$$

(see [68], Chap. 9, Theorem 15; [64]).

Finally we suppose that $t = 2$ and

$$A + 2B = -X - Y \ ,$$

where $X, Y \in \mathcal{T}$, $X \neq Y$, $A \neq 0$. We now find that

$$a = x + y\ , \quad (b + a)^2 = xy\ ,$$

and so x and y are distinct roots of the equation

$$u^2 + au + (a^2 + b^2) = 0\ .$$

A necessary and sufficient condition for this equation to have distinct roots is that

$$tr\left(\frac{a^2 + b^2}{a^2}\right) = tr\left(1 + \frac{b}{a}\right) = 1 + tr\left(\frac{b}{a}\right) = 0\ .$$

4.2.1.4 Quaternary Reed-Muller Codes

In this section we define quaternary Reed-Muller code, $QRM(r, m)$, whose image under the map α is $RM(r, m)$ for all r, and which includes the Kerdock and 'Preparata' codes as special cases.

Definition. Let $QRM(0, m)$ be the quaternary repetition code of length $n = 2^m$, and for $1 \leq r \leq m$ let $QRM(r, m)$ be generated by $QRM(0, m)$ together with all vectors of the form

$$(0, T(\lambda_j), T(\lambda_j \xi^j), T(\lambda_j \xi^{2j}), \ldots, T(\lambda_j \xi^{(n-1)j}))$$

where j ranges over all representatives of cyclotomic cosets mod $2^m - 1$ for which $wt(j) \leq r$, and λ_j ranges over R. Then $QRM(r, m)$ is a quaternary code of length $n = 2^m$ and type 4^k, where

$$k = 1 + \binom{m}{1} + \cdots + \binom{m}{r}\ .$$

Theorem 4.11.

$$\begin{aligned} QRM(1, m) &= \mathcal{K}\ , & (4.51)\\ QRM(m-2, m) &= \mathcal{P}\ , & (4.52)\\ \alpha(QRM(r, m)) &= RM(r, m)\ , & (4.53)\\ QRM(r, m)^{\perp} &= QRM(m - r - 1, m)\ . & (4.54) \end{aligned}$$

Proof. Equation (4.51) follows from Theorem 11(b)[48], (4.52) from (4.54), and (4.53) from [68, Chap. 13, Section 5]. It remains to prove (4.54). This follows from the transform domain characterization of $QRM(r, m)$ as the set of vectors a for which $\widehat{a}(\lambda) = 0$ whenever $wt(\lambda) \leq m - 1 - r$, and $QRM(r, m)^{\perp}$ as the set of vectors for which $\widehat{a}(\lambda) = 0$ whenever $wt(\lambda) \leq r$. (Equivalently, we consider the cyclic codes obtained by deleting the first coordinate, and use the fact that the zeros of a code are the reciprocals of the nonzeros of the dual code.) ■

4.2.1.5 Automorphism Groups.

Consider any system Ω of linear equations over $\mathbb{Z}_4$, in the variables c_x, $x \in \mathcal{T}$ (see (4.27)), that includes

$$\sum_{x\in\mathcal{T}} c_x = 0 , \tag{4.55}$$

$$\sum_{x\in\mathcal{T}} c_x x = 0 , \tag{4.56}$$

together with equations of the form

$$2\left(\sum_{x\in\mathcal{T}} c_x x^{2^j+1}\right) = 0 . \tag{4.57}$$

Theorem 4.12. *The linear system Ω is invariant under the doubly transitive group G of 'affine' permutations of the form*

$$x \to (ax+b)^{2^m} ,$$

where $a, b \in \mathcal{T}$ and $a \neq 0$. The order of G is $2^m(2^m-1)$.

Proof. Repeated application of the Frobenius automorphism (4.34) to (4.56) gives

$$\sum_{x\in\mathcal{T}} c_x x^{2^j} = 0 , \tag{4.58}$$

for all j. It now follows from (4.55), (4.56) and (4.58) that

$$\begin{aligned}\sum_{x\in\mathcal{T}} c_x(ax+b)^{2^m} &= \sum_{x\in\mathcal{T}} c_x(a^{2^m}x^{2^m} + b^{2^m} + 2a^{2^{m-1}}b^{2^{m-1}}x^{2^{m-1}}) \\ &= a\sum_{x\in\mathcal{T}} c_x x + b\sum_{x\in\mathcal{T}} c_x + 2a^{2^{m-1}}b^{2^{m-1}}\sum_{x\in\mathcal{T}} x^{2^{m-1}} = 0 .\end{aligned}$$

Finally

$$\begin{aligned}2\sum_{x\in\mathcal{T}} c_x[(ax+b)^{2^m}]^{2^j+1} &= 2\sum_{x\in\mathcal{T}} c_x(a^{2^m}x^{2^m} + b^{2^m})^{2^j+1} \\ &= 2\sum_{x\in\mathcal{T}} c_x(ax+b)^{2^j+1} \\ &= 2\sum_{x\in\mathcal{T}} c_x(a^{2^j+1}x^{2^j+1} + b^{2^j+1} + a^{2^j}bx^{2^j} + b^{2^j}ax) \\ &= 0 .\end{aligned}$$

Hence Ω is invariant under G. ■

Corollary. *Our quaternary Kerdock, 'Preparata' codes are invariant under a doubly transitive group of order* $2^{m+1}(2^m-1)m$ *generated by G, negation, and the Frobenius map (4.34) acting on* $\mathcal{T}$.

Proof. The presence of negation follows from $\mathbb{Z}_4$-linearity, the action of G from Theorem 4.12 and the Frobenius map from (4.58). ■

Remarks. By the automorphism group Aut(C) of a binary nonlinear code C we will mean the set of all coordinate permutations that preserve the code. It is easy to see that if $C = \phi(\mathcal{C})$ is the binary image of a linear quaternary code $\mathcal{C}$, then Aut$(\mathcal{C})$ is isomorphic to a subgroup of Aut(C).

The automorphism groups of the binary Nordstrom-Robinson, Kerdock, classical Preparata, and Delsarte-Goethals codes are known (Berlekamp [9], Carlet [23–25], Kantor [53,54]). For odd $m \geq 5$ these groups have the same orders as those in the Corollary.

We conclude that, *for odd* $m \geq 5$, *the groups mentioned in the Corollary are the full automorphism groups of these quaternary codes.* (For the 'Preparata' codes we use the fact that they have the same automorphism group as their duals.)

The case $m = 3$ is exceptional. The quaternary octacode has an automorphism group of order 1344 (Conway and Sloane [28]), whereas the group of the binary Nordstrom-Robinson code has order 80640 (Berlekamp [9], see also Conway and Sloane [27]).

4.2.1.6 A New Family of Distance Regular Graphs of Diameter 4.

As before, $P = \phi(\mathcal{P}) = \phi(\mathcal{K}^\perp)$ denotes our 'Preparata' code of length $N = 2^{m+1}$, with m odd ≥ 3.

Definition. A $\mathbb{Z}_4$-*coset* of P is the image under ϕ of a coset of $\mathcal{P}$ in $\mathbb{Z}_4^{N/2}$. We construct a graph Γ_m on the $\mathbb{Z}_4$-cosets of P by joining two cosets by an edge if they are the images of cosets $x+\mathcal{P}$, $y+\mathcal{P}$ such that $x-y+\mathcal{P}$ has minimal Lee weight 1.

Let Π denote the partition of $\mathbb{Z}_2^N$ into $\mathbb{Z}_4$-cosets of P. Then Γ_n can be thought of as the quotient graph ([1, Section 11.1.B]) of the N-hypercube by the partition Π.

The aim of this section is to show that Γ_m is distance regular and to compute its distance distribution diagram and eigenmatrix $\mathbf{P}$. For this purpose, we need certain regularity properties of P and Π.

If C is a binary code of length N, its outer distribution matrix $B = (B_{x,j})$ is the $2^N \times (N+1)$ matrix with typical entry

$$B_{x,j} = |\{y \in C : \ d(x,y) = j\}|$$

(Delsarte [34]). In other words, the rows of B are the weight distributions of the translates of C.

A *code* C of covering radius r is said to be *completely regular* [34] if B contains exactly $r+1$ distinct rows. A *partition* Π of $\mathbb{Z}_2^N$ into cosets is said to be *completely regular* if all members of the partition are completely regular with the same matrix.

Lemma 4.4. *The covering radius of P is 4.*

Proof. In the previous section, we saw that it is at most 4. But P is contained in a code with the same weight distribution as an extended Hamming code (Theorem 4.7), and so by the supercode lemma [26] the covering radius is at least 4. ■

Lemma 4.5. *The codewords of weight 6 in P form a $3-(2^{m+1}, 6, (2^{m+1}-4)/3)$ design.*

Proof. The proof of Theorem 33 of [68], Chapter 15 can be used, since it depends only on the annihilator polynomial of P. ■

Theorem 4.13. *The 'Preparata' code P is completely regular.*

Proof. The well-known recurrence relation between the columns of B ([34,68]) has order 4, by Lemma 4.4, and so it is sufficient to check that $B_{x,j}$ can take at most five different values for fixed $x \in \mathbb{Z}_2^N$ and $0 \leq j \leq 4$. If $d(x, P) \leq 2$, the fact that P has minimal distance 6 shows that $B_{x,j}$ is either 0 or 1. If $d(x, P) = 3$ then Lemma 4.5 shows that $B_{x,3} = (N-1)/3$. Clearly $B_{x,0} = B_{x,1} = B_{x,2} = 0$. Finally if $d(x, P) = 4$ then $B_{x,0} = B_{x,1} = B_{x,2} = B_{x,3} = 0$. ■

As in [29] it will be noticed that P is neither linear, perfect, nor uniformly packed, and so (in the notation of Levenshtein [63]) is not a design of Delsarte type (i.e. $d \geq 2s' - 1$); P is a highly nontrivial example of a completely regular code. Furthermore the $\mathbb{Z}_4$-linearity of P and the properties of ϕ show that each $\mathbb{Z}_4$-coset of P is completely regular with the same outer distribution matrix. Hence Π is completely regular. The next result follows immediately from Theorems 11.1.6 and 11.1.5 of [1].

Theorem 4.14. *The graph Γ_m is distance regular on N^2 vertices with diameter 4 and degree N.*

We now proceed to a more detailed study of the parameters of Γ_m. Recall that the valencies v_j are the numbers of points at distance j from a given point. The intersection numbers a_j, b_j, c_j are defined in Chapter 4 of [1].

Lemma 4.6. *Γ_m is bipartite.*

Proof. Let us take a parity check matrix H of the form (4.40) for $\mathcal{P}$. For a coset $x + \mathcal{P}$ let Hx' be the associated syndrome and let $\nu(x)$ be the leading bit of Hx'. Then ν is a map from the vertices of Γ_m onto $\{0, 1\}$. Let X_j be the set $\nu^{-1}(j)$, $j = 0, 1$. Since $\nu(x) = 1$ if x has weight 1, two cosets with the same image under ν cannot have adjacent images in Γ_m. ■

Lemma 4.7. *If $x \in \mathbb{Z}_2^n$ is at distance 4 from P, then $B_{x,4} = N(N-1)/12$.*

Proof. From [87] and the fact that P has size $2^N/N^2$ and four nonzero dual distances d_1', d_2', d_3', d_4' we obtain

$$B_{x,4} = \frac{2^4}{4!N^2} \prod_{j=1}^{4} d_j' .$$

The desired result then follows from $d_1' = (N-\sqrt{N})/2$, $d_2' = N/2$, $d_3' = (N+\sqrt{N})/2$, $d_4' = N$. ■

Theorem 4.15. *The valencies of* Γ_m *are* $v_0 = 1$, $v_1 = N$, $v_2 = \binom{N}{2}$, $v_3 = \frac{N(N-2)}{2}$, $v_4 = \frac{N-2}{2}$. *The intersection numbers of* Γ_m *are* $b_0 = N$, $c_1 = 1$, $b_1 = N-1$, $c_2 = 2$, $b_2 = N-2$, $c_3 = N-1$, $b_3 = 1$, $c_4 = N$. *Furthermore* $a_j = 0$ *for* $j = 0, 1, 2, 3, 4$.

Proof. By Lemma 4.6, Γ_m is bipartite, hence without circuits of odd length. Therefore $a_j = 0$ for $0 \leq j \leq 4$.

The intersection numbers add up to the degree, so $N = b_j + c_j$ for $0 \leq j \leq 4$, and it only remains to calculate the c_j. The values of c_1 and c_2 are clear from the double-error-correcting character of $\mathcal{P}$. Finally c_3 and c_4 are computed from the formula of Theorem 11.1.8 of [1] by observing that $e_{l,j} = B_{x,j}$ if $d(x, P) = l$. Moreover $e_{3,3} = (N-1)/3$ by Lemma 4.5 and $e_{4,4} = N(N-1)/12$ by Lemma 4.7. The intersection numbers of the N-cube are well known to be $a_j = 0$, $b_j = N - j$, $c_j = j$. ■

Corollary. *The eigenmatrix* **P** *for* Γ_m *is*

$$P = \begin{pmatrix} 1 & N & \binom{N}{2} & \frac{N(N-2)}{2} & \frac{N-2}{2} \\ 1 & \sqrt{N} & 0 & -\sqrt{N} & -1 \\ 1 & 0 & -\dfrac{N}{2} & 0 & 1-\dfrac{N}{2} \\ 1 & -\sqrt{N} & 0 & \sqrt{N} & -1 \\ 1 & -N & \binom{N}{2} & -\frac{N(N-2)}{2} & \frac{N}{2}-1 \end{pmatrix}$$

Proof. See [1], Section 4.1.2, or [29], Proposition 3.17. ■

Remarks. 1) Let R_j denote the jth class of the association scheme corresponding to Γ_m. Then $R_1 + R_3$ has only three eigenvalues, N^2, 0, $-N^2$, and is a strongly regular graph isomorphic to the complete bipartite graph $K_{N^2/2,\, N^2/2}$. It would be interesting to see if R_3 is also distance regular.
2) Π is a 4-partition design in the sense of [22,29].

4.2.2 Codes over $\mathbb{Z}_{2^k}$

4.2.2.1 Coding Theory

Definition 4.5. *An* (n, M, d) *binary code is a cardinality* M *subset of* $\mathbb{Z}_2^n = \mathbb{Z}_2 \times \cdots \times \mathbb{Z}_2$ *with Hamming distance at least* d *between every two different elements. An*

$(n, 2n, n/2)$ binary code is called a Hadamard code. *An $(n-1, 2^{n-1}/n, 3)$ binary code is called a one-perfect code.*

In [62], Krotov introduced a new way to generalize the Gray map, defining a map

$$\Phi : \mathbb{Z}_{2m}^n \to 2^{\mathbb{Z}_2^{mn}}$$
$$\Phi(x_1, \ldots, x_n) = H_{x_1} \times \cdots \times H_{x_n},$$

where $\{H_{x_1}, \ldots, H_{x_n}\}$ is a partition of $\mathbb{Z}_2^m$ into extended one-perfect $(m, \frac{2^m}{2m}, 4)$ codes. If $C \subset \mathbb{Z}_{2m}^n$ then $\Phi(C) = \bigcup_{x \in C} \Phi(x)$. Also the weight function $wt^{\diamondsuit} : \mathbb{Z}_{2^k} \to \mathbb{R}^+$ was defined as

$$wt^{\diamondsuit}(x) = \begin{cases} 0 & \text{if } x = 0, \\ 1 & \text{if } x \text{ is odd}, \\ 2 & \text{if } x \text{ is even}. \end{cases}$$

Now, the *Krotov distance* $d^{\diamondsuit}$ on $\mathbb{Z}_{2^k}^n$ is defined by the standard way from the above weight

$$d^{\diamondsuit}((x_1, \ldots, x_n), (y_1, \ldots, y_n)) = \sum_{i=1}^{n} wt^{\diamondsuit}(y_i - x_i).$$

(In [81], this distance was defined for the more general case of codes over $\mathbb{Z}_{p^k}$; further, it was generalized in [37], were it got the name *overweight distance.*) Note that in the special case $k = 2$, the Krotov distance coincides with the Lee distance. A code C is called an $(n, M, d)^{\diamondsuit}$-code if $C \subseteq Z_{2m}^n$, $\mid C \mid= M$, and the $d^{\diamondsuit}$-distance between any two different elements of C is not less than d.

We can extend w to a weight function on R^n in the obvious way

$$w(x_1, \ldots, x_n) = \sum_{i=1}^{n} w(x_i).$$

In the seminal paper [20], the so-called homogeneous weight was introduced for integer residue rings $\mathbb{Z}_{2m}$. In the case of m a power of 2 it becomes

$$wt^* : \mathbb{Z}_{2m} \to \mathcal{Q},$$

$$wt^*(x) = \begin{cases} 0 & \text{if } x = 0, \\ m & \text{if } x = m, \\ m/2 & \text{otherwise}. \end{cases}$$

Now, the distance d^* is defined by the standard way. A code C is called an $(n, M, d)^*$-code if $C \subseteq Z_{2m}^n$, $\mid C \mid= M$, and the d^*-distance between any two different elements of C is not less than d.

A linear code C over the ring R of length n is an R-submodule of R^n. The homogenous weight enumerator $A_C(z)$ of C is given by

$$A_C(z) = \sum_{c \in C} z^{w(c)}.$$

A *character* of a finite abelian group G is a map $\chi : G \longrightarrow C^*$ such that $\chi(x+y) = \chi(x)\chi(y)$. The next theorem describes the homogeneous weight in terms of sum of characters. We refer to [3,14] for definitions and background on Frobenius rings and generating character. Roughly speaking all additive characters in a Frobenius ring can be deduced from the generating character by scaling of the argument.

Theorem 4.16 ([43]). *Let R be a finite Frobenius ring with generating character χ and set of units $R^\times$. Then the homogeneous weights on R are precisely the functions $w : R \longrightarrow \mathbb{R}$ with*

$$x \mapsto \gamma\Big(1 - \frac{1}{|R^\times|} \sum_{u \in R^\times} \chi(xu)\Big),$$

for some $\gamma > 0$.

Proposition 4.1 ([3], Corollary 16). *Let C be a proper, regular, projective two-weight code with nonzero integral weights $w_1 < w_2$. Then there exists a positive integer d, a divisor of $|C|$, and a positive integer u such that $w_1 = du$ and $w_2 = d(u+1)$.*

4.2.2.2 $\mathbb{Z}_{2^k}$-codes and syndrome graphs

In [19] Carlet introduced a generalization to $\mathbb{Z}_{2^k}$ of the Gray map defined originally for $\mathbb{Z}_4$ in [48]. Denote by $\mathbb{F}_2$ the field with two elements.

Definition 4.6. *Let k be any positive integer, u any element of $\mathbb{Z}_{2^k}$, and $\sum_{i=1}^{k} 2^{i-1}u_i$ its binary expansion ($u_i = 0, 1$). The image of u under the generalized Gray map ϕ is the following Boolean function on $\mathbb{F}_2^{k-1}$*

$$\phi(u) : (y_1, \ldots, y_{k-1}) \mapsto u_k + \sum_{i=1}^{k-1} u_i y_i.$$

In this general case, we can naturally identify any Boolean function on $\mathbb{F}_2^{k-1}$ to a binary word of length 2^{k-1} by listing all its values in arbitrary order.

The generalized Gray map can be extended, in the obvious way, into a mapping from $(\mathbb{Z}_{2^k})^n$ to $\mathbb{F}_2^{2^{k-1}n}$. The homogeneous weights of the elements of $(\mathbb{Z}_{2^k})^n$ are the Hamming weights of their images under the generalized Gray map. This homogeneous weight w_G on $\mathbb{Z}_{2^k}$ is defined by its values $0, 2^{k-2}$, and 2^{k-1} respectively on $\{0\}$, $\mathbb{Z}_{2^k} - 2^{k-1}\mathbb{Z}_{2^k}$ and $2^{k-1}\mathbb{Z}_{2^k} - \{0\}$.

It is well-known that for the ring $\mathbb{Z}_n$, a generating character is the map $\chi(a) = \omega^a$, where $\omega^n = 1$. As the Hamming weight is a homogeneous weight, by Definition 2.6 (ii) [77] $\gamma = 2^{k-2}$ for $R = \mathbb{Z}_{2^k}$. So by Theorem 2.6 [77], the following lemma is immediate.

Lemma 4.8. *For $x \in \mathbb{Z}_{2^k}$, we have*

$$w_G(x) = 2^{k-2} - \frac{1}{2} \sum_{u \in \mathbb{Z}_{2^k}^\times} \chi(xu).$$

From now on, we call $\mathbb{Z}_{2^k}$*−code* a linear code over the ring $\mathbb{Z}_{2^k}$ for some integer $k \geq 2$. If C is a $\mathbb{Z}_{2^k}$-code of length n, we define a graph $\Gamma(C)$ on the column space T of its parity check matrix, two such columns being connected if they differ by a column of this matrix. This graph is a Cayley graph on the additive group of T with set of generators $S = \{uh_i : u \in \mathbb{Z}_{2^k}^\times, \text{for } i = 1, \dots, n\}$, where $H = [h_1 \mid \cdots \mid h_n]$ is the parity check matrix of C. Also the zero column $1_T \notin S$ and $S^{-1} \subseteq S$, so this graph is simple. The graph $\Gamma(C)$ is regular with degree $|S| = 2^{k-1}n$ on $\frac{2^{kn}}{|C|}$ vertices. It is well-known that there is a one-to-one correspondence between syndromes of a code and its cosets, that maps $x + C$ to Hz, where z is any element of $x + C$. So this graph can be defined on the cosets of C, and two cosets are adjacent if and only if their difference is a coset with an element like $(x_1, \dots, x_n) + C$ where there exists i such that $x_i = u \in \mathbb{Z}_{2^k}^\times$ and for each $j \neq i$, $x_j = 0$.

Lemma 4.9. *If C is a proper, regular, projective two weight $\mathbb{Z}_{2^k}$-code with homogeneous weight $w_1 < w_2$. Then there are integers u and t such that $w_1 = u2^t$ and $w_2 = (u+1)2^t$.*

Proof. See Proposition 4.1. □

Theorem 4.17. *Suppose that C is a $\mathbb{Z}_{2^k}$-code with homogeneous weight enumerator for the dual code $A_{C^\perp}(z) = \sum_{i=0}^{2^{k-1}n} m_i z^i$. Then the eigenvalues of $\Gamma(C)$ are $n2^{k-1} - 2i$ with multiplicity m_i, for $i = 0, \dots, 2^{k-1}n$.*

Proof. For any $x \in C^\perp$ we construct the eigenvector e_x of the adjacency matrix A of $\Gamma(C)$ by $(e_x)_y = \chi_x(y)$ where $x = (x_1, \dots, x_n)$ and

$$\chi_x : \frac{\mathbb{Z}_{2^k}^n}{C} \longrightarrow \mathcal{C},$$

$$\chi_x((a_1, \dots, a_n) + C) = \omega^{(x_1,\dots,x_n)(a_1,\dots,a_n)}.$$

As $A(e_x)_y = \sum_{s\in S} \chi_x(s)(e_x)_y$, the corresponding eigenvalue is $\sum_{s\in S} \chi_x(s)$ ([2], Page 11). Now, by using Lemma 4.8, we have:

$$w_G(x) = \sum_{i=1}^{n} w(x_i) = \sum_{i=1}^{n} (2^{k-2} - \frac{1}{2} \sum_{u \in \mathbb{Z}_{2^k}^\times} \chi(x_i u)) = n2^{k-2} - \frac{1}{2} \sum_{s\in S} \chi_x(s).$$

So $\sum_{s\in S} \chi_x(s) = n2^{k-1} - 2w_G(x)$. □

Theorem 4.18. *Suppose that C is a two-weight $\mathbb{Z}_{2^k}$-code with weights w_1 and w_2. The syndrome graph $\Gamma(C^\perp)$ of $C^\perp$ is a strongly regular graph with degree $2^{k-1}n$ and eigenvalues $K(w_1)$ and $K(w_2)$, where $K(x) = 2^{k-1}n - 2x$. For this graph $\lambda > 0$ yields $d^\diamond(C^\perp) \leq 3$.*

Proof. In the proof of Theorem 4.17, for each weight of the dual code we have an eigenvalue in the form of $2^{k-1}n - 2w_i$. Since C is a two-weight code, the graph

$\Gamma(C^\perp)$ has exactly two eigenvalues $K(w_i) = 2^{k-1}n - 2w_i$ for $i = 1, 2$. Now, by applying Theorem 9.1.2 of [2], this graph is a strongly regular graph.

If $\lambda > 0$, there is at least one triangle in the graph $\Gamma(C^\perp)$. Assume that the vertices of this triangle are x, y, z. By definition of adjacency $x - y$, $x - z$, and $y - z$ are in S, so there exist $u_1, u_2, u_3 \in \mathbb{Z}_{2^k}^\times$ such that $x - y = u_1 g_i$, $x - z = u_2 g_j$ and $y - z = u_3 g_k$, where $G = [g_1 \mid \cdots \mid g_n]$ is the generator matrix of C. Consider $x = (x_1, \ldots, x_n)$ such that

$$x_i = u_1,\ x_j = -u_2,\ x_k = -u_3,\ x_m = 0 :\ m \neq i, j, k$$

is an element of $C^\perp$ with $d^\diamond(x) = 3$, we see that $d^\diamond(C^\perp) \leq 3$. □

Theorem 4.19. *If C is a proper, regular, projective two-weight $\mathbb{Z}_{2^k}$-code with weights $w_1 < w_2$ of dual Krotov distance $d^\diamond$ at least 4, length $n > 3$, then there is an integer t such that*

$$w_1 = 2^t, \qquad w_2 = 2^{t+1}, \qquad n = 2^{t-k+2}.$$

Proof. By Theorem 4.18, suppose that $r_i = 2^{k-1}n - 2w_i$ for $i = 1, 2$ are eigenvalues of $\Gamma(C^\perp)$. The condition on the dual distance implies, by Theorem 4.18, that $\lambda = 0$. If we set $N = 2^{k-1}n$ and apply Theorem 2.3 [77], the following relations are obtained:

$$\mu = N + (N - 2w_1)(N - 2w_2), \tag{4.59}$$

$$\mu = 2(w_1 + w_2) - 2N, \tag{4.60}$$

$$4(w_2 - w_1)^2 = \mu^2 + 4(N - \mu). \tag{4.61}$$

Now, we use Lemma 4.9 and put the value of w_i in (4.60), so we get the following equation

$$2^{2t+2} = \mu^2 + 4(N - \mu). \tag{4.62}$$

As $\mu \leq N$, (4.62) yields that $\mu \leq 2^{t+1}$. Now, we obtain the value of N from (4.59) and put this value in (4.62). Now, we get the following equation:

$$2^{2t+2} = \mu^2 - 6\mu + (2u + 1)2^{t+2}. \tag{4.63}$$

This equation entails that

$$2^{t+2} | \mu(\mu - 6), \tag{4.64}$$

and $\mu \neq 6$. So there exists an integer b such that $\mu(\mu - 6) = 2^{t+2}b$. Suppose that $\mu = 2^\ell a$, where a is an odd positive integer. Therefore

$$2^{\ell+1}a(2^{\ell-1}a - 3) = 2^{t+2}b. \tag{4.65}$$

First we show that b is odd. If b is even, then $2^{t+3} | \mu(\mu - 6)$. In (4.63), we divide the two sides of the equation by 2^{t+2} and we get:

$$2^t = \frac{\mu(\mu - 6)}{2^{t+2}} + (2u + 1).$$

As $2u + 1$ is odd and 2^t (when $t \neq 0$) is even. This yields that $\frac{\mu(\mu-6)}{2^{t+2}}$ is odd, which means that $2^{t+3} \nmid \mu(\mu - 6)$. On the other hand, we have already shown that $\mu \leq 2^{t+1}$, so when $t = 0$ it is clear that $\mu = 0, 1,$ or 2. By plugging these values in (4.62) we see that $N \leq 2$, but we consider $n > 3$ and so $N \geq 6$ by considering $N = 2^{k-1}n$. Therefore b is odd.

By (4.65), since a and $(2^{\ell-1}a - 3)$ are odd, we can conclude that $\ell = t + 1$. So $\mu = 2^{t+1}a$, but before we have shown that $\mu \leq 2^{t+1}$. Therefore $a = 1$ and $\mu = 2^{t+1}$. Now, put $\mu = 2^{t+1}$ in (4.59), we see that 2^{t+1} divides $N(N + 1)$. Since $N + 1$ is odd, 2^{t+1} divides N. Let $x = \frac{N}{2^{t+1}} \geq 1$, and rewrite (4.59), we have the equation

$$x - 1 = (x - u)((u + 1) - x)2^{t+1}.$$

Since $x \geq 1$, by considering sign in two sides of this equation we obtain $x = u = 1$. Therefore $N = 2^{t+1}$, $n = 2^{t-k+2}$ and also by Lemma 4.9, $w_1 = 2^t$, and $w_2 = 2^{t+1}$. □

Example 4.6. *Suppose that H is a linear $\mathbb{Z}_{2^k}^n$ code with parameters $(n, n2^k, n2^{k-2})^*$. By Lemma 5.1 in [62], $H^\perp$ is a linear code over $\mathbb{Z}_{2^k}^n$ with $d^\diamond \geq 4$. So one can apply Theorem 4.19 and find the parameters of the code $(n, n2^k, n2^{k-2})^* = (2^{t-k+2}, 2^{t+2}, 2^t)$ and its weights, namely 2^t, 2^{t+1}.*

Example 4.7. *One can apply Theorem 4.19 to obtain the parameters of the dual codes of the codes in ([62] Example 4.2), since the Krotov distance of these codes is 4.*
If $k = 3$ and $n = 16$ by Lemma 4.19, $t = 5$ and hence $w_1 = 32$, $w_2 = 64$.
If $k = 3$ and $n = 32$, then $t = 6$. So $w_1 = 64$, $w_2 = 128$.

4.2.2.3 Two-Weight $\mathbb{Z}_4$-Codes

Let C be a quaternary linear code. Since C is a subgroup of $\mathbb{Z}_4^n$, it is isomorphic to an abelian group of the form $\mathbb{Z}_4^{k_1} \times \mathbb{Z}_2^{k_2}$. Therefore C is of type $4^{k_1}2^{k_2}$ for some nonegative integers k_1, k_2 satisfying $1 \leq 2k_1 + k_2 \leq 2n$, and such that $|C| = 2^{2k_1+k_2}$.

Proposition 4.2. *If C is a two-Lee weight projective $\mathbb{Z}_4$-code of length n, with weights* w_1 and w_2. *Then*

$$n(2n + 1) - 2n(w_1 + w_2) + 2w_1w_2(1 - \frac{1}{|C|}) = 0.$$

Proof. See [76, Proposition 5.1 (12)]. □

Now, we characterize certain two-weight $\mathbb{Z}_4$-codes by their parameters.

Theorem 4.20. *If C is a regular two-Lee weight $\mathbb{Z}_4$-code of dual Lee distance at least 4, length $n > 3$, and size 2^κ, with weights $w_1 < w_2$, then there is an integer $t > 1$ such that*

$$\kappa = t + 2, \qquad n = 2^t,$$
$$w_1 = 2^t, \qquad w_2 = 2^{t+1}.$$

Proof. By using Theorem 4.19, we solve the equation of Proposition 4.2 for $|C|$ and we obtain that $\kappa = t + 2$. □

In other words, the Gray image of such a code is a Hadamard code in the sense of [72]. These $\mathbb{Z}_4$-linear Hadamard codes are classified up to equivalence in [72, Section III] and [61]. For each possible length $n = 2^t$, there are exactly $\lfloor \frac{t+1}{2} \rfloor$ non-equivalence classes of such codes with $t + 1 = k_1 + 2k_2$, where the code is of type $4^{k_1}2^{k_2}$ with a generator matrix consisting of all column vectors of the form

$$(2\mathbb{Z}_4)^{k_2} \times 1 \times (\mathbb{Z}_4)^{k_1 - 1}.$$

4.2.2.4 Two Infinite Families of Two-Weight $\mathbb{Z}_4$-Codes

If $k = 1$, let $Y_k = (1)$, when $k \geq 2$, define

$$Y_k = \begin{pmatrix} Y_{k-1} & Y_{k-1} & Y_{k-1} & Y_{k-1} & B_{k-1} \\ \mathbf{0} & \mathbf{1} & \mathbf{2} & \mathbf{3} & \underbrace{1\,1 \cdots 1}_{2^{k-1}} \end{pmatrix} \tag{4.66}$$

where B_{k-1} is a $(k-1) \times 2^{k-1}$ matrix over $2\mathbb{Z}_4$ with all different columns and $\mathbf{i}$ is the row vector $(i, i, \cdots, i)$ of length $2^{k-2}(2^{k-1} - 1)$, $i \in \mathbb{Z}_4$.

Obviously, Y_k is a $k \times n$ matrix over $\mathbb{Z}_4$ where $n = 2^{k-1}(2^k - 1)$. Let $\mathbf{x}_i$ and $\mathbf{x}_j$ be two columns of Y_k, then it is easy to check that $\mathbf{x}_i \neq m\mathbf{x}_j$, where $m = 1, 2, 3$ and $1 \leq i, j \leq n$.

Besides, if we delete $\mathbf{0}$ (a column of B_k) from B_k, then we get a $k \times (2^k - 1)$ matrix B'_k over $2\mathbb{Z}_4$. Let C' be the code generated by B'_k. It's easy to check that C' is a special case of the code obtained in Theorem 4.5 in [76], when $k_1 = 0$, $k_2 = k$, then C' is a one-Lee weight linear code with the nonzero Lee weight $w' = 2^k$.

Proposition 4.3. *If C is the code generated by Y_k, then C is a two-Lee weight code of length $n = 2^{k-1}(2^k - 1)$, with the nonzero Lee weight $w_1 = 2^{k-1}(2^k - 1)$, $w_2 = 2^{2k-1}$. Moreover, $A_{w_1} = 4^k - 2^k$ and $A_{w_2} = 2^k - 1$, respectively.*

Proof. Let

$$G_k = \begin{pmatrix} Y_k & 3Y_k & B_k \end{pmatrix}, \tag{4.67}$$

then G_k is a $k \times 4^k$ matrix over $\mathbb{Z}_4$. Since B_k is a $k \times 2^k$ matrix over $2\mathbb{Z}_4$ with all different columns, then all columns of G_k are pairwise different. If we delete the $\mathbf{0}$ from G_k (in fact we delete $\mathbf{0}$ from B_k), then we get a $k \times (4^k - 1)$ matrix G'_k. Therefore, G'_k is a matrix with all nonzero different columns over $\mathbb{Z}_4$.

Let C'_k be the code generated by G'_k. It's easy to check that C'_k is a special case of the code obtained in Theorem 4.5 in [76], when $k_1 = k$, $k_2 = 0$, then C'_k is a one-Lee weight linear code with the nonzero Lee weight $w' = 4^k$.

Let $\mathbf{c} = (\mathbf{x}, 3\mathbf{x}, \mathbf{y}) \in C'$ and the lengths of $\mathbf{x}$ and $\mathbf{y}$ are $2^{k-1}(2^k - 1)$ and $2^k - 1$, respectively. Obviously, $W_L(\mathbf{y}) = 2^k$ or 0, and

$$W_L(\mathbf{c}) = W_L(\mathbf{x}) + W_L(3\mathbf{x}) + W_L(\mathbf{y}) = 2W_L(\mathbf{x}) + W_L(\mathbf{y}),$$

then, we have

$$W_L(\mathbf{x}) = \frac{W_L(\mathbf{c}) - W_L(\mathbf{y})}{2} = \frac{4^k - W_L(\mathbf{y})}{2}. \tag{4.68}$$

If $W_L(\mathbf{y}) \neq 0$, then $W_L(\mathbf{x}) = w_1 = 2^{k-1}(2^k - 1)$. Otherwise, $W_L(\mathbf{x}) = w_2 = 2^{2k-1}$. It is not difficult to get $A_{w_2} = 2^k - 1$ when $W_L(\mathbf{y}) = \mathbf{0}$. □

Corollary 4.1. *Let C be the code generated by Y_k, $\mathbf{c} \in C$ and $\mathbf{c} \neq \mathbf{0}$, then*

$$W_L(\mathbf{c}) = \begin{cases} 2^{2k-1} & \textit{if the order of } \mathbf{c} \textit{ is } 2, \\ 2^{k-1}(2^k - 1) & \textit{if the order of } \mathbf{c} \textit{ is } 4. \end{cases}$$

Proof. Let $\mathbf{r}_1, \mathbf{r}_2, \cdots, \mathbf{r}_k$ be the rows of Y_k. From the proof of 4.3, $W_L(\mathbf{r}_i) = w_1 = 2^{k-1}(2^k - 1)$, but $W_L(2\mathbf{r}_i) = w_2 = 2^{2k-1}$.

Step 1. Let

$$\mathbf{c} = p_1\mathbf{r}_1 + p_2\mathbf{r}_2 + \cdots + p_k\mathbf{r}_k,$$

where $p_i \in \mathbb{Z}_4$, $1 \le i \le k$ and there exists $p_i \neq 0$. We can prove that $p_i = 0$ or 2 for all $1 \le i \le k$, if the order of $\mathbf{c}$ is 2. By (4.66), let

$$\mathbf{r}_k = (\mathbf{0\ 1\ 2\ 3}\ \overbrace{1\,1\ \cdots\ 1}^{2^{k-1}}),$$

where $\mathbf{i}$ is the row vector $(i, i, \cdots, i)$ of length $2^{k-2}(2^{k-1} - 1)$, $i \in \mathbb{Z}_4$. If the order of $\mathbf{c}$ is 2, and $p_k = 1$ or 3, then $\mathbf{c}$ contains at least a 1 or a 3 according to the definition of Y_k in (4.66), contradiction. Therefore, $p_k = 0$ or 2 and $p_k\mathbf{r}_k = \mathbf{0}$ or $2\mathbf{r}_k$. So the order of $p_1\mathbf{r}_1 + p_2\mathbf{r}_2 + \cdots + p_{k-1}\mathbf{r}_{k-1}$ is 2. Similarly, by (4.66), we can get $p_{k-1} = 0$ or 2. Then we can prove $p_i = 0$ or 2 for all $1 \le i \le k$.

Step 2. By (4.68), we can get if $p_i = 2$ or 0, then $W_L(p_i\mathbf{r}_i) = 2^{2k-1}$ or 0, $1 \le i \le k$. Similarly,

$$W_L(p_1\mathbf{r}_1 + p_2\mathbf{r}_2 + \cdots + p_k\mathbf{r}_k) = 2^{2k-1} = w_2,$$

where $p_i = 0$ or 2, $1 \le i \le k$ and there exists $p_i \neq 0$. By **Step 1**, these are all the codewords of order 2. Therefore, if the order of $\mathbf{c}$ is 2, then $W_L(\mathbf{c}) = 2^{2k-1} = w_2$. Since $A_{w_2} = 2^k - 1$ by 4.3 and C is a two-Lee weight code, we can get if the order of $\mathbf{c}$ is 4, then $W_L(\mathbf{c}) = 2^{k-1}(2^k - 1)$. □

When $k \geq 2$, let

$$X_k = \begin{pmatrix} Y_{k-1} & Y_{k-1} & Y_{k-1} & Y_{k-1} \\ \mathbf{0} & \mathbf{1} & \mathbf{2} & \mathbf{3} \end{pmatrix}, \tag{4.69}$$

where $\mathbf{i}$ is the row vector $(i, i, \cdots, i)$ of length $2^{k-2}(2^{k-1}-1)$, $i \in \mathbb{Z}_4$.

Similar to 4.3, we have

Proposition 4.4. *If C is the code generated by X_k, then C is a two-Lee weight code of length $n = 2^k(2^{k-1}-1)$ with the nonzero Lee weight $w_1 = 2^k(2^{k-1}-1)$, and $w_2 = 2^{2k-1}$. Moreover, $A_{w_1} = 4^k - 2^{k-1}$ and $A_{w_2} = 2^{k-1} - 1$, respectively.*

Proof. From the structure of X_k and 4.3, we know there are $2^{k-1}-1$ codewords of weight 2^{2k-1} generated by the first $k-1$ rows. Let $\mathbf{r}_1, \mathbf{r}_2, \cdots, \mathbf{r}_k$ be the rows of X_k, and denote the codeword generated by the first $k-1$ rows by $\mathbf{c}$. Without loss of generality, by (4.69), let

$$\mathbf{r}_k = (\mathbf{0}\ \mathbf{1}\ \mathbf{2}\ \mathbf{3}),$$

where $\mathbf{i}$ is the row vector $(i, i, \cdots, i)$ of length $2^{k-2}(2^{k-1}-1)$, $i \in \mathbb{Z}_4$.

It is sufficient to prove $W_L(\mathbf{r}_k + \mathbf{c}) = W_L(2\mathbf{r}_k + \mathbf{c}) = 2^k(2^{k-1}-1)$. It is easy to check that $W_L(\mathbf{r}_k) = W_L(2\mathbf{r}_k) = 2^k(2^{k-1}-1)$. Let $\mathbf{c} = (\mathbf{c}', \mathbf{c}', \mathbf{c}', \mathbf{c}')$, n_i be the number of i in $\mathbf{c}'$ and n' be the length of $\mathbf{c}'$. So $n = 4n' = 2^k(2^{k-1}-1)$.

1. If the order of $\mathbf{c}$ is 2, according to Corollary 4.1, $W_L(\mathbf{c}) = 2^{2k-1}$, $n' = n_0 + n_2$, and

$$\begin{aligned} W_L(\mathbf{r}_k + \mathbf{c}) &= W_L(\mathbf{c}') + W_L(\mathbf{c}' + \mathbf{1}) + W_L(\mathbf{c}' + \mathbf{2}) + W_L(\mathbf{c}' + \mathbf{3}) \\ &= 2n_2 + n' + 2(n' - n_2) + n' = 4n' = 2^k(2^{k-1}-1). \end{aligned}$$

2. If the order of $\mathbf{c}$ is 4, $n' = n_0 + n_1 + n_2 + n_3$, and

$$\begin{aligned} W_L(\mathbf{r}_k + \mathbf{c}) &= W_L(\mathbf{c}') + W_L(\mathbf{c}' + \mathbf{1}) + W_L(\mathbf{c}' + \mathbf{2}) + W_L(\mathbf{c}' + \mathbf{3}) \\ &= (n_1 + 2n_2 + n_3) + (n_0 + 2n_1 + n_2) \\ &\quad + (2n_0 + n_1 + n_3) + (n_0 + n_2 + 2n_3) \\ &= 4(n_1 + n_3 + n_2 + n_0) = 4n' = 2^k(2^{k-1}-1). \end{aligned}$$

Similarly, we can also prove $W_L(2\mathbf{r}_k + \mathbf{c}) = 2^k(2^{k-1}-1) = w_1$. Therefore, $A_{w_2} = 2^{k-1} - 1$, and $A_{w_1} = 4^k - 2^{k-1}$. □

Example 4.8. *If $k = 3$, then $n = 24, w_1 = 24, w_2 = 32$, then, according to Proposition 4.4, there is a two-Lee weight code with the generator matrix:*

$$\begin{pmatrix} 111102 & 111102 & 111102 & 111102 \\ 012311 & 012311 & 012311 & 012311 \\ 000000 & 111111 & 222222 & 333333 \end{pmatrix}. \tag{4.70}$$

Next, we will give some structures of two-Lee weight projective codes over $\mathbb{Z}_4$.

Lemma 4.10. *Let C be a linear code over $\mathbb{Z}_4$ with type $4^{k_1}2^{k_2}$ of length n. If $k_2 = 1$ and the columns of the generator matrix $G_{(k_1+1)\times n}$ are all distinct nonzero vectors $(c_1, c_2, \cdots, c_{k_1}, c_{k_1+1})^T$, where $(c_1, c_2, \cdots, c_{k_1})^T$ is one column of Y_{k_1}, $c_{k_1+1} = 0$ or 2, then C is a two-Lee weight code of length $n = 2^{k_1}(2^{k_1} - 1)$ with the nonzero Lee weights $w_1 = 2^{k_1}(2^{k_1} - 1)$ and $w_2 = 2^{2k_1}$. Moreover, $A_{w_1} = 2^{k_1}(2^{k_1+1} - 1)$ and $A_{w_2} = 2^{k_1} - 1$, respectively.*

Proof. Without loss of generality, let

$$G_{(k_1+1)\times n} = \begin{pmatrix} Y_{k_1} & Y_{k_1} \\ \mathbf{0} & \mathbf{2} \end{pmatrix}, \tag{4.71}$$

where $\mathbf{i}$ is the row vector $(i, i, \cdots, i)$ of length $2^{k_1-1}(2^{k_1} - 1)$, $i \in 2\mathbb{Z}_4$ and $\mathbf{r}_1, \mathbf{r}_2, \cdots, \mathbf{r}_k, \mathbf{r}_{k_1+1}$ be the rows of $G_{(k_1+1)\times n}$. Since the columns of $G_{(k_1+1)\times n}$ are all nonzero distinct vectors, it follows that $n = 2^{k_1}(2^{k_1} - 1)$. Let $\mathbf{c} = (\mathbf{x}, \mathbf{x}) \in C$, where $\mathbf{x}$ is generated by Y_{k_1}. According to Proposition 4.3, then $W_L(\mathbf{c}) = 2W_L(\mathbf{x}) = 2^{k_1}(2^{k_1} - 1)$ or 2^{2k_1}. Denote the number of i in $\mathbf{x}$ by n_i, and the length of $\mathbf{x}$ by n'. Obviously, $n = 2n' = 2^{k_1}(2^{k_1} - 1)$.

1. If the order of $\mathbf{x}$ is 4, $n' = n_0 + n_1 + n_2 + n_3$, then we have

$$\begin{aligned} W_L(\mathbf{r}_{k_1+1} + \mathbf{c}) &= W_L(\mathbf{x}) + W_L(\mathbf{x} + \mathbf{2}) = n_1 + 2n_2 + n_3 + 2n_0 + n_1 + n_3 \\ &= 2(n_0 + n_1 + n_2 + n_3) = 2n' = 2^{k_1}(2^{k_1} - 1). \end{aligned}$$

2. If the order of $\mathbf{x}$ is 2, $n' = n_0 + n_2$, $n_1 = n_3 = 0$, then we have

$$\begin{aligned} W_L(\mathbf{r}_{k_1+1} + \mathbf{c}) &= W_L(\mathbf{x}) + W_L(\mathbf{x} + \mathbf{2}) = 2n_2 + 2n_0 \\ &= 2n' = 2^{k_1}(2^{k_1} - 1). \end{aligned}$$

Hence, C is a two-Lee weight code with the nonzero Lee weights $w_1 = 2^{k_1}(2^{k_1} - 1)$, and $w_2 = 2^{2k_1}$. Obviously, A_{w_2} does not change comparing with Y_{k_1}, *i.e.* $A_{w_2} = 2^{k_1} - 1$, thus, $A_{w_1} = 2^{k_1}(2^{k_1+1} - 1)$. □

Theorem 4.21. *Let C be a linear code over $\mathbb{Z}_4$ with type $4^{k_1}2^{k_2}$ of length n. If the columns of the generator matrix $G_{(k_1+k_2)\times n}$ are all distinct nonzero vectors $(c_1, c_2, \cdots, c_{k_1}, c_{k_1+1}, \cdots, c_{k_1+k_2})^T$, where $(c_1, c_2, \cdots, c_{k_1})^T$ is one column of Y_{k_1}, $c_i = 0$ or 2, $k_1 + 1 \le i \le k_1 + k_2$, then C is a two-Lee weight code of length $n = 2^{k_1+k_2-1}(2^{k_1} - 1)$ with the nonzero Lee weights $w_1 = 2^{k_1+k_2-1}(2^{k_1} - 1)$ and $w_2 = 2^{2k_1+k_2-1}$. Moreover, $A_{w_1} = 2^{k_1}(2^{k_1+k_2} - 1)$ and $A_{w_2} = 2^{k_1} - 1$, respectively.*

Proof. If $k_2 = 1$, the result follows from Lemma 4.10. Assume the result is valid for any $k_2 = m$, we denote the generator matrix and the nonzero weights by $G_{k_1,m}$ and by $w_1' = 2^{k_1+m-1}(2^{k_1} - 1)$, $w_2' = 2^{2k_1+m-1}$, respectively, then for $k_2 = m + 1$, without loss of generality, the generator matrix $G_{k_1,m+1}$ of C can be expressed as

$$G_{k_1,m+1} = \begin{pmatrix} G_{k_1,m} & G_{k_1,m} \\ \mathbf{0} & \mathbf{2} \end{pmatrix}, \tag{4.72}$$

where $\mathbf{i}$ is the row vector $(i, i, \cdots, i)$ of length $2^{k_1+m-1}(2^{k_1}-1)$, $i \in 2\mathbb{Z}_4$. It's easy to verify that C is a two-Lee weight code with $w_1 = 2w_1' = 2^{k_1+m}(2^{k_1}-1)$ and $w_2 = 2w_2' = 2^{2k_1+m}$. Similar to the proof of Lemma 4.10, we can get $A_{w_2} = 2^{k_1}-1$ and $A_{w_1} = 2^{k_1}(2^{k_1+k_2}-1)$. By induction hypothesis, the results are valid. □

Example 4.9. *If $k_1 = 2, k_2 = 1$, then $n = 12, w_1 = 12, w_2 = 16$, then, according to Theorem 4.21, there is a two-Lee weight code with the generator matrix:*

$$G_{2,1} = \left(\begin{array}{c|c} 111102 & 111102 \\ 012311 & 012311 \\ 222222 & 000000 \end{array}\right). \tag{4.73}$$

Example 4.10. *If $k_1 = 2, k_2 = 2$, then $n = 24, w_1 = 24, w_2 = 32$, then, according to Theorem 4.21, there is a two-Lee weight code with the generator matrix:*

$$G_{2,2} = \left(\begin{array}{c|c} 111102111102 & 111102111102 \\ 012311012311 & 012311012311 \\ 222222000000 & 222222000000 \\ 222222222222 & 000000000000 \end{array}\right). \tag{4.74}$$

Corollary 4.2. *Let C be a linear code over $\mathbb{Z}_4$ with type $4^{k_1}2^{k_2}$ of length n. When $k_1 \geq 2$, the columns of the generator matrix $G'_{(k_1+k_2)\times n}$ are all distinct nonzero vectors $(c_1, c_2, \cdots, c_{k_1}, c_{k_1+1}, \cdots, c_{k_1+k_2})^T$, where $(c_1, c_2, \cdots, c_{k_1})^T$ is one column of X_{k_1} defined above, c_i = 0 or 2, $k_1 + 1 \leq i \leq k_1 + k_2$, then C is a two-Lee weight code of length $n = 2^{k_1+k_2}(2^{k_1-1}-1)$ with the nonzero Lee weights $w_1 = 2^{k_1+k_2}(2^{k_1-1}-1)$, $w_2 = 2^{2k_1+k_2-1}$. Moreover, $A_{w_1} = 4^{k_1}2^{k_2} - 2^{k_1-1}$ and $A_{w_2} = 2^{k_1-1} - 1$, respectively.*

Proof. Assume that $k_1 \geq 2$, then according to Theorem 4.21, we can construct a code C' with type $4^{k_1-1}2^{k_2}$, and denote its generator matrix by G'. By the definition of X_k in (4.69), without loss of generality, let

$$G'_{(k_1+k_2)\times n} = \begin{pmatrix} \mathbf{0} & \mathbf{1} & \mathbf{2} & \mathbf{3} \\ G' & G' & G' & G' \end{pmatrix}, \tag{4.75}$$

where $\mathbf{i}$ is the row vector $(i, i, \cdots, i)$ of length $2^{k_1+k_2-2}(2^{k_1-1}-1)$, $i \in \mathbb{Z}_4$. It is easily seen that C is a two-Lee weight code with the nonzero weight $w_1 = 2^{k_1+k_2}(2^{k_1-1}-1)$ and $w_2 = 2^{2k_1+k_2-1}$. By induction hypothesis, C is a two-Lee weight code over $\mathbb{Z}_4$. □

Example 4.11. *If $k_1 = 2, k_2 = 2$, then $n = 16, w_1 = 32, w_2 = 16$, according to Corollary 4.2, there is a two-Lee weight code with the generator matrix:*

$$G'_{2,2} = \begin{pmatrix} 1111 & 1111 & 1111 & 1111 \\ 0123 & 0123 & 0123 & 0123 \\ 0000 & 0000 & 2222 & 2222 \\ 0000 & 2222 & 0000 & 2222 \end{pmatrix}. \tag{4.76}$$

Example 4.12. *If $k_1 = 3, k_2 = 1$, then $n = 48, w_1 = 48, w_2 = 64$, according to Corollary 4.2, there is a two-Lee weight code with the generator matrix $G'_{3,1}$:*

$$\left(\begin{array}{cccc|cccc} 111102 & 111102 & 111102 & 111102 & 111102 & 111102 & 111102 & 111102 \\ 012311 & 012311 & 012311 & 012311 & 012311 & 012311 & 012311 & 012311 \\ 000000 & 111111 & 222222 & 333333 & 000000 & 111111 & 222222 & 333333 \\ 222222 & 222222 & 222222 & 222222 & 000000 & 000000 & 000000 & 000000 \end{array}\right). \quad (4.77)$$

Next, we are ready to give the sufficient and necessary conditions for a two-Lee weight linear code to be projective over $\mathbb{Z}_4$.

Lemma 4.11. *Let C be a linear code over $\mathbb{Z}_4$ with type $4^{k_1}2^{k_2}$ of length n and with the generator matrix G. Then C is projective if and only if the following conditions hold:*

1. *every column of G contains 1 or 3;*
2. *any two columns of G are not multiple of each other by ± 1.*

Proof. Let $\mathbf{x}_1, \mathbf{x}_2, \cdots, \mathbf{x}_n$ be the columns of G. Let us first prove the sufficiency. Let $C^\perp$ be the dual code of C. If $\mathbf{c} \in C^\perp$, then we only need to prove $W_L(\mathbf{c}) \geq 3$. Obviously, $W_L(\mathbf{c})$ cannot be 1 because all columns are nonzero. Assume $W_L(\mathbf{c}) = 2$. Obviously, $\mathbf{c}$ cannot be written as $\mathbf{c} = (0, \cdots, 0, 2, 0, \cdots, 0)$ due to (1). Then we have two cases:

Case 1. Without loss of generality, let $\mathbf{c} = (0, \cdots, 0, 1, 0, \cdots, 0, 1, 0, \cdots, 0)$, where $c_i = c_j = 1, i < j$. Then $\mathbf{x}_i + \mathbf{x}_j = \mathbf{0}$. Thus, $\mathbf{x}_j = 3\mathbf{x}_i$, $\mathbf{x}_i$ and $\mathbf{x}_j$ are in proportion by the unit 3, contradiction.

Case 2. Without loss of generality, let $\mathbf{c} = (0, \cdots, 0, 1, 0, \cdots, 0, 3, 0, \cdots, 0)$, where $c_i = 1, c_j = 3, i < j$. Then $\mathbf{x}_i + \mathbf{x}_j = \mathbf{0}$. Thus, $\mathbf{x}_j = \mathbf{x}_i$, contradiction.

Therefore, $W_L(\mathbf{c}) \geq 3$. By definition, C is projective.

Now we begin to prove the necessity. If (1) doesn't hold, let $\mathbf{x}_j \in 2\mathbb{Z}_4^{k_1+k_2}$, then the codeword $\mathbf{c} = (0, \cdots, 0, 2, 0, \cdots, 0) \in C^\perp$, where $c_j = 2$. If (2) doesn't hold, $\mathbf{x}_i = \pm\mathbf{x}_j$ for some pair of indices $i < j$.

Case 1. If $\mathbf{x}_i = \mathbf{x}_j$, we can construct $\mathbf{c}' = (0, \cdots, 0, 1, 0, \cdots, 0, 3, 0, \cdots, 0)$, where $c'_i = 1, c'_j = 3, i < j$ and $\mathbf{c}' \in C^\perp$.

Case 2. If $\mathbf{x}_i = 3\mathbf{x}_j$, then $\mathbf{c}'' = (0, \cdots, 0, 1, 0, \cdots, 0, 1, 0, \cdots, 0) \in C^\perp$, where $c''_i = c''_j = 1$.

But the Lee weight of $\mathbf{c}$, $\mathbf{c}'$ and $\mathbf{c}''$ is 2. Contradiction with $w_L(C^\perp) = 3$. □

Theorem 4.22. *Let C be a linear code over $\mathbb{Z}_4$ with type $4^{k_1}2^{k_2}$ of length n obtained in Theorem 4.21, then C is projective.*

Proof. Denote the generator matrix of C by G_{k_1,k_2}, then it is sufficient to prove that G_{k_1,k_2} satisfies Lemma 4.11.

Step 1. Prove Y_{k_1} defined in (4.66) can make it. The result is obvious.

Step 2. Use induction for k_2. If $k_2 = 1$, we know the generator matrix has the form (4.71) in Lemma 4.10. And by Step 1 and Lemma 4.11, we can easily prove that the code is projective. Assume the result is valid for any $k_2 = m$, then for $k_2 = m + 1$, according to (4.72) in Theorem 4.21, without loss of generality, the generator matrix $G_{k_1,m+1}$ of C is

$$G_{k_1,m+1} = \begin{pmatrix} G_{k_1,m} & G_{k_1,m} \\ \mathbf{0} & \mathbf{2} \end{pmatrix}, \tag{4.78}$$

where $\mathbf{i}$ is the row vector $(i, i, \cdots, i)$ of length $2^{k_1+m-1}(2^{k_1} - 1)$, $i \in 2\mathbb{Z}_4$. So, all the columns of $G_{k_1,m+1}$ contain a 1 or a 3. If there are two columns in proportion by ± 1, then there are two columns of $G_{k_1,m}$ in proportion by ± 1, a contradiction. So the result is valid for $k_2 = m+1$. By induction hypothesis, G_{k_1,k_2} satisfies Lemma 4.11 for any $k_1, k_2 \geq 1$, then C is projective by Lemma 4.11. □

Similar to the proof of Theorem 4.22, we have the following corollary.

Corollary 4.3. *The linear code obtained in Corollary 4.2 is projective.*

For $\mathbf{x} = (x_1, x_2, \cdots, x_n) \in \mathbb{Z}_4^n$, define $\alpha(\mathbf{x}) = (\alpha(x_1), \alpha(x_2), \cdots, \alpha(x_n))$, where $\alpha(0) = \alpha(2) = 0$ and $\alpha(1) = \alpha(3) = 1$.

Lemma 4.12 (*See [48]*). *Let C be a linear code over $\mathbb{Z}_4$, $\mathbf{x_1}, \mathbf{x_2}, \cdots, \mathbf{x_m}$ be a set of generators of C, then $\Phi(C)$ is linear if and only if $2\alpha(\mathbf{x_i}) * \alpha(\mathbf{x_j}) \in C$ for all i, j, satisfying $1 \leq i \leq j \leq m$.*

Let $q = p^m$, where p is prime and m is a positive integer. According to [15], an $[n, k]$ code C over $GF(q)$ is a k-dimensional subspace of $GF(q)^n$. Next proposition is from the classical paper in [15] about the survey of two weight codes over finite fields.

Proposition 4.5 (*.See [15]*). *Example SU 1: Let Ω be the complement of a t-dimensional subspace of $GF(q)^k$ where $1 \leq t \leq k - 1$. If v is a non-zero vector, then $|v^{\perp} \cap \Omega| = q^{k-1} - q^t$ or $q^{k-1} - q^{t-1}$.*

For a two-weight code, if the field is $GF(q)$, its parameters are

$$n = \frac{q^l - q^t}{q - 1}, k = l, w_1 = q^{l-1} - q^{t-1}, w_2 = q^{l-1}, A_{w_1} = q^l - q^{l-t}, A_{w_2} = q^{l-t} - 1.$$

where $1 \leq t \leq l - 1$.

Proposition 4.6. *The Gray images of the codes of Theorem 4.21 and Corollary 4.2 are binary distance invariant codes with the weights given $w_1 = q^{l-1} - q^{t-1}$ and $w_2 = q^{l-1}$.*

Note that we do not claim these binary codes to be linear.

Table 4.1 is a table about the parameters of the Gray images $\Phi(C_1)$, $\Phi(C_2)$ and $\Phi(C_3)$, where C_1 and C_2 are the codes obtained in Theorem 4.21 and Corollary 4.2 in this paper, respectively, and C_3 is the code obtained in [76, Theorem 5.6].

TABLE 4.1 The Parameters of $\Phi(C_1)$, $\Phi(C_2)$ and $\Phi(C_3)$

	n	w_1	w_2	A_{w_1}	A_{w_2}
SU 1	$\frac{q^l-q^t}{q-1}$	$q^{l-1}-q^{t-1}$	q^{l-1}	q^l-q^{l-t}	$q^{l-t}-1$
$\Phi(C_1)$	$2^{k_1+k_2}(2^{k_1}-1)$	$2^{k_1+k_2-1}(2^{k_1}-1)$	$2^{2k_1+k_2-1}$	$4^{k_1}2^{k_2}-2^{k_1}$	$2^{k_1}-1$
$\Phi(C_2)$	$2^{k_1+k_2}(2^{k_1}-2)$	$2^{k_1+k_2}(2^{k_1-1}-1)$	$2^{2k_1+k_2-1}$	$4^{k_1}2^{k_2}-2^{k_1-1}$	$2^{k_1-1}-1$
$\Phi(C_3)$	$2^{2k_1+k_2-1}$	$2^{2k_1+k_2-2}$	$2^{2k_1+k_2-1}$	$4^{k_1}2^{k_2}-2$	1

Remark 4.1. *If $k_1 = 1$, $\Phi(C_1)$ and $\Phi(C_3)$ have the same parameters. Besides, due to [76, Corollary 6.8], when $k_1 = 1$, $\Phi(C_1)$ reaches the Plotkin bound and Griesmer bound.*

Remark 4.2. *It can be seen that the values of l for $\Phi(C_1)$, $\Phi(C_2)$ and $\Phi(C_3)$ are all $2k_1+k_2$. But the values of t are k_1+k_2, k_1+k_2+1, $2k_1+k_2-1$ for $\Phi(C_1)$, $\Phi(C_2)$ and $\Phi(C_3)$, respectively. Therefore, these three families of codes lead to non-equivalent two-Lee weight linear codes.*

Finally, we will consider the linearity of the image codes in Theorem 4.21 and Corollary 4.2.

Theorem 4.23. *If the columns of the generator matrix of C satisfy the conditions of Theorem 4.21, then $\Phi(C)$ is linear when $k_1 = 1$ and nonlinear when $k_1 \geq 2$.*

Proof. When $k_1 = 1$, the result can be obtained in [76, Corollary 6.8]. When $k_1 \geq 2$, denote the generator matrix of C by G_{k_1,k_2} and let $\mathbf{r}_1, \mathbf{r}_2, \cdots, \mathbf{r}_{k_1}, \cdots, \mathbf{r}_{k_1+k_2}$ be the rows of G_{k_1,k_2}, where $\mathbf{r}_j$ is a vector over $2\mathbb{Z}_4$, $k_1+1 \leq j \leq k_1+k_2$. By Theorem 4.21,

$$G_{k_1+1,k_2} = \begin{pmatrix} G_{k_1,k_2} & G_{k_1,k_2} \\ \mathbf{0} & \mathbf{2} \end{pmatrix}, \tag{4.79}$$

where $\mathbf{i}$ is the row vector $(i, i, \cdots, i)$ of length $2^{k_1+k_2-1}(2^{k_1}-1)$, $i \in 2\mathbb{Z}_4$. Obviously, if either $\mathbf{r}_1$ or $\mathbf{r}_2$ is from the last k_2 rows, it's easy to check that $2\alpha(\mathbf{r}_1) * \alpha(\mathbf{r}_2) = \mathbf{0} \in C$.

(1) If $k_1 = k+1 \geq 2$, $k_2 = 0$, then $\mathbf{r}_1, \mathbf{r}_2, \cdots, \mathbf{r}_k$ and $\mathbf{r}_{k+1}$ become the rows of Y_{k+1}. By the definition of Y_k, we can get

$$Y_{k+1} = \begin{pmatrix} Y_k & Y_k & Y_k & Y_k & B_k \\ \mathbf{0} & \mathbf{1} & \mathbf{2} & \mathbf{3} & \mathbf{1}' \end{pmatrix}$$

where $k \geq 1$, B_k is a $k \times 2^k$ matrix over $2\mathbb{Z}_4$ with all different columns, $\mathbf{i}$ is a row vector $(i, i, \cdots, i)$ of length $2^{k-1}(2^k-1)$, $i \in \mathbb{Z}_4$ and $\mathbf{1}'$ is the row vector $(1, 1, \cdots, 1)$ of length 2^k.

Without loss of generality, let

$$\mathbf{r}_1 = (\overbrace{0\,0\,\cdots\,0}^{2^{k-1}(2^k-1)}\,\overbrace{1\,1\,\cdots\,1}^{2^{k-1}(2^k-1)}\,\overbrace{2\,2\,\cdots\,2}^{2^{k-1}(2^k-1)}\,\overbrace{3\,3\,\cdots\,3}^{2^{k-1}(2^k-1)}\,\overbrace{1\,1\,\cdots\,1}^{2^k}) \tag{4.80}$$

be the last row of Y_{k+1}, and

$$\mathbf{r}_2 = (\mathbf{x}\ \mathbf{x}\ \mathbf{x}\ \mathbf{x}\ \mathbf{b}),$$

where $\mathbf{x}$ is a row of Y_k, and $\mathbf{b}$ is a row of B_k. Then

$$\begin{aligned} 2\alpha(\mathbf{r}_1) * \alpha(\mathbf{r}_2) &= 2(\alpha(\mathbf{0})\ \alpha(\mathbf{1})\ \alpha(\mathbf{2})\ \alpha(\mathbf{3})\ \alpha(\mathbf{1}')) * (\alpha(\mathbf{x})\ \alpha(\mathbf{x})\ \alpha(\mathbf{x})\ \alpha(\mathbf{x})\ \alpha(\mathbf{b})) \\ &= 2(\mathbf{0}\ \alpha(\mathbf{x})\ \mathbf{0}\ \alpha(\mathbf{x})\ \mathbf{0}), \end{aligned}$$

we get

$$W_L(2\alpha(\mathbf{r}_1) * \alpha(\mathbf{r}_2)) = 4W_L(\alpha(\mathbf{x})).$$

Since the length of $\mathbf{x}$ is $2^{k-1}(2^k - 1)$, then

$$4W_L(\alpha(\mathbf{x})) \leq 4 \times 2^{k-1}(2^k - 1) = 2^{k+1}(2^k - 1).$$

By Proposition 4.3, C is a two-Lee weight linear code with the nonzero Lee weights $2^k(2^{k+1}-1)$, and 2^{2k+1}. And the nonzero smallest Lee weight of C is $2^k(2^{k+1}-1) > 2^{k+1}(2^k - 1)$. Thus, $2\alpha(\mathbf{r}_1) * \alpha(\mathbf{r}_2) \notin C$. By Lemma 4.12, $\Phi(C)$ is nonlinear when $k_1 \geq 2$, $k_2 = 0$.

(2) If $k_2 \geq 1$, then the first k_1 rows of G_{k_1,k_2} can be written as:

$$\underbrace{Y_{k_1}\ Y_{k_1}\ \cdots Y_{k_1}}_{2^{k_2}},$$

and let

$$\mathbf{r}_i' = \underbrace{(\mathbf{r}_i\ \mathbf{r}_i\ \cdots\ \mathbf{r}_i)}_{2^{k_2}},$$

where $i = 1$ or 2. Since $k_1 = k + 1$, we have

$$\begin{aligned} W_L(2\alpha(\mathbf{r}_1') * \alpha(\mathbf{r}_2')) &= 2^{k_2} W_L(2\alpha(\mathbf{r}_1) * \alpha(\mathbf{r}_2)) = 2^{k_2+2} W_L(\alpha(\mathbf{x})) \\ &\leq 2^{k_2+2} \times 2^{k-1}(2^k - 1) = 2^{k_1+k_2}(2^{k_1-1} - 1). \end{aligned}$$

By Theorem 4.21, C is a two-Lee weight code with the nonzero Lee weights $2^{k_1+k_2-1}(2^{k_1} - 1)$ and $2^{2k_1+k_2-1}$. And the nonzero smallest Lee weight is $2^{k_1+k_2-1}(2^{k_1} - 1) > 2^{k_1+k_2}(2^{k_1-1} - 1)$. Thus, $2\alpha(\mathbf{r}_1') * \alpha(\mathbf{r}_2') \notin C$. By Lemma 4.12, $\Phi(C)$ is nonlinear when $k_1 \geq 2$, $k_2 \geq 0$. □

Example 4.13. *Consider the code defined in Example 4.9. Let* $\mathbf{r}_1$ *and* $\mathbf{r}_2$ *denote the first and second rows of* $G_{2,1}$ *in* (4.73)*, respectively. Then the Lee weight of*

$$\begin{aligned} 2\alpha(\mathbf{r}_1) * \alpha(\mathbf{r}_2) &= 2\alpha(111102111102) * \alpha(012311012311) \\ &= 2(111100111100) * (010111010111) \\ &= (020200020200) \end{aligned}$$

is 8 which can not be obtained in C *since* C *is a two-Lee weight code with* $w_1 = 12$, $w_2 = 16$.

Theorem 4.24. *If the columns of the generator matrix of C satisfy the conditions of Corollary 4.2, then $\Phi(C)$ is linear when $k_1 = 2$ and nonlinear when $k_1 \geq 3$.*

Proof. Denote the generator matrix of C by G_{k_1,k_2}. Let $\mathbf{r}_1, \mathbf{r}_2, \cdots, \mathbf{r}_{k_1}, \cdots, \mathbf{r}_{k_1+k_2}$ be the rows of G_{k_1,k_2}, where $\mathbf{r}_j$ is a vector over $2\mathbb{Z}_4$, $k_1 + 1 \leq j \leq k_1 + k_2$.

(1) By Corollary 4.2, if $k_1 = 2$, $k_2 = 1$, the generator matrix is:

$$G_{2,1} = \begin{pmatrix} 1111 & 1111 \\ 0123 & 0123 \\ 2222 & 0000 \end{pmatrix}.$$

So, if $\mathbf{r}_1$, $\mathbf{r}_2$, $\mathbf{r}_j$ are the rows of G_{2,k_2}, $3 \leq j \leq k_2 + 2$, we can get $\mathbf{r}_1$ is all 1 vector. Therefore, $\alpha(\mathbf{r}_1) = \mathbf{1}$ and $\alpha(\mathbf{r}_j) = \mathbf{0}$. Besides,

$$\begin{aligned} 2\alpha(\mathbf{r}_1) * \alpha(\mathbf{r}_2) = 2\mathbf{r}_2 \in C, 2\alpha(\mathbf{r}_1) * \alpha(\mathbf{r}_j) &= \mathbf{0} \in C, \\ 2\alpha(\mathbf{r}_2) * \alpha(\mathbf{r}_j) = \mathbf{0} \in C, \quad 2\alpha(\mathbf{r}_j) * \alpha(\mathbf{r}_k) &= \mathbf{0} \in C, \end{aligned}$$

where $3 \leq j, k \leq 2 + k_2$. By Lemma 4.12, $\Phi(C)$ is linear when $k_1 = 2$, $k_2 \geq 0$.

(2) If $k_1 \geq 3$, by (4.75) in the proof of Corollary 4.2,

$$G_{k_1,k_2} = \begin{pmatrix} \mathbf{0} & \mathbf{1} & \mathbf{2} & \mathbf{3} \\ G'_{k_1-1,k_2} & G'_{k_1-1,k_2} & G'_{k_1-1,k_2} & G'_{k_1-1,k_2} \end{pmatrix},$$

where G'_{k_1-1,k_2} is the generator matrix of the code C' obtained in Theorem 4.21, and $\mathbf{i}$ is the row vector $(i, i, \cdots, i)$ of length $2^{k_1+k_2-2}(2^{k_1-1} - 1)$, $i \in \mathbb{Z}_4$. Without loss of generality, let

$$\mathbf{r}_1 = (\mathbf{0}\ \mathbf{1}\ \mathbf{2}\ \mathbf{3}) \text{ and } \mathbf{r}_2 = (\mathbf{r}_0\ \mathbf{r}_0\ \mathbf{r}_0\ \mathbf{r}_0)$$

where $\mathbf{r}_0$ is one row of G'_{k_1-1,k_2}. Then

$$2\alpha(\mathbf{r}_1) * \alpha(\mathbf{r}_2) = 2(\mathbf{0}\ \alpha(\mathbf{r}_0)\ \mathbf{0}\ \alpha(\mathbf{r}_0)).$$

Since the length of $\mathbf{r}_0$ is $2^{k_1+k_2-2}(2^{k_1-1} - 1)$,

$$W_L(2\alpha(\mathbf{r}_1) * \alpha(\mathbf{r}_2)) = 4W_L(\alpha(\mathbf{r}_0)) \leq 2^{k_1+k_2}(2^{k_1-1} - 1). \tag{4.81}$$

If $\mathbf{0} \neq 2\alpha(\mathbf{r}_1) * \alpha(\mathbf{r}_2) \in C$, since C is a two-Lee weight code by Corollary 4.2, $W_L(2\alpha(\mathbf{r}_1) * \alpha(\mathbf{r}_2)) = 2^{k_1+k_2}(2^{k_1-1} - 1)$ or $2^{2k_1+k_2-1}$, Thus

$$2^{k_1+k_2}(2^{k_1-1} - 1) \leq W_L(2\alpha(\mathbf{r}_1) * \alpha(\mathbf{r}_2)) = 4W_L(\alpha(\mathbf{r}_0)). \tag{4.82}$$

Therefore, combining (4.81) and (4.82), we have

$$W_L(2\alpha(\mathbf{r}_1) * \alpha(\mathbf{r}_2)) = 4W_L(\alpha(\mathbf{r}_0)) = 2^{k_1+k_2}(2^{k_1-1} - 1),$$

which forces $\alpha(\mathbf{r}_0)$ to be all 1 vector. Thus we have $W_L(\mathbf{r}_0) = 2^{k_1+k_2-2}(2^{k_1-1} - 1)$, $W_L(2\mathbf{r}_0) = 2^{k_1+k_2-1}(2^{k_1-1} - 1)$ and then $W_L(2\mathbf{r}_0) = 2W_L(\mathbf{r}_0)$. However, $\mathbf{r}_0$ is one row of G'_{k_1-1,k_2}, the generator matrix obtained in Theorem 4.21, and the two Lee weights w_1 and w_2 can not satisfy $w_2 = 2w_1$, contradiction. Thus there are at least 2 rows $\mathbf{r}_1$, $\mathbf{r}_2$ satisfying $2\alpha(\mathbf{r}_1) * \alpha(\mathbf{r}_2) \notin C$. Namely, $\Phi(C)$ is nonlinear when $k_1 \geq 3$, $k_2 \geq 0$. □

Example 4.14. *Consider the code defined in Example 4.12. Let* $\mathbf{r}_1$ *annd* $\mathbf{r}_2$ *denote the first and second rows of* $G'_{3,1}$ *in (4.77), respectively. Then the Lee weight of*

$$\begin{aligned}2\alpha(\mathbf{r}_1) * \alpha(\mathbf{r}_2) &= 2\alpha(11110211110211110211110211110211110211110211110 2) \\ &* \ \alpha(012311012311012311012311012311012311012311012311) \\ &= (020200020200020200020200020200020200020200020200)\end{aligned}$$

is 32 which can not be obtained in C *since* C *is a two-Lee weight code with* $w_1 = 48$, $w_2 = 64$.

4.2.3 The Punctured Dodecacode over $\mathbb{F}_4$

4.2.3.1 Graphs

All graphs in this note are finite, undirected, connected, without loops or multiple edges. The neighborhood $\Gamma(x)$ is the set of vertices connected to x. The *degree* of a vertex x is the size of $\Gamma(x)$. A graph is *regular* if every vertex has the same degree. The i-neighborhood $\Gamma_i(x)$ is the set of vertices at geodetic distance i to x. A graph is *distance regular* (DR) if for every pair or vertices u and v at distance i apart the quantities

$$\begin{aligned} a_i &= |\Gamma_{i+1}(u) \cap \Gamma(v)| \\ c_i &= |\Gamma_{i-1}(u) \cap \Gamma(v)| \end{aligned}$$

solely depend on i and not on the special choice of the pair (u, v). In that situation the graphs Γ_i are regular of degree v_i and we will refer to the v_is as the *successive degrees* of Γ. The *automorphism group* of a graph is the set of permutations of the vertices that preserve adjacency. A graph is *distance-transitive* if its automorphism group is transitive on its vertices and on each of the sets $\Gamma_i(x)$ for any i and any fixed x. A DR graph of diameter 2 is called a *Strongly Regular Graph* (SRG). Its parameters $(\nu, \kappa, \lambda, \mu)$ denote the number of vertices, the degree, the number of common neighbors of a pair of connected vertices, the number of common neighbors of a pair of disconnected vertices. The *spectrum* of a graph is the set of distinct eigenvalues of its adjacency matrix. It is denoted by $\{\lambda_1^{m_1}, \lambda_2^{m_2}, \dots\}$, where m_i stands for the multiplicity of the eigenvalue λ_i. The *Hamming graph* $H(n, q)$ is the DR graph on $\mathbb{F}_q^n$, two vectors being connected if they are at Hamming distance one.

4.2.3.2 Quaternary Codes

The dodecacode D is a code of length 12 over $\mathbb{F}_4 = \mathbb{F}_2(w)$ of parameters $(12, 4^6, 6)$ that is additive but not linear [31]. It can be defined as a cyclic code with one generator. In the notation of [31] we have $D = (w10100100101)$. Thus, puncturing at any of the twelve coordinates gives an equivalent code D^- of parameters $(11, 4^6, 5)$, of generator matrix given in Table 4.2. Its primal and dual weight distributions and automorphism groups are easily computed in Magma [6] and Sage [90].

TABLE 4.2 Generator Matrix of D^-

$$\begin{pmatrix}
0 & 0 & 0 & 0 & 0 & w^2 & w^2 & 0 & w & 1 & w \\
0 & 0 & 0 & 0 & 0 & w & 0 & w & w & w & 1 \\
1 & 0 & 0 & 0 & 0 & 1 & 0 & 1 & w^2 & w^2 & 1 \\
w & 0 & 0 & 0 & 0 & 0 & w & 1 & w & w & w \\
0 & 1 & 0 & 0 & 0 & 0 & 1 & 1 & 1 & w^2 & w^2 \\
0 & w & 0 & 0 & 0 & w^2 & 1 & w^2 & w & w & 0 \\
0 & 0 & 1 & 0 & 0 & w & w & 1 & 1 & 0 & 1 \\
0 & 0 & w & 0 & 0 & w & 1 & w & w^2 & w^2 & 0 \\
0 & 0 & 0 & 1 & 0 & w & w & w & 0 & 1 & w \\
0 & 0 & 0 & w & 0 & w^2 & 1 & 1 & w^2 & 0 & w \\
0 & 0 & 0 & 0 & 1 & w^2 & w^2 & w^2 & 1 & 0 & w^2 \\
0 & 0 & 0 & 0 & w & w & 1 & 0 & 1 & w & w
\end{pmatrix}$$

Theorem 4.25. *The weight distribution and the dual weight distribution of D^- are*

$$[\langle 0,1\rangle, \langle 5,198\rangle, \langle 6,198\rangle, \langle 7,990\rangle, \langle 8,495\rangle, \langle 9,1650\rangle, \langle 10,330\rangle, \langle 11,234\rangle]$$
$$\text{and} \quad [\langle 0,1\rangle, \langle 6,198\rangle, \langle 8,495\rangle, \langle 10,330\rangle],$$

respectively, where $\langle i, A_i\rangle$ means that there are A_i codewords of weight i.

Theorem 4.26. *The monomial automorphism group of D^- is generated by the following two monomial matrices:*

$$\begin{bmatrix}
0&0&0&0&0&w&0&0&0&0&0\\
1&0&0&0&0&0&0&0&0&0&0\\
0&0&w^2&0&0&0&0&0&0&0&0\\
0&0&0&0&0&0&0&w^2&0&0&0\\
0&0&0&0&0&0&0&0&0&1&0\\
0&0&0&0&w&0&0&0&0&0&0\\
0&0&0&0&0&0&0&0&0&0&w\\
0&0&0&1&0&0&0&0&0&0&0\\
0&w^2&0&0&0&0&0&0&0&0&0\\
0&0&0&0&0&0&0&0&w^2&0&0\\
0&0&0&0&0&0&1&0&0&0&0
\end{bmatrix}, \quad
\begin{bmatrix}
1&0&0&0&0&0&0&0&0&0&0\\
0&0&0&0&0&0&0&0&0&0&w^2\\
0&0&0&0&0&w^2&0&0&0&0&0\\
0&0&0&0&0&0&0&w^2&0&0&0\\
0&0&0&0&0&0&0&0&w&0&0\\
0&0&w&0&0&0&0&0&0&0&0\\
0&0&0&0&0&0&0&0&0&1&0\\
0&0&0&w&0&0&0&0&0&0&0\\
0&0&0&0&w^2&0&0&0&0&0&0\\
0&0&0&0&0&0&1&0&0&0&0\\
0&w&0&0&0&0&0&0&0&0&0
\end{bmatrix}.$$

It is isomorphic to a subgroup of S_{33} of order $54 = 2 \times 3^3$ with generators (1 4 27 29 14 18)(2 5 25 30 15 16)(3 6 26 28 13 17)(7 9 8)(10 22 12 24 11 23)(19 31 20 32 21 33) *and* (4 32)(5 33)(6 31)(7 17)(8 18)(9 16)(10 23)(11 24)(12 22)(13, 27)(14 25)(15 26)(19 28)(20 29)(21 30) *in disjoint cycles product notation (this subgroup reflects the action of the automorphism group on the words of weight* 1 *in the order* 10...0, w0...0, $w^2$0...0, 010...0, ..., 0...0w^2*). The complete automorphism group of D^- is the product of the monomial automorphism group with the group of translations and has the order $54 \cdot 2^{12}$.*

Remark 4.3. *Since 54 is not divisible by 33 the code D^- cannot be completely transitive [89].*

To compute the parameters λ and μ, we require the following lemma, of independent interest.

Lemma 4.13. *If C is a uniformly packed code of distance $2e+1$, length n, and weight distribution A_i over $\mathbb{F}_q$, then we have*

$$\lambda(q-1)^e\binom{n}{e} = A_{2e+1}\binom{2e+1}{e}$$

$$A_{2e+1}\binom{2e+1}{e}(e+1)(q-2)+A_{2e+2}\binom{2e+2}{e+1} = (\lambda-\mu)A_{2e+1}\binom{2e+1}{e} + (\mu-1)(q-1)^{e+1}\binom{n}{e+1}.$$

Proof. Double counting the number of pairs (x,y) such that $\mathrm{wt}(x)=e$, $\mathrm{wt}(y)=2e+1$, $\mathrm{wt}(y-x)=e+1$, $y\in C$ relates λ and A_{2e+1}. This yields the first relation.

Double counting the number of pairs (x,y) such that $\mathrm{wt}(x)=e+1$, $\mathrm{wt}(y)\in\{2e+1,2e+2\}$, $\mathrm{wt}(y-x)=e+1$, $y\in C$ (note that among the $(q-1)^{e+1}\binom{n}{e+1}$ words of weight $e+1$, exactly $A_{2e+1}\binom{2e+1}{e}$ are at distance e from the code, and the other are at distance $e+1$ from the code) relates λ, μ, A_{2e+1}, and A_{2e+2}. This gives us the second relation. □

We are now in a position to state and prove the main result of this subsection.

Theorem 4.27. *The code D^- is uniformly packed with $(\lambda,\mu)=(4,5)$.*

Proof. We know the code is uniformly packed since its minimum distance equals twice its external distance minus one [1, Cor. 11.1.2]. To compute λ and μ we specialize the above lemma to $q=4$, $n=11$, $e=2$, and $A_5=A_6=198$ that is known from Theorem 4.25. The first relation yields $\lambda=4$. Reporting into the second gives $\mu=5$. □

Denote by D^{3-} any $(9,4^6)$ code obtained by puncturing D^- on any pair of coordinates. The dual of D^{3-} is an additive two-weight code. The following result comes from an easy Magma computation [6].

Theorem 4.28. *The dual weight distribution of D^{3-} is $[\langle 0,1\rangle,\langle 6,36\rangle,\langle 8,27\rangle]$.*

4.2.3.3 Binary Codes

We aim to construct a binary code with a coset graph isomorphic to that of D^-. To that end, we define a universal correspondence between quaternary and binary codes that preserves coset graphs, up to isomorphism.

Define the concatenation map of a quaternary code of length n with the zero sum code $R_3^\perp = \{000, 011, 110, 101\}$ as follows. For binary scalars a and b, let $\phi(a + bw) = (b, b + a, a)$. Equivalently, $\phi(c) = (\mathrm{Tr}(c), \mathrm{Tr}(wc), \mathrm{Tr}(w^2c))$. The action of ϕ is extended to the quaternary vectors coordinatewise.

Theorem 4.29. *Assume that Q is a quaternary additive code of length n and $B = \phi(Q^\perp)^\perp$. The following assertions hold:*

i. *if Q is an $(n, 4^k)$ code, then B is a linear $[3n, n + 2k]$ binary code;*

ii. *if w is a weight of $Q^\perp$ with frequency A_w, then $2w$ is a weight of $B^\perp$ with frequency A_w, and all the weights of $B^\perp$ arise in this way;*

iii. *the coset graphs Γ_Q and Γ_B are isomorphic.*

Proof. (i) Trivially, the length of $B^\perp = \phi(Q^\perp)$ is $3n$ and the size is the same as the size of $Q^\perp$, i.e., 4^{n-k}. Hence, the size of B is $2^{3n}/4^{n-k} = 2^{n+2k}$.

(ii) From the concatenation, we readily see that $\mathrm{wt}(\phi(c)) = 2\mathrm{wt}(c)$.

(iii) We will show that the graphs Γ_Q and Γ_B built on the syndromes of the check matrices P and $\phi(P)$ coincide. To see this, we first note that both graphs are built on the binary columns of height $2n - 2k$. Next, we consider the connecting syndromes that correspond to the weight-1 vectors. Denote by P_1 the first column of the check matrix. The syndromes corresponding to the weight-1 quaternary vectors $(1, 0, ..., 0)$, $(w, 0, ..., 0)$, and $(w^2, 0, ..., 0)$ are $\mathrm{Tr}(P_1)$, $\mathrm{Tr}(w^2P_1)$, and $\mathrm{Tr}(wP_1)$, where the trace map acts on the column component-wise. On the other hand, by the definition of ϕ, the first three columns of $\phi(P)$ are $\mathrm{Tr}(P_1)$, $\mathrm{Tr}(wP_1)$, and $\mathrm{Tr}(w^2P_1)$. Therefore, the syndromes corresponding to the binary 1-weight vectors $(1, 0, ..., 0)$, $(0, 1, 0, ..., 0)$, $(0, 0, 1, 0, ..., 0)$ are $\mathrm{Tr}(P_1)$, $\mathrm{Tr}(wP_1)$, $\mathrm{Tr}(w^2P_1)$ again. Considering in a similar way every other column of P, we find that the set of syndromes corresponding to the weight-1 vectors is the same for P and $\phi(P)$. Hence, $\Gamma_Q = \Gamma_B$. □

Remark 4.4. *It can be shown that Q and B are related by the following correspondence (denoted by ψ) that associates 2^n codewords of B to any codeword of Q. For simplicity, we write it for $n = 1$.*

$$0 \xrightarrow{\psi} 000, 111; \qquad 1 \xrightarrow{\psi} 100, 011; \qquad w \xrightarrow{\psi} 001, 110; \qquad w^2 \xrightarrow{\psi} 010, 101.$$

Thus the images of any of the four symbols form an antipodal pair of vertices in the three-cube. We have the following commutative diagram where the down arrow means "dual".

$$\begin{array}{ccc} Q & \longrightarrow & \psi(Q) \\ \downarrow & & \downarrow \\ Q^\perp & \longrightarrow & \phi(Q^\perp) = \psi(Q)^\perp \end{array}$$

Thus ψ is the pullback of ϕ in this diagram.

Corollary 4.4. *Let $B^- = \phi(D^{-\perp})^\perp$. The binary code B^- is a completely regular code of parameters* $[33, 23, 3]$*, with dual weight distribution*

$$[\langle 0,1\rangle, \langle 12,198\rangle, \langle 16,495\rangle, \langle 20,330\rangle],$$

and weight distribution

$$\begin{aligned}[\langle 0,1\rangle, &\langle 3,11\rangle, \langle 5,198\rangle, \langle 6,1243\rangle, \langle 7,4158\rangle, \langle 8,13563\rangle, \langle 9,38445\rangle,\\ &\langle 10,88638\rangle, \langle 11,185397\rangle, \langle 12,352902\rangle, \langle 13,568788\rangle, \langle 14,786885\rangle,\\ &\langle 15,998052\rangle, \langle 16,1156023\rangle, \langle 17,1156023\rangle, \langle 18,998052\rangle, \langle 19,786885\rangle,\\ &\langle 20,568788\rangle, \langle 21,352902\rangle, \langle 22,185397\rangle, \langle 23,88638\rangle, \langle 24,38445\rangle,\\ &\langle 25,13563\rangle, \langle 26,4158\rangle, \langle 27,1243\rangle, \langle 28,198\rangle, \langle 30,11\rangle, \langle 33,1\rangle].\end{aligned}$$

Proof. The dual weight distribution follows by the preceding theorem combined with Theorem 4.25. The weight distribution is then computed by MacWilliams transform. □

Remark 4.5. *The code B^- is a nontrivial example of a completely regular code since its minimum distance is* 3*, and its external distance is* 3*. Thus it is neither perfect nor uniformly packed.*

Corollary 4.5. *The code $\phi(D^{3-\perp})$ is a binary two-weight* $[27, 6; 12, 16]$ *code.*

Proof. The weight distribution follows by Theorem 4.29, where $n = 9$, $k = 6$, $Q = D^{3-}$, and $B = \phi(D^{3-\perp})^\perp$, combined with Theorem 4.28. □

Remark 4.6. *Binary completely regular codes with the last parameters are known (see e.g. Chapter 7); all such codes are related to bent functions, see Theorem 12.12 and the following paragraph in [15]. Puncturing different coordinates, we obtain three nonequivalent codes with these parameters.*

4.2.3.4 Graphs

In this section, we study the coset graphs of D^-, and of D^{3-}.

Theorem 4.30. *The graph Γ_{D^-} is distance-regular of diameter* 3*, of spectrum*

$$\{27^1, 9^{198}, 1^{495}, (-7)^{330}\}.$$

Its successive degrees are $(1, 33, 495, 495)$.

Proof. The spectrum of Γ_{D^-} is easily computed from the weight distribution of Theorem 4.25 upon applying Theorem 4.4. The distance-regularity follows from the fact that D^- is completely regular, being uniformly packed in the sense of [40], as having minimum distance $5 = 2 \times 3 - 1$ and external distance 3. In particular, it is completely regular as per [40, Thm 7]. The fact that the error correcting capacity is two implies that the first three degrees are $1, 33 = 3 \times 11, 495 = \binom{11}{2}3^2$. The last degree is followed by $2^{10} - 1 - 33 - 495 = 495$. □

More structural information on this graph is as follows.

Theorem 4.31. *The intersection array of the graph* Γ_{D^-} *is* $\{33, 30, 15; 1, 2, 15\}$.

Proof. We sketch the three steps of the proof as follows. We know λ and μ from Theorem 4.27. The outer distribution matrix B of an uniformly packed code is uniquely determined by n, e, q and these two parameters [1, Cor. 11.1.2]. From the known intersection array of $H(n, 4)$, and this data, the result follows by [1, Th. 11.1.8], with $\Gamma = H(n, 4)$, and Π being the completely regular partition induced by the cosets of D^-. □

Theorem 4.32. *The automorphism group of the graph has order* $2^{10} \cdot 54$, *acts transitively on the vertices, and has two orbits on the edges, of size* $2^9 \cdot 6$ *and* $2^9 \cdot 27$. *The stabilizer of a vertex has a structure of type* $(C_9 \rtimes C_3) \rtimes C_2$. *The induced subgraph fixed by an element of the automorphism group, depending on the order* p *of the element, can be the following, up to isomorphism:*

i. *the whole graph,* $p = 1$;

ii. *the null graph,* $p = 2$;

iii. *the disjoint union of eight complete graphs of order* 4, $p = 2$;

iv *the edge-free graph on four vertices,* $p = 3$;

v. *the Hamming graph* $H(2, 4)$, $p = 3$;

vi. *the edge-free graph on two vertices,* $p = 6$;

vii. *the one-vertex graph,* $p = 9$;

viii. *the complete graph of order* 4, $p = 9$.

Proof. Since the graph is a Cayley graph, it is vertex transitive. The other assertions were established by Sage [90] computations. □

The possible induced subgraphs fixed by an element of a prime order of the automorphism group of a distance-regular graph with intersection array $\{33, 30, 15; 1, 2, 15\}$ were studied in [69]. The case (iv) of the theorem above is missed in the main Theorem of [69], which should be completed by the subcase where every two vertices of the fixed subgraph are at distance three [70].

The following strongly regular graph is found by standard spectral techniques, as indicated in [1, p. 428]. For the next two results, we assume that the reader is familiar with the theory of duality in association schemes [1,4,33].

Theorem 4.33. *The graph* $(\Gamma_{D^-})_2$ *is strongly regular with parameters* (2^{10}, 495, 238, 240).

Proof. The P-matrix of the association scheme underlying Γ_{D^-} can be computed by the formulas in [30, pp. 135–136] or [33, Th. 5.25], as

$$\begin{pmatrix} 1 & 33 & 495 & 495 \\ 1 & 9 & 15 & -25 \\ 1 & 1 & -17 & 15 \\ 1 & -7 & 15 & -9 \end{pmatrix}.$$

The second column is the spectrum of $(\Gamma_{D^-})_2$, discounting multiplicities. The result follows by the spectral characterization of strongly regular graphs [2]. □

Strongly regular graphs with these parameters are known [7]. We do not know if any of them is isomorphic to $(\Gamma_{D^-})_2$.

There is a Q-analogue of the preceding results.

Theorem 4.34. *The Delsarte dual of the underlying association scheme of Γ_{D^-} is Q-polynomial with Krein array $\{33, 30, 15; 1, 2, 15\}$, multiplicities 1, 33, 495, 495 and valencies 1, 198, 495, 330. The second relation of that association scheme is a SRG of parameters $(2^{10}, 495, 238, 240)$.*

Proof. Recall that the Delsarte dual of a coset graph is an association scheme on the codewords of the dual code [33], called hereafter the *distance scheme.* If the weights are numbered

$$w_1 = 6 < w_2 = 8 < w_3 = 10,$$

then for $x, y \in D^{-\perp}$, we define the relation R_i as xR_iy iff $\mathrm{wt}(x + y) = w_i$. By convention xR_0y iff $x = y$. This is exactly the situation of [30] with $e = 2$, up to the fact that D^- is not linear but only additive. By Delsarte duality, the multiplicities of this scheme are the degrees of Γ_{D^-}. Similarly, its valencies are the multiplicities of the spectrum of Γ_{D^-}. The Krein array coincides with the intersection array of the coset graph. The fact that the second relation of the distance scheme is a SRG comes from computation of the P-matrix which can be done as in [30, pp. 135–136].

$$\begin{pmatrix} 1 & 198 & 495 & 330 \\ 1 & 54 & 15 & -70 \\ 1 & 6 & -17 & 10 \\ 1 & -10 & 15 & -6 \end{pmatrix}.$$

□

Remark 4.7. *By Delsarte duality, the P-matrix of the coset scheme is the Q-matrix of the distance scheme and conversely.*

Remark 4.8. *This Q-polynomial scheme appears as an open problem in the table of Jason Williford [94]. We thank Bill Martin for pointing this out.*

Remark 4.9. *We do not know if the SRG constructed in that theorem is isomorphic with $(\Gamma_{D^-})_2$.*

We also obtain a strongly regular graph on the cosets of D^{3-}.

Theorem 4.35. *The graph* Γ_{D^3-} *is a SRG of parameters* $(64, 27, 10, 12)$. *The spectrum of* Γ_{D^3-} *is* $\{27^1, 3^{36}, (-5)^{27}\}$.

Proof. The spectrum of Γ_{D^3-} is easily computed from the weight distribution of Theorem 4.28 upon applying [1, 11.1.11]. The strong regularity follows then by the spectral characterization of SRG's [2]. The parameters are followed by the data of the spectrum [2, Th. 9.1.3]. □

Remark 4.10. *Depending on the punctured coordinates we obtain three non-isomorphic SRG*$(64, 27, 10, 12)$. *One of them is isomorphic to the SRG corresponding to a linear* $[9, 3]$ *two-weight quaternary code. All the codes* D^{3-} *are, however, nonlinear.*

4.2.4 Three Weight Codes over Rings

4.2.4.1 *Triple Sum and s-Sum Sets*

We extend the definition of triple sum sets given in [18] from finite fields to finite commutative rings R with a unit group $R^\times$. The set $\Omega \subseteq R^k$ is a *triple sum set* (TSS) if it is stable by scalar multiplication by units $\in R^\times$ and if there are integer constants σ_0 and σ_1 such that a non-zero $h \in R^k$ can be written as

$$h = x + y + z, \text{ with } x, y, z \in \Omega,$$

- σ_0 times if $h \in \Omega$,
- σ_1 times if $h \in R^k \setminus \Omega$.

This can generalized immediately to the notion of an *s-sum set* by replacing the equation $h = x + y + z$ by $h = \sum_{i=1}^{s} x_i$ where each $x_i \in \Omega$.

4.2.4.2 *Syndrome Graph*

An *eigenvalue* of a graph Γ (i.e., an eigenvalue of its adjacency matrix) is called a *restricted eigenvalue* if there is a corresponding eigenvector which is not a multiple of the all-one vector $\mathbf{1}$. Note that for an η-regular connected graph, the restricted eigenvalues are simply the eigenvalues different from η.

Definition 4.7. *Let* T *be a finite abelian group and* $S \subseteq T$ *a subset satisfying* $S = -S$ *and* $0_T \notin S$. *The corresponding* Cayley graph $\mathrm{C}(T, S)$ *has vertex set equal to* T*; two vertices* $g, h \in T$ *are adjacent in* $\mathrm{C}(T, S)$ *iff* $g - h \in S$.

The conditions imposed on S ensure that $\mathrm{C}(T, S)$ is a simple graph. The graph $\mathrm{C}(T, S)$ is regular of degree $|S|$, and it is connected iff S generates the group T. We use the same idea as in [15] to associate with any linear code C over $\mathbb{Z}_{p^m}$ a certain Cayley graph, called the *syndrome graph* of C and denoted by $\Gamma(C)$.

Let $H = [h_1 \mid \cdots \mid h_n] \in (\mathbb{Z}_{p^m})^{\ell\times n}$ be a parity-check matrix of C. The vertex set of $\Gamma(C)$ is $V = \{Hx; x \in (\mathbb{Z}_{p^m})^n\}$, the column space of H, which is isomorphic to the dual code $C^\perp$. The elements of V are called *syndromes* of C. Two syndromes Hx, Hy are adjacent in $\Gamma(C)$ if they differ by a unit-multiple of a column of H:

$$Hx \sim Hy \;:\Longleftrightarrow\; H(x-y) = uh_i \quad \text{for some } 1 \le i \le n \text{ and } u \in \mathbb{Z}_{p^m}^\times. \tag{4.83}$$

It is obvious that $\Gamma(C)$ is the Cayley graph of V corresponding to the generating $S = \{uh_i; u \in \mathbb{Z}_{p^m}^\times, 1 \le i \le n\}$. If C is regular then $0 \notin S$ and $-S \subseteq S$, implying that $\Gamma(C)$ is simple. Moreover, if C is projective, $\Gamma(C)$ is regular of degree $|S| = p^{m-1}(p-1)n$ and has $\frac{p^{mn}}{|C|}$ vertices. The latter follows from $|C|\,|C^\perp| = |(\mathbb{Z}_{p^m})^n| = p^{mn}$.

As defined, $\Gamma(C)$ depends on the particular choice of H. The terminology "syndrome graph of C" is justified, however, since $\Gamma(C)$ has another representation as a "coset graph" of C, which shows that its isomorphism type is determined by C. It is well-known that there is a one-to-one correspondence between syndromes of a code and its cosets [66], given by $x + C \mapsto Hx$. The condition in (4.83) can be restated as $H(x - y - ue_i) = 0$, where e_i denotes the ith standard unit vector in $(\mathbb{Z}_{p^m})^n$. Thus $\Gamma(C)$ is isomorphic to the graph whose vertices are the cosets of C and in which two vertices $x + C$, $y + C$ are adjacent if their difference is a coset of the form $ue_i + C$. This is the Cayley graph of the quotient group $(\mathbb{Z}_{p^m})^n/C$ with generating set

$$S^* = \cup_{i=1}^n \cup_{u\in\mathbb{Z}_{p^m}^\times} (ue_i + C).$$

Now, we recall the relation between the weight distribution of a linear code over $\mathbb{Z}_{p^m}$ and the eigenvalues of the syndrome graph of its dual code. This extension of Lemma 3.4 in [15] was derived in [80] Theorem 4.1 and its proof is omitted here.

Theorem 4.36. *Suppose that C is a regular, projective linear code over $\mathbb{Z}_{p^m}$ with homogeneous weights w_i and corresponding weight distribution $A_i = |\{x \in C; \mathrm{w_{hom}}(x) = w_i\}|$. Then the eigenvalues of $\Gamma(C^\perp)$ are $n(p-1)p^{m-1} - pw_i$ with multiplicity A_i.*

The *coset graph with b loops* $\Gamma(C)^b$ of a code C is obtained from $\Gamma(C)$ by adding b loops around every vertex. The following result is immediate from the previous theorem. The proof is omitted. (In general, we call dual weight distribution of a code the weight distribution of the dual code).

Corollary 4.6. *If C is a $\mathbb{Z}_{p^m}$-code, with dual weight distribution $[\langle w_i, A_i\rangle]$, such that $C^\perp$ is regular and projective, then the spectrum of $\Gamma(C)^b$ is $\{(b+n(p-1)p^{m-1} - pw_i)^{A_i}\}$. Thus A_i is the frequency of the homogeneous weight w_i in $C^\perp$ and the multiplicity of the eigenvalue $b + n(p-1)p^{m-1} - pw_i$.*

4.2.4.3 Spectral Graph Theory

Let $\overline{R^k}$ denote the set of all regular vectors in R^k, i.e.,

$$\overline{R^k} = \{\omega \in R^k \,:\, \{\omega \cdot x \,:\, x \in R^k\} = R\}.$$

If $\Omega \subseteq \overline{R^k}$, we denote by $C(\Omega)$ the regular and projective code of length $n = \frac{|\Omega|}{|R^\times|}$ obtained as the kernel of the $k \times n$ matrix H with columns the projectively non-equivalent elements of Ω. Thus H is the check matrix of $C(\Omega)$. To simplify notation, let $\Gamma_{C(\Omega)} = \Gamma(\Omega)$ and $\Gamma^b_{C(\Omega)} = \Gamma(\Omega)^b$, where we also write $\Gamma(\Omega)^0$ for $\Gamma(\Omega)$.

The following result is immediate from the definitions in the introduction and will be used repeatedly and implicitly in the rest of this chapter.

Theorem 4.37. *Keep the above notation. Assume that $C(\Omega)$ is regular and projective. The following are equivalent*

- *Ω (resp $\Omega \cup 0$) is an s-sum set,*
- *$\Gamma(\Omega)$ (resp. $\Gamma(\Omega)^1$) is an s-SWRG.*

Proof. Observe that $\Gamma(\Omega)$ is a Cayley graph on the group $(R^k, +)$ with generators the columns of H. Thus the definition of an s-sum set can be regarded as a statement on the number of paths of length s between 0 and a linear combination of s columns of H. (By translation, we can assume that 0 is the first endpoint of the path). □

The following result comes from the well-known connection between SRGs and two-weight codes. It is described for codes over fields in [1,2], and for codes over rings in [13].

Theorem 4.38. *If $C(\Omega)^\perp$ is a two-weight code, then both $\Omega \cup 0$ and Ω are s-sum sets for all $s > 1$.*

Proof. If $C(\Omega)^\perp$ is a two-weight code, it is well-known that $\Gamma(\Omega)$ is an SRG [13]. The result follows by Theorem 4.37, and the observation, made in the introduction, that SRGs are s-SWRG for all $s > 1$. □

The following result was observed in [41] Theorem III.6 for the Hamming weight.

Theorem 4.39. *If Ω or $\Omega \cup 0$ is an s-sum set, then $C(\Omega)^\perp$ has at most three nonzero weights. Furthermore, if s is even, then $C(\Omega)^\perp$ is a two-weight code.*

Proof. This is a consequence of [35, Theorem 3.4] applied to the graph $\Gamma(\Omega)$, and based on Theorem 4.37. □

We come to the coding-theoretic characterization of 3-sum sets which is contained for finite fields in [18, Th. 2.1]. The consideration of syndrome graphs with loops might seem artificial but is justified by the existence of several non-trivial examples, as evidenced by Section 4.4 in [75].

The relation with the 3-SWRG property and the eigenvalues of a connected regular graph was obtained in [35] Proposition 4.1.

Theorem 4.40. *Let Γ be a connected regular graph with four distinct eigenvalues $k > \theta_1 > \theta_2 > \theta_3$. Then Γ is a 3-SWRG iff $\theta_1 + \theta_2 + \theta_3 = 0$.*

In our setting the formulas for the eigenvalues in Theorem 4.36 and Corollary 4.6 give:

Theorem 4.41. *Assume that $C(\Omega)^\perp$ is of length n and has three nonzero weights $w_1 < w_2 < w_3$. Let b be any non-negative integer.*

- *Ω is a TSS iff $w_1 + w_2 + w_3 = 3n(p-1)p^{m-2}$;*
- *$\Omega \cup 0$ is a TSS iff $w_1 + w_2 + w_3 = 3\frac{1+n(p-1)p^{m-1}}{p}$;*
- *$\Gamma^b_{C(\Omega)}$ is a 3-SWRG iff $w_1 + w_2 + w_3 = 3\frac{b+n(p-1)p^{m-1}}{p}$;*

Proof. For the first two cases we apply Theorem 4.37 to conclude the equivalence to $\Gamma^b_{C(\Omega)}$ being a 3-SWRG for $b = 0$ and $b = 1$, respectively. Combining Corollary 4.6 with Theorem 4.40 gives the third case, which contains the first two cases for $b = 0$ and $b = 1$, respectively. □

Note that if $m \geq 2$ and $p \neq 3$, then the second case of Theorem 4.41 is impossible. More generally, for the third case of Theorem 4.41 the assumptions $m \geq 2$ and $p \neq 3$ imply that b has to be divisible by p. Then, the condition may be rewritten as $w_1 + w_2 + w_3 \geq 3n(p-1)p^{m-2}$ and $w_1 + w_2 + w_3 \equiv 0 \pmod 3$.

Example 1: If $p = m = 2$ and $n = 4$, then the possible triples of weights in the first case of Theorem 4.41 are, with simplified notation, and by increasing the value of w_2 equal to

$$129,\ 138,\ 237,\ 147,\ 246,\ 345,\ 156,\ 255.$$

We will see in Section 4.2 in [75] that none of these possibilities can be attained by a $\mathbb{Z}_4$-code and the smallest possible length for $p = m = 2$ and $b = 0$ is indeed $n = 6$.

Proposition 4.3 in [35] gives an explicit diophantine equation bearing on the eigenvalues of Γ to be an s-SWRG:

Theorem 4.42. *Let Γ be a connected regular graph with four distinct eigenvalues $k > \theta_1 > \theta_2 > \theta_3$ and $s \geq 3$. Then Γ is an s-SWRG iff*

$$(\theta_2 - \theta_3)\theta_1^s + (\theta_3 - \theta_1)\theta_2^s + (\theta_1 - \theta_2)\theta_3^s = 0. \tag{4.84}$$

We remark that for $s = 3$ (4.84) is equivalent to $(\theta_1 - \theta_2)(\theta_1 - \theta_3)(\theta_2 - \theta_3)(\theta_1 + \theta_2 + \theta_3) = 0$, i.e., Theorem 4.40 is contained as a special case. Similar to the proof of Theorem 4.41, we can use the formulas for the eigenvalues in Theorem 4.36 and Corollary 4.6 to transfer [35, Proposition 4.3] to s-sum sets:

Theorem 4.43. *Assume that $C(\Omega)^\perp$ is of length n and has three nonzero weights $w_1 < w_2 < w_3$. Let $b \geq 0$ and $s \geq 3$ be integers.*

- *Ω is an s-sum set iff $(w_3 - w_2)(n(p-1)p^{m-2} - pw_1)^s + (w_1 - w_3)(n(p-1)p^{m-2} - pw_2)^s + (w_2 - w_1)(n(p-1)p^{m-2} - pw_3)^s = 0$.*
- *$\Omega \cup 0$ is an s-sum set iff $(w_3 - w_2)(1 + n(p-1)p^{m-2} - pw_1)^s + (w_1 - w_3)(1 + n(p-1)p^{m-2} - pw_2)^s + (w_2 - w_1)(1 + n(p-1)p^{m-2} - pw_2)^s = 0$.*

- $\Gamma^b_{C(\Omega)}$ *is an s-SWRG iff* $(w_3 - w_2)(b + n(p-1)p^{m-2} - pw_1)^s + (w_1 - w_3)(b + n(p-1)p^{m-2} - pw_2)^s + (w_2 - w_1)(b + n(p-1)p^{m-2} - pw_2)^s = 0$.

Example 4: If $n = 12$, $p = m = 2$ the only triples (w_1, w_2, w_3) satisfying the first condition of Theorem 4.43 for $s = 5$ are $\{11 - i, 12, 13 + i\}$, for $i = 0, 1, \dots, 10$.

Note that $w_2 = 12 = n$ in all cases. This can be partially explained by the fact that (4.84) admits the solution $\theta_2 = 0$, $\theta_1 = -\theta_3$ for all odd s. In other words, a graph Γ with these eigenvalues is an s-SWRG for all odd $s > 1$, cf. [35, Proposition 4.2]. Using Theorem 4.37 and Theorem 4.36, this observation translates into a result that is new in the context of s-sum sets:

Theorem 4.44. *Assume that* $C(\Omega)^\perp$ *is of length n and has three nonzero weights* $w_1 < w_2 = n(p-1)p^{m-2} < w_3$, *with* $w_1 + w_3 = 2n(p-1)p^{m-2}$. *Then* Ω *is an s-sum set for all odd* $s > 1$.

In [35, Theorem 4.4] it is essentially proven that if (4.84) admits a solution $(\theta_1, \theta_2, \theta_3)$ with $\theta_1 > \theta_2 > \theta_3$ for two different values of $s > 1$, then $\theta_2 = 0$ and $\theta_1 = -\theta_3$. This directly leads to:

Theorem 4.45. *Assume that* $C(\Omega)^\perp$ *is of length n and has three nonzero weights* $w_1 < w_2 < w_3$. *If* $w_2 \neq n(p-1)p^{m-2}$, *or* $w_1 + w_3 \neq 2n(p-1)p^{m-2}$, *then there is at most one* $s > 1$ *such that* Ω *is an s-sum set.*

If we are not in the situation of Theorem 4.44, then any given code can yield an s-sum set for at most one value of s. There are other solutions of (4.84), e.g. $(5, \frac{1}{2}(-3 + \sqrt{281}), \frac{1}{2}(-3 - \sqrt{281}))$ for $s = 5$. However, the found solutions are non-integral and no graph that is an s-SWRG and attains these eigenvalues is known, see [35, p. 808]. Since the eigenvalues corresponding to an s-sum set are integral, we state:

Conjecture 4.1. Ω *is an s-sum set for some* $s > 3$ *iff*

$$(w_1, w_2, w_3) = (w_1, n(p-1)p^{m-2}, 2n(p-1)p^{m-2} - w_1),$$

for some $0 < w_1 < n(p-1)p^{m-2}$.

At least for the special case $p = m = 2$, i.e., $\mathbb{Z}_4$-codes, Conjecture 4.1 seems to be true also for $s = 3$, see Conjecture 4.2. In [17] it has been shown that all s-sum sets of a special type also have to be 3-sum sets.

$\mathbb{Z}_4$-codes

Here we consider $\mathbb{Z}_4$ linear codes and assume that the reader is familiar with [48]. When we say that C is an $[n, k]$ code, where $k = (k_1, k_2)$ is the *shape*, this means that C is a linear code over $\mathbb{Z}_4$ of length n, which admits a generator matrix of the form

$$G = \begin{pmatrix} I_{k_1} & A & B \\ 0 & 2I_{k_2} & 2D \end{pmatrix}, \tag{4.85}$$

where I_{k_1} and I_{k_2} denote the $k_1 \times k_1$ and $k_2 \times k_2$ identity matrices, respectively, A and D are 0, 1 valued matrices , and B is a $\mathbb{Z}_4$ matrix. We assume that the stated generator matrix consists of n columns (and k_1+k_2 rows). We will say that the length n is *effective* to mean that the generator matrix does not contain an all-zero vector $\mathbf{0}$ as a column. The code size, i.e., the number of codewords, is given by $\#C = 4^{k_1}\cdot 2^{k_2} = 2^{2k_1+k_2}$. For $\mathbb{Z}_4$-codes we note that the homogeneous weight coincides with the Lee weight.

We note the following compatibility between a $\mathbb{Z}_4$-code and its Gray image.

Proposition 4.7. *If C is a $\mathbb{Z}_4$-code with three Lee weights satisfying the hypotheses of Theorem 4.41, then its binary image C^G satisfies the hypotheses of Theorem 4.41.*

Proof. As is well-known, the Hamming weights of C^G are the Lee weights of C. If n is the length of C then C^G is of length $2n$. The third condition of Theorem 4.41 becomes with $p=m=2$,

$$w_1+w_2+w_3 = 3\frac{b+n(p-1)p^{m-1}}{p} = 3\frac{b+2n}{2},$$

which is the same with $p=2, m=1$ and length $2n$. □

By A_i we denote the number of codewords of C of homogeneous weight i and by B_i we denote the number of codewords of homogeneous weight i of its dual code $C^\perp$. The cardinality of the dual code is $\#C^\perp = 2^{2n-2k_1-k_2}$. Since we assume that a corresponding generator matrix does not have a zero column, we have $B_1=0$ in our context. As mentioned at the end of Section 2.1 in [75], every $\mathbb{Z}_4$ code has minimum dual distance $d^\perp \geq 3$ iff it is regular and projective. The numbers A_i can be encoded in a weight enumerator of C:

$$\mathrm{Hom}_C(X,Y) = \sum_{c\in C} X^{2n-\mathrm{w_{hom}}(c)}Y^{\mathrm{w_{hom}}(c)} = \sum_{i=0}^{2n} A_i X^{2n-w_i}Y^{w_i}. \tag{4.86}$$

The corresponding MacWilliams identity is given by $\mathrm{Hom}_C(X,Y) = \frac{1}{\#C^\perp}\cdot \mathrm{Hom}_{C^\perp}(X+Y,X-Y)$, see e.g. [48]. Note that this formula equals the Hamming weight enumerator for a linear code over $\mathbb{F}_2$ with cardinality $\#C=2^k$ if we replace n by $2n$ and k by $2k_1+k_2$. As mentioned in the introduction and used in the proof of Lemma 4.7, the same conclusion can be obtained via the Gray map. Thus, we can easily rewrite the classical *MacWilliams identities* [68] or the *Pless power moments* [71] for the Hamming weight over $\mathbb{F}_2$. Using $A_0=B_0=1$ and $B_1=0$ the first four *(Pless) power moments* for the $\mathrm{w_{hom}}$ are given by:

$$\sum_{i>0} A_i = 2^{2k_1+k_2}-1, \tag{4.87}$$

$$\sum_{i\geq 0} iA_i = 2^{2k_1+k_2}n, \tag{4.88}$$

$$\sum_{i\geq 0} i^2A_i = 2^{2k_1+k_2-1}(B_2+n(2n+1)), \tag{4.89}$$

$$\sum_{i\geq 0} i^3A_i = 2^{2k_1+k_2-2}(3(B_2n-B_3)+2n^2(2n+3)). \tag{4.90}$$

We remark, that the first three equations (including B_1) can also be found in [85, Theorem 3.1].

In this section, we are interested in structural results for l-weight codes where l is small. Mostly we will restrict our attention to projective codes only. In some cases, we will require additional constraints on the sum $\sum_{i=1}^{l} w_i$ of weights (which are motivated by strongly walk regular graphs, see e.g. [35] and see also [83], which studies the field case).

4.2.4.4 Theoretical Results for $\mathbb{Z}_4$-Codes with Few Weights

In this section, we give only very few results that are of interest for the remaining part of this paper. For more information, we refer the interested reader to e.g. [84,85], where one-Lee weight codes and projective two-Lee weight codes are studied. For every pair (k_1, k_2) of non-negative integers there exists exactly one non-isomorphic one-weight $[n,(k_1,k_2)]$ code (for a unique length n depending on k_1 and k_2), see [16].

Lemma 4.14. *Let C be an $[n,(k_1,k_2)]$ code. By C_2 we denote the subcode of C spanned by the codewords of even weight. The cardinality of C_2 is either $2^{2k_1+k_2-1}$ or $2^{2k_1+k_2}$ and all codewords of C_2 have an even weight.*

Proof. Let E be the code consisting of codewords of C of even Lee weight. This code is linear as the dual of the code spanned by the all-2 vector, so that $C_2 = E \cap C$. Thus C_2 is linear and its size depends on whether $2 \cdot \mathbf{1} \in C^\perp$ or not. □

We also call C_2 the even-weight subcode of C.

Lemma 4.15. *Let C be an $[n,(k_1,k_2)]$ code with $t \geq 2$ different nonzero weights. Then, at most $t-1$ weights can be odd.*

Proof. Let $c \in C$ be a codeword of odd weight, then $2c \in C$ is a codeword of even weight. □

Lemma 4.16. *Let C be an $[n,(k_1,k_2)]$ three-weight code with $d^\perp \geq 3$ and weights w_1, w_2, and w_3 that occur A_1, A_2, and A_3 times, respectively. Using the abbreviation $y = 2^{2k_1+k_2-1}$ we have*

$$A_1 = \frac{y \cdot (2n^2 - 2nw_2 - 2nw_3 + 2w_2w_3 + n) - w_2w_3}{(w_2 - w_1)(w_3 - w_1)} \tag{4.91}$$

$$A_2 = \frac{y \cdot (2n^2 - 2nw_1 - 2nw_3 + 2w_1w_3 + n) - w_1w_3}{(w_2 - w_3)(w_2 - w_1)} \tag{4.92}$$

$$A_3 = \frac{y \cdot (2n^2 - 2nw_1 - 2nw_2 + 2w_1w_2 + n) - w_1w_2}{(w_3 - w_1)(w_3 - w_2)} \tag{4.93}$$

and

$$\begin{aligned} 3B_3 &= 2n^2(2n+3) - (w_1 + w_2 + w_3)\,2n(2n+1) - 4w_1w_2w_3 \\ &\quad +2\,(w_1w_2 + w_1w_3 + w_2w_3)\,2n + \frac{2w_1w_2w_3}{y} \end{aligned}$$

for the number of dual codewords of weight 3.

Proof. Solving the first three power moment equations (4.87)–(4.89) for A_1, A_2, and A_3 gives the first statement. Plugging into the fourth power moment equation (4.90) and solving for $3B_3$ gives the last equation of the statement. □

We remark that the integrality of B_3 implies that the product of the three weights $w_1w_2w_3$ is divisible by $y/2 = 2^{2k_1-k_2-2}$. Now let us look at the first case of Theorem 4.41, which is $w_1 + w_2 + w_3 = 3n$ in our situation. Our first observation is that the length n has to be even.

Lemma 4.17. *Let C be an $[n,(k_1,k_2)]$ three-weight code with $d^{\perp} \geq 3$ and weights w_1, w_2, and w_3 satisfying $w_1 + w_2 + w_3 = 3n$. Then $n \equiv 0 \pmod 2$.*

Proof. Assume that n is odd. Then, also $w_1+w_2+w_3 = 3n$ is odd. Using Lemma 4.15 we conclude that exactly one weight – say w_2 – is odd. We consider the even weight subcode C_2 of C. Due to Lemma 4.14 C_2 only contains codewords of even weight and has cardinality $\#C_2 = 2^{2k_1+k_2-1}$, i.e., $A_2 = 2^{2k_1+k_2-1}$. Since C is projective C_2 has effective length $n' \in \{n-1, n\}$. Equation (4.88) gives

$$w_1A_1 + w_2A_2 + w_3A_3 = 2^{2k_1+k_2}n$$

for C and

$$w_1A_1 + w_3A_3 = 2^{2k_1+k_2-1}n'$$

for C_2. Taking the difference and dividing by $A_2 = 2^{2k_1+k_2-1}$ yields $w_2 = n + (n-n')$. Since w_2, n are odd and $n - n' \in \{0,1\}$, we have $n' = n$ and $w_2 = n$. So, let us write $w_1 = n - t$ and $w_3 = n + t$ for some odd integer t (observe that w_1 and w_3 have to be even). With this, equation (4.92) and $A_2 = 2^{2k_1+k_2-1} = y$ gives $(y-1)t^2 = (y-n)n$. Note that $t = 1$ gives $y = n+1$ or $n = 1$. In the first case, we have $A_1 = 0$ and $w_1 = 0$ in the second, so that we can assume $t \geq 3$. Solving the above equations for n gives

$$n = \frac{y}{2} \pm \frac{\sqrt{-4t^2y + 4t^2 + y^2}}{2}.$$

So, we consider the diophantine equation

$$x^2 = -4t^2y + 4t^2 + y^2, \tag{4.94}$$

where y is a power of 2 and all integers are positive.[1]

In our situation we have $t \geq 3$, so that $y \geq 64 = 2^6$, since x^2 would be negative for $t \geq 3$ and $y \leq 32$. Now let us set $x = 2z$ and $y = 2^{r+1}$, i.e., $r \geq 5$. With this we can rewrite (4.94) to

$$(z-t)(z+t) = 2^{2r} - 2^{r+1}t^2. \tag{4.95}$$

Since t is odd, z has to be odd too. We conclude that either $z - t$ or $z + t$ has to be divisible by 2^r from the fact that the greatest common divisor of $z - t$ and $z + t$ divides $2t$ and $r \geq 5$.

Assume that $z = s2^r - t$ for some nonzero integer s (this is the case that 2^r divides z+t). Plugging in into (4.95), dividing by 2^{r+1}, and simplifying gives $(2^{r-1}s - t) \cdot s = 2^{r-1} - t^2$, so that

$$2^{r-1} \cdot (s^2 - 1) = t \cdot (s - t).$$

Note that $s^2 \geq 1$. If $s^2 = 1$, then $s = t$ or $t = 0$, which contradicts $t \geq 3$. Thus, the left-hand side is positive and $s > t \geq 3$. This implies $t(s-t) \leq s(s-t) < 2^{r-1}(s^2-1)$ – contradiction.

Assume that $z = s2^r + t$ for some nonzero integer s (this is the case that 2^r divides z-t). Plugging in into (4.95), dividing by 2^{r+1}, and simplifying gives $(2^{r-1}s + t) \cdot s = 2^{r-1} - t^2$, so that

$$2^{r-1} \cdot (s^2 - 1) = t \cdot (-s - t).$$

Note that $s^2 \geq 1$. If $s^2 = 1$, then $s = -t$ or $t = 0$, which contradicts $t \geq 3$. Thus, the left hand side is positive and $-s > t \geq 3$. This implies $t(s-t) \leq s(s-t) < 2^{r-1}(s^2-1)$ – contradiction. □

We remark that in the case where also C_2 is projective the proof can be shortened significantly since [3, Corollary 16] implies that $2t$ is a power of two. An example where Lemma 4.17 excludes parameters of a code is given by $n = 29$, $2k_1 + k_2 = 8$, $w = (24, 31, 32)$, and weight distribution $0^1 24^{76} 31^{128} 32^{51}$. Applying the MacWilliams transform gives $B_0 = 1$, $B_1 = B_2 = 0$, $B_3 = 164$, and indeed all B_i are non-negative integers. Moreover, the number of even-weight codewords is 128, i.e., half the size of the code.

Next we show that under the assumption of Theorem 4.45 the length and the three weights all are divisible by large powers of 2 if the length is suitably large.

Proposition 4.8. *Let C be an $[n, (k_1, k_2)]$ three-weight code with $d^\perp \geq 3$ and weights satisfying $w_1 + w_2 + w_3 = 3n$ and $w_2 = n$. For each positive integer r there exists an integer $N(r)$ such that $n \geq N(r)$ implies that 2^r divides n and 2^{r-1} divides all three weights w_1, w_2, w_3.*

Proof. Let t be a positive integer with $w_1 = n - t$ and $w_3 = n + t$. With this (and $y = 2^{2k_1+k_2-1}$) the equations of Lemma 4.16 are equivalent to

$$A_1 = \frac{n(y-n-t)}{2t^2} \tag{4.96}$$

$$A_2 = \frac{t^2(2y-1) - n(y-n)}{t^2} \tag{4.97}$$

$$A_3 = \frac{n(y-n+t)}{2t^2} \tag{4.98}$$

and

$$3B_3 = \frac{2n(n-t)(n+t)}{y}. \tag{4.99}$$

Since $A_3 - A_1 = \frac{n}{t}$ the effective length n has to be divisible by t. From $A_2 \in \mathbb{N}$ we conclude that t^2 divides $n(n-y)$. So, if p^l divides t for some odd prime p, then p^{2l} has to divide n since y is a power of 2. As an abbreviation, we set $k = 2k_1 + k_2$. Now let us try to parameterize $t = 2^u \cdot v$ and $n = 2^x \cdot v^2 \cdot z$ for odd positive integers v, z and non-negative integers u, x. Plugging in and simplifying gives

$$A_1 = \frac{z \cdot (2^{k-u-1} - 2^{x-u}v^2 z - v)}{2^{u+1-x}} \tag{4.100}$$

$$A_2 = 2^{2(x-u)}v^2 z^2 + 2^k - 2^{x+k-2u-1}z - 1 \tag{4.101}$$

$$A_3 = \frac{z \cdot (2^{k-u-1} - 2^{x-u}v^2 z + v)}{2^{u+1-x}} \tag{4.102}$$

$$3B_3 = \frac{v^4 z \cdot (2^{x-u}vz - 1) \cdot (2^{x-u}vz + 1)}{2^{k-x-2u-2}}, \tag{4.103}$$

where $u \le x$ (t divides n) and $x \le k-1$ ($n \le 2^k - 1$ since C is projective).

If $k - x - 2u - 2 \ge 1$ then $B_3 \in \mathbb{N}$ and $v, z \equiv 1 \pmod 2$ imply $u = x$. Since $\gcd(vz-1, vz+1) = 2$, we have that 2^{k-3u-3} either divides $vz - 1$ or $vz + 1$. So, we use the parameterization $vz = s \cdot \left(2^{k-3u-3}\right) + \alpha$ for some positive integer s and $\alpha \in \{-1, 1\}$. With this $A_1 > 0$ gives

$$2^{k-u-1} - v\left(s \cdot \left(2^{k-3u-3}\right) + \alpha + 1\right) > 0,$$

so that $vs < 2^{2u+2}$, i.e., $sv \le 2^{2u+2} - 1$. Now $A_2 > 0$ gives $v^2 z^2 + 2^k > 2^{k-u-1}z$, which is equivalent to

$$sv(vz)^2 + 2^k sv > s2^{k-u-1}vz = s^2 2^{2k-4u-4} + \alpha s 2^{k-u-1} > s^2 2^{2k-4u-4} - s2^{k+2u+2}. \tag{4.104}$$

Since $sv \le 2^{2u+2} - 1$ the left hand side is at most

$$\begin{aligned}
&s^2 2^{2k-4u-4} - s^2 2^{2k-6u-6} + s\alpha 2^{k-u} - s\alpha 2^{k-3u-2} \\
&+ \alpha^2 2^{2u+2} - \alpha^2 + 2^{k+2u+2} - 2^k \\
\le& s^2 2^{2k-4u-4} - s^2 2^{2k-6u-6} + 2s2^{k-u} + 2^{2u+2} + 2^{k+2u+2} \\
\le& s^2 2^{2k-4u-4} - s^2 2^{2k-6u-6} + 3s2^{k+2u+2}.
\end{aligned}$$

Thus

$$4s \cdot 2^{k+2u+2} > s^2 2^{2k-6u-6} \tag{4.105}$$

has to be satisfied, so that $k \le 8u + 9$ and

$$x = u \ge \frac{k-9}{8}. \tag{4.106}$$

Otherwise we have $k \le x + 2u + 2$. From $A_1 > 0$ we conclude $y - n > 0$. Since $y - n$ is an integer and both y and n are divisible by 2^x we have $y - n \ge 2^x$. Now $A_2 > 0$, (4.97), and $2y - 1 \le 2^k$ imply

$$v^2 2^{2u} \cdot 2^k - 2^x v^2 z \cdot 2^x > 0,$$

so that $k + 2u > 2x$, i.e.,

$$k \geq 2x - 2u + 1. \tag{4.107}$$

Combined with $k \leq x + 2u + 2$ we obtain $x \leq 4u + 1$ and $k \leq 6u + 3$, i.e.,

$$x \geq u \geq \frac{k-3}{6}. \tag{4.108}$$

In both cases we can conclude $x \geq u \geq \frac{k-9}{8}$, so that the result follows from $n < 2^k$. □

Conjecture 4.2. *Let C be an $[n, (k_1, k_2)]$ three-weight code with $d^\perp \geq 3$ and weights satisfying $w_1 + w_2 + w_3 = 3n$ and $w_1 < w_2 < w_3$. Then, $w_2 = n$.*

We remark that in order to prove Conjecture 4.2 the Diophantine conditions of Lemma 4.16, i.e., $A_1, A_2, A_3 \in \mathbb{N}_{>0}$ and $B_3 \in \mathbb{N}_{\geq 0}$ are not sufficient. Up to $n = 50$ we have the following exceptional tuples $(n, w_1, w_2, w_3, y, A_1, A_2, A_3, B_3)$:

- $(29, 24, 31, 32, 64, 76, 128, 51, 164)$
- $(33, 29, 32, 38, 64, 64, 111, 80, 157)$
- $(34, 30, 32, 40, 128, 64, 299, 148, 36)$
- $(50, 46, 48, 56, 64, 32, 145, 78, 580)$

So, from Lemma 4.17 we can conclude that Conjecture 4.2 is true for all $n < 34$. We remark that the examples for $n \in \{29, 34, 50\}$ satisfy $B_i \in \mathbb{N}_{\geq 0}$ for all $1 \leq i \leq n$, i.e., the MacWilliams identities are not sufficient in order to conclude the non-existence of the corresponding code.

4.2.4.5 Classification of Three-Weights $\mathbb{Z}_4$-Codes with $d^\perp \geq 3$ and Short Length

Having Theorem 4.41 in mind we aim to classify three-weights $\mathbb{Z}_4$-codes with $d^\perp \geq 3$ and short length. Optimal linear codes over $\mathbb{Z}_4$ with length $n \leq 7$ were classified in [93]. Without the restriction to optimal codes, all linear codes over $\mathbb{Z}_4$ with length $n \leq 10$ and $2k_1 + k_2 \leq 8$ were classified in [36] for the homogeneous weight. The corresponding tables at `http://www.algorithm.uni-bayreuth.de/en/research/Coding_Theory/CanonicalForm/Classification/index.html` contain the complete counts for $n \leq 10$. E.g. for $n = 10$ and $(k_1, k_2) = (5, 0)$ there are exactly 8 848 026 non-isomorphic codes. For each given n our approach is to loop over all $1 \leq w_1 < w_2 < w_3 \leq 2n$ such that A_1, A_2, A_3, B_3, as specified in Lemma 4.16, are integers with $A_1, A_2, A_3 \geq 1$ and $B_3 \geq 0$. Since $2^{2k_1+k_2-2}$ has to divide $w_1 w_2 w_3$, which is due to $B_3 \in \mathbb{N}$, we obtain a finite list of possibilities for each length n. Note that $d^\perp \geq 3$ implies $k_1 \geq 1$. With respect to Theorem 4.41 we consider all remaining tuples where $w_1 + w_2 + w_3 \geq 3n$ and $w_1 + w_2 + w_3 \equiv 0 \pmod 3$. In Table 4.3 we list the parameters that can be attained by a $\mathbb{Z}_4$ codes and refer to a corresponding generator matrix. As abbreviations we

TABLE 4.3 Attainable Parameters for $S \equiv 0 \pmod 3$

n	w	A	$(k_1,k_2) \to G$
6	(4,6,8)	(6,16,9)	$(2,1) \to G^1_{6,1}$
6	(4,6,8)	(18,24,21)	$(3,0) \to G^1_{6,2}$, $(2,2) \to G^1_{6,3}$
8	(4,8,12)	(1,27,3)	$(2,1) \to G^1_{8,1}$
8	(4,8,12)	(5,51,7)	$(3,0) \to G^1_{8,2}$, $(2,2) \to G^1_{8,3}$
3	(2,4,6)	(15,15,1)	$(2,1) \to G^2_{3,1}$
5	(4,6,8)	(16,12,3)	$(2,1) \to G^2_{5,1}$
7	(6,8,10)	(42,7,14)	$(3,0) \to G^2_{7,1}$
9	(8,10,12)	(15,12,4)	$(2,1) \to G^2_{9,1}$
10	(8,12,16)	(62,64,1)	$(3,1) \to G^2_{10,1}$
10	(8,12,16)	(130,120,5)	$(4,0) \to G^2_{10,2}$

TABLE 4.4 Computationally Excluded Cases for $S = 3n$

n	$2k_1 + k_2$	w	A	(k_1,k_2)
2	3	(1,2,3)	(1,3,3)	(1,1)
4	4	(2,4,6)	(1,11,3)	(2,0), (1,2)
4	5	(2,4,6)	(5,19,7)	(2,1), (1,3)
4	6	(2,4,6)	(13,35,15)	(3,0), (2,2), (1,4)
6	5	(4,6,8)	(6,16,9)	(1,3)
6	6	(4,6,8)	(18,24,21)	(1,4)
8	5	(6,8,10)	(6,15,10)	(2,1), (1,3)
8	6	(6,8,10)	(22,15,26)	(3,0), (2,2), (1,4)
8	7	(6,8,10)	(54,15,58)	(3,1), (2,3), (1,5)
8	5	(4,8,12)	(1,27,3)	(1,3)
8	6	(4,8,12)	(5,51,7)	(1,4)
8	7	(4,8,12)	(13,99,15)	(3,1), (2,3), (1,5)
10	5	(8,10,12)	(5,16,10)	(2,1), (1,3)
10	6	(8,10,12)	(25,8,30)	(3,0), (2,2), (1,4)

use $w = (w_1, w_2, w_3)$, $S = w_1 + w_2 + w_3$, and $A = (A_1, A_2, A_3)$. The parameters of computationally excluded codes are listed in Tables 4.4 and 4.5. For the latter we perform an exhaustive search using the given restrictions on the weights w_i, the length n, and the shape (k_1, k_2).

$$G^1_{6,1} = \begin{pmatrix} 101112 \\ 010331 \\ 002200 \end{pmatrix}, \quad G^1_{6,2} = \begin{pmatrix} 100122 \\ 010212 \\ 001331 \end{pmatrix}, \quad G^1_{6,3} = \begin{pmatrix} 101112 \\ 011121 \\ 002002 \\ 000220 \end{pmatrix},$$

$$G^1_{8,1} = \begin{pmatrix} 10101122 \\ 01112311 \\ 00220202 \end{pmatrix}, \quad G^1_{8,2} = \begin{pmatrix} 10001222 \\ 01022111 \\ 00110013 \end{pmatrix}, \quad G^1_{8,3} = \begin{pmatrix} 10111112 \\ 01112331 \\ 00200020 \\ 00020020 \end{pmatrix},$$

TABLE 4.5 Computationally Excluded Cases for $S > 3n$, $S \equiv 0 \pmod 3$

n	$2k_1 + k_2$	w	A	(k_1,k_2)
3	5	(2,4,6)	(15,15,1)	(1,3)
5	5	(4,6,8)	(16,12,3)	(1,3)
7	5	(6,8,10)	(16,11,4)	(2,1), (1,3)
7	6	(6,8,10)	(42,7,14)	(2,2), (1,4)
7	7	(4,8,12)	(31,95,1)	(3,1), (2,3), (1,5)
7	8	(4,8,12)	(65,187,3)	(4,0), (3,2), (2,4), (1,6)
9	5	(8,10,12)	(15,12,4)	(1,3)
9	6	(8,11,14)	(43,16,4)	(3,0), (2,2), (1,4)
10	7	(8,12,16)	(62,64,1)	(2,3), (1,5)
10	8	(8,12,16)	(130,120,5)	(3,2), (2,4), (1,6)

$$G_{3,1}^2 = \begin{pmatrix} 101 \\ 011 \\ 002 \end{pmatrix}, \quad G_{5,1}^2 = \begin{pmatrix} 10122 \\ 01111 \\ 00202 \end{pmatrix}, \quad G_{7,1}^2 = \begin{pmatrix} 1001112 \\ 0101231 \\ 0012133 \end{pmatrix},$$

$$G_{9,1}^2 = \begin{pmatrix} 101111122 \\ 011223311 \\ 002020202 \end{pmatrix}, \quad G_{10,1}^2 = \begin{pmatrix} 1000211111 \\ 0101223010 \\ 0010111233 \\ 0002220222 \end{pmatrix}, \quad G_{10,2}^2 = \begin{pmatrix} 1000212110 \\ 0100031232 \\ 0010132012 \\ 0001210231 \end{pmatrix}.$$

We remark that more examples can be found for the parameters:

- $n = 15$, $k = 9$, $w = (12, 16, 20)$, $A = (190, 255, 66)$;
- $n = 18$, $k = 8$, $w = (16, 20, 24)$, $A = (153, 72, 30)$;
- $n = 22$, $k = 7$, $w = (20, 24, 28)$, $A = (71, 43, 13)$.

Cyclic Kerdock Codes

The Kerdock code K of length 2^s with $s \geq 3$ odd over $\mathbb{Z}_4$ can be expressed as

$$K = \mathbb{Z}_4\mathbf{j} \oplus Q,$$

where $\mathbf{j}$ denotes the all-one vector. The code Q is a free code of size 4^s with weight spectrum

$$2^s, 2^s \pm 2^{\frac{s-1}{2}}.$$

By puncturing on the first coordinate we obtain a cyclic code of length $n = 2^s - 1$ that was used in [88] to generate low correlation quadriphase sequences. This code is called K^- in [48] and has the same weights as Q. It can be shown to be projective by the same type of arguments as in [48, Section 3.3]. The sum of these weights being $3 \cdot 2^s = 3(n+1)$, we see that, by Theorem 4.41 with $b = 2$, the syndrome graph of K^- with a double loop at each vertex is a 3-SWRG. It can be easily checked that the parameters of K^- go in line with our results in the previous subsection.

We observe that the binary images have a length congruent to 2 (mod 4), which is enough to show they are not linear, since by [59, Corollary 1], we should have a length multiple of 4. This is an alternative proof that Kerdock codes cannot be linear. The classical proof builds on the fact that the Preparata code, or any code with the same weight distribution is nonlinear [38].

4.2.4.6 Extension to Chain Rings

In this section, we briefly indicate how the preceding work extends when replacing $\mathbb{Z}_p^m$ by a finite chain ring R of depth $r \geq 2$, say, and residue field $\mathbb{F}_q$, for some prime power q. Recall that such a ring is a local ring whose lattice of ideals form a chain of length r. In particular, the maximal ideal is principal, generated by an element γ, a nilpotent element of nilpotency index r [79, Chap. 3]. An example is $\mathbb{Z}_p^m$, when $r = m$, and $q = p$. The definition of the homogeneous weight becomes

$$\mathrm{w}_{\mathrm{hom}}(x) = \begin{cases} 0 & \text{if } x = 0, \\ q^{r-1} & \text{if } 0 \neq x \in (\gamma^{e-1}), \\ (q-1)q^{r-2} & \text{otherwise.} \end{cases}$$

The definition of the syndrome graph in Section 2.3 in [75] to this generalized situation only needs obvious modifications. Also the proof of [80, Theorem 4.1] can be generalized to conclude the connection between coset graph eigenvalues and weights:

Theorem 4.46. *Suppose that C is a regular, projective linear code over R with homogeneous weights w_i and corresponding weight distribution $A_i = |\{x \in C; \mathrm{w}_{\mathrm{hom}}(x) = w_i\}|$. Then the eigenvalues of $\Gamma(C^\perp)$ are $n(q-1)q^{r-1} - qw_i$ with multiplicity A_i.*

In particular, for the coset graph with loops, we have the following result.

Corollary 4.7. *If C is an R-code, of minimum distance at least three, with dual weight distribution $[\langle w_i, A_i \rangle]$, then the spectrum of $\Gamma(C)^b$ is $\{(b + n(q-1)q^{r-1} - qw_i)^{A_i}\}$. Thus A_i is the frequency of the homogeneous weight w_i in $C^\perp$ and the multiplicity of the eigenvalue $b + n(q-1)q^{r-1} - qw_i$.*

From there, upon observing that Theorem 4.37 is valid for any finite commutative ring, the main condition for the existence of a TSS becomes

Theorem 4.47. *Assume that $C(\Omega)^\perp$ is of length n and has three nonzero weights $w_1 < w_2 < w_3$. Let b be any integer.*

- *Ω is a TSS iff $w_1 + w_2 + w_3 = 3n(q-1)q^{r-2}$;*
- *$\Omega \cup 0$ is a TSS iff $w_1 + w_2 + w_3 = 3\frac{1+n(q-1)q^{r-1}}{q}$;*
- *$\Gamma^b_{C(\Omega)}$ is a 3-SWRG iff $w_1 + w_2 + w_3 = 3\frac{b+n(q-1)q^{r-1}}{q}$;*

We remark that $r \geq 2$ and q not being divisible by 3 in the second case is impossible and b has to be divisible by q in the third case.

4.2.4.7 Trace Codes over Chain Rings

Consider the chain ring $\mathbb{F}_p + u\mathbb{F}_p$, with $u^2 = 0$. This is a chain ring with $r = 2$ and $q = p$. The codes $C(m,p)$ of length $\frac{p^{2m}-p^m}{2}$ with m singly even in [82, Section 5.1] are defined as

$$\{(Tr(ax))_{x\in L} \mid a \in \mathcal{R}\},$$

where $L = Q + u\mathbb{F}_p^m$, Q is the set of squares of $\mathbb{F}_{p^m}$, and $\mathcal{R} = \mathbb{F}_{p^m} + u\mathbb{F}_{p^m}$. Note that, since m is even, we have $\mathbb{F}_p^\times \subseteq Q$. This implies that these codes are replications by a factor $p-1$ of projective codes $P(m,p)$ say of length $n = \frac{p^{2m}-p^m}{2(p-1)}$. These latter codes have weights

$$p^{2m-1} - p^{m-1},\, p^{2m-1} - p^{m-1} - p^{m-1}(p^{m/2}+1),\, p^{2m-1} - p^{m-1} + p^{m-1}(p^{m/2}-1),$$

and so satisfy the relation

$$w_1 + w_2 + w_3 = 3(1-1/p)n,$$

for the weight of the Gray map in [82], normalized by a factor 2. This shows, by the first condition of Theorem 4.47, that the coset graph $\Gamma(C)$ is a 3-SWRG, or, equivalently, that the Ω such that $C(\Omega) = C$ is a TSS.

4.2.4.8 Generalized Teichmüller Codes

Let $q = 2^r$ and $R = \mathrm{GR}(4,r)$ the Galois ring of characteristic 4 and residue field $\mathbb{F}_{2^r}$. The order of R is q^2. For integers $k \geq 2$ and

$$s \in \begin{cases} \{0,2,4,\ldots,(k-1)r\} & \text{if } k \text{ odd;} \\ \{r, r+2, r+4, \ldots, (k-1)r\} & \text{if } k \text{ even,} \end{cases}$$

the *generalized Teichmüller codes* $\mathcal{T}_{q,k,s}$ are constructed in [57, Sec. 3.1], [58], generalizing the *Teichmüller codes* in [60]. They are regular projective R-linear codes of length $n = 2^s \cdot \frac{q^k-1}{q-1}$ and with three non-zero homogeneous weights, scaled by the factor $1/q^{r-2}$,[2]

$$w = \left(2^s q^k - 2^{s/2} q^{(k-1)/2},\, 2^s q^k,\, 2^s q^k + 2^{s/2} q^{(k-1)/2}\right)$$

of frequencies

$$A = \left(\frac{1}{2}(q^k-1)(q^k + 2^{s/2}q^{(k+1)/2}),\, q^k - 1,\, \frac{1}{2}(q^k-1)(q^k - 2^{s/2}q^{(k+1)/2})\right).$$

From the length we get $2^s q^k = n(q-1)$. Thus, the sum of the weights is $S = 3 \cdot 2^s q^k q^{r-2} = 3(n(q-1) + 2^s)q^{r-2} = \frac{3}{q}(b + n(q-1)q^{r-1})$ for $b = 2^s q^{r-1}$. By Theorem 4.47 we see that the coset graph with b loops on each vertex is a 3-SWRG.

We remark that in [60] and [57], more projective three-weight-codes over Galois rings are found, among them the dualized extended Kerdock codes $\hat{\mathcal{K}}_{k+1}^*$ and the dualized generalized Teichmüller codes $\mathcal{T}_{q,k,s}^*$. However, they are either not regular or the sum of the weights does not satisfy the divisibility conditions needed in the context of this paper.

4.2.4.9 Classification of Three-Weights Codes over $\mathbb{F}_2 + u\mathbb{F}_2$ with $d^{\perp} \geq 3$ and Short Length

In a similar manner as what we did over $\mathbb{Z}_4$, we classify three-weight codes with dual weight $d^{\perp} \geq 3$ over the other proper chain ring of order 4, which is the ring $R = \mathbb{F}_2 + u\mathbb{F}_2$ of dual numbers over $\mathbb{F}_2$. The Gray image of such a code is always a linear binary code, and a linear binary code is the Gray image of an R-linear code if and only if it has a fixed-point free automorphism of order 2 [51]. We classify the codes in the same range of parameters as in the $\mathbb{Z}_4$-case, that is $n \leq 10$, $S \geq 3n$ and $S \equiv 0 \mod 3$. Since the MacWilliams identities for R-linear codes are the same as for $\mathbb{Z}_4$-linear codes (and up to doubling the length, the same as the MacWillimans identities for the linear Gray image), we can start with the same list of feasible parameters we computed for $\mathbb{Z}_4$-linear codes.

The feasible parameters, together with the weight distribution and the value $S-3n \geq 0$, are listed in Table 4.6. The column (k_1, k_2) lists the shapes of the realizable codes. An entry "–" indicates that no R-linear code with the given parameters does exist.

A list of suitable generator matrices for the realizable parameters is given in Table 4.7.

TABLE 4.6 Classification for $R = \mathbb{F}_2[X]/(X^2)$, $S \geq 3n$, $n \leq 10$

n	$2k_1 + k_2$	Weights	S–$3n$	$(k_1 + k_2)$
2	3	$1^1 2^3 3^3$	0	–
4	4	$2^1 4^{11} 6^3$	0	–
4	5	$2^5 4^{19} 6^7$	0	–
4	6	$2^{13} 4^{35} 6^{15}$	0	–
6	5	$4^6 6^{16} 8^9$	0	(2,1)
6	6	$4^{18} 6^{24} 8^{21}$	0	(3,0), (2,2)
8	5	$6^6 8^{15} 10^{10}$	0	–
8	6	$6^{22} 8^{15} 10^{26}$	0	–
8	7	$6^{54} 8^{15} 10^{58}$	0	–
8	5	$4^1 8^{27} 12^3$	0	(2,1)
8	6	$4^5 8^{51} 12^7$	0	(3,0), (2,2)
8	7	$4^{13} 8^{99} 12^{15}$	0	–
10	5	$8^5 10^{16} 12^{10}$	0	–
10	6	$8^{25} 10^8 12^{30}$	0	–
3	5	$2^{15} 4^{15} 6^1$	3	(2,1)
5	5	$4^{16} 6^{12} 8^3$	3	(2,1)
7	5	$6^{16} 8^{11} 10^4$	3	–
7	6	$6^{42} 8^7 10^{14}$	3	–
7	7	$4^{31} 8^{95} 12^1$	3	–
7	8	$4^{65} 8^{187} 12^3$	3	–
9	5	$8^{15} 10^{12} 12^4$	3	(2,1)
9	6	$8^{43} 11^{16} 14^4$	6	–
10	7	$8^{62} 12^{64} 16^1$	6	(3,1)
10	8	$8^{130} 12^{120} 16^5$	6	(4,0)

TABLE 4.7 Generator Matrices for $R = \mathbb{F}_2[X]/(X^2)$, $S \geq 3n$, $n \leq 10$

n	k	Weights	S–$3n$	$(k_1 + k_2)$	G
3	5	$2^{15}4^{15}6^1$	3	(2,1)	$\begin{pmatrix} 1&0&1 \\ 0&1&1 \\ 0&0&X \end{pmatrix}$
5	5	$4^{16}6^{12}8^3$	3	(2,1)	$\begin{pmatrix} 1&0&1&0&X \\ 0&1&1&1&1 \\ 0&0&0&X&X \end{pmatrix}$
6	5	$4^6 6^{16} 8^9$	0	(2,1)	$\begin{pmatrix} 1&0&X&X+1&1&1 \\ 0&1&1&1&1&X \\ 0&0&0&0&X&X \end{pmatrix}$
6	6	$4^{18}6^{24}8^{21}$	0	(3,0)	$\begin{pmatrix} 1&0&0&1&1&X+1 \\ 0&1&0&1&X&0 \\ 0&0&1&1&1&1 \end{pmatrix}$
6	6	$4^{18}6^{24}8^{21}$	0	(2,2)	$\begin{pmatrix} 1&0&1&1&0&1 \\ 0&1&1&X&1&X+1 \\ 0&0&X&X&0&0 \\ 0&0&0&0&X&X \end{pmatrix}$
8	5	$4^1 8^{27} 12^3$	0	(2,1)	$\begin{pmatrix} 1&0&X&1&0&1&X&1 \\ 0&1&1&X&1&1&1&X+1 \\ 0&0&0&0&X&X&X&X \end{pmatrix}$
8	6	$4^5 8^{51} 12^7$	0	(3,0)	$\begin{pmatrix} 1&0&0&X&X&0&1&X \\ 0&1&0&0&1&1&0&1 \\ 0&0&1&1&1&X&X&X+1 \end{pmatrix}$
8	6	$4^5 8^{51} 12^7$	0	(2,2)	$\begin{pmatrix} 1&0&X&1&1&1&1&X+1 \\ 0&1&1&1&0&0&0&1 \\ 0&0&0&X&X&0&X&X \\ 0&0&0&0&0&X&X&0 \end{pmatrix}$
9	5	$8^{15}10^{12}12^4$	3	(2,1)	$\begin{pmatrix} 1&0&1&X&0&1&1&1&X+1 \\ 0&1&1&1&1&X&1&0&1 \\ 0&0&0&0&X&X&X&X&0 \end{pmatrix}$
10	7	$8^{62}12^{64}16^1$	6	(3,1)	$\begin{pmatrix} 1&0&0&1&X&1&X&0&1&0 \\ 0&1&0&1&X&0&1&1&X+1&X+1 \\ 0&0&1&1&1&0&1&X&1&1 \\ 0&0&0&X&X&X&X&X&0&X \end{pmatrix}$
10	8	$8^{130}12^{120}16^5$	6	(4,0)	$\begin{pmatrix} 1&0&0&0&1&0&X&X&X+1&1 \\ 0&1&0&0&X&X&0&1&X+1&1 \\ 0&0&1&0&0&X&1&X&1&1 \\ 0&0&0&1&X&1&X&0&1&1 \end{pmatrix}$

We remark that we found further examples of R-linear three-weight codes with $d^{\perp} \geq 3$, $S - 3n \geq 0$ and $3 \mid S$. Their parameters are displayed in Table 4.8

TABLE 4.8 Further Projective Three-Weight-Codes over $R = \mathbb{F}_2[X]/(X^2)$ with $S \geq 3n$

n	k	Weights	S–$3n$
11	9	$8^{162}12^{312}16^{37}$	3
11	10	$8^{330}12^{616}16^{77}$	3
13	6	$12^{45}16^{17}20^1$	9
15	9	$12^{190}16^{255}20^{66}$	3
18	8	$16^{153}20^{72}24^{30}$	6
26	8	$24^{172}28^{32}32^{51}$	6
42	8	$40^{189}48^{63}56^3$	18
49	7	$48^{91}52^{28}56^8$	9
54	11	$48^{724}56^{1104}64^{219}$	6
61	11	$56^{980}64^{847}72^{220}$	9
63	13	$56^{2556}64^{4095}72^{1540}$	3

Notes

1. Note that (4.94) has infinitely many solutions for $t \in \{-1, 0, 1\}$. Also if y is not a power of 2 there are lots of solutions, e.g. $(x, t, y) \in \{(1, 2, 15), (11, 2, 21), (14, 3, 40), (17, 6, 145)\}$.
2. The scaling of the homogeneous weight in [57] and [60] differs by ours by the factor q^{r-2}.

REFERENCES

[1] A. E. Brouwer, A. M. Cohen and A. Neumaier, *Distance-Regular Graphs*, Springer Verlag, Berlin, 1989.

[2] A. E. Brouwer and W. H. Haemers, *Spectra of Graphs*, Springer, New York, 2012.

[3] E. Byrne, M. Kiermaier and A. Sneyd, Properties of codes with two homogeneous weights, *Finite Fields Appl.*, 18(4): 711–727, 2012.

[4] E. Bannai and T. Ito, *Algebraic Combinatorics I: Association Schemes*, Benjamin/Cummings, Menlo Park, CA, 1984.

[5] J. Borges, J. Rifà and V. A. Zinoviev, Nonexistence of completely transitive codes with error-correcting capability $e > 3$, *IEEE Trans. Inf. Theory*, 47(4): 1619–1621, 2001.

[6] W. Bosma, J. Cannon and C. Playoust, The Magma algebra system. The user language. *J. Symbolic Comput.* 24(3–4): 235–265, 1997.

[7] A. E. Brouwer and H. Van Maldeghem, *Strongly Regular Graphs*, Ser. *Encycl. Math. Appl.*, Vol. 182, Cambridge University Press, Cambridge, 2022.

[8] R. D. Baker, J. H. van Lint and R. M. Wilson, On the Preparata and Goethals codes, *IEEE Trans. Inf. Theory*, 29: 342–345, 1983.

[9] E. R. Berlekamp, Coding theory and the Mathieu groups, *Inf. Control*, 18: 40–64, 1971.

[10] S. Boztaş, Near-optimal 4ϕ (4-phase) sequences and optimal binary sequences for CDMA, Ph. D. dissertation, Univ. Southern Calif., Los Angeles, 1990.

[11] S. Boztaş, A. R. Hammons, Jr. and P. V. Kumar, 4-Phase sequences with near-optimum correlation properties, *IEEE Trans. Inf. Theory*, 38: 1101–1113, 1992.

[12] A. E. Brouwer and L. M. G. M. Tolhuizen, A sharpening of the Johnson bound for binary linear codes and the nonexistence of linear codes with the Preparata properties, preprint, 1992.

[13] E. Byrne, M. Greferath and T. Honold, Ring geometries, two-weight codes and strongly regular graphs, *Des. Codes Cryptogr.*, 48(1): 1–16, 2008.

[14] E. Byrne, M. Kiermaier and A. Sneyd, Properties of codes with two homogeneous weights, *Finite Fields Appl.*, 18(4): 711–727, 2012.

[15] R. Calderbank and W. M. Kantor, The geometry of two-weight codes, *Bull. London Math. Soc.*, 18(2): 97–122, 1986.

[16] C. Carlet, One-weight $\mathbb{Z}_4$-linear codes, in *Coding Theory, Cryptography and Related Areas: Proceedings of an International Conference on Coding Theory, Cryptography and Related Areas, held in Guanajuato, Mexico, in April, 57–72, 1998*, pp. 57–72. Springer, 2000,

[17] J. R. Coma, All s-sum sets of type 2 or 3 are triple-sum sets, *Comput. Math. Appl.*, 23(1): 35–43, 1992.

[18] B. Courteau and J. Wolfmann, On triple sum sets and two or three-weight codes, *Discr. Math.*, 50(2–3): 179–191, 1984.

[19] C. Carlet, $\mathbb{Z}_{2^k}$-linear codes, *IEEE Trans. Inf. Theory*, 44: 1543–1547, 1998.

[20] I. Constantinescu and W. Heise, A metric for codes over residue class rings of integers, *Probl. Inf. Transm*, 33: 208–213, 1997.

[21] A. R. Calderbank, A. R. Hammons, Jr., P. V. Kumar, N. J. A. Sloane and P. Solé, A linear construction for certain Kerdock and Preparata codes, *Bull. Amer. Math. Soc.*, 29: 218–222, 1993.

[22] P. Camion, B. Courteau and P. Delsarte, On r-partition designs in Hamming space, *AAECC*, 2: 147–162, 1992.

[23] C. Carlet, The automorphism groups of the Kerdock codes, *J. Inf. Optim. Sci.*, 12(3): 387–400, 1991.

[24] C. Carlet, Les groupes dáutomorphism des codes de Delsarte-Goethals, *C. R. Acad. Sci. Paris Série I*, 315: 475–478, 1992.

[25] C. Carlet, The automorphism groups of the Delsarte-Goethals codes, *Des. Codes Cryptogr.*, 3: 237–249, 1993.

[26] G. D. Cohen, M. R. Karpovsky, H. F. Mattson, Jr. and J. R. Schatz, Covering radius-survey and recent results, *IEEE Trans. Inf. Theory*, 31: 328–343, 1985.

[27] J. H. Conway and N. J. A. Sloane, Orbit and coset analysis of the Golay and related codes, *IEEE Trans. Inf. Theory*, 36: 1038–1050, 1990.

[28] J. H. Conway and N. J. A. Sloane, Self-dual codes over the integers modulo 4, *J. Comb. Theory Ser. A*, 62: 30–45, 1993.

[29] B. Courteau and A. Montpetit, A class of codes admitting at most three nonzero dual distances, *Discrete Math.*, 90: 267–280, 1991.

[30] A. R. Calderbank and J. M. Goethals, On a pair of dual subschemes of the Hamming scheme $H_n(q)$. *Eur. J. Comb.* 6(2): 133–147, 1985.

[31] A. R. Calderbank, E. M. Rains, P. W. Shor and N. J. A. Sloane, Quantum error correction via codes over $\mathbb{F}_4$, *IEEE Trans. Inf. Theory*, 44(4): 1369–1387, 1998.

[32] P. J. Cameron and J. H. van Lint, *Designs, Graphs, Codes and their Links, Number 22 in London Mathematical Society Student Texts*, Cambridge University Press, Cambridge, 1991.

[33] P. Delsarte, *An Algebraic Approach to Association Schemes of Coding Theory*, Philips Res. Rep. Supplement, Vol. 10, Philips, Eindhoven, 1973, viii+97 pp.

[34] P. Delsarte, Four fundamental parameters of a code and their combinatorial significance, *Inf. Control*, 23: 407–438, 1973.

[35] E. R. van Dam and G. R. Omidi, Strongly walk-regular graphs, *J. Comb. Theory Ser. A*, 120(4): 803–810, 2013.

[36] T. Feulner, Canonization of linear codes over $\mathbb{Z}_4$, *Adv. Math. Commun.*, 5(2): 245–266, 2011.

[37] N. Gassner, M. Greferath, J. Rosenthal, V. Weger, Bounds for coding theory over rings, *Entropy*, 24(10): 1473(1–16), 2022.

[38] J.-M. Goethals, Nonlinear codes defined by quadratic forms over $\mathbb{F}_2$, *Inf. Control*, 31: 43–74, 1976.

[39] M. Giudici and C. E. Praeger: Completely transitive codes in Hamming graphs, *Eur. J. Comb.*, 20(7): 647–662, 1999.

[40] J. M. Goethals and H. C. A. van Tilborg, Uniformly packed codes, *Philips Res. Rep.*, 30: 9–36, 1975.

[41] M. Griera, J. Rifa, and L. Hughet, On s-sum sets and projective codes, *Lect. Notes Comp. Sci.*, 229: 135–142, 1986.

[42] M. Greferath and S. E. Schmidt, Finite-ring combinatorics and MacWilliams' equivalence theorem, *J. Comb. Theory Ser. A*, 92(1): 17–28, 2000.

[43] T. Honold, Characterization of finite Frobenius rings, *Arch. Math.*, 76(6): 406–415, 2001.

[44] T. Honold and I. Landjev, Linearly representable codes over chain rings, *Abhandlungen aus dem mathematischen Seminar der Universität Hamburg*, 69: 187–203, 1999.

[45] T. Honold and I. Landjev, Linear codes over finite chain rings, *Electron. J. Comb.*, 7: 11–22, 2000.

[46] T. Honold and I. Landjev, Linear codes over finite chain rings and projective Hjelmslev geometries, in *Codes Over Rings, Proceedings of the CIMPA Summer School, Ankara, Turkey, 18–29 August, 2008, Series on Coding Theory and Cryptology*, vol. 6, pp. 60–123. World Scientific, 2009.

[47] T. Honold and I. Landjev, Codes over rings and ring geometries, *Curr. Res. Top.*, 7: 159–184, 2011.

[48] A. R. Hammons, P. V. Kumar, A. R. Calderbank, N. J. A. Sloane and P. Solé, The $\mathbb{Z}_4$-linearity of Kerdock, Preparata, Goethals and related codes, *IEEE Trans. Inf. Theory*, 40: 301–319, 1994.

[49] W. C. Huffman and V. Pless, *Fundamentals of Error-Correcting Codes*, Cambridge University Press, Cambridge, 2003.

[50] T. W. Hungerford, *Algebra*, Springer-Verlag, New York, 1974.

[51] T. Honold and I. Landjev, All Reed-Muller codes are linearily representable over the ring of dual numbers over $\mathbb{Z}_2$, *IEEE Trans. Inf. Theory*, 45(2): 700–701, 1999.

[52] T. Honold and A. A. Nechaev, Weighted modules and representations of codes, *Prob. Inf. Transm.*, 35(3): 205–223, 1999.

[53] W. M. Kantor, Spreads, translation planes and Kerdock sets, *SIAM J. Alg. Discr. Math.*, 3: 151–165 and 308–318, 1982.

[54] W. M. Kantor, On the inequivalence of generalized Preparata codes, *IEEE Trans. Inform. Theory*, 29: 345–348, 1983.

[55] M. Klemm, Über die Identität von MacWilliams für die Gewichtsfunktion von Codes, *Arch. Math.*, 49: 400–406, 1987.

[56] M. Klemm, Selbstduale codes über dem Ring der ganzen Zahlen modulo 4, *Arch. Math.*, 53: 201–207, 1989.

[57] M. Kiermaier, Geometrische Konstruktionen linearer Codes über Galois-Ringen der Charakteristik 4 von hoher homogener Minimaldistanz, Dissertation, Universität Bayreuth, 2012, `https://epub.uni-bayreuth.de/id/eprint/195`.

[58] M. Kiermaier, Generalized Teichmüller codes, in preparation.

[59] M. Kiermaier, S. Kurz and P. Solé, A. Wassermann, On strongly walk regular graphs, triple sum sets and their codes, *Des. Codes Cryptogr.*, 91(2): 645–675, 2023.

[60] M. Kiermaier and J. Zwanzger, New ring-linear codes from dualization in projective Hjelmslev geometries, *Des. Codes Cryptogr.*, 66(1–3): 39–55, 2013.

[61] D. Krotov, $\mathbb{Z}_4$-linear Hadamard and extended perfect codes, *Electron. Notes Discrete Math*, 6: 107–112, 2001.

[62] D. Krotov, On $\mathbb{Z}_{2^k}$-dual binary codes, *IEEE Trans. Inf. Theory*, 53: 1532–1537, 2007.

[63] V. I. Levenshtein, Designs as maximum codes in polynomial metric spaces, *Acta Appl. Math.*, 29: 1–82, 1992.

[64] R. Lidl and H. Niederreiter, *Finite Fields, 2nd ed.*, Ser. *Encyclopedia of Mathematics and Its Applications*, vol. 20, Cambridge University Press, Cambridge, 1996.

[65] R. A. Liebler and R. A. Mena, Certain distance-regular digraphs and related rings of characteristic 4, *J. Combin. Theory Ser. A*, 47: 111–123, 1988.

[66] S. Ling and C. P. Xing, *Coding Theory A First Course*, Cambridge University Press, Cambridge, 2004.

[67] B. R. MacDonald, *Finite Rings with Identity*, Marcel Dekker, New York, 1974.

[68] F. J. MacWilliams and N. J. A. Sloane, *The Theory of Error-Correcting Codes*, North-Holland, Amsterdam, 1977.

[69] A. A. Makhnev, On automorphisms of a distance-regular graph with intersection array $\{33, 30, 15; 1, 2, 15\}$, *Inst. Math. Mech.*, 84(3): 235-C241, 2012.

[70] A. A. Makhnev, Private communucation, 2018.

[71] V. Pless, Power moment identities on weight distributions in error correcting codes, *Inf. Control*, 6(2): 147–152, 1963.

[72] K. T. Phelps, J. Rifà and M. Villanueva, On the additive ($\mathbb{Z}_4$−linear and non-$\mathbb{Z}_4$−linear) Hadamard codes: rank and kernel, *IEEE Trans. Inf. Theory*, 52: 316–319, 2006.

[73] N. V. Semakov, V. A. Zinoviev and G. V. Zaitsev, Uniformly packed codes, *Problemy Peradachi Informatsii*, 7(1): 38–50, 1971.

[74] P. Shankar, On BCH codes over arbitrary integer rings, *IEEE Trans. Inf. Theory*, 25: 480–483, 1979.

[75] M. Shi, M. Kiermaier, S. Kurz and P. Solé. Three-weight codes over rings and strongly walk regular graphs. *Graphs Comb.*, 38(3): 56, 2022.

[76] M. Shi and Y. Wang, Optimal binary codes from one-Lee weight codes and two-Lee weight projective codes over $\mathbb{Z}_4$, *J. Syst. Sci. Complex.*, 27(4): 795–810, 2014.

[77] M. Shi, Z. Sepasdar, A. Alahmadi and P. Solé. On two-weight $\mathbb{Z}_{2^k}$-codes. *Des. Codes Cryptogr.*, 86(6): 1201–1209, 2018.

[78] M. Shi , D. S. Krotov and P. Solé. A new distance-regular graph of diameter 3 on 1024 vertices. *Des. Codes Cryptogr.*, 87: 2091–2101, 2019.

[79] M. Shi, A. Alahmadi and P. Solé, *Codes and Rings: Theory and Practice*, Academic Press, Amsterdam, 2017.

[80] M. Shi, T. Honold, Y. Qiu, Z. Sepasdar, R. Wu and P. Solé, The geometry of two-weight codes over $\mathbb{Z}_{p^m}$, *IEEE Trans. Inf. Theory*, 67(12): 7769–7781, 2021.

[81] M. Shi, R. Wu and D. S. Krotov, On $Z_pZ_{p^k}$-additive codes and their duality, *IEEE Trans. Inf. Theory*, 65(6): 3841–3847, 2019.

[82] M. Shi, R. Wu, Y. Liu and P. Solé, Two and three weight codes over $\mathbb{F}_p + u\mathbb{F}_p$, *Cryptogr. Commun.*, 9(5): 637–646, 2017.

[83] M. Shi and P. Solé, Three weight codes, triple sum sets and strongly walk regular graphs, *Des. Codes Cryptogr.*, 87(10): 2395–2404, 2019.

[84] M. Shi, L. Xu and G. Yang, A note on one weight and two weight projective $\mathbb{Z}_4$-codes, *IEEE Trans. Inf. Theory*, 63(1): 177–182, 2017.

[85] M. Shi and Y. Wang, A note on one weight and two weight projective $\mathbb{Z}_4$-codes, *J. Syst. Sci. Complex.*, 27(4): 795–810, 2014.

[86] P. Solé, A quaternary cyclic code, and a family of quadriphase sequences with low correlation properties, *Lect. Notes Comput. Sci.*, 388: 193–201, 1989.

[87] P. Solé, An inversion formula for Krawtchouk polynomials with application to coding theory, *J. Inf. Optim. Sci.*, 11: 207–213, 1990.

[88] P. Solé, A quaternary cyclic code, and a family of quadriphase sequences with low correlation properties, in *Coding Theory and Applications: 3rd International Colloquium, Toulon, France, November 2-4, 1988. Proceedings*, vol. 88, pp. 193–201, 1989.

[89] P. Solé, Completely regular codes and completely transitive codes, *Discrete Math.*, 81(2): 193–201, 1990.

[90] W. A. Stein, et al, *Sage Mathematics Software (Version 6.9)*, 2015, Online: `http://www.sagemath.org`.

[91] J. V. Uspensky, *Theory of Equations*, McGraw-Hill, New York, 1948.

[92] P. Udaya and M. U. Siddiqi, Large linear complexity sequences over $\mathbb{Z}_4$ for quadriphase modulated communication systems having good correlation properties, in *IEEE Internat. Symp. Inform. Theory*, Budapest, Hungary, 1991.

[93] J. N. C. Wong, Classification of small optimal linear codes over $\mathbb{Z}_4$, Master thesis, University of Victoria, 2002.

[94] J. S. Williford, Table of feasible parameter sets for primitive 3-class Q-polynomial association schemes. Online: `https://jaanos.github.io/tables/qpoly/qprim3_table.html`.

[95] J. A. Wood, Duality for modules over finite rings and applications to coding theory, *Am. J. Math.*, 121(3): 555–575, 1999.

[96] M. Yamada, Distance-regular digraphs of girth 4 over an extension ring of $Z/4Z$, *Graphs Combin.*, 6: 381–394, 1990.

[97] G. V. Zaitzev, V. A. Zinoviev and N. V. Semakov, Interrelation of Preparata and Hamming codes and extension of Hamming codes to new double-error correcting codes, in *Proceedings of the Second International Symposium on Information Theory, Tsahkadsor, Armenia, 1971*; Akadémiai Kiadó, Budapest, pp. 257–263, 1973.

[98] O. Zariski and P. Samuel, *Commutative Algebra*, Van Nostrand, Princeton, NJ, 2, 1960.

CHAPTER 5

Group Actions on Codes in Graphs

Daniel R. Hawtin and Cheryl E. Praeger

5.1 INTRODUCTION: CODES IN GRAPHS

Traditionally error-correcting codes are modelled as subsets C of vectors in a finite vector space $V = \mathbb{F}_q^n$ over a field of order q, where the vectors are represented as n-dimensional row vectors with entries from $\mathbb{F}_q$, and distance between two vectors is the number of entries where they differ. The *minimum distance* δ of C is the smallest distance between two codewords, and the code is called *perfect* if the balls of radius $\lfloor(\delta - 1)/2\rfloor$ partition the space V. Much work was devoted by many researchers to understanding and finding new perfect codes, and eventually, Tietäväinen [89], and Zinoviev and Leontiev [92], independently showed that the only non-trivial perfect codes over finite fields are 'Hamming-like' codes (perfect single-error correcting codes), and the famous Golay codes in $\mathbb{F}_2^{23}$ and $\mathbb{F}_3^{11}$, (see for example, the 1975 survey on perfect codes by Van Lint [90]). Larger families of codes with desirable properties were sought. For example, in 1971, Semakov et al. [81] introduced a family of codes, properly containing the perfect codes, which they called uniformly packed codes, and which retained strong regularity properties for 'packing' codewords in the vector space. Then independently, around 1973, Norman Biggs [13,14] and Philippe Delsarte [28,29] suggested a complete change of focus. Biggs introduced the concept of a perfect code in a graph, while Delsarte developed a general theory of association schemes in coding theory. This led to the notion of *codes in graphs*, which is the theme of this chapter.

This chapter provides an overview of recent work on codes in graphs, which are neighbour-transitive or have stronger symmetry properties. Most of the graphs considered are distance-regular, and our particular focus is codes in Hamming graphs since the current work on neighbour-transitive codes in Johnson graphs (a particular study advocated by Delsarte) has been recently covered in the 2021 survey by the second author [74].[1] In Section 5.3, we give a brief historical account of completely

 DOI: 10.1201/9781003393931-5

transitive codes in Hamming graphs and Johnson graphs which motivated the more general developments on neighbour-transitivity, particularly in Hamming graphs, surveyed in Section 5.5. We also summarise very recent work on codes in other graphs, such as Kneser graphs and the incidence graphs of generalised quadrangles, as well as state some results for Grassmann graphs and bilinear forms graphs (the q-analogues of the Johnson and Hamming graphs).

5.1.1 Codes in Graphs: Neighbour-Transitivity and Complete Transitivity

A *graph* $\Gamma = (V, E)$ consists of a set V of *vertices* and a set E of *edges* (unordered pairs of vertices), and a *code* $\mathcal{C}$ in Γ is defined as a subset of V. *Distance* between distinct *codewords* (elements of $\mathcal{C}$) is given by the length of a shortest path between them in the graph Γ, and the *minimum distance* $\delta(\mathcal{C})$ is the minimum distance between distinct codewords. This viewpoint also suggests a natural measure for the symmetry of a code $\mathcal{C}$ in Γ. Symmetry of the graph Γ is measured by its *automorphism group* $\mathrm{Aut}(\Gamma)$, namely the subgroup of permutations of V which leave invariant the edge-set E. We take as the *automorphism group* $\mathrm{Aut}(\mathcal{C})$ of the code $\mathcal{C}$ the subgroup of $\mathrm{Aut}(\Gamma)$ consisting of all elements that leave $\mathcal{C}$ invariant (setwise). Delsarte suggested choosing Γ to be a *distance-regular graph*: that is to say, for any (possibly equal) vertices v and w, the number of vertices at distance j from v and at distance k from w depends only upon j, k, and the distance between v and w. He defined a special type of code, now called a completely-regular code, 'which enjoys combinatorial (and often algebraic) symmetry akin to that observed for perfect codes' (see [68, p. 1]). Again, disappointingly, not many examples of completely regular codes were found with good error-correcting properties – that is to say, with large distance between distinct codewords (see the comments and references in [74, Section 1.1]).

A useful notion for exploring a code $\mathcal{C}$ in a graph Γ is the *distance partition* of $\mathcal{C}$. This is the partition:

> $\{\mathcal{C}_0, \mathcal{C}_1, \ldots, \mathcal{C}_\rho\}$ of V such that $\mathcal{C}_0 = \mathcal{C}$ and, for $i \geq 1$, $\mathcal{C}_i$ is the set of all vertices γ such that the minimum distance between γ and a codeword is equal to i. The largest integer i such that $\mathcal{C}_i \neq \emptyset$ is called the *covering radius* of $\mathcal{C}$, and is denoted ρ and sometimes $\rho(\mathcal{C})$. (5.1)

Since graph automorphisms preserve distance, the automorphism group $\mathrm{Aut}(\mathcal{C})$ fixes each of the subsets $\mathcal{C}_i$ setwise, and so each $\mathcal{C}_i$ is a union of $\mathrm{Aut}(\mathcal{C})$-orbits. A natural, but very strong, symmetry condition to place on a code is to require that each $\mathcal{C}_i$ is a single orbit of $\mathrm{Aut}(\mathcal{C})$. Codes with this property are called *completely transitive*, and this family of codes was among the first family of codes to be studied in various families of graphs. We discuss some of the results about completely transitive codes in Section 5.3.

Most recent studies of codes in graphs have focused on strictly larger families than the completely regular codes or the completely transitive codes, where their strict regularity or symmetry conditions have been replaced by more 'local' conditions. This new approach perhaps dates back to a discussion between the second author and (Bob) Liebler in 2005 (leading to [66]). In the context of codes in Johnson graphs, Bob suggested that the stringent regularity conditions imposed for complete regularity could be replaced by a 'local transitivity' property. This led to the notion of a *neighbour-transitive code* in an arbitrary graph Γ, that is to say, a code C such that $\mathrm{Aut}(C)$ is transitive both on C and on the set C_1 of the distance partition. The vertices in C_1 are called *code-neighbours* (the non-codewords that are adjacent in Γ to some codeword). More recently, for any positive integer $s \leq \rho(C)$, a code C in Γ is called *s-neighbour-transitive* if each of $C_0, C_1, \ldots, C_s$ is an $\mathrm{Aut}(C)$-orbit. Thus, the neighour-transitive codes are 1-neighbour-transitive, and the completely transitive codes are $\rho(C)$-neighbour-transitive.

5.2 FUNDAMENTAL CONCEPTS: NEIGHBOUR-TRANSITIVE CODES

In this section, we present some general concepts and results regarding codes in graphs. We begin with a discussion in arbitrary graphs before specialising to the Hamming graphs and other specific graph families.

5.2.1 Parameters and Regularity Properties for Codes in Graphs

Let $\Gamma = (V, E)$ be a graph. Note that we will always assume that Γ is simple, finite, undirected, and connected. Let $\alpha, \beta \in V$. Denote by $d(\alpha, \beta)$ the *distance* in Γ between α and β, that is, the length of the shortest path between α and β. Define $\Gamma_i(\alpha) = \{\gamma \in V \mid d(\alpha, \gamma) = i\}$. Furthermore, the *ball* $B_i(\alpha)$ of radius i centered at α is defined to be $\bigcup_{j=0}^{i} \Gamma_j(\alpha)$.

Let C be a code in a graph $\Gamma = (V, E)$. We refer to elements of C as *codewords*. A code C such that $|C| \leq 1$ or $C = V(\Gamma)$ is called *trivial*, and we will often be assuming without statement that C is non-trivial. If C' is a subset of C then we say that C' is a *subcode* of C. The *minimum distance* δ of C is the smallest distance between a pair of distinct elements of C. The *error-correction capacity* e of C is defined to be the largest value of i for which, given distinct $\alpha, \beta \in C$, the balls $B_i(\alpha)$ and $B_i(\beta)$ are disjoint. These two parameters are related by $e = \lfloor(\delta - 1)/2\rfloor$. The *covering radius* ρ of C is the smallest value of i for which $\bigcup_{\alpha \in C} B_i(\alpha) = V$.

Lemma 5.1. *Let C be a code with error-correction capacity e in a graph Γ and let $i \leq e$. Then the following hold.*

1. *For each $\gamma \in C_i$ there exists a unique $\alpha \in C$ such that $\gamma \in \Gamma_i(\alpha)$.*

2. *C_i is the disjoint union $\bigcup_{\alpha \in C} \Gamma_i(\alpha)$.*

3. *$\Gamma_i(\alpha) \neq \emptyset$ for all $\alpha \in C$.*

Proof. All the assertions hold for a trivial code C, with $i = e = 0$, so we may assume that C is non-trivial. Then its minimum distance $\delta = \delta(C)$ is a positive integer. Since every element of C_i is at distance i from some element of C we have

$C_i \subseteq \bigcup_{\alpha \in C} \Gamma_i(\alpha)$. Let $\alpha \in C$ and $\gamma \in \Gamma_i(\alpha)$. Consider a codeword $\beta \in C$ such that $\alpha \neq \beta$. If $d(\beta, \gamma) \leq i$, then $\gamma \in B_i(\alpha) \cap B_i(\beta)$ and hence $d(\alpha, \beta) \leq 2i \leq 2e < \delta$, which is a contradiction. Thus $d(\beta, \gamma) > i$. In particular, α is the unique codeword in C at distance i from γ, and part (1) holds. This implies moreover that $\gamma \in C_i$. Thus $\Gamma_i(\alpha) \subseteq C_i$ for each $\alpha \in C$, and hence $\bigcup_{\alpha \in C} \Gamma_i(\alpha) = C_i$. Now, if β_1 and β_2 are distinct codewords in C, then the assumption $i \leq e$ implies that $B_i(\alpha) \cap B_i(\beta) = \emptyset$ so that $\Gamma_i(\alpha) \cap \Gamma_i(\beta) = \emptyset$. Thus the union is disjoint, and part (2) is proved.

Consider distinct codewords $\alpha, \beta \in C$, and let $(\gamma_0, \gamma_1, \ldots, \gamma_r)$ be a path in Γ of length $r = d(\alpha, \beta)$ from $\alpha = \gamma_0$ to $\gamma_r = \beta$. It follows from the minimality in the definition of $d(\alpha, \beta)$ that $d(\alpha, \gamma_j) = j$ and $d(\beta, \gamma_{r-j}) = j$ whenever $0 \leq j \leq r$. If $r \leq i$, then taking $j = r$ we have $\beta \in B_i(\alpha) \cap B_i(\beta)$, which is a contradiction, since $i \leq e$. Hence $r > i$ and $\gamma_i \in \Gamma_i(\alpha)$, proving part (3). □

Definition 5.1. *For a code C in a graph Γ and a non-negative integer s at most the covering radius ρ, C is said to be* s-regular *if for each $i \in \{0, 1, \ldots, s\}$ there exist non-negative integers a_i, b_i (if $s < \rho$), and c_i (if $i > 0$), such that for each vertex $\alpha \in C_i$ there are precisely:*

1. *a_i vertices in $\Gamma_1(\alpha) \cap C_i$,*
2. *b_i vertices in $\Gamma_1(\alpha) \cap C_{i+1}$, and,*
3. *c_i vertices in $\Gamma_1(\alpha) \cap C_{i-1}$,*

and a_i, b_i, c_i depend only on i, and not on the particular choice of α. If $s = \rho$ then C is said to be completely regular.

It is worth mentioning that a graph Γ is distance-regular, according to the usual definition, if and only if, for every vertex α of Γ, the singleton set $\{\alpha\}$ is a completely regular code in Γ.

5.2.2 Symmetry of Codes in Graphs

The *symmetric group* $\mathrm{Sym}(V)$ is the group of all permutations of V, and, for $\alpha \in V$ and $g \in G$, we write α^g for the image of α under g. Each element $g \in \mathrm{Sym}(V)$ permutes subsets of V in a natural way, namely for $U \subseteq V$, the image U^g of U under g is the set $\{\alpha^g \mid \alpha \in U\}$ of images for the elements of U. The *setwise stabiliser* of U is the set $\mathrm{Sym}(V)_U := \{g \in \mathrm{Sym}(V) \mid U^g = U\}$, and this is a subgroup of $\mathrm{Sym}(V)$. In particular, for each $i < |V|$, g permutes the i-element subsets of V among themselves and, for a set E of i-element subsets, we say that g *leaves E invariant* if $U^g \in E$ for all $U \in E$. We often deal with *permutation groups on V*, that is, subgroups $G \leq \mathrm{Sym}(V)$. A permutation group is *transitive* on V if, for all $u, v \in V$, there exists $g \in G$ such that $u^g = v$. For $1 < i < v$, we say that G is *i-transitive* on V if G is transitive on V and, for $v \in V$, the stabiliser G_v is $(i-1)$-transitive on $V \setminus \{v\}$. Also G is said to be *i-homogeneous* on V if G is transitive on the set of i-element subsets of V.

Let $\Gamma = (V, E)$ be a graph. Then an *automorphism* of Γ is a permutation $g \in \mathrm{Sym}(V)$ such that g leaves the edge set E invariant. The set of all automorphisms of Γ forms a subgroup $\mathrm{Aut}(\Gamma)$ called the *automorphism group* of Γ.

If $\mathcal{C}$ is a code in Γ then, as introduced in Section 5.1, the *automorphism group* of $\mathcal{C}$ is the setwise stabiliser $\mathrm{Aut}(\mathcal{C})$ of $\mathcal{C}$ in $\mathrm{Aut}(\Gamma)$, that is to say, $\mathrm{Aut}(\mathcal{C}) = \mathrm{Sym}(V)_{\mathcal{C}} \cap \mathrm{Aut}(\Gamma)$. (5.2)

Two codes are *equivalent* if there exists an automorphism of Γ mapping one to the other. Note that equivalent codes have many of the same properties, for instance, the same minimum distance, the same covering radius and isomorphic automorphism groups. Hence, we will often be interested in codes only up to equivalence.

The following concepts are the main focus of this chapter and may be viewed as algebraic analogues of Definition 5.1.

Definition 5.2. *Let $\mathcal{C}$ be a code with covering radius ρ in a graph Γ, let $G \leq \mathrm{Aut}(\mathcal{C})$, and let $s \in \{1, \ldots, \rho\}$. Then we make the following definitions.*

1. *$\mathcal{C}$ is* (G, s)-neighbour-transitive *if G acts transitively on each of the sets $\mathcal{C}, \mathcal{C}_1, \ldots, \mathcal{C}_s$.*

2. *$\mathcal{C}$ is* G-neighbour-transitive *if $\mathcal{C}$ is $(G, 1)$-neighbour-transitive.*

3. *$\mathcal{C}$ is* G-completely transitive *if $\mathcal{C}$ is (G, ρ)-neighbour-transitive.*

Moreover, we say that $\mathcal{C}$ is neighbour-transitive, s-neighbour-transitive, *or* completely transitive, *respectively, if $\mathcal{C}$ is $\mathrm{Aut}(\mathcal{C})$-neighbour-transitive, $(\mathrm{Aut}(\mathcal{C}), s)$-neighbour-transitive, or $\mathrm{Aut}(\mathcal{C})$-completely transitive, respectively.*

It turns out that many famous codes have these symmetry properties. We mention several of them in Example 5.1. We also give in Example 5.2 a simple example of an explicit infinite family of completely transitive codes.

Example 5.1. *The following well-known codes are completely transitive codes in the Hamming graphs (see Section 5.4.1 for introductory material relating to the Hamming graphs and see [18, Sections 5.1 and 5.2] for more examples):*

1. *The perfect binary Golay code is a 12-dimensional vector-subspace of $\mathbb{F}_2^{23}$ with minimum distance 7 and covering radius 3, and its extended code is a 12-dimensional vector-subspace of $\mathbb{F}_2^{24}$ with minimum distance 8 and covering radius 4. These codes are G-completely transitive for $G = T \rtimes \mathrm{M}_{23}$ and $G = T \rtimes \mathrm{M}_{24}$, respectively, where T is the group of translations by codewords in each case; see [84, p. 199]. This extended Golay code was used by NASA's Voyager spacecraft to send back to earth hundreds of colour pictures of Jupiter and Saturn in their 1979, 1980, and 1981 fly-bys. Error correction was vital to data transmission since memory constraints dictated offloading data virtually instantly leaving no second chances, and the data needed to be transmitted within a constrained telecommunications bandwidth.*[2]

2. *The perfect ternary Golay code is a 6-dimensional vector-subspace of* $\mathbb{F}_3^{11}$ *with minimum distance* 5 *and covering radius* 2 *and its extended code is a* 6*-dimensional vector-subspace of* $\mathbb{F}_3^{12}$ *with minimum distance* 6 *and covering radius* 3*. These codes are* G*-completely transitive for* $G = T \rtimes (2.\mathrm{M}_{11})$ *and* $G = T \rtimes (2.\mathrm{M}_{12})$*, respectively, where* T *is the group of translations by codewords in each case; see [46, p. 653].*

3. *The Nordstrom–Robinson code is a non-linear code consisting of* 256 *codewords with minimum distance* 5 *and covering radius* 3 *in* $\mathbb{F}_2^{15}$ *and its extended code is a non-linear code consisting of* 256 *codewords with minimum distance* 6 *and covering radius* 4 *in* $\mathbb{F}_2^{16}$*. These codes are* G*-completely transitive for* $G \cong 2^5 \rtimes \mathrm{A}_8 \cong 2^5 \rtimes \mathrm{GL}_4(2)$ *and* $G \cong 2^5 \rtimes \mathrm{AGL}_4(2)$*, respectively; see [43].*

Example 5.2. *Let* Γ *be the graph with vertex set* $\mathbb{Z}_{2n}$ *and vertices* i, j *defined to be adjacent if* $i - j = \pm 1$ *(i.e.,* Γ *is a cycle of length* $2n$*) and let* $C = \{0, n\}$*. Then* $Aut(C)$ *is generated by the rotation* $i \mapsto i+n$ *(for* $i \in \mathbb{Z}_{2n}$*), and the reflection* $i \mapsto -i$ *(for* $i \in \mathbb{Z}_{2n}$*). Hence* $Aut(C) \cong C_2^2$ *and* $Aut(C)$ *acts transitively on* C*. We also have* $C_i = \{\pm i, n \pm i\}$ *for each* i *satisfying* $1 \leq i \leq n/2$*. Note that when* $1 \leq i < n/2$ *the cardinality* $|C_i| = 4$*, but when* n *is even we have* $|C_{n/2}| = 2$*. In all cases* $Aut(C)$ *acts transitively on* C_i *for each* i *satisfying* $1 \leq i \leq n/2$*, and thus* C *is completely transitive.*

Lemma 5.2 gives two additional equivalent conditions for (G, s)-neighbour-transitivity for integers s at most the error-correction capacity. It is most useful for s-neighbour-transitive codes with 'large' minimum distance δ, namely $\delta \geq 2s + 1$. Applications of this result are two-fold. Firstly, it is often simpler to prove s-neighbour-transitivity of a code in terms of the 'local action' of the stabiliser of a codeword α on the ball $B_s(\alpha)$ (Lemma 5.2(2)). Secondly, Lemma 5.2 often allows, in the case of a specific graph, for us to prove structural results. We will see examples of such structural results in later sections.

Lemma 5.2. *Let* C *be a code in a graph* Γ *such that* C *has error-correction capacity* $e \geq 1$*, let* $G \leq \mathrm{Aut}(C)$*, let* $\alpha \in C$*, and let* s *be an integer such that* $1 \leq s \leq e$*. Then the following are equivalent.*

1. C *is* (G, s)*-neighbour-transitive.*

2. G *acts transitively on* C *and, for each* $i \in \{1, \ldots, s\}$*, the stabiliser* G_α *is transitive on* $\Gamma_i(\alpha)$*.*

3. *For each* $i \in \{1, \ldots, s\}$*,* G *acts transitively on the set* $\{(\beta, \gamma) \mid \beta \in C, \gamma \in \Gamma_i(\beta)\}$*.*

Proof. Suppose that part (1) holds. Then, by Definition 5.2, G acts transitively on C and transitively on C_i, for each $i \in \{1, \ldots, s\}$. Let $\gamma_1, \gamma_2 \in \Gamma_i(\alpha)$. Then, by Lemma 5.1(2), $\gamma_1, \gamma_2 \in C_i$ and so there exists $g \in G$ such that $\gamma_1^g = \gamma_2$. By Lemma 5.1(1), α is the unique element of C such that $d(\alpha, \gamma_1) = d(\alpha, \gamma_2) = i$, and hence $\alpha^g = \alpha$, that is, $g \in G_\alpha$. Thus G_α acts transitively on $\Gamma_i(\alpha)$ and so part (1) implies part (2).

Suppose that part (2) holds. Let $1 \leq i \leq s$ and let $\beta \in \mathcal{C}$ and $\gamma \in \Gamma_i(\beta)$. Since G acts transitively on $\mathcal{C}$, there exists $g_1 \in G$ such that $(\beta, \gamma)^{g_1} = (\alpha, \gamma')$ for some $\gamma' \in \Gamma_i(\alpha)$. Furthermore, G_α acts transitively on $\Gamma_i(\alpha)$, and hence there exists $g_2 \in G_\alpha$ such that $(\gamma')^{g_2} = \gamma_1$. Thus, $(\beta, \gamma)^{g_1 g_2} = (\alpha, \gamma_1)$ and so part (2) implies part (3).

Finally, suppose that part (3) holds. Let $i \in \{1, \ldots, s\}$, let $\nu_1, \nu_2 \in \mathcal{C}_i$ and, as in Lemma 5.1, let β_1, β_2 be the unique elements of $\mathcal{C}$ such that $d(\beta_1, \nu_1) = d(\beta_2, \nu_2) = i$. Then (β_j, ν_j) (for $j = 1, 2$) lies in the set of pairs in part (3), and so there exists an element $h \in G$ such that $(\beta_1, \nu_1)^h = (\beta_2, \nu_2)$. Thus $\nu_1^h = \nu_2$ and G acts transitively on $\mathcal{C}_i$. Also by Lemma 5.1 part (3), for $\beta_1, \beta_2 \in \mathcal{C}$, there exist ν_1, ν_2 such that, for each $j = 1, 2$, $\nu_j \in \Gamma_i(\beta_j)$ so (β_j, ν_j) lies in the set of pairs in part (3). Thus by part (3), we have $(\beta_1, \nu_1)^h = (\beta_2, \nu_2)$ for some $h \in G$, and it follows that G is transitive also on $\mathcal{C}$. Thus part (3) implies part (1). □

The case $s = 1$ of Lemma 5.2 follows from [66, Theorem 1.2], noting that $\delta(\mathcal{C}) \geq 3$ is equivalent to error capacity $e \geq 1$. A G-neighbour-transitive code $\mathcal{C}$ with the property of Lemma 5.2(2) is called *strongly-incidence-transitive*, and the theory of strongly incidence transitive codes in Johnson graphs is developed in [66]. The expository chapter [74] gives a recent account focusing especially on links between such codes and a family of combinatorial designs called Delandtsheer designs.

In the following proposition, we assume the conclusion Lemma 5.2(2) holds with $s = 2$.

Proposition 5.1. *Let $\mathcal{C}$ be a non-trivial code with covering radius ρ and minimum distance δ in a connected graph Γ. Suppose that $G \leq \mathrm{Aut}(\mathcal{C})$ such that G acts transitively on $\mathcal{C}$ and, for $\alpha \in \mathcal{C}$, G_α acts transitively on each of $\Gamma_1(\alpha)$ and $\Gamma_2(\alpha)$. Then one of the following holds:*

1. $\rho \geq 2$;

2. $\rho = 1$, $\delta = 3$ *and $\mathcal{C}$ is a perfect code.*

3. $\rho = 1$, $\delta = 2$, *and either*

    a. *Γ is bipartite and $\mathcal{C}$ is one of the biparts; or*

    b. *for every pair $\mu, \nu \in \mathcal{C}_1$ with $d(\mu, \nu) = 1$ we have that $\Gamma_1(\mu) \cap \mathcal{C} = \Gamma_1(\nu) \cap \mathcal{C}$.*

Proof. If $\rho = 0$ then $\mathcal{C} = V(\Gamma)$ is a trivial code, but since this is not the case we have $\rho \geq 1$. If $\rho \geq 2$ then part (1) holds. Hence, we may assume that $\rho = 1$. Since Γ is connected, G acts transitively on $\mathcal{C}$ and G_α acts transitively on $\Gamma_1(\alpha)$, it follows that $\mathcal{C}_1 = \bigcup_{\beta \in \mathcal{C}} \Gamma_1(\beta)$. In particular $\delta \geq 2$. If $\delta \geq 3$, then this union is disjoint, and since $\rho = 1$, $|\mathcal{C}| \geq 2$ and Γ is connected, there must be an edge between some vertex of $\Gamma_1(\beta)$ and some vertex of $\Gamma_1(\beta')$ for some distinct codewords β and β', and hence $d(\beta,\beta')=3$, so $\delta=3$. This implies that $\mathcal{C}$ has error-correction capacity $e=1$, and any pair of balls of radius 1 centered at distinct codewords is disjoint. Moreover, since $\rho=1$, the vertex set $V(\Gamma)=\mathcal{C}\cup\mathcal{C}_1$, and hence the set of balls of radius 1 centered at the codewords of $\mathcal{C}$ partitions $V(\Gamma)$. Thus $\mathcal{C}$ is perfect, as in part (2).

Thus we may assume that $\delta=2$. Then, since G acts transitively on $\mathcal{C}$ and G_α acts transitively on $\Gamma_2(\alpha)$, it follows, for each codeword $\beta\in\mathcal{C}$, that $\Gamma_2(\beta)$ is contained in $\mathcal{C}$. If there are no edges between distinct vertices of $\mathcal{C}_1$, then all edges of Γ are incident with a vertex of $\mathcal{C}$ and a vertex of $\mathcal{C}_1$, that is to say, Γ is bipartite and $\mathcal{C},\mathcal{C}_1$ form a bipartition as in part (3)(a). Hence we may assume that there exists some pair $\mu,\nu\in\mathcal{C}_1$ such that $d(\mu,\nu)=1$. For any such pair μ,ν, suppose that $\gamma\in\Gamma_1(\mu)\cap\mathcal{C}$. Then (γ,μ,ν) is a path of length 2 in Γ so $d(\gamma,\nu)\leq 2$. We have shown that $\Gamma_2(\gamma)\subseteq\mathcal{C}$, and since $\nu\in\mathcal{C}_1$ this implies that $d(\gamma,\nu)=1$, that is, $\gamma\in\Gamma_1(\nu)\cap\mathcal{C}$ also. A similar argument holds with μ and ν interchanged, and hence $\Gamma_1(\mu)\cap\mathcal{C}=\Gamma_1(\nu)\cap\mathcal{C}$ and part (3)(b) holds. □

In Section 5.5.5 we apply Proposition 5.1 to codes in Hamming graphs (see Remark 5.4). In that case, Proposition 5.1(3)(b) never holds, so we are able to make much stronger conclusions. We pose the following problem related to this.

Problem 5.1. *Investigate codes satisfying Proposition 5.1(3)(b).*

5.2.3 Elusive Codes

In this subsection, we make a short commentary on the concept of neighbour-transitivity. Let $\mathcal{C}$ be a code in a connected graph Γ such that the minimum distance $\delta(\mathcal{C})\geq 3$ so, by Lemma 5.1, the set of code neighbours is the disjoint union $\mathcal{C}_1=\cup_{\alpha\in\mathcal{C}}\Gamma_1(\alpha)$. It turns out that $\mathcal{C}_1$ determines the code $\mathcal{C}$ if $\delta(\mathcal{C})$ is large enough and if the graph Γ is *reduced*, that is, if:

$$\Gamma_1(\alpha)=\Gamma_1(\alpha') \text{ if and only if } \alpha=\alpha'. \tag{5.3}$$

We note that all the graphs we consider in this chapter are reduced in this sense.

Lemma 5.3. *Let $\mathcal{C}_1$ be the set of code-neighbours of a non-trivial code in a connected regular reduced graph Γ. If $\delta(\mathcal{C})\geq 5$ then $\mathcal{C}=\{\alpha\in V(\Gamma)\mid\Gamma_1(\alpha)\subseteq\mathcal{C}_1\}$, and hence $\mathcal{C}_1$ determines $\mathcal{C}$.*

Proof. Let $X:=\{\alpha\in V(\Gamma)\mid\Gamma_1(\alpha)\subseteq\mathcal{C}_1\}$. Since $\delta(\mathcal{C})\geq 5$, $\mathcal{C}_1$ contains $\Gamma_1(\alpha)$ for each codeword α and hence $\mathcal{C}\subseteq X$. We claim that equality holds. Suppose to the contrary that $\alpha\in X\setminus\mathcal{C}$, and let β,β' be distinct vertices in $\Gamma_1(\alpha)\subseteq\mathcal{C}_1$. By the definition of $\mathcal{C}_1$ there are codewords $\gamma,\gamma'\in\mathcal{C}$ such that $(\gamma,\beta,\alpha,\beta',\gamma')$ is a path in Γ of length 4. Since the minimum distance $\delta(\mathcal{C})\geq 5$, it follows that $\gamma=\gamma'$. For fixed α,β and γ, letting β' range over $\Gamma_1(\alpha)$, we see that $\Gamma_1(\alpha)=\Gamma_1(\gamma)$, which is a contradiction since Γ is reduced. Thus $X=\mathcal{C}$. □

It follows from Lemma 5.3 that, under the conditions of that lemma, the setwise stabilisers in $\mathrm{Aut}(\Gamma)$ of $\mathcal{C}$ and of $\mathcal{C}_1$ are equal, that is to say, $\mathrm{Aut}(\mathcal{C}) = \mathrm{Aut}(\mathcal{C}_1)$. Then since $\mathcal{C}_1$ is the disjoint union $\cup_{\alpha\in\mathcal{C}}\Gamma_1(\alpha)$, if $\mathrm{Aut}(\mathcal{C}_1)$ is transitive on $\mathcal{C}_1$ then it must also be transitive on $\mathcal{C}$. Thus $\mathcal{C}$ is neighbour-transitive if and only if $\mathrm{Aut}(\mathcal{C}_1)$ is transitive on $\mathcal{C}_1$, and this would be a simplification of Definition 5.2 of neighbour-transitivity.

However for smaller $\delta(\mathcal{C})$, it is possible for $\mathcal{C}_1$ to be the set of code neighbours of more than one code. Such a code $\mathcal{C}$ is said to be *elusive*. An infinite family of elusive neighbour-transitive codes in binary Hamming graphs (see Definition 5.4) was described by Gillespie and the second author in [40, Section 5], and the smallest code in the family is given in Example 5.3.

Example 5.3. *Let $\Gamma = H(4,2)$ (see Definition 5.4) and write the vertex set as $V\Gamma = F \times F$ with $F = \mathbb{F}_2^2$. Define*

$$\mathcal{C} := \{(0,0,0,0),(1,1,1,1)\}$$
$$\mathcal{C}' := \{(0,0,0,0),(1,0,1,0),(0,1,0,1),(1,1,1,1)\}$$
$$\mathcal{X} := \{(\beta,\beta'),(\beta',\beta) \mid \beta \in \{(0,0),(1,1)\}, \beta' \in \{(0,1),(1,0)\}\}.$$

Then $\mathcal{C}$ and $\mathcal{C}'$ are both linear codes in Γ, with $\delta(\mathcal{C}) = 4$ and $\delta(\mathcal{C}') = 2$. Further, $\mathcal{X}$ is the set of code neighbours of both of the codes $\mathcal{C}$ and $\mathcal{C}'$; and the code $\mathcal{C}$ is neighbour-transitive.

Two new constructions for infinite families of elusive neighbour-transitive codes in $H(n,q)$ were given by Gillespie and the authors in [55, Sections 3.1 and 3.2], this time with minimum distance 3. The constructions produce *elusive pairs* $(\mathcal{C}, X)$, where $\mathcal{C}$ is an elusive code and $X \leq \mathrm{Aut}(\mathcal{C}_1)$ such that X does not fix $\mathcal{C}$ setwise. For these examples there were precisely two distinct images of $\mathcal{C}$ under elements of X, but very recent constructions by the first author in [52, Theorem 1.1 and 1.2] show that the number of images can be arbitrarily large, answering [55, Question 1.4].

A *spherical bitrade* in a graph Γ is a pair $(\mathcal{C}, \mathcal{C}')$ of codes in Γ with the property that for any $\alpha \in V(\Gamma)$ we have $|\Gamma_1(\alpha) \cap \mathcal{C}| = |\Gamma_1(\alpha) \cap \mathcal{C}'| \in \{0,1\}$. Spherical bitrades are investigated in [69] in relation to a type of switching construction for obtaining new perfect codes from known ones. If $\mathcal{C}$ is an elusive code with $\delta(\mathcal{C}) \geq 3$, and $g \in \mathrm{Aut}(\mathcal{C}_1) \setminus \mathrm{Aut}(\mathcal{C})$, then $(\mathcal{C}, \mathcal{C}^g)$ is a spherical bitrade. As is pointed out in [69], the examples of spherical bitrades in $H(q,q)$ given in [69, Theorem 2] were first constructed as elusive codes in [55, Example 1]. Moreover, one application of the product construction of [69, Theorem 1] is to produce spherical bitrades in $H(kq,q)$ for arbitrary k; examples of elusive codes in $H(kq,q)$ were first given in [55, Lemma 3.9]. It is worth noting also that if elusive codes are used as input for [69, Theorem 1] then the resulting codes are also elusive.

An extension of the concept of an elusive code relevant to s-neighbour-transitive codes was introduced in [52]: a code $\mathcal{C}$ in a graph Γ is said to be *s-elusive* if there exists a code $\mathcal{C}'$ distinct from $\mathcal{C}$, but equal to the image of $\mathcal{C}$ under an element of $\mathrm{Aut}(\Gamma)$, such that the sets of s-neighbours of $\mathcal{C}$ and $\mathcal{C}'$ are the same. Interesting new examples were found in [52, Theorem 1.1] for $s = 1, 2, 3$ developed from Reed–Muller codes, Preparata codes, and binary Golay codes. It would be interesting to know of examples of s-elusive codes in graphs other than the Hamming graphs.

Problem 5.2. *Find s-elusive, neighbour-transitive codes ($s \geq 1$) in other interesting distance-regular graphs, or show that none exist.*

5.2.4 s-Neighbour-Transitive Codes and s-Distance-Transitive Graphs

There is a beautiful link between (G, s)-neighbour-transitive codes and (G, s)-*distance-transitive graphs*, namely connected graphs $\Gamma = (V, E)$ for which $G \leq \mathrm{Aut}(\Gamma)$ is transitive on the distance sets $\Gamma_i := \{(\alpha, \beta) \mid d(\alpha, \beta) = i\}$, for $i = 0, 1, \ldots, s$. The link involves a *quotient graph* Γ_N modulo a normal subgroup $N \trianglelefteq G$: the quotient Γ_N is the graph with vertices the N-orbits in V such that distinct N-orbits U, U' form an edge of Γ_N provided there exist $\alpha \in U$ and $\beta \in U'$ such that $\{\alpha, \beta\} \in E$. For a vertex $\alpha \in V$, we denote the N-orbit containing α by $\alpha^N := \{\alpha^x \mid x \in N\}$.

Proposition 5.2. *Let $\Gamma = (V, E)$ be a graph and $G \leq \mathrm{Aut}(\Gamma)$ such that G acts transitively on V, and let $N \lhd G$ be intransitive on V. Suppose further that, for some $\alpha \in V$, the set $\mathcal{C} = \alpha^N$ is a $(G_\mathcal{C}, s)$-neighbour-transitive code in Γ. Then the quotient graph Γ_N is $(G/N, s)$-distance-transitive.*

Proof. Since $N \lhd G$ and G is transitive on V, it follows that G acts transitively on the set of N-orbits in V, that is to say, the vertex set $V(\Gamma_N)$ of Γ_N, see [75, Lemma 2.20]. As N leaves each of its orbits invariant, N is contained in the kernel of this G-action and the quotient group G/N acts vertex-transitively on Γ_N. Recall that $\mathcal{C} = \alpha^N$ is one of the N-orbits, and as a code in Γ, it is $(G_\mathcal{C}, s)$-neighbour-transitive, by assumption.

Let U, U' be adjacent vertices of Γ_N, and let $\{\alpha_1, \alpha_2\} \in E$ such that $\alpha_1 \in U$ and $\alpha_2 \in U'$. Since N acts transitively on U and on U', it follows that every vertex of U is adjacent in Γ to some vertex in U', and every vertex of U' is adjacent in Γ to some vertex in U.

Let X, Y be vertices of Γ_N at distance i in Γ_N from $\mathcal{C}$, where $1 \leq i \leq s$. Then, arguing as in the preceding paragraph, there exist vertices $\beta \in X$ and $\gamma \in Y$ such that there are paths of length i in Γ from α to each of β and γ. Suppose that there exists a codeword $\alpha' \in \mathcal{C}$ such that there is a path of length j from α' to β with $0 < j < i$. Then $\mathcal{C}$ and X would be at distance less than i in Γ_N, which is not the case. Hence each codeword of $\mathcal{C}$ has distance at least i from β, and it follows that $\beta \in \mathcal{C}_i$. By an identical argument $\gamma \in \mathcal{C}_i$. Since $\mathcal{C}$ is $(G_\mathcal{C}, s)$-neighbour-transitive, there exists an element $g \in G_\mathcal{C}$ such that $\beta^g = \gamma$. Since G permutes the N-orbits among themselves, this means that g maps the N-orbit X containing β to the N-orbit Y containing γ. Thus $G_\mathcal{C}$ induces a transitive action on the vertices of Γ_N at distance i from $\mathcal{C}$. Since we have already shown that Γ_N is vertex-transitive, and since N is in the kernel of the G-action on Γ_N, it follows that G/N acts transitively on the set of ordered pairs of Γ_N-vertices at distance i. Finally since this holds for all $i \leq s$, the result follows. □

On the one hand, this observation can be used to produce s-distance-transitive graphs from certain s-neighbour-transitive codes in Hamming graphs:

Example 5.4. *Let C be an s-neighbour-transitive and linear code in the Hamming graph $H(n,q)$ (see Section 5.4.1 for the definitions of 'linear' and $H(n,q)$). Moreover, let T_C be the group of translations by codewords of C and let $H = \mathrm{Aut}(C)_\mathbf{0}$ be the stabiliser of the codeword $\mathbf{0} \in C$. Then we may apply Proposition 5.2 with $N = T_C$, so that $C = \mathbf{0}^N$, and $G = T_{V(\Gamma)} \rtimes H$, where $T_{V(\Gamma)} \cong \mathbb{F}_q^n$ is the group of translations by all vectors in $V(\Gamma)$, to obtain a $(G/N, s)$-distance-transitive graph Γ_N. For examples of such codes C, with $s = 2$, see Theorem 5.7(3) and Propositions 5.9–5.13.*

On the other hand, there are many (G, s)-distance-transitive graphs Γ known for which the group G has a nontrivial vertex-intransitive normal subgroup N, and the question arising here is whether an N-orbit provides an example of an s-neighbour-transitive code:

Example 5.5. *(1) Let Γ be a $(G,2)$-arc-transitive graph[3] and let N be a normal subgroup of G with at least 3 orbits on $V(\Gamma)$. It was shown by the second author in [72, Theorem 4.1] that the quotient Γ_N, as defined above, is $(G/N,2)$-arc-transitive and the graph Γ is a cover of Γ_N. This means that, for each $\alpha \in V(\Gamma)$, the code $C = \alpha^N$ in Γ has minimum distance at least the girth of Γ_N. Note that*

$$\rho(C) = 1 \iff \Gamma_N \textit{ is a complete graph} \iff \delta(C) = 3.$$

In all other cases Γ_N has girth at least 4, so $\delta(C) \geq 4$, $\rho(C) \geq 2$, and C is $(G_C, 2)$-neighbour-transitive.

(2) Let Γ be an antipodal G-distance-transitive graph with diameter δ. Then an antipodal block C (that is, a maximal set of vertices mutually at distance δ) is a code in Γ with minimum distance δ, and usually C is an orbit of a normal subgroup of G. For example, $(G_C, 1)$-neighbour-transitive codes C with $\delta = 3$ may be obtained in this way from antipodal distance-transitive covers of complete graphs, and these graphs have been classified in [47]. Similarly $(G_C, 2)$-neighbour-transitive codes C with $\delta = 4$ arise in this way from antipodal distance-transitive covers of complete bipartite graphs, which are classified in [58].

Regarding the examples in Example 5.5(1), we observe that for a $(G,2)$-arc-transitive graph Γ, a vertex-stabiliser G_α is 2-transitive on $\Gamma_1(\alpha)$ [73, Lemma 9.4] and in particular is primitive, so Γ is G-locally-primitive. The 2-arc-transitivity condition can be weakened to local primitivity, and each code of the form $C = \alpha^N$, for an intransitive normal subgroup N of G, will be neighbour-transitive with $\delta \geq 3$ (since Γ covers Γ_N by [73, Theorem 10.2]). Weakening the condition further to Γ being G-locally-quasiprimitive will yield neighbour-transitive codes $\mathcal{C} = \alpha^N$ with $\delta \geq 2$, see [64, Theorem 1.3].

In seeking examples, if we start with a (G, s)-distance-transitive graph Γ and a normal subgroup $N \lhd G$ with at least three vertex-orbits, then the normal quotient Γ_N is always $(G/N, s)$-distance-transitive [33, Theorem 1.1]. It would be interesting to have examples for Proposition 5.2 where the graph Γ is not itself (G, s)-distance-transitive.

Problem 5.3. *Find graph-group pairs (Γ, G) such that $G \leq Aut(\Gamma)$ is vertex-transitive, Γ is not (G, s)-distance-transitive, and for some vertex-intransitive subgroup $N \lhd G$ the code $\mathcal{C} = \alpha^N$ is $(G_{\mathcal{C}}, s)$-neighbour-transitive – thus giving a $(G/N, s)$-distance-transitive quotient graph Γ_N.*

5.2.5 Further Techniques for Analysing Codes

We now define a concept that has proved useful for keeping track of code invariants, especially for analysing s-neighbour-transitive codes.

Definition 5.3. *Let G be a group acting on a set Ω and let $\iota : \Omega \to S$, for some set S. If for all $\alpha \in \Omega$ and $g \in G$ the equality $\iota(\alpha) = \iota(\alpha^g)$ holds, then ι is called* G-invariant. *The elements of S are called* types, *namely if $\alpha \in \Omega$, then we say that α has* type $\iota(\alpha)$.

The archetypal example of a G-invariant map $\iota : \Omega \to S$ is to take S as the set of G-orbits in Ω and define $\iota(\alpha) = \alpha^G$, for each $\alpha \in \Omega$. We consider another example below for the Johnson graphs which are defined in Definition 5.6. Note that this example will be typical for us, in that Ω will often be the vertex set of a graph containing a code of interest.

Example 5.6. *Let $\mathcal{C}$ be a G-neighbour-transitive code in the Johnson graph $J(v, k)$ (see Section 5.4.2) with underlying set $\mathcal{V}$ such that G fixes setwise some subset $U \subseteq \mathcal{V}$ with $0 < |U| < v$. For a vertex α of $J(v, k)$ define $\iota(\alpha) = |\alpha \cap U|$. Let $g \in G$. Then, since $U^g = U$, it follows that $\iota(\alpha^g) = |\alpha^g \cap U| = |(\alpha \cap U)^g| = \iota(\alpha)$. Hence ι is G-invariant.*

In Example 5.6, since G acts transitively on $\mathcal{C}$, every element of $\mathcal{C}$ has the same type. The next result expands on this observation.

Lemma 5.4. *Let $\mathcal{C}$ be a (G, s)-neighbour-transitive code with minimum distance δ in the connected graph $\Gamma = (V, E)$, let ι be a G-invariant map on V, let $\alpha \in \mathcal{C}$ and let $i \in \{0, 1, \ldots, s\}$. Then the following hold.*

1. *All vertices in $\mathcal{C}_i$ have the same type.*
2. *$B_i(\alpha)$ contains at most $i + 1$ different types of vertices.*
3. *If $\delta \geq 2i$ then all vertices in $\Gamma_i(\alpha)$ have the same type.*

Proof. Let $\nu_1, \nu_2 \in \mathcal{C}_i$. By Definition 5.2, G acts transitively on $\mathcal{C}_i$ and hence there exists $g \in G$ such that $\nu_1^g = \nu_2$. Since ι is G-invariant, $\iota(\nu_1) = \iota(\nu_1^g) = \iota(\nu_2)$. This proves part (1). Next, $B_i(\alpha) \subseteq \mathcal{C} \cup \mathcal{C}_1 \cup \cdots \cup \mathcal{C}_i$. By part (1), the type of any vertex $\mu \in B_i(\alpha)$ is the same as the type of a vertex in $\mathcal{C}_j$, for some $j \in \{0, 1, \ldots, i\}$. Hence there are at most $i+1$ possibilities for $\iota(\mu)$, and part (2) is proved. Finally, if $\beta \in \Gamma_i(\alpha)$ and $\delta \geq 2i$ then, by the triangle inequality, $d(\beta, \gamma) \geq i$ for all $\gamma \in \mathcal{C} \setminus \{\alpha\}$, and hence $\beta \in \mathcal{C}_i$. Thus $\Gamma_i(\alpha) \subseteq \mathcal{C}_i$, and so, by part (1), all vertices in $\Gamma_i(\alpha)$ have the same type, proving part (3). □

Lemma 5.4 was used implicitly in [66] to classify the codes in Example 5.6, that is, G-neighbour-transitive codes in $J(v, k)$ (see Section 5.4.2) where G acts intransitively on $\mathcal{V}$. We include a different proof of this result, below, in order to illustrate how Lemma 5.4 may be applied. For a set U and integer $j \leq |U|$, we denote by $\binom{U}{j}$ the set of all j-element subsets of U.

Theorem 5.1 ([66, Proposition 3.3]). *Let $\mathcal{C}$ be a non-trivial G-neighbour-transitive code in $\Gamma = J(v, k)$, where $2 \leq k < v$, and suppose that $U \subseteq \mathcal{V}$ is such that G fixes U setwise and $0 < |U| < v$. Then $\mathcal{C}$ has minimum distance 1 and, possibly replacing U by $\mathcal{V} \setminus U$, one of the following holds.*

1. $k < |U|$ *and* $\mathcal{C} = \{\alpha \in V(\Gamma) \mid \alpha \subseteq U\}$;
2. $k > |U|$ *and* $\mathcal{C} = \{\alpha \in V(\Gamma) \mid U \subseteq \alpha\}$.

Proof. We consider the type $\iota(\alpha) = |\alpha \cap U|$ of a vertex α of Γ, as in Example 5.6. By Lemma 5.4, all vertices in $\mathcal{C}$ have the same type, and all vertices in $\mathcal{C}_1$ have the same type. Moreover, for every $\alpha \in \mathcal{C}$, $\Gamma_1(\alpha) \subseteq \mathcal{C} \cup \mathcal{C}_1$, and hence either every vertex of $\Gamma_1(\alpha)$ has the same type, or $\delta(\mathcal{C}) = 1$ and the vertices in $\Gamma_1(\alpha)$ have one of precisely two different types. Note that $\iota(\alpha) \leq |\alpha| = k$, and equality holds if and only if $\alpha \subseteq U$.

First, suppose that $\iota(\alpha) = k$ for all $\alpha \in \mathcal{C}$. Then $\bigcup_{\alpha \in \mathcal{C}} \alpha \subseteq U$ and since $\mathcal{C}$ is non-trivial (*i.e.*, $|\mathcal{C}| \geq 2$) we have $|U| > k$. Let $\alpha \in \mathcal{C}$. Then $\alpha \subseteq U$ and $\Gamma_1(\alpha)$ consists of the set of all vertices $\nu \cup \{a\}$, where $\nu \in \binom{\alpha}{k-1}$ and $a \in \mathcal{V} \setminus \alpha$. Moreover, $\iota(\nu \cup \{a\}) = k$ if $a \in U$ (there are $k(|U| - k) > 0$ such pairs (ν, a)) and $\iota(\nu \cup \{a\}) = k - 1$ if $a \notin U$ (there are $k(v - |U|) > 0$ such pairs (ν, a)). Hence $\Gamma_1(\alpha)$ contains precisely two types of vertices and so, by Lemma 5.4, $\delta(\mathcal{C}) = 1$ and, for every codeword $\alpha \in \mathcal{C}$, each element of $\Gamma_1(\alpha)$ having type k is again a codeword. Observe that the induced subgraph of Γ on the set $\binom{U}{k}$ is isomorphic to $J(|U|, k)$ and is thus connected, and hence the set $\binom{U}{k}$ is precisely the set of vertices of type k; the previous sentence implies that $\binom{U}{k} \subseteq \mathcal{C}$. Hence $\mathcal{C} = \binom{U}{k}$ and part (2) holds.

Thus we may assume that codewords α have (constant) type $\iota(\alpha) = \ell$ strictly less than k. Suppose next that $\ell = |U|$. This implies that $|U| < k$. Let $\alpha \in \mathcal{C}$. Then $U \subset \alpha$ and $\Gamma_1(\alpha)$ consists of the set of all vertices $\nu \cup \{a\}$, where $\nu \in \binom{\alpha}{k-1}$ and $a \in \mathcal{V} \setminus \alpha$. Moreover, $\iota(\nu \cup \{a\}) = \ell$ if $\nu \cap U = \alpha \cap U = U$ (there are $(k - \ell)(v - k) > 0$ such pairs (ν, a)) and $\iota(\nu \cup \{a\}) = \ell - 1$ if $\nu \cap U \neq \alpha \cap U$ (there are $\ell(v - k) > 0$

such pairs (ν, a)). Hence $\Gamma_1(\alpha)$ contains precisely two types of vertices, and by Lemma 5.4, each element of $\Gamma_1(\alpha)$ of type ℓ is again a codeword. Now the induced subgraph of Γ on the set of vertices $\{\alpha \in V(\Gamma) \mid U \subseteq \alpha\}$ is isomorphic to $J(v-\ell, k-\ell)$ with vertex set $\binom{\mathcal{V}\setminus U}{k-\ell}$. Since $J(v-\ell, k-\ell)$ is connected, we deduce that $\{\alpha \in V(\Gamma) \mid U \subseteq \alpha\} \subseteq \mathcal{C}$ and in fact equality holds as in part (1).

This leaves the case where $0 \leq \ell < \min\{|U|, k\}$. If $\ell = 0$, then all codewords α are contained in $\mathcal{V} \setminus U$, and replacing U with $\mathcal{V} \setminus U$, the argument in the previous paragraph shows that part (1) holds. Similarly, if $v - |U| = k - \ell$ then all codewords α contain $\mathcal{V} \setminus U$, and replacing U with $\mathcal{V} \setminus U$, the argument in paragraph two of the proof shows that part (2) holds. Thus we may assume in addition that $0 < \ell$ and $v - |U| > k - \ell$. However in this case a codeword α is adjacent to vertices of types $\ell - 1, \ell$ and $\ell + 1$. This contradicts Lemma 5.4(2), completing the proof. □

5.3 COMMENTARY ON THE ORIGINS OF COMPLETELY TRANSITIVE CODES

In a 1987 preprint [83, Section 7], Patrick Solé introduced the term 'completely transitive code' in the context of binary linear codes. His concept is different in a number of respects from that introduced in Definition 5.2(3), and we comment more below. To the best of our knowledge, the notion of a 'completely-transitive code in a graph' dates back to 1988, when Chris Godsil suggested to the second author the study of completely transitive codes in Johnson graphs (see Definition 5.6). Since a subset of vertices in a Johnson graph $J(v, k)$ is a set of k-subsets of the underlying v-set $\mathcal{V}$, each code $\mathcal{C}$ in $J(v, k)$ has a natural interpretation as a 'design' with point-set $\mathcal{V}$ and with constant block size k. Thus, completely transitive codes in Johnson graphs were first called *completely transitive designs*. In 1988, Bill Martin was a PhD student of Chris Godsil studying completely regular designs (completely regular codes in $J(v, k)$) and Chris's hope was that by studying the more restricted family of completely transitive designs we might discover new examples and characterisations. An account of this investigation, dating from 1997, can be found in [48]. It was never published as a journal article, but was posted on the arXiv in 2014 because of repeated requests for copies.

When regarded as codes in the Hamming graph $H(n, 2)$ (Definition 5.4), the 'completely transitive codes' defined by Solé [83, Section 7], or see [84] for the 1990 published version, are precisely those binary linear codes $\mathcal{C} \subseteq \mathbb{F}_2^n$ that are G-completely transitive in the sense of Definition 5.2(3) for the subgroup $G = T \rtimes G_0$ of affine transformations, where T is the group of translations of $\mathbb{F}_2^n$ by codewords in $\mathcal{C}$, and G_0 is the subgroup of permutation matrices which leave $\mathcal{C}$ invariant. Note that G_0 is the group traditionally regarded as the automorphism group of a linear code. In Ref. [84, Section 7], Solé gave examples of such codes, and also a necessary condition, and a sufficient condition (separate conditions), for complete transitivity in terms of the natural action of G_0 as a permutation group on a set of n points. Shortly afterwards, in 1991, Rifà and Pujol [76] (or see [18, Proposition 13]) showed that, for a binary linear completely transitive code $\mathcal{C}$ of length n, the natural coset graph on the set $\mathbb{F}_2^n/C$ of additive cosets of C in $\mathbb{F}_2^n$ is distance-transitive. Solé's concept of complete transitivity extends naturally to linear codes in $H(n,q)$ for arbitrary prime powers q, as does the link with distance transitivity of the quotient graph $\mathbb{F}_q^n/C$ (see Proposition 5.2 with $s = \rho(C)$). For this reason, in order to distinguish Solé's completely transitive linear codes from the more general family of completely transitive codes in Hamming graphs, Solé's notion is sometimes called *coset-complete-transitivity*, see [45,46]. For linear codes in $H(n,q)$, where q is a prime power, the concepts of complete transitivity and coset-complete transitivity are equivalent if $q \leq 3$ and definitely not equivalent if $q = 7$ or $q \geq 9$, see [46, Theorem 1.3 and Example 3.1].

Giudici, in his masters thesis [45] and a joint paper with the second author [46], introduced the notion given in Definition 5.2 of a G-completely transitive code in a Hamming graph $H(n,q)$ for arbitrary n and q, and made a general study of the structure of such graphs. He gave families of examples, and significant structural descriptions, which pointed to the importance of those codes in $H(n,q)$ for which the automorphism group G induces a transitive action on the set of n entries of codewords, see [46, Sections 4, 5 and 6]. A recent survey article of Borges, Rifa and Zinoviev contains an overview of this and more recent work, see in particular [18, Section 3] on coset-completely transitive codes.

The error-correcting capacity e of a binary coset-completely transitive code was shown by Borges, Rifà and Zinoviev to be no larger than three, [16,17] or see [18, Theorem 12], but there are families of examples with unbounded covering radius (with $e = 1$) [78]. We will see that similar statements can be made about the more general family of s-neighbour-transitive codes. In particular, we will prove the following result in Section 5.5.4.

Theorem 5.2. *Suppose that C is an s-neighbour-transitive code in the Hamming graph $H(n,q)$ such that C has error-correcting capacity e with* $\min\{e,s\} \geq 4$. *Then* $n \geq 9$, $q = 2$, $e = \lfloor \frac{n-1}{2} \rfloor$, $s = \lceil \frac{n-1}{2} \rceil$, *and C is equivalent to the binary repetition code* $\mathrm{Rep}_n(2)$ *(see Definition* 5.5(3)*).*

Borges, Rifà and Zinoviev showed further that coset-completely transitive codes may be obtained via some innovative combinatorial methods, such as a Kronecker product construction ([79] or see [18, Section 5.5]), or a lifting procedure ([80] or see [18, Section 5.2]), while Gill, Gillespie and Semeraro [37, Theorem C] constructed new families from the incidence matrices of certain designs. Also, sometimes a well-known family of codes may contain only a few codes which are completely transitive. For example, a Preparata code of length n, or its extension, is completely transitive if and only if $n = 15$, ([43] or see [18, Theorem 15]); these exceptional completely transitive codes are the Nordstrom–Robinson codes.

In this chapter, we will see that a similar situation arises for the larger families of neighbour-transitive, and 2-neighbour-transitive codes, where substructures in graphs, and in finite geometries such as generalised quadrangles, have been used in constructions, and where group theory, sometimes relying on the finite simple group classification, has been applied to achieve classifications.

5.4 GRAPHS OF INTEREST

In this section, we introduce several families of graphs. For each family, we provide at least one example of a code arising from an interesting combinatorial or algebraic structure. Most of these families are analysed further in later sections. We have already informally met the first family of graphs, the Hamming graphs. They will also receive the most attention throughout this chapter.

5.4.1 Hamming Graphs

Definition 5.4. *Let $\mathcal{N}$ be a set of size n and $\mathcal{Q}$ a set of size q, where $n, q \geq 2$. We define the* Hamming graph $H(n, q)$ *in the following two equivalent, but dual, ways:*

1. *The vertex set of $H(n, q)$ is the set of all n-tuples $(a_1, \ldots, a_n)$, where $a_i \in \mathcal{Q}$ for each $i \in \mathcal{N}$ and we have identified $\mathcal{N}$ with $\{1, \ldots, n\}$. Two such n-tuples form an edge if and only if they differ in precisely one position.*

2. *The vertex set of $H(n, q)$ is the set of all functions $\alpha : \mathcal{N} \to \mathcal{Q}$ with an edge between vertices α and β when there exists a unique $i \in \mathcal{N}$ such that $\alpha(i) \neq \beta(i)$.*

The set $\mathcal{N}$ is called the set of entries, *and the set $\mathcal{Q}$ the* alphabet, *of $H(n, q)$. We may use the notation $H(\mathcal{N}, q)$, $H(n, \mathcal{Q})$ or $H(\mathcal{N}, \mathcal{Q})$ if we wish to indicate either or both of the sets $\mathcal{N}$ or $\mathcal{Q}$ explicitly.*

To see that the two definitions given above are equivalent: let $\mathcal{N} = \{1, \ldots, n\}$ and consider functions $\alpha, \beta : \mathcal{N} \to \mathcal{Q}$. The n-tuples $(\alpha(1), \ldots, \alpha(n))$ and $(\beta(1), \ldots, \beta(n))$ differ in precisely one entry if and only if there exists a unique $i \in \mathcal{N}$ such that $\alpha(i) \neq \beta(i)$. We will use whichever is the most convenient notation for each particular context, functions or n-tuples. The following example illustrates this further.

Example 5.7. *Let $\mathcal{N} = \mathbb{F}_2^3$, let $\mathcal{Q} = \mathbb{F}_2$, let $\mathbb{F}_2[x_1, x_2, x_3]$ denote the polynomial ring over $\mathbb{F}_2$ in the three variables x_1, x_2, x_3, and let C be the $\mathbb{F}_2$-vector space*

$$C = \langle 1, x_1, x_2, x_3 \rangle \subseteq \mathbb{F}_2[x_1, x_2, x_3].$$

Considering each polynomial in C as a function from $\mathcal{N}$ to $\mathcal{Q}$, the set C defines a linear code in $H(8, 2)$. The elements 0 and 1 in C correspond to the 8-tuples $(0, \ldots, 0)$ and $(1, \ldots, 1)$, respectively. The remaining elements of C are precisely the linear polynomials in $\mathbb{F}_2[x_1, x_2, x_3]$ and, upon evaluation on the elements of $\mathcal{N}$, these correspond to the characteristic vectors of the 2-flats (the affine 2-spaces) of the affine geometry $\mathrm{AG}_3(2)$. If the entries are labelled by the elements of $\mathcal{N}$ as follows

$$(0,\, e_1,\, e_2,\, e_1 + e_2,\, e_3,\, e_1 + e_3,\, e_2 + e_3,\, e_1 + e_2 + e_3),$$

then we find that $\mathcal{C}$ consists of the following 8-tuples:

$$\begin{array}{ll}
(1,1,1,1,1,1,1,1) & (0,0,0,0,0,0,0,0)\\
(1,1,1,1,0,0,0,0) & (0,0,0,0,1,1,1,1)\\
(1,1,0,0,1,1,0,0) & (0,0,1,1,0,0,1,1)\\
(1,1,0,0,0,0,1,1) & (0,0,1,1,1,1,0,0)\\
(1,0,1,0,1,0,1,0) & (0,1,0,1,0,1,0,1)\\
(1,0,1,0,0,1,0,1) & (0,1,0,1,1,0,1,0)\\
(1,0,0,1,1,0,0,1) & (0,1,1,0,0,1,1,0)\\
(1,0,0,1,0,1,1,0) & (0,1,1,0,1,0,0,1).
\end{array}$$

Note that $\mathcal{C}$ is an extended Hamming code of length 8 [67, Example, p. 27], but is also known as the Reed–Muller code $\mathcal{RM}_2(1,3)$ (see Definition 5.15).

The full automorphism group of the Hamming graph $\Gamma = H(n,q)$ is the semi-direct product $\mathrm{Aut}(\Gamma) = B \rtimes L$, where the *base group* B is $\mathrm{Sym}(\mathcal{Q})^n$ and the *top group* L is $\mathrm{Sym}(\mathcal{N})$ (see [20, Theorem 9.2.1]). Let $x = h\sigma \in \mathrm{Aut}(\Gamma)$, where $h = (h_1,\dots,h_n) \in B$ and $\sigma \in L$. Then h, σ and x act on an n-tuple $\alpha = (a_1,\dots,a_n)$ via

$$\alpha^h = \left(a_1^{h_1},\dots,a_n^{h_n}\right), \quad \alpha^\sigma = \left(a_{(1^{\sigma^{-1}})},\dots,a_{(n^{\sigma^{-1}})}\right),$$
$$\text{and} \quad \alpha^x = \left(a_{1^{\sigma^{-1}}}^{h_{1^{\sigma^{-1}}}},\dots,a_{n^{\sigma^{-1}}}^{h_{n^{\sigma^{-1}}}}\right). \tag{5.4}$$

For example, $(a_1,a_2,a_3,a_4)^{(1\,2\,3)} = (a_3,a_1,a_2,a_4)$. If instead we consider a vertex α to be a function $\mathcal{N} \to \mathcal{Q}$ then h, σ and x act on α via

$$\alpha^h(i) = (\alpha(i))^{h_i}, \quad \alpha^\sigma(i) = \alpha\left(i^{\sigma^{-1}}\right), \quad \text{and} \quad \alpha^x(i) = \left(\alpha(i^{\sigma^{-1}})\right)^{h_{i^{\sigma^{-1}}}}. \tag{5.5}$$

Let $\mathcal{C}$ be a code in $H(n,q)$, $G \leq \mathrm{Aut}(\mathcal{C})$ and let $\mathcal{M}$ be a subset of $\mathcal{N}$ with $m = |\mathcal{M}|$. Then we define

a. the *projection* $\pi_{\mathcal{M}}(\mathcal{C})$ of $\mathcal{C}$ with respect to $\mathcal{M}$ to be the code in $H(\mathcal{M},q)$ consisting of the functions $\alpha|_{\mathcal{M}} : \mathcal{M} \to \mathcal{Q}$ given by restricting α to $\mathcal{M}$, for each $\alpha \in \mathcal{C}$. Moreover, we define

b. $\chi_{\mathcal{M}}(G)$ to be the subgroup of $\mathrm{Aut}(H(\mathcal{M},q))$ induced by the setwise stabiliser $G_{\mathcal{M}}$ in its action on the restrictions $\alpha|_{\mathcal{M}}$ to $\mathcal{M}$ of functions $\mathcal{N} \to \mathcal{Q}$.

c. We say that a code $\mathcal{C}$ is *linear*, or $\mathbb{F}_q$*-linear*, if $\mathcal{Q} = \mathbb{F}_q$ and $\mathcal{C}$ is an $\mathbb{F}_q$-vector subspace of $\mathbb{F}_q^n$.

d. If $\mathcal{Q} = \mathbb{F}_q$ then it makes sense to define the translation $t_\alpha : \beta \mapsto \alpha + \beta$ by a vertex α of $H(n,q)$, and if $\mathcal{C}$ is linear then $\mathrm{Aut}(\mathcal{C})$ contains the subgroup $T_{\mathcal{C}} = \{t_\alpha \mid \alpha \in \mathcal{C}\}$. If 0 is a distinguished element of $\mathcal{Q}$ and α is a vertex in $H(n,q)$ then $\mathrm{supp}(\alpha) = \{i \in \mathcal{N} \mid \alpha(i) \neq 0\}$ and $\mathrm{wt}(\alpha) = |\mathrm{supp}(\alpha)|$.

Example 5.8. *Consider the code* $\mathcal{C} = \langle 1, x_1, x_2, x_3 \rangle$ *(an* $\mathbb{F}_2$ *vector subspace) in* $H(8,2)$ *with* $\mathcal{N} = \mathbb{F}_2^3$ *and* $\mathcal{Q} = \mathbb{F}_2$*, as in Example 5.7. Then for each* $\alpha \in \mathcal{C}$*,* $\mathrm{Aut}(\mathcal{C})$ *contains the translation* $t_\alpha : \beta \to \alpha + \beta$*, and moreover each* $\sigma \in \mathrm{AGL}_3(2) \leq \mathrm{Sym}(\mathcal{N})$ *defines an automorphism of* $\mathcal{C}$*. Hence* $\mathrm{Aut}(\mathcal{C})$ *contains* $G = T_{\mathcal{C}}.\mathrm{AGL}_3(2)$*. Let* $\mathcal{M} = \langle e_1, e_2 \rangle$*, where* e_i *corresponds to* x_i *in Example 5.7. Then the projection code* $\mathcal{C}' = \pi_{\mathcal{M}}(\mathcal{C}) = \langle 1, x_1, x_2 \rangle$ *and we may express the codewords of* $\pi_{\mathcal{M}}(\mathcal{C})$ *as 4-tuples by taking the first four entries of each 8-tuple from Example 5.7:*

$$\begin{array}{ll} (1,1,1,1) & (0,0,0,0) \\ (1,1,0,0) & (0,0,1,1) \\ (1,0,1,0) & (0,1,0,1) \\ (1,0,0,1) & (0,1,1,0). \end{array}$$

Also, the projection $\chi_{\mathcal{M}}(G) = T_{\mathcal{C}'}.\mathrm{AGL}_2(2)$ *is a subgroup of* $\mathrm{Aut}(\mathcal{C}')$*.*

Definition 5.5. *Let* $\mathcal{C}$ *be a code in* $H(\mathcal{N}, \mathcal{Q})$ *and let* $\mathcal{M} = \mathcal{N} \times \{1, \ldots, k\}$*, from which we introduce the following definitions related to* $H(\mathcal{M}, \mathcal{Q})$*.*

1. *If* $\alpha_1, \ldots, \alpha_k$ *are vertices of* $H(\mathcal{N}, \mathcal{Q})$ *(that is, functions* $\mathcal{N} \to \mathcal{Q}$*) then we define a vertex of* $H(\mathcal{M}, \mathcal{Q})$*,* $\beta_{\alpha_1,\ldots,\alpha_k} : \mathcal{M} \to \mathcal{Q}$*, by*

$$\beta_{\alpha_1,\ldots,\alpha_k} : (i,j) \mapsto \alpha_j(i).$$

2. *The* k-fold product $\mathrm{Prod}_k(\mathcal{C})$ *is the code in* $H(\mathcal{M}, \mathcal{Q})$ *given by*

$$\mathrm{Prod}_k(\mathcal{C}) = \{\beta_{\alpha_1,\ldots,\alpha_k} \mid \alpha_1, \ldots, \alpha_k \in \mathcal{C}\}.$$

3. *The* k-fold repetition $\mathrm{Rep}_k(\mathcal{C})$ *in* $H(\mathcal{M}, \mathcal{Q})$ *is*

$$\mathrm{Rep}_k(\mathcal{C}) = \{\beta_{\alpha,\ldots,\alpha} \mid \alpha \in \mathcal{C}\};$$

and if $|\mathcal{N}| = 1$ *so* $H(\mathcal{N}, \mathcal{Q})$ *is degenerate (a complete graph on* q *vertices), then this code is the usual repetition code, and we write* $\mathrm{Rep}_k(q)$ *if* $\mathcal{C} = H(1, \mathcal{Q})$*.*

4. *For* $g_1, \ldots, g_k \in \mathrm{Sym}(\mathcal{N})$*, we define* $\sigma_{g_1,\ldots,g_k}$ *to be the automorphism of* $H(\mathcal{M}, \mathcal{Q})$ *that maps the vertex*

$$\gamma : (i,j) \mapsto \gamma(i,j) \quad to \quad \gamma^{\sigma_{g_1,\ldots,g_k}} : (i,j) \mapsto \gamma\left(i^{g_j^{-1}}, j\right).$$

5. *For* $h_1, \ldots, h_k \in \mathrm{Sym}(\mathcal{Q})$*, we define* $x_{h_1,\ldots,h_k}$ *to be the automorphism of* $H(\mathcal{M}, \mathcal{Q})$ *that maps the vertex*

$$\gamma : (i,j) \mapsto \gamma(i,j) \quad to \quad \gamma^{x_{h_1,\ldots,h_k}} : (i,j) \mapsto (\gamma(i,j))^{h_j}.$$

Note that the automorphism group of $\mathrm{Prod}_k(\mathcal{C})$ *contains the wreath product* $\mathrm{Aut}(\mathcal{C}) \wr \mathrm{S}_k$ *and the automorphism group of* $\mathrm{Rep}_k(\mathcal{C})$ *contains* $\mathrm{Aut}(\mathcal{C}) \times \mathrm{S}_k$*.*

5.4.2 Johnson and Kneser Graphs

Recall that $\binom{\mathcal{V}}{k}$ denotes the set of all k-element subsets of a set $\mathcal{V}$; this set forms the vertex set of both the Johnson and Kneser graphs

Definition 5.6. *Let $\mathcal{V}$ be a set of size v and let k be an integer with $2 \leq k \leq v-1$. The* Johnson graph *$J(v,k)$ has vertex set $\binom{\mathcal{V}}{k}$ and a pair of distinct vertices are defined to be adjacent if they intersect in $(k-1)$-subset. We say that $\mathcal{V}$ is the* underlying set *of $J(v,k)$.*

If $v \neq 2k$ then $\mathrm{Aut}(J(v,k)) = \mathrm{Sym}(\mathcal{V})$ and if $v = 2k$ then $\mathrm{Aut}(J(v,k)) = \mathrm{Sym}(\mathcal{V}) \times C_2$ [20, Theorem 9.1.2]. As mentioned in the introduction, we refer the reader to [74] for a recent survey regarding neighbour-transitive codes in Johnson graphs.

Definition 5.7. *Let $\mathcal{V}$ be a set of size v and let k be an integer with $2 \leq k \leq (v-1)/2$. The* Kneser graph *$K(v,k)$ has vertex set $\binom{\mathcal{V}}{k}$ and a pair of distinct vertices are defined to be adjacent if they are disjoint. If $v = 2k+1$ then $K(v,k)$ is called the* odd graph *O_{k+1}. We say that $\mathcal{V}$ is the* underlying set *of $K(v,k)$ (or O_{k+1}).*

The automorphism group of $K(v,k)$ is $\mathrm{Sym}(\mathcal{V})$ [49, Corollary 7.8.2]. The next example explores the relationship between codes in Johnson graphs and codes in Kneser graphs.

Example 5.9. *Let $\mathcal{V}$ be a set of size v, let k satisfy $2 \leq k \leq (v-1)/2$, and let $J(v,k)$ and $K(v,k)$ be the Johnson and Kneser graphs, respectively, with underlying set $\mathcal{V}$. Since both graphs have the same vertex set, namely $\binom{\mathcal{V}}{k}$, a code in one is also a code in the other, generally with different parameters. For instance:*

a. Let $U \subseteq \mathcal{V}$ such that $k < |U| < v$ and let $C = \{\alpha \in \binom{\mathcal{V}}{k} \mid \alpha \subset U\}$. By Theorem 5.1(1), as a code in the Johnson graph, C has minimum distance 1 and is neighbour-transitive. We now consider C to be a code in the Kneser graph. In this case, since there exists a pair of disjoint k-subsets of U if and only if $|U| \geq 2k$, it follows that C has minimum distance 1 in the Kneser graph if and only if $|U| \geq 2k$. Moreover, since there always exists a pair of k-subsets of U at distance 2 in $K(v,k)$ (in particular, $|U| \geq k+1$ implies there exists such a pair intersecting in a $(k-1)$-subset of U), we deduce that C has minimum distance 2 in the Kneser graph when $|U| < 2k$. Moreover, it turns out that C is neighbour-transitive in the Kneser graph if and only if $|U| = v-1$ (see Theorem 5.10(1) and Example 5.16).

b. Let $U \subseteq \mathcal{V}$ such that $0 < |U| < k$ and let $C = \{\alpha \in \binom{\mathcal{V}}{k} \mid U \subset \alpha\}$. Theorem 5.1(2) again tells us that, as a code in the Johnson graph, C has minimum distance 1 and is neighbour-transitive. However, since every codeword contains U, no pair of elements of C are disjoint. Hence C has minimum distance at least 2 in the Kneser graph, and it is not difficult to see that the *minimum distance equals 2 since $v \geq 2k+1$. Moreover, it turns out that $\mathcal{C}$ is always neighbour-transitive in the Kneser graph (see Theorem 5.10(3) and Example 5.16).*

5.4.3 q-Analogues of Various Graphs

Roughly speaking, if a combinatorial definition is phrased in terms of sets then, for a prime power q, the *q-analogue* of this definition is obtained by rephrasing the original definition in terms of $\mathbb{F}_q$-vector spaces. The next definition gives the q-analogues of the Hamming graphs. Codes in these graphs are known as *rank-metric* codes and have received interest lately due to their application in network coding, see [12] for a recent survey.

Definition 5.8. *Let $X \cong \mathbb{F}_q^m$ and $Y \cong \mathbb{F}_q^n$ with $m, n \geq 2$. The* bilinear forms graph *$H_q(m,n)$ is the graph with vertex set the vector space of all linear maps $\alpha : X \to Y$ where two vertices α and β are adjacent if* $\mathrm{rank}(\alpha - \beta) = 1$.

Example 5.10. *Delsarte [30] and Gabidulin [36] independently introduced the following class of codes. Let n, k, s be positive integers such that* $\gcd(n,s) = 1$. *Identify X and Y with $\mathbb{F}_{q^n}$, considered as an n-dimensional $\mathbb{F}_q$-vector space, and note that each of the polynomials $x^{q^{is}}$ defines an $\mathbb{F}_q$-linear map $X \to Y$.*

Define $\mathcal{C}$ to be the code in $\Gamma = H_q(n,n)$ given by the polynomials in the $\mathbb{F}_{q^n}$-vector space

$$\mathcal{C} = \left\langle x, x^{q^s}, \ldots, x^{q^{s(k-1)}} \right\rangle_{\mathbb{F}_{q^n}}.$$

Note that each polynomial in $\mathcal{C}$ is a linearised *polynomial (see [65, Section 3.4]) and thus defines an $\mathbb{F}_q$-linear map from X to Y. Also, $\mathcal{C}$ has minimum distance $\delta = n-k+1$. Hence, as long as $k \leq n-1$, the set $\Gamma_1(0)$ comprises all rank 1 linear maps. The trace function* $\mathrm{tr}(x) = x^{q^{n-1}} + \cdots + x^q + x$ *is one such rank 1 linear map and all others may be written as $f_{a,b}(x) = a \cdot \mathrm{tr}(bx)$ for some $a, b \in \mathbb{F}_{q^n}^\times$. Since $\mathcal{C}$ is an $\mathbb{F}_{q^n}$-vector space, it follows that Aut($\mathcal{C}$) contains all translations by elements of $\mathcal{C}$. Moreover, Aut($\mathcal{C}$) contains a subgroup isomorphic to $(\mathbb{F}_{q^n}^\times \times \mathbb{F}_{q^n}^\times)/\mathbb{F}_q^\times$ given by the maps $f(x) \mapsto a \cdot f(bx)$ for each $a, b \in \mathbb{F}_{q^n}^\times$. Thus Aut($\mathcal{C}$) acts transitively on $\mathcal{C}$ and the stabiliser in Aut($\mathcal{C}$) of the zero map acts transitively on the set of all rank 1 linear maps. Hence, by Lemma 5.2, $\mathcal{C}$ is neighbour-transitive.*

The above example suggests the following open problem.

Problem 5.4. *Find more examples and characterise families of s-neighbour-transitive codes in $H_q(m,n)$.*

Next we introduce the q-analogues of the Johnson graphs. For $X = \mathbb{F}_q^d$, we write $\binom{X}{k}_q$ for the set of k-dimensional subspaces of X.

Definition 5.9. *Let $X = \mathbb{F}_q^d$. The* Grassmann graph *$J_q(d,k)$ has vertex set $\binom{X}{k}_q$ and vertices α and β are adjacent when $\alpha \cap \beta$ has dimension $k-1$.*

For an integer $k \geq 2$, a *k-spread* of a vector space V is set of k-dimensional subspaces of V such that each non-zero vector of V is contained in precisely one spread element. In the setting of projective geometry, a spread is a partition of the points of $\mathrm{PG}_{d-1}(q)$ such that each part forms a $(k-1)$-dimensional projective subspace. A *regulus* of $\mathrm{PG}_{2k-1}(q)$ is a set $\mathcal{R}$ of size $q+1$ consisting of pairwise disjoint $(k-1)$-dimensional projective subspaces such that any line meeting at least 3 elements of $\mathcal{R}$ meets every element of $\mathcal{R}$; see [32, Section 5.1]. Three pairwise disjoint $(k-1)$-dimensional projective subspaces contained in a $(2k-1)$-subspace determine a unique regulus containing all three of them. A spread $\mathcal{S}$ is called *regular* (or sometimes *Desarguesian*) if $\mathcal{S}$ contains all the elements of the regulus determined by any three pairwise distinct elements of $\mathcal{S}$ contained in a projective $(2k-1)$-subspace. Equivalently, a regular spread is one obtained via *field reduction*, see [63]. The next example uses field reduction to show that a regular spread is a neighbour-transitive code in a Grassmann graph.

Example 5.11. *Let k be a positive integer dividing d, let $Y = \mathbb{F}_{q^k}^{d/k}$ and $X = \mathbb{F}_q^d$. Then X and Y are isomorphic as $\mathbb{F}_q$-vector spaces. Fix such an $\mathbb{F}_q$-vector space isomorphism $\phi : Y \to X$. Let C be the image under ϕ of the set of all one-dimensional $\mathbb{F}_{q^k}$-subspaces of Y. Then C is a regular spread of X and C is a code in $J_q(d,k)$ with minimum distance $\delta = k$. The stabiliser of C in $\Gamma\mathrm{L}_d(q)$ is $\Gamma\mathrm{L}_{d/k}(q^k)$, and hence $\mathrm{Aut}(C) \cong \Gamma\mathrm{L}_{d/k}(q^k)/Z$, where Z is the center of $\mathrm{GL}_d(q)$. Thus $\mathrm{Aut}(C)$ acts transitively on C. Moreover, if $\alpha \in C$ then $\mathrm{Aut}(C)_\alpha$ contains $(q^{d-k} : ((q^k-1) \times \mathrm{GL}_{d/k-1}(q^k)))/Z$, which acts transitively on the set of k-dimensional $\mathbb{F}_q$-subspaces of X that intersect α in a $(k-1)$-dimensional $\mathbb{F}_q$-subspace. Hence, by Lemma 5.2(2), C is neighbour-transitive.*

5.4.4 Two Graphs Related to Incidence Structures

An *incidence structure* (see [32]) is a triple $\mathcal{S} = (\mathcal{P}, \mathcal{L}, \mathrm{I})$, where $\mathcal{P}$ and $\mathcal{L}$ are disjoint non-empty sets of elements called *points* and *lines*, respectively, and $\mathrm{I} \subseteq \mathcal{P} \times \mathcal{L}$ is a point-line incidence relation. If $(p, \ell) \in \mathrm{I}$ then we say that p *lies on* ℓ, p is *incident* with ℓ, or ℓ is *incident* with p. The elements of I are called *flags* and if $p \in \mathcal{P}$, $\ell \in \mathcal{L}$ and $(p, \ell) \notin \mathrm{I}$ then (p, ℓ) is called an *antiflag*. If there exists an $\ell \in \mathcal{L}$ such that $(p_1, \ell), (p_2, \ell) \in \mathrm{I}$ then p_1 and p_2 are *collinear*. The *dual* incidence structure $\mathcal{S}^D = (\mathcal{L}, \mathcal{P}, \mathrm{I}^D)$ is obtained from $\mathcal{S}$ by interchanging the sets of points and lines, with incidence I^D defined by $(\ell, p) \in \mathrm{I}^D$ if $(p, \ell) \in \mathrm{I}$.

Definition 5.10. *Let $\mathcal{S} = (\mathcal{P}, \mathcal{L}, \mathrm{I})$ be an incidence structure. Then the* incidence graph *of $\mathcal{S}$ is the graph with vertex set $\mathcal{P} \cup \mathcal{L}$ and an edge $\{p, \ell\}$ whenever $(p, \ell) \in \mathrm{I}$ (and no further edges). The* collinearity graph *of $\mathcal{S}$ is the graph with vertex set $\mathcal{P}$ and an edge $\{p_1, p_2\}$ whenever p_1 and p_2 are collinear.*

Lemma 5.5. *Let Γ be the incidence graph of an incidence structure $\mathcal{S} = (\mathcal{P}, \mathcal{L}, \mathrm{I})$ and let C be an s-neighbour-transitive code in Γ such that $C \subseteq \mathcal{P}$. Then C is an $\lfloor s/2 \rfloor$-neighbour-transitive code in the collinearity graph of $\mathcal{S}$.*

Proof. This follows from applying Lemma 5.2 to $\mathcal{C}$ as a vertex subset of both the incidence graph of $\mathcal{S}$ and the collinearity graph of $\mathcal{S}$. □

Example 5.12. *Let $\mathcal{S} = (\mathcal{P}, \mathcal{L}, \mathrm{I})$ be the incidence structure obtained by letting $\mathcal{P}$ be the set of points, and $\mathcal{L}$ the set of lines, of the projective geometry $\mathrm{PG}_3(q)$, with the usual point-line incidence. Let Γ be the incidence graph of $\mathcal{S}$, and let $\mathcal{C}$ be a regular spread of $\mathrm{PG}_3(q)$, as defined in Example 5.11 (with $d = 4$, $k = 2$), but considered now as a code in the incidence graph Γ.*

a. *Then as a code in Γ, $\mathcal{C}$ is completely transitive (and 2-neighbour-transitive) with minimum distance 4 and covering radius 2; and*

b. *as a code in the collinearity graph of $\mathcal{S}^D$, which is isomorphic to $J_q(4,2)$, $\mathcal{C}$ is completely transitive (and neighbour-transitive) with minimum distance 2 and covering radius 1.*

Lemma 5.6. *All the claims of Example 5.12 are valid.*

Proof. Let $X = \mathbb{F}_q^4$. It follows from Definition 5.9 that the collinearity graph of $\mathcal{S}^D$ is $J_q(4,2)$, with vertex set $\binom{X}{2}_q$. Also, from Example 5.11, we see that $\mathrm{Aut}(\mathcal{C})$ contains $\mathrm{GL}_2(q^2)/Z$, where Z is the scalar subgroup of $\mathrm{GL}_4(q)$, and as a code in $J_q(4,2)$, $\mathcal{C}$ has minimum distance 2, covering radius 1 and $\mathrm{Aut}(\mathcal{C})$ is transitive on both $\mathcal{C}$ and $\mathcal{L}\backslash\mathcal{C}$. Thus all the assertions of part (b) are valid. Now we consider $\mathcal{C}$ as a code in the incidence graph Γ of $\mathcal{S}$. The set of code neighbours comprises the whole of the point set $\mathcal{P}$, since from the definition of a spread each projective point lies on exactly one line in the spread. Thus $\mathcal{C}$ has covering radius 2. Also the subgroup $\mathrm{GL}_2(q^2)/Z$ of $\mathrm{Aut}(\mathcal{C})$ is transitive on $\mathcal{P}$, and it follows that $\mathcal{C}$ is 2-neighbour-transitive and hence also completely transitive. Finally, any two codewords, that is lines ℓ, ℓ' in $\mathcal{C}$, have no point in common and so are at distance strictly greater than 2 in Γ. Since Γ is bipartite this means that ℓ, ℓ' are at distance at least 4. In fact their distance is equal to 4, since choosing points p, p' on ℓ, ℓ', respectively, and letting ℓ'' be the line containing p and p', we have a path $(\ell, p, \ell'', p', \ell')$ of length 4 in Γ. Thus all the assertions of part (a) are also valid. □

5.5 CODES IN HAMMING GRAPHS

The next three propositions are applications of Lemma 5.2 and are the starting point for our analysis of s-neighbour-transitive codes in Hamming graphs. We include here a proof of the first in order to illustrate the general idea. Recall the notions of i-homogeneous and i-transitive group actions from Section 5.2.2.

Proposition 5.3. *[38, Proposition 2.5] Let C be a (G, s)-neighbour-transitive code, with error-correction capacity $e \geq 1$, in $\Gamma = H(n,q) = H(\mathcal{N}, q)$ and let $\alpha \in C$. Then, for each $i \leq \min\{e, s\}$, G_α acts i-homogeneously on $\mathcal{N}$.*

Proof. Since the automorphism group of the Hamming graph is vertex-transitive, we may assume (replacing $\mathcal{C}$ if necessary by $\mathcal{C}^g$, for some $g \in \mathrm{Aut}(H(n,q)) = \mathrm{Sym}(q) \wr \mathrm{Sym}(n)$) that $\alpha = (0,\ldots,0) \in \mathcal{C}$ for some distinguished element $0 \in \mathcal{Q}$. By Lemma 5.2, the stabiliser G_α acts transitively on $\Gamma_i(\alpha)$ for each $i \leq \min\{e,s\}$. Now $\Gamma_i(\alpha)$ is the set of all weight i vertices of $H(n,q)$. Let $I, J \in \binom{\mathcal{N}}{i}$ and $\beta_1, \beta_2 \in \Gamma_1(\alpha)$ such that $\mathrm{supp}(\beta_1) = I$ and $\mathrm{supp}(\beta_2) = J$. Then there exists $g \in G$ such that $\beta_1^g = \beta_2$. Since $i \leq e$, the codeword α is the only vertex of $\mathcal{C}$ that has distance i from β_1 and β_2, and thus $g \in G_\alpha$. This means that g is of the form $g = (h_1,\ldots,h_n)\sigma$, where $\sigma \in L$, $(h_1,\ldots,h_n) \in B$, and $0^{h_i} = 0$ for each $i = 1,\ldots,n$. It follows that $I^\sigma = J$, and the result is proved. □

Proposition 5.3 implies in particular that, for each G-neighbour-transitive code $\mathcal{C}$ in $H(\mathcal{N},q)$ with error correction capacity at least 1, the group G induces a transitive action on $\mathcal{N}$. This action may be imprimitive, that is to say, G may leave invariant some non-trivial partition of $\mathcal{N}$. It turns out that, in this case, the projection $\pi_J(\mathcal{C})$, for any part J of such a partition, is also a neighbour-transitive code in the smaller Hamming graph $H(J,q)$. (Recall the projection maps π_J and χ_J from Section 5.4.)

Proposition 5.4. *Suppose that $\mathcal{C}$ is a G-neighbour-transitive code in $H(\mathcal{N},q)$ with minimum distance $\delta \geq 3$, and suppose further that $\mathcal{J}$ is a non-trivial G-invariant partition of $\mathcal{N}$, and $J \in \mathcal{J}$ is such that $\pi_J(\mathcal{C})$ is not the complete code $H(J,q)$. Then the following hold:*

1. *$\pi_J(\mathcal{C})$ is $\chi_J(G)$-neighbour-transitive ([41, Proposition 3.4]).*
2. *$\pi_J(\mathcal{C})$ has minimum distance at least 2 ([41, Corollary 3.7]).*

Now $\mathrm{Aut}(H(\mathcal{N},\mathcal{Q})) = \mathrm{Sym}(\mathcal{Q}) \wr \mathrm{Sym}(\mathcal{N}) = B \rtimes L$. For a subgroup $G \leq B \rtimes L$, and $i \in \mathcal{N}$, let G_i denote the subgroup consisting of all elements $x = h\sigma \in B \rtimes L$ lying in G such that $i^\sigma = i$. By (5.4), such an element $x = h\sigma$ maps each vertex $(\alpha_1,\ldots,\alpha_n)$ to a tuple with i^{th} entry $\alpha_i^{h_i}$. Thus G_i induces an action on the set $\mathcal{Q}_i$ of ith entries of vertices of $H(\mathcal{N},\mathcal{Q})$. (Of course $\mathcal{Q}_i$ is a copy of $\mathcal{Q}$.) A more intricate argument than that in the proof of Proposition 5.3 yields the following.

Proposition 5.5. *[38, Proposition 2.7] Let $\mathcal{C}$ be a G-neighbour-transitive code, with minimum distance $\delta \geq 3$, in $\Gamma = H(\mathcal{N},\mathcal{Q}) = H(n,q)$, and let $i \in \mathcal{N}$. Then $G_i^{\mathcal{Q}_i}$ acts 2-transitively on $\mathcal{Q}_i$.*

By an old theorem of Burnside ([21, Section 154], or see [75, Theorem 3.21]) every finite 2-transitive group is either a group of affine transformations of a finite vector space, or is an almost-simple group. Thus, Proposition 5.5 implies that every $(G,2)$-neighbour-transitive code satisfies precisely one of the conditions in Definition 5.11 below.

Definition 5.11. *Let $\mathcal{C}$ be a G-neighbour-transitive code in $H(\mathcal{N},q)$, let K be the kernel of the action of G on $\mathcal{N}$, and let $i \in \mathcal{N}$. Then precisely one of the following holds for $(\mathcal{C},G)$; i.e. $\mathcal{C}$ is*

1. G-entry-faithful *if G acts faithfully on $\mathcal{N}$, that is, $K = 1$;*

2. G-alphabet-almost-simple *if $K \neq 1$, G acts transitively on $\mathcal{N}$, and $G_i^{Q_i}$ is a 2-transitive almost-simple group; and*

3. G-alphabet-affine *if $K \neq 1$, G acts transitively on $\mathcal{N}$, and $G_i^{Q_i}$ is a 2-transitive affine group.*

We use descriptors such as G-entry-faithful for arbitrary code-group pairs $(\mathcal{C}, G)$ in Hamming graphs.

Codes that are G-entry-faithful and $(G, 2)$-neighbour-transitive were considered by the authors with Gillespie and Giudici [38].

Theorem 5.3 ([38, Theorem 1.1]). *Suppose that $\mathcal{C}$ is a code in $H(n, q)$, with $|\mathcal{C}| \geq 2$ and minimum distance $\delta \geq 5$. Then $\mathcal{C}$ is G-entry-faithful and $(G, 2)$-neighbour-transitive if and only if $\mathcal{C}$ is equivalent to either:*

1. *a binary repetition code with $\delta = n$, or,*

2. *the even weight subcode of the punctured Hadamard code of length 12, $G \cong \mathrm{M}_{11}$, and $\delta = 6$.*

5.5.1 Alphabet-Almost-Simple Codes in Hamming Graphs

It turns out that permutation codes (see [6,15] and Definition 5.12) provide useful building blocks for constructing a wide range of neighbour-transitive codes, many of which are alphabet-almost-simple, as defined in Definition 5.11.

Definition 5.12. *Let $\mathcal{N} = \mathcal{Q} = \{1, \ldots, q\}$ and let $T \subseteq \mathrm{Sym}(\mathcal{Q})$. Then, using the 'function' Definition 5.4(2) for the Hamming graph $H(\mathcal{N}, \mathcal{Q})$, the* permutation code *$\mathcal{C}(T)$ is the code comprising the elements of T considered as functions from $\mathcal{N}$ to $\mathcal{Q}$, that is to say, $t \in T$ corresponds to the codeword $\alpha_t : i \to i^t$ (for $i \in \mathcal{N}$). In terms of the 'q-tuple' Definition 5.4(1) for $H(\mathcal{N}, \mathcal{Q})$, the codeword α_t in $\mathcal{C}(T)$ corresponding to $t \in T$ is $\alpha_t = (1^t, 2^t, \ldots, q^t)$, that is to say, $\alpha_t(i) = i^t$.*

Here is a simple example of a permutation code in $H(4, 4)$.

Example 5.13. *Letting $T = \mathrm{A}_4$, the permutation code $\mathcal{C}(T)$ consists of the following 4-tuples:*

$$\begin{array}{cccc} (1,2,3,4) & (2,1,4,3) & (3,4,1,2) & (4,3,2,1) \\ (3,1,2,4) & (4,1,3,2) & (4,2,1,3) & (1,4,2,3) \\ (2,3,1,4) & (2,4,3,1) & (3,2,4,1) & (1,3,4,2). \end{array}$$

Next we identify some automorphisms of $\mathcal{C}(T)$ among the following elements of $\mathrm{Aut}(H(q, q)) = B \rtimes L$, by examining their actions on 'permutation' type vertices, that is, vertices with pairwise distinct entries, and hence of the form α_t for some $t \in \mathrm{Sym}(\mathcal{Q})$. For an element $g \in \mathrm{Sym}(\mathcal{Q})$, we write σ_g for the corresponding element of the top group $L = \mathrm{Sym}(\mathcal{N})$, we write $x_g = (g, g, \ldots, g)$ for the corresponding 'diagonal' element of the base group $B = \mathrm{Sym}(\mathcal{Q})^q$. The actions of these elements on $H(\mathcal{N}, \mathcal{Q})$ are given in (5.4) and (5.5). For any Hamming graph $H(\mathcal{N}, \mathcal{Q}) = H(n, q)$ and any subgroup $H \leq \mathrm{Sym}(\mathcal{Q})$, we define the *diagonal subgroup* $\mathrm{Diag}_n(H)$ of the base group $B = \mathrm{Sym}(\mathcal{Q})^n$ of $\mathrm{Aut}(H(n, q))$ by

$$\mathrm{Diag}_n(H) = \{x_g = (g, g, \ldots, g) \in B \mid g \in H\}. \tag{5.6}$$

Proofs of the following assertions are given in [41, Lemma 8], and are easily derived.

1. For $g \in \mathrm{Sym}(\mathcal{Q})$, the element σ_g maps $\alpha_t \in \mathcal{C}(T)$ to $\alpha_{tg^{-1}}$.

2. For $g \in \mathrm{Sym}(\mathcal{Q})$, the element x_g maps $\alpha_t \in \mathcal{C}(T)$ to α_{tg}.

3. For $g \in \mathrm{Sym}(\mathcal{Q})$, the product $x_g \sigma_g$ maps $\alpha_t \in \mathcal{C}(T)$ to $\alpha_{g^{-1}tg}$.

Now suppose that T is a subgroup of $\mathrm{Sym}(\mathcal{Q})$. Then choosing g to lie in T in parts (1) and (2) we see that $\mathrm{Aut}(\mathcal{C}(T))$ contains $\mathrm{Diag}_q(T) \times \{\sigma_g \mid g \in T\} \cong T \times T$. Moreover, choosing g in the normaliser $N_{\mathrm{Sym}(\mathcal{Q})}(T)$ in part (3), we obtain the larger subgroup $\mathrm{Diag}_q(T) \rtimes A(T)$ of $\mathrm{Aut}(\mathcal{C}(T))$, where $A(T) := \{x_g \sigma_g \mid g \in N_{\mathrm{Sym}(\mathcal{Q})}(T)\} \cong N_{\mathrm{Sym}(\mathcal{Q})}(T)$. This group is called the holomorph of T, [75, Section 3.3], and we note that $\mathrm{Diag}_q(T) \rtimes A(T) \leq \mathrm{Diag}_q(\mathrm{Sym}(q)) \rtimes L$ (recalling that $L = \mathrm{Sym}(q)$ is the top group).

Definition 5.13. *A code $\mathcal{C}$ in $H(q, q)$ is* diagonally s-neighbour-transitive *if $\mathcal{C}$ is (G, s)-neighbour-transitive for some $G \leq \mathrm{Diag}_n(\mathrm{Sym}(\mathcal{Q})) \rtimes L$. If $s = 1$ we say simply that $\mathcal{C}$ is diagonally neighbour-transitive.*

In [42], necessary and sufficient conditions on T were obtained for the code $\mathcal{C}(T)$ to be diagonally neighbour-transitive.

Theorem 5.4 ([42, Theorem 2]). *Let $\mathcal{N} = \mathcal{Q} = \{1, 2, \ldots, q\}$, let T be a subgroup of* $\mathrm{Sym}(\mathcal{Q})$, *and let $\mathcal{C}(T)$ be the permutation code as in Definition 5.12. Then $\mathcal{C}(T)$ is diagonally neighbour-transitive if and and only if $N_{\mathrm{Sym}(\mathcal{Q})}(T)$ is 2-transitive.*

Examples for which this 2-transitivity condition holds include elementary abelian groups T acting regularly, as well as almost simple 2-transitive groups. We note that all finite 2-transitive groups are known explicitly, see for example [22], and that this classification depends on the finite simple group classification. The diagonally neighbour-transitive codes $\mathcal{C}(T)$ are building blocks for constructing *frequency permutation arrays* (codes in $H(mq, q)$ where each element of $\mathcal{Q}$ occurs m times as an entry in each codeword; introduced in [57]). It was shown in [42, Theorem 2] that, for each diagonally neighbour-transitive permutation code $\mathcal{C}(T)$ and each positive integer m, the repetition code $\mathrm{Rep}_m(\mathcal{C}(T))$ (see Definiiton 5.5) is a diagonally neighbour-transitive frequency permutation array in $H(mq, q)$. In particular, when $N_{\mathrm{Sym}(\mathcal{Q})}(T)$ is an almost simple 2-transitive group, all of these diagonally neighbour-transitive codes, $\mathcal{C}(T)$ and $\mathrm{Rep}_m(\mathcal{C}(T))$, are alphabet-almost-simple as in Definition 5.11. In fact, these may be regarded as archetypical examples for alphabet-almost-simple codes with minimum distance at least three. In the following result, part (1) was proved first in [41, Section 7] and more succinctly in [39, Proposition 3.3]; whereas part (2) was proved in [41, Theorem 1.1].

Theorem 5.5. *Let $\mathcal{C}$ be a code in $H(\mathcal{N},\mathcal{Q}) = H(n,q)$ which is G-alphabet-almost-simple and G-neighbour-transitive with $\delta \geq 3$. Then*

1. *there is a G-invariant partition $\mathcal{J}$ of $\mathcal{N}$ such that, for $J \in \mathcal{J}$, the projection $\pi_J(\mathcal{C})$ is equivalent to a diagonally $\chi_J(G)$-neighbour-transitive code with minimum distance $\delta(\pi_J(\mathcal{C})) \geq 2$;*

2. *$\mathcal{C}$ has a sub-code $\mathcal{S}$ which is a neighbour-transitive frequency permutation array, and $\mathcal{C}$ is a disjoint union of G-images of $\mathcal{S}$.*

We make some comments about the links between the two parts of Theorem 5.5, and the structure of the sub-code $\mathcal{S}$ in part (2).

Remark 5.1. *(a) It follows from [42, Theorems 1 and 3] that the projected codes $\pi_J(\mathcal{C})$ in part (1) of Theorem 5.5 are either frequency permutation arrays or repetition codes.*

b. Some explicit information about the possibilities for the neighbour-transitive sub-code $\mathcal{S}$ in part (2) of Theorem 5.5 is provided by [41, Theorem 1.1] as follows: $\mathcal{S}$ is equivalent to a code of the form $\mathrm{Prod}_\ell(\mathrm{Rep}_k(\mathcal{C}'))$ (with $\ell, k \geq 1$), where Prod_ℓ and Rep_k are the product and repetition constructions defined in Definition 5.5. Moreover, the small code $\mathcal{C}'$ itself has one of three specific forms described in [41, (7,4), (7.5), (7.6)], namely the trivial code $\mathcal{Q}$ of length 1, or a permutation code $\mathcal{C}(T)$ as in Definition 5.12, or a twisted version $\mathcal{C}(T, T^\sigma)$ of a permutation code defined and studied in [3,44].

c. The code $\mathcal{C}$ in Theorem 5.5 may have minimum distance strictly larger than that of the projected codes $\pi_J(\mathcal{C})$ in part (1). An insightful example was given in [41, Example 9.1], where $\mathcal{S} = \mathrm{Prod}_\ell(\mathcal{C}(A_q))$, and $\mathcal{C}$ is a disjoint union of two G-translates of $\mathcal{S}$. Each of the codes $\mathcal{C}$ and $\mathcal{S}$ has minimum distance 3, while $\pi_J(\mathcal{C})$ is the permutation code $\mathcal{C}(S_q)$ with minimum distance 2.

We give some explicit constructions of alphabet-almost-simple neighbour-transitive codes, the first built from permutation codes.

Example 5.14. *Let $\mathcal{N} = \mathcal{Q} = \{1, \ldots, q\}$, and let $T \leq \mathrm{S}_q$ be such that the permutation code $\mathcal{C} = \mathcal{C}(T)$ in $H(q,q)$ is diagonally neighbour-transitive (see Theorem 5.4). Then, letting $\mathcal{M} = \mathcal{N} \times \{1, \ldots, k\}$, as in Definition 5.5, the product code $\mathrm{Prod}_k(\mathcal{C})$ and the repetition code $\mathrm{Rep}_k(\mathcal{C})$ are neighbour-transitive codes in $H(\mathcal{M},\mathcal{Q})$. Moreover, for each of these codes, and for each $i = 1, \ldots, k$, the projection with respect to $\mathcal{N} \times \{i\}$ is equal to $\mathcal{C}(T)$. The following codes are additional examples of codes that again project to $\mathcal{C}(T)$ with respect to $\mathcal{N} \times \{i\}$, and properly lie between $\mathrm{Rep}_k(\mathcal{C})$*

and $\mathrm{Prod}_k(\mathcal{C})$. *Recall from Definition 5.5 that if* $g_1, \dots, g_k$ *are functions* $\mathcal{N} \to \mathcal{Q}$ *then* $\beta_{g_1,\dots,g_k}$ *is a vertex of* $H(\mathcal{M}, \mathcal{Q})$, *and note that if* $g \in T$ *then* g *is a function* $\mathcal{Q} \to \mathcal{Q}$.

1. *Let* $T = \mathrm{S}_q$ *and consider the code consisting of all* $\beta_{t_1,\dots,t_k}$ *such that* $t_1, \dots, t_k \in T$ *and* $t_1 t_2 \cdots t_k \in \mathrm{A}_n$. *(If* k *is even then this is one of the codes corresponding to the 'elusive' code* $\mathcal{C}(q, k)$ *in [55, Lemma 3.9], from which it follows that* $\mathrm{Prod}_k(\mathcal{C})$ *is neighbour-transitive.)*

2. *Let* $1 \neq H \lhd T$ *and let*
$$\mathrm{Prod}(T, k, H) = \{\beta_{h_1 t,\dots,h_k t} \mid h_1, \dots, h_k \in H, t \in T\}.$$
We state below that $\mathrm{Prod}(T, k, H)$ *is neighbour-transitive as long as* $N_{\mathrm{S}_q}(H)$ *is 2-transitive. Note that several explicit examples are given, and precise requirements on* H *and* T *related to the neighbour-transitivity of* $\mathrm{Prod}(T, k, H)$ *are discussed further, in [54].*

Proposition 5.6 ([54]). *Let* k *be a positive integer, let* $T \leq \mathrm{S}_q$ *and let* $1 \neq H \lhd T$ *such that* $N_{\mathrm{S}_q}(H)$ *is 2-transitive. Then the code* $\mathrm{Prod}(T, k, H)$ *in Example 5.14(2) is neighbour-transitive.*

Problem 5.5. *Investigate further the codes* $\mathrm{Prod}(T, k, H)$ *in Example 5.14.*

The second construction is a general version of the twisted permutation codes $\mathcal{C}(T, T^{\sigma})$ from Remark 5.1(b). Not all of them will arise as the building block $\mathcal{C}'$ for alphabet-almost-simple codes. More details of this construction are available in [3,44].

Definition 5.14. *Let* T *be a group with (not necessarily distinct) permutation representations* $\rho_1, \dots, \rho_k : G \to \mathrm{S}_q$, *let* $\mathcal{Q} = \{1, \dots, q\}$ *and let* $\mathcal{N} = \mathcal{Q} \times \{1, \dots, k\}$. *We define the* twisted permutation code $\mathcal{C}(T; \rho_1, \dots, \rho_k)$ *in* $H(kq, q)$ *as the code consisting of the functions* $\mathcal{N} \to \mathcal{Q}$ *given by* $(i, j) \mapsto i^{\rho_j(t)}$, *for each* $t \in T$.

One benefit of a twisted permutation code $\mathcal{C}(T; \rho_1, \dots, \rho_k)$ in $H(kq, q)$ over the usual repetition code $\mathrm{Rep}_k(\mathcal{C}(T))$ in the same graph $H(kq, q)$ is that the minimum distance of the twisted version may be greater than for the untwisted code $\mathrm{Rep}_k(\mathcal{C}(T))$. Some examples are given in Table 5.1. While it is only the examples in Lines 1–3 which arise in connection with alphabet-almost-simple codes, examples in the other lines, especially Lines 5–6 demonstrate that the differences between the minimum distances of the twisted and untwisted codes can be unbounded. The groups G_t in line 6 are defined and studied in [3, Section 3.1]; G_t is a certain subgroup of $\mathrm{AGL}_t(p)$, it contains the translation group and has order p^{t+1}.

Among this diverse family of alphabet-almost-simple, neighbour-transitive codes, very few are 2-neighbour-transitive. In fact we have the following classification of such codes.

Theorem 5.6 ([39, Theorem 1.1]). *Suppose that* $\mathcal{C}$ *is an alphabet-almost-simple, 2-neighbour-transitive code in* $H(n, q)$ *with minimum distance* $\delta \geq 3$. *Then* $n = \delta = 3$, $q \geq 5$, *and* $\mathcal{C}$ *is equivalent to the repetition code* $\mathrm{Rep}_3(q)$.

TABLE 5.1 Groups T Giving Twisted Permutation Codes in $H(kq, q)$ with Minimum Distance δ_{tw} Strictly Greater than the Minimum Distance δ_{rep} of the k-Fold Repetition of $\mathcal{C}(T)$

Line	T	q	k	δ_{tw}	δ_{rep}
1	S_6	6	2	8	4
2	A_6	6	2	8	6
3	$\mathrm{ASL}_3(2)$	8	2	12	8
4	S_6	60	4	≤ 224	176–192
5	$\mathrm{Sp}_4(2^n)$	$2^{3n}+2^{2n}+2^n+1$	2	$2^{3n+1}+2^{2n}$	2^{3n+1}
6	G_t	p^t	p	$p^{t+1}-p$	$p^{t+1}-p^2$

Lines 1–4 can be found in [44] while lines 5 and 6 are in [3].

5.5.2 Alphabet-Affine Codes in Hamming Graphs

In this section, we consider codes that are alphabet-affine (Definition 5.11), that is, codes $H(n, q)$ having an automorphism group not acting faithfully on entries, and giving rise to a 2-transitive affine group in the action induced on the alphabet. A natural family of examples of such codes are the cyclic codes: a code $\mathcal{C}$ in $H(\mathcal{N}, \mathbb{F}_q)$

$$\text{is called } \textit{cyclic} \text{ if } \mathcal{C} \text{ is linear and there exists an } n\text{-cycle } \sigma \in L \cap \mathrm{Aut}(\mathcal{C}). \tag{5.7}$$

In particular each cyclic code $\mathcal{C}$ is alphabet-affine since $\mathrm{Aut}(\mathcal{C})$ contains the subgroup of translations by elements of $\mathcal{C}$ (acting transitively on $\mathcal{C}$). Moreover, a cyclic code is neighbour-transitive since $\mathrm{Aut}(\mathcal{C})$ also contains both the subgroup of scalars and an n-cycle from the top group L. On the other hand, if $x = h\sigma$ with $h \in B$ and $\sigma \in L$, then $x \in \mathrm{Aut}(\mathcal{C})$ does not necessarily imply that $\sigma \in \mathrm{Aut}(\mathcal{C})$. That is to say, a linear code $\mathcal{C}$ may not be cyclic even if the induced group $\mathrm{Aut}(\mathcal{C})^{\mathcal{N}}$ on entries contains an n-cycle. We exhibit a small non-cyclic linear code with this property in Example 5.15. In fact, as we see below, the class of codes that are alphabet-affine and neighbour-transitive is strictly larger than the class of cyclic codes.

Proposition 5.7. *Let $\mathcal{C}$ be a linear code with minimum distance $\delta \geq 3$ in $H(\mathcal{N}, \mathbb{F}_q)$. Then $\mathcal{C}$ is neighbour-transitive if and only if $\mathrm{Aut}(\mathcal{C})^{\mathcal{N}}$ is transitive. In particular, if $\mathcal{C}$ is a cyclic code then $\mathcal{C}$ is neighbour-transitive.*

Proof. If $\mathcal{C}$ is neighbour-transitive, then, by Proposition 5.3, $\mathrm{Aut}(\mathcal{C})^{\mathcal{N}}$ is transitive. Suppose now that $\mathrm{Aut}(\mathcal{C})^{\mathcal{N}}$ is transitive. Since $\mathcal{C}$ is linear, $\mathrm{Aut}(\mathcal{C})$ contains $T_{\mathcal{C}}$, which acts transitively on $\mathcal{C}$. Moreover, this implies that $\mathrm{Aut}(\mathcal{C}) = T_{\mathcal{C}}.\mathrm{Aut}(\mathcal{C})_{\mathbf{0}}$. In particular, $\mathrm{Aut}(\mathcal{C})_{\mathbf{0}}^{\mathcal{N}}$ is transitive. For each $i \in \mathcal{N}$, let $e_i : \mathcal{N} \to \mathbb{F}_q$ such that $e_i : i \to 1$ and $e_i : k \to 0$ if $k \neq i$. Then $\Gamma_1(\mathbf{0}) = \{ae_i \mid a \in \mathbb{F}_q^\times\}$. Let $i, j \in \mathcal{N}$ and $a_i, a_j \in \mathbb{F}_q^\times$ so that $a_ie_i, a_je_j \in \Gamma_1(\mathbf{0})$. Then, since $\mathrm{Aut}(\mathcal{C})_{\mathbf{0}}$ induces a transitive action on $\mathcal{N}$, there exists an $x \in \mathrm{Aut}(\mathcal{C})_{\mathbf{0}}$ such that $(a_ie_i)^x = be_j$, for some $b \in \mathbb{F}_q^\times$. Using the linearity of $\mathcal{C}$ again, scalar multiplication by a_jb^{-1} gives an element of $\mathrm{Aut}(\mathcal{C})_{\mathbf{0}}$, and hence we can map a_ie_i to a_je_j. It then follows from Lemma 5.2 that $\mathcal{C}$ is neighbour-transitive. The last sentence holds since $\mathcal{C}$ being cyclic implies that $\mathrm{Aut}(\mathcal{C})^{\mathcal{N}}$ is transitive. □

Example 5.15. *Consider the linear code $\mathcal{C}$ in $H(4, \mathbb{F}_3)$ consisting of the codeword $(0,0,0,0)$ as well as the following 8 non-zero codewords:*

$$\begin{array}{cc} (0,1,1,1) & (0,2,2,2) \\ (1,0,1,2) & (2,0,2,1) \\ (2,1,0,2) & (1,2,0,1) \\ (2,2,1,0) & (1,1,2,0). \end{array}$$

The automorphism $x : (a,b,c,d) \mapsto (d,a,b,2c)$ of $\mathcal{C}$ fixes $(0,0,0,0)$ and cycles through the 8 non-zero codewords of $\mathcal{C}$; x has order 8 and projects to the 4-cycle $(1\,2\,3\,4)$ in the top group $L \cong \mathrm{S}_4$. However $\mathcal{C}$ is not cyclic, for if σ is a 4-cycle in L then, since the only codeword consisting of one 0 and three 1s is $(0,1,1,1)$, the image $(0,1,1,1)^\sigma$ cannot lie in $\mathcal{C}$. On the other hand, $\mathcal{C}$ has minimum distance $\delta = 3$ and satisfies all the hypotheses of Proposition 5.7, so it is neighbour-transitive. Moreover, $\mathcal{C}$ is not equivalent to any cyclic code (one way to see this is to observe that the subspace spanned by $\{\alpha^{\sigma^k} \mid k = 0,1,2,3\}$ contains a weight 2 vector, for each of $\alpha = (0,1,1,1), (0,1,1,2), (0,1,2,1)$ and where $\sigma = (1\,2\,3\,4)$). Note also that $\mathcal{C}$ is the projective Reed–Muller code $\mathcal{PRM}_3(1,2)$, which we meet later on in Definition 5.16.

The next result suggests investigating the submodule structure of certain modules, which we consider further in subsequent sections. For a group G and a prime p, $O_p(G)$ denotes the largest normal p-subgroup of G.

Proposition 5.8. *[56, Proposition 3.5] Let $\mathcal{C}$ be a code in the Hamming graph $H(n,q)$, with $q = p^d$ for a prime p, such that $\mathcal{C}$ is G-alphabet-affine and $(G,2)$-neighbour-transitive, with $\delta \geq 5$, and suppose that $\underline{0} \in C$. Then C contains a subcode $\mathcal{S}$ such that $\mathcal{S}$ is the code formed by the orbit of $\underline{0}$ under $O_p(K)$, where $K = G \cap B$. Moreover, it follows that:*

1. *$\mathcal{S}$ is a block of imprimitivity for the action of G on $\mathcal{C}$, and $G_{\mathcal{S}} = O_p(K) \rtimes G_{\mathbf{0}}$,*
2. *$\mathcal{S}$ is $G_{\mathcal{S}}$-alphabet-affine and $(G_{\mathcal{S}},2)$-neighbour-transitive with minimum distance $\delta_{\mathcal{S}} \geq \delta$,*
3. *$\mathcal{S}$ is an $\mathbb{F}_p G_{\mathbf{0}}$-module, and if $\mathcal{S} \neq \mathrm{Rep}_n(2)$ then q^2 divides $|\mathcal{S}|$.*

The following result gives further, fairly strong, restrictions on the structure of the automorphism group of a 2-neighbour-transitive code.

Lemma 5.7 ([53]). *Let $\mathcal{C}$ be a code in $H(n,q) = H(\mathcal{N}, \mathcal{Q})$ that is $(G,2)$-neighbour-transitive with minimum distance $\delta \geq 5$.*

1. *If in addition $\mathcal{C}$ is G-alphabet-affine, then $G_{\mathbf{0},i}^{\mathcal{Q}_i^\times} \leq \Gamma\mathrm{L}_1(q)$.*
2. *Alternatively, if $\mathbf{0} \in \mathcal{C}$ and $K = B \cap G$, then $K_{\mathbf{0}} \cong \mathrm{Diag}_n(H)$, where H acts semi-regularly on $\mathcal{Q}_i^\times$ for all $i \in \mathcal{N}$.*

5.5.3 Codes in Binary Hamming Graphs

When we restrict the alphabet $\mathcal{Q}$ to be the field $\mathbb{F}_2$ of order 2, then we obtain very tight descriptions of the possibilities for (G, s)-neighbour-transitive codes with minimal distance not too small. Recall from Proposition 5.3 that for such codes, the stabiliser in G of a codeword induces an s-homogeneous action on entries. Also, for a code $\mathcal{C} \leq \mathbb{F}_2^n$, $T_{\mathcal{C}}$ denotes the set of all translations by elements of $\mathcal{C}$; if $T_{\mathcal{C}}$ is a subgroup of $\mathrm{Aut}(\mathcal{C})$ then $\mathcal{C}$ is additively closed. The next result gives a nearly complete description of the binary 2-neighbour-transitive codes with δ at least 5.

Theorem 5.7 ([56, Theorem 1.2]). *Let $\mathcal{C}$ be a code in $H(n, 2)$ with minimum distance $\delta \geq 5$. Then $\mathcal{C}$ is 2-neighbour-transitive if and only if one of the following holds:*

1. *$\mathcal{C}$ is the binary repetition code $\mathrm{Rep}_n(2)$ with $\delta = n$.*
2. *$\mathcal{C}$ is one of the following codes (see [38, Definition 4.1]):*
   a. *the Hadamard code with $n = 12$ and $\delta = 6$;*
   b. *the punctured Hadamard code with $n = 11$ and $\delta = 5$;*
   c. *the even weight subcode of the punctured Hadamard code with $n = 11$ and $\delta = 6$.*
3. *There exists a linear subcode $\mathcal{S} \subseteq \mathcal{C}$ with dimension k and minimum distance δ, and a subgroup $G_{\mathbf{0}} \leq Aut(\mathcal{C})_{\mathbf{0}}$, where $\mathcal{S}$, $G_{\mathbf{0}}$, n, δ and k are as in Table 5.2, such that*
   a. *$\mathcal{S}$ is $(G, 2)$-neighbour-transitive, where $G = T_{\mathcal{S}} \rtimes G_{\mathbf{0}} \leq Aut(\mathcal{C})$, and,*
   b. *$\mathcal{C}$ is the union of a set Δ of cosets of $\mathcal{S}$, and $Aut(\mathcal{C})$ acts transitively on Δ.*

Remark 5.2. *Table 5.2 gives the possibilities for the linear subcode $\mathcal{S}$ of $\mathcal{C}$ in Theorem 5.7(3)(a). For each line of Table 5.2, the code $\mathcal{C}$ in Theorem 5.7(3)(b) is identified in the relevant part of the proof of [56, Theorem 4.5]. We note that the minimum distance δ of $\mathcal{C}$ satisfies $5 \leq \delta < n$ in all lines except possibly line 9. In both of the lines 8 and 9 of Table 5.2 we have $\mathrm{soc}(G_{\mathbf{0}}) = \mathrm{PSU}_3(r)$, and it follows from the proof of [56, Theorem 4.5] that $\mathcal{S}$ is self-orthogonal (that is, $\mathcal{S} \subseteq \mathcal{S}^{\perp}$) if $r \equiv 3 \pmod 4$ but not if $r \equiv 1 \pmod 4$. Self-orthogonality ensures that only the example in line 8 of a minimal $(G, 2)$-neighbour-transitive code arises when $r \equiv 3 \pmod 4$, while for $r \equiv 1 \pmod 4$ we have two different examples, one for each of lines 8 and 9 of Table 5.2. It was proved, see [56, Remark 1.3], that the codes in line 9 have minimum distance $\delta \geq 4$, and it is an open problem to determine precisely which values of $r \equiv 1 \pmod 4$ correspond to a code with $\delta \geq 5$.*

In Theorem 5.7(3), the linear subcode $\mathcal{S}$ (see Proposition 5.8) is a submodule of the permutation module over $\mathbb{F}_2$ of a 2-homogeneous permutation group. The next lemma explores the properties of such subcodes.

TABLE 5.2 Parameters for the $(G, 2)$-Neighbour-Transitive Code $\mathcal{S}$ in Theorem 5.7(3)

Line	soc(G_0)	n	δ	k	Conditions
1	$\mathbb{Z}_p^d$	$r = p^d$	$\geq (r-1)^{1/2} + 1$	$\frac{r-1}{2}$	$23 \leq r \equiv 7 \pmod 8$ 2-hom. not 2-trans.
2	$\mathbb{Z}_2^t$	2^t	2^{t-1}	$t+1$	$t \geq 4$, 2-trans.
3	$\mathrm{PSL}_t(2^a)$	$\frac{2^{at}-1}{2^a-1}$	$\geq \frac{2^{a(t-1)}-1}{2^a-1} + 1$	t^a	$t \geq 3$, $(a,t) \neq (1,3)$
4	A_7	15	8	4	–
5	$\mathrm{PSL}_2(r)$	$r+1$	$\geq r^{1/2} + 1$	$\frac{r+1}{2}$	$23 \leq r \equiv \pm 1 \pmod 8$ not 3-trans.
6	$\mathrm{Sp}_{2t}(2)$	$2^{2t-1} - 2^{t-1}$	$2^{2t-2} - 2^{t-1}$	$2t+1$	$t \geq 3$
7	$\mathrm{Sp}_{2t}(2)$	$2^{2t-1} + 2^{t-1}$	2^{2t-2}	$2t+1$	$t \geq 3$
8	$\mathrm{PSU}_3(r)$	r^3+1	$\geq r^2+1$	$r^2 - r + 1$	r is odd
9	$\mathrm{PSU}_3(r)$	r^3+1	≥ 4	$r^3 - r^2 + r$	$r \equiv 1 \pmod 4$
10	$\mathrm{Ree}(r)$	r^3+1	$\geq r^2+1$	$r^2 - r + 1$	$r \geq 3$
11	M_{22}	22	8	10	–
12	M_{23}	23	8	11	–
13	M_{24}	24	8	12	–
14	HS	176	≥ 50	21	–
15	Co_3	276	100	23	–

See Remark 5.2 for more information.

Lemma 5.8 ([56, Lemma 4.3]). *Let H act 2-homogeneously on a set $\mathcal{N}$ of size $n \geq 5$, let $V \cong \mathbb{F}_2^n$ be the permutation module for the action of H on $\mathcal{N}$, and let $\mathcal{C}$ be a proper $\mathbb{F}_2H$-submodule of V. Then $\mathcal{C}$ is a code in $H(\mathcal{N}, \mathbb{F}_2)$ with minimum distance δ and the group $G := T_{\mathcal{C}} \rtimes H \leq \mathrm{Aut}(\mathcal{C})$, where precisely one of the following holds:*

1. *$\mathcal{C}, \delta, G$ satisfy one of the lines of Table 5.3;*
2. *$\delta = 3$, $\mathcal{C}$ is a perfect code in $H(\mathcal{N}, \mathbb{F}_2)$, and $\mathcal{C}$ is G-neighbour-transitive; or*
3. *$4 \leq \delta < n$ and $\mathcal{C}$ is $(G, 2)$-neighbour-transitive.*

Every perfect code with minimum distance 3 in $H(n, 2)$ has the same parameters as a Hamming code, see [90]. Thus, in a sense, Lemma 5.8 part (2) is well understood, as is part (1). In addition, those codes in Lemma 5.8 part (3) having minimum distance at least 5 are described in Theorem 5.7. Thus, the codes in Lemma 5.8 for which least is known are those in part (3) with $\delta = 4$. These are the subject of Problem 5.6.

TABLE 5.3 Codes Arising in Lemma 5.8(1) and Their Properties

$\mathcal{C}$	δ	Properties
$\mathrm{Rep}_n(2)$	n	$(G, 2)$-neighbour-transitive
$\mathrm{Rep}_n(2)^\perp$	2	G-completely transitive

Problem 5.6. *Determine all codes with minimum distance $\delta = 4$ corresponding to submodules of the permutation module over $\mathbb{F}_2$ of a 2-homogeneous permutation group.*

Note that a completely transitive code $\mathcal{C}$ in $H(n,2) = H(\mathcal{N},2)$ with minimum distance $\delta \geq 5$ has covering radius $\rho \geq 2$, and hence is 2-neighbour-transitive so that Theorem 5.7 may be applied. In particular, if $\mathcal{C} \neq \mathrm{Rep}_n(2)$, then the induced subgroup $S \cong \mathrm{soc}(\mathrm{Aut}(\mathcal{C})_{\mathbf{0}}^{\mathcal{N}})$ is as in Theorem 5.7(2) or (3), so S is either a small Mathieu group $\mathrm{M}_{11}, \mathrm{M}_{12}$, or S is one of the groups in the column 'soc($G_{\mathbf{0}}$)' of Table 5.2. This observation was exploited in [7] to obtain a partial classification, stated in Theorem 5.8, of binary completely transitive codes with minimum distance at least 5.

Theorem 5.8 ([7, Theorem 1.3]). *Let $\mathcal{C}$ be a non-trivial completely transitive code in $H(n,2)$ with minimum distance $\delta \geq 5$, and suppose that the socle S of the group induced by* $\mathrm{Aut}(\mathcal{C})_{\mathbf{0}}$ *on $\mathcal{N}$ is as in one of the lines of Table* 5.4. *Then $\mathcal{C}$ is equivalent to one of the codes in Table* 5.5. *Moreover, each code in Table 5.5 is completely transitive.*

Note that, of the groups S occurring in Theorem 5.7(2) or (3), the only ones not considered in Theorem 5.8 are the following groups 'soc($G_{\mathbf{0}}$)' of Table 5.2: i) $\mathrm{PSL}_t(2^a)$ unless $(t,a) = (3,2)$, (ii) $\mathrm{Sp}_{2t}(2)$ for $t \geq 3$, (iii) $\mathbb{Z}_2^t$ for $t \geq 3$, or iv) $\mathbb{Z}_p^d$ where $r = p^d \equiv 7 \pmod 8$. Thus the following problem is still open.

Problem 5.7. *Classify the binary completely transitive codes $\mathcal{C}$ in $H(n,2)$ with minimum distance at least* 5, *for which* $\mathrm{soc}(\mathrm{Aut}(\mathcal{C})_{\mathbf{0}}^{\mathcal{N}})$ *is one of i)* $\mathrm{PSL}_t(2^a)$ *with $(t,a) \neq (3,2)$, ii)* $\mathrm{Sp}_{2t}(2)$ *for $t \geq 3$, iii) $\mathbb{Z}_2^t$ for $t \geq 3$, or iv) $\mathbb{Z}_p^d$ where $23 \leq n = p^d \equiv 7 \pmod 8$.*

5.5.4 A Non-Existence Result: Proof of Theorem 5.2

We now have enough information about G-alphabet-affine and (G,s)-neighbour-transitive codes to prove Theorem 5.2 which was stated in Section 5.3.

TABLE 5.4 Some Groups S for Which There Exists a Non-Trivial Binary 2-Neighbour-Transitive Code $\mathcal{C}$ in $H(n,2)$ such that $\mathcal{C} \neq \mathrm{Rep}_n(2)$ and $S \cong \mathrm{soc}(\mathrm{Aut}(\mathcal{C})_{\mathbf{0}}^{\mathcal{N}})$

S	n	**Conditions**
$\mathrm{PSL}_3(4)$	21	–
A_7	15	–
$\mathrm{PSL}_2(r)$	$r+1$	$23 \leq r \equiv \pm 1 \pmod 8$
$\mathrm{PSU}_3(r)$	r^3+1	r is odd
$\mathrm{Ree}(r)$	r^3+1	$r \geq 3$
M_m	$11, 12, 22, 23, 24$	–
HS	176	–
Co_3	276	–

TABLE 5.5 Non-Trivial Binary Completely Transitive Codes $\mathcal{C}$ with Minimum Distance $\delta \geq 5$ and Where $\mathrm{Aut}(\mathcal{C})_{\mathbf{0}}^{\mathcal{N}}$ Has as Socle One of the Groups S in Table 5.4

Line	$\mathcal{C}$	$\mathrm{Aut}(\mathcal{C})$	**Parameters**
1	$\mathcal{H}$	$2\mathrm{M}_{12}$	$(12, 24, 6; 3)$
2	$\mathcal{PH}$	$2 \rtimes \mathrm{M}_{11}$	$(11, 24, 5; 3)$
3	$\mathcal{NR}$	$2^5 \rtimes \mathrm{A}_8$	$(15, 256, 5; 3)$
4	$\langle \mathcal{L}, \Delta_1 \rangle \cup \langle \mathcal{L}, \Delta_2 \rangle$	$T_{\mathcal{L}} \rtimes \mathrm{P\Gamma L}_3(4)$	$(21, 2^{10} \cdot 3, 5; 6)$
5	$\mathcal{P}^{\perp}$	$T_{\mathcal{C}} \rtimes \mathrm{P\Gamma L}_3(4)$	$[21, 12, 5; 3]$
6	$\langle \mathcal{L}, \Delta_1 \rangle$	$T_{\mathcal{C}} \rtimes \mathrm{P\Sigma L}_3(4)$	$[21, 11, 5; 6]$
7	$\mathcal{L}$	$T_{\mathcal{C}} \rtimes \mathrm{P\Gamma L}_3(4)$	$[21, 10, 5; 6]$
8	$\mathcal{G}_{24}$	$T_{\mathcal{C}} \rtimes \mathrm{M}_{24}$	$[24, 12, 8; 4]$
9	$\mathcal{G}_{23}$	$T_{\mathcal{C}} \rtimes \mathrm{M}_{23}$	$[23, 12, 7; 3]$
10	$\mathcal{G}_{23}^{\perp}$	$T_{\mathcal{C}} \rtimes \mathrm{M}_{23}$	$[23, 11, 8; 7]$
11	$\mathcal{G}_{22}$	$T_{\mathcal{C}} \rtimes (\mathrm{M}_{22} : 2)$	$[22, 12, 6; 3]$
12	$\mathcal{EG}_{22}$	$T_{\mathcal{C}} \rtimes (\mathrm{M}_{22} : 2)$	$[22, 11, 6; 7]$
13	$\mathcal{SG}_{22}$	$T_{\mathcal{C}} \rtimes \mathrm{M}_{22}$	$[22, 11, 7; 6]$

See [7, Section 3] for the definitions of these codes. The codes in Lines 1–4 are non-linear with parameters $(n, |\mathcal{C}|, \delta; \rho)$, while the remaining codes are linear with parameters $[n, k, \delta; \rho]$, where ρ is the covering radius of $\mathcal{C}$ and k is the dimension.

Proof of Theorem 5.2. If $n \geq 9$ and $\mathcal{C} = \mathrm{Rep}_n(2)$ in $H(n, 2)$ then, by Theorem 5.3, $\mathcal{C}$ is completely transitive with minimum distance $\delta = n$, and hence error-correction capacity $e = \lfloor \frac{n-1}{2} \rfloor$. Since $\Gamma_i(\mathbf{0})$ consists precisely of the weight i vertices of $H(n, 2)$ and $\Gamma_i(\mathbf{0})$ is contained in $\mathcal{C}_i$ if and only if $i \leq n/2$, the largest value for s for which $\mathcal{C}$ is s-neighbour-transitive is $s = \lceil \frac{n-1}{2} \rceil$. Thus $\min\{e, s\} \geq 4$ and all the conditions of Theorem 5.2 hold.

Suppose from now on that $\mathcal{C}$ is an s-neighbour-transitive code in $H(n, q)$ with error capacity e such that $\min\{e, s\} \geq 4$, and that $\mathcal{C} \neq \mathrm{Rep}_n(2)$. Let $G \leq \mathrm{Aut}(\mathcal{C})$ such that $\mathcal{C}$ is (G, s)-neighbour-transitive. Since $e \geq 4$ we have $\delta \geq 9$, which also implies that $n \geq 9$. It now follows from Theorem 5.3 that, if $\mathcal{C}$ is G-entry-faithful then $\mathcal{C}$ is equivalent to $\mathrm{Rep}_n(2)$. Let us assume now that $\mathcal{C}$ is not G-entry-faithful, and that $\mathcal{C}$ is not equivalent to $\mathrm{Rep}_n(2)$. Since $\delta \geq 9$, it follows from Theorem 5.6 that $\mathcal{C}$ is not G-alphabet-almost-simple and thus, by Proposition 5.5 and Definition 5.11, $\mathcal{C}$ is G-alphabet-affine.

Replacing $\mathcal{C}$ by an equivalent code we may assume that $\underline{0} \in \mathcal{C}$. By Proposition 5.3, $G_{\mathbf{0}}$ acts 4-homogeneously on $\mathcal{N}$. Hence, by [22, Table 7.4] and [60], $G_{\mathbf{0}}^{\mathcal{N}}$ is one of the groups in Table 5.6 of degree n. Also, by Proposition 5.8, there exists a subcode $\mathcal{S}$ of $\mathcal{C}$ such that $\mathcal{S}$ is an $\mathbb{F}_p G_{\mathbf{0}}$-module. If $q = 2$, then Theorem 5.7 eliminates each possibility for $G_{\mathbf{0}}^{\mathcal{N}}$ (recalling where necessary that $\delta \geq 9$). Thus $q \geq 3$.

Let $I \subseteq \mathcal{N}$ with $|I| = 4$. Now $\mathcal{C}$ is $(G, 4)$-neighbour-transitive (as $s \geq 4$), so $G_{\mathbf{0}}$ is transitive on the set of all weight 4 vertices of $H(n, q)$. Also, as $G_{\mathbf{0}}$ is 4-homogeneous on $\mathcal{N}$, the setwise stabiliser $G_{\mathbf{0},I}$ acts transitively on the set of

TABLE 5.6 Four-Homogeneous Groups for the Proof of Theorem 5.2

$G_{\mathbf{0}}^{\mathcal{N}}$	n	Sol	$q{\geq}3$
A_n	n	$2^3 \cdot 3$	3
S_n	n	$2^4 \cdot 3$	3
M_{11}	11	$2^3 \cdot 3$	3
M_{12}	12	$2^6 \cdot 3$	3 or 5
M_{23}	23	$2^7 \cdot 3^2$	3 or 5
M_{24}	24	$2^9 \cdot 3^2$	3, 5 or 9
$PSL_2(8)$	9	$2^2 \cdot 3$	–
$P\Gamma L_2(8)$	9	$2^2 \cdot 3$	–
$P\Gamma L_2(32)$	33	2^2	–

weight 4 vertices having support I. Hence $(q-1)^4$ divides the order of $G_{\mathbf{0},I}^{H(I,q)}$. Now by Lemma 5.7(1) the induced group $G_{\mathbf{0},I}^{H(I,q)}$ is a subgroup of $\Gamma L_1(q) \wr S_4$, and in particular is soluble. Also, by Lemma 5.7(2), the group $K_{\mathbf{0}} := B \cap G_{\mathbf{0}}$ has order dividing $q-1$ and $K_{\mathbf{0}} \cong K_{\mathbf{0}}^{H(I,q)}$. By the definition of $K_{\mathbf{0}}$ we have $G_{\mathbf{0,I}}^{\mathcal{N}} \cong G_{\mathbf{0,I}}/K_{\mathbf{0}}$, and hence $G_{\mathbf{0},I}^{H(I,q)}/K_{\mathbf{0}}^{H(I,q)}$ is a soluble quotient of $G_{\mathbf{0},I}^{\mathcal{N}}$ with order divisible by $(q-1)^3$.

Recall that $n \geq 9$, and let Sol be the order of the largest soluble normal subgroup of $G_{\mathbf{0},I}^{\mathcal{N}}$. For example, if $G_{\mathbf{0}}^{\mathcal{N}} = S_n$ then $G_{\mathbf{0},I}^{\mathcal{N}} = S_4 \times S_{n-4}$ so Sol $= 2^4 \cdot 3$. Since $(q-1)^3$ divides Sol it follows that the possibilities for q are as in Table 5.6. In particular, the last three lines of Table 5.6 are ruled out as there are no possibilities for q. If q is 3 or 5, then $K_{\mathbf{0}}$ consists of scalars and the subcode $\mathcal{S}$ of $\mathcal{C}$ mentioned above is an $\mathbb{F}_q G_{\mathbf{0}}^{\mathcal{N}}$-module, of dimension k, say. In particular $\mathcal{S}$ is a linear $[n, k, \delta']$ code with $\delta' \geq \delta \geq 9$, so by the Singleton bound [67, Theorem 11, Chapter 1], $k \leq n - \delta + 1 \leq n - 8$. This gives an immediate contradiction for the first four lines of Table 5.6: for A_n and S_n by [61, Proposition 5.3.7], for M_{11} and M_{12} by [61, Proposition 5.3.8]. The remaining cases are $G_{\mathbf{0}}^{\mathcal{N}} = M_{23}$ or M_{24}. A similar argument to that in the previous paragraph, for a subset $J \subseteq \mathcal{N}$ with $|J| = 2$ shows that $q-1$ divides the order of a soluble quotient of $G_{\mathbf{0},J}^{\mathcal{N}}$. However, for $G_{\mathbf{0}}^{\mathcal{N}} = M_{23}$ or M_{24}, the largest soluble quotient of $G_{\mathbf{0},J}^{\mathcal{N}}$ has order 2. Thus $q = 3$, and again the subcode $\mathcal{S}$ of $\mathcal{C}$ is an $\mathbb{F}_q G_{\mathbf{0}}^{\mathcal{N}}$-module, of dimension k, say. This implies, by [59], that $k \geq 22$, whereas the Singleton bound requires $k \leq n - \delta + 1 \leq 24 - 8 = 16$. □

5.5.5 Codes in Hamming Graphs from Permutation Modules

In this section, we present examples of linear codes that are s-neighbour-transitive, for $s \geq 2$, arising from permutation modules. We construct these modules via polynomial algebras. Historically, polynomial algebras have been used to construct many interesting examples of codes, such as the generalised Reed–Muller codes and the projective Reed–Muller codes; see Definitions 5.15 and 5.16 below. We present a richer family of examples that may eventually lead to a classification of alphabet-affine, 2-neighbour-transitive codes in $H(n, q)$ with minimum distance at least five: such a classification would be a huge strengthening of Theorem 5.2.

Throughout this section let $R = \mathbb{F}_q[x_1, \ldots, x_t]$, the ring of polynomials with coefficients in $\mathbb{F}_q$ in the variables $x_1, \ldots, x_t$. Each element of R may be viewed as a function $\mathbb{F}_q^t \to \mathbb{F}_q$ and conversely, by Lagrange interpolation (see [65, Theorem 1.7.1]), every such function may be represented (in at least one way) by an element of R. A monomial $x_1^{a_1} \cdots x_t^{a_t}$ is said to have *degree* $a_1 + \cdots + a_t$ and the degree of a polynomial is the maximum value of the degrees of its constituent monomials. The generalised Reed–Muller codes arise as follows as subspaces of R; see [31].

Definition 5.15. *Let k be an integer with $0 \le k \le t(q-1)$. The* k-th order q-ary generalised Reed–Muller code $\mathcal{RM}_q(k,t)$ *in $H(\mathbb{F}_q^t, \mathbb{F}_q)$ is the subspace of R consisting of all polynomials of degree at most k. When $q = 2$ these are simply called* Reed–Muller *codes.*

The parameters of the generalised Reed–Muller codes are given in [5, Theorem 5.4.1 and Corollary 5.5.4]. Let $d \in \{1, 2, \ldots, q-1\}$ and let $R[d]$ be the subspace of R consisting of all polynomials f such that $f(ax_1, \ldots, ax_t) = a^d f(x_1, \ldots, x_t)$, for each $a \in \mathbb{F}_q$. In other words, $R[d]$ consists of all polynomials f of R such that each monomial of f has non-zero degree equivalent to d modulo $q-1$. This leads to a similar construction to that in the previous definition, but now involving $R[d]$. See [85] for more details. Note that we are taking $\mathcal{N}$ to be a set of representatives for the one-dimensional subspaces of $\mathbb{F}_q^t$ and then regarding the elements of $R[d]$ as functions $\mathcal{N} \to \mathbb{F}_q$.

Definition 5.16. *Let k be an integer with $0 \le k \le t(q-1)$, let $\mathcal{N}$ be a fixed set of representatives for the one-dimensional subspaces of $\mathbb{F}_q^t$ and let $d \in \{1, 2, \ldots, q-1\}$ such that $d \equiv k \pmod{q-1}$. The* k-th order q-ary projective Reed–Muller *code* $\mathcal{PRM}_q(k,t)$ *in $H(\mathcal{N}, \mathbb{F}_q)$ is the subspace of $R[d]$ consisting of all polynomials having degree at most k.*

The $\mathbb{F}_q\mathrm{AGL}_t(q)$-submodule structure of R was determined by Sin [82], while the $\mathbb{F}_q\mathrm{GL}_t(q)$-submodule structure of R (and, in particular, of each $R[d]$) is determined in [11]. Any submodule of R determines a code in $H(\mathbb{F}_q^t, \mathbb{F}_q)$ and any submodule of $R[d]$ determines a code in $H(\mathcal{N}, \mathbb{F}_q)$, where $\mathcal{N}$ is a set of representatives for the one-dimensional subspaces of $\mathbb{F}_q^t$. Indeed, the generalised Reed–Muller code $\mathcal{RM}_q(k,t)$ is an $\mathbb{F}_q\mathrm{AGL}_t(q)$-submodule of R and the projective Reed–Muller code $\mathcal{PRM}_q(k,t)$ is an $\mathbb{F}_q\mathrm{GL}_t(q)$-submodule of $R[d]$, where $d \in \{1, 2 \ldots, q-1\}$ such that $d \equiv k \pmod{q-1}$. However, the submodule lattices of R and $R[d]$ are, in general, considerably more complicated than the chains of submodules given by varying the parameter k in $\mathcal{RM}_q(k,t)$ and $\mathcal{PRM}_q(k,t)$, respectively. In particular, if r is such that $q = p^r$ for some p prime then the submodule lattices of R and $R[d]$ are parameterised by r-tuples of integers satisfying certain restrictions, see

[11, Theorems A and C]. The next two results show that, under certain conditions on the parameters q, d, many submodules of R and $R[d]$ (including the generalised and the projective Reed–Muller codes and their duals) give examples of 2-neighbour-transitive codes. Note that in Proposition 5.9, since $q = 2$ is prime, the codes arising are precisely the Reed–Muller codes and their duals. Recall that we denote by $T_{\mathcal{C}}$ the group of translations by elements of a linear code $\mathcal{C}$. In all the propositions in this section, we assume that the covering radius is at least 2; in Remark 5.4, we discuss the situation when this does not hold.

Proposition 5.9 ([53] or see [51, Proposition 9.1.8]). *Let $q = 2$ and $\mathcal{C}$ be an $\mathbb{F}_2\mathrm{AGL}_t(2)$-submodule of R such that $\mathcal{C}$ is a code with covering radius $\rho \geq 2$ in $H(\mathbb{F}_2^t, \mathbb{F}_2)$. Then $\mathcal{C}$ is $(G, 2)$-neighbour-transitive, where $G = T_{\mathcal{C}} \rtimes \mathrm{AGL}_t(2)$.*

Proposition 5.10 ([53] or see [51, Proposition 9.2.3]). *Let $d \in \{1, 2, \ldots, q-1\}$ with $\gcd(d, q-1) = 1$, let $\mathcal{N}$ be a set of representatives for the one-dimensional subspaces of $\mathbb{F}_q^t$, and let $\mathcal{C}$ be an $\mathbb{F}_q\mathrm{GL}_t(q)$-submodule of $R[d]$ such that $\mathcal{C}$ is a code with covering radius $\rho \geq 2$ in $H(\mathcal{N}, \mathbb{F}_q)$. Then $\mathcal{C}$ is $(G, 2)$-neighbour-transitive, where $G = T_{\mathcal{C}} \rtimes \mathrm{GL}_t(q)$.*

Remark 5.3. *The conclusion of Proposition 5.9 is generally false if we instead consider $q \geq 3$, as is the conclusion of Proposition 5.10 when $\gcd(d, q-1) \neq 1$; see [53], or [51, Proposition 9.1.9]. Note also that the proofs of each of the propositions stated in this section proceed identically to the proofs of the respective references from [51], though the statements here are more general than there.*

Recalling from Definition 5.16 that elements of $R[d]$ are polynomial functions, we can obtain further examples of 2-neighbour-transitive codes by appropriate restrictions of the domain of these functions. For the first examples, we embed an affine space into a projective space and restrict to the one-dimensional subspaces corresponding to the affine points.

Proposition 5.11 ([53] or see [51, Proposition 9.3.3]). *Let $d \in \{0, 1, \ldots, q-1\}$ with $\gcd(d, q-1) = 1$, and fix an embedding of $\mathrm{AG}_{t-1}(q)$ into $\mathrm{PG}_{t-1}(q)$, the points of which are the one-dimensional subspaces of $\mathbb{F}_q^t$. Let $\mathcal{N}$ be a set of representatives for the one-dimensional subspaces corresponding to the points of $\mathrm{AG}_{t-1}(q)$ and let $\mathcal{C}$ be an $\mathbb{F}_q\mathrm{GL}_t(q)$-submodule of $R[d]$ such that $\mathcal{C}$ is a code with covering radius $\rho \geq 2$ in $H(\mathcal{N}, \mathbb{F}_q)$. Then $\mathcal{C}$ is $(G, 2)$-neighbour-transitive, where $G = T_{\mathcal{C}} \rtimes (\mathbb{F}_q^\times \times \mathrm{AGL}_{t-1}(q))$.*

It is worth briefly comparing the automorphism groups of the codes in Definition 5.15 with the codes occurring in Proposition 5.11. There is a subgroup $\mathrm{AGL}_{t-1}(q)$ appearing in the automorphism group of the code $\mathcal{RM}_q(k, t-1)$ in $H(q^{t-1}, q)$, and also inside the automorphism group of any code in $H(q^{t-1}, q)$ arising as an $\mathbb{F}_q\mathrm{GL}_t(q)$-submodule of $R[d]$ as in Proposition 5.11. Moreover, $\mathrm{AGL}_{t-1}(q)$ acts faithfully on the entries of the Hamming graph in each case. However, the actions are not the same: in the former case $\mathrm{AGL}_{t-1}(q)$ occurs as a subgroup of the top group L, while this is not true in the latter case. This is the key difference that allows Proposition 5.11 to be proved for more general values of q, while $q = 2$ in Proposition 5.9.

Next, we present two further infinite families of 2-neighbour-transitive codes, the first arising from the Suzuki–Tits ovoids and the second from classical unitals. The Suzuki group $\mathrm{Sz}(q)$, where $q = 2^{2f+1}$ for some positive integer f, acts 2-transitively on the Suzuki–Tits ovoid consisting of q^2+1 points of the projective space $\mathrm{PG}_3(q)$, no three of which are collinear; see [34, p. 250]. The unitary group $\mathrm{PGU}_3(q)$ acts 2-transitively on the unital consisting of the q^3+1 isotropic points of $\mathrm{PG}_2(q^2)$ under a non-degenerate Hermitian form; see [34, p. 248].

Proposition 5.12 ([53] or see [51, Proposition 9.4.6]). *Let $q = 2^{2f+1}$, let $d \in \{0, 1, \ldots, q-1\}$ with $\gcd(d, q-1) = 1$, and let $\mathcal{N}$ be a set of representatives in $\mathbb{F}_q^4$ for the one-dimensional subspaces corresponding to the points of the Suzuki–Tits ovoid in $\mathrm{PG}_3(q)$. Furthermore, let $\mathcal{C}$ be an $\mathbb{F}_q\mathrm{GL}_t(q)$-submodule of $R[d]$ such that $\mathcal{C}$ is a code with covering radius $\rho \geq 2$ in $H(\mathcal{N}, \mathbb{F}_q)$. Then $\mathcal{C}$ is $(G, 2)$-neighbour-transitive, where $G = T_{\mathcal{C}} \rtimes (\mathbb{F}_q^\times \rtimes \mathrm{Sz}(q))$.*

Proposition 5.13 ([53] or see [51, Proposition 9.4.8]). *Let $q = 2^f$, let $d \in \{0, 1, \ldots, q-1\}$ with $\gcd(d, q-1) = 1$, and let $\mathcal{N}$ be a set of representatives in $\mathbb{F}_{q^2}^3$ for the one-dimensional subspaces corresponding to the points of the classical unital in $\mathrm{PG}_2(q^2)$. Furthermore, let $\mathcal{C}$ be an $\mathbb{F}_q\mathrm{GL}_t(q)$-submodule of $R[d]$ such that $\mathcal{C}$ is a code with covering radius $\rho \geq 2$ in $H(\mathcal{N}, \mathbb{F}_q)$. Then $\mathcal{C}$ is $(G, 2)$-neighbour-transitive, where $G = T_{\mathcal{C}} \rtimes (\mathbb{F}_q^\times \rtimes \mathrm{PGU}_3(q))$.*

Remark 5.4. *Note that for a code to be 2-neighbour-transitive it must have covering radius at least 2, hence this is an assumption in all the propositions of this section. However, since in each of the Propositions 5.9–5.13 the group G acts transitively on both $\Gamma_1(\mathbf{0})$ and $\Gamma_2(\mathbf{0})$, it follows from Proposition 5.1 that any non-trivial code $\mathcal{C}$ arising from the respective submodule in the relevant Hamming graph, but having covering radius $\rho \leq 1$, is either perfect with $(\rho, \delta) = (1, 3)$, or has $\rho = 1$ and $\delta = 2$. Moreover, if $\Gamma = H(n, q)$ then either $\Gamma_1(\mathbf{0})$ and $\Gamma_2(\mathbf{0})$ both contain no pairs of adjacent vertices (when $q = 2$) or each contains a pair of adjacent vertices (when $q \geq 3$). This implies that Proposition 5.1(3)(b) does not occur. In particular, any relevant (linear) code with covering radius $\rho \leq 1$ is either a perfect Hamming code*[4] *or is the binary repetition code, and is thus known.*

Assmus and Key [5, Section 5.7] construct and analyse the *subfield subcodes* of the generalised and projective Reed–Muller codes. Further examples of 2-neighbour-transitive codes may be obtained in a similar manner from submodules of R and $R[d]$; see [53] or [51, Section 9.5].

The assumption in Propositions 5.9–5.13 that $\mathcal{C}$ is an $\mathbb{F}_q\mathrm{GL}_t(q)$-submodule of $R[d]$ is more restrictive than necessary. However we have stated the results in this way because, although the lattice of $\mathbb{F}_q\mathrm{GL}_t(q)$-submodules of $R[d]$ is known (see [11]), we wonder whether there may be additional subspaces invariant under the subgroups of $\mathrm{GL}_t(q)$ occurring in these propositions, which may yield new interesting codes.

Problem 5.8. *Determine more information about the G_0-submodule structure of $R[d]$, where $G_0 = \mathbb{F}_q^\times \times \mathrm{AGL}_{t-1}(q)$, $\mathbb{F}_q^\times \rtimes \mathrm{Sz}(q)$ or $\mathbb{F}_q^\times \rtimes \mathrm{PGU}_3(q)$ in Proposition 5.12, 5.11 or 5.13, respectively.*

As alluded to earlier, a reasonable amount is known about the minimum distance of each of the generalised and projective Reed-Muller codes, see [5, Section 5.5]. It would be nice to have similar results for the other codes discussed in this section arising from submodules.

Problem 5.9. *Find the minimum distances of the codes from submodules of R and $R[d]$ in Propositions 5.11–5.13.*

It should be remarked also that some of the codes in this section, and their subfield subcodes, are related to codes arising from incidences between subspaces of differing dimensions in projective and affine geometries; see [5, Section 5.6] and [11, Section 8]. Interesting work has been done concerning the geometric structure of codewords of low weight in some of these cases; see, for example, [2,27,62].

5.6 CODES IN KNESER GRAPHS

Neighbour-transitive and 2-neighbour-transitive codes have recently been studied in the Kneser graphs [25]. The next result is a classification of 2-neighbour-transitive codes $\mathcal{C}$ in Kneser graphs with minimum distance $\delta(\mathcal{C}) \geq 5$. Recall from Definition 5.7 that in a Kneser graph $K(\mathcal{V}, k)$, the cardinality $v = |\mathcal{V}|$ is at least $2k+1$. Note also that, for $|\mathcal{V}| = 23$, by an *endecad* we mean a subset of $\mathcal{V}$ such that its characteristic vector corresponds to a weight 11 codeword of the perfect binary Golay code in the Hamming graph $H(\mathcal{V}, 2)$ (see [23, Page 71]).

Theorem 5.9 ([25, Theorem 1.2]). *Let $\mathcal{C}$ be a 2-neighbour-transitive code in $\Gamma = K(v,k)$ with minimum distance $\delta \geq 5$. Then $v = 2k+1$, and hence Γ is the odd graph O_{k+1}, and one of the following holds.*

1. $\mathrm{Aut}(\mathcal{C}) \cong \mathrm{M}_{23}$ *with $v = 23$ and $\mathcal{C}$ consists of the endecads.*

2. $\mathrm{Aut}(\mathcal{C}) \cong \mathrm{P\Gamma L}_d(2)$, *where $d \geq 5$, Ω is the set of all points, and $\mathcal{C}$ is the set of all hyperplanes, in $\mathrm{PG}_{d-1}(2)$.*

Recall from Example 5.9 that any code in a Kneser graph, and hence also any code in an odd graph, may be regarded as a code in a Johnson graph. A similar situation exists for codes in Johnson graphs which can also be viewed as codes in binary Hamming graphs, and in this case, there is a clear connection between the minimum distances of the codes, see [70, Section 1.3]. The following lemma gives a similar relationship between the minimum distances for codes in Kneser graphs and the same codes in the corresponding Johnson graphs. This lemma is key to proving Theorem 5.9, where the proof proceeds by reducing to the case of odd graphs (see [25, Theorem 3.1]) and then applying results about neighbour-transitive codes in Johnson graphs [74] along with the following Lemma 5.9.

TABLE 5.7 Conditions on the Parameters of $\mathcal{C}_{\rm int}(a,b;c,d)$ in Example 5.16 Ensuring It Is Neighbour-Transitive with Minimum Distance δ

Line	a	b	c	d	δ
1	1	$v-1$	0	k	1
2	$2e$	$2f+1$	e	f	1
3	$< k$	$v-a$	a	$k-a$	2

Lemma 5.9 ([25, Lemma 3.2]). *Let $\mathcal{C}$ be a 2-neighbour-transitive code in O_{k+1} with minimum distance $\delta \geq 5$. Then $\mathcal{C}$ is also a code in the Johnson graph $J(2k+1,k)$, with the same vertex set as O_{k+1}, and $\mathcal{C}$ is neighbour-transitive in $J(2k+1,k)$ with minimum distance $\delta' \geq 3$.*

Given the classification in Theorem 5.9, the paper [25] then considers neighbour-transitive codes $\mathcal{C}$ in Kneser graphs $K(\mathcal{V},k)$ in general, roughly separating the analysis into cases where the action of Aut$(\mathcal{C})$ on $\mathcal{V}$ is: intransitive, transitive but imprimitive, and primitive. The following example and theorem deal with the intransitive case, and provide a useful application of the concept of types (see Definition 5.3) and Lemma 5.4.

Example 5.16. *Let $\Gamma = K(\mathcal{V},k)$, let a,b,c and d be non-negative integers such that $a \geq c$, $b \geq d$, $a+b = v = |\mathcal{V}|$ and $c+d = k$. Let $\mathcal{V}$ be the disjoint union $A \cup B$ with $|A| = a$ and $|B| = b$. For $\alpha \in V(\Gamma)$ let $\iota(\alpha) = (|\alpha \cap A|, |\alpha \cap B|)$. Define*

$$\mathcal{C}_{\rm int}(a,b;c,d) = \{\alpha \in V(\Gamma) \mid \iota(\alpha) = (c,d)\}.$$

Then $\mathcal{C}_{\rm int}(a,b;c,d)$ is neighbour-transitive if a,b,c,d are as in one of the lines of Table 5.7 (see [25, Lemma 4.2]).

Theorem 5.10 ([25, Theorem 1.3]). *Let $\mathcal{C}$ be a non-trivial neighbour-transitive code in $K(\mathcal{V},k)$ with minimum distance δ and suppose that Aut$(\mathcal{C})$ acts intransitively on $\mathcal{V}$. Then Aut$(\mathcal{C})$ has precisely two non-empty orbits on $\mathcal{V}$, say A and B, and $\mathcal{C}$ is equivalent to a subcode of one of the codes in Example 5.16.*

The following example shows that there may indeed be proper subcodes of the codes given in Example 5.16 which are neighbour-transitive and have automorphism groups intransitive on $\mathcal{V}$.

Example 5.17 ([25, Example 4.6]). *Let $\mathcal{V}$ be the disjoint union $A \cup B$, where A is the set of points of the affine geometry* AG$_3(2)$ *and $|B| = 5$. Furthermore, let $\mathcal{T}$ be the set of all* tetrahedrons *of A, where a tetrahedron is set of 4 points of* AG$_3(2)$ *that do not form an affine plane, and let $\mathcal{C}$ be the code in $O_7 = K(13,6)$ consisting of all vertices α such that $\alpha \cap A \in \mathcal{T}$ and $|\alpha \cap B| = 2$. Note that $\mathcal{C}$ is a proper subcode of $\mathcal{C}_{int}(8,5;4,2)$ and is neighbour-transitive (see [25, Lemma 4.7]).*

The next theorem concerns the case where $\mathcal{C}$ is neighbour-transitive and Aut$(\mathcal{C})$ acts transitively on $\mathcal{V}$. Note that a 2-homogeneous permutation group is primitive, [75, Lemma 2.30].

Theorem 5.11 ([25, Theorem 1.7]). *Let $\mathcal{C}$ be a neighbour-transitive code in $K(v,k)$ with minimum distance $\delta \geq 3$.*

1. *If* $\text{Aut}(\mathcal{C})$ *is transitive and imprimitive on* $\mathcal{V}$ *then* $v = 2k+1$ *so* $K(v,k)$ *is the odd graph* O_{k+1}*; and*

2. *if* $\text{Aut}(\mathcal{C})$ *is primitive on* $\mathcal{V}$*, then then either* $v = 2k+1$ *and* $K(v,k) = O_{k+1}$*, or* $\text{Aut}(\mathcal{C})$ *is* 2*-homogeneous on* $\mathcal{V}$.

Noting that the 2-homogeneous groups are classified (see [34, Section 7.7 and Theorem 9.4B]), we pose the following problems.

Problem 5.10. *[25, Problem 1.5] Classify the neighbour-transitive codes $\mathcal{C}$ in $K(\mathcal{V},k)$ such that $\delta(\mathcal{C}) \geq 3$ and* $\text{Aut}(\mathcal{C})$ *acts 2-homogeneously on* $\mathcal{V}$.

Problem 5.11. *Find examples of neighbour-transitive codes $\mathcal{C}$ in $K(2k+1,k) = O_{k+1}$ such that $\delta(\mathcal{C}) \geq 3$, and* $\text{Aut}(\mathcal{C})$ *is primitive on* $\mathcal{V}$ *but not* 2*-homogeneous (see also [25, Problem 1.7]).*

Given the above results and open problems, we should comment briefly on neighbour-transitive codes $\mathcal{C}$ in odd graphs where $\text{Aut}(\mathcal{C})$ is imprimitive on $\mathcal{V}$. The next example and theorem are again applications of types and Lemma 5.3. Given a multiset M we write $M = \{b_1^{a_1}, \ldots, b_s^{a_s}\}$ where each b_i is an element of M that occurs with multiplicity a_i, for $i = 1, \ldots, s$. For example, the multiset $\{0,1,1,2,2,2\}$ could be written as $\{0^1, 1^2, 2^3\}$.

Example 5.18. *Let $\Gamma = K(2k+1,k) = O_{k+1}$ and let $\mathcal{B} = \{B_1, \ldots, B_a\}$ be a partition of $\mathcal{V}$ into a blocks each having size b. For a vertex $\alpha \in V(\Gamma)$, let $\iota(\alpha)$ be the multiset $\{\alpha \cap B_1, \ldots, \alpha \cap B_a\}$. For a multiset M, define*

$$\mathcal{C}_{\text{imp}}(a,b;M) = \{\alpha \in V(\Gamma) \mid \iota(\alpha) = M\}.$$

Then, by [25, Lemma 5.2], $\mathcal{C}_{\text{imp}}(a,b;M)$ is neighbour-transitive if and only if M is as in one of the lines of Table 5.8.

Theorem 5.12 ([25, Theorem 1.6]). *Let $\mathcal{C}$ be a non-trivial neighbour-transitive code in $K(2k+1,k) = O_{k+1}$ such that* $\text{Aut}(\mathcal{C})$ *acts transitively but imprimitively on $\mathcal{V}$. Then $\mathcal{C}$ is equivalent to a subcode of one of the codes in Example 5.18.*

TABLE 5.8 Multisets M for which $\mathcal{C}_{\text{imp}}(a,b;M)$, as in Example 5.16, Is Neighbour-Transitive with Minimum Distance δ

Line	M	δ
1	$\{((b-1)/2)^{(a+1)/2}, ((b+1)/2)^{(a-1)/2}\}$	1
2	$\{0^{(a-1)/2}, (b-1)/2, b^{(a-1)/2}\}$	1
3	$\{b^{a_0}, b_1^{a_1}\}$	2

Note that there are currently no known examples of codes $\mathcal{C}$ where $\mathrm{Aut}(\mathcal{C})$ acts imprimitively on $\mathcal{V}$ and $\mathcal{C}$ is a proper subcode of a code from Example 5.18. Hence we finish this section with the following research problem.

Problem 5.12 ([25, Problem 1.6]). *Find new examples of codes satisfying Theorem 5.12, or prove that all examples are equivalent to a code in Example 5.18.*

5.7 CODES IN INCIDENCE GRAPHS OF GENERALISED QUADRANGLES

A *generalised quadrangle* is an incidence structure[5] $\mathcal{Q} = (\mathcal{P}, \mathcal{L}, \mathrm{I})$ such that:

1. Each point is incident with $t+1$ lines ($t \geq 1$) and two distinct points are incident with at most one line.

2. Each line is incident with $s+1$ points ($s \geq 1$) and two distinct lines are incident with at most one point.

3. If p is a point and L is a line not incident with p, then there is a unique pair $(q, M) \in \mathcal{P} \times \mathcal{L}$ for which $p \,\mathrm{I}\, M \,\mathrm{I}\, q \,\mathrm{I}\, L$.

A generalised quadrangle $\mathcal{Q}$ satisfying the above axioms is said to have *order* (s, t) and has $(s+1)(st+1)$ points and $(t+1)(st+1)$ lines; and $\mathcal{Q}$ is called *thick* if both $s, t \geq 2$. The dual of a generalised quadrangle of order (s, t) is a generalised quadrangle of order (t, s). For further background on generalised quadrangles see [71].

Let $\mathcal{Q}$ be a generalised quadrangle and let Γ be its incidence graph (see Definition 5.10). Then Γ is bipartite, has degrees $s+1$ and $t+1$, diameter 4 and girth 8. Note that, since Γ has diameter 4, a code $\mathcal{C}$ in Γ has minimum distance $\delta(\mathcal{C}) \leq 4$.

An *ovoid* (respectively, a *partial ovoid*) of a generalised quadrangle $\mathcal{Q} = (\mathcal{P}, \mathcal{L}, \mathrm{I})$ is a subset $\mathcal{O}$ of $\mathcal{P}$ such that each line $\ell \in \mathcal{L}$ is incident with exactly one (respectively, at most one) point of $\mathcal{O}$. Dually, a *spread* (respectively, a *partial spread*) of a generalised quadrangle $\mathcal{Q} = (\mathcal{P}, \mathcal{L}, \mathrm{I})$ is a subset $\mathcal{S}$ of $\mathcal{L}$ such that each point $p \in \mathcal{P}$ is incident with exactly one (respectively, at most one) line of $\mathcal{S}$. A partial ovoid (respectively, spread) is called *maximal* if there is no partial ovoid (spread) properly containing it. In particular, an ovoid (spread) is a maximal partial ovoid (spread). These geometric conditions can be reformulated in the language of coding theory as follows.

Lemma 5.10 ([26, Lemma 3.6]). *Let $\mathcal{C}$ be a code in a generalised quadrangle $\mathcal{Q}$ with minimum distance $\delta = 4$ and covering radius ρ. Then the following hold:*

1. *$\mathcal{C}$ is a partial ovoid or a partial spread of $\mathcal{Q}$.*

2. *$\mathcal{C}$ is a maximal partial ovoid or a maximal partial spread of $\mathcal{Q}$ if and only if $\rho \leq 3$.*

3. *$\mathcal{C}$ is an ovoid or spread of $\mathcal{Q}$ if and only if $\rho = 2$.*

Now we consider *classical* generalised quadrangles, which are associated with certain classical groups; see [71, Chapter 3] for their constructions and [91, Sections 3.5.6 and 3.6.4] for more about their automorphism groups. Those that arise in our next result are the *symplectic generalised quadrangle* $\mathsf{W}_3(q)$ and the *hermitian generalised quadrangle* $H_3(q^2)$, defined as follows: let V be the underlying vector space of the projective geometry $\mathrm{PG}_3(q^\tau)$ equipped with a non-singular symplectic or hermitian form f, where $\tau = 1$ or 2, respectively. The points are the totally isotropic one-dimensional subspaces, and the lines are the totally isotropic 2-dimensional subspaces of V, with incidence given by symmetrised inclusion. If q is a square then a regular spread of $\mathsf{W}_3(q)$ can obtained by embedding $\mathsf{W}_1(q^2)$ into it, see [10, Section 3.2]; and a classical ovoid of $\mathsf{H}_3(q^2)$ can be constructed by taking the absolute points of a non-degenerate unitary polarity, see [10, Section 3.1]. The associated codes were shown to be neighbour-transitive in [26, Lemmas 4.2 and 4.4].

Theorem 5.13 ([26, Theorem 4.5]). *Let C be a neighbour-transitive code with minimum distance* 4 *and covering radius $\rho = 2$ in the incidence graph of a thick classical generalised quadrangle $\mathcal{Q}$ and assume that* Aut(C) *is insoluble. Then C is equivalent to one of the following:*

1. *A regular spread of $\mathsf{W}_3(q)$, where q is a square.*
2. *A classical ovoid of $\mathsf{H}_3(q^2)$.*

Some sporadic examples of maximal partial spreads in $\mathsf{W}_3(q)$ are given in Example 5.19. We note that, if q is even then $\mathsf{W}_3(q)$ is self dual, while if q is odd then the dual of $\mathsf{W}_3(q)$ is the classical generalised quadrangle $\mathsf{Q}_4(q)$. Thus for odd q, a maximal partial spread in $W_3(q)$ is a maximal partial ovoid in $\mathsf{Q}_4(q)$. It is a conjecture of Thas [88, Conjecture, p. 13] that when q is sufficiently large then there are no maximal partial ovoids of size $q^2 - 1$ in $\mathsf{Q}_4(q)$ (the dual of $\mathsf{W}_3(q)$).

Example 5.19 ([26, Example 5.4]). *Let $V \cong \mathbb{F}_q^4$ with symplectic form f such that $f(x, y) = x_1y_2 - x_2y_1 - x_3y_4 + x_4y_3$. Let q, G be as in one of the rows of Table 5.9, so $G \leq SL_2(q)$ (represented as 2×2 matrices) and G is sharply transitive on the non-zero vectors of $\mathbb{F}_q^2$. Let C be the following set of 2-dimensional subspaces of V, where each is represented as the row-space of a 2×4 matrix:*

$$C = \{[I\ A] \mid A \in G\},$$

where I is the 2×2 identity matrix. Letting x be the first row of $[I\ A]$ and y be the second row, we have $f(x, y) = \det I - \det A$. Thus the row-space of $[I\ A]$

TABLE 5.9 Subgroups of $\mathrm{SL}_2(q)$ of Order $q^2 - 1$ (see [86, Chapter 3, Section 6])

q	2	3	5	7	11
G	$\mathrm{GL}_1(4)$	Q_8	$2.\mathrm{A}_4$	$2.\mathrm{S}_4$	$\mathrm{SL}_2(5)$

TABLE 5.10 Results Table for Theorem 5.14

	Conditions	Conclusions
1.	$\rho = 2$	C is equivalent to a regular spread
2.	$\|C\| = q^2$	$\rho = 4$, and C can be extended to a spread or an ovoid
3.	$\|C\| = q^2 - 1$ and $\rho = 3$	$q \in \{2, 3, 5, 7, 11\}$ and C is equivalent to a code in Example 5.19
4.	$\|C\| = q + 1$ and $\rho = 3$	C is equivalent to the set of points on a hyperbolic line
5.	$q = \rho = 3$	C as in line (3) or (4), or C is equivalent to the sporadic code given in [26, Example 5.3] with $\|C\| = 5$

is an isotropic 2-space if and only if $\det I - \det A = 0$, *that is,* $\det A = 1$. *Since* $A \in SL_2(q)$, *the code* C *is indeed a subset of lines of* $\mathsf{W}_3(q)$. *By [26, Lemma 5.5],* C *is a neighbour-transitive maximal partial ovoid of* $\mathsf{W}_3(q)$. *Note that this is the dual of a construction for maximal partial spreads given in [24].*

We collect together, in Theorem 5.14, information about neighbour-transitive partial ovoids and spreads in $\mathsf{W}_3(q)$, focusing mainly on maximal partial ovoids and spreads. Examples for each case can be found in [26].

Theorem 5.14 ([26, Theorem 1.2])**.** *Let* C *be a neighbour-transitive code with minimum distance 4 and covering radius* ρ *in the incidence graph of the generalised quadrangle* $\mathsf{W}_3(q)$. *Then, for each line of Table 5.10, if the 'Conditions' hold then the 'Conclusions' also hold. Moreover, in line (1), the converse assertion is also valid.*

Remark 5.5. *The statement of Theorem 5.14 differs slightly from that of [26, Theorem 1.2]: in lines (3) and (5) of Table 5.10 we have* $\rho = 3$ *as a hypothesis rather than a conclusion (correcting a mistake in [26, Theorem 1.2]), and the conclusion in line (5) of Table 5.10 is stronger than in [26, Theorem 1.2], reflecting the discussion preceding [26, Conjecture 1.4].*

Generalised quadrangles are examples of a broader class of incidence structures called *polar spaces.* We pose the following open problem.

Problem 5.13. *Investigate* s*-neighbour-transitive codes in the incidence graphs of other classical generalised quadrangles and, more generally, in other classical polar spaces.*

5.8 CODA: FINAL REFLECTIONS AND SUMMARY OF OPEN PROBLEMS

Our aim in this chapter has been to outline the state-of-the-art regarding our understanding of s-neighbour-transitive codes in various graphs. This is an area of

TABLE 5.11 References for Open Problems Stated in This Chapter and a Rough Description of Each

Problem	Topic
5.1	Codes with covering radius 1
5.2	s-Elusive codes
5.3	s-Distance-transitive quotient graphs
5.4	Codes in bilinear forms graphs
5.5	Frequency permutation arrays
5.6	Codes in $H(n, 2)$ with minimum distance 4
5.7	Binary completely transitive codes
5.8	2-Neighbour-transitive codes from submodules
5.9	Parameters of codes from submodules
5.10	2-Homogeneous actions and Kneser graphs
5.11	Primitive actions and odd graphs
5.12	Imprimitive actions and odd graphs
5.13	s-Neighbour-transitive codes in polar spaces

active research and we have posed several open problems throughout this chapter; these are summarised for reference in Table 5.11. In the remainder of this section, we briefly reflect on the most significant achievements and major open problems we have covered. We also mention a few interesting related results and areas of research which for reasons of space we could not discuss in detail.

Regarding codes in Hamming graphs, Theorem 5.2 effectively gives an upper bound on $\min\{e, s\}$, where e is the error-correction capacity of an s-neighbour-transitive code. Furthermore, progress has been made classifying completely transitive codes with minimum distance at least 5 in the binary case (see Section 5.5.3), though there is still significant work to be done here (see Problem 5.7). The results of Sections 5.5.2 and 5.5.5 pave the way towards a deeper understanding of 2-neighbour-transitive codes with minimum distance at least 5 in $H(n, q)$ when $q \geq 3$; these may lead in the future towards classification results for completely transitive codes with alphabet size larger than 2.

Turning to other graphs, classification results have been obtained for neighbour-transitive codes in Johnson graphs (see [74]) and progress has been made on neighbour-transitive codes in Kneser graphs (see Section 5.6). There are still open problems related to codes in each of these families of graphs, as there are for numerous other families of distance-regular and distance-transitive graphs (see, for instance, Sections 5.4.3 and 5.7).

Many of the results we have stated for codes in Hamming graphs assume some small lower bound (typically 5) on the minimum distance of a code. That is not to say that codes with smaller minimum distances are not interesting. For example, recently Borges, Rifà and Zinoviev [19] investigated the complete transitivity of certain "supplementary" codes in $H(n, q)$. These are constructed via a concatenation method previously introduced by the same authors. They find several infinite families of codes with minimum distance 3 and covering radius 1 or 2. They also conjecture that these are all the completely transitive codes that may be obtained via their construction – their conjecture is informed by computational results on the sizes of the automorphism groups of some of the codes.

A *maximum distance separable* (MDS) code is a code $\mathcal{C}$ in the Hamming graph $H(n,q)$ that meets the Singleton bound [67, Theorem 11, Chapter 1], that is, if $|\mathcal{C}| = q^k$ and δ is the minimum distance of $\mathcal{C}$ then $n = \delta + k - 1$ (see [87]). By [87, Theorem 3], a linear MDS code $\mathcal{C}$ is "equivalent" to an *n-arc* in the projective space $\mathrm{PG}_{k-1}(q)$, that is, the column vectors of a generator matrix for $\mathcal{C}$ are representatives for a set of n points in $\mathrm{PG}_{k-1}(q)$, with $n \geq k$, such that no subset of k points is contained in a hyperplane. The archetypal example of an MDS code is a Reed–Solomon code $\mathcal{C}$ in $H(q+1,q)$, which corresponds to a geometric object known as a *normal rational curve* (see [87, Section 2]). In fact, such a Reed–Solomon code is equivalent to the projective Reed–Muller code $\mathcal{C} = \mathcal{PRM}_q(k-1,2)$ and, by Proposition 5.10, is 2-neighbour-transitive when $\gcd(k-1, q-1) = 1$ with automorphism group $T_{\mathcal{C}} \rtimes \Gamma\mathrm{L}_2(q)$ (see also [35], noting that a different notion of automorphism group is used there). The *MDS conjecture* states that if $\mathcal{C}$ is an MDS code of size q^k in $H(n,q)$ and $4 \leq k \leq q-3$ then $n \leq q+1$. Ball [8] proved the MDS conjecture when q is a prime; but it is still open for non-prime q. More recently, additive MDS codes have been classified over small fields [9] and certain additive MDS codes have been shown to be equivalent to linear codes [1].

A code $\mathcal{C}$ in a graph is called *propelinear*[6] if $\mathrm{Aut}(\mathcal{C})$ contains a subgroup H such that H acts regularly on $\mathcal{C}$ (that is H is transitive on $\mathcal{C}$ and codeword stabilisers fix $\mathcal{C}$ pointwise). In particular, the automorphism group of any linear code $\mathcal{C}$ in $H(n,q)$ contains the group of translations by codewords; this group acts regularly on $\mathcal{C}$, and thus each linear code is propelinear. Propelinear codes have primarily been studied in the Hamming graphs: for example, Rifà and Pujol [77] studied a subclass of propelinear codes, known as *translation-invariant* propelinear codes. In addition, many interesting codes in $H(2k,q)$ have been constructed as additive codes in $\mathbb{Z}_4^k$, and are thus propelinear (see [50]). This is an active research area, with new interesting examples of propelinear codes still being discovered (see, for instance, [4]). It would be interesting to study propelinear codes in other distance-regular graphs.

Notes

1. This was the Clay Lecture at the British Combinatorial Conference 2021, and a version of the survey may be downloaded from https://www.claymath.org/events/british-combinatorial-conference-2021/
2. See en.wikipedia.org/wiki/Binary_Golay_code.
3. That is, G acts transitively on the set of all paths (u,v,w) in Γ with $u \neq w$.
4. To see this: by [67, Theorem 37, Chapter 6] a perfect linear code $\mathcal{C}$ with covering radius 1 in $H(n,\mathbb{F}_q)$ necessarily has length $n = (q^k-1)/(q-1)$, dimension k and minimum distance 3. The condition 'minimum distance 3' implies that each column of a parity-check matrix H for $\mathcal{C}$ is non-zero, and no pair of columns of H is linearly dependent. This implies that the columns of H are a set of representatives for the one-dimensional subspaces of $\mathbb{F}_q^k$, *i.e.*, that $\mathcal{C}$ is a Hamming code.

5. Recall the definition of an incidence structure from Section 5.4.4.
6. The reader should note that we have stated this definition in the language of this chapter. Much of the literature regarding propelinear codes uses a different, but equivalent, definition.

REFERENCES

[1] S. Adriaensen and S. Ball. On additive MDS codes with linear projections. *Finite Fields Appl.*, 91:102255, 2023.

[2] S. Adriaensen and L. Denaux. Small weight codewords of projective geometric codes. *J. Combin. Theory Ser. A*, 180:105395, 2021.

[3] M. Akbari, N.I. Gillespie, and C.E. Praeger. Increasing the minimum distance of codes by twisting. *Electron. J. Comb.*, 25(3):3–36, 2018.

[4] J.A. Armario, I. Bailera, and R. Egan. Butson full propelinear codes. *Des. Codes Cryptogr.*, 91(2):333–351, 2023.

[5] E.F. Assmus and J.D. Key. *Designs and Their Codes*, Cambridge Tracts in Mathematics, Vol. 103. Cambridge University Press, Cambridge, 1994.

[6] R.F. Bailey. Error-correcting codes from permutation groups. *Discrete Math.*, 309(13):4253–65, 2009.

[7] R.F. Bailey and D.R. Hawtin. On the classification of binary completely transitive codes with almost-simple top-group. *Euro. J. Comb.*, 107:103604, 2022.

[8] S. Ball. On sets of vectors of a finite vector space in which every subset of basis size is a basis. *J. Eur. Math. Soc.*, 14(3):733–748, 2012.

[9] S. Ball, G. Gamboa, and M. Lavrauw. On additive MDS codes over small fields. *Adv. Math. Commun.*, 17(4):828–844, 2023.

[10] J. Bamberg and T. Penttila. A classification of transitive ovoids, spreads, and m-systems of polar spaces. *Forum Math.*, 21(2):181–216, 2009.

[11] M. Bardoe and P. Sin. The Permutation modules for $\mathrm{GL}(n+1, \mathbb{F}_q)$ acting on $\mathbb{P}^n(\mathbb{F}_q)$ and $\mathbb{F}_q^{n+1}$. *J. Lond. Math. Soc.*, 61(1):58–80, 2000.

[12] H. Bartz, L. Holzbaur, H. Liu, S. Puchinger, J. Renner, and A. Wachter-Zeh. Rank-Metric Codes and Their Applications. *Foundations and Trends in Communications and Information Theory*, 19(3):390–546, 2022. (`arXiv:2203.12384`).

[13] N. Biggs. Perfect codes in graphs. *J. Combin. Theory Ser. B*, 15(3):289–296, 1973.

[14] N. Biggs. Perfect codes and distance transitive graphs, in *Combinatorics, Proc. of the Third British Comb. Conf., Aberystwyth 1973*, McDonough, F.P. and Mavron, V.C., Eds., London Mathematical Society Lecture Notes Series, Vol. 13, pp. 1–8, 1973.

[15] I.F. Blake, G. Cohen, and M. Deza. Coding with permutations. *Inf. Control*, 43(1):1–9, 1979.

[16] J. Borges and J. Rifà. On the nonexistence of completely transitive codes, *IEEE Trans. Inf. Theory*, 46(1):279–280, 2000.

[17] J. Borges, J. Rifà, and V.A. Zinoviev. Nonexistence of completely transitive codes with error-correcting capability $e > 3$. *IEEE Trans. Inf. Theory*, 47(4):1619–1621, 2001.

[18] J. Borges, J. Rifà, and V.A. Zinoviev. On completely regular codes. *Probl. Inf. Transm.*, 55(1):1–45, 2019.

[19] J. Borges, J. Rifà, and V.A. Zinoviev. On new infinite families of completely regular and completely transitive codes. Preprint, (`arXiv:2303.11190`), 2023.

[20] A.E. Brouwer, A.M. Cohen, and A. Neumaier. *Distance-Regular Graphs*. Springer-Verlag, Berlin, 1989.

[21] W. Burnside. *Theory of Groups of Finite Order*, 2nd ed. Cambridge University Press, Cambridge, 1911; Reprint by photo offset Dover Publications, New York, 1955.

[22] P.J. Cameron. *Permutation Groups*, London Mathematical Society Student Texts, Vol. 45. Cambridge University Press, Cambridge, 1999.

[23] J.H. Conway. *Atlas of Finite Groups: Maximal Subgroups and Ordinary Characters for Simple Groups*. Clarendon Press, Oxford, 1985.

[24] K. Coolsaet, J. De Beule, and A. Siciliano. The known maximal partial ovoids of size $q^2 - 1$ of $Q(4, q)$. *J. Combin. Des.*, 21(3):89–100, 2013.

[25] D. Crnković, D.R. Hawtin, N. Mostarac, and A. Švob. Neighbour-transitive codes in Kneser graphs. Preprint, (`arXiv:2307.09752`), 2023.

[26] D. Crnković, D.R. Hawtin, and A. Švob. Neighbour-transitive codes and partial spreads in generalised quadrangles. *Des. Codes Cryptogr.*, 90:1521–1533, 2022.

[27] M. De Boeck. Small weight codewords in the dual code of points and hyperplanes in PG(n, q), q even. *Des. Codes Cryptogr.*, 63(2):171–182, 2012.

[28] P. Delsarte. An algebraic approach to the association schemes of coding theory. *Philips Res. Rep. Suppl.*, 10:vi–97, 1973.

[29] P. Delsarte. The association schemes of coding theory, in *Combinatorics: Proceedings of the NATO Advanced Study Institute held at Nijenrode Castle, Breukelen, The Netherlands 8–20 July 1974*. Springer Netherlands, Dordrecht, pp. 143–161, 1975.

[30] P. Delsarte. Bilinear forms over a finite field, with applications to coding theory. *J. Combin. Theory Ser. A*, 25(3):226–241, 1978.

[31] P. Delsarte, J.-M. Goethals, and F.J. Mac Williams. On generalized Reed-Muller codes and their relatives. *Inf. Control*, 16(5):403–442, 1970.

[32] P. Dembowski. *Finite Geometries*, Vol. 44. Springer Science & Business Media, New York, 1997.

[33] A. Devillers, M. Giudici, C. H. Li, and C. E. Praeger. Locally s-distance transitive graphs. *J. Graph Theory*, 69:176–197, 2012.

[34] J.D. Dixon and B. Mortimer. *Permutation Groups*. Springer-Verlag, New York, 1996.

[35] A. Dur. The automorphism groups of Reed-Solomon codes. *J. Combin. Theory Ser. A*, 44(1):69–82, 1987.

[36] E.M. Gabidulin. Theory of codes with maximum rank distance. *Problemy peredachi informatsii*, 21(1):3–16, 1985.

[37] N. Gill, N.I. Gillespie, and J. Semeraro. Conway groupoids and completely transitive codes. *Combinatorica*, 38:399–442, 2018.

[38] N.I. Gillespie, M. Giudici, D.R. Hawtin, and C.E. Praeger. Entry-faithful 2-neighbour transitive codes. *Des. Codes Cryptogr.*, 79(3):549–564, 2016.

[39] N.I. Gillespie and D.R. Hawtin. Alphabet-almost-simple 2-neighbour-transitive codes. *Ars Math. Contemp.*, 14(2):345–357, 2017.

[40] N.I. Gillespie and C.E. Praeger. Neighbour transitivity on codes in Hamming graphs. *Des. Codes Cryptogr.*, 67(3):385–393, 2013.

[41] N.I. Gillespie and C.E. Praeger. Characterisation of a family of neighbour transitive codes. Preprint (`arXiv:1405.5427`), 2014.

[42] N.I. Gillespie and C.E. Praeger. Diagonally neighbour transitive codes and frequency permutation arrays. *J. Algebraic Combin.*, 39(3):733–747, 2014.

[43] N.I. Gillespie and C.E. Praeger. New characterisations of the Nordstrom–Robinson codes. *Bull. London Math. Soc.*, 49(2):320–330, 2017.

[44] N.I. Gillespie, C.E. Praeger, and P. Spiga. Twisted permutation codes. *J. Group Theory*, 18(3):407–433, 2015.

[45] M. Giudici. *Completely Transitive Codes in Hamming Graphs.* (Master's thesis, The University of Western Australia, Perth, Australia, 1998).

[46] M. Giudici and C.E. Praeger. Completely transitive codes in Hamming graphs. *Euro. J. Comb.*, 20(7):647–662, 1999.

[47] C.D. Godsil, R.A. Liebler, and C.E. Praeger. Antipodal distance transitive covers of complete graphs. *Euro. J. Comb.*, 19:455–478, 1998.

[48] C. Godsil and C.E. Praeger. Completely transitive designs. Manuscript from 1997, uploaded to the math arXiv May 2014. Preprint, (`arXiv:1405.2176`), 2014.

[49] C. Godsil and G.F. Royle. *Algebraic Graph Theory.* Springer Science & Business Media, New York, 2013.

[50] A.R. Hammons, P.V. Kumar, A.R. Calderbank, N.J. Sloane, and P. Solé. The $\mathbb{Z}_4$-linearity of Kerdock, Preparata, Goethals, and related codes. *IEEE Trans. Inf. Theory*, 40(2):301–319, 1994.

[51] D.R. Hawtin. *Algebraic Symmetry of Codes in Hamming Graphs.* (PhD thesis, The University of Western Australia, Perth, Australia, 2017.) `doi: 10.4225/23/5a5808f48f2d0`

[52] D.R. Hawtin. s-Elusive codes in Hamming graphs. *Des. Codes Cryptogr.*, 89:1211–1220, 2021.

[53] D.R. Hawtin. Alphabet-affine 2-neighbour-transitive codes. Preprint, (`arXiv:2411.08351`), 2024.

[54] D.R. Hawtin. A new construction for frequency permutation arrays related to permutation groups. In preparation.

[55] D.R. Hawtin, N.I. Gillespie, and C.E. Praeger. Elusive codes in Hamming graphs. *Bull. Austral. Math. Soc.*, 88:286–296, 2013.

[56] D.R. Hawtin and C.E. Praeger. Minimal binary 2-neighbour-transitive codes. *J. Comb. Theory Ser. A*, 171:105–173, 2020.

[57] S. Huczynska and G.L. Mullen. Frequency permutation arrays. *J. Comb. Des.*, 14(6):463–478, 2006.

[58] A.A. Ivanov, R. Liebler, T. Penttila, and C.E. Praeger. Antipodal distance transitive covers of complete bipartite graphs. *Euro. J. Comb.* 18:11–34, 1997.

[59] C. Jansen, K. Lux, R. Parker, and R. Wilson. *An Atlas of Brauer Characters.* Clarendon Press, Oxford, 1995.

[60] W.M. Kantor. k-Homogeneous groups. *Math. Z.*, 124(4):261–265, 1972.

[61] P.B. Kleidman and M. Liebeck. *The Subgroup Structure of the Finite Classical Groups.* Cambridge University Press, Cambridge, 1990.

[62] M. Lavrauw, L. Storme, and G. Van de Voorde. On the code generated by the incidence matrix of points and k-spaces in $\mathrm{PG}(n,q)$ and its dual. *Finite Fields Appl.*, 14(4):1020–1038, 2008.

[63] M. Lavrauw and G. Van de Voorde. Field reduction and linear sets in finite geometry. *Top. Finite Fields*, 632:271–293, 2015.

[64] C.H. Li, C.E. Praeger, A. Venkatesh, and S. Zhou. Finite locally-quasiprimitive graphs. *Discrete Math.*, 246:197–218, 2002.

[65] R. Lidl and H. Niederreiter. *Finite Fields.* Encyclopedia of Mathematics and Its Applications, Vol. 20. Cambridge University Press, Cambridge, 1997.

[66] R.A. Liebler and C.E. Praeger. Neighbour-transitive codes in Johnson graphs. *Des. Codes Cryptogr.*, 73(1):1–25, 2014.

[67] F.J. MacWilliams and N.J.A. Sloane. *The Theory of Error Correcting Codes.* North-Holland Mathematical Library, North-Holland, Amsterdam, 1978.

[68] W.J. Martin. Completely regular codes: A viewpoint and some problems, in *Proceedings* Com^{2}MaC *Workshop on Distance-Regular Graphs and Finite Geometry, Busan, Korea, July 24–26, 2004*, pp. 43–56, 2004.

[69] I.Y. Mogilnykh and F.I. Solov'eva. On existence of perfect bitrades in Hamming graphs. *Discrete Math.*, 343(12):112–128, 2020.

[70] M. Neunhöffer and C.E. Praeger. Sporadic neighbour-transitive codes in Johnson graphs. *Des. Codes Cryptogr.*, 72(1):141–152, 2014.

[71] S.E. Payne and J.A. Thas. *Finite Generalized Quadrangles.* EMS Press, European Mathematical Society, Zurich, 2009.

[72] C.E. Praeger. An O'Nan–Scott theorem for finite quasiprimitive permutation groups, and an application to 2-arc transitive graphs. *J. London Math. Soc.*, 47:227–239, 1993.

[73] C.E. Praeger (with the assistance of Cai Heng Li and Alice C. Niemeyer). Finite transitive permutation groups and finite vertex-transitive graphs, in *Graph Symmetry: Algebraic Methods and Applications*, G. Hahn, G. Sabidussi, Eds., NATO ASI, Ser.C, Vol. 497. Kluwer, Dordrecht/Boston/London, pp. 277–318, 1997.

[74] C.E. Praeger. Codes and designs in Johnson graphs with high symmetry, in *Surveys in Combinatorics 2021*, K.K. Dabrowski, M. Gadouleau, N. Georgiou, M. Johnson, G.B. Mertzios, D. Paulusma, Eds., Vol. 470. Cambridge University Press, Cambridge, pp. 321–342, 2021.

[75] C.E. Praeger and C. Schneider. *Permutation Groups and Cartesian Decompositions.* Cambridge University Press, Cambridge, 2018.

[76] J. Rifà and J. Pujol. Completely transitive codes and distance transitive graphs, in *Applied Algebra, Algebraic Algorithms and Error-Correcting Codes (Proc. 9th Int. Sympos. AAECC-9, New Orleans, USA, Oct. 7–11, 1991)*, Mattson, H.F., Mora, T., and Rao, T.R.N., Eds., Lecture Notes in Computer Science, Vol. 539. Springer, Berlin, pp. 360–367, 1991.

[77] J. Rifà and J. Pujol. Translation-invariant propelinear codes. *IEEE Trans. Inf. Theory*, 43(2):590–598, 1997.

[78] J. Rifà and V.A. Zinoviev. On a class of binary linear completely transitive codes with arbitrary covering radius. *Discrete Math.*, 309:5011–5016, 2009.

[79] J. Rifà and V.A. Zinoviev. New completely regular q-ary codes based on Kronecker products. *IEEE Trans. Inf. Theory*, 56:266–272, 2010.

[80] J. Rifà and V.A. Zinoviev. On lifting perfect codes. *IEEE Trans. Inf. Theory*, 57:5918–5925, 2011.

[81] N.V. Semakov, V.A. Zinoviev, and G.V. Zaitsev. Uniformly packed codes. *Probl. Inf. Transm.*, 7(1):30–39, 1971.

[82] P. Sin. On codes that are invariant under the affine group. *Electron. J. Comb.*, 19(4):P20, 2012.

[83] P. Solé. Completely regular codes and completely transitive codes. RR-0727, *INRIA*. 1987. inria-00075825

[84] P. Solé. Completely regular codes and completely transitive codes. *Discrete Math.*, 81(2):193–201, 1990.

[85] A.B. Sorensen. Projective Reed–Muller codes. *IEEE Trans. Inf. Theory*, 37(6):1567–1576, 1991.

[86] M. Suzuki. *Group Theory I.* Grundlehren der mathematischen Wissenschaften. Springer, Berlin, Heidelberg, 1982.

[87] J.A. Thas. MDS codes and arcs in projective spaces: A survey. *Matematiche.* 47(2):315–328, 1992.

[88] K. Thas. *Symmetry in Finite Generalized Quadrangles.* Birkhäuser Verlag, Basel, 2004.

[89] A. Tietäväinen. On the nonexistence of perfect codes over finite fields. *SIAM J. Appl. Math.*, 24(1):88–96, 1973.

[90] J.H. Van Lint. A survey of perfect codes. *Rocky Mountain J. Math.*, 5:199–224, 1975.

[91] R. Wilson. *The Finite Simple Groups*, Graduate Texts in Mathematics, Vol. 251. Springer Science & Business Media, Germany, 2009.

[92] V.A. Zinoviev and V.K. Leontiev. The nonexistence of perfect codes over Galois fields (Engl. Transl.). *Probl. Control Inf. Theory*, 2(2):16–24, 1973.

CHAPTER 6

Some Completely Regular Codes in Doob Graphs

Evgeny A. Bespalov and Denis S. Krotov

6.1 INTRODUCTION

In algebraic graph theory, Doob graphs $D(m,n)$ are known as distance-regular graphs with the same intersection array as the Hamming graph $H(N,4)$, $N = 2m+n$. The Doob graph $D(m,n)$ is the Cartesian product of $m > 0$ copies of the Shrikhande graph, which is the srg$(16,6,2,2)$ different from $K_4 \square K_4$, and $n \geq 0$ copies of the complete graph K_4. The Doob graph can be naturally represented as a Cayley graph on the module $\mathbb{E}_4^m \times \mathbb{F}_4^n$ over $\mathbb{E}_4 = \mathrm{GR}(4^2)$ or, alternatively, on the module $\mathbb{Z}_4^{2m} \times \mathbb{Z}_2^{2n'} \times \mathbb{Z}_4^{n''}$ over $\mathbb{Z}_4$, where $n' + n'' = n$. With this representation, we can treat linear and additive codes, as well as dual codes in the sense of a given inner product. In Section 6.2, we prove the MacWilliams identities connecting the weight distributions of a linear or additive code and its dual. In particular, for each case, we determine the dual scheme, on the same set but with different metrics, such that the weight distribution of an additive code C in the Doob scheme $D(m, n' + n'')$ is related by the MacWilliams identities with the weight distribution of the dual code $C^\perp$ in the dual scheme. In the other sections of this chapter, we consider special classes of codes in Doob graphs most of them are related to completely regular codes with covering radius 1. In Section 6.3, we characterize the parameters of perfect and extended perfect codes in Doob graphs, including the characterization of parameters of additive and linear such codes (there are Doob graphs in which perfect codes exist but linear or additive perfect codes do not exist). Sections 6.4 and 6.5 contain the characterization of MDS codes in Doob graphs, separately for code distance more than 2 and code distance 2. In particular, Section 6.5 contains elements of the theory of unrestricted distance-2 MDS codes in Hamming graphs (equivalently, multary quasigroups) and the characterization theorem [43] for such codes in $D(m,n)$ including the case $m = 0$ corresponding to $H(n,4)$ [45]. Some attention has been paid to properties of other completely regular codes with covering radius 1 and minimum eigenvalue, which can be treated as multifold MDS codes. Section 6.6 is

DOI: 10.1201/9781003393931-6

devoted to the construction and the study of parameters of completely regular codes with covering radius 1, or, equivalently, equitable 2-partitions, in Doob graphs. The technique developed there includes some novel tools for the Cartesian product of graphs and extends the results obtained in [12] for Hamming graphs. In the end of Section 6.6, one can find the characterization of parameters of multifold one-perfect codes in Doob graphs.

6.1.1 General Definitions

We imply that the reader is familiar with the following concepts, defined in Chapter 1:

- *graphs* (multiple edges and loops are allowed), *vertices*, *edges*, *adjacent* vertices, *neighborhood*, *degree* of a vertex, *regular graph*, *induced subgraph* (Definition 1.1);
- *connected graph*, *graph distance* (Definition 1.2);
- *sum* of graphs, *complement graph* (Definition 1.8);
- *clique* (k-*clique*), *coclique*, *independent set* (Definition 1.12);
- *automorphism* of a graph, *automorphism group* $\mathrm{Aut}(\Gamma)$ of a graph Γ, *isomorphism* between two graphs (Definition 1.6);
- *Cartesian product* of graphs (Definition 1.9);
- *Cayley graph* $\mathrm{Cay}(G, S)$ on a group G with *connecting* set $S \subset G$, $S = S^{-1}$ (Definition 1.10);
- *complete graph*, *Hamming graph*, *hypercube* (n-*cube*) (Definition 1.11);
- *adjacency matrix*, *eigenfunctions*, and *eigenvalues* of a graph (Definition 1.14);
- *perfect coloring*, *equitable partition*, *quotient matrix* (Definition 1.17);
- *completely regular code*, *covering radius*, *intersection array*, *completely regular partition* (Definition 1.21);
- *covering* of a *target* graph by a *cover* graph (Definition 1.22);
- *strongly regular graph* with parameters $\mathrm{srg}(v, k, \lambda, \mu)$ (Definition 1.23);
- *distance-regular graph* with *intersection array* $\{b_0, \ldots, b_{r-1}; 1 = c_1, \ldots, c_r\}$ (Definition 1.24);
- *code* in a graph, *code distance*, *automorphism group* of a code, *equivalent* codes (Definition 1.31);
- *retract*, i-*retract* (Definition 1.7).

Definition 6.1 (tensor product of matrices and graphs). *For given $r_1 \times t_1$-matrix $A = (a_{i,j})$ and $r_2 \times t_2$-matrix $B = (b_{i,j})$, their* Kronecker product *(also known as* tensor product*) $A \otimes B$ is the $(r_1 r_2 \times t_1 t_2)$-matrix is defined as follows:*

$$A \otimes B = \begin{pmatrix} a_{1,1}B & a_{1,2}B & \dots & a_{1,t_1}B \\ a_{2,1}B & a_{2,2}B & \dots & a_{2,t_1}B \\ \dots & \dots & \dots & \dots \\ a_{r_1,1}B & a_{r_1,2}B & \dots & a_{r_1,t_1}B \end{pmatrix}.$$

For given graphs $G = (V_G, E_G)$ and $H = (V_H, E_H)$ with adjacency matrices A_G and A_H, respectively, their tensor product *$G \otimes H$ is the graph on the vertex set $V_G \times V_H$ with adjacency matrix $A_G \otimes A_H$.*

6.1.2 Shrikhande Graph

Definition 6.2 (Shrikhande graph). *The* Shrikhande graph Sh *is the Cayley graph on the group $\mathbb{Z}_4^{2+}$ with connecting set $\{01, 03, 10, 30, 11, 33\} = \{\pm 01, \pm 10, \pm 11\}$. Naturally,* Sh *is a quotient graph of the infinite triangular grid* $\mathrm{Cay}(\mathbb{Z}^{2+}, \{\pm 01, \pm 10, \pm 11\})$*, where the cosets of $4\mathbb{Z}^2$ are identified to the vertices of* Sh*. Keeping in mind this representation, it is often convenient to draw the Shrikhande graph as a fragment from 16 vertices of the infinite triangular grid, not necessarily of rhombic shape, see e.g. Figure 6.8. The Shrikhande graph is strongly regular with parameters $(16, 6, 2, 2)$ the only (up to isomorphism) other strongly regular graph with the same parameters is the Hamming graph $H(2, 4) = K_4 \square K_4$.*

Lemma 6.1 (4-cocliques in Sh). *In the Shrikhande graph, there are exactly two 4-cocliques up to equivalence: $\{00, 02, 20, 22\}$ (Figure 6.1a) and $\{00, 02, 21, 23\}$ (Figure 6.1b); totally sixteen 4-cocliques (see also Figure 6.6). Each 4-coclique in* Sh *is a completely regular code with quotient matrix $\begin{pmatrix} 0 & 6 \\ 2 & 4 \end{pmatrix}$. Each vertex of the Shrikhande graph belongs to exactly four different 4-cocliques. There are exactly three equivalence classes of partitions of the vertex set of the Shrikhande graph into 4-cocliques, see Figure 6.2a–c.*

Lemma 6.2 ([11, Lemma 8], [9]). *Let us be given two edge-disjoint (i.e., $E' \cap E'' = \emptyset$) graphs $\Gamma' = (V, E')$ and $\Gamma'' = (V, E'')$ on the same vertex set V, each of them being isomorphic to either* Sh *or $K_4 \square K_4$. Then their complement $\Gamma = \overline{\Gamma' \cup \Gamma''}$ is the sum $K_4 + K_4 + K_4 + K_4$ of four vertex-disjoint graphs K_4.*

(a) 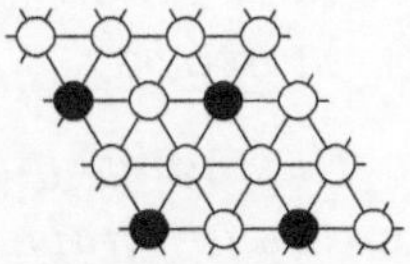 (b) 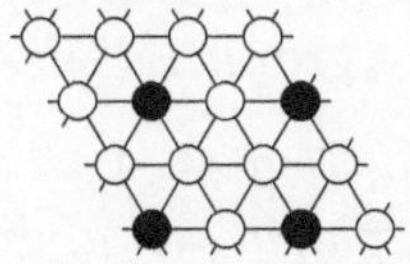

Figure 6.1 All 4-cocliques in Sh up to equivalence. (a) represents an equivalence class of size 4; (b) represents an equivalence class of size 12.

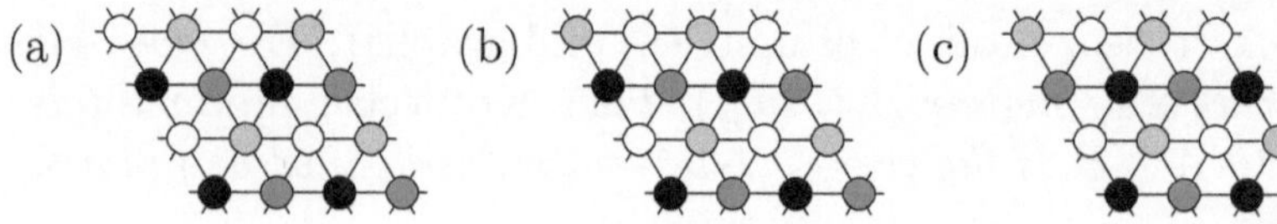

Figure 6.2 All partitions of the vertex set of Sh into 4-cocliques, up to equivalence. (a) represents an equivalence class of size 1; (b) represents an equivalence class of size 3; (c) represents an equivalence class of size 6.

Each of them is an equitable partition with quotient matrix $\begin{pmatrix} 0 & 2 & 2 & 2 \\ 2 & 0 & 2 & 2 \\ 2 & 2 & 0 & 2 \\ 2 & 2 & 2 & 0 \end{pmatrix}$.

The proof below inherits the idea of the proof that K_{10} cannot be partitioned into three Petersen graphs, which can be found in Peter Cameron's blog https://cameroncounts.wordpress.com/2013/07/24/partitions-into-petersens/. The proof in [11, Lemma 8] is different and contains a gap, which was corrected in [9].

Proof. Each of the graphs Γ', Γ'' has three eigenvalues 6, 2, and -2, with multiplicities 1, 6, and 9, respectively. Let A', A'', and A be the adjacency matrices of Γ', Γ'', and Γ, respectively, and let J be the all-1 16×16 matrix. We have

$$A' + A'' + A + I = J$$

The nine-dimensional eigenspaces of A' and A'' lie inside the 15-dimensional space orthogonal to the all-1 vector; so they have intersection of dimension at least 3. For each vector v in this eigenspace, we have $A'v = A''v = -2v$, $Iv = v$, and $Jv = 0$. It follows that $Av = 3v$. Additionally, the all-1 vector is also an eigenvector of A with eigenvalue 3. We see that the regular graph Γ of valency 3 has multiplicity at least 4 for the eigenvalue 3. Hence it has at least 4 connected components. The only case is $K_4 + K_4 + K_4 + K_4$. □

6.1.3 Doob Graphs and Covering by Hypercubes

Definition 6.3. *Denote by $D(m,n)$ the Cartesian product of m copies of the Shrikhande graph* Sh *and n copies of the complete graph K_4 of order 4. If $m = 0$, then $D(m,n)$ is the Hamming graph $H(n,4)$. If $m \geq 1$, then $D(m,n)$ is called a* Doob graph.

The Doob graph $D(m,n)$ is a distance regular graph with the same intersection array as the Hamming graph $H(2m+n,4)$. Moreover,

Theorem 6.1 ([20]). *If $q \neq 4$, then the Hamming graphs $H(N,q)$ is uniquely defined up to isomorphism as a distance-regular graph with intersection array $\{N(q-1),(N-1)(q-1),\dots,(q-1);1,2,\dots,N\}$. If $q = 4$, then there exists (up to isomorphism) exactly $\lfloor N/2 \rfloor$ other distance-regular graphs with the same intersection array, namely, the Doob graphs $D(m,n)$, $2m+n = N$, $m \geq 1$.*

Since K_4 can be represented as a Cayley graph over or $\mathbb{Z}_2^{2+}$ or $\mathbb{Z}_4^+$, the Doob graph can also be represented as a Cayley graph over the additive group of $(\mathbb{Z}_4^2)^m \times (\mathbb{Z}_2^2)^{n'} \times \mathbb{Z}_4^{n''}$ for every decomposition $n' + n''$ of n. In Section 6.2, we will consider a more advanced algebraic structure on the set of vertices the Doob graph, the structure of a module over the ring GR(4^2) of order 16.

An important fact is existence of a covering of the Doob graph by a hypercube. In particular, the following proposition answers an inexplicit question in [Remark 3.8.A][34], where the authors wrote "We do not know any example where Γ is a Hamming graph and Π is a completely regular partition such that Γ/Π is a Doob graph."

Proposition 6.1. *For every pair (m, n) of nonnegative integers, there is a covering of $D(m, n)$ by $H(6m + 3n, 2)$.*

Proof. It is well known that the structure of a Cayley graph over $\mathbb{Z}_2^N$ is not the only representation of $H(N, 2)$ as a Cayley graph. In particular, $H(6m + 3n, 2)$ is isomorphic to $\mathrm{Cay}(\mathbb{Z}_4^{3m} \times \mathbb{Z}_2^{3n}, \{\pm \boldsymbol{e}_1, \ldots, \pm \boldsymbol{e}_{3m}, \boldsymbol{e}_{3m+1}, \ldots, \boldsymbol{e}_{3m+3n}\})$, where $\boldsymbol{e}_i$ is a $(3m + 3n)$-tuple with 1 in the ith position and 0 in the others. The isomorphism with the classical representation of the hypercube is given by replacing each $\mathbb{Z}_4$-coordinate by two $\mathbb{Z}_2$-coordinates with the help of the map $0 \to 00$, $1 \to 01$, $2 \to 11$, $3 \to 10$, but for convenience, we will use the same notation $H(6m + 3n, 2)$ for this Cayley graph. The Doob graph $D(m, n)$ is

$$\mathrm{Cay}(\mathbb{Z}_4^{2m} \times \mathbb{Z}_2^{2n}, \{\pm \boldsymbol{e}'_1, \pm \boldsymbol{e}''_1, \pm(\boldsymbol{e}'_1 + \boldsymbol{e}''_1), \ldots, \pm \boldsymbol{e}'_m, \pm \boldsymbol{e}''_m, \pm(\boldsymbol{e}'_m + \boldsymbol{e}''_m),$$
$$\boldsymbol{e}'_{m+1}, \boldsymbol{e}''_{m+1}, \boldsymbol{e}'_{m+1} + \boldsymbol{e}''_{m+1}, \ldots, \boldsymbol{e}'_{m+n}, \boldsymbol{e}''_{m+n}, \boldsymbol{e}'_{m+n} + \boldsymbol{e}''_{m+n}, \}).$$

where $\boldsymbol{e}'_i$ and $\boldsymbol{e}''_i$ are the vectors with $1, 0$ and $0, 1$, respectively, in the ith pair of positions and zeros in the other positions. We now consider the linear map ϕ from $\mathbb{Z}_4^{3m} \times \mathbb{Z}_2^{3n}$ onto $\mathbb{Z}_4^{2m} \times \mathbb{Z}_2^{2n}$ such that

$$\phi(\boldsymbol{e}_{3i-2}) = \boldsymbol{e}'_i, \quad \phi(\boldsymbol{e}_{3i-1}) = \boldsymbol{e}''_i, \quad \phi(\boldsymbol{e}_{3i}) = \boldsymbol{e}'_i + \boldsymbol{e}''_i$$

(the other values of ϕ are obtained by linearity). It is straightforward that ϕ is a covering of $D(m, n)$ by $H(6m + 3n, 2)$. Indeed, if a vertex $\boldsymbol{v}$ of $H(6m + 3n, 2)$ has ϕ-color $\boldsymbol{u}$, i.e., $\phi(\boldsymbol{v}) = \boldsymbol{u}$, then the neighbors $\boldsymbol{u} + \boldsymbol{e}$ (where $\boldsymbol{e}$ runs over the connecting set of $H(6m + 3n, 2)$) of $\boldsymbol{u}$ have ϕ-colors $\boldsymbol{u} + \phi(\boldsymbol{e})$, where $\phi(\boldsymbol{e})$ runs over the connecting set of $D(m, n)$ and hence $\boldsymbol{u} + \phi(\boldsymbol{e})$ runs over the neighborhood of $\boldsymbol{u}$. Hence, ϕ is a perfect coloring with quotient matrix being the adjacency matrix of $D(m, n)$. □

Taking into account Proposition 1.36(i) in Chapter 1, Proposition 6.1 means that every equitable partition (in particular, every completely regular code) in a Doob graph corresponds to an equitable partition (completely regular code) of a hypercube with the same quotient matrix.

Example 6.1. *For example, the equitable partition*

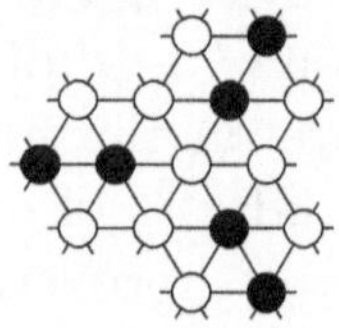

of the Shrikhande graph with quotient matrix $(\begin{smallmatrix}1&5\\3&3\end{smallmatrix})$ *corresponds to a known equitable partition of the cover* 6*-cube graph* $H(6,2)$*, which generates a known infinite series [64] of unbalanced Boolean functions with the best correlation immunity* $2n/3-1$ *(see the bound in [22] and also the discussion in Section 1.4.3.3 of Chapter 1 of this book).*

Example 6.2. *Another nice picture*

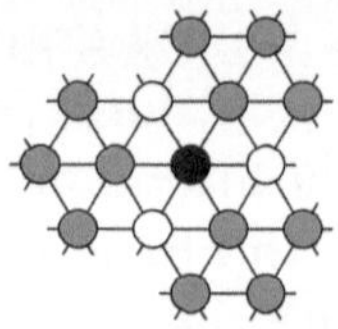

illustrates (taking into account the covering of Sh *by* $H(6,2)$*) the "perfect semicoloring" of the* 6*-cube used as an intermediate step in the construction [24] of a perfect coloring of* $H(12,2)$ *with quotient matrix* $(\begin{smallmatrix}3&9\\7&5\end{smallmatrix})$*, which, in its turn, induces another infinite class of unbalanced Boolean functions with the best correlation immunity. The black and white colors are treated as the vectors* $(1,0)$ *and* $(0,1)$*, respectively, while the gray it treated as* $(\frac{1}{2},\frac{1}{2})$*. The corresponding* $|V|\times 2$ *coloring matrix* F *then satisfies the matrix equation* $AF = FS$ *(see equation (1.2) in Chapter 1), where* A *is the adjacency matrix of the graph (*Sh *or* $H(6,2)$*, for the coloring induced by the covering) and* $S = \frac{1}{2}\begin{pmatrix}3 & 9\\7 & 5\end{pmatrix}$*. An additional requirement for such a coloring in the constriction [24] is that the gray vertices induce* 4*-cycles in* $H(6,2)$*; it can be straightforwardly checked that the covering preimage of the three gray rhombuses in the picture consists of twelve* 4*-cycles.*

In the rest of this section, we list all nonequivalent equitable partitions of the Shrikhande graph (Table 6.1).

Up to equivalence, there are seven completely regular codes with covering radius 1 in the Shrikhande graph:

- one with quotient matrix $\begin{pmatrix}1 & 5\\3 & 3\end{pmatrix}$ (see Section 6.5);
- one with quotient matrix $\begin{pmatrix}4 & 2\\2 & 4\end{pmatrix}$;

TABLE 6.1 Perfect Colorings of the Shrikhande Graph.

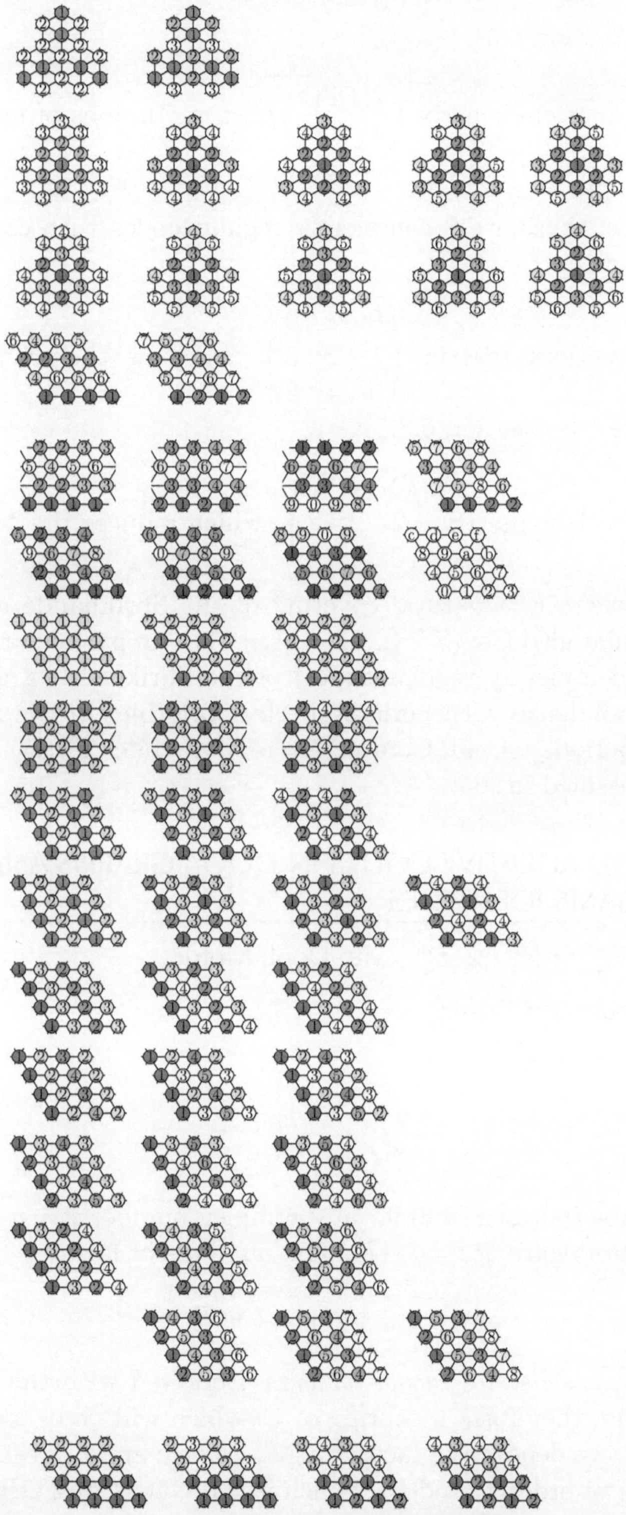

- two with quotient matrix $\begin{pmatrix} 0 & 6 \\ 2 & 4 \end{pmatrix}$, which are maximum cocliques or MDS codes (see Section 6.5);
- three with quotient matrix $\begin{pmatrix} 2 & 4 \\ 4 & 2 \end{pmatrix}$, which are unions of maximum cocliques or two-fold MDS codes (see Section 6.5).

Up to equivalence, there are 2 completely regular codes with covering radius 2 in the Shrikhande graph:

- one with quotient matrix $\begin{pmatrix} 0 & 6 & 0 \\ 1 & 2 & 3 \\ 0 & 4 & 2 \end{pmatrix}$, which is the trivial extended one-perfect code (see Section 6.3.5);
- one with quotient matrix $\begin{pmatrix} 2 & 4 & 0 \\ 2 & 2 & 2 \\ 0 & 4 & 2 \end{pmatrix}$, which induces the 4-cycle.

Note that there is a natural covering of the Shrikhande graph by the infinite triangular grid $\mathrm{Cay}(\mathbb{Z}^2, \{\pm 01, \pm 11, \pm 11\})$. In particular, perfect colorings (similarly, completely regular codes) of the Shrikhande graph are in one-to-one correspondence with periodic perfect colorings of the infinite triangular grid with periods 04, 40. Completely regular codes in the infinite triangular grid are classified in [65].

6.2 LINEAR AND ADDITIVE CODES IN DOOB GRAPHS AND MACWILLIAMS IDENTITIES

6.2.1 Codes in $\mathbb{E}_4^m \times \mathbb{F}_4^{n'} \times \mathbb{Z}_4^{n''}$ with Doob Metric

6.2.1.1 Linear Codes and Duality

Let i be a square root of -1, let

$$\omega := \frac{-1 + i\sqrt{3}}{2}$$

be a primitive cube root of 1, and for any complex number $x = a + bi \in \mathbb{C}$, $a, b \in \mathbb{R}$, let $\bar{x}$ denote its conjugate, $\bar{x} := a - bi$. The numbers of form

$$a + b\omega, \quad a, b \in \mathbb{Z}$$

are known as the *Eisenstein–Jacobi integers*. Endowed with the complex addition and multiplication, they form a subring of $\mathbb{C}$, which will be denoted by $\mathbb{E}$.

By $\mathbb{E}_p$, $p \in \mathbb{E}$, we denote the factor ring $\mathbb{E}/p\mathbb{E}$; we are interested in $\mathbb{E}_2$, which is the finite field $\mathbb{F}_4$ of order 4, and $\mathbb{E}_4$, which is the Galois ring $\mathrm{GR}(4^2)$ (Figure 6.3). The set of tuples from the Cartesian product $V_{m,n} := \mathbb{E}_4^m \times \mathbb{F}_4^n$ of m copies of $\mathbb{E}_4$ and

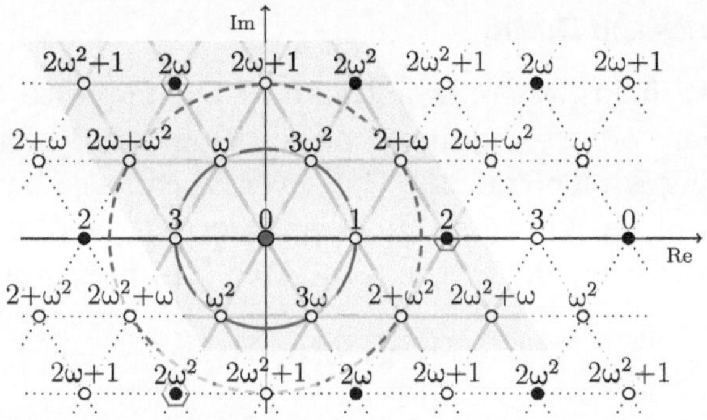

Figure 6.3 Representation of the Shrikhande graph as a Cayley graph of $\mathbb{E}/4\mathbb{E} \simeq \mathrm{GR}(4^2)$. The solid circle indicates the set Ω; the dashed circle indicates the coset $\psi\Omega$; the three small hexagons indicate the coset 2Ω.

n copies of $\mathbb{F}_4$ with the component-wise addition and multiplication by a constant (modulo 4 in the first m positions, modulo 2 in the last n positions) forms a module over $\mathbb{E}_4$.

In $\mathbb{E}_4$, we consider the set of powers of $-\omega$ and call it Ω:

$$\Omega := \{\pm 1, \pm\omega, \pm\bar{\omega}\}.$$

It forms a multiplicative subgroup of $\mathbb{E}_4$, and $\mathbb{E}_4$ is partitioned into the multiplicative cosets 0Ω, Ω, 2Ω, and $\psi\Omega$, where

$$\psi := \omega - 1.$$

Any nonempty subset of $V_{m,n}$ closed under addition and multiplication by a constant is called a *linear code* or a *linear* $\mathbb{E}_4\mathbb{F}_4$ *code*. So, the linear codes correspond to the submodules of $V_{m,n}$.

Any nonempty subset of $V_{m,n}$ closed under addition is called an *additive code* or an *additive* $\mathbb{E}_4\mathbb{F}_4$ *code*.

For $\boldsymbol{x} = (x_1, \ldots, x_k)$ and $\boldsymbol{y} = (y_1, \ldots, y_k)$ from $\mathbb{E}_4^k$ or $\mathbb{F}_4^k$, the standard inner product $\langle \boldsymbol{x}, \boldsymbol{y} \rangle$ is defined as usual:

$$\langle \boldsymbol{x}, \boldsymbol{y} \rangle := x_1 y_1 + \cdots + x_k y_k. \tag{6.1}$$

For $\boldsymbol{x} = (\boldsymbol{x}^*, \boldsymbol{x}')$ and $\boldsymbol{y} = (\boldsymbol{y}^*, \boldsymbol{y}')$, where $\boldsymbol{x}^*, \boldsymbol{y}^* \in \mathbb{E}_4^m$, $m > 0$, and $\boldsymbol{x}', \boldsymbol{y}' \in \mathbb{F}_4^n$, the inner product is defined as

$$\langle \boldsymbol{x}, \boldsymbol{y} \rangle := \langle \boldsymbol{x}^*, \boldsymbol{y}^* \rangle + 2 \cdot \langle \boldsymbol{x}', \boldsymbol{y}' \rangle, \tag{6.2}$$

where $2\cdot$ is the group homomorphism $z + 2\mathbb{E} \to 2z + 4\mathbb{E}$ from $\mathbb{F}_4^+$ to $\mathbb{E}_4^+$.

Two vectors $\boldsymbol{x}$ and $\boldsymbol{y}$ are said to be *orthogonal* if $\langle \boldsymbol{x}, \boldsymbol{y} \rangle = 0$. For a linear code C in $V_{m,n}$, its *dual*, denoted $C^\perp$, is defined as the set of all vectors from $V_{m,n}$ that are orthogonal to every vector from C. The dual $C^\perp$ is also a linear code and one has $C^{\perp\perp} = C$ and $|C| \cdot |C^\perp| = |V_{m,n}|$.

6.2.1.2 *Additive Codes and Duality*

For codes over finite fields, there is a standard technique to extend the concept of duality from linear codes to additive codes; below, we describe the approach in application to the ring $\mathbb{E}_4$. The module $V_{m,n}$ over $\mathbb{E}_4$ can also be viewed as a module over the subring $\mathbb{Z}_4$ of $\mathbb{E}_4$. Given an inner product $\langle\cdot,\cdot\rangle$ and a surjective $\mathbb{Z}_4$-linear map L from $\mathbb{E}_4$ onto $\mathbb{Z}_4$, we define the associated $\mathbb{Z}_4$-inner product $[\cdot,\cdot]_L$ as

$$[\cdot,\cdot]_L := L(\langle\cdot,\cdot\rangle).$$

Two vectors $\boldsymbol{x}$ and $\boldsymbol{y}$ are said to be *L-orthogonal* if $[\boldsymbol{x},\boldsymbol{y}]_L = 0$. Given an additive code C in $V_{m,n}$, its *L-dual* is defined as the set of all vectors from $V_{m,n}$ that are L-orthogonal to every vector from C. The following known and easy fact shows that the L-duality is an extension of the duality to the additive codes.

Proposition 6.2. *For every surjective $\mathbb{Z}_4$-linear map L from $\mathbb{E}_4$ onto $\mathbb{Z}_4$, two linear codes are L-dual if and only if they are dual.*

Proof. Let C be a linear code. Obviously, the orthogonality of two vectors implies their L-orthogonality. So, any vector from $C^\perp$ also belongs to the L-dual of C. Let us consider a vector $\boldsymbol{v}$ from the L-dual. By the definition,

$$L(\langle\boldsymbol{v},\boldsymbol{x}\rangle) = 0 \qquad \text{for all } \boldsymbol{x} \in C.$$

Since C is linear,

$$L(a\langle\boldsymbol{v},\boldsymbol{x}\rangle) = L(\langle\boldsymbol{v},a\boldsymbol{x}\rangle) = 0 \quad \text{for all } \boldsymbol{x} \in C,\ a \in \mathbb{E}_4. \tag{6.3}$$

Seeking a contradiction, assume that $b := \langle\boldsymbol{v},\boldsymbol{x}\rangle \neq 0$ for some $\boldsymbol{x}$ from C. Take $d \in \mathbb{E}_4$ such that $L(d) = 1$. There should be a such that $ab = d$ or $ab = 2d$. For this a, we have $L(a\langle\boldsymbol{v},\boldsymbol{x}\rangle) \in \{1,2\}$, contradicting (6.3). Hence, for all $\boldsymbol{x}$ from C we have $\langle\boldsymbol{v},\boldsymbol{x}\rangle = 0$, and $\boldsymbol{v}$ belongs to $C^\perp$. □

A standard linear map from $\mathbb{E}_4$ to $\mathbb{Z}_4$ (as well as from a finite field to its prime-order subfield) used for different purposes is the *trace* Tr:

$$\mathrm{Tr}(z) := z + \bar{z}, \qquad z \in \mathbb{E}_4$$

(for a Galois ring in general, $\mathrm{Tr}(z) := \sum \sigma(z)$, where the sum is over all automorphisms σ of the ring). Below, for brevity, the Tr-duality will be referred to as duality (by Proposition 6.2, it is in agreement with the duality of linear codes), and the Tr-inner product $[\cdot,\cdot]_{\mathrm{Tr}}$ will be denoted as $[\cdot,\cdot]$. The notation $C^\perp$ is also extended to additive codes C.

6.2.1.3 Extending the Concept of Additive Codes

We can further extend the definition of additive codes, the concept of duality, and the corresponding notation to the codes in

$$V_{m,n',n''} := \mathbb{E}_4^m \times \mathbb{F}_4^{n'} \times \mathbb{Z}_4^{n''}$$

by defining the following $\mathbb{Z}_4$-inner product:

$$[\boldsymbol{x}, \boldsymbol{y}] := \mathrm{Tr}\langle \boldsymbol{x}^*, \boldsymbol{y}^* \rangle + 2 \cdot \mathrm{Tr}\langle \boldsymbol{x}', \boldsymbol{y}' \rangle + \langle \boldsymbol{x}'', \boldsymbol{y}'' \rangle.$$

Here and below, for any vector $\boldsymbol{z}$ from $V_{m,n',n''}$, the notations $\boldsymbol{z}^*$, $\boldsymbol{z}'$, and $\boldsymbol{z}''$ denote the $\mathbb{E}_4$-, $\mathbb{F}_4$-, and $\mathbb{Z}_4$-parts of $\boldsymbol{z}$, respectively; z_i^*, z_i', and z_i'' refer to the i-th coordinate of $\boldsymbol{z}^*$, $\boldsymbol{z}'$, or $\boldsymbol{z}''$, respectively. I.e.,

$$\boldsymbol{z} = (\boldsymbol{z}^*, \boldsymbol{z}', \boldsymbol{z}'') = (z_1^*, \ldots, z_m^*, z_1', \ldots, z_{n'}', z_1'', \ldots, z_{n''}'').$$

The motivation for this extension lies in the extended possibility to construct codes with different parameters in the same metric space (as we will see in the next subsection, the coordinates from the last two groups are metrically equivalent in the considered scheme). Examples of exploiting this possibility can be found in the construction of additive one-perfect codes [59].

6.2.1.4 The Doob Metric

We define two weight function on $V_{m,n',n''}$:

$$\mathrm{wt}(\boldsymbol{x}) := \sum_{j=1}^{m} \mathrm{wt}(x_j^*) + \sum_{j=1}^{n'} \mathrm{wt}(x_j') + \sum_{j=1}^{n''} \mathrm{wt}(x_j''),$$

$$\widetilde{\mathrm{wt}}(\boldsymbol{x}) := \sum_{j=1}^{m} \widetilde{\mathrm{wt}}(x_j^*) + \sum_{j=1}^{n'} \mathrm{wt}(x_j') + \sum_{j=1}^{n''} \mathrm{wt}(x_j''),$$

called the *weight* and the *coweight* of $\boldsymbol{x}$, where

$$\mathrm{wt}(x_j^*) = \begin{cases} 0 & \text{if } x_j^* = 0 \\ 1 & \text{if } x_j^* \in \Omega \\ 2 & \text{otherwise,} \end{cases} \qquad \widetilde{\mathrm{wt}}(x_j^*) = \begin{cases} 0 & \text{if } x_j^* = 0 \\ 1 & \text{if } x_j^* \in \psi\Omega \\ 2 & \text{otherwise,} \end{cases}$$

$$\mathrm{wt}(x_j') = \begin{cases} 0 & \text{if } x_j' = 0 \\ 1 & \text{otherwise,} \end{cases} \qquad \mathrm{wt}(x_j'') = \begin{cases} 0 & \text{if } x_j'' = 0 \\ 1 & \text{otherwise.} \end{cases}$$

Naturally, the weight function wt defines the distance

$$d(\boldsymbol{x}, \boldsymbol{y}) := \mathrm{wt}(\boldsymbol{y} - \boldsymbol{x}) \tag{6.4}$$

and the distance-1 graph $D(m, n' + n'')$ on $V_{m,n',n''}$, where two vectors are adjacent if and only if they are at distance 1. If $m > 0$, then $D(m, n)$ is called a *Doob graph* (the case $m = 0$ corresponds to the quaternary Hamming graph $H(n, 4)$). The graph $D(m, n' + n'')$ is the Cartesian product of m copies of the *Shrikhande graph* $D(1, 0)$

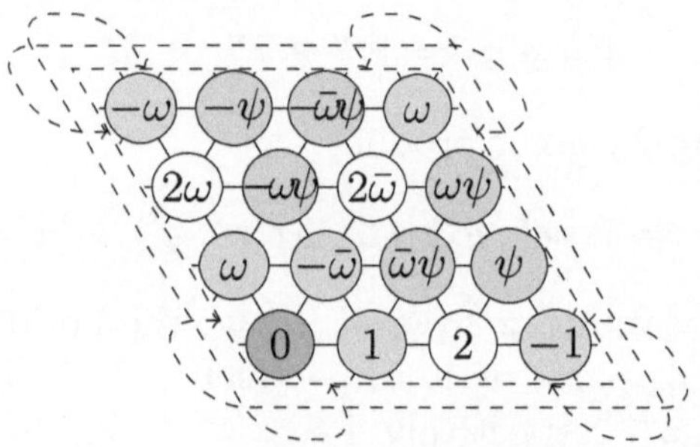

and $n' + n''$ copies of the complete graph of order 4 (n' copies of the complete graph with the vertex set $\mathbb{F}_4$ and n'' copies of the complete graph with the vertex set $\mathbb{Z}_4$).

It is a distance-regular graph, and the corresponding metric induces an association scheme, a *Doob scheme.* From the metrical point of view, there is no difference between the $n = n' + n''$ last coordinates; there are isometries of the metric space that permute these coordinates in an arbitrary manner. The difference between these two groups becomes important only when we consider the additive structure, for example, when constructing additive codes.

In a similar way, $\widetilde{\mathrm{wt}}$ defines another graph on $V_{m,n',n''}$, say $\widetilde{D}(m, n' + n'')$. It is isomorphic to $D(m, n' + n'')$ (an isomorphism is given by multiplying the first m coordinates of the vectors by ψ). From the theory below, one can see that the graphs $D(m, n' + n'')$ and $\widetilde{D}(m, n' + n'')$ are in some sense dual to each other.

6.2.2 MacWilliams Identities for Codes in $\mathbb{E}_4^m \times \mathbb{F}_4^{n'} \times \mathbb{Z}_4^{n''}$

For any complex-valued function f on $V_{m,n',n''}$, define its *Fourier transform* $\widehat{f}\colon V_{m,n',n''} \to \mathbb{C}$:

$$\widehat{f}(\boldsymbol{u}) := \sum_{\boldsymbol{v} \in V_{m,n',n''}} i^{[\boldsymbol{u},\boldsymbol{v}]} f(\boldsymbol{v}).$$

Lemma 6.3. *Let C be an additive code in $V_{m,n',n''}$, and let $C^\perp$ be its dual. Then for every complex-valued function f on $V_{m,n',n''}$,*

$$\sum_{\boldsymbol{z} \in C^\perp} f(\boldsymbol{z}) = \frac{1}{|C|} \sum_{\boldsymbol{u} \in C} \widehat{f}(\boldsymbol{u}). \tag{6.5}$$

Proof. The proof is standard for such kind of identities. In the right part of (6.5), we substitute the definition of the Fourier transform and change the summation order:

$$\sum_{\boldsymbol{u} \in C} \widehat{f}(\boldsymbol{u}) = \sum_{\boldsymbol{u} \in C} \sum_{\boldsymbol{v} \in V_{m,n',n''}} i^{[\boldsymbol{u},\boldsymbol{v}]} f(\boldsymbol{v}) = \sum_{\boldsymbol{v} \in V_{m,n',n''}} f(\boldsymbol{v}) \sum_{\boldsymbol{u} \in C} i^{[\boldsymbol{u},\boldsymbol{v}]}. \tag{6.6}$$

If $\boldsymbol{v} \in C^\perp$, then $[\boldsymbol{u},\boldsymbol{v}] = 0$ for all $\boldsymbol{u}$ from C. Hence,

$$\sum_{\boldsymbol{u}\in C} i^{[\boldsymbol{u},\boldsymbol{v}]} = |C| \qquad \text{if } \boldsymbol{v} \in C^\perp. \tag{6.7}$$

If $\boldsymbol{v} \notin C^\perp$, then there exists $\boldsymbol{z}$ in C such that $[\boldsymbol{z},\boldsymbol{v}] \neq 0$. Since C is additive, $\boldsymbol{u} \in C$ is equivalent to $\boldsymbol{z}+\boldsymbol{u} \in C$, and we have

$$\sum_{\boldsymbol{u}\in C} i^{[\boldsymbol{u},\boldsymbol{v}]} = \sum_{\boldsymbol{z}+\boldsymbol{u}\in C} i^{[\boldsymbol{z}+\boldsymbol{u},\boldsymbol{v}]} = \sum_{\boldsymbol{u}\in C} i^{[\boldsymbol{z}+\boldsymbol{u},\boldsymbol{v}]} = i^{[\boldsymbol{z},\boldsymbol{v}]} \sum_{\boldsymbol{u}\in C} i^{[\boldsymbol{u},\boldsymbol{v}]}.$$

Since $[\boldsymbol{z},\boldsymbol{v}] \neq 0$, it holds $i^{[\boldsymbol{z},\boldsymbol{v}]} \neq 1$ and hence

$$\sum_{\boldsymbol{u}\in C} i^{[\boldsymbol{u},\boldsymbol{v}]} = 0 \qquad \text{if } \boldsymbol{v} \notin C^\perp. \tag{6.8}$$

From (6.7) and (6.8), we see that the right expression in (6.6) equals $\sum_{\boldsymbol{v}\in C^\perp} f(\boldsymbol{v})|C|$. □

Consider the function

$$f(\boldsymbol{v}) := \prod_{j=1}^{m} X_{v_j^*} \prod_{j=1}^{n'} Y_{v_j'} \prod_{j=1}^{n''} Z_{v_j''}$$

which also depends on the formal variables X_a, $a \in \mathbb{E}_4$, Y_b, $b \in \mathbb{F}_4$, Z_0, Z_1, Z_2, Z_3. The sum of f over some code C will be denoted by

$$W_C(X_a, a \in \mathbb{E}_4; Y_b, b \in \mathbb{F}_4; Z_c, c \in \mathbb{Z}_4), \text{ or, for short,} W_C(\boldsymbol{X};\boldsymbol{Y};\boldsymbol{Z}),$$

and called the *complete weight enumerator* of C. The *weight* and *coweight enumerators* of C

$$\begin{aligned} W_C(A,B) &:= \sum_{\boldsymbol{z}\in C} A^{N-\mathrm{wt}(\boldsymbol{z})} B^{\mathrm{wt}(\boldsymbol{z})} \quad \text{and} \\ \widetilde{W}_C(\mathcal{A},\mathcal{B}) &:= \sum_{\boldsymbol{z}\in C} \mathcal{A}^{N-\widetilde{\mathrm{wt}}(\boldsymbol{z})} \mathcal{B}^{\widetilde{\mathrm{wt}}(\boldsymbol{z})}, \quad N := 2m+n'+n'', \end{aligned} \tag{6.9}$$

respectively, are obtained from $W_C(\boldsymbol{X};\boldsymbol{Y};\boldsymbol{Z})$ by identifying

$$\begin{aligned} X_a &= A^{2-\mathrm{wt}(a)} B^{\mathrm{wt}(a)} = \mathcal{A}^{2-\widetilde{\mathrm{wt}}(a)} \mathcal{B}^{\widetilde{\mathrm{wt}}(a)}, \\ Y_0 &= Z_0 = A = \mathcal{A}, \quad Y_1 = Y_\omega = Y_{\bar\omega} = Z_1 = Z_2 = Z_3 = B = \mathcal{B}. \end{aligned} \tag{6.10}$$

From the definitions of f and the Fourier transform, we find

$$\begin{aligned} \widehat{f}(\boldsymbol{u}) &= \sum_{\boldsymbol{v}\in V_{m,n',n''}} i^{\mathrm{Tr}[\boldsymbol{u},\boldsymbol{v}]} \prod_{j=1}^{m} X_{v_j^*} \prod_{j=1}^{n'} Y_{v_j'} \prod_{j=1}^{n''} Z_{v_j''} \\ &= \prod_{j=1}^{m} \sum_{v\in\mathbb{E}_4} i^{\mathrm{Tr}(u_j^* v)} X_v \cdot \prod_{j=1}^{n'} \sum_{v\in\mathbb{F}_4} i^{2\cdot\mathrm{Tr}(u_j' v)} Y_v \cdot \prod_{j=1}^{n''} \sum_{v\in\mathbb{Z}_4} i^{u_j'' v} Z_v. \end{aligned}$$

After expressing (6.5), we get

$$W_{C^\perp}(\boldsymbol{X};\boldsymbol{Y};\boldsymbol{Z}) \tag{6.11}$$
$$= \frac{1}{|C|}W_C\Big(\sum_{v\in\mathbb{E}_4} i^{\mathrm{Tr}(av)}X_v, a\in\mathbb{E}_4;\ \sum_{v\in\mathbb{F}_4}(-1)^{\mathrm{Tr}(bv)}Y_v, b\in\mathbb{F}_4;\ \sum_{v\in\mathbb{Z}_4} i^{cv}Z_v, c\in\mathbb{Z}_4\Big).$$

We now identify the variables to get $W_{C^\perp}(A,B)$ in the left side of (6.11). By straightforward calculations using the character table (Figure 6.4), we can find the sum $\sum_{v\in\mathbb{E}_4} i^{\mathrm{Tr}(av)}A^{2-\mathrm{wt}(v)}B^{\mathrm{wt}(v)}$ for every a in $\mathbb{E}_4$. For example, if $a=\bar{\omega}\psi$, then the sum is

$$\begin{aligned}
&(-i)\cdot AB+(-1)\cdot B^2+\ (i)\cdot B^2+\ (1)\cdot AB\\
&+(-1)\cdot B^2+\ (i)\cdot B^2+\ (1)\cdot B^2+(-i)\cdot B^2\\
&+\ (i)\cdot AB+\ (1)\cdot AB+(-i)\cdot B^2+(-1)\cdot B^2\\
&+\ (1)\cdot A^2+(-i)\cdot AB+(-1)\cdot B^2+\ (i)\cdot AB
\end{aligned}$$

which is $A^2+2AB-3B^2=(A+3B)(A-B)$. Similarly, for every a in $\mathbb{E}_4$ we have

$$\sum_{v\in\mathbb{E}_4} i^{\mathrm{Tr}(av)}A^{2-\mathrm{wt}(v)}B^{\mathrm{wt}(v)} = (A+3B)^{2-\widehat{\mathrm{wt}}(a)}(A-B)^{\widehat{\mathrm{wt}}(a)}. \tag{6.12}$$

Also, directly,

$$\sum_{v\in\mathbb{F}_4}(-1)^{\mathrm{Tr}(bv)}A^{1-\mathrm{wt}(v)}B^{\mathrm{wt}(v)} = (A+3B)^{1-\mathrm{wt}(b)}(A-B)^{\mathrm{wt}(b)},$$

$$\sum_{v\in\mathbb{Z}_4} i^{cv}A^{1-\mathrm{wt}(v)}B^{\mathrm{wt}(v)} = (A+3B)^{1-\mathrm{wt}(c)}(A-B)^{\mathrm{wt}(c)}.$$

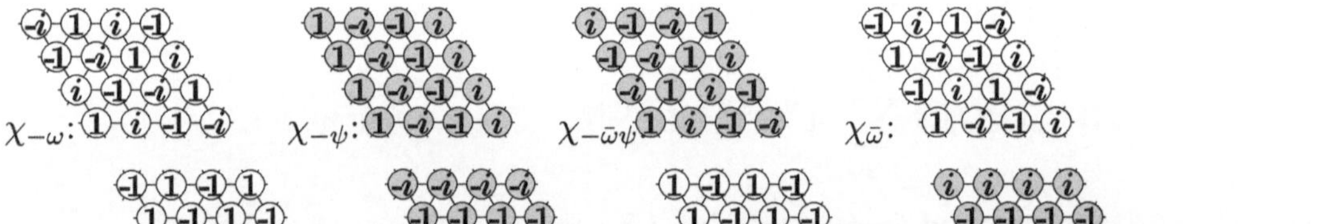

Figure 6.4 Table of characters $\chi_v(u) := i^{\mathrm{Tr}(uv)}$, $u,v\in\mathbb{E}_4$.

Now, one can see that identifying the variables as in (6.10) results in the first formula in the following theorem; the second formula is similar.

Theorem 6.2. *Let C be an additive code in $\mathbb{E}_4^m \times \mathbb{F}_4^{n'} \times \mathbb{Z}_4^{n''}$. The Doob weight and coweight enumerators* (6.9) *of C and its dual $C^\perp$ are connected by the MacWilliams identities*

$$W_{C^\perp}(A,B) = \frac{1}{|C|}\widetilde{W}_C(A+3B, A-B), \qquad (6.13)$$

$$W_C(A,B) = \frac{1}{|C^\perp|}\widetilde{W}_{C^\perp}(A+3B, A-B). \qquad (6.14)$$

Finally we note that in the case $n'' = 0$, each vector $\boldsymbol{x}$ from $V_{m,n',n''}$ satisfies $\mathrm{wt}(\boldsymbol{x}) = \widetilde{\mathrm{wt}}(\psi\boldsymbol{x})$. On the other hand, for a linear code C, $\boldsymbol{x} \in C$ is equivalent to $\psi\boldsymbol{x} \in C$. Hence, $W_C(A,B) = \widetilde{W}_C(A,B)$ for linear codes, which is not true for additive codes in general. We conclude the following.

Theorem 6.3. *Let C be a linear code in $\mathbb{E}_4^m \times \mathbb{F}_4^{n'}$. The Doob weight enumerators* (6.9) *of C and its dual $C^\perp$ satisfy*

$$W_{C^\perp}(A,B) = \frac{1}{|C|}W_C(A+3B, A-B). \qquad (6.15)$$

For additive codes in general, the claim of Theorem 6.3 is not always true. For example, the additive codes $C = \{0, 2, 2\omega, 2\bar{\omega}\}$ and $D = \{0, \psi, 2\psi, -\psi\}$ have the same weight enumerator $A^2 + 3B^2$, but their duals $C^\perp = \{0, 2, 2\omega, 2\bar{\omega}\}$ and $D^\perp = \{0, \omega, 2\omega, -\omega\}$ have different weight enumerators $A^2 + 3B^2$ and $A^2 + 2AB + B^2$ (but the same coweight enumerator $\mathcal{A}^2 + 3\mathcal{B}^2$). However, we can modify the inner product, replacing Tr by another function, to make the weight distributions of an additive code C and its dual (in the sense of the modified inner product) be connected my the MacWilliams identity.

Theorem 6.4. *Let C be an additive code in $\mathbb{E}_4^m \times \mathbb{F}_4^{n'} \times \mathbb{Z}_4^{n''}$. If $C^{\perp(\psi)}$ is the dual of C in the sense of the $\mathbb{Z}_4$-inner product*

$$[\boldsymbol{x}, \boldsymbol{y}]_{(\psi)} := \mathrm{Tr}(\psi\langle \boldsymbol{x}^*, \boldsymbol{y}^*\rangle) + \mathrm{Tr}\langle \boldsymbol{x}', \boldsymbol{y}'\rangle + \langle \boldsymbol{x}'', \boldsymbol{y}''\rangle,$$

then the Doob weight enumerators (6.9) *of C and $C^{\perp(\psi)}$ satisfy the MacWilliams identity*

$$W_{C^{\perp(\psi)}}(A,B) = \frac{1}{|C|}W_C(A+3B, A-B). \qquad (6.16)$$

Proof. Consider the map σ that transforms $(\boldsymbol{x}^*, \boldsymbol{x}', \boldsymbol{x}'')$ to $(\psi\boldsymbol{x}^*, \boldsymbol{x}', \boldsymbol{x}'')$. Clearly, $[\boldsymbol{x}, \boldsymbol{y}]_{(\psi)} = [\sigma(\boldsymbol{x}), \boldsymbol{y}]$ and hence $\sigma(C^{\perp(\psi)}) = C^\perp$. On the other hand, σ is obviously an isometry between the metric spaces over $\mathbb{E}_4^m \times \mathbb{F}_4^{n'} \times \mathbb{Z}_4^{n''}$ with weight and coweight metrics. Hence, $\widetilde{W}_{C^\perp} = \widetilde{W}_{\sigma(C^{\perp(\psi)})} = W_{C^{\perp(\psi)}}$. Now, (6.16) follows from Theorem 6.2. □

Remark 6.1. *In the theory of self-dual and quantum codes, an important role is played by the Hermitian inner product defined as*

$$\langle \boldsymbol{x}, \boldsymbol{y}\rangle_{\mathrm{Herm}} := x_1\bar{y}_1 + \cdots + x_k\bar{y}_k$$

for $\boldsymbol{x} = (x_1, \ldots, x_k)$ *and* $\boldsymbol{y} = (y_1, \ldots, y_k)$ *from* $\mathbb{E}_4^k$ *or* $\mathbb{F}_4^k$ *(or, in general, for any other ring admitting an automorphisms* $y \to \bar{y}$ *of order 2), see e.g. [18]. We note that the theory above remains valid if we replace the standard inner product* $\langle\cdot,\cdot\rangle$ *(6.1) by* $\langle\cdot,\cdot\rangle_{\mathrm{Herm}}$*. Indeed, obviously, the conjugacy does not change the weight; so, the dual code* $C^\perp$ *and the Hermitian dual* $\overline{C}^\perp$ *have the same weight (or coweight) distribution.*

6.2.3 Codes in $\mathbb{Z}_4^{2m} \times \mathbb{Z}_2^{2n'} \times \mathbb{Z}_4^{n''}$ and MacWilliams Identities

As was mentioned above, $\mathbb{E}_4^m \times \mathbb{F}_4^{n'} \times \mathbb{Z}_4^{n''}$ can be considered as a $\mathbb{Z}_4$-module, and in this sense it is isomorphic to the structure $\mathbb{Z}_4^{2m} \times \mathbb{Z}_2^{2n'} \times \mathbb{Z}_4^{n''}$, consisting of $(2m + 2n' + n'')$-tuples with the first $2m$ and the last n'' elements from $\mathbb{Z}_4$ and the "middle" $2n'$ elements from $\mathbb{Z}_2$. From this point of view, one can find convenient to consider the Doob scheme directly on $\mathbb{Z}_4^{2m} \times \mathbb{Z}_2^{2n'} \times \mathbb{Z}_4^{n''}$ and consider the duality in a usual manner, as for $\mathbb{Z}_2\mathbb{Z}_4$-codes (see e.g. [13]).

For convenience, we treat the elements of $(\mathbb{Z}_4^2)^m \times (\mathbb{Z}_2^2)^{n'} \times \mathbb{Z}_4^{n''}$ as $(m+n'+n'')$-tuples, where the first m elements are pairs (we write like $a{:}b$, omitting ":" between numbers, e.g. $21 = 2{:}1$) from $\mathbb{Z}_4^2$, the next n' elements are pairs from $\mathbb{Z}_2^2$, and the last n'' elements are from $\mathbb{Z}_4$. For vectors

$$\boldsymbol{x} := (x_1{:}y_1, \ldots, x_m{:}y_m, x'_1{:}y'_1, \ldots, x'_{n'}{:}y'_{n'}, x''_1, \ldots, x''_{n''}),$$
$$\boldsymbol{u} := (u_1{:}v_1, \ldots, u_m{:}v_m, u'_1{:}v'_1, \ldots, u'_{n'}{:}v'_{n'}, u''_1, \ldots, u''_{n''}),$$

the *inner product* $\langle \boldsymbol{u}, \boldsymbol{x}\rangle$ is defined as

$$\langle \boldsymbol{u}, \boldsymbol{x}\rangle := \sum_{j=1}^{m}(u_jx_j+v_jy_j) + 2\cdot\sum_{j=1}^{n'}(u'_jx'_j+v'_jy'_j) + \sum_{j=1}^{n''}u''_jx''_j; \tag{6.17}$$

the *weight* and *coweight* are defined as

$$\mathrm{wt}(\boldsymbol{x}) := \sum_{j=1}^{m}\mathrm{wt}(x_j{:}y_j) + \sum_{j=1}^{n'}\mathrm{wt}(x'_j{:}y'_j) + \sum_{j=1}^{n''}\mathrm{wt}(x''_j),$$
$$\widetilde{\mathrm{wt}}(\boldsymbol{x}) := \sum_{j=1}^{m}\widetilde{\mathrm{wt}}(x_j{:}y_j) + \sum_{j=1}^{n'}\mathrm{wt}(x'_j{:}y'_j) + \sum_{j=1}^{n''}\mathrm{wt}(x''_j),$$

where wt on $\mathbb{Z}_4^2$ is defined by $\mathrm{wt}(00) = 0$, $\mathrm{wt}(a{:}b) = 1$ for $a{:}b \in \{01, 03, 10, 30, 11, 33\}$, $\mathrm{wt}(a{:}b) = 2$ in the other cases; similarly, $\widetilde{\mathrm{wt}}$ on $\mathbb{Z}_4^2$ is defined by $\widetilde{\mathrm{wt}}(00) := 0$, $\widetilde{\mathrm{wt}}(a{:}b) := 1$ for $a{:}b \in \{01, 03, 10, 30, 13, 31\}$, $\widetilde{\mathrm{wt}}(a{:}b) := 2$ in the other cases; on $\mathbb{Z}_2^2$ or $\mathbb{Z}_4$ (in the last two groups of coordinates), $\mathrm{wt}(00) = \mathrm{wt}(0) := 0$ and $\mathrm{wt}(01) = \mathrm{wt}(10) = \mathrm{wt}(11) = \mathrm{wt}(1) = \mathrm{wt}(2) = \mathrm{wt}(3) := 1$. As in the previous section, the corresponding metrics and graphs are defined. The Shrikhande graph $D(1,0)$ and its dual $\widetilde{D}(1,0)$, with respect to the considered inner product, are as in the following picture:

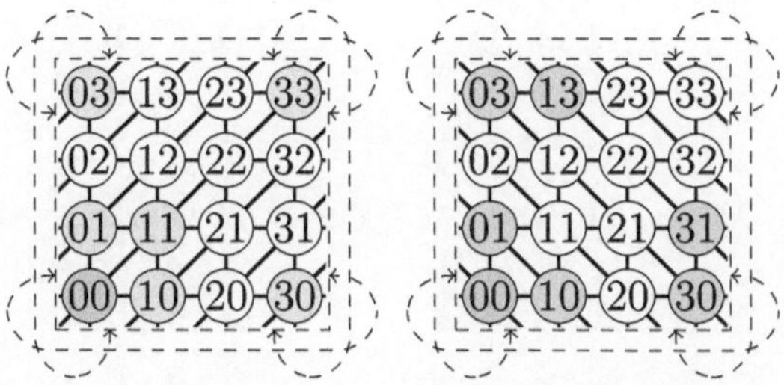

In the notation presented in the current section, the weight and coweight enumerators are defined as before:

$$W_C(A,B) := \sum_{z\in C} A^{N-\mathrm{wt}(z)} B^{\mathrm{wt}(z)} \quad \text{and}$$

$$\widetilde{W}_C(\mathcal{A},\mathcal{B}) := \sum_{z\in C} \mathcal{A}^{N-\widetilde{\mathrm{wt}}(z)} \mathcal{B}^{\widetilde{\mathrm{wt}}(z)}, \quad N := 2m + n' + n''.$$

Moreover, the MacWilliams identities are also written as previously (equations 6.13 and 6.14), in the new notation, see the following theorem.

Theorem 6.5. *Let C be an additive code in $(\mathbb{Z}_4^2)^m \times (\mathbb{Z}_2^2)^{n'} \times \mathbb{Z}_4^{n''}$ with the weight (coweight) functions* (6.17). *The Doob weight and coweight enumerators* (6.18) *of C and its dual $C^\perp$ are connected by the MacWilliams identities*

$$W_{C^\perp}(A,B) = \frac{1}{|C|}\widetilde{W}_C(A+3B, A-B),$$

$$W_C(A,B) = \frac{1}{|C^\perp|}\widetilde{W}_{C^\perp}(A+3B, A-B).$$

The technique to prove Theorem 6.5 is similar to that for Theorem 6.2 in the previous section. To check the identity

$$\sum_{v\in\mathbb{Z}_4^2} i^{\langle a,v\rangle} A^{2-\mathrm{wt}(v)} B^{\mathrm{wt}(v)} = (A+3B)^{2-\widetilde{\mathrm{wt}}(a)} (A-B)^{\widetilde{\mathrm{wt}}(a)},$$

an analog of (6.12), the character table (Figure 6.5) is useful.

Again, to make the weight distribution of an additive code and its dual connected in the same metric, we can modify the duality by modifying the inner product.

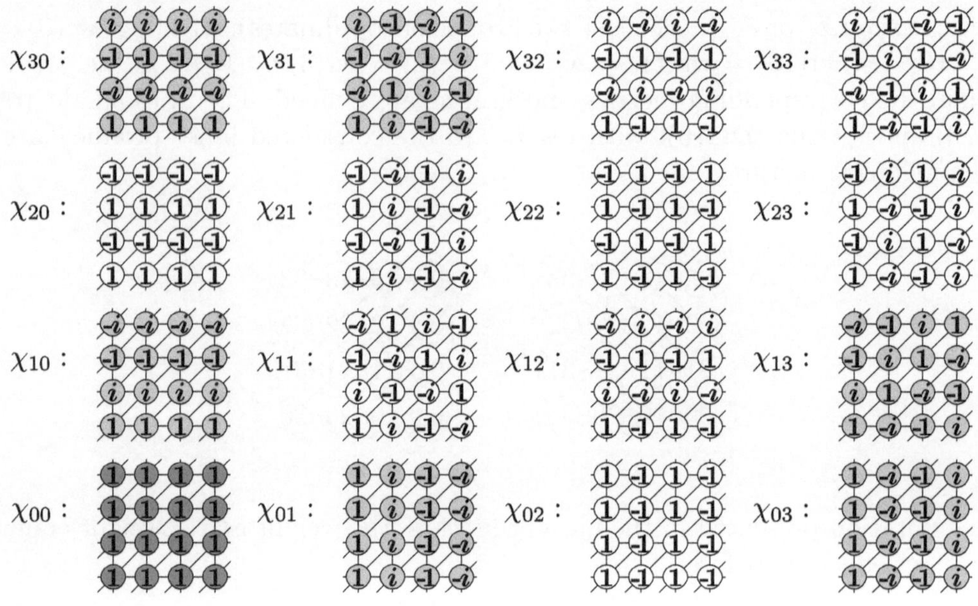

Figure 6.5 Table of characters $\chi_v(u) := i^{\langle u,v\rangle}$, $u, v \in \mathbb{Z}_4^2$.

Theorem 6.6. *Let C be an additive code in $(\mathbb{Z}_4^2)^m \times (\mathbb{Z}_2^2)^{n'} \times \mathbb{Z}_4^{n''}$. If $C_-^\perp$ is the dual of C in the sense of the modified inner product*

$$\langle \boldsymbol{u}, \boldsymbol{x}\rangle_- := \sum_{j=1}^{m}(u_j x_j - v_j y_j) + 2\cdot\sum_{j=1}^{n'}(u'_j x'_j + v'_j y'_j) + \sum_{j=1}^{n''} u''_j x''_j \tag{6.18}$$

(instead of (6.17)), then the Doob weight enumerators (6.9) of C and $C_-^\perp$ satisfy

$$W_{C_-^\perp}(A, B) = \frac{1}{|C|} W_C(A + 3B, A - B). \tag{6.19}$$

Proof. Consider the map σ that transforms

$$(u_1{:}v_1, \ldots, u_m{:}v_m, u'_1{:}v'_1, \ldots, u'_{n'}{:}v'_{n'}, u''_1, \ldots, u''_{n''})$$

to

$$(u_1{:}-v_1, \ldots, u_m{:}-v_m, u'_1{:}v'_1, \ldots, u'_{n'}{:}v'_{n'}, u''_1, \ldots, u''_{n''}).$$

By the definitions of the inner products (6.17) and (6.18) we have $\langle \boldsymbol{u}, \boldsymbol{x}\rangle_- = \langle \sigma(\boldsymbol{u}), \boldsymbol{x}\rangle$ and hence $\sigma(C_-^\perp) = C^\perp$. On the other hand, σ is obviously an isometry between the metric spaces over $\mathbb{Z}_4^{2m} \times \mathbb{Z}_2^{2n'} \times \mathbb{Z}_4^{n''}$ with weight and coweight metrics. Hence, $\widetilde{W}_{C^\perp} = \widetilde{W}_{\sigma(C_-^\perp)} = W_{C_-^\perp}$. Now, (6.19) follows from Theorem 6.5. □

6.2.4 Conclusion

We conclude that to connect the weight distributions of an additive code and its dual in a Doob scheme by the MacWilliams identities, one should consider their weights in different (but isomorphic) Doob graphs. An alternative approach is to consider the weights in the same metric, but to treat duality in a modified way, with a special inner product, see Theorems 6.4 and 6.6. The last way can be recommended if a code is defined by a check matrix which is treated as a generator matrix of the dual code.

In the following section, classes of linear and additive one-perfect codes in Doob graphs are constructed. By Theorem 6.3, the code generated by a check matrix of a linear one-perfect code in $D(m,n)$ has the parameters of the code dual to the linear 1-perfect quaternary Hamming code of length $2m+n$ (the dual code is a simplex code, or a tight 2-design, see e.g. [35]). According to Theorem 6.6, to get a generator matrix of a tight 2-design from a check matrix of an additive (linear over $\mathbb{Z}_4$) one-perfect code in a Doob graph $D(m,n'+n'')$, one should multiply by -1 the first m columns of the matrix with even numbers.

6.3 PERFECT CODES AND EXTENDED PERFECT CODES

In any graph, an e-perfect code is defined as a set of vertices such that every ball of radius e contains exactly one code vertex. Straightforwardly, every e-perfect code in a distance-regular graph is completely regular with covering radius e. Moreover, one-perfect codes are completely regular in any regular graph.

The history of the study of perfect codes in Doob graphs started from the paper [35], where it was shown that nontrivial e-perfect codes in $D(m,n)$ can only exist when $e=1$ and $2m+n=(4^k-1)/3$ for some integer k and two one-perfect codes, in $D(2,1)$ and $D(1,3)$, were constructed. In [40], infinite series of perfect codes in Doob graphs were obtained. In particular, it was shown that the necessary condition $2m+n=(4^k-1)/3$ is sufficient if $m<n-o(2m+n)$; the class of linear perfect codes was completely characterized; a class of additive perfect codes was constructed and necessary conditions on m, n', n'' for the existence of additive perfect codes in $D(m,n'+n'')$ were obtained. In [59], it was shown that those conditions are also sufficient; so, the parameters of additive perfect codes were characterized. In the recent paper [60], an infinite series of quasicyclic additive one-perfect codes in $D((2^\Delta-1)(2^\Delta-2)/6, 2^\Delta-1)$, Δ odd, was constructed. In [41], it was shown that the condition $2m+n=(4^k-1)/3$ is not only necessary but also sufficient for the existence of unrestricted one-perfect codes in $D(m,n)$. To compare, unrestricted one-perfect codes exist in all 10 Doob graphs of diameter $2m+n=21$, a linear one-perfect code exists in $D(8,5)$ only; additive one-perfect codes exist in $D(7,7)$ and $D(8,5)$. Further, unrestricted one-perfect codes exist in all 42 Doob graphs of diameter $2m+n=85$, a linear one-perfect code exists in $D(32,21)$ and $D(40,5)$; additive one-perfect codes exist in $D(31,23)$, $D(32,21)$, $D(35,15)$, $D(36,13)$, $D(37,11)$, $D(38,9)$, $D(39,7)$, and $D(40,5)$.

The codes formally dual to the one-perfect codes are known as simplex codes or tight 2-designs (the formal duality of two codes means that the MacWilliams transform of the distance distribution of one code gives the distance distribution of the other code). A *simplex code* in a Hamming graph $H(N,q)$ or a Doob graph

$D(m,n)$ has $N(q-1)+1$ codewords at mutual distance $(N(q-1)+1)/q$ from each other (for $D(m,n)$, we put $N = 2m+n$ and $q = 4$). In every $D(m,n)$ such that $2m+n = (4^k-1)/3$ for some k, simplex codes were constructed in [35]. So, it is safe to say that for every one-perfect code C in a Doob graph there is a simplex code (tight 2-design) that is formally dual to C. However, to treat duality in the usual sense, as a duality between two submodules, we need additive codes (note that the duality should be defined in a special way to be compatible with the MacWilliams transform, see Section 6.2). Since additive one-perfect codes (and hence, additive simplex codes) exist only if the parameters satisfy the additional conditions (6.20)–(6.22), for any other parameters meeting $2m+n = (4^k-1)/3$ the class of one-perfect codes is connected with the class of simplex codes only in the way of formal duality (there is still a challenge in finding a more strict connection, as was done, for example, in [26] for the formally dual classes of Preparata-like codes and Kerdock-like codes). It should also be noted that the parameters of simplex codes (tight 2-designs) are not proven to be restricted by the case $2m+n = (4^k-1)/3$ only. The problem of existence of simplex codes for other parameters is open for Doob graphs, as well as for Hamming graphs, where most attempts focus on the binary case (see the Hadamard conjecture).

Extended perfect codes form another nice class related to perfect codes. In Hamming graphs, an extended perfect code is defined as an even-distance code whose projection along one of coordinates is a perfect code. In Doob graphs, we can define them by parameters. As we will see in Section 6.3.5, extended codes in a Doob graph $D(m,n)$, $m > 0$, are completely regular and exist only if $(m,n) = (1,0)$, $(2,2)$.

6.3.1 Linear One-Perfect Codes

Theorem 6.7 ([40]). *Linear one-perfect codes in $\mathbb{E}_4^m \times \mathbb{E}_2^n$ with $D(m,n)$-metric exist if and only if*

$$n = (4^{\gamma+\delta} - 1)/3 \qquad \textit{and} \qquad m = (4^{\gamma+2\delta} - 4^{\gamma+\delta})/6$$

for some nonnegative integer γ and δ.

Note that $\delta = 0$ corresponds to the Hamming graph $D(0,n) = H(n,4)$, while for Doob graphs $m > 0$ implies $\delta > 0$. The necessity of the conditions in the theorem is a special case of the necessity in Theorem 6.9 below, for additive codes. In the following, we consider a construction of a check matrix of such a code.

Let $\gamma \geq 0$ and $\delta > 0$ be integers. Consider two matrices $A^* = A^*_{\gamma,\delta}$ and $A' = A'_{\gamma,\delta}$. The matrix A^* consists of all columns from $\mathbb{E}_4^{\gamma+\delta}$ satisfying the following:

(*) the order of the column is 4;

(**) the first regular (order-4) element of the column is either 1 or $\psi = 2+\omega$;

(***) the last γ elements of the column are zero divisors.

The number of such columns is $(16^\delta 4^\gamma - 4^{\delta+\gamma})/6$, which will be denoted by m; so, A^* is a $(\gamma+\delta) \times m$ matrix over $\mathbb{E}_4$.

The matrix A' consists of all $n = (4^{\delta+\gamma} - 1)/3$ nonzero columns from $\mathbb{E}_2^{\gamma+\delta}$ whose first nonzero element is 1.

We now merge the matrices A^* and A' into the matrix $A = A_{\gamma,\delta} = A^*|A'$ of size $(\gamma+\delta) \times (m+n)$ and define the multiplication Az^{T} for $z = (x|y) \in \mathbb{E}_4^m \times \mathbb{E}_2^n$ as $A^*x^{\mathrm{T}} + 2A'y^{\mathrm{T}}$ (here, the result of the multiplication by 2 is considered as a column-vector over $\mathbb{E}_4$). For example,

$$A_{0,2} = \left(\begin{array}{cccccccccccccccccc} 0 & 0 & 2 & 2 & 2\omega & 2\omega & 2\omega^2 & 2\omega^2 & 1 & 1 & 1 & 1 & 1 & 1 & 1 & 1 & 1 & 1 \\ 1 & \psi & 1 & \psi & 1 & \psi & 1 & \psi & 0 & 2 & 2\omega & 2\omega^2 & 1 & -\omega^2 & \omega & -1 & \omega^2 & -\omega \\ \hline \end{array} \right.$$

$$\left. \begin{array}{ccccccccc|ccccc} 1 & 1 & 1 & 1 & 1 & 1 & \psi & \dots & \psi & 0 & 1 & 1 & 1 & 1 \\ 2{+}\omega & 2\omega{+}1 & 2\omega{+}\omega^2 & 2\omega^2{+}\omega & 2\omega^2{+}1 & 2{+}\omega^2 & 0 & \dots & 2{+}\omega^2 & 1 & 0 & 1 & \omega & \omega^2 \\ \hline \end{array} \right),$$

$$A_{1,1} = \left(\begin{array}{cccccccc|ccccc} 1 & 1 & 1 & 1 & \psi & \psi & \psi & \psi & 0 & 1 & 1 & 1 & 1 \\ \hline 0 & 2 & 2\omega & 2\omega^2 & 0 & 2 & 2\omega & 2\omega^2 & 1 & 0 & 1 & \omega & \omega^2 \end{array} \right),$$

$$A_{0,1} = \left(\begin{array}{cc|c} 1 & \psi & 1 \\ \hline \end{array} \right).$$

Lemma 6.4 ([40]). *Let the matrix $A = A_{\gamma,\delta}$ be constructed as above. The set $C = C_{\gamma,\delta} = \{c \in \mathbb{E}_4^m \times \mathbb{E}_2^n : \ Ac^{\mathrm{T}} = \overline{0}^{\mathrm{T}}\}$ is a linear one-perfect code in the Doob graph $D(m,n)$.*

The proof is standard for linear One-perfect codes represented by a check matrix in different translation-invariant metrics and consists of the check that the multiset $\{Ae^{\mathrm{T}} : \ \mathrm{wt}(e) = 1\}$ contains each admissible vector (syndrome) with multiplicity 1.

Two codes in $\mathbb{E}_4^m \times \mathbb{E}_2^n$ with $D(m,n)$-metric are said to be *monomially equivalent* if one code can be obtained from the other with the following operations, applied simultaneously to all codewords: permutation of the first m coordinates; permutation of the last n coordinates; multiplying the value in the ith coordinate by an element a_i from Ω, $i = 1, \dots, m$; multiplying the value in the jth coordinate by a non-zero element b_j from $\mathbb{F}_4$, $j = m+1, \dots, m+n$. It is not difficult to see the uniqueness of linear one-perfect codes in Doob graphs up to monomial equivalence.

Theorem 6.8. *Any two linear one-perfect codes in $\mathbb{E}_4^m \times \mathbb{E}_2^n$ with $D(m,n)$-metric are monomially equivalent.*

Combining the results in the current section with Theorem 6.3, we get the following.

Corollary 6.1. *A linear distance-$(3(2m+n)+1)/4$ code of cardinality $(3(2m+n)+1)$ (linear simplex code, linear tight 2-design) in $D(m,n)$ exists if and only if $n = (4^{\gamma+\delta} - 1)/3$ and $m = (4^{\gamma+2\delta} - 4^{\gamma+\delta})/6$ for some nonnegative integer γ and δ. For each such m and n, such code is unique up to monomial equivalence and a generator matrix of the code is a check matrix of a linear one-perfect code in the same graph.*

6.3.2 Additive One-Perfect Codes

Theorem 6.9 ([40,59]). *An additive one-perfect code in the Doob graph $D(m, n' + n'')$ with the structure of $\mathbb{Z}_4^{2m} \times \mathbb{Z}_2^{2n'} \times \mathbb{Z}_4^{n''}$ exists if and only if for some even $\Gamma \geq 0$ and integer $\Delta \geq 2$,*

$$2m + n' + n'' = (2^{\Gamma+2\Delta} - 1)/3, \quad (6.20)$$
$$3n' + n'' = 2^{\Gamma+\Delta} - 1, \quad (6.21)$$
$$1 \neq n'' \leq 2^{\Delta} - 1. \quad (6.22)$$

Note that the case $\Delta = 0$ has sense but is excluded from the theorem because $D(0, n')$ is not named a Doob graph; it is the Hamming graph $H(n', 4)$. Before we consider the proof of Theorem 6.9, we mention the connection with tight 2-designs, codes with parameters dual to 1-perfect. Combining Theorems 6.9 and 6.5, we get the following.

Corollary 6.2. *An additive distance-$(3(2m + n' + n'') + 1)/4$ code of cardinality $3(2m + n' + n'') + 1$ (additive simplex code, additive tight 2-design) in $\mathbb{Z}_4^{2m} \times \mathbb{Z}_2^{2n'} \times \mathbb{Z}_4^{n''}$ with $D(m, n' + n'')$-metric exists if and only if (6.20)–(6.22) hold for for some even nonnegative Γ and nonnegative integer Δ different from 1 (the case $\Delta = 0$ corresponds to the Hamming graph $D(0, (2^\Gamma - 1)/3) = H((2^\Gamma - 1)/3, 4)$). A generator matrix of such a code is obtained from a check matrix, over $\mathbb{Z}_4$, of an additive one-perfect code by changing the sign of the first m columns with even indices.*

Proof of Theorem 6.9, "only if". The proof inherits some ideas from the proof of [14, Theorem 24]. Assume $C \subset \mathbb{Z}_4^{2m} \times \mathbb{Z}_2^{2n'} \times \mathbb{Z}_4^{n''}$ is an additive one-perfect code in $D(m, n' + n'')$. Then C contains the all-zero word $\overline{0}$ and does not contain elements of weight 1 or 2. For every weight-1 vertex e, the set $[e] = e + C$ is also a one-perfect code. As follows from the definition of one-perfect code, all such $[e]$, together with C itself, form a partition of the module; hence, they form the factor-group $(\mathbb{Z}_4^{2m} \times \mathbb{Z}_2^{2n'} \times \mathbb{Z}_4^{n''})^+/C$. This group is isomorphic to $(\mathbb{Z}_2^\Gamma \times \mathbb{Z}_4^\Delta)^+$ for some nonnegative integers Γ and Δ. The number of elements of order 2 in this group is $2^{\Gamma+\Delta} - 1$. On the other hand, the number of order-2 tuples of weight 1 is $3n' + n''$ (namely, these are the weight-1 tuples that have 01, 11, or 10 in a $\mathbb{Z}_2^2$-coordinate or 2 in a $\mathbb{Z}_4$-coordinate). Moreover, if e is an order-4 tuple of weight 1, then $e + e$ does not coincide with $\overline{0}$, is adjacent to e, and thus has weight 1 or 2. The last means that $e + e$ cannot belong to C; hence $[e]$ has order 4 in the factorgroup as well. So, $3n' + n''$ is also the number of elements of order 2 in the factorgroup, and (6.21) holds. Additionally, as the order of the factorgroup coincides with the number of weight-1 vertices plus one, we get $2^{\Gamma+2\Delta} = 6m + 3(n' + n'') + 1$, i.e. (6.20); we also note that Γ cannot be odd because $2^{\Gamma+2\Delta} - 1$ is not a multiple of 3 in this case.

To prove the right inequality in (6.22), we note that n'' weight-1 vertices (namely, the weight-1 vertices that have 2 in one of the last n coordinates) are multiples of 2, i.e., have the form $2e$ for some e; hence, the same is true for the corresponding n'' cosets. But the number of such nonzero multiples of 2 in the group $(\mathbb{Z}_2^\Gamma \times \mathbb{Z}_4^\Delta)^+$ is $2^\Delta - 1$; so, n'' cannot exceed this value.

Let us prove that $n'' \neq 1$. Assume the contrary, $n'' = 1$. Consider the subgroup of the factorgroup that consists of the $2^{\Gamma+\Delta}$ elements of order at most 2. This subgroup is isomorphic to $(\mathbb{Z}_2^{\Gamma+\Delta})^+$. Its nonzero elements are divided into n' triples $[e_{2m+2i-1}]$, $[e_{2m+2i}]$, $[e_{2m+2i-1} + e_{2m+2i}]$, $i = 1, \ldots, n'$, and one additional element $[2e_{2m+2n'+1}]$, where e_j is the tuple with one in the jth position and zeros in the others. We see that the sum of all elements of the considered subgroup is $[2e_{2m+2n'+1}]$, i.e., non-zero, which is obviously impossible if $\Gamma + \Delta > 1$. The case $\Gamma + \Delta = 1$ is degenerate and yields $m = 0$, which is not allowed by the definition of a Doob graph.

From equation (6.20) and inequality (6.22) we find that $\Delta = 0$ also implies $m = 0$.

It remains to prove that $\Delta \neq 1$. Indeed, if $\Delta = 1$, then (6.22) yields $n'' = 0$ and we see from equations (6.20) and (6.21) that both $2^{\Gamma+2} - 1$ and $2^{\Gamma+1} - 1$ are multiples of 3, a contradiction. □

The proof of the "if" part of Theorem 6.9 consists of several constructions. They are based on the representation of additive codes in Doob graph $D(m, n)$ as submodules of $\mathbb{Z}_4^{2m} \times \mathbb{Z}_2^{2n'} \times \mathbb{Z}_4^{n''}$ (see Section 6.2.3) with the inner product (6.17), which is used to represent a code via a check matrix as follows.

A matrix $(B|B'|B'')$ (where B and B'' are matrices over $\mathbb{Z}_4$, while B' is over $\mathbb{Z}_2$) is a *check matrix* of an additive code C, $C \subset \mathbb{Z}_4^{2m} \times \mathbb{Z}_2^{2n'} \times \mathbb{Z}_4^{n''}$, if

$$C = \{c \in \mathbb{Z}_4^{2m} \times \mathbb{Z}_2^{2n'} \times \mathbb{Z}_4^{n''} : (B|B'|B'')c^{\mathrm{T}} = \bar{0}^{\mathrm{T}}\},$$

where

$$(B|B'|B'')(c^*|c'|c'')^{\mathrm{T}} = Bc^{*\mathrm{T}} + 2 \cdot B'c'^{\mathrm{T}} + B''c''^{\mathrm{T}}.$$

For a tuple $z \in \mathbb{Z}_4^{2m} \times \mathbb{Z}_2^{2n'} \times \mathbb{Z}_4^{n''}$, the value $(B|B'|B'')z^{\mathrm{T}}$ is called the *syndrome* of z. We will say that s is *covered* by the coordinate i with $2(m + n') + 1 \leq i \leq 2(m + n') + n''$ or by the pair of coordinates $2i - 1$, $2i$ with $1 \leq i \leq m + n'$, if it is the syndrome of some e of weight 1 with the only non-zero value in the position i or the only non-zero values in the positions $2i - 1$, $2i$, respectively (in the last case, only one of the two values if positions $2i - 1$, $2i$ is nonzero or both are nonzero and equal to each other, according to the definition of the Doob weight). We also make an agreement that by a *pair* of coordinates (or of columns of a check matrix) we will always mean a pair of coordinates of form $2i - 1$, $2i$, where $1 \leq i \leq m + n'$, i.e., a pair that corresponds to the same Sh or K_4 component in the representation of $D(m, n' + n'')$ as the Cartesian product of m copies of the Shrikhande graph Sh and $n' + n''$ copies of the complete graph K_4 of order 4.

The following lemma is a straightforward reformulation of the definition of one-perfect codes in terms of check matrices. It is a variant of a general principle [28] of recognizing additive one-perfect codes in abelian groups with different translation-invariant metrics: if the number of nonzero syndromes equals the number of weight-1 words, then it is sufficient to check that every nonzero syndrome is covered at least once (where "covered" means that it is a syndrome of a weight-1 word).

Lemma 6.5. *An additive code in* $\mathbb{Z}_4^{2m} \times \mathbb{Z}_2^{2n'} \times \mathbb{Z}_4^{n''}$ *with a check matrix* $(B|B'|B'')$ *is 1-perfect in* $D(m, n'+n'')$ *if and only if the matrix does not have all-zero columns and every nonzero syndrome from* $\{(B|B'|B'')z \mid z \in \mathbb{Z}_4^{2m} \times \mathbb{Z}_2^{2n'} \times \mathbb{Z}_4^{n''}\}$ *is uniquely covered by one of the first* $m+n'$ *pairs of coordinates or the last* n'' *coordinates.*

In the five subsections below, we describe constructions that cover all admissible values of Δ, Γ, and n'' that satisfy the conditions of Theorem 6.9. These constructions prove the sufficiency of these conditions for existence of additive one-perfect codes in Doob graphs. Section 6.3.2.1 follows the construction in [40, Section 5], while the remaining subsections inherit the main ideas from [59], slightly rearranging the details (in particular, we now can increase Δ by 2, Γ by 2, and n'' by 3 in an arbitrary order).

6.3.2.1 Construction for even Δ

The construction of linear one-perfect codes in Section 6.3.1 provides additive one-perfect codes satisfying the conditions of Theorem 6.9 with $n'' = 0$ and any even $\Delta > 0$ and $\Gamma \geq 0$. More explicitly, we have the following:

Lemma 6.6 (Corollary 2 in [40]). *Let* $\widetilde{\ }$ *be the homomorphism from* $\mathbb{E}_4$ *(from* $\mathbb{F}_4$*) into the ring of* 2×2 *matrices over* $\mathbb{Z}_4$ *(over* $\mathbb{Z}_2$*, respectively) induced by* $\omega \to \widetilde{\omega} = \left(\begin{smallmatrix}3&1\\3&0\end{smallmatrix}\right)$*. Let the matrix* $B = B^*|B'$ *be obtained from the matrix* $A = A_{\gamma,\delta} = (A^*|A')$ *constructed in Section 6.3.1 by replacing every item* x *of* A^* *by the* 2×2 *matrix* $\widetilde{x}$ *over* $\mathbb{Z}_4$ *and every item* y *of* A' *by the* 2×2 *matrix* $\widetilde{y}$ *over* $\mathbb{Z}_2$*. The set* $\widehat{C} = \{c \in \mathbb{Z}_4^{2m} \times \mathbb{Z}_2^{2n} : Bc^{\mathrm{T}} = \overline{0}^{\mathrm{T}}\}$*, where* $B(x|y)^{\mathrm{T}} = B^*x^{\mathrm{T}} + 2B'y^{\mathrm{T}}$*, is an additive one-perfect code in the Doob graph* $D(m,n)$*,* $n = (4^{\gamma+\delta}-1)/3$*,* $m = (4^{\gamma+2\delta} - 4^{\gamma+\delta})/6$*.*

Remark 6.2. *Alternatively, one-perfect codes with even* Δ *can also be found using the recursive constructions below and starting with suitable small case.*

The next observation (essentially, Theorem 3 in [40]) allows to modify the check matrix above to construct additive one-perfect codes with even Γ and Δ and any admissible n''. In the matrix $A_{\gamma,\delta} = (A^*|A')$ constructed in Section 6.3.1, the $\mathbb{F}_4$-submatrix A' contains $4^\delta - 1 = 2^\Delta - 1$ columns with zeros in the last γ positions. Each of these columns, say a', is equal modulo 2 to some of the columns, say a^*, of the $\mathbb{E}_4$-submatrix A^*. By applying the transformation $\widetilde{\ }$ as in Lemma 6.6, the columns a^* and a' are mapped to four columns b_1^*, b_2^* and b_1', b_2' of the submatrices B^* and B', respectively, of the matrix B. We now see that these four columns cover the nine syndromes b^*, $2b^*$, $3b^*$, $b \in \{b_1^*, b_2^*, b_1^* + b_2^*\}$. If we remove these four columns (two in B^* and two in B') and add the three columns b_1^*, b_2^*, $b_1^* + b_2^*$ into the new third part B'' of the check matrix, then the set of covered syndromes will not change and n'' will increase by 3. This is illustrated by the following example ($\gamma = 1$, $\delta = 2$):

$$\left(\begin{array}{c|c} \ldots 1 \ldots & \ldots 1 \ldots \\ \ldots \omega^2 \ldots & \ldots \omega^2 \ldots \\ \hline \ldots 0 \ldots & \ldots 0 \ldots \end{array}\right) \xrightarrow{\sim} \left(\begin{array}{c|c} \ldots 10 \ldots & \ldots 10 \ldots \\ \ldots 01 \ldots & \ldots 01 \ldots \\ \ldots 03 \ldots & \ldots 01 \ldots \\ \ldots 13 \ldots & \ldots 11 \ldots \\ \hline \ldots 00 \ldots & \ldots 00 \ldots \\ \ldots 00 \ldots & \ldots 00 \ldots \end{array}\right) \longrightarrow \left(\begin{array}{c|c|c} \ldots & \ldots & \ldots 0 \;\; 1 \;\; 3 \ldots \\ \ldots & \ldots & \ldots 1 \;\; 0 \;\; 3 \ldots \\ \ldots & \ldots & \ldots 3 \;\; 0 \;\; 1 \ldots \\ \ldots & \ldots & \ldots 3 \;\; 1 \;\; 0 \ldots \\ \hline \ldots & \ldots & \ldots 0 \;\; 0 \;\; 0 \ldots \\ \ldots & \ldots & \ldots 0 \;\; 0 \;\; 0 \ldots \end{array}\right)$$

Since we can repeat this substitution up to $(2^\Delta - 1)/3$ times, it covers all admissible values of n'' (note that due to (6.21) n'' is divisible by 3 if Δ is even).

6.3.2.2 Construction for Odd Δ: The Starting Point

The construction of additive one-perfect codes with odd $\Delta \geq 3$ starts with the following explicit example of a check matrix of an additive one-perfect code in $D(8, 1+4)$ ($\Delta = 3$, $\Gamma = 0$, $n'' = 4$):

$$\left(\begin{array}{c|c|c} 10\;22\;01\;12\;21\;23\;11\;\mathbf{10} & \mathbf{10} & 1\;0\;0\;1 \\ 01\;10\;22\;23\;12\;21\;21\;\mathbf{31} & \mathbf{11} & 0\;1\;0\;1 \\ 22\;01\;10\;21\;23\;12\;12\;\mathbf{03} & \mathbf{01} & 0\;0\;1\;1 \end{array}\right). \tag{6.23}$$

6.3.2.3 Construction for Odd Δ: Increasing Δ

The next recursive construction allows to increase Δ by 2 if $\Gamma = 0$ (n'' remains the same). Assume that $B = (B^*|B'|B'')$ is a check $(\Delta+\Gamma) \times (m + n' + n'')$ matrix of a one-perfect code in $D(m, n' + n'')$, i.e., each nonzero syndrome from $\mathbb{Z}_4^\Delta \times \mathbb{Z}_2^\Gamma$ is uniquely covered by a pair of columns in B^* or B' or a column in B''. If we add two zero rows to the top of B, then we obtain a matrix that covers all nonzero syndromes from $\mathbb{Z}_4^{\Delta+2} \times \mathbb{Z}_2^\Gamma$ with the first two elements equal 0. We are going to add columns that cover syndromes with a non-zero pair of the first symbols.

Let α and β be two invertible $(\Delta+\Gamma) \times (\Delta+\Gamma)$ matrices over $\mathbb{Z}_4$ that map $\mathbb{Z}_4^\Delta \times \mathbb{Z}_2^\Gamma$ onto itself such that the matrix $\alpha + \beta$ is also invertible. For example,

$$\alpha = \begin{pmatrix} \alpha_\Delta & \mathbf{0} \\ \mathbf{0} & \alpha_\Gamma \end{pmatrix}, \quad \beta = \begin{pmatrix} \beta_\Delta & \mathbf{0} \\ \mathbf{0} & \beta_\Gamma \end{pmatrix}, \quad \alpha_i = \begin{pmatrix} 1 & 0 & 0 & 0 \\ 0 & 1 & 0 & 0 \\ 0 & 0 & 1 & 0 \\ 1 & 0 & 0 & 1 \end{pmatrix}, \quad \beta_i = \begin{pmatrix} 3 & 1 & 0 & 0 \\ 0 & 3 & 1 & 0 \\ 0 & 0 & 3 & 1 \\ 0 & 0 & 0 & 3 \end{pmatrix}.$$

Let $c_1, \ldots, c_s$, $s = 4^\Delta 2^\Gamma$, be all columns from $\mathbb{Z}_4^\Delta \times \mathbb{Z}_2^\Gamma$ such that the first t, $t = 4^\Delta 2^\Gamma - 2^{\Delta+\Gamma}$, columns $c_1, \ldots, c_t$ are of order 4 (the last $s - t$ are of order 2 or 1) and there are no two opposite columns among $c_1, \ldots, c_{t/2}$. We define the matrices U^*, W^*, E^*, V^* (over $\mathbb{Z}_4$), D', E' (over $\mathbb{Z}_2$), E'' (over $\mathbb{Z}_4$) as follows ($\overline{0}$ denotes the all-zero row):

$$U^* = \begin{pmatrix} 0 & 2 & 0 & 2 & \cdots & 0 & 2 \\ 2 & 0 & 2 & 0 & \cdots & 2 & 0 \\ \alpha c_1 & \beta c_1 & \alpha c_2 & \beta c_2 & \cdots & \alpha c_{\frac{t}{2}} & \beta c_{\frac{t}{2}} \end{pmatrix},$$

$$W^* = \begin{pmatrix} 0 & 1 & 0 & 1 & \cdots & 0 & 1 \\ 1 & 0 & 1 & 0 & \cdots & 1 & 0 \\ \alpha c_1 & \beta c_1 & \alpha c_2 & \beta c_2 & \cdots & \alpha c_s & \beta c_s \end{pmatrix},$$

$$V^* = \begin{pmatrix} 1 & 1 & 1 & 1 & \cdots & 1 & 1 \\ 2 & 3 & 2 & 3 & \cdots & 2 & 3 \\ \alpha c_1 & \beta c_1 & \alpha c_2 & \beta c_2 & \cdots & \alpha c_s & \beta c_s \end{pmatrix},$$

$$D' = \frac{1}{2}\begin{pmatrix} 0 & 2 & 0 & 2 & \cdots & 0 & 2 \\ 2 & 0 & 2 & 0 & \cdots & 2 & 0 \\ \alpha c_{t+1} & \beta c_{t+1} & \alpha c_{t+2} & \beta c_{t+2} & \cdots & \alpha c_s & \beta c_s \end{pmatrix},$$

$$E^* = \begin{pmatrix} \overline{0} \\ \overline{0} \\ B^* \end{pmatrix}, \quad E' = \begin{pmatrix} \overline{0} \\ \overline{0} \\ B' \end{pmatrix}, \quad E'' = \begin{pmatrix} \overline{0} \\ \overline{0} \\ B'' \end{pmatrix}.$$

In [59, Proposition 1], the following claim (in a slightly different form) was proven for $\Gamma = 0$.

Lemma 6.7. *The matrix $(U^*\, W^*\, V^*\, E^* \mid D'\, E' \mid E'')$ is a check matrix (with $\Delta + 2$ rows of order 4 and Γ rows of order 2) of an additive one-perfect code in $D(t/2 + 2s + m, (s - t + n') + n'')$.*

Indeed, $(E^*|E'|E'')$ covers all nonzero syndromes beginning with $(0,0)$; U^* covers all order-4 syndromes beginning with $(0,2)$, $(2,0)$, or $(2,2)$; W^* covers all syndromes beginning with $(0,1)$, $(1,0)$, $(1,1)$, $(0,3)$, $(3,0)$, $(3,3)$; V^* covers all syndromes beginning with $(1,2)$, $(1,3)$, $(2,1)$, $(3,2)$, $(3,1)$, $(2,3)$; D' covers all order-2 syndromes beginning with $(0,2)$, $(2,0)$, or $(2,2)$.

Thus, starting from the check matrix (6.23), we can recursively construct check matrices for one-perfect codes corresponding to $\Gamma = 0$, $n'' = 4$, and all odd $\Delta \geq 3$.

6.3.2.4 *Construction for Odd Δ: Increasing Γ*

Here, we describe a recursive way to increase Γ by 2 (in [59, Section 3.4], Γ was increased from 0 in one step). We make the same assumption as in Section 6.3.2.3 about the matrices $B = (B^*|B'|B'')$, α, β and column vectors c_1, …, c_s. Construct the following matrices:

$$U^* = \begin{pmatrix} \alpha c_1 & \beta c_1 & \alpha c_2 & \beta c_2 & \cdots & \alpha c_{\frac{t}{2}} & \beta c_{\frac{t}{2}} \\ 0 & 2 & 0 & 2 & \cdots & 0 & 2 \\ 2 & 0 & 2 & 0 & \cdots & 2 & 0 \end{pmatrix},$$

$$D' = \frac{1}{2}\begin{pmatrix} \alpha c_{t+1} & \beta c_{t+1} & \alpha c_{t+2} & \beta c_{t+2} & \cdots & \alpha c_s & \beta c_s \\ 0 & 2 & 0 & 2 & \cdots & 0 & 2 \\ 2 & 0 & 2 & 0 & \cdots & 2 & 0 \end{pmatrix},$$

$$E^* = \begin{pmatrix} B^* \\ \overline{0} \\ \overline{0} \end{pmatrix}, \quad E' = \begin{pmatrix} B' \\ \overline{0} \\ \overline{0} \end{pmatrix}, \quad E'' = \begin{pmatrix} B'' \\ \overline{0} \\ \overline{0} \end{pmatrix}.$$

Lemma 6.8. *The matrix* $(U^* \, E^* \,|\, D' \, E' \,|\, E'')$ *is a check matrix (with* Δ *rows of order* 4 *and* $\Gamma + 2$ *rows of order* 2*) of an additive one-perfect code in* $D(t/2 + 2s + m, (s - t + n') + n'')$.

Indeed, $(E^*|E'|E'')$ covers all nonzero syndromes ending with $(0,0)$; U^* covers all order-4 syndromes ending with $(0,2)$, $(2,0)$, or $(2,2)$; D' covers all order-2 syndromes ending with $(0,2)$, $(2,0)$, or $(2,2)$.

Thus, starting from the check matrix (6.23), we can recursively construct check matrices for one-perfect codes corresponding to $n'' = 4$, and all odd $\Delta \geq 3$ and even $\Gamma \geq 0$.

6.3.2.5 Construction for Odd Δ: Increasing n''

To increase n'', we can utilize the substitution trick from Section 6.3.2.1. To apply it up to $2^\Delta - 5$ times, it is sufficient to observe the following:

Lemma 6.9. *If* $(B^*|B'|B'')$ *is the matrix in* (6.23) *or obtained from it with the recursions described in Sections 6.3.2.3 and 6.3.2.4, then for every pair* (b'_1, b'_2) *of columns in* B' *with last* Γ *elements equal to* 0*, there is a pair* (b^*_1, b^*_2) *of columns in* B' *such that* $2b^*_1 = 2 \cdot b'_1$ *and* $2b^*_2 = 2 \cdot b'_2$.

Proof. For the matrix (6.23), it is straightforward (see the bold four columns in (6.23)). The induction step $\Delta \to \Delta + 2$ follows by observing that $2 \cdot D'$ is a submatrix of $2W^*$ is Section 6.3.2.3. The induction step $\Gamma \to \Gamma + 2$ follows by observing that all syndromes with zeros in the Γ-part are covered by the submatrix $(E^* \,|\, E' \,|\, E'')$ of the matrix $(U^* \, E^* \,|\, D' \, E' \,|\, E'')$. □

So, the substitution described is Section 6.3.2.1 can be applied up to $(2^\Delta - 5)/3$ times (the number of pairs of columns in B'), each step increasing n'' by 3. The maximum possible value of n'' (for odd Δ), $2^\Delta - 1$, implies $n' = 0$ if $\Gamma = 0$. This special subcase (Δ odd, $\Gamma = 0$, $n' = 0$, $n'' = 2^\Delta - 1$) will be considered also in the nest section, where existence of additive quasicyclic codes in Doob graph is considered.

6.3.3 Quasicyclic Additive 1-Perfect Codes

In this subsection, we assume that (in the notation of Theorem 6.9) Δ is odd, $\Delta \geq 3$, $\Gamma = 0$, and $n'' = 2^\Delta - 1$; so, $n' = 0$ and $m = (2^\Delta - 1)(2^\Delta - 2)/6$. The columns of a check matrix are associated with elements of the Galois ring $\mathrm{GR}(4^\Delta) = \mathbb{Z}_4[z]/\langle h(z) \rangle$, where $h(z)$ is an basic primitive polynomial of degree Δ with a primitive root ξ.

Theorem 6.10 (Corollary 1 in [60]). *For odd* $\Delta \geq 3$*, there exists a vector* $H = (h_1, \ldots, h_{2m+n''})$ *of length* $2m + n''$ *over* $\mathrm{GR}(4^\Delta)$ *(which can be treated as a* $\Delta \times (2m + n'')$*-matrix over* $\mathbb{Z}_4$*), where* $m = (2^\Delta - 1)(2^\Delta - 2)/6$ *and* $n'' = 2^\Delta - 1$*, such that* *the code C in $\mathbb{Z}^{2m+n''}$ defined by*

$$C = \{c \in \mathbb{Z}^{2m+n''} : \langle H, c\rangle = 0\}$$

is a one-perfect code in $D(m, n'')$ and

the set of pairs $\{(h_{2i-1}, h_{2i})\}_{i=1}^{m}$ and the set $\{h_{2m+j}\}_{j=1}^{n''}$ are invariant under the multiplication by ξ of $\mathrm{GR}(4^\Delta)$ and, if Δ is not divisible by 3, under the action of the automorphism group of the ring $\mathrm{GR}(4^\Delta)$.

In particular, the code C is invariant under a quasicyclic coordinate-permutation automorphism of order $2^\Delta - 1$ (induced by the multiplication by ξ) and, if Δ is not divisible by 3, by an additional coordinate-permutation automorphism of order Δ (induced by an automorphism of $\mathrm{GR}(4^\Delta)$).

Theorem 6.11 (Proposition 2 in [60]). *A quasicyclic code C from Theorem 6.10 is not equivalent to a code constructed recursively in Sections 6.3.2.3 and 6.3.2.5.*

The invariant that proves the nonequivalence is the number of codewords of weight 3 with all three nonzero positions in the right part of the codeword. Every additive 1-perfect code in $D((2^\Delta-1)\frac{2^\Delta-2}{6}, 0+(2^\Delta-1))$ has $(2^\Delta-1)(2^\Delta-2)/6$ such codewords of order 2 (their supports form a Steiner triple system on the last $(2^\Delta-1)$ coordinates as points). The quasicyclic codes have no such codewords or order 4, while every code constructed recursively in Section 6.3.2 has such codewords.

6.3.4 Unrestricted 1-Perfect Codes

In this section, we show how to construct a one-perfect code in an arbitrary Doob graph $D(m, n)$ such that where $2m + n = \frac{4^k-1}{3}$ for some integer k. One of the approaches was shown in [41], where, starting with the Hamming code $\mathcal{H}$ over $\mathbb{F}_4$ in $H(2m+n, 4)$ and replacing subwords of length 4 corresponding to the positions $4i-3, 4i-2, 4i-1, 4i$ of the codewords by subwords of length 4 over $\mathbb{Z}_4$ from some special codes in $D(2, 0)$ or $D(1, 2)$, one can construct a one-perfect code in $D(2, 2m+n-4)$, then in $D(4, 2m+n-8)$, and so on, finally coming to $D(m, n)$ in $\lfloor m/2\rfloor$ steps. That approach is partially similar to that of [35] for tight 2-designs (the codes formally dual to 1-perfect), differing in details and metrical properties of the used mappings. In this section, we present another representation of the construction, using the ideas of the generalized concatenation, see [67] for general principles and [57] for the construction of nonbinary one-perfect codes. The construction below coincides with the one in [41] essentially (but not in description) if in Theorem 6.12 we take the linear Hamming code for the role of the first "outer" code $\mathcal{H}$ and the linear distance-2 MDS code

$$\{(y_1, \ldots, y_s, y_0) \in \mathbb{F}_4^{s+1} : y_1 + \cdots + y_n + y_0 = 0\} \tag{6.24}$$

for the role of the second "outer" code M.

6.3.4.1 *Completely Regular Codes in $D(1,2)$, $D(2,0)$, and $H(4,4)$*

We start with constructing examples of small codes that will be used as 'inner' codes for the concatenation construction in the next subsection. In each of the graphs $D(1,2)$, $D(2,0)$, $H(4,4)$, we need to construct a partition of the vertex set into four distance-2 codes, each of which, in its turn, is partitioned into four distance-3 codes. We will construct such partitions as partitions into cosets of additive codes (actually, completely regular MDS codes, see Section 6.4).

Denote

$$\begin{aligned} \bar{x} = (0,1,2,3),\ \bar{y} &= (1,0,1,2) \in \mathbb{Z}_4^4; \\ \bar{z} &= (0,0,1,1) \in \mathbb{Z}_4^4; \\ \bar{u} = (0,0,0,2),\ \bar{v} &= (0,0,2,0) \in \mathbb{Z}_4^4; \\ \bar{x}' = (1,1,1,1),\ \bar{y}' &= (0,1,\xi,\xi^2) \in \mathbb{F}_4^4; \\ \bar{z}' &= (0,0,1,1) \in \mathbb{F}_4^4. \end{aligned}$$

Define

$$\begin{aligned} C'' &= \langle \bar{x}, \bar{y} \rangle, & C' &= \langle \bar{x}, \bar{y}, \bar{z} \rangle; \\ D'' &= \langle \bar{x}, \bar{y} \rangle, & D' &= \langle \bar{x}, \bar{y}, \bar{u}, \bar{v} \rangle; \\ E'' &= \langle \bar{x}', \bar{y}' \rangle, & E' &= \langle \bar{x}', \bar{y}', \bar{z}' \rangle. \end{aligned}$$

Lemma 6.10. *The following assertions hold:*

a. $C'' \subset C'$, $D'' \subset D'$, $E'' \subset E'$;

b. C', D', E' *are distance-*2 *codes of cardinality* 64 *in* $D(1,2)$, $D(2,0)$, $H(4,4)$, *respectively;*

c. C'', D'', E'' *are distance-*3 *codes of cardinality* 16 *in* $D(1,2)$, $D(2,0)$, $H(4,4)$, *respectively.*

Next, we denote

- by $\{C^{(z)}\}_{z\in\mathbb{F}_4}$ the partition of $\mathbb{Z}_4^4$ into the cosets of C',
- by $\{D^{(z)}\}_{z\in\mathbb{F}_4}$ the partition of $\mathbb{Z}_4^4$ into the cosets of D',
- by $\{E^{(z)}\}_{z\in\mathbb{F}_4}$ the partition of $\mathbb{F}_4^4$ into the cosets of E'

(the indices z from $\mathbb{F}_4$ are assigned to the four elements of the partition in an arbitrary order), and for $z \in \mathbb{F}_4$ denote

- by $\{C^{(z,y)}\}_{y\in\mathbb{F}_4}$ the partition of $C_i^{(z)}$ into cosets of C'',
- by $\{C^{(z,y)}\}_{y\in\mathbb{F}_4}$ the partition of $D_i^{(z)}$ into cosets of D'',
- by $\{E^{(z,y)}\}_{y\in\mathbb{F}_4}$ the partition of $E_i^{(z)}$ into cosets of E''.

Remark 6.3. *The partition* $\{C^{(z,y)}\}_{z,y\in\mathbb{F}_4}$ *of* $\mathbb{Z}_4^4$ *(partition* $\{D^{(z,y)}\}_{z,y\in\mathbb{F}_4}$ *of* $\mathbb{Z}_4^4$*, partition* $\{E^{(z,y)}\}_{z,y\in\mathbb{F}_4}$ *of* $\mathbb{F}_4^4$*) is an equitable partition of* $D(1,2)$ *(respectively,* $D(2,0)$*,* $H(4,4)$*) with quotient matrix*

$$\begin{pmatrix} 0_4 & J_4 & J_4 & J_4 \\ J_4 & 0_4 & J_4 & J_4 \\ J_4 & J_4 & 0_4 & J_4 \\ J_4 & J_4 & J_4 & 0_4 \end{pmatrix}, \quad 0_4 = \begin{pmatrix} 0 & 0 & 0 & 0 \\ 0 & 0 & 0 & 0 \\ 0 & 0 & 0 & 0 \\ 0 & 0 & 0 & 0 \end{pmatrix}, \quad J_4 = \begin{pmatrix} 1 & 1 & 1 & 1 \\ 1 & 1 & 1 & 1 \\ 1 & 1 & 1 & 1 \\ 1 & 1 & 1 & 1 \end{pmatrix}.$$

6.3.4.2 Concatenation Construction of 1-Perfect Codes in Doob Graphs

Theorem 6.12. *Let* $\mathcal{H}$ *be an arbitrary one-perfect code in* $H(s,4)$*,* $s = (4^{k-1}-1)/3$ *(for example, the Hamming code), and let* $M \subset \mathbb{F}_4^{s+1}$ *be an arbitrary* $(s+1, 4^s, 2)$ *MDS code (distance-2 code of cardinality* 4^s*, not necessarily linear) in* $H(s+1,4)$ *(for example, (6.24)). Assume* $m \in \{0,\dots,2s\}$*. For* $i \in \{1,\dots,s\}$*, we define*

- $G_i^{(a,b)} = D^{(a,b)}$*,* $a,b \in \mathbb{F}_4$ *if* $i \le \frac{m}{2}$*,*
- $G_i^{(a,b)} = C^{(a,b)}$*,* $a,b \in \mathbb{F}_4$ *if* m *is odd and* $i = \frac{m+1}{2}$*,*
- $G_i^{(a,b)} = E^{(a,b)}$*,* $a,b \in \mathbb{F}_4$ *if* $i \ge \frac{m}{2}+1$*.*

Under the notation above, the code

$$C = \{(\bar{x}_1,\dots,\bar{x}_s,y_0) : \ \bar{x}_i \in G_i^{(z_i,y_i)}, \ (z_1,\dots,z_s) \in \mathcal{H}, \ (y_1,\dots,y_s,y_0) \in M\}$$

is a one-perfect code in $D(m,n)$*,* $n = 4s+1-2m$*.*

Remark 6.4. *In the concatenation construction above, there is a possibility to choose the second "outer" code* M *and the partitions* $\{G_i^{(a,b)}\}$ *into "inner" codes independently for each codeword* $(z_1,\dots,z_s)$ *of the first "outer" code* $\mathcal{H}$*. This possibility is well known in the theory of non-linear one-perfect codes and allows to construct a huge amount of inequivalent one-perfect codes (which grows double-exponentially in the code length), but is out of the scope of the current survey. Another way to generalize is to use a larger length of the "inner" codes, as in [57] for perfect codes in nonbinary Hamming spaces.*

Corollary 6.3. *The Doob graph* $D(m,n)$ *has a non-trivial e-perfect code if and only if* $e = 1$ *and there is a positive integer* k *such that* $2m+n = (4^k-1)/3$*.*

Proof. The "if" and "only if" parts of the statement come from Theorem 6.12 and [35, Theorem 3], respectively. □

6.3.5 Nonexistence of Extended 1-Perfect Codes

Definition 6.4. *A code* C *in* $D(m,n)$ *is called an* extended one-perfect code *if* $2m+n = (4^l-1)/3+1$*,* $|C| = 4^{2m+n-l-1}$ *for some positive integer* l *and any two different codewords are at distance at least 4 from each other.*

If $n > 0$, then the projection of an extended one-perfect code along one of the last n coordinates results in a one-perfect code (this operation, projection, also known as puncturing in coding theory, is inverse to extension, which motivates the name of the considered codes). However, for $n = 0$, we have no such treatment.

Codes of cardinality 1 in $D(1, 0)$ and $D(0, 2)$ can be called trivial extended one-perfect codes. If $2m + n = 6$ extended one-perfect codes are also MDS codes with code distance 4. Such codes exist in $D(0, 6)$ and in $D(2, 2)$, while in cases $D(1, 4)$ and $D(3, 0)$ there are no such codes. See Section 6.4.2 for more details. In Theorem 6.16 below, we show that there are no other extended one-perfect codes. Moreover, this implies the nonexistence of completely regular codes with corresponding intersection arrays.

6.3.5.1 Extended 1-Perfect Codes Are Completely Regular

Theorem 6.13. *Let C be a code in $D(m, n)$ with code distance 4, then*

$$|C| \le \frac{4^{2m+n-1}}{6m + 3n - 2}.$$

Moreover, in the case of equality, C is a completely regular code with quotient matrix

$$\begin{pmatrix} 0 & 6m+3n & 0 \\ 1 & 2 & 6m+3n-3 \\ 0 & 2m+n & 4m+2n \end{pmatrix}.$$

Proof. Let C be a code in $D(m, n)$ with code distance 4. We split the vertex set of $D(m, n)$ into the following three subsets: $C^{(0)} = C$, $C^{(1)} = \{x : d(x, C) = 1\}$, and $C^{(2)} = \{x : d(x, C) \ge 2\}$.

For $i, j \in \{0, 1, 2\}$, we define the function $s^i_j : C^{(i)} \to \mathbb{N}$, whose value $s^i_j(x)$ on a vertex x in $C^{(i)}$ is the number of vertices in $C^{(j)}$ neighbor to x.

Obviously, $s^0_0 \equiv 0$ (as the code distance is 4), $s^0_2 \equiv 0$, and $s^2_0 \equiv 0$ (by the definition). We have $s^1_0 \equiv 1$ (indeed, for any $x \in C^{(1)}$ we have $s^1_0 \ge 1$ by the definition and $s^1_0 \le 1$ by the code distance). For any vertex x in $C^{(1)}$, we denote by $o(x)$ its unique neighbor in $C^{(0)}$. If some vertices y and y' in $C^{(1)}$ are adjacent, then $o(y) = o(y')$ (indeed, $d(o(y), o(y')) \le 3$, while the code distance is 4); therefore, all $C^{(1)}$-neighbors of y belong to the neighborhood of $o(y)$. The number of common neighbors of two adjacent vertices in $D(m, n)$ is 2; considering the vertices $y \in C^{(1)}$ and $o(y) \in C^{(0)}$, we see that $s^1_1(y) = 2$; hence, $s^1_1 \equiv 2$. For each vertex x of $D(m, n)$, we have $s^i_0(x) + s^i_1(x) + s^i_2(x) = 6m + 3n$, where i is the color of x. Therefore, $s^0_1 \equiv 6m + 3n$ and $s^1_2 \equiv 6m + 3n - 3$.

It remains to consider s^2_1. At first, let us prove that $s^2_1(x) \le 2m + n$ for any $x \in C^{(2)}$.

Let u be an arbitrary vertex in $D(m, n)$. For $j \in \{1, \dots, n\}$, denote by $h_{j,u}$ the K_4-subgraph of $D(m, n)$ induced by u and the vertices that differ from u in the $(m + j)$-th position only. For $i \in \{1, \dots, m\}$, denote by $d_{i,u}$ the Shrikhande-subgraph of $D(m, n)$ induced by u and the vertices that differ from u in the i-th position only.

We call an edge $\{w, u\}$ of $D(m, n)$ an (i, j)-*edge* if $w \in C^{(i)}$ and $u \in C^{(j)}$ or vice versa. For any vertex v in $C^{(2)}$, denote by $\alpha_{i,v}$ ($\beta_{j,v}$) the number of $(1,2)$-edges in $d_{i,v}$ (respectively, in $h_{j,v}$) divided by $|d_{i,v} \cap C^{(2)}|$ (respectively, by $|h_{j,v} \cap C^{(2)}|$). Let us prove the following:

(*) *for any* $i \in \{1, \dots, m\}$, $j \in \{1, \dots, n\}$, *and* $v \in C^{(2)}$, *we have* $\alpha_{i,v} \leq 2$ *and* $\beta_{j,v} \leq 1$; *moreover,*
if $\alpha_{i,v} = 2$, *then any vertex of* $d_{i,v} \cap C^{(2)}$ *has exactly two neighbors in* $d_{i,v} \cap C^{(1)}$;
if $\beta_{j,v} = 1$, *then any vertex of* $h_{i,v} \cap C^{(2)}$ *has exactly one neighbor in* $h_{i,v} \cap C^{(1)}$.

Let $i \in \{1, \dots, m\}$ and $v \in C^{(2)}$. Consider two cases. If $d_{i,v}$ contains a vertex u of C, then all neighbors of u are in $C^{(1)}$ and the remaining 9 vertices of the Shrikhande graph $d_{i,v}$ are in $C^{(2)}$; it is straightforward to see that the conclusion of (*) about $\alpha_{i,v}$ is satisfied in this case. In the second case, $d_{i,v}$ has no vertices of C. In this case $d_{i,v} \cap C^{(1)}$ must be an independent set (indeed, if some vertices $u, w \in d_{i,v} \cap C^{(1)}$ are adjacent, then $o(u) \in C^{(0)}$ is their common neighbor, but these vertices have only two common neighbors, both in $d_{i,v}$). So $\alpha_{i,v} = \frac{6x}{16-x}$, where $x = |d_{i,v} \cap C^{(1)}| \leq 4$ (a maximum independent set in the Shrikhande graph has cardinality 4). If $x < 4$, then $\alpha_{i,v} < 2$. If $x = 4$, then $\alpha_{i,v} = 2$, and there are 16 (two equivalence classes) independent sets of size 4 in the Shrikhande graph (see Lemma 6.1 and Figure 6.6); for each of them, the conclusion holds.

Similarly, for any $j \in \{1, \dots, n\}$ and $v \in C^{(2)}$ we see that $h_{j,v}$ has three vertices in $C^{(2)}$ and one in $C^{(1)}$. The claim for $\beta_{j,v}$ is straightforward, which concludes the proof of (*).

Denote by α the value $\sum_{x \in C^{(2)}} s_1^2(x)$, i.e. the number of $(1,2)$-edges in $D(m, n)$. Obviously,

$$\alpha = \sum_{v \in C^{(2)}} \Big(\sum_{i \in \{1,\dots,m\}} \alpha_{i,v} + \sum_{j \in \{1,\dots,n\}} \beta_{j,v} \Big).$$

From (*), we have $\alpha \leq |C^{(2)}|(2m+n)$. Moreover, if $\alpha = (2m+n)|C^{(2)}|$ then $\alpha_{i,v} = 2$ and $\beta_{j,v} = 1$ for any $i \in \{1, \dots, m\}$, $j \in \{1, \dots, n\}$, and $v \in C^{(2)}$, which would imply $s_1^2(v) = 2m + n$ for any $v \in C^{(2)}$.

Let us evaluate the cardinalities of $C^{(1)}$ and $C^{(2)}$. By double-counting the number of $(0,1)$-edges, we see $|C^{(1)}| = (6m+3n)|C|$. Since $s_2^1 \equiv 6m+3n-3$, the number α of $(1,2)$-edges equals $3(2m+n-1)|C^{(1)}|$. On the other hand, since $s_1^2(x) \leq 2m+n$ for any $x \in C^{(2)}$ we have

$$3(2m+n-1)(6m+3n)|C| = \alpha \leq |C^{(2)}|(2m+n) = (4^{2m+n} - |C|(1+6m+3n))(2m+n).$$

This implies required bound

$$|C| \leq \frac{4^{2m+n-1}}{6m + 3n - 2}. \tag{6.25}$$

In the case of equality, we have $s_1^2(x) = 2m + n$ for any $x \in f^{-1}(2)$. Hence, there are no vertices at distance more than 2 from C, and $s_1^2 \equiv 2m + n$. Therefore, C is a completely regular code with required quotient matrix. □

Checking straightforwardly the code distance and the cardinality, we also have the following.

Theorem 6.14. *If C is a completely regular code in $D(m,n)$ with quotient matrix*

$$\begin{pmatrix} 0 & 6m+3n & 0 \\ 1 & 2 & 6m+3n-3 \\ 0 & 2m+n & 4m+2n \end{pmatrix},$$

then C is an extended one-perfect code.

6.3.5.2 Nonexistence of Extended Perfect Codes

To prove the nonexistence of extended one-perfect codes, we use the following theorem.

Theorem 6.15. *Let f be a perfect k-coloring of $D(m,n)$ with quotient matrix S that has eigenvalues $\lambda_0 > \lambda_1 > \ldots > \lambda_l$. Let i be a color of f, and let $s^t_{i,i}$ be the (i,i)-th element of S^t, $t = 1, \ldots, l-1$. Then the system of linear equations*

$$\begin{pmatrix} 1 & 1 & \ldots & 1 \\ \lambda_1 & \lambda_2 & \ldots & \lambda_l \\ \lambda_1^2 & \lambda_2^2 & \ldots & \lambda_l^2 \\ \ldots & \ldots & \ldots & \ldots \\ \lambda_1^{l-1} & \lambda_2^{l-1} & \ldots & \lambda_l^{l-1} \end{pmatrix} \begin{pmatrix} x_1 \\ x_2 \\ x_2 \\ \ldots \\ x_l \end{pmatrix} = |f^{-1}(i)| \begin{pmatrix} 1 \\ s^1_{i,i} \\ s^2_{i,i} \\ \ldots \\ s^{l-1}_{i,i} \end{pmatrix} - \frac{|f^{-1}(i)|^2}{4^{2m+n}} \begin{pmatrix} 1 \\ \lambda_0 \\ \lambda_0^2 \\ \ldots \\ \lambda_0^{l-1} \end{pmatrix}$$

has a unique solution $(a_1, \ldots, a_l)$. Moreover, $a_j \cdot 4^{2m+n}$ is a non-negative integer for $j = 1, \ldots, l$.

If $m = 0$, this theorem was proved in [6]. The proof uses the fact that there is an orthogonal basis from 4^{2m+n} eigenfunctions with the values from $\{\pm 1, \pm\sqrt{-1}\}$. Such a basis exists for $m > 0$ as well (see Figure 6.4 or 6.5 for $D(1,0)$; for arbitrary m and n, the basis eigenfunctions for $D(m,n)$ are constructed from those for $D(1,0)$ and $D(0,1)$ using the tensor product). Using this fact, the proof of Theorem 6.15 for $m = 0$ in [6] can be applied to arbitrary m.

The following lemma can be checked directly.

Lemma 6.11. *The matrix*

$$\begin{pmatrix} 0 & 6m+3n & 0 \\ 1 & 2 & 6m+3n-3 \\ 0 & 2m+n & 4m+2n \end{pmatrix}$$

has the following eigenvalues: $\lambda_0 = 6m+3n$, $\lambda_1 = 2$, $\lambda_2 = -2m-n$.

Now we can apply Theorem 6.15 to prove the non-existence of extended one-perfect codes in Doob graphs.

Proposition 6.3. *Let C be an extended one-perfect code in $D(m,n)$ (including the case $D(0,n) = H(n,4)$), where $2m+n = \frac{4^l+2}{3}$, $l \in \mathbb{N}$. Then $l \leq 3$.*

Proof. Let C be an extended one-perfect code in $D(m,n)$, where $2m+n = \frac{4^l+2}{3}$ for some positive integer l.

In this case, we have the following system of equations

$$\begin{cases} a_1 + a_2 = 4^{2m+n-l-1} - 4^{2m+n-2l-2} \\ 2a_1 - (2m+n)a_2 = -3(2m+n)4^{2m+n-2l-2}. \end{cases}$$

From this system we have

$$a_2 \cdot 4^{2m+n} = \frac{4^{4m+2n-2l-2}(2 \cdot 4^{l+1} - 2 + 4^l + 2)}{\frac{4^l+8}{3}} = \frac{4^{4m+2n-l-3}3^3}{4^{l-1}+2}.$$

By Theorem 6.15 the number $a_2 \cdot 4^{2m+n}$ is integer. If $l = 1$, then $4^2 \cdot a_2 = 9$. Let $l > 1$. Since the greatest common divisor of the denominator $4^{l-1}+2$ and $4^{4m+2n-l-3}$ equals 2, it follows that $2 \cdot 4^{l-2} + 1$ divides 27. This implies $l \in \{2,3\}$. □

The following proposition solves the remaining cases in $D(m,n)$.

Proposition 6.4. *There are no extended one-perfect codes in $D(m,n)$, where $2m+n = 22$.*

Proof. Since $D(m,n)$, $2m+n = 22$, is a distance-regular graph with intersection array $\{66, 63, \ldots, 3; 1, 2, \ldots, 22\}$, any equitable partition of $D(m,n)$ with quotient matrix S is an equitable partition of its distance-i graph with quotient matrix $S^{(i)}$, $i = 1, 2, \ldots, 22$, where

$$S^{(i)}S = (69-3i)S^{(i-1)} + 2iS^{(i)} + (i+1)S^{(i+1)}, \quad S^{(0)} = \mathrm{Id}, \quad S^{(1)} = S.$$

For extented one-perfect codes, we find

$$S^{(1)} = \begin{pmatrix} 0 & 66 & 0 \\ 1 & 2 & 63 \\ 0 & 22 & 44 \end{pmatrix}, \quad S^{(2)} = \begin{pmatrix} 0 & 0 & 2079 \\ 0 & 693 & 1386 \\ 11 & 484 & 1584 \end{pmatrix}, \quad S^{(3)} = \begin{pmatrix} 0 & 13860 & 27720 \\ 210 & 9660 & 31710 \\ \frac{440}{3} & \frac{33220}{3} & 30360 \end{pmatrix}$$

and see that $S^{(3)}$ is not integer, a contradiction. □

The idea of seeking a noninteger coefficient in a distance quotient matrix of a putative completely regular code provides an alternative approach of proving the nonexistence of extended one-perfect codes. It was developed by Vorob'ev [66] who showed the nonexistence of extended one-perfect codes in $H(n,q)$ for $n > q+2$ and prime power $q \geq 5$ (for $q = 3, 4$, this was proved earlier in [5]).

Remind that formally the singleton from any vertex in $D(0,2)$ or $D(1,0)$ is an extended one-perfect code, called trivial. Also all extended one-perfect codes in $D(m,n)$ with $2m+n = 6$ are characterized in [1] ($m = 0$) and [11] ($m > 0$, see Section 6.4.2). In this case, extended one-perfect codes are also MDS codes. From Propositions 6.3 and 6.4, we have the following.

Theorem 6.16. *An extended one-perfect code in $D(m,n)$ (including the case $D(0,n) = H(n,4)$) exists if and only if (m,n) is one of $(0,2)$, $(1,0)$, $(0,6)$, $(2,2)$.*

Corollary 6.4. *A completely regular code with quotient matrix*

$$\begin{pmatrix} 0 & 6m+3n & 0 \\ 1 & 2 & 6m+3n-3 \\ 0 & 2m+n & 4m+2n \end{pmatrix}$$

exists in $D(m,n)$ if and only if (m,n) is one of $(0,2)$, $(1,0)$, $(0,6)$, $(2,2)$.

6.4 MDS CODES WITH DISTANCE MORE THAN 2

In coding theory, MDS codes are codes attaining the Singleton bound [62]. We should emphasize that we do not imply that an MDS code is linear, or additive, or having any other additional structure. The Singleton bound holds for all unrestricted codes, and, following [62], any unrestricted code attaining it is called MDS. The classical Singleton bound for Hamming graphs says that the cardinality of any distance-d code C in $H(N,q)$, $d \le N$, is not greater than q^{N-d+1}. It is not difficult to see that the Singleton bound for $H(N,4)$ also holds for codes in any Doob graph $D(m,n)$ of the same diameter $N = 2m+n$:

Lemma 6.12 (Singleton bound for codes in the Doob graphs). *A distance-d code C in $D(m,n)$, $d \le 2m+n$, satisfy*

$$|C| \le 4^{2m+n-d+1}.$$

Proof. We consider two cases.

In first case, suppose that $D(m,n)$ contains a subgraph D' of diameter $d-1$ isomorphic to $D(m',n')$, where $m' \le m$ and $n' \le n$. Such subgraph contains at most one codeword of C because its diameter is smaller than the code distance. Since the vertex set of $D(m,n)$ is covered by $16^{m-m'} \cdot 4^{n-n'}$ translations of D', the result follows. (Similarly, the Singleton bound is proved in the Hamming case.)

In second case, suppose that $D(m,n)$ does not contain a subgraph of diameter $d-1$ isomorphic to a Doob graph. This can only happen if $n = 0$ and d is even. In this case, $D(m,n)$ contains a subgraph D' isomorphic to $D(d/2,0)$. For the argument similar to the previous case to work, it remains to show that $D(d/2,0)$ contains no more than 4 codewords. Indeed, for the distance between two words $(x_1,\ldots,x_{d/2})$ and $(y_1,\ldots,y_{d/2})$ to be d, the distance between x_i and y_i in the Shrikhande graph must equal 2, $i = 1,\ldots,d/2$. It follows that the ith components of the codewords form a coclique in Sh. Hence, the number of codewords is not greater than the maximum size of a coclique in Sh, that is, 4. □

An *MDS code* with parameters $((m,n),4^k,d)$ in the graph $D(m,n)$ is a code of cardinality 4^k with code distance $d = 2m+n-k+1$, i.e., a code attaining the Singleton bound.

In this section, we consider characterization theorems for MDS codes in Doob graphs. We separate the material into two parts according to the code distance, larger than 2 (Section 6.4) and 2 (Section 6.5).

In the rest of this introduction, we briefly survey known classification results for MDS codes in Hamming graphs.

There are only linear MDS codes in $H(n,2)$ and $H(n,3)$ (for the case of distance 2 and $q = 3$, see e.g. [49, Corollary 13.25, Exercise 13.11]); apart of the distance-2 and distance-n codes there is only a distance-3 MDS code in $H(4,3)$. MDS codes with distance at least 3 in Hamming graphs $H(n,4)$ are classified in [1]. Quaternary MDS codes of distance 2 are characterized in [45]; we will consider the characterization theorem for this class in Section 6.5 because it is a part of characterization of distance-2 MDS codes in Doob graphs. The computer-aided classification of MDS codes with distance at least 3 and alphabet size up to 8 can be found in [31–33]. Distance-2 MDS codes in $H(n,4)$ and $H(n,5)$, $n \leq 6$ and $H(n,6)$, $n = 4$, are classifies up to equivalence in [53]. Representatives of nonequivalent $(3,q^2,2)_q$ MDS codes (essentially, latin squares, see also [52] and references there) and $(4,q^2,3)_q$ MDS codes (essentially, pairs of orthogonal latin squares, or graeco-latin squares) with $q \leq 8$ can be found at Brendan McKay's web-page [50]. It remains to note that distance-n MDS code in $H(n,q)$ is unique up to equivalence for any n and q

It is worth noting that for MDS codes in the Hamming graph $H(N,q)$ there is a well-known conjecture that MDS codes with distance greater than 2 and less than N exist only for $N \leq q+2$ (for $q = 4$, this is 6), with equality only for even q. For linear codes over finite fields (prime power q), this conjecture is proved in the case of a prime q; for nonprime values of q, important results are obtained, which are close to a final solution (see [3,4]). The things are more difficult for non-prime-power q; for example, for many cases the maximum length n of a known $(n,q^2,n-1)_q$ MDS code is very small (such a code is equivalent to a system of $n-2$ mutually orthogonal latin squares, MOLS; for example, existence of 3 mutually orthogonal latin squares of order $q = 10$ is a known problem, see e.g. [51]), MDS codes with parameters $(5,q^3,3)_q$ are not known for many q, see e.g. [29,55].

Finally, in the context of this book, we note that not all MDS codes are completely regular. Distance-2 MDS codes and (1-perfect) distance-3 MDS codes in $H(q+1,q)$ are completely regular with covering radius 1 and quotient matrices $\begin{pmatrix} 0 & (q-1)n \\ n & (q-2)n \end{pmatrix}$ and $\begin{pmatrix} 0 & q^2-1 \\ 1 & q^2-2 \end{pmatrix}$, respectively; distance-3 MDS codes of length less than $q+1$ are completely regular with covering radius 2 (see, e.g., [31, Corollary 6]; it also follows from Delsarte's theory [14] because the dual distance distribution of such code has only two nonzero weights) and quotient matrix

$$\begin{pmatrix} 0 & (q-1)n & 0 \\ 1 & n^2+q-3n & (n-1)(q+1-n) \\ 0 & n(n-1) & n(q-n) \end{pmatrix};$$

extended 1-perfect distance-4 MDS codes in $H(q+2,q)$ (they can only exist for even q [6]) are completely regular with covering radius 2 and quotient matrix

$$\begin{pmatrix} 0 & (q+2)(q-1) & 0 \\ 1 & q-2 & q^2-1 \\ 0 & q+2 & q^2-4 \end{pmatrix}.$$

In Section 6.4, we will see that not all distance-3 MDS codes in Doob graphs are completely regular; for each power-of-two q, there is a distance-4 MDS code in $H(q+1,q)$ (see e.g. [15, (F.5)]) that is completely regular with quotient matrix $\begin{pmatrix} 0 & q^2-1 & 0 & 0 \\ 1 & q-2 & q^2-q & 0 \\ 0 & q & q^2-q-2 & 1 \\ 0 & 0 & q^2-1 & 0 \end{pmatrix}$; however, it is not clear if all MDS codes with such parameters are completely regular (for $q=4$, such code is unique [1]; for $q=8$, there are two nonequivalent such codes, both are linear [33] and completely regular). From Section 6.4.2, we will see that similar distance-4 MDS codes in $D(1,3)$ are not completely regular (but the one in $D(2,1)$ is).

6.4.1 Distance-$(2m+n)$ MDS Codes

We start with considering MDS codes of distance $2m+n$ in $D(m,n)$. Each such code consists of four codewords $\boldsymbol{z}^j = (x_1^j, \ldots, x_m^j, y_1^j, \ldots, y_n^j)$, $j = 1,2,3,4$, where x_i^j belongs to the set of vertices of Sh, say $\mathbb{Z}_4^2$, and y_i^j belongs to the set of vertices of K_4, say $\{0,1,2,3\}$. For each i from $\{1,\ldots,m\}$, the set $\{x_i^1, x_i^2, x_i^3, x_i^4\}$ is a 4-coclique in Sh. There are 16 different 4-cocliques (Lemma 6.1) and $4! \times 16 = 384$ ways to choose the ordered 4-tuple $(x_i^1, x_i^2, x_i^3, x_i^4)$. Thus, there are $384^m \times 24^n$ ways to choose an ordered 4-tuple $(\boldsymbol{z}^1, \boldsymbol{z}^2, \boldsymbol{z}^3, \boldsymbol{z}^4)$ of codewords of a distance-$(2m+n)$ MDS code in $D(m,n)$. Dividing by 4!, we get the following.

Proposition 6.5. *There are exactly* $384^m \times 24^{n-1}$ *different distance-*$(2m+n)$ *MDS codes in* $D(m,n)$.

To evaluate the number of nonequivalent codes, we need the following technical lemma. By *ordered MDS code* we mean an ordered quadruple $(\boldsymbol{z}^1, \boldsymbol{z}^2, \boldsymbol{z}^3, \boldsymbol{z}^4)$ of vertices of a Doob graph such that the set $\{\boldsymbol{z}^1, \boldsymbol{z}^2, \boldsymbol{z}^3, \boldsymbol{z}^4\}$ is an MDS code.

Lemma 6.13. *In* Sh, *there are exactly* 4 *equivalence classes (under automorphisms of* Sh*) of ordered MDS codes, with representatives* (00, 21, 02, 23), (00, 21, 23, 02), (00, 02, 21, 23) *(corresponding to Figure 6.1b), and* (00, 02, 20, 22) *(corresponding to Figure 6.1a. In* K_4*, there is only one ordered MDS code, up to equivalence.*

Now, it is easy to count the number of nonequivalent ordered MDS codes in $D(m,n)$. Each equivalence class is indexed by 4 nonnegative integers m_1, m_2, m_3, m_4 with $m_1+m_2+m_3+m_4 = m$, where m_s denotes the number of positions i such that the quadruple of elements $(x_i^1, x_i^2, x_i^3, x_i^4)$ (the elements in the ith positions of the codewords) belongs to the sth equivalence class from Lemma 6.13.

Proposition 6.6. *There are exactly* $\binom{m+3}{3}$ *nonequivalent (under automorphisms of* $D(m,n)$*) ordered MDS codes in* $D(m,n)$*.*

To count the number of equivalence classes on usual (unordered) MDS codes, we should take into account that changing the order of codewords can change the equivalence class of ordered MDS codes the 3 equivalence classes of ordered MDS codes in Sh corresponding to Figure 6.1b can turn be permuted in 3! ways. So, the number of equivalence classes of distance-$(2m+n)$ MDS codes is asymptotically $\binom{m+3}{3}/3!$. More accurately, the following holds.

Theorem 6.17 (Proposition 1 in [11]). *The number* $F(m)$ *of equivalence classes of MDS codes of size 4 in* $D(m,n)$ *is*

$$\frac{m^3}{36}+\frac{7m^2}{24}+\frac{11m}{12}+1-\frac{m \bmod 2}{8}-\frac{m \bmod 3}{9}.$$

m	0	1	2	3	4	5	6	7	8	9	10	11	12	13	14	15	16	$\cdots$
$F(m)$	1	2	4	7	11	16	23	31	41	53	67	83	102	123	147	174	204	$\cdots$

In the context of the book on completely regular codes, it is worthy to note the following.

Proposition 6.7. *All distance-*$(2m+n)$ *MDS codes in Doob graphs* $D(m,n)$, $d=2m+n\geq 4$, *are not completely regular.*

Proof. Indeed, if $\boldsymbol{x}=(x_1,\ldots)$, $\boldsymbol{y}=(y_1,\ldots)$, $\boldsymbol{z}=(z_1,\ldots)$, $\boldsymbol{w}=(w_1,\ldots)$ are the codewords, then we consider the word $\boldsymbol{v}$ obtained from $\boldsymbol{x}$ by changing the first element x_1 to y_1 and the word $\boldsymbol{u}$ obtained from $\boldsymbol{x}$ by changing the first element x_1 to one of the 6 vertices of Sh at distance 2 from x_1 and different from y_1, z_1, w_1. Both $\boldsymbol{v}$ and $\boldsymbol{u}$ are at distance 2 from the code and from $\boldsymbol{x}$, but $\boldsymbol{v}$ is also at distance $d-2$ from $\boldsymbol{y}$, while the distance from $\boldsymbol{u}$ to each of $\boldsymbol{y}$, $\boldsymbol{z}$, $\boldsymbol{w}$ is at least $d-1$. According to Delsarte's definition, the code is not completely regular. □

6.4.2 Distance More Than 2 and Less Than $2m+n$

According to the main result of [11], there are exactly ten nonequivalent MDS codes of distance more than 2 and less than $2m+n$ in Doob graphs $D(m,n)$:

- 2 distance-3 size-4^2 MDS codes in $D(2,0)$, one linear $\langle(1,\omega-1)\rangle$ and one non-additive (Table 6.2a);
- 1 distance-3 size-4^2 MDS code in $D(1,2)$, with additive structure $(\mathbb{Z}_4^2)^1\times\mathbb{Z}_4^2$: $\langle(21,1,0),(12,0,1)\rangle$;
- 2 distance-4 size-4^2 MDS codes in $D(2,1)$, one linear $\langle(1,2\omega+1;1)\rangle$ [35] and one non-additive (Table 6.2b);
- 1 distance-4 size-4^2 MDS code in $D(1,3)$, with additive structure $(\mathbb{Z}_4^2)^1\times\mathbb{Z}_4^3$: $\langle(21,1,0,1),(12,0,1,1)\rangle$ [35];

- 2 distance-3 size-4^3 MDS codes in $D(2,1)$ (one-perfect codes), one linear $\langle(1,\omega-1;\omega)\rangle^{\perp}$ [35] and one non-additive (Table 6.2c);
- 1 distance-3 size-4^3 MDS code in $D(1,3)$ (one-perfect code), with additive structure $(\mathbb{Z}_4^2)^1 \times \mathbb{Z}_4^3$: $\langle(21,1,0,1),(32,0,1,1)\rangle^{\perp}$ [35];
- 1 distance-4 size-4^3 MDS code in $D(2,2)$, linear extended one-perfect code $\langle(1,2\omega+1;1,0),(2,0;1,1)\rangle$.

The code $\langle(1,\omega-1)\rangle$ in $H(2,0)$ is self-dual with respect to the Hermitian inner product (see Remark 6.1); the codes $\langle(1,2\omega+1;1)\rangle$ in $H(2,1)$ and $\langle(1,2\omega+1;1,0),(2,0;1,1)\rangle$ in $H(2,2)$ are self-orthogonal with respect to the inner product (6.2).

Straightforward calculations [7] show that among the codes above, only nonlinear distance-4 MDS codes in $D(2,1)$ and $D(1,3)$ are not completely regular. The 3 distance-3 MDS codes in $D(2,0)$ and $D(1,2)$ are completely regular with quotient matrix $\begin{pmatrix}0 & 12 & 0\\ 1 & 8 & 3\\ 0 & 12 & 0\end{pmatrix}$. The 3 distance-4 MDS codes in $D(2,1)$ and $D(1,3)$ are actually simplex codes, or tight 2-designs; the distance between every two distinct codewords is always 4. However, the linear distance-4 MDS code in $D(2,1)$ is completely regular with quotient matrix $\begin{pmatrix}0 & 15 & 0 & 0\\ 1 & 2 & 12 & 0\\ 0 & 4 & 10 & 1\\ 0 & 0 & 15 & 0\end{pmatrix}$, while the other two distance-4 MDS codes, the non-additive one in $D(2,1)$ and the additive one in $D(1,3)$, are not completely regular. The 3 distance-3 MDS codes in $D(2,1)$ and $D(1,3)$ are 1-perfect and hence completely regular with quotient matrix $\begin{pmatrix}0 & 15\\ 1 & 14\end{pmatrix}$.

TABLE 6.2 MDS Codes That Are Not Equivalent to any Linear or Additive Code: (a) Distance-3 Code in $D(2,0)$, (b) Distance-4 Simplex MDS Code in $D(2,1)$, (c) 1-Perfect Distance-3 Code in $D(2,1)$.

(a)	00 00,	02 02,	20 23,	22 21,	01 22,	03 20,	21 01,	23 03,
	10 32,	12 30,	31 10,	33 12,	11 13,	13 11,	30 31,	32 33
(b)	00 00 0,	02 02 0,	20 23 0,	22 21 0,	01 22 1,	03 20 1,	21 01 1,	23 03 1,
	10 32 2,	12 30 2,	31 10 2,	33 12 2,	11 13 3,	13 11 3,	30 31 3,	32 33 3
(c)	00 00 0,	02 00 1,	21 00 2,	23 00 3,	02 02 0,	00 02 1,	23 02 2,	21 02 3,
	23 20 0,	21 20 1,	00 20 2,	02 20 3,	21 22 0,	23 22 1,	02 22 2,	00 22 3,
	22 01 0,	20 01 1,	01 01 2,	03 01 3,	20 03 0,	22 03 1,	03 03 2,	01 03 3,
	01 21 0,	03 21 1,	20 21 2,	22 21 3,	03 23 0,	01 23 1,	22 23 2,	20 23 3,
	32 10 0,	30 10 1,	12 10 2,	10 10 3,	30 12 0,	32 12 1,	10 12 2,	12 12 3,
	10 31 0,	12 31 1,	32 31 2,	30 31 3,	12 33 0,	10 33 1,	30 33 2,	32 33 3,
	13 11 0,	11 11 1,	33 11 2,	31 11 3,	11 13 0,	13 13 1,	31 13 2,	33 13 3,
	31 30 0,	33 30 1,	13 30 2,	11 30 3,	33 32 0,	31 32 1,	11 32 2,	13 32 3

The distance-4 MDS codes in $D(2,2)$ are extended one-perfect and hence completely regular with quotient matrix $\begin{pmatrix} 0 & 18 & 0 \\ 1 & 2 & 15 \\ 0 & 6 & 12 \end{pmatrix}$.

6.5 MDS CODES WITH DISTANCE 2 AND MULTIFOLD MDS CODES

In this section with an unpalatable title, we describe a series of very important results concerning completely regular codes of covering radius 1 with the smallest eigenvalue in Doob and Hamming graphs, which is $-2m-n$ for the Doob graph $D(m,n)$ and $-N$ in the Hamming $H(N,q)$ case. The theory of such codes for Doob graphs is essentially based on the one for Hamming graphs, so it is very natural to present here all related results for these two classes of graphs. For convenience and brevity, we will use a unified notation $DH(m,n,q)$, called a *DH graph*, for the Cartesian product of m copies of Shrikhande graph and n copies of K_q, always implying that $m=0$ or $q=4$ (or both), so $DH(m,n,q)$ is always the Doob graph $D(m,n)=DH(m,n,4)$ or the Hamming graph $H(n,q)=DH(0,n,q)$. The vertex set of $DH(m,n,q)$ is implied to be $\mathbb{E}_4^m \times \mathbb{Z}_q^n$, where $\mathbb{E}_4$ is also used for brevity as the ring structure discussed in Section 6.2 does not play any role.

Distance-2 MDS codes in DH graphs can be also characterized as maximum independent sets (attaining the Hoffman bound [17, Propostion 1.3.2(i)]), algebraic $(2m+n-1)$-designs of minimum size (a set of vertices is an *algebraic t-design* if the decomposition of its characteristic function into the sum of eigenfunctions of the graph does not include eigenfunctions corresponding to the t largest non-main eigenvalues of the graph), in the case of Hamming graphs, equivalent to orthogonal arrays $\mathrm{OA}(q^{n-1},n,q,n-1)$ (see a.g. [27]), $(n-1)$-ary quasigroups of order q, latin $(n-1)$-cubes of order q (see Definition 6.5 in Section 6.5.2). Multifold MDS codes are equivalent to algebraic $(2m+n-1)$-designs, completely regular codes with covering radius 1 and eigenvalue $-2m-n$ (the minimum eigenvalue for $DH(m,n,q)$), simple orthogonal arrays of strength $n-1$ (in the Hamming case) [25].

In the context of the this chapter, the main result in this section is the constructive characterization of MDS codes in $D(m,n)$ and $H(n,4)$ (Theorem 6.21), satisfying that every such code is semilinear or reducible. However, we also consider several important results for MDS and multifold MDS codes in DH graphs in general, both used in the proof of the main theorem and having an independent value.

The quotient matrix of a completely regular code of covering radius 1 with the minimum eigenvalue in $D(m,n)$ or $H(N,q)$ has the form (we assume that $N=2m+n$ and $q=4$ for $D(m,n)$)

$$\begin{pmatrix} N(l-1) & N(q-l) \\ Nl & N(q-l-1) \end{pmatrix} \tag{6.26}$$

for some l (it is easy to derive this from the row sum, which must equal to the degree of the graph, and the eigenvalue $-N$). If $l=1$, then the code is an MDS code with distance 2; for brevity, we will omit mentioning the distance of MDS codes in this section because it is always 2. Such MDS codes are independent sets that attain the Hoffman bound [17, Propostion 1.3.2(i)], and hence they are all completely regular. The codes with arbitrary l will be called *l-fold MDS codes*, or *multifold MDS codes*, because a code with such parameters (quotient matrix) can be obtained (if l is integer, see below) as the union of l disjoint MDS codes. In Doob graphs $D(m,n)$ with $n \geq 1$ and Hamming graphs $H(N,q)$, there are cliques of size q (4 for $D(m,n)$) (Delsarte cliques). Such a clique intersects with every l-fold MDS codes in exactly l points (see Corollary 6.5 below), and hence l is integer. However, it is not true for $D(m,n)$ and another value of l, 1.5, is possible. In general, we have the following.

Lemma 6.14. *Assume the graph Γ is the Cartesian product of s graphs Γ_i, $i = 1,\ldots,s$. A function f on the vertices of Γ is an eigenfunction of Γ with minimum eigenvalue if and only if for every $i \in \{1,\ldots,s\}$ the retract of f on every copy Γ' of Γ_i obtained as a subgraph by fixing elements in all non-i'th components of the product is an eigenfunction of Γ' with the minimum eigenvalue.*

Proof (sketch). The space of functions on the vertices of Γ has a basis from functions

$$\phi(x_1,\ldots,x_s) = \phi_1(x_1)\ldots\phi_s(x_s), \tag{6.27}$$

where ϕ_i is a basis eigenfunction of Γ_i. The eigenvalue corresponding to such ϕ is the product of the eigenvalues corresponding to ϕ_i. If f is an eigenfunction corresponding to the minimum eigenvalue of Γ, then it is the sum of basis eigenfunctions of form (6.27), where each component ϕ_i is an eigenfunction corresponding to the minimum eigenvalue of Γ_i; hence, we have such an eigenfuction in all retracts of the considered form.

If f is *not* an eigenfunction corresponding to the minimum eigenvalue of Γ, then it is the sum of basis eigenfunctions of form (6.27), where at least one component ϕ_j in one summand is an eigenfunction corresponding to a non-minimum eigenvalue of Γ_i. In this case, we can choose constants a_i, $i \in \{1,\ldots,s\}\backslash\{j\}$, such that $\phi_i(a_i) \neq 0$ in this summand. The corresponding retract of f is then not an eigenfunction of Γ_j with minimum eigenvalue. □

Corollary 6.5. *A set of vertices of $D(m,n)$ ($H(N,q)$) is an l-fold MDS code if and only if its retract in every subgraph isomorphic to* Sh *or K_4 (K_q for $H(N,q)$) is an l-fold MDS code.*

Proof. It is known and easy to see that in any graph, equitable 2-partitions with quotient matrix $\left(\begin{smallmatrix} a & b \\ c & d \end{smallmatrix}\right)$ are in one-to-one correspondence with two-valued eigenfunctions with values, say, 1 in the vertices of the first cell and $-b/c$ in the vertices of the second cell of the partition. Now, the claim is straightforward from Lemma 6.14. □

Proposition 6.8. *In $D(m,0)$, any completely regular code with covering radius* 1 *and minimum eigenvalue has one of the following quotient matrices:*

$$\begin{pmatrix} 0 & 6m \\ 2m & 4m \end{pmatrix}, \begin{pmatrix} m & 5m \\ 3m & 3m \end{pmatrix}, \begin{pmatrix} 2m & 4m \\ 4m & 2m \end{pmatrix}, \begin{pmatrix} 3m & 3m \\ 5m & m \end{pmatrix}, \begin{pmatrix} 2m & m \\ 3m & 0 \end{pmatrix}.$$

In $H(N,q)$ and $D(m,n)$, $n \geq 1$, any completely regular code with covering radius 1 *and minimum eigenvalue has quotient matrix of form* (6.26), *where l is integer and $N = 2m+n$, $q = 4$ for $D(m,n)$.*

Proof. The claim follows from Corollary 6.5. In K_q, an l-fold MDS code is just a set of any l vertices, so l is necessarily integer for Hamming graphs and Doob graphs $D(m,n)$ with $n > 0$. However, $D(m,0)$ does not contain a 4-clique and this condition is not necessary. By Corollary 6.5, it is sufficient to show the statement for the Shrikhande graph $\mathrm{Sh} = D(1,0)$. From the matrix (6.26), we see that for $N = 2m = 2$ all coefficients are integer and nonnegative only if $l \in \{1, \frac{3}{2}, 2, \frac{5}{2}, 3\}$. □

6.5.1 An Injective Map

We first describe a standalone result that shows that the number of distance-2 MDS codes (maximum independent sets) in $D(m,n)$, $m > 0$, is smaller than that number in $D(m-1, n+2)$ (and, by induction, in $D(0, n+2m) = H(2m+n, 4)$) by constructing an injective map from the first class (we denote it by $\mathrm{MDS}_{m,n}$) to the second. At first, we define the map ξ from $\mathrm{MDS}_{1,0}$ into $\mathrm{MDS}_{0,2}$, see Figure 6.6.

For arbitrary m, $n \geq 0$, the action of $\kappa : \mathrm{MDS}_{m+1,n} \to \mathrm{MDS}_{m,n+2}$ is defined as follows:

$$\kappa M \stackrel{\mathrm{def}}{=} \big\{(x_1, \dots, x_m, z_1, z_2, y_1, \dots, y_n) \in D(m, n+2) : (z_1, z_2) \in \xi M_{x_1,\dots,x_m,y_1,\dots,y_n}\big\},$$

where

$$M_{x_1,\dots,x_m,y_1,\dots,y_n} \stackrel{\mathrm{def}}{=} \{v \in \mathrm{Sh} : (x_1, \dots, x_m, v, y_1, \dots, y_n) \in M\}.$$

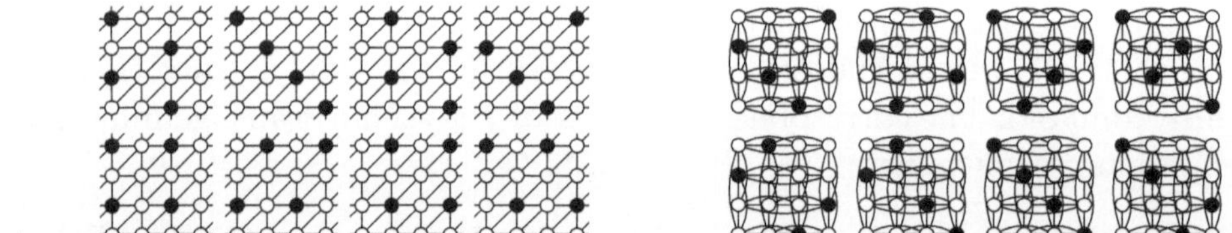

Figure 6.6 The 16 MDS codes in Sh and the corresponding MDS codes in K_4^2 (the last row shows the 8 MDS codes in K_4^2 that have no preimages under ξ).

Lemma 6.15 ([39]). *For every distance-2 MDS code M in $D(m+1,n)$, the set κM is a distance-2 MDS code in $D(m,n+2)$. Moreover, for two different distance-2 MDS codes M_1 and M_2 in $D(m+1,n)$, the MDS codes κM_1 and κM_2 are also different.*

Corollary 6.6. *The number of distance-2 MDS codes M in $D(m+1,n)$ is strictly less than the number of distance-2 MDS codes M in $D(m,n+2)$.*

The strictness follows from the fact that ξ is not surjective (see Figure 6.6).

6.5.2 Latin Colorings and Reducibility

Definition 6.5 (latin coloring, latin hypercube, latin square). *We say that a $\mathbb{Z}_q$-valued function f on the vertices of the Hamming $H(N,q)$ or Doob $D(m,n)$ (where $N=2m+n$ and $q=4$) graph is a* latin coloring *if it is a perfect q-coloring with quotient matrix $N(J_q-I_q)$, where J_q and I_q are the all-one and identity matrices, respectively, of size $q\times q$. In the case of the Hamming graph $H(n,q)$ latin colorings are known as* latin hypercubes *of order q and dimension n (in the case $n=3$,* latin cubes*; in the case $n=2$,* latin squares*).*

Lemma 6.16. (a) *In a latin coloring of $H(N,q)$ or $D(m,n)$ ($q=4$), each q-clique contains exactly one vertex of each color from $\mathbb{Z}_q$. Moreover, the induced coloring on each Shrikhande subgraph (in the case of the Doob graph) is perfect with quotient matrix $\begin{pmatrix} 0&2&2&2\\ 2&0&2&2\\ 2&2&0&2\\ 2&2&2&0 \end{pmatrix}$ and hence belongs to one of the three equivalence classes illustrated in Figure 6.2.*

(b) *Latin colorings are equivalent to partitions of the vertices of the graph into q MDS codes; namely, a function f is a latin coloring if and only if $f^{-1}(i)$ is an MDS code for every i in $\mathbb{Z}_q$.*

(c) *Latin colorings of $H(N,q)$ or $D(m,n)$ are equivalent to (distance-2) MDS codes in $\Gamma=(V,E)$, where $\Gamma=H(N+1,q)$ or $\Gamma=D(m,n+1)$, respectively; namely, a function f is a latin coloring if and only if $\{(\boldsymbol{x},f(\boldsymbol{x})) : \boldsymbol{x}\in V\}$ is an MDS code.*

Proof (sketch). (a) If we have a $\mathbb{Z}_q$-valued latin coloring f, then the function $\phi(\cdot)=\sigma_i(f(\cdot))$, where $\sigma_i(i)=1$ and $\sigma_i(j)=-1/(q-1)$ for $j\in\mathbb{Z}_q\backslash\{i\}$, is an eigenfunction with eigenvalue $-N$, which is the minimum eigenvalue of the considered graph (for $D(m,n)$, $N=2m+n$ and $q=4$). By Lemma 6.14, we have a minimum-value eigenfunction in every K_q- or Sh-retract. It is easy to see that it possesses 1 in exactly 1 vertex of K_q and exactly 4 vertices of Sh. The claim is now straightforward.

(b) and (c) are straightforward. □

So, one can study MDS codes in terms of latin colorings, taking into account Lemma 6.16(c), with the only exception of MDS codes in $D(m,n)$ each of both representations (MDS codes and latin colorings) has its advantages and disadvantages. For example, in the latin-coloring representation of MDS codes, one of coordinates is distinguished and considered as the value of the coloring, while the other coordinates correspond to independent arguments. This makes this representation less symmetric and sometimes one need to consider the dependent and independent coordinates separately, while in the MDS representation they have "the same rights." On the other hand, latin colorings are convenient to combine in the composition, constructing new latin colorings of a larger Hamming or Doob graph. For this reason, it is convenient to combine both these representations in the same claims or formulas.

Notation agreement. We will associate a code with its characteristic function, naming it by the same letter; for example, if C is an MDS code, then $C(\boldsymbol{x})$ equals 1 if $\boldsymbol{x} \in C$ and equals 0 otherwise.

The following fact is straightforward. Actually, all three claims are the same, taking into account Lemma 6.16(c).

Lemma 6.17. (a) *If F is an MDS code in $DH(m, n+1, q)$ and g is a latin coloring of $DH(m', n', q)$, then the composition*

$$F(x_1, \ldots, x_m, y_1, \ldots, y_{i-1}, g(\boldsymbol{v}), y_i, \ldots, y_n),$$
$$x_1, \ldots, x_m \in \mathbb{Z}_4^2, \quad y_1, \ldots, y_n \in \mathbb{Z}_q, \quad \boldsymbol{v} \in \mathbb{E}_4^{m'} \times \mathbb{Z}_q^{n'} \tag{6.28}$$

is an MDS code in $DH(m+m', n+n', q)$.

(b) *If f and g are latin colorings of $DH(m, n, q)$ and $DH(m', n', q)$, respectively, then the composition*

$$C = \{(\boldsymbol{u}, \boldsymbol{v}) \in (\mathbb{E}_4^m \times \mathbb{Z}_q^n) \times (\mathbb{E}_4^{m'} \times \mathbb{Z}_q^{n'}) : \ f(\boldsymbol{u}) = g(\boldsymbol{v})\} \tag{6.29}$$

is an MDS code in $DH(m+m', n+n', q)$.

(c) *If f and g are latin colorings of $DH(m, n+1, q)$ and $DH(m', n', q)$, respectively, then the composition*

$$f(x_1, \ldots, x_m, y_1, \ldots, y_{i-1}, g(\boldsymbol{v}), y_i, \ldots, y_n),$$
$$x_1, \ldots, x_m \in \mathbb{Z}_4^2, \quad y_1, \ldots, y_n \in \mathbb{Z}_q, \quad \boldsymbol{v} \in \mathbb{E}_4^{m'} \times \mathbb{Z}_q^{n'} \tag{6.30}$$

is a latin coloring of $DH(m+m', n+n', q)$.

Keeping in mind Lemma 6.16(c), we see that (6.28) allows to combine MDS codes in $DH(m, n+1, q)$ and $DH(m', n'+1, q)$ into a new MDS code in $DH(m+m', n+n', q)$.

Definition 6.6 (reducible MDS codes). *An MDS code in $DH(M, N, q)$ is called* reducible *if it can be, up to permutation of coordinates, represented in form* (6.28) *for some m, m', n, n' such that $M = m+m'$, $N = n+n'$, $2m+n > 1$, $2m'+n' > 1$, some MDS code M in $DH(m, n+1, q)$ and some latin coloring f of $DH(m', n', q)$ (equivalently, in form* (6.29) *for some latin colorings f and g of $DH(m, n, q)$ and $DH(m', n', q)$, respectively).*

The condition $2m+n > 1$, $2m'+n' > 1$ excludes trivial cases of decomposition and implies that we combine two MDS codes in Doob or Hamming graphs to obtain an MDS code in a DH graph of larger diameter. In particular, and MDS code in a DH graph of diameter less than 4 cannot be reducible according to the definition. The following theorem, proved in [46], gives a test for reducibility of MDS codes in terms of reducibility of their retracts. (The case $6 \leq i < n$ in the first part of the theorem was considered in [38]; the results in [38,46] are formulated in terms of n-ary quasigroups.)

Theorem 6.18 (Theorem 1 in [46]). *Assume M is an MDS code in $H(n,q)$, $n \geq 5$. If for some i, $i \leq n$, all i- and all $(i-1)$-retracts of M are reducible, then M is reducible. Moreover, if q is prime and for some i all i-retracts are reducible, then M is reducible.*

Currently, there is no known version of this result for MDS codes in Doob graphs. Theorem 6.18 can be treated as follows.

Corollary 6.7. *An irreducible MDS code in $H(n,q)$ has an irreducible $(n-1)$- or $(n-2)$-retract; if q is prime, then it has an irreducible $(n-1)$-retract.*

In [37] (see also [10]) it was shown that the last "prime-q" claims of Theorem 6.18 and Corollary 6.7 cannot be generalized to an arbitrary q:

Proposition 6.9. *For every odd m, $m \geq 5$, and every q divisible by 4 there exists an irreducible MDS code in $H(n,q)$ such that all its $(n-1)$-retracts are reducible.*

Corollary 6.7 can be very helpful for the characterization of different retract-closed classes of MDS codes (latin hypercubes, multiary quasigroups): arguing inductively and proving some hypothesis about the structure of irreducible MDS codes, we can assume that the MDS code has an $(n-1)$- or $(n-2)$-retract (in the case of prime q, $(n-1)$-retract) that satisfies the hypothesis; this information can help to analyze the structure of the MDS code and to prove that the hypothesis is true for n. This strategy was applied for the characterization of MDS codes of order 4 in [45], see Theorem 6.21 below. In the simplest case, this method can be used to prove that some classes of MDS codes do not contain irreducible codes:

Proposition 6.10 (MDS treatment of Corollary 2 in [46] for n-ary quasigroups). *Every MDS code in $H(n,5)$, $n \geq 5$, or $H(n,7)$, $n \geq 4$, such that every its 3-retract is equivalent to a linear MDS code $\{(x,y,z) \in \mathbb{Z}_q^3 : x+y+z=0\}$ is reducible.*

The proof is by induction, utilizing computations and Corollary 6.7 for the induction base and step. It would be very interesting to find a proof for other values of q.

6.5.3 Decomposability of $q/2$-Fold MDS Codes

In this subsection, we consider another test for reducibility of MDS codes in graphs $D(m,n)$, including $H(n,4)$, suitable only for $q=4$. It includes a criterion on decomposability (see the definition below) of $q/2$-fold MDS codes.

Definition 6.7 (decomposable multifold MDS codes). *A $q/2$-fold MDS code (q even) is called* decomposable (indecomposable) *if its characteristic function can (cannot) be represented as the modulo-2 sum of two $\{0,1\}$-functions in disjoint nonempty collections of variables.*

Definition 6.8 (component, multifold MDS codes). *A connected component of a subgraph of $DH(m,n,q)$ induced by a multifold MDS code M will be called a* component *(of M). A multifold MDS code is called* connected *if it is a component itself. A two-fold MDS code in $DH(m,n,q)$ is called* bipartite *if it is the union of two MDS codes.*

It follows that these functions are the characteristic functions of $q/2$-fold MDS codes in DH graphs, which, in their turns, can be decomposable or not. As a result, the characteristic function of any $q/2$-fold MDS code is the modulo-2 sum of the characteristic functions of one, two, or more (up to $m+n$) indecomposable $q/2$-fold MDS codes. With some natural assumptions, such a representation is unique:

Proposition 6.11 (on a unique decomposition [36, Theorem 4.1(a), Remark 4.3], [43, Lemma 2]). *For any $q/2$-fold MDS code M in $DH(m,n,q)$, there is a unique collection $\{M_1,\ldots,M_k\}$ of indecomposable $q/2$-fold MDS codes not containing the all-zero tuple, such that*

$$M(\alpha(\boldsymbol{x}_1,\ldots,\boldsymbol{x}_k),\beta(\boldsymbol{y}_1,\ldots,\boldsymbol{y}_k)) = M_1(\boldsymbol{x}_1,\boldsymbol{y}_1)+\cdots+M_k(\boldsymbol{x}_k,\boldsymbol{y}_k)+\sigma \bmod 2,$$
$$\boldsymbol{x}_i\in\mathbb{E}_4^{m_i},\quad \boldsymbol{y}_i\in\mathbb{Z}_q^{n_i},\quad (6.31)$$

where $n=n_1+\cdots+n_k$, $m=m_1+\cdots+m_k$, for all $i\in\{1,\ldots,k\}$ it holds $m_i+n_i\geq 1$, α and β are some permutations of m and n coordinates, respectively, and $\sigma\in\{0,1\}$.

The main fact about decomposability of $q/2$ MDS codes, and also one of the key facts in the proof of the characterization of MDS codes in $D(m,n)$ is the following theorem, proved in [36, Theorem 4.1(b), Remark 4.3] for $m=0$ and in [43, Proposition 1] for $m>0$:

Theorem 6.19. *A $q/2$-fold MDS code in $DH(m,n,q)$, $(m,n)\neq(1,0)$, is decomposable if and only if it is not connected.*

Definition 6.9 (linear two-fold MDS code, semilinear MDS code). *We will say that a two-fold MDS code in $D(m,n)$ is* linear *if $k=m+n$ in the representation (6.31) (i.e., all M_i are two-fold MDS codes in* Sh *or K_4) and, moreover, the ones that are in* Sh *are not connected (Figure 6.7a).*

The next simple lemma means that the class of semilinear MDS codes in $D(m,n)$ is constructive and well understood.

Lemma 6.18. *A linear MDS code in $D(m,n)$ is unique up to equivalence, it contains exactly 2^{2m+n-1} different subsets that are MDS codes. The number of semilinear MDS codes in $D(m,n)$ is $2\cdot 3^{m+n}\cdot 2^{2^{2m+n-1}}(1+o(1))$ as $m+n\to\infty$.*

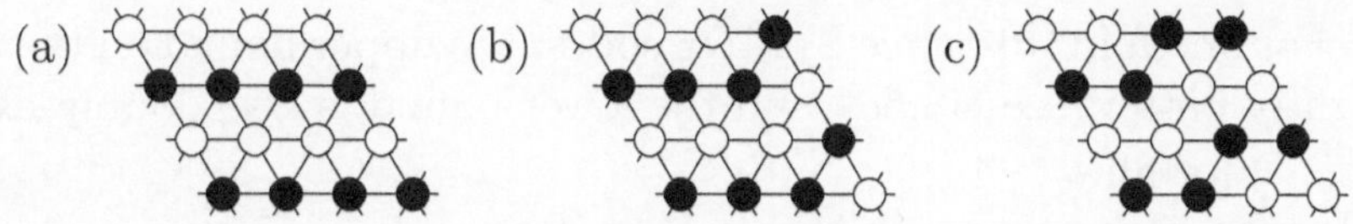

Figure 6.7 All two-fold MDS codes in Sh, up to equivalence; (a) is not connected, (b) and (c) are connected.

Indeed, for any bipartite two-fold MDS code (not only in $D(m,n)$ but also in $H(n,q)$ for any q), an MDS-code is just a bipartite part of an induced subgraph of $DH(m,n,q)$. The number to choose such a part if 2^r, where r is the number of components.

The next test for reducibility was proved in [36] for $m = 0$ and generalized in [43].

Theorem 6.20. *Assume that an MDS code S in $D(m,n)$ is a subset of a two-fold MDS code M, which has a canonical decomposition* (6.31). *If S is irreducible, then one of the following three cases takes place:*

- *M is linear;*
- *$k = 1$;*
- *$k = 2$ and either $(m_1, n_1) = (0,1)$ or $(m_2, n_2) = (0,1)$.*

In all other cases, S is reducible.

6.5.4 Characterization of MDS Codes in $D(m,n)$

The following theorem was proved in [45] for $H(n,4)$ (in terms of multiary quasigroups of order 4) and generalized to $D(m,n)$ in [43].

Theorem 6.21. *Every MDS code in $D(m,n)$ or $H(n,4)$ is semilinear (Definition 6.9) or reducible (Definition 6.6).*

The prove uses tests on reducibility in Theorems 6.18 and 6.20 and analyzes the remaining cases. As a result, the complete proof includes results in several papers. The problem of finding a shorter proof of this theorem (even in the case $m = 0$) looks intriguing.

Much less is known about two-fold MDS codes in $D(m,n)$ (except the bipartite case); the most interesting result is Theorem 6.22 below. In [61], an upper bound $2^{2.925^n}$ on their number for large n was proved in the Hamming case $m = 0$ (but the asymptotic of the double-logarithm of this bound does not achieve the similar asymptotic for the known lower bound $2^{2^{n(1+o(1))}}$) and the two-fold MDS codes in $H(4,4)$ were computationally classified up to equivalence. In particular, it was found that a two-fold MDS code can be equivalent to its complement. Recently [47], the upper bound from [61] was generalized to multifold MDS codes in $H(n,q)$ for arbitrary q. For one-fold (distance-2) MDS codes, the upper bound on the number of codes is slightly better, see [48,56], and the lower bound is also doubly-exponential (of form $2^{2^{\Omega(n)}}$) for all $q \geq 4$.

The following theorem was proved in [54]; the proof uses Theorem 6.21 ($m = 0$) and is surprisingly long, and existence of a shorter proof is also an intriguing question. An analog of this result for Doob graphs is not yet known.

Theorem 6.22. *If a two-fold MDS code in $H(n, 4)$ is bipartite, then its complement is also bipartite.*

For larger q, the question of existence of multifold MDS codes that cannot be split into MDS codes was also studied, see e.g. [30,44].

6.5.5 $1\frac{1}{2}$-Fold MDS Codes

Under this intriguing notion, $1\frac{1}{2}$-fold MDS codes, we only mean a completely regular code in $D(m, 0)$ whose quotient matrix $\left(\begin{smallmatrix} 1m & 5m \\ 3m & 3m \end{smallmatrix}\right)$ is the arithmetic mean of the quotient matrices $\left(\begin{smallmatrix} 0m & 6m \\ 2m & 4m \end{smallmatrix}\right)$ and $\left(\begin{smallmatrix} 2m & 4m \\ 4m & 2m \end{smallmatrix}\right)$ of a one-fold and two-fold MDS code. Such a code exists for all m and can be constructed as $C = \{(x_1, \ldots, x_m) : x_1 + \cdots + x_m \in C'\}$, where C' is a completely regular code in the Shrikhande graph with quotient matrix $\left(\begin{smallmatrix} 1 & 5 \\ 3 & 3 \end{smallmatrix}\right)$, unique up to equivalence (see Figure 6.8). Below we show that by slightly modifying this construction, we can construct doubly exponential number of completely regular codes with the same parameters.

Proposition 6.12. *Let $\lambda : \mathbb{Z}_4^{m-1} \to \{0, 1\} \subset \mathbb{Z}_4$ be an arbitrary function. Define the following two sets of vertices of $D(m, 0)$:*

$$C_0 = \{(x_1^*, \ldots, x_m^*) \in (\mathbb{Z}_4^2)^m : x_1^* + \cdots + x_m^* \in \{01, 02, 10, 33\}\},$$

$$C_1 = \{((y_1, z_1), \ldots, (y_m, z_m)) \in (\mathbb{Z}_4^2)^m : y_1 + \cdots + y_m = 2, \\ z_1 + \ldots + z_m - \lambda(y_1, \cdots, y_{m-1}) \in \{0, 2\}\}.$$

Then $C_0 \cup C_1$ is a completely regular code in $D(m, 0)$ with quotient matrix $\left(\begin{smallmatrix} 1m & 5m \\ 3m & 3m \end{smallmatrix}\right)$.

Proof. Fix all positions except the ith one and consider the retract of the constructed code in the corresponding Shrikhande subgraph. Denote $y = -y_1 - \ldots - y_{i-1} - y_{i+1} - \cdots - y_m$ and $z = -z_1 - \cdots - z_{i-1} - z_{i+1} - \cdots - z_m$. The vertices

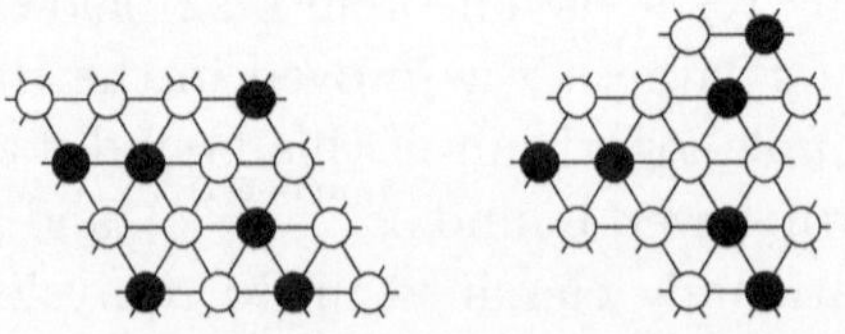

Figure 6.8 An equitable partition of Sh with quotient matrix [[1, 5], [3, 3]].

that have the first color form either the set $\{01, 02, 10, 33, 20, 22\} + (y, z)$ or the set $\{01, 02, 10, 33, 21, 23\} + (y, z)$ of vertices in this subgraph, depending on the value $\lambda(y_1, \ldots, y_{m-1})$, where $y_i = y + 2$. Each of these two sets is a completely regular code in the Shrikhande graph with quotient matrix $\left(\begin{smallmatrix} 1 & 5 \\ 3 & 3 \end{smallmatrix}\right)$. So, each code vertex has exactly 1 code neighbor in each of the m Shrikhande subgraphs it belongs and each noncode vertex has exactly 3 code neighbors in each of the m Shrikhande subgraphs it belongs. □

Corollary 6.8. *For any m there at least $2^{4^{m-1}}$ different perfect colorings in $D(m, 0)$ with quotient matrix $\left(\begin{smallmatrix} 1m & 5m \\ 3m & 3m \end{smallmatrix}\right)$. The number of equivalence classes of such colorings is at least $2^{4^{m-1}}/\mathrm{Aut}(D(m, n)) = 2^{4^{m-1} - O(n \ln n)}$.*

6.6 COMPLETELY REGULAR CODES WITH COVERING RADIUS 1

In this section, we consider objects that are equivalent to completely regular codes with covering radius 1, namely perfect 2-colorings. We review the results of [8] and describe the constructions which form the basis of these results. We focus on the case of two colors but some perfect colorings with more than two colors also emerge in connection with constructions of 2-colorings. Particularly, we describe a series of completely regular codes in Doob graphs with covering radius 2.

Some completely regular codes with covering radius 1 in Doob graphs are well known, for example, one-perfect codes and distance-2 MDS codes. The union of disjoint one-perfect codes or distance-2 MDS codes is also a completely regular code with covering radius 1.

We say that a vertex coloring f of a graph is a *$\{b; c\}$-coloring* if it is a perfect 2-coloring with quotient matrix $\begin{pmatrix} a & b \\ c & d \end{pmatrix}$. By default, we will assume that the first color of a $\{b; c\}$-coloring is 0 and the second is 1. The problem of the existence of $\{b; c\}$-colorings in Doob graphs can be divided into two problems. The first is classification of all pairs (b, c) for which there is a $\{b; c\}$-coloring in some Doob graph, see more details in the next subsection. The second is for a fixed pair (b, c) describe all m, n for which $\{b; c\}$-colorings exist in $D(m, n)$. Here we describe results related to the first problem and lower bounds for the second problem.

Here we continue the ideas invented in [23] and generalized in [12] where perfect 2-colorings in Hamming graphs have been considered. Some constructions in these works are based on covering and splitting which follows from the existence of partitions of distance-2 MDS codes in $H(q, q)$ into codes with code distance 3. We consider some generalizations of these ideas by involving multifold tilings and some partitions of distance-2 MDS codes in Doob graphs into codes with code distance at least 3.

6.6.1 Admissibility of Perfect 2-Colorings

Proposition 6.13. *For any nonnegative integers m, n, m', n' there is a covering of $D(m,n)+(6m'+3n')$ by $D(m+m',n+n')$, where $D(m,n)+(6m'+3n')$ is the graph obtained from $D(m,n)$ by adding $6m'+3n'$ loops at every vertex.*

This covering can be defined as

$$\phi(x_1^*,\ldots,x_{m+m'}^*;x_1',\ldots,x_{n+n'}') = (x_1^*,\ldots,x_m^*;x_1',\ldots,x_n').$$

Corollary 6.9. *If there is a perfect coloring in $D(m,n)$ with quotient matrix S, then for any $m',n' \geq 0$ there is a perfect coloring in $D(m+m',n+n')$ with quotient matrix $S+(6m'+3n')I$.*

In view of Corollary 6.9, we can divide the problem of existence of $\{b;c\}$-colorings into two problems. The first problem is the characterization of all pairs (b,c) such that there is a $\{b;c\}$-coloring in some Doob graph. We call a pair (b,c) *completely admissible* if for some D a $\{b;c\}$-coloring in $D(m,n)$ exists for all m, n such that $2m+n \geq D$. Since for fixed D there are few Doob graphs of diameter D, we also define a-admissible and ∞-admissible pairs. A pair (b,c) is called *a-admissible*, where a is some nonnegative integer, if for some D a $\{b;c\}$-coloring in $D(m,n)$ exists for all m, n such that $2m+n \geq D$ and $n \geq a$. For example, the pair $(4^k-1,1)$ is 1-admissible (it follows from the existence of one-perfect codes) but not 0-admissible (see Proposition 6.14) and hence not completely admissible. A pair (b,c) is called *∞-admissible* if for some N a $\{b;c\}$-coloring exists in $D(0,N)=H(N,4)$.

There are standard necessary conditions on the existence of a $\{b;c\}$-coloring in $D(m,n)$.

Proposition 6.14. *If there exists a $\{b;c\}$-coloring in $D(m,n)$, then*

1. $\dfrac{b+c}{\gcd(b,c)} = 2^k$ *for some positive k;*
2. $b+c=4i$ *for some $i \in \{1,\ldots,2m+n\}$;*
3. *if $c=1$ ($b=1$), then $b=4^l-1$ (respectively, $c=4^l-1$) for some integer l; moreover, $n \neq 0$.*

Proof. Claim 1 is valid because the cardinality of the first color is integer and equals $\frac{c}{b+c} \cdot 4^{2m+n}$; Claim 2 is essentially the Lloyd condition [25, Lemma 2.2(c)].

For the first part of Claim 3, it is sufficient to show that $c=1$ implies that b is divisible by 3 (indeed, by Claim 1, we have $b+c=2^k$ for some integer k). Consider a vertex x of first color. The neighborhood of x is divided into n groups of 3 vertices and m groups of 6 vertices, where the neighbors from the same group together with x belong to the same induced K_4- or Sh-subgraph. It is easy to see that if a group contains both first- and second-color vertices, then some of those second-color vertices are adjacent with at least two vertices of first color, including x. It follows that $b = 2^k - 1$ is divisible by 3, which implies that $k = 2l$ for some integer l. Moreover, if $n = 0$, then b is divisible by 6, which is also a contradiction because $2^k - 1$ is odd. □

Claim 3 is an analog of [12, Theorem 3.4] for Hamming graphs.

Corollary 6.10. *Let a pair (b, c) be completely admissible, ∞-admissible, or a-admissible for some a. Then*

1. $\dfrac{b+c}{\gcd(b,c)} = 2^k$ *for some positive k;*
2. *$b + c$ is divisible by 4;*
3. *if $c = 1$ $(b = 1)$ then $b = 4^k - 1$ $(c = 4^l - 1)$ for some l.*

As we will see in Section 6.6.6, that the inverse is also true, i.e., any pair (b, c) that satisfies the conclusion of Corollary 6.10 is ∞-admissible and completely admissible or a-admissible for some nonnegative a, see Theorem 6.30.

6.6.2 Multifold Tilings

The concept of tiling below differs from how this name is used in most of the literature. Often, a tiling means a partition of a group onto translations of the same set. For our purposes, the condition of being translations of the same set is not necessary (although in the most of cases it is possible to construct such a tiling with required parameters, we do not focus on this property), and we call by tiling an arbitrary partition of the vertex set of a graph into codes with the same parameters (in our case, completely regular codes). This definition is in agreement with definition of tiling in [21, p. 43]; however, we suggest to consider our definition as a local terminology.

Definition 6.10 (Multifold tilings). *Let $f_1, \ldots, f_m$ be $\{b; c\}$-colorings of a given graph G. We call the collection $(f_1, \ldots, f_m)$ a* λ-fold $\{b; c\}$-tiling *if any vertex x of G has the first color in exactly λ colorings from the collection, i.e. $|\{j \in \{1, \ldots, m\} : f_j(x) = 1\}| = \lambda$. A one-fold $\{b; c\}$-tiling is called just a* $\{b; c\}$-tiling. *In terms of completely regular codes, this is defined in the following way: a collection $(C_1, \ldots, C_m)$ of completely regular codes of a graph G with intersection array $\{b; c\}$ is called a* λ-fold $\{b; c\}$-tiling *if any vertex of G belongs to exactly λ codes from the collection.*

This definition can be generalized to perfect k-colorings. Let $f_1, \ldots, f_m$ be perfect k-colorings of a given graph G with quotient matrix S. We call the collection $(f_1, \ldots, f_m)$ a $\overline{\lambda}$-fold tiling, *where $\overline{\lambda} = (\lambda_0, \ldots, \lambda_{k-1})$ and $\lambda_0, \ldots, \lambda_{k-1}$ are positive* *integers, if for any vertex x of G and any color i from $\{0,\dots,k-1\}$, the vertex x has color i in exactly λ_i colorings from the collection, i.e., $|\{j \in \{1,\dots,m\} : f_j(x) = i\}| = \lambda_i$. However, we will mostly consider the case of two colors.*

Denote by I_l the identity $l \times l$ matrix. Denote by $J_{r,s}$ the $r \times s$ all-one matrix and by J_l the matrix $J_{l,l}$. We will omit the indices if they are clear from context.

Note that the first colors of colorings in $\{b;c\}$-tiling in a given graph G form an equitable partition with quotient matrix $c(J-I)+aI$, where a is uniquely determined by parameters. In other hand, any equitable partition with quotient matrix represented in form $c(J-I)+aI$ correspond to $\{b;c\}$-tiling for uniquely determined b. So, tilings are in one-to-one correspondence with perfect colorings with quotient matrices of type $c(J-I)+aI$.

The existence of a $\{b;c\}$-coloring f in $D(m,n)$ implies the existence of λ-fold $\{b;c\}$-tiling in this graph for some λ. Indeed, if we take all 4^{2m+n} translations of f we obtain a $|f^{-1}(1)|$-fold $\{b;c\}$-tiling. However, in general $|f^{-1}(1)|$ is not always the minimum possible multiplicity.

The Direct Product construction allows to construct new tilings.

Definition 6.11. *Let $\mathbb{G}$ be a finite group in additive notation. A $|\mathbb{G}| \times |\mathbb{G}|$ matrix $A = (a_{i,j})$ whose rows and column are indexed by elements of $\mathbb{G}$ is called $\mathbb{G}$-*circulant *(if $\mathbb{G} = \mathbb{Z}_k^+$, just* circulant*) if $a_{i,j} = a_{i+s,j+s}$ for any $i,j,s \in \mathbb{G}$.*

Theorem 6.23. *Let $\mathbb{G}$ be a finite abelian group. Let g and h be $\mathbb{G}$-valued perfect colorings of graphs $G = (V_G, E_G)$ and $H = (V_H, E_H)$ with $\mathbb{G}$-circulant quotient matrices A and B, respectively. Then*

$$f(x,y) = g(x) + h(y), \qquad x \in V_G,\ y \in V_H,$$

is a perfect k-coloring of the Cartesian product $G \square H$ with quotient matrix $A+B$, which is also $\mathbb{G}$-circulant.

Corollary 6.11. *Let $(g_1,\dots,g_k)$ be a $\{b_1;c_1\}$-tiling in a graph G and $(h_1,\dots,h_k)$ be a $\{b_2;c_2\}$-tiling in a graph H. Then there is $\{b_1+b_2;c_1+c_2\}$-tiling in $G \square H$.*

Proof. Since a $\{b;c\}$-tiling of a d-regular graph is an equitable partition with circulant quotient matrix $c(J-I)+(d-b)I$ (actually, it is $\mathbb{G}$-circulant for any group $\mathbb{G}$ of order k), the claim follows from Theorem 6.23. □

Sometimes, the quotient matrix of a perfect coloring is not circulant but the colors can be divided into blocks such that construction based on the direct product also works.

Theorem 6.24. *Let g be a perfect k-coloring of a graph G with quotient matrix*

$$S = \begin{pmatrix} A_{1,1} & A_{1,2} & \dots & A_{1,l} \\ A_{2,1} & A_{2,2} & \dots & A_{2,l} \\ \dots & \dots & \dots & \dots \\ A_{l,1} & A_{l,2} & \dots & A_{l,l} \end{pmatrix}$$

where $A_{i,i}$ is a $\mathbb{Z}_{k_i}$-circulant matrix for any i, $A_{i,j}$ is the matrix $a_{i,j}J_{k_i,k_j}$, and $k_1+\cdots+k_l = k$. Let h_i be a perfect k_i-coloring of a graph H with quotient matrix B_i which is $\mathbb{Z}_{k_i}$-circulant. Then there is a perfect coloring f of the graph $F = G\Box H$ with quotient matrix

$$\begin{pmatrix} A_{1,1}+B_1 & A_{1,2} & \dots & A_{1,l} \\ A_{2,1} & A_{2,2}+B_2 & \dots & A_{2,l} \\ \dots & \dots & \dots & \dots \\ A_{l,1} & A_{l,2} & \dots & A_{l,l}+B_l \end{pmatrix}.$$

The required coloring can be obtained as

$$f(x,y) = g(x) + (0, h_r(y)),$$

where the colors are represented as pairs (i, j), $i \in \mathbb{Z}_l$ and $j \in \mathbb{Z}_{k_i}$, $g(x) = (r, t)$.

Tilings can be used in constructions of perfect 2-colorings. At first, by the definition elements of tilings are perfect colorings themselves. Also the union of colors in perfect coloring with quotient matrix $c(J - I) + aI$ is also a perfect 2-coloring.

Proposition 6.15. *Let there be a perfect l-coloring of a graph G with quotient matrix $c(J - I) + aI$. Then for any $k \in \{1, \dots, l-1\}$ there is a $\{kc; (l-k)c\}$-coloring in G. Moreover, there is k-fold $\{kc; (l-k)c\}$-tiling of G.*

Corollary 6.11 also works if one of tiling is multifold.

Proposition 6.16. *Let $(g_1, \dots, g_l)$ be a $\{b_1; c_1\}$-tiling of a graph G; let $(h_1, \dots, h_l)$ be a λ-fold $\{b_2; c_2\}$-tiling of a graph H. Then $(f_1, \dots, f_m)$, where*

$$f_i(x,y) = h_{g(x)+i}(y),$$

is a λ-fold $\{b_2 + (m-\lambda)c_1; c_2 + \lambda c_1\}$-tiling of the Cartesian product graph $G\Box H$.

Proposition 6.17. *Let g be a perfect $(l+1)$-coloring of a ck-regular graph G with quotient matrix*

$$\begin{pmatrix} 0 & 0 & 0 & \dots & 0 & k \\ 0 & 0 & 0 & \dots & 0 & k \\ \dots & \dots & \dots & \dots & \dots & \dots \\ 0 & 0 & 0 & \dots & 0 & k \\ r & r & r & \dots & r & k-lr \end{pmatrix}$$

Let $(h_1, \dots, h_l)$ be a λ-fold $\{b; c\}$-tiling in a graph H, where $c = \lambda r$. Then

$$f(x,y) = \begin{cases} h_{g(x)}(y) & \text{if } g(x) \in \{1, \dots, l\} \\ 1 & \text{if } g(x) = l+1 \end{cases}$$

is a $\{b+k; c\}$-coloring in the Cartesian product $G\Box H$.

Proposition 6.18. *Let g be a perfect kl-coloring of a graph G with quotient matrix $(c(J_k - I_k) + aI_k) \otimes J_l$. Let $\mu_1, \ldots, \mu_k$ be such that μ_i equals 0 or l or there is a μ_i-fold $\{b_i; c_i\}$-tiling $(h_1^i, \ldots, h_l^i)$ of a graph H, where $b_i + c_i = l(c - a)$. If $\mu_i = 0$ ($\mu_i = l$) let $h_1^i, \ldots, h_l^i$ be constant functions with value 1 (value 0). Then $(f^{(0,0)}, \ldots, f^{(k-1,l-1)})$, where*

$$f^{(i,j)}(x, y) = h_{s+j}^{r+i}(y), \qquad \text{where } g(x) = (r, s),$$

is a $(\mu_1 + \cdots + \mu_k)$-fold $\{ckl - c(\mu_1 + \cdots + \mu_k); c(\mu_1 + \cdots + \mu_k)\}$-tiling of $G \Box H$.

Propositions 6.17 and 6.18, where l is power of 2, will be used below to construct perfect colorings in Doob graphs.

It is not a surprise that constructions based on the product of graphs have a big potential for generalizations and it makes sense to keep searching for new constructions of perfect colorings, but here we stay on those ones that are used in the current study.

6.6.3 Splittable Coverings

For any m and n the graph $D(m, n)$ can be partitioned into 4 disjoint MDS codes. Hence, the graph $D(m, n)$ covers the graph $(2m+n)K_4$ (each MDS code is mapped to its own vertex of K_4). The Shrikhande graph $mD(1, 0)$ with multiple edges also can be covered by $D(m, 0)$ (but not by $D(m, n)$ with $n > 0$); the covering is $f(x_1^*, \ldots, x_m^*) = (x_1^* + \cdots + x_m^*)$. The two coverings above can be combined with the following observation.

Lemma 6.19. *Let a graph G_1 cover a graph H_1 and a graph G_2 cover a graph H_2. Then the graph $G_1 \Box G_2$ covers the graph $H_1 \Box H_2$.*

As a result, we have the following.

Proposition 6.19. *Let k, m, n, m' and n' be nonnegative integers such that $2m' + n' = kn$. If k is even or k is odd and $n' \geq n$, then the graph $D(mk + m', n') = D(m', n') \Box D(mk, 0)$ covers $kD(m, n) = (kK_4)^n \Box kD(m, 0)$.*

This implies the following way to construct a new perfect coloring.

Corollary 6.12. *Let there be a perfect coloring in $D(m, n)$ with quotient matrix S. Then*

1. *for any even k and any m', n' such that $2m' + n' = kn$, there is a perfect coloring in $D(mk + m', n')$ with quotient matrix kS;*

2. *for any odd k and any m', n' such that $2m' + n' = kn \leq kn'$, there is a perfect coloring in $D(mk + m', n')$ with quotient matrix kS.*

In the construction above, densities of colors do not change. In particular, if for the original $\{b; c\}$-coloring we have $\dfrac{b + c}{\gcd(b, c)} = 2^l$, then for the resulting $\{kb; kc\}$-coloring the value $\dfrac{kb + kc}{\gcd(kb, kc)}$ is 2^l with the same l. To construct colorings with another l, we will define another series of coverings between Doob graphs.

Definition 6.12 (l-splittable covering, $G[l]$). *For a given graph G with adjacency matrix A, denote by $G[l]$ the graph with adjacency matrix $A \otimes J_l$, i.e., the tensor product of G and K_l+1, the complete graph on l vertices with one loop added to each vertex. Each vertex of $G[l]$ has the form (v, i), where v is a vertex of G and $i \in \{0, 1, \dots, l-1\}$. For graphs G and H, the covering of $H[l]$ by G is called an* l-splittable covering *of H by G. Note that in this case G covers lH.*

Proposition 6.20. *If $G_1, \dots, G_k$ are simple graphs and $G = G_1 \square \dots \square G_k$, then*

$$\phi((v_1, i_1), \dots, (v_k, i_k)) = (v_1, \dots, v_k, (i_1 + \dots + i_k \bmod l))$$

is a covering of $G[l]$ by $G_1[l] \square \dots \square G_k[l]$.

Proposition 6.21 ([63, Property 2.3]). *Let there be perfect colorings of graphs G and H with quotient matrices S and T, respectively. Then in the tensor product of G and H there is a perfect coloring with quotient matrix $S \otimes T$.*

Corollary 6.13. *If there is a perfect coloring of a graph H with quotient matrix S, then there is a perfect coloring of $H[l]$ with quotient matrix $S \otimes J_l$.*

As a consequence of Proposition 6.20 and Corollary 6.13, we have the following.

Proposition 6.22. *Let $G_1, \dots, G_t$ be some graphs, and let there be a perfect coloring of $G_1 \square \dots \square G_t$ with quotient matrix S. Then for any positive integer l there is a perfect coloring of the graph $G_1[l] \square \dots \square G_t[l]$ with quotient matrix $S \otimes J_l$.*

Corollary 6.14. *Let there be a perfect coloring of the graph $G_1 \square \dots \square G_t$ with quotient matrix S. If for every $i = 1, \dots, t$ there is an l-splittable covering of the graph G_i by a graph H_i, then there is a perfect coloring of $H_1 \square \dots \square H_t$ with quotient matrix $S \otimes J_l$.*

In the next section, we describe some splittable coverings and discuss some constructions based on them.

6.6.4 Partitions of MDS Codes

Here we describe an important series of perfect colorings obtained from partitions of MDS codes. We will use the special case $q = 4$ of the following fact.

Proposition 6.23. *If $n = p^k$ and $q = p^m$, where p is prime and $m \le k$, then there are q pairwise disjoint MDS codes $C_0, \dots, C_{q-1}$ in $H(n, q)$ such that for each i the code C_i is partitioned into n codes $C_{i,0}, \dots, C_{i,n-1}$ with distance at least 3.*

Proof. We first note that the statement is true for the graph $H(n, n)$ on the vertex set $\mathbb{F}_n^n$. Indeed, each of the cosets of the linear MDS code C with check matrix

$$H = (1, \dots, 1),$$

can be partitioned into n cosets of the linear distance-3 code D with check matrix

$$H' = \begin{pmatrix} 1 & 1 & \dots & 1 & 1 \\ \alpha_1 & \alpha_2 & \dots & \alpha_{n-1} & 0 \end{pmatrix}$$

where $\mathbb{F}_n = \{\alpha_0 = 0, \alpha_1, \dots, \alpha_{n-1}\}$.

Let $T = \{\beta_0 = 0, \beta_1, \dots, \beta_{q-1}\}$ be an arbitrary m-dimensional $\mathbb{F}_p$-subspace of $\mathbb{F}_n$. The subgraph of $H(n,n)$ induced by $T^n = T \times \dots \times T$ is isomorphic to $H(n,q)$. The corresponding retract C_0 of C is an additive MDS code in $H(n,q)$. Moreover, the p^m nonempty retracts of cosets of C are also MDS in $H(n,q)$, cosets of C_0; we denote them C_i, where $C_i = \{x \in T^n : \; Hx^{\mathrm{T}} = \beta_i\}$, $i = 0, 1, \dots, q-1$. Each C_i is partitioned into $C_{i,j}$, where $C_{i,j} = \{x \in T^n : \; H'x^{\mathrm{T}} = (\beta_i, \alpha_j)^{\mathrm{T}}\}$, $j = 0, 1, \dots, n-1$. Each $C_{i,j}$ is a coset of $C_{0,0}$, which is a retract of D and hence has distance at least 3. □

The obtained retract codes C_0 and $C_{0,0}$ are additive. The code $C_{0,0}$ is linear only if T is the order-q subfield of $\mathbb{F}_n$, which is only possible if m divides k; in this case, $C_{0,0}$ is essentially the generalized Reed–Muller code (see, e.g., [2, Section 5.4]) $\mathcal{R}_q((q-1)t-2, t)$ of order $(q-1)t-2$, $t = k/m$, whose dual is the generalized Reed–Muller code $\mathcal{R}_q(1,t)$ of order 1. In [16,58], a series of additive two-weight codes in $H(n = p^k, q = p^m)$, $m \le k$, was obtained, whose dual are completely regular codes with intersection array $\{n(q-1), n-1; 1, n(q-1)\}$. That dual is essentially our code $C_{0,0}$.

Below, we will often mention coverings of $K_4[2^k]$ (i.e., 2^k-splittable coverings of $K_4 = H(1,4) = D(0,1)$) by $D(m,n)$, where $2m + n = 2^k$. From the definition we see that actually $K_4[2^k]$ in the complete 4-partite graph $K_{2^k,2^k,2^k,2^k}$ on the vertex set is $\{(i,j) : i \in \{0, \dots, 3\}, \, j \in \{0, \dots, 2^k-1\}\}$, where (i,j) and (i',j') are adjacent if and only if $i \ne i'$; the adjacency matrix is $M_k = (J_4 - I_4) \otimes J_{2^k}$, which is the quotient matrix of any covering of $K_4[2^k]$. We also note that if we have a coloring of $K_4[2^k]$ by $D(m,n)$, then the vertices of colors $(i,0)$, …, $(i, 2^k-1)$ for fixed i form an MDS code; moreover, for any i, j, i', j', $i \ne i'$, the vertices of colors (i,j) and (i',j') induce a 1-regular subgraph (matching).

Corollary 6.15. *For every $k \ge 2$, there is a covering of $K_4[2^k]$ by $H(2^k, 4)$.*

Proof. The equitable partition $(C_{i,j})_{i=0}^{3}{}_{j=0}^{2^k-1}$ corresponding to the required covering is constructed in Proposition 6.23. Indeed, each vertex x from C_i has exactly $3 \cdot 2^k$ neighbors, none of which is in C_i. On the other hand, x cannot have more than one neighbor in any $C_{i',j'}$, $i \in \{0,1,2,3\} \setminus \{i\}$, $j = 0, \dots, 2^k - 1$, because the code distance of $C_{i',j'}$ is more than 2. We conclude that x has exactly one neighbor in each such $C_{i',j'}$. □

Some variation of the construction in the proof of Proposition 6.23 (namely, removing the last column in each of the matrices H, H') leads to the first equitable partition in the following corollary. The second equitable partition is obtained from the first one by joining the colors $(0,j)$, $(1,j)$, $(2,j)$, $(3,j)$ for each j.

Corollary 6.16. *There is a perfect 2^k-coloring of graph $H(2^k-1,4)$ with quotient matrix $(J_4-I_4)\otimes J_{2^k}$. There is a perfect 2^k-coloring of graph $H(2^k-1,4)$ with quotient matrix $3(J_{2^k}-I_{2^k})$.*

With Corollary 6.14, the covering from Corollary 6.15 gives the following.

Proposition 6.24. *Let there be a perfect coloring in $H(n,4)$ with quotient matrix S. Then for any $k\geq 2$ there is a perfect coloring of $H(n2^k,4)$ with quotient matrix $S\otimes J_{2^k}$.*

As we will see below, the results similar to Corollaries 6.15 and 6.16 and Proposition 6.24 are also true for Doob graphs. At first it is true for case $k=2$.

Proposition 6.25. *There are coverings of $K_4[2^2]$ by $D(2,0)$ and by $D(1,2)$.*

Corollary 6.17. *If there is a perfect coloring of $H(N,4)$ with quotient matrix S, then there is a perfect coloring of $D(m,n)$, where $2m+n=4N$, with quotient matrix $S\otimes J_4$.*

Applying Corollary 6.17 to to the colorings of Hamming graphs constructed in Corollaries 6.15 and 6.16 and Proposition 6.24 gives Corollaries 6.18–6.20 below.

Corollary 6.18. *If $2m+n=2^k$, where $k\geq 2$ and $k=3$ implies $m=0$, then there is a covering of $K_4[2^k]$ by $D(m,n)$.*

Corollary 6.19.

1. *For any m and n such that $2m+n=2^k-1$, $k=2$ or $k\geq 4$, there is a perfect 2^k-coloring of $D(m,n)$ with quotient matrix $3(J_{2^k}-I_{2^k})$.*
2. *For any m and n such that $2m+n=2^k$, $k\geq 2$, there is a perfect 2^k-coloring of $D(m,n)$ with quotient matrix $3J_{2^k}$.*

Note that the case $k=3$ in the second item does not follow from Corollary 6.17. It is obtained from colorings with quotient matrices $\begin{pmatrix}1&2\\2&1\end{pmatrix}\otimes J_4$, $\begin{pmatrix}2&1\\1&2\end{pmatrix}\otimes J_4$, found explicitly, and Theorem 6.23, where $\mathbb{G}=\mathbb{Z}_2^+\times\mathbb{Z}_4^+$.

Corollary 6.20. *If $2m+n=2^k$, where $k\geq 2$ and $k=3$ implies $m=0$, then $D(m,n)$ covers $K_4[2^k]$.*

Additionally, we have the following.

Proposition 6.26. *$D(2^k,0)$ covers $D(1,0)[2^k]$.*

Proof (a sketch). We divide the elements of $\mathbb{Z}_4^2$ into 4 subsets $A_0=\{00,02,20,22\}$, $A_1=\{01,03,10,30\}$, $A_\alpha=\{10,12,30,32\}$, and $A_{\alpha^2}=\{11,13,31,33\}$, where $\mathbb{F}_4=\{0,1,\alpha,\alpha^2\}$. For any vertex $x=(x_1^*,\ldots,x_{2^k}^*)$ of $D(2^k,0)$, denote by $y(x)$ the vertex $(y_1,\ldots,y_{2^k})$ of $H(2^k,4)$, where $x_i^*\in A_{y_i}$. The vertex $y(x)$ has some color (i,j) in the covering of $K_4[2^k]$ by $D(0,n)$. We define the required covering as $((x_1^*+\cdots+x_{2^k}^*),j)$. □

From Corollary 6.20, Proposition 6.26, Lemma 6.19 and Proposition 6.20 we get the following.

Proposition 6.27. *If $2m' + n' = 2^k n$, then $D(2^k m + m', n')$ covers $D(m, n)[2^k]$.*

Corollary 6.21. *If $2m' + n' = 2^k n$, then the existence of a perfect coloring of $D(m, n)$ with quotient matrix S implies the existence of a perfect coloring of $D(2^k m + m', n')$ with quotient matrix $S \otimes J_{2^k}$.*

From Theorem 6.23 with $\mathbb{G} = \mathbb{Z}_4^+ \times \mathbb{Z}_{2^k}^+$, we have the following.

Corollary 6.22. *If $2m + n = \gamma \cdot 2^k$, where $k \geq 2$ and $k = 3$ implies $m = 0$, then there is a perfect 2^{k+2}-coloring of $D(m, n)$ with quotient matrix $\gamma(J_4 - I_4) \otimes J_{2^k}$.*

If we unite last $6m + 3n$ colors in the covering of $K_4[2^k]$ from Corollary 6.22, then we obtain the following perfect $(2^k + 1)$-coloring.

Corollary 6.23. *Let $2m + n = \gamma \cdot 2^k$, where $\gamma \geq 1$, $k \geq 2$, and $k = 3$ implies $m = 0$. Then there is a perfect $(2^k + 1)$-colorings f of $D(m, n)$ with quotient matrix*

$$S = \begin{pmatrix} 0 & 0 & 0 & \dots & 0 & (6m+3n) \\ 0 & 0 & 0 & \dots & 0 & (6m+3n) \\ \dots & \dots & \dots & \dots & \dots & \\ 0 & 0 & 0 & \dots & 0 & (6m+3n) \\ \gamma & \gamma & \gamma & \dots & \gamma & (4m+2n) \end{pmatrix}.$$

The first color of such coloring is a completely regular code. So, we obtain an infinite series of completely regular codes with covering radius 2 in Doob graphs.

Corollary 6.24. *Let $2m + n = \gamma \cdot 2^k$ for some odd γ and $k \geq 4$. Then there is a completely regular code C in $D(m, n)$ with quotient matrix equals*

$$\begin{pmatrix} 0 & 6m+3n & 0 \\ c & 4m+2n & 2m+n-c \\ 0 & 6m+3n & 0 \end{pmatrix},$$

where $c = \gamma \cdot b$ for some $b \in \{1, 2^k - 1\}$.

Corollary 6.25. *Let there be a perfect 2-coloring g of $D(m, n)$ with quotient matrix $S = \begin{pmatrix} a & b \\ c & d \end{pmatrix}$. If the eigenvalue $a - c$ is not positive, then there is a $\{t_1 c + t_2 b, (4 - t_1)c + (4 - t_2)b\}$-coloring f of $D(m', n')$, where $2m' + n' = 2m + n + (b + c)$, $m' \geq 4m$ and $t_1, t_2 \in \{0, 1, \dots, 4\}$, $t_1 + t_2 \notin \{0, 8\}$.*

Proof. By Corollary 6.21, there is a perfect coloring of $D(m_1, n_1)$, where $2m_1 + n_1 = 4(2m+n)$, with quotient matrix $S \otimes J_4$. Also we have a perfect coloring of $D(m_2, n_2)$, where $2m_2 + n_2 = c - a = b + c - 6m - 3n$, with quotient matrix $(c - a)(J_4 - I_4)$

(a partition into MDS codes). Therefore by Theorem 6.24, there is a perfect coloring of $D(m', n')$ with quotient matrix

$$\begin{pmatrix} a & c & c & c & b & b & b & b \\ c & a & c & c & b & b & b & b \\ c & c & a & c & b & b & b & b \\ c & c & c & a & b & b & b & b \\ c & c & c & c & d & b & b & b \\ c & c & c & c & b & d & b & b \\ c & c & c & c & b & b & d & b \\ c & c & c & c & b & b & b & d \end{pmatrix}.$$

The union of t_1 colors from the first block and t_2 colors from the second block gives the requirement partition. □

A similar construction works also for Hamming graphs over prime power alphabet, see [12, Theorem 4.13].

Remark 6.5. *Note that the statement similar to Corollary 6.25 with 2^k instead of 4 is true if there is an appropriate tiling corresponding to a perfect coloring with quotient matrix $(c-a)(J_{2^k} - I_{2^k}) + rI_{2^k}$ for some r.*

Since by Corollary 6.18 for any $k \geq 3$ there is a covering of $K_4[2^2]$ by $D(m', n')$, where $2m' + n' = 2^k$ (if $k = 3$ then $m' = 0$), as a special case of Proposition 6.18 we have the following construction.

Corollary 6.26. *Assume that for each $i \in \{1, 2, 3, 4\}$ we have a λ_i-fold $\{b_i; c_i\}$-tiling $(h^i_1, \ldots, h^i_{2^l})$ of $D(m, n)$, where $b_i + c_i = 2^l$. Let $2m' + n' = 2^k$, where $k \geq 2$ and $k = 3$ implies $m' = 0$. Then for each μ_i in $\{0, \lambda_i, 2^l\}$, $i = 1, 2, 3, 4$, there is a $(\mu_1 + \cdots + \mu_4)$-fold $\{2^{k+2} - (\mu_1 + \cdots + \mu_k); (\mu_1 + \cdots + \mu_k)\}$-tiling $(f^{(0,0)}, \ldots, f^{(3,2^l-1)})$ of $D(m + m', n + n')$.*

Consider the difference between constructions from Corollaries 6.25 and 6.26. Let there be a $\{b; c\}$-coloring of $D(m, n)$, where b and c are relatively prime and $b + c = 2^l$ and $b + c \geq 6m + 3n$ (i.e., the second eigenvalue of the quotient matrix is not positive). Then by Corollary 6.25 there is a $\{4b + 3c; c\}$-coloring in some Doob graph of diameter $2m + n + b + c$. On the other hand, suppose there is not only a $\{b; c\}$-coloring of $D(m, n)$ but also a c-fold $\{b; c\}$-tiling of $D(m, n)$. In this case, by Corollary 6.26 there is a $\{4b + 3c; c\}$-coloring of some Doob graph of diameter $2m + n + (b + c)$. So, in both cases, we obtain the same quotient matrices of graphs of the same diameter but maybe for different number of Shrikhande components. Since often for fixed quotient matrix the existence of perfect coloring with this quotient matrix does not depend on m and n, the different number of Shrikhande components is not always important. The advantage of the first construction is that we do not need any tiling. The advantage of the second construction is that we can take colorings with positive second eigenvalue. So, both constructions make sense and can be useful. In Section 6.6.6 we show how the second construction can be used. Note also that in both cases if we have a $\{b;c\}$-coloring (multifold $\{b;c\}$-tiling) that satisfies the condition, then the resulting coloring (tiling) also satisfies the conditions, so both constructions can be applied recursively.

6.6.5 Existence of Tilings

In this subsection, we consider just simple tilings, i.e. one-fold tilings. Tilings of $D(m,n)$ correspond to perfect colorings with quotient matrix of type $c(J_l - I_l) + aI_l$. Note that the cardinalities of colors are same; hence l is a power of 2.

Recall that there are one-perfect codes in $D(m,n)$, where $2m+n = \frac{4^l-1}{3}$ (see Corollary 6.3). A partition of the vertex set into one-perfect codes (which can be easily obtained by the translation of a one-perfect code with vectors of weight at most 1) is an equitable partition with quotient matrix $(J_{4^l} - I_{4^l})$; joining pairs of the codes in such a partition we get an equitable partition with quotient matrix $2(J_{2^{2l-1}} - I_{2^{2l-1}}) + I_{2^{2l-1}}$. Also, by Corollary 6.19, we have perfect 2^l-colorings of Doob graphs with quotient matrices $3(J_{2^l} - I_{2^l})$ or $3J_{2^l}$.

All results in the current subsection can be obtained by using these colorings and constructions from previous subsections.

Proposition 6.28. *We have the following.*

1. *For any integer $l \geq 1$ and $s \geq 1$, there is a perfect 4^l-coloring of $D(0,n)$, $n = \dfrac{s(4^l-1)}{3}$, with quotient matrix $s(J-I)$.*

2. *For any integer $l \geq 2$, $s \geq 2$, and $a \in \{0,1,2\}$ such that $a+s \equiv 0 \bmod 3$, there is a perfect 2^{2l-1}-coloring of $D(0,n)$, $n = \dfrac{s(2^{2l-1}-1)+a}{3}$, with quotient matrix $s(J-I)+aI$.*

3. *For any integer $l \geq 1$, there is a perfect 2^{2l-1}-coloring of $D(m,n)$, $2m+n = \dfrac{(2^{2l}-1)}{3}$, with quotient matrix $2(J-I)+I$.*

4. *For any positive integer $l \geq 1$, there is a perfect 2^{2l-1}-coloring of $D(m,0)$, where $2m = \dfrac{2(2^{2l-1}+1)}{3}$, with quotient matrix $2(J-I)+4I$.*

Proposition 6.29. *We have the following.*

1. *For any $l \geq 1$ and $s \geq 1$, there is a perfect 4^l-coloring in $D(m,n)$, where $2m+n = \dfrac{s(4^l-1)}{3}$, with quotient matrix $s(J-I)$.*

2. *For any $l \geq 1$ and odd $s \geq 3$, there is a perfect 4^l-coloring in $D(m,0)$, where $2m = \dfrac{s(4^l-1)+3}{3}$, with quotient matrix $s(J-I)+3I$.*

Proposition 6.30. *We have the following.*

1. *For any* $l \geq 3$*, even* $s \geq 2$*, and* $a \in \{0, 2, 4\}$ *such that* $s + a \equiv 0 \bmod 6$*, there is a perfect* 2^{2l-1}*-coloring in* $D(m, n)$*, where* $2m + n = \frac{s(2^{2l-1}-1)+a}{3}$*, with quotient matrix* $s(J - I) + aI$*.*

2. *For any* $l \geq 3$*, odd* $s \geq 3$*, and* $a \in \{0, 2, 4\}$ *such that* $s - 3 + a \equiv 0 \bmod 6$*, there is a perfect* 2^{2l-1}*-coloring in* $D(m, n)$*, where* $2m + n = \frac{s(2^{2l-1}-1)+a}{3}$*, with quotient matrix* $s(J - I) + aI$*.*

3. *For any* $l \geq 3$*, odd* $s \geq 3$*, and* $a \in \{3, 5, 7\}$ *such that* $s + a \equiv 0 \bmod 6$*, there is a perfect* 2^{2l-1}*-coloring in* $D(m, 0)$*, where* $2m = \frac{s(2^{2l-1}-1)+a}{3}$*, with quotient matrix* $s(J - I) + aI$*.*

4. *For any odd* $s \geq 3$ *and* $a \in \{3, 5, 7\}$ *such that* $s + a \equiv 0 \bmod 6$*, there is a perfect* 8*-coloring in* $D(m, n)$*, where* $2m + n = \frac{7s+a}{3}$*, with quotient matrix* $s(J - I) + aI$*.*

Proposition 6.31. *There is a* $\{b; b\}$*-coloring in* $D(m, n)$ *if and only if* $b = 2i$ *for some* $i \in \{1, \ldots, 2m + n\}$*.*

The existence of tilings with the following parameters remains open.

Does there exist a perfect 2^{2l-1}-coloring with quotient matrix $s(J - I) + I$ in $D(m, n)$, where $2m + n = \frac{s(2^{2l-1}-1)+1}{3}$, for $l \geq 2$ and $s \equiv 2, 5 \bmod 6$?

Does there exist a perfect 8-coloring with quotient matrix $s(J - I) + aI$ in $D(m, n)$, where $2m+n = \frac{7s+a}{3}$, for odd s, $a \in \{0, 2, 4\}$, $a+s-3 = 0 \equiv \bmod 6$?

6.6.6 Existence of $\{b, c\}$-Colorings

The following coloring is well known; the first color is a c-fold one-perfect code (for example, the union of c disjoint one-perfect codes).

Theorem 6.25. *Let* $b + c = 4^l$*. Then there is a perfect coloring in* $D(m, n)$*, where* $2m + n = \frac{4^l-1}{3}$*, with the quotient matrix* $\begin{pmatrix} c-1 & b \\ c & b-1 \end{pmatrix}$*.*

The following three theorems are straightforward consequences from the tilings obtained in previous sections and Proposition 6.15.

Theorem 6.26. *Let* $\gcd(b, c) \geq 2$ *and* $\dfrac{b+c}{\gcd(b, c)} = 4^l$*. Then*

1. *There is a perfect coloring in* $D(m, n)$*, where* $2m + n = \dfrac{\gcd(b, c) \cdot (4^l - 1)}{3}$*, with quotient matrix*

$$\begin{pmatrix} c - \gcd(b, c) & b \\ c & b - \gcd(b, c) \end{pmatrix}.$$

2. *If* $\gcd(b,c)$ *is odd then there is a perfect coloring in* $D(m,0)$, *where* $2m = \frac{\gcd(b,c)(4^l-1)+3}{3}$, *with quotient matrix*

$$\begin{pmatrix} c-\gcd(b,c)+3 & b \\ c & b-\gcd(b,c)+3 \end{pmatrix}.$$

Theorem 6.27. *Let* $\gcd(b,c)=2$ *and* $\frac{b+c}{\gcd(b,c)} = 2^{2l-1}$ *for some* $l \geq 2$. *Then*

1. *There is a perfect coloring in* $D(m,n)$, *where* $2m+n = \frac{2^{2l}-1}{3}$, *with quotient matrix* $\begin{pmatrix} c-1 & b \\ c & b-1 \end{pmatrix}$.

2. *There is a perfect coloring in* $D(m,0)$, *where* $2m = \frac{2^{2l}+2}{3}$, *with quotient matrix* $\begin{pmatrix} c+2 & b \\ c & b+2 \end{pmatrix}$.

Theorem 6.28. *Let* $\gcd(b,c) > 2$ *and* $\frac{b+c}{\gcd(b,c)} = 2^{2l-1}$ *for some* $l \geq 2$. *Then*

1. *If* $\gcd(b,c)$ *is even and* $a \in \{0,2,4\}$ *such that* $\gcd(b,c)+a \equiv 0 \bmod 6$, *then there is a perfect coloring in* $D(m,n)$, *where* $2m+n = \frac{\gcd(b,c)(2^{2l-1}-1)+a}{3}$, *with quotient matrix*

$$\begin{pmatrix} c-\gcd(b,c)+a & b \\ c & b-\gcd(b,c)+a \end{pmatrix}.$$

2. *If* $\gcd(b,c)$ *is odd,* $l=2$ *implies* $m=0$, *and* $a \in \{0,2,4\}$ *such that* $\gcd(b,c)-3+a \equiv 0 \bmod 6$, *then there is a perfect coloring in* $D(m,n)$, *where* $2m+n = \frac{\gcd(b,c)(2^{2l-1}-1)+a}{3}$, *with quotient matrix*

$$\begin{pmatrix} c-\gcd(b,c)+a & b \\ c & b-\gcd(b,c)+a \end{pmatrix}.$$

3. *If* $\gcd(b,c)$ *is odd and* $a \in \{3,5,7\}$ *is such that* $\gcd(b,c)+a \equiv 0 \bmod 6$, *then there is a perfect coloring in* $D(m,0)$, *where* $2m = \frac{\gcd(b,c)(2^{2l-1}-1)+a}{3}$, *with quotient matrix*

$$\begin{pmatrix} c-\gcd(b,c)+a & b \\ c & b-\gcd(b,c)+a \end{pmatrix}.$$

4. If $\gcd(b,c)$ *is odd and* $a \in \{3,5,7\}$ *is such that* $\gcd(b,c)+a \equiv 0 \bmod 6$, *then there is a perfect 8-coloring in* $D(m,n)$, *where* $2m+n = \dfrac{7\gcd(b,c)+a}{3}$, *with quotient matrix*

$$\begin{pmatrix} c-\gcd(b,c)+a & b \\ c & b-\gcd(b,c)+a \end{pmatrix}.$$

Consider the case when $(b+c)$ is odd power of 2 and b and c are relatively prime. Here we need Corollary 6.26 for the step of induction and the following multifold tilings in $D(1,0)$ and $D(0,3)$ for the basis of induction.

Lemma 6.20. *For any* $c \in \{2,\ldots,6\}$,

1. *there is a c-fold* $\{8-c,c\}$*-tiling* $(g_1,\ldots,g_8)$ *of* $D(1,0)$;
2. *there is a c-fold* $\{8-c;c\}$*-tiling* $(h_1,\ldots,h_8)$ *of* $H(3,4)$.

The first series of tilings is illustrated in Figure 6.9. For example, the four black sets and the four grey sets are completely regular codes with intersection array $\{5;3\}$, forming a 3-fold $\{5;3\}$-tiling (the complements form a 5-fold $\{3;5\}$-tiling). The four white sets form a one-fold $\{6;2\}$-tiling (the complements form a 3-fold $\{6;2\}$-tiling); by repeating each set twice we get a two-fold $\{6;2\}$-tiling. Any equitable partition with quotient matrix $[[2,4],[4,2]]$ is a one-fold $\{4;4\}$-tiling.

Similarly, the second series of tilings is illustrated in Figure 6.10.

Starting with tilings from Lemma 6.20 and applying Corollary 6.26 to these tilings and coverings of $K_4[2^k]$, we get the following.

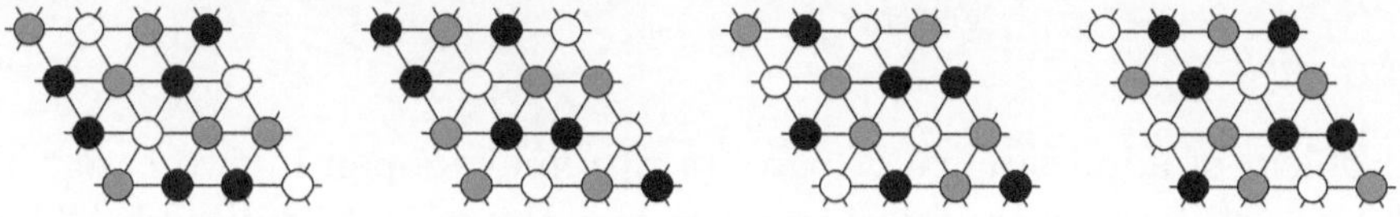

Figure 6.9 Illustration to Lemma 6.20, item 1.

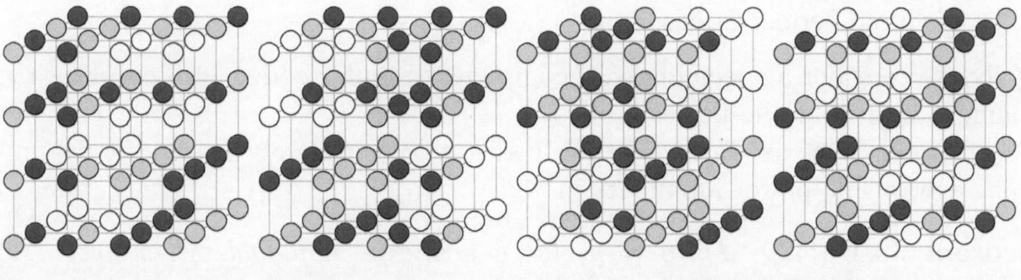

Figure 6.10 Illustration to Lemma 6.20, item 2 (four graphs H(3,4)).

Theorem 6.29. *If* $\gcd(b, c) = 1$, $b > 1$, $c > 1$, *and* $b + c = 2^{2l-1}$, $l \geq 3$, *then*

1. *there is a perfect coloring in* $D(m, n)$, *where* $2m + n = \frac{b+c-2}{3}$, $n \geq 8$, $m \geq 1$, *with quotient matrix* $\begin{pmatrix} c-2 & b \\ c & b-2 \end{pmatrix}$;
2. *there is a perfect coloring in* $H(n, 4)$, *where* $n = \frac{b+c+1}{3}$, *with quotient matrix* $\begin{pmatrix} c+1 & b \\ c & b+1 \end{pmatrix}$.

Summarizing the statements of this subsection, we have the following.

Theorem 6.30. *We have the following:*

1. *A pair* $(b, 1)$ *is* ∞*-admissible if and only if* $b = 4^l - 1$ *for some* l.
2. *A pair* (b, c), *where* $1 \notin \{b, c\}$, *is* ∞*-admissible if and only if* $\dfrac{b+c}{\gcd(b,c)} = 2^l$ *for some* l.
3. *A pair* $(b, 1)$ *is* 1*-admissible if and only if* $b = 4^l - 1$ *for some* l.
4. *If* $\gcd(b, c) \geq 2$ *and* $\dfrac{b+c}{\gcd(b,c)} = 2^l$ *for some* l, *then the pair* (b, c) *is* 0*-admissible.*
5. *If* $\gcd(b, c) = 1$ *and* $\dfrac{b+c}{\gcd(b,c)} = 4^l$ *for some* l, *then the pair* (b, c) *is* 1*-admissible.*
6. *If* $\gcd(b, c) = 1$ *and* $\dfrac{b+c}{\gcd(b,c)} = 2^{2l-1}$ *for some* l, *then the pair* (b, c) *is* 8*-admissible.*

In the problem of admissibility of pairs (b, c) remains open in two cases. The first case is $b + c = 4^l$, b and c are relatively prime: is this pair 0-admissible? The second case is $b + c = 2^{2l-1}$, b and c are relatively prime: for which $a \in \{0, \ldots, 7\}$ the pair (b, c) is a-admissible?

6.6.7 Multifold Perfect Codes

Finally, we consider a special case of completely regular codes with covering radius 1, namely, multifold one-perfect codes.

The problem of existence of such codes was completely solved for Hamming graphs over prime power alphabets in [42].

Theorem 6.31 ([42]). *Let* $q = p^t$ *for a prime* p. *A* μ*-fold one-perfect code in* $H(n, q)$ *exists if and only if*

i. $\mu < n(q-1) + 1$,

ii. $\dfrac{\mu q^n}{n(q-1)+1}$ *is integer, and*

iii. $n \equiv 1 \mod q$.

For the Doob graphs the same result is also true.

Theorem 6.32. *There is a c-fold perfect code in $D(m,n)$ if and only if*

1. $2m + n \equiv 1 \bmod 4$ *and*

2. $c = \alpha l$, *where α is such that* $6m+3n+1 = \alpha 2^k$, *α is odd and* $l \in \{1, \ldots, 2^k-1\}$.

A sketch of proof. Let us briefly discuss the idea of the proof. At first, for given m and n satisfying the conditions we can construct not only α-fold one-perfect codes but also a tiling from such codes. Further, if we represent the diameter $D = 2m+n$ of $D(m,n)$ as $2^k + D'$, $k = \lfloor \log_2 D \rfloor$, then by induction we can construct a tiling consisting from multifold perfect codes using a $\{3(2^k-1);3\}$-tiling in the Doob graph of diameter 2^k and a tiling consisting from multifold perfect codes in the Doob graph of diameter D', which exists by the induction hypothesis. The only case for which this does not work is $D = \frac{4^l-1}{3}$. This case corresponds to the parameters for which one-perfect code exists and it is a solved problem. □

Note that μ-fold one-perfect code exist in $D(m,n)$ if and only if it exists in $H(2m+n,4)$, i.e., the existence does not depend on the number of Shrikhande components. Also note that for the existence of multifold one-perfect codes in the Doob graphs, as well as for all prime-power-ary Hamming graphs, the standard necessary conditions (the cardinality is integer and Lloyd's condition holds) are sufficient.

The characterization of parameters of additive multifold one-perfect codes needs further study.

Acknowledgments

The work on the results of Sections 6.5.5 and 6.6 was carried out in the Sobolev Institute of Mathematics at the expense of the Russian Science Foundation (grant 22-11-00266), https://rscf.ru/project/22-11-00266/. The other sections contain results supported in part by the Russian Science Foundation (grant 18-11-00136), and within the framework of the state contract of the Sobolev Institute of Mathematics (Project FWNF-2022-0017).

REFERENCES

[1] T. L. Alderson. (6, 3)-MDS codes over an alphabet of size 4. *Des. Codes Cryptogr.*, 38(1):11–40, 2006. https://doi.org/10.1007/s10623-004-5659-4.

[2] E. F. Assmus, Jr. and J. D. Key. *Designs and Their Codes.* Cambridge University Press, Cambridge, 1992. https://doi.org/10.1017/CBO9781316529836.

[3] S. Ball. On sets of vectors of a finite vector space in which every subset of basis size is a basis. *J. Eur. Math. Soc.*, 14(3):733–748, 2012. https://doi.org/10.4171/JEMS/316.

[4] S. Ball and J. De Beule. On sets of vectors of a finite vector space in which every subset of basis size is a basis II. *Des. Codes Cryptogr.*, 65(1–2):5–14, 2012. https://doi.org/10.1007/s10623-012-9658-6.

[5] E. Bespalov. On the non-existence of extended perfect codes and some perfect colorings. E-print 2008.13260, arXiv.org, 2020. Available at http://arxiv.org/abs/2008.13260.

[6] E. Bespalov. On the non-existence of extended 1-perfect codes and MDS codes. *J. Comb. Theory Ser. A*, 189:105607(1–11), 2002. https://doi.org/10.1016/j.jcta.2022.105607.

[7] E. Bespalov. Equitable partitions in the Doob graphs. In *Proceedings of Thirteenth International Workshop on Algebraic and Combinatorial Coding Theory ACCT 2018*, pp. 179–182, Svetlogorsk, Russia, September 2018. Online: https://www.dropbox.com/s/h7u89lh8vyirww9/Proceedings%20final.pdf.

[8] E. Bespalov. On multifold perfect codes and some other completely regular codes in the Doob graphs and quaternary Hamming graphs. E-print 2312.07109, arXiv.org, 2023. https://doi.org/10.48550/arXiv.2312.07109.

[9] E. A. Bespalov. Methods of algebraic graph theory in the study of MDS codes. PhD thesis, Novosibirsk State University, 2018. In Russian.

[10] E. A. Bespalov and D. S. Krotov. On one test for the switching separability of graphs modulo q. *Sib. Math. J.*, 57(1):7–17, 2016. https://doi.org/10.1134/S003744661601002X, translated from *Sib. Mat. Zh.*, 57(1):10–24, 2017.

[11] E. A. Bespalov and D. S. Krotov. MDS codes in Doob graphs. *Probl. Inf. Trans.*, 53(2):136–154, 2017. https://doi.org/10.1134/S003294601702003X, translated from *Probl. Peredachi Inf.*, 53(2):40–59, 2017.

[12] E. A. Bespalov, D. S. Krotov, A. A. Matiushev, A. A. Taranenko, and K. V. Vorob'ev. Perfect 2-colorings of Hamming graphs. *J. Comb. Des.*, 29(6):367–396, 2021. https://doi.org/10.1002/jcd.21771.

[13] J. Borges, C. Fernández-Córdoba, J. Pujol, J. Rifà, and M. Villanueva. Z_2Z_4-linear codes: Generator matrices and duality. *Des. Codes Cryptogr.*, 54(2):167–179, 2010. https://doi.org/10.1007/s10623-009-9316-9.

[14] J. Borges and J. Rifà. A characterization of 1-perfect additive codes. *IEEE Trans. Inf. Theory*, 45(5):1688–1697, 1999. https://doi.org/10.1109/18.771247.

[15] J. Borges, J. Rifà, and V. A. Zinoviev. On completely regular codes. *Probl. Inf. Trans.*, 55(1):1–45, 2019. https://doi.org/10.1134/S0032946019010010.

[16] J. Borges, D. V. Zinoviev, and V. A. Zinoviev. On the classification of completely regular codes with covering radius two and antipodal dual. E-print 2302.10556, ArXiv.org, 2023. https://doi.org/10.48550/arXiv.2302.10556.

[17] A. E. Brouwer, A. M. Cohen, and A. Neumaier. *Distance-Regular Graphs.* Springer-Verlag, Berlin, 1989. https://doi.org/10.1007/978-3-642-74341-2.

[18] A. R. Calderbank, E. M. Rains, P. M. Shor, and N. J. A. Sloane. Quantum error correction via codes over GF(4). *IEEE Trans. Inf. Theory*, 44(4):1369–1387, 1998. https://doi.org/10.1109/18.681315.

[19] P. Delsarte. *An Algebraic Approach to Association Schemes of Coding Theory*, volume 10 of Philips Research Reports, Supplement. N.V. Philips' Gloeilampenfabrieken, Eindhoven, 1973.

[20] Y. Egawa. Characterization of $H(n, q)$ by the parameters. *J. Comb. Theory Ser. A*, 31(2):108–125, 1981. https://doi.org/10.1016/0097-3165(81)90007-8.

[21] T. Etzion. *Perfect Codes and Related Structures*. World Scientific, Singapore, 2022. https://doi.org/10.1142/12823.

[22] D. G. Fon-Der-Flaass. A bound on correlation immunity. *Sib. Èlektron. Mat. Izv.*, 4:133–135, 2007. Online: http://mi.mathnet.ru/eng/semr149.

[23] D. G. Fon-Der-Flaass. Perfect 2-colorings of a hypercube. *Sib. Math. J.*, 48(4):740–745, 2007. https://doi.org/10.1007/s11202-007-0075-4 translated from *Sib. Mat. Zh.* 48(4):923–930, 2007.

[24] D. G. Fon-Der-Flaass. Perfect colorings of the 12-cube that attain the bound on correlation immunity. *Sib. Èlektron. Mat. Izv.*, 4:292–295, 2007. In Russian. English translation: https://arxiv.org/abs/1403.8091.

[25] C. D. Godsil. *Algebraic Combinatorics*. Chapman and Hall, New York, 1993.

[26] A. R. Hammons, Jr, P. V. Kumar, A. R. Calderbank, N. J. A. Sloane, and P. Solé. The Z_4-linearity of Kerdock, Preparata, Goethals, and related codes. *IEEE Trans. Inf. Theory*, 40(2):301–319, 1994. https://doi.org/10.1109/18.312154.

[27] A. S. Hedayat, N. J. A. Sloane, and J. Stufken. *Orthogonal Arrays: Theory and Applications*. Springer Series in Statistics. Springer, New York, NY, 1999. https://doi.org/10.1007/978-1-4612-1478-6.

[28] O. Heden and M. Güzeltepe. On perfect 1-$\mathcal{E}$-error-correcting codes. *Math. Commun.*, 20(1):23–35, 2015.

[29] L. Ji and J. Yin. Constructions of new orthogonal arrays and covering arrays of strength three. *J. Comb. Theory Ser. A*, 117(3):236–247, 2010. https://doi.org/10.1016/j.jcta.2009.06.002.

[30] M. Kochol. Non-extendible latin parallelepipeds. *Inf. Process. Lett.*, 112(24):942–943, 2012. https://doi.org/10.1016/j.ipl.2012.08.014.

[31] J. I. Kokkala, D. S. Krotov, and P. R. J. Östergård. On the classification of MDS codes. *IEEE Trans. Inf. Theory*, 61(12):6485–6492, 2015. https://doi.org/10.1109/TIT.2015.2488659.

[32] J. I. Kokkala and P. R. J. Östergård. Classification of Graeco–Latin cubes. *J. Comb. Des.*, 23(12):509–521, 2015. https://doi.org/10.1002/jcd.21400.

[33] J. I. Kokkala and P. R. J. Östergård. Further results on the classification of MDS codes. *Adv. Math. Commun.*, 10(3):489–498, 2016. https://doi.org/10.3934/amc.2016020.

[34] J. H. Koolen, W. S. Lee, W. J. Martin, and H. Tanaka. Arithmetic completely regular codes. *Dis. Math. Theor. Comput. Sci.*, 17(3):59–76, 2016. https://doi.org/10.46298/dmtcs.2150.

[35] J. H. Koolen and A. Munemasa. Tight 2-designs and perfect 1-codes in Doob graphs. *J. Stat. Plann. Inference*, 86(2):505–513, 2000. https://doi.org/10.1016/S0378-3758(99)00126-3.

[36] D. S. Krotov. On decomposability of 4-ary distance 2 MDS codes, double-codes, and n-quasigroups of order 4. *Dis. Math.*, 308(15):3322–3334, 2008. https://doi.org/10.1016/j.disc.2007.06.038.

[37] D. S. Krotov. On irreducible n-ary quasigroups with reducible retracts. *Eur. J. Comb.*, 29(2):507–513, 2008. https://doi.org/10.1016/j.ejc.2007.01.005.

[38] D. S. Krotov. On reducibility of n-ary quasigroups. *Dis. Math.*, 308(22):5289–5297, 2008. https://doi.org/10.1016/j.disc.2007.08.099.

[39] D. S. Krotov. On the number of maximum independent sets in Doob graphs. *Sib. Èlektron. Mat. Izv.*, 12:508–812, 2015. https://doi.org/10.17377/semi.2015.12.043.

[40] D. S. Krotov. Perfect codes in Doob graphs. *Des. Codes Cryptogr.*, 80(1):91–102, 2016. https://doi.org/10.1007/s10623-015-0066-6.

[41] D. S. Krotov. The existence of perfect codes in Doob graphs. *IEEE Trans. Inf. Theory*, 66(3):1423–1427, 2020. https://doi.org/10.1109/TIT.2019.2946612.

[42] D. S. Krotov. Multifold 1-perfect codes. E-print 2212.03644, arXiv.org, 2022. Available at http://arxiv.org/abs/2212.03644.

[43] D. S. Krotov and E. A. Bespalov. Distance-2 MDS codes and latin colorings in the Doob graphs. *Graphs Comb.*, 34(5):1001–1017, 2018. https://doi.org/10.1007/s00373-018-1926-4.

[44] D. S. Krotov and V. N. Potapov. On multifold MDS and perfect codes that are not splittable into onefold codes. *Probl. Inf. Trans.*, 40(1):5–12, 2004. https://doi.org/10.1023/B:PRIT.0000024875.79605.fc translated from *Probl. Peredachi Inf.* 40(1):6–14, 2004.

[45] D. S. Krotov and V. N. Potapov. n-Ary quasigroups of order 4. *SIAM J. Dis. Math.*, 23(2):561–570, 2009. https://doi.org/10.1137/070697331.

[46] D. S. Krotov and V. N. Potapov. On connection between reducibility of an n-ary quasigroup and that of its retracts. *Dis. Math.*, 311(1):58–66, 2011. https://doi.org/10.1016/j.disc.2010.09.023.

[47] D. S. Krotov and V. N. Potapov. An upper bound on the number of frequency hypercubes. *Dis. Math.*, 347(1):113657(1–9), 2024. https://doi.org/10.1016/j.disc.2023.113657.

[48] D. S. Krotov, V. N. Potapov, and P. V. Sokolova. On reconstructing reducible n-ary quasigroups and switching subquasigroups. *Quasigroups Relat. Syst.*, 16(1):55–67, 2008.

[49] C. F. Laywine and G. L. Mullen. *Discrete Mathematics Using Latin Squares.* Wiley, New York, 1998.

[50] B. D. McKay. Combinatorial data. Latin squares. https://cs.anu.edu.au/people/Brendan.McKay/data/latin.html.

[51] B. D. McKay, A. Meynert, and W. Myrvold. Small Latin squares, quasigroups, and loops. *J. Comb. Des.*, 15(2):98–119, 2007. https://doi.org/10.1002/jcd.20105.

[52] B. D. McKay and I. M. Wanless. On the number of Latin squares. *Ann. Comb.*, 9(3):335–344, 2005. https://doi.org/10.1007/s00026-005-0261-7.

[53] B. D. McKay and I. M. Wanless. A census of small Latin hypercubes. *SIAM J. Dis. Math.*, 22(2):719–736, 2008. https://doi.org/10.1137/070693874.

[54] V. N. Potapov. On extensions of partial n-quasigroups of order 4. *Sib. Adv. Math.*, 22(2):135–151, 2012. https://doi.org/10.3103/S1055134412020058 translated from *Mat. Tr.* 14(2):147–172, 2011.

[55] V. N. Potapov. Constructions of pairs of orthogonal latin cubes. *J. Comb. Des.*, 28(8):604–613, 2020. https://doi.org/10.1002/jcd.21718.

[56] V. N. Potapov and D. S. Krotov. On the number of n-ary quasigroups of finite order. *Dis. Math. Appl.*, 21(5–6):575–585, 2011. DOI: 10.1515/dma.2011.035, translated from *Diskretn. Mat.* 24(1):60–69, 2012.

[57] A. M. Romanov. On perfect and Reed–Muller codes over finite fields. *Probl. Inf. Trans.*, 57(3):199–211, 2021. https://doi.org/10.1134/S0032946021030017, translated from *Probl. Peredachi Inf.* 57(3):3–16, 2021.

[58] N. V. Semakov, V. A. Zinoviev, and V. G. Zaitsev. Class of maximal equidistant codes. *Probl. Inf. Trans.*, 5(2):65–68, 1969. Translated from *Probl. Peredachi Inf.*, 5(2):84–87, 1969.

[59] M. Shi, D. Huang, and D. S. Krotov. Additive perfect codes in Doob graphs. *Des. Codes Cryptogr.*, 87(8):1857–1869, 2019. https://doi.org/10.1007/s10623-018-0586-y.

[60] M. Shi, X. Li, D. S. Krotov, and F. Özbudak. Quasi-cyclic perfect codes in Doob graphs and special partitions of Galois rings. *IEEE Trans. Inf. Theory*, 69(9):5597–5603, 2023. https://doi.org/10.1109/TIT.2023.3272566.

[61] M. J. Shi, S. K. Wang, X. X. Li, and D. S. Krotov. On the number of frequency hypercubes $\mathrm{F}^n(4;2,2)$. *Sib. Math. J.*, 62(5):951–962, 2021. https://doi.org/10.33048/smzh.2021.62.516, translated from *Sib. Mat. Zh.*, 62(5):1173–1187, 2021.

[62] R. Singleton. Maximum distance q-nary codes. *IEEE Trans. Inf. Theory*, 10(2):116–118, 1964. https://doi.org/10.1109/TIT.1964.1053661.

[63] A. A. Taranenko. Algebraic properties of perfect structures. *Linear Algebra Appl.*, 607:286–306, 2020. https://doi.org/10.1016/j.laa.2020.08.012.

[64] Yu. Tarannikov. On resilient Boolean functions with maximal possible nonlinearity. Cryptology ePrint Archive 2000/005, 2000. https://eprint.iacr.org/2000/005.

[65] A. Yu. Vasil'eva. Completely regular codes in the triangular grid. *Proc. MIPT*, 14(2):14–42, 2022. In Russian.

[66] K. Vorob'ev. On extended perfect codes. E-print 2403.10992, arXiv.org, 2024. Available at http://arxiv.org/abs/2403.10992.

[67] V. A. Zinoviev. Generalized concatenated codes. *Probl. Inf. Trans.*, 12(1):2–9, 1976. Translated from *Probl. Peredachi Inf.*, 12(1):5–15, 1976.

CHAPTER 7

Completely Regular Codes

Tables of Small Parameters for Binary and Ternary Hamming Graphs

Jack H. Koolen, Denis S. Krotov, and William J. Martin

IN THIS CHAPTER, WE PRESENT tables of small parameters of completely regular (CR) codes in Hamming graphs $H(n, q)$, $q = 2, 3$. We list all putative intersection arrays with intersection numbers up to some limit (depending on the covering radius) that pass some basic tests (see Sections 7.1.1, 7.1.2, 7.1.5, 7.1.4) and, for each array A, show the best known lower and upper (if exists) bounds on the smallest n such that an A-CR code exists in $H(n, q)$.

The tables are preceded by two auxiliary sections, where we describe known necessary conditions and constructions that are sufficient to build the tables. We should note that some nice families of linear CR codes are not represented in the tables because of large parameters; moreover, first questionable parameters of linear CR codes are rather large (see [40]). So, we can say that our tables are mainly focused on unrestricted CR codes and cannot be considered as a good reference for linear CR codes.

The following concepts defined in Chapter 1 are used in this chapter: *Hamming graph* $H(n, q)$ (*hypercube* $H(n, 2)$), *s-face* and its free directions, *weight* of a vertex in $H(n, q)$, *equitable partition* (and the corresponding *quotient matrix*), *perfect coloring*, graph *covering*, *completely regular (CR) code* (and the corresponding *intersection array*), *covering radius*, *linear code*, *coset graph*, *Cayley graph.*

DOI: 10.1201/9781003393931-7

7.1 NECESSARY CONDITIONS

7.1.1 Monotonicity

The intersection array of any CR code in a Hamming graph satisfies the following well-known property (see Section 1.2.2.4 in Chapter 1 for the discussion of more general classes of graphs in this context).

Lemma 7.1. *If there exists a* $\{b_0,\ldots,b_{\rho-1};c_1,\ldots,c_\rho\}$*-CR code* C *in* $H(n,q)$*, then* $b_{i-1} > b_i$ *and* $c_i < c_{i+1}$*,* $i = 1,\ldots,\rho-1$*.*

Proof. Consider two neighbor vertices $x \in C^{(i-1)}$ and $y \in C^{(i)}$. For each of b_i neighbors z in $C^{(i+1)}$, the vertices x and z have two common neighbors, both in $C^{(i)}$; one of them is y, and the other is some $v(z)$. It is easy to see that $v(z)$ are different for different z. Hence, $b_{i-1} \geq b_i$. Since x has at least one more neighbor y in $C^{(i)}$, we get $b_{i-1} > b_i$. Similarly, $c_i < c_{i+1}$. □

7.1.2 Cardinalities

The integrality of the cardinalities of cells found from the quotient matrix is one of the well-known basic necessary conditions for the existence of equitable partitions. For CR codes, we have the following.

Lemma 7.2. *For a* $\{b_0,\ldots,b_{\rho-1};c_1,\ldots,c_\rho\}$*-CR code* C*, the cardinalities of the cells of the distance partition* $(C = C^{(0)},C^{(0)},\ldots,C^{(\rho)})$ *can be found from the following relations:* $b_i \cdot |C^{(i)}| = c_{i+1} \cdot |C^{(i+1)}|$*,* $i = 0,1,\ldots,\rho-1$*, and the number of vertices of the graph. In* $H(n,q)$*, if the proportion between the cell cardinalities is* $s_0 : s_1 : \ldots : s_\rho$*, then* $(s_0+s_1+\cdots+s_\rho)/\gcd(s_0,s_1,\ldots,s_\rho)$ *must divide a power of* q*.*

7.1.3 Orthogonal Arrays

It is known [14] that a CR code in $H(n,q)$ (as well as every cell from the distance partition) with quotient matrix eigenvalues $\theta_0 > \theta_{i_1} > \ldots > \theta_{i_\rho}$, where $\theta_i = n(q-1)-qi$, is a simple orthogonal array of strength $t = i_1 - 1$, i.e., have the same number of codewords in every subgraph of $H(n,q)$ isomorphic to $H(n-t,q)$. This "same" number can be found from the parameters and of course must be integer.

For example, consider the quotient matrix $\begin{pmatrix} 0 & 14 & 0 & 0 \\ 10 & 1 & 3 & 0 \\ 0 & 12 & 0 & 2 \\ 0 & 0 & 13 & 1 \end{pmatrix}$ of a putative CR code C in $H(7,3)$ (see Table 7.10). The cardinalities of the cells are $260 \cdot 3$, $364 \cdot 3$, $91 \cdot 3$, and $14 \cdot 3$. The eigenvalues are $\theta_0 = 14$, $\theta_3 = 5$, $\theta_5 = -1$, $\theta_6 = -4$; the orthogonal-array strength is 2. Hence, every $H(5,3)$-subgraph of $H(7,3)$ must contain $260 \cdot 3/3^2$ codewords of C. This number is not integer; hence, such a code does not exist. We then have a lower bound 8 on the minimum n of $H(n,3)$ where a $\{14,3,2;10,12,13\}$-CR code can exist.

Remark 7.1. *After inventing the new test shown in Lemma 7.6, all arrays for which the orthogonal array property increased the lower bound on n (including the example above) disappeared from the small-parameters tables. However, the orthogonal-array test remains to be potentially useful. For example, it shows the nonexistence of one-perfect codes in $H(19,6)$ [23], which are CR with intersection array $\{95;1\}$. In Chapter 1, Remark 1.15, a generalization of the orthogonal-array test to the orthogonality of CR codes is discussed, which is also a promising tool to prove nonexistence.*

Lemma 7.3(i) below is a special case of the fact [16,25] that orthogonal arrays of size less than 2^{n-1} in $H(n,2)$ cannot have strength larger than $2n/3-1$ (see also Section 1.2.1.2 in Chapter 1). It is also a special case of Proposition 7.1, proved in the next subsection.

Lemma 7.3. *For a $\{b;c\}$-CR code in $H(n,2)$, $b \neq c$, we have the following:*

(i) $b+c \leq 4n/3$ [16];

(ii) *moreover, if $b+c = 4n/3$, then $b/\gcd(b,c)$ or $c/\gcd(b,c)$ is divisible by* 3 [31].

Claim (ii) follows the key fact proved in [31] that the number of edges of a given direction containing exactly one codeword does not depend on the choice of the direction. The nonexistence in $H(12,2)$ of a $\{11;5\}$-CR code, satisfying the hypothesis of Lemma 7.3(ii), was proved earlier in [18].

7.1.4 Eigenvalues and Eigenvectors

The following fact is often referred to as the Lloyd theorem, because of the famous work [32], where the corresponding fact was proved for a very special case, perfect codes.

Lemma 7.4 (see e.g. [19, Chapter 5, Lemma 2.2]). *All eigenvalues of the quotient matrix S of an equitable partition of a graph Γ are eigenvalues of Γ. In particular, if $\Gamma = H(n,q)$, then all eigenvalues of S are in $\{-n, -n+q, \ldots, (q-1)n\}$.*

The following lemma is a special case of [26, Corollary 2.7].

Lemma 7.5 ([26]). *The difference between the two largest eigenvalues of the quotient matrix S of a CR code in $H(n,q)$ must be the largest difference between two neighbor eigenvalues of S.*

The next property of quotient matrices requires calculating the eigenvectors. Theorems 1.2 and 1.3 in Chapter 1 narrowed to Hamming graphs look as follows.

Lemma 7.6 (Triangle inequality). *The basis eigenvectors $\varphi_{i_0}, \varphi_{i_1}, \ldots, \varphi_{i_\rho}$ of the quotient matrix S of an equitable partition of $H(n,q)$, where φ_{i_j} is an eigenvector with eigenvalue $\theta_{i_j} = (q-1)n - q \cdot i_j$, satisfy the following property. For any $j,l \in \{0,\ldots,\rho\}$, let*

$$\varphi_{i_j} \bullet \varphi_{i_l} = \sum_{h=0}^{k} \beta^h_{j,l} \varphi_{i_h}.$$

be the decomposition of the pointwise product $\varphi_{i_j} \bullet \varphi_{i_j}$ of φ_{i_j} and φ_{i_j} into a linear combination of the eigenvectors.

(i) *If $\beta^h_{j,l} \neq 0$, then $|i_j - i_l| \leq i_h \leq i_j + i_l$.*

(ii) *Moreover, if $q = 2$, then $\beta^h_{j,l} \neq 0$ also implies $i_h \leq 2n - i_j - i_l$ and $i_h + i_j + i_l \equiv 0 \bmod 2$.*

Remark 7.2. *By linear algebra, $\beta^h_{j,l} = 0$ is equivalent to $(\varphi_{i_j} \bullet \varphi_{i_l})D\varphi_{i_h} = 0$, where D is a diagonal matrix such that DS is a symmetric matrix.*

Now we can prove a generalization of Lemma 7.3 to CR-codes of arbitrary covering radius whose intersection numbers form arithmetic progressions.

Proposition 7.1 (correlation-immunity bound for arithmetic CR codes). *If there exists a $\{\rho b, \ldots, 2b, b; c, 2c, \ldots, \rho c\}$-CR code C in $H(n, 2)$, where $b \neq c$, then*

$$n \geq \frac{3(b+c)}{4}\rho.$$

Proof. The eigenvalues of the quotient matrix S of C form the arithmetic progression θ_0, $\theta_{\frac{b+c}{2}}$, ..., $\theta_{\rho\cdot\frac{b+c}{2}}$ (see [27, Lemma 3.3]), where $\theta_i = n - 2i$, $i = 0, \ldots, n$, are the eigenvalues of $H(n, 2)$. The $\theta_{\rho\cdot\frac{b+c}{2}}$-eigenvector of S is $\phi_\rho = \left((-\frac{c}{b})^0, (-\frac{c}{b})^1, \ldots, (-\frac{c}{b})^\rho\right)$, up to a multiplicative constant. The cardinalities of cells $C^{(i)}$ of the distance partition with respect to C are

$$|C^{(i)}| = \frac{2^n c^\rho}{(b+c)^\rho} \cdot \binom{\rho}{i} \cdot \frac{b^i}{c^i}, \quad i = 0 \ldots, \rho.$$

Denote by $D = (D_{i,j})^\rho_{i,j=0}$ the diagonal $(\rho+1) \times (\rho+1)$ matrix with $D_{i,i} = |C^{(i)}|$. With all above, we see that

$$(\phi_\rho \bullet \phi_\rho) \cdot D \cdot \phi_\rho = \sum_{i=0}^{\rho} \frac{2^n c^\rho}{(b+c)^\rho} \cdot \binom{\rho}{i} \cdot \frac{b^i}{c^i} \cdot \left(-\frac{c}{b}\right)^{3i} = \frac{2^n c^\rho (b-c)^\rho}{b^{2\rho}(b+c)^\rho} \neq 0.$$

By Lemma 7.6(ii) and Remark 7.2, we have $\rho\frac{b+c}{2} \leq 2n - \rho\frac{b+c}{2} - \rho\frac{b+c}{2}$, which gives the required inequality. □

By Lemma 7.15, the existence of a $\{b; c\}$-CR code attaining the correlation-immunity bound implies the existence of a $\{\rho b, \ldots, 2b, b; c, 2c, \ldots, \rho c\}$-CR code attaining this bound for arbitrary ρ. At the moment, there are three known families of such CR codes. The first one, with $b = 3s$, $c = s$, starts from the one-perfect code $\{000, 111\}$ in $H(3, 2)$, corresponding to $s = 1$, $\rho = 1$; the second family, with $b = 5s$, $c = 3s$, starts from the $\{5; 3\}$-CR code in $H(6, 2)$ with the characteristic function g_6 from [42, p.17]; the third family, with $b = 9s$, $c = 7s$, starts from the $\{9; 7\}$-CR code in $H(12, 2)$ constructed in [18]. The first open parameters $\{b; c\}$ attaining the correlation-immunity bound are $\{23; 9\}$, $\{21; 11\}$, and $\{17; 15\}$ in $H(24, 2)$.

7.1.5 Weight Distribution

An equitable partition with quotient matrix S of a distance-regular graph Γ is an equitable partition of the corresponding distance-w graph for every w up to the diameter of Γ. The quotient matrix of this equitable partition can be found from S, w, and the parameters of the graph. This matrix must be integer and non-negative. This test for feasibility of a matrix is stronger than the previous one (of course, if for each w the matrix is integer, then the formula for the cardinality of each cell also gives integer values). We need the following polynomial of degree w, known as the Krawtchouk polynomial:

$$K_{w;n,q}(x) = \sum_{j=0}^{w} (-1)^j (q-1)^{w-j} \binom{x}{j} \binom{n-x}{w-j}; \tag{7.1}$$

in particular,

$$K_{1;n,q}^{-1}(y) = \frac{(q-1)n - y}{q}. \tag{7.2}$$

Lemma 7.7. *If $K_{w;n,q}(K_{1;n,q}^{-1}(S))$ is not an integer matrix for some $w \in \{0, 1, \ldots, n\}$, then for every nonnegative integer k, equitable partitions with quotient matrix $S + k(q-1)\mathrm{Id}$ do not exist in $H(n+k, q)$.*

Proof. If C is an equitable partition of $H(n+k, q)$ with quotient matrix S, then $S_w = K_{w;n,q}(K_{1;n,q}^{-1}(S))$ is the quotient matrix of C in the distance-w graph $\Gamma^{(w)}$ of $H(n+k, q)$ (two vertices are adjacent in $\Gamma^{(w)}$ if the distance between them in $H(n+k, q)$ is w), see e.g. [28]. Hence, S_w is integer and the claim holds for $k = 0$.

For $k = 1$, the claim follows from the identity

$$\begin{aligned} K_{w;n+1,q}\big(K_{1;n+1,q}^{-1}(x + (q-1))\big) &= K_{w;n+1,q}\big(K_{1;n,q}^{-1}(x)\big) \\ &= K_{w;n,q}\big(K_{1;n,q}^{-1}(x)\big) + (q-1) K_{w-1;n,q}\big(K_{1;n,q}^{-1}(x)\big) \end{aligned} \tag{7.3}$$

(which is straightforward from equation $K_{1;n+1,q}^{-1}(x + (q-1)) = K_{1;n,q}^{-1}(x)$ and $\binom{n+1-x}{w-j} = \binom{n-x}{w-j} + \binom{n+1-x}{w-1-j}$). Indeed, from (7.3), we see that for the quotient matrix $S' = S + (q-1)\mathrm{Id}$ in $H(n+1, q)$, the corresponding quotient matrix in the distance-w graph of $H(n+1, q)$ is $S'_w = S_w + (q-1)S_{w-1}$. From this expression, we see that if w_0 is the smallest value of w for which S_w is not an integer matrix, then it is also the smallest value of w for which S'_w is not an integer matrix.

Applying the same argument inductively, we obtain the claim for any $k \geq 0$. □

Lemma 7.8. (i) *If for some $w \in \{0, 1, \ldots, n\}$ the matrix $K_{w;n,q}(K_{1;n,q}^{-1}(S))$ has a negative entry, then equitable partitions with quotient matrix S do not exist in $H(n, q)$.*

(ii) *If $K_{n;n,q}(K_{1;n,q}^{-1}(S))$ has a negative entry and S has a complete collection of eigenvalues that are eigenvalues of $H(n, q)$ (i.e., of form $n(q-1) - iq$, $i \in \{0, \ldots, n\}$), then for every nonnegative integer k, equitable partitions with quotient matrix $S + k(q-1)\mathrm{Id}$ do not exist in $H(n+k, q)$.*

Proof. The first claim is proved similarly to Lemma 7.7, $k = 0$.

To prove the second claim, we start with two observations. At first, as follows from (7.2) and the hypothesis on the eigenvalues of S, the eigenvalues of $K^{-1}_{1;n,q}(S)$ are all in $\{0, \dots, n\}$. At second, for $x \in \{0, \dots, n\}$, we have $K_{n+1;n,q}(x) = 0$. From these two facts, we get $K_{n+1;n,q}(K^{-1}_{1;n,q}(S)) = 0$. Therefore, substituting $x = S$ to (7.3), we get

$$K_{n+1;n+1,q}\big(K^{-1}_{1;n+1,q}(S + (q-1)\mathrm{Id})\big) = (q-1)K_{n;n,q}\big(K^{-1}_{1;n,q}(S)\big).$$

By the hypothesis, the matrix in the right part is not nonnegative. Hence, the matrix in the left part is not nonnegative, and $S + (q-1)\mathrm{Id}$ also satisfies the hypothesis. By induction $S + k(q-1)\mathrm{Id}$ satisfies the hypothesis for any nonnegative integer k, and by the first part of the lemma equitable partitions with quotient matrix $S + k(q-1)\mathrm{Id}$ do not exist in $H(n+k, q)$. □

7.1.6 Interweight Distribution

Lemma 7.9 ([44]). *For an equitable partition $(C_0, \dots, C_\rho)$ of $H(n, 2)$, and a vertex x in C_i the number of pairs (y, z), $y \in C_j$, $z \in C_k$, such that $d(x, y) = a$, $d(x, z) = b$, and $d(x, z) = c$ is uniquely determined by n, i, j, k, a, b, c, and the quotient matrix.*

Formulas for finding all these *interweight* numbers (for all feasible i, j, k, a, b, c) can be found in [29]. Obviously, if, for a putative intersection array, the formulas give negative or noninteger interweight numbers, then the corresponding equitable partitions do not exist. In our experience, if the array successfully passes the test for integrity of the weight distribution, then the interweight numbers are also integer. In contrast, the non-negativity of the interweight numbers is a strong test that can reject parameters that pass all other tests. For some putative intersection arrays, it establishes a new lower bound on the dimension of the hypercube $H(n, 2)$ in which an equitable partition can exist. Empirically, for CR-1 codes in $H(n, 2)$, the interweight-distribution test rejects the same parameters as Lemma 7.3(i); this fact needs a theoretical explanation.

7.1.7 There Are No $\{22, 20, 15, 2, 1; 1, 2, 12, 20, 22\}$-CR Codes in $H(n, 3)$

Up to now, $A = \{22, 20, 15, 2, 1; 1, 2, 12, 20, 22\}$ is the only intersection array which satisfies all known general necessary conditions but has its own proof of asymptotical (for all n) nonexistence. The corresponding quotient matrix is

$$S = \begin{pmatrix} 0 & 22 & 0 & 0 & 0 & 0 \\ 1 & 1 & 20 & 0 & 0 & 0 \\ 0 & 2 & 5 & 15 & 0 & 0 \\ 0 & 0 & 12 & 8 & 2 & 0 \\ 0 & 0 & 0 & 20 & 1 & 1 \\ 0 & 0 & 0 & 0 & 22 & 0 \end{pmatrix} \qquad (7.4)$$

(for $n = 11$; for arbitrary n all diagonal entries are increased by $2(n - 11)$). The first observation is that the union of of an A-code C and its metrical antipode $C^{(5)}$ is a $\{22, 20; 1, 2\}$-CR code, which is the ternary Golay code if $n = 11$.

Lemma 7.10. *There are no $\{22, 20, 15, 2, 1; 1, 2, 12, 20, 22\}$-CR codes in $H(11, 3)$.*

Proof. Without loss of generality, assume $\bar{0} \in C$. From the distance invariance of equitable partitions in distance-regular graphs, we have that $(C, C^{(1)}, \ldots, C^{(5)})$ is an equitable partition of the distance-3 graph of $H(n, 3)$. In particular, from the corresponding quotient matrix, we see that there is exactly one codeword of C at distance 3 from every vertex in $C^{(2)}$ (actually, this can be also observed by analyzing the original quotient matrix (7.4)). In particular, the weight-5 codewords of C form a 2-design in the following meaning: every weight-2 word of $H(11, 3)$ is at distance 3 from exactly 1 weight-5 codeword of C. It is computationally checked that the ternary Golay code does not include such a configuration. □

Lemma 7.11. *For each $s \geq 0$, there is a unique perfect coloring of $H(11 + s, 3)$ with quotient matrix $S_s = \begin{pmatrix} 2s & 22 & 0 \\ 1 & 1+2s & 20 \\ 0 & 2 & 20+2s \end{pmatrix}$, up to equivalence. Moreover, this coloring does not essentially depend on s coordinates.*

Proof. Since for $s = 0$ the claim follows from the uniqueness of the Golay code in $H(11, 3)$ [33, 20.8], it is sufficient to prove only the second claim. Let $(D, D^{(1)}, D^{(2)})$ be an equitable partition of $H(11 + s, 3)$ with quotient matrix S_s. Because of 1 in the $(1, 0)$th entry of the quotient matrix, it is not difficult to conclude the following (see [30, Lemma 2]):

(*) *D is the union of 729 s-faces at distance at least 3 from each other.*

Without loss of generality, we assume that D includes the face F consisting of all vertices with 0 in the first 11 positions. Denote by $F^{\perp}$ the face consisting of all vertices with 0 in the last s position. We claim the following:

(**) *The weight distribution of $D \cap F^{\perp}$ is* $[1, 0, 0, 0, 0, 132, 132, 0, 330, 110, 0, 24]$.

This claim is straightforward from the following two facts.

- *The weight distributions of $D \cap F^{\perp}$, $D^{(1)} \cap F^{\perp}$, and $D^{(2)} \cap F^{\perp}$ are completely determined by the weight distributions of $D \cap F$, $D^{(1)} \cap F$, $D^{(2)} \cap F$ and are independent of the choice of a concrete equitable partition $D \cap F$, $D^{(1)} \cap F$, $D^{(2)} \cap F$ with considered parameters.* (See [45] for general formulas for equitable partitions in $H(n, q)$.)

- *There is at least one equitable partition with considered parameters for which (**) is true.* (Indeed, the weight distribution in (**) is the weight distribution of the Golay code in $H(11, 3)$. So, we have the fact for $s = 0$. Adding nonessential coordinates, we get it for arbitrary s.)

Next, since the intersection of two non-disjoint faces is a face, from (*) we have that $D \cap F^{\perp}$ is the union of faces. On the other hand, from the weight distribution in (**) we see that $D \cap F^{\perp}$ does not include 2-faces or larger (indeed, any 2-face contains vertices of three different subsequent weights, while the sequence in (**) has no three subsequent nonzeros). Since every 1-face consists of 3 words of weights i, $i+1$, $i+1$ for some i, from (**) we see that that i can only be 5 or 8, and there are at most $\frac{132}{2} + \frac{110}{2} = 121$ 1-faces in $D \cap F^{\perp}$. It follows that $D \cap F^{\perp}$ has at least $729 - 3 \cdot 121 = 366$ isolated vertices and D includes at least 366 s-faces having the same fixed coordinates as F. Since 366 is more than half of 729, the same is true for all s-faces in D (otherwise, applying the same arguments to one of the remaining s-faces, we get a contradiction). We see that D does not depend on the last s coordinates. It follows that the same is true for the distance partition with respect to D (and the corresponding perfect coloring). □

Theorem 7.1. *There are no* $\{22, 20, 15, 2, 1; 1, 2, 12, 20, 22\}$*-CR codes in* $H(n, 3)$.

Proof. The union of a $\{22, 20, 15, 2, 1; 1, 2, 12, 20, 22\}$-CR code C and $C^{(5)}$ is a $\{22, 20; 20, 22\}$-CR code. By Lemma 7.11, it consists if 729 parallel $(n-11)$-faces. Obviously, vertices of both C and $C^{(5)}$ cannot meet in one of those faces. It follows that C is the union of parallel $(n-11)$-faces and is obtained from a $\{22, 20, 15, 2, 1; 1, 2, 12, 20, 22\}$-CR code C_0 in $H(11, 3)$ by direct product with an $(n-11)$-face. But by Lemma 7.10 such C_0 does not exist. □

7.2 CONSTRUCTIONS

7.2.1 Linear Codes

Assume that we have a subset M of the vector space $\mathbb{F}_q^n$ over the finite field $\mathbb{F}_q$ of order q. We assume that M spans $\mathbb{F}_q^n$ and consists of pairwise independent vectors (however, M can be redundant as a generating set as linear dependencies involving three or more vectors are allowed), which means "nonzero and distinct" in the binary case $q = 2$. Consider the Cayley graph $\Gamma_M = \mathrm{Cay}(\mathbb{F}_q^{n+}, \mathbb{F}_q^* M)$ on the additive group $\mathbb{F}_q^{n+}$ of $\mathbb{F}_q^n$ with connecting set $\mathbb{F}_q^* M = \{az : a \in \mathbb{F}_q, a \neq 0, z \in M\}$, where two vectors are adjacent if and only if their difference belongs to aM for some nonzero $a \in \mathbb{F}_q$.

Let H_M be a matrix with column set M. It is straightforward that $x \to H_M x$ is a covering of the graph Γ_M by $H(|M|, q)$. As a corollary, we have the following.

Proposition 7.2. *Assume that we have an* $n \times N$ *matrix* H_M *of rank* n *over* $\mathbb{F}_q$ *without collinear columns and* M *is the column set of* H_M. *Denote by* Γ_M *the Cayley graph* $\mathrm{Cay}(\mathbb{F}_q^{n+}, \mathbb{F}_q^* M)$.

a. *A coloring* f *of the Cayley graph* Γ_M *is perfect with quotient matrix* Q *if and only if* $F(x) = f(H_M \cdot x)$ *is a perfect coloring of* $H(|M|, q)$ *with the same quotient matrix* Q.

b. *A code B in Γ_M is Q-CR if and only if $H_M B = \{H_M \cdot x : x \in B\}$ is a Q-CR code.*

c. *A code B in Γ_M is linear if and only if $H_M B = \{H_M \cdot x : x \in B\}$ is a linear code.*

d. *[8, 11.1.C] the graph Γ_M is distance-regular with intersection array A if and only if the linear code $C = \{x : H_M \cdot x = 0\}$ is A-CR (the code distance of C is at least 3).*

So, there is a correspondence between distance-regular Cayley graphs on vector spaces over $\mathbb{F}_q$ with simple connecting set of form $\mathbb{F}_q^* M$ and linear q-ary CR codes with distance at least 3. Note that linear CR codes with distances 1 and 2 are reduced to linear CR codes with distance at least 3, see Corollary 1.3 in Chapter 1.

There are many different constructions of linear CR codes in Hamming graphs, a survey for the case of covering radius 2 can be found in [12], see also the lemma below. Examples of distance-regular Cayley graphs of large diameter are halved hypercubes and folded halved hypercubes (the corresponding CR codes were considered in [35,37]), bilinear forms graphs (the corresponding CR codes were considered in [36]), and the other graphs described in [8, Sections 9.5 and 9.6]. An important fact is that $H(n, q^s)$ is a Cayley graph on $\mathbb{F}_q^{ns}$. So, any CR code (linear or nonlinear) in $H(n, q^s)$ corresponds to a (linear or nonlinear, respectively) CR code in $H(n(q^s - 1)/(q - 1), q)$ with the same quotient matrix.

A survey of different constructions of linear CR-2 codes can be found in [12].

Lemma 7.12 ([12]). *Assume that we have an $n \times N$ matrix H_M of rank n over $\mathbb{F}_q$ without collinear columns and M is the column set of H_M. Let the code C be the kernel of H_M, i.e., $C = \{x : H_M \cdot x = 0\}$, and its dual $C^\perp$ the span of rows of H_M.*

Let C be a linear code with distance 3, let $C^\perp$ be its dual, and let M be the column set of the check matrix of C (generator matrix of $C^\perp$).

The following assertions are equivalent:

1. *C is a CR-2.*
2. *$C^\perp$ is a projective two-weight code.*
3. *$\mathbb{F}_q^* M$ is a (λ_1, λ_2) difference set for some positive integer λ_1 and λ_2 (that is, for every element v in $\mathbb{F}_q^* M$ it holds $\lambda_1 = |\{(x, y) : x - y = v;\ x, y \in \mathbb{F}_q^* M\}|$ and for every nonzero v not in $\mathbb{F}_q^* M$ it holds $\lambda_2 = |\{(x, y) : x - y = v;\ x, y \in \mathbb{F}_q^* M\}|$.*
4. *the Cayley graph with connecting set $\mathbb{F}_q^* M$ is strongly regular (i.e., distance regular of radius 2; this graph can also be considered at the coset graph of C).*

7.2.2 Construction "+1"

The following simple fact is very important for the study of feasible parameters of CR codes in Hamming graphs. It means that if, for a given intersection array A, a A-CR code exist in $H(n_0, q)$, then A-CR codes exist in $H(n, q)$ for all $n \geq n_0$.

Lemma 7.13. *If* $(C^{(0)}, C^{(1)}, \ldots, C^{(\rho)})$ *is an equitable partition of* $H(n, q)$ *with quotient matrix* S*, then* $(D^{(0)}, D^{(1)}, \ldots, D^{(\rho)})$*, where*

$$D^{(i)} = \{(x, a) : \; x \in C^{(i)}, \; a \in \{0, \ldots, q-1\}\}$$

is an equitable partition of $H(n+1, q)$ *with quotient matrix* $S + (q-1)\mathrm{Id}$.

7.2.3 Construction "$\times t$"

Lemma 7.14. *If* $(C^{(0)}, C^{(1)}, \ldots, C^{(\rho)})$ *is an equitable partition of* $H(n, q)$ *with quotient matrix* S *and* $t \geq 2$*, then* $(D^{(0)}, D^{(1)}, \ldots, D^{(\rho)})$*, where*

$$D^{(i)} = \{(x_1, \ldots, x_t) \in (\{0, \ldots, q-1\}^n)^t : \; x_1 + \ldots + x_t \in C^{(i)}\},$$

is an equitable partition of $H(tn, q)$ *with quotient matrix* tS*. Here,* $+$ *is an arbitrary Abelian group operation on* $\{0, \ldots, q-1\}$ *(for example, the module-*q *addition), which acts coordinate-wise on* n*-tuples.*

7.2.4 Construction "$\{b; c\}^t$"

The following lemma was proved in [27] (the special case where C is a Hamming code was considered earlier in [41], for t=2, in [2]).

Lemma 7.15. *If* C *is a CR code with intersection array* $\{b; c\}$ *in a graph* Γ*, then its* t^{th} *Cartesian power*

$$C^t = \{(x_1, \ldots, x_t) : \; x_i \in C\}$$

is a CR code with intersection array $\{tb, (t-1)b, \ldots, b; c, 2c, \ldots, tc\}$ *(an arithmetic CR code, see [27]) in the Cartesian product of* t *copies of* Γ *(if* $\Gamma = H(n, q)$*, then the product is* $H(tn, q)$*).*

Actually, we have even a more general fact.

Lemma 7.16. *If* C' *and* C'' *are CR codes with intersection arrays* $\{t'b, (t'-1)b, \ldots, b; c, 2c, \ldots, t'c\}$*,* $\{t''b, (t''-1)b, \ldots, b; c, 2c, \ldots, t''c\}$ *in* Γ' *and* Γ''*, respectively, then the Cartesian product*

$$C' \times C'' = \{(x', x'') : \; v' \in V(\Gamma'), \; x'' \in V(\Gamma'')\}$$

is a CR code with intersection array $\{tb, (t-1)b, \ldots, b; c, 2c, \ldots, tc\}$*,* $t = t' + t''$ *in* $\Gamma' \times \Gamma''$.

7.2.5 Golay Family

The binary Golay family of linear CR codes consists of the 3-perfect binary Golay $[23, 12, 7]$ code G_{23}, the extended binary Golay $[24, 12, 8]$ code $\overline{G}_{24}$ (actually, G_{23} is the $\{0, 1\}$-shortened $\overline{G}_{24}$), and several shortened Golay codes:

- $\{23, 22, 21; 1, 2, 3\}$-CR binary Golay code G_{23}, see [8, 11.3E];
- $\{22, 21, 20; 1, 2, 6\}$-CR $\{0, 1\}$-shortened G_{23}, see [8, 11.3F];
- $\{21, 20, 16; 1, 2, 12\}$-CR $\{00, 01, 10, 11\}$-shortened G_{23}, see [8, 11.3G];
- $\{24, 23, 22, 21; 1, 2, 3, 24\}$-CR extended binary Golay code $\overline{G}_{24}$, see [8, 11.3D];
- $\{21, 20, 16, 6, 2, 1; 1, 2, 6, 16, 20, 21\}$-CR $\{00, 11\}$-shortened G_{23}, see [8, 11.3H];
- $\{21, 20, 16, 9, 2, 1; 1, 2, 3, 16, 20, 21\}$-CR $\{000, 111\}$-shortened $\overline{G}_{24}$, see [8, 11.3H];
- $\{22, 21, 20, 3, 2, 1; 1, 2, 3, 20, 21, 22\}$-CR $\{0\}$-shortened G_{23}, see [8, 11.3H];
- $\{22, 21, 20, 16, 6, 2, 1; 1, 2, 6, 16, 20, 21, 22\}$-CR $\{00, 11\}$-shortened $\overline{G}_{24}$, see [6];
- $\{23, 22, 21, 20, 3, 2, 1; 1, 2, 3, 20, 21, 22, 23\}$-CR $\{0\}$-shortened $\overline{G}_{24}$, see [6].

The last two arrays where known as the intersection arrays of the bipartite doubles of the graphs of cosets of G_{23} and shortened G_{23}, respectively, see [8, 11.3EF]. These graphs are clearly Cayley graphs of $\mathbb{F}_2^{12}$; however, the fact that there are CR codes with these intersection arrays was unnoticed for some time. Observing this fact [6] refined the known Neumaier conjecture: the $\overline{G}_{24}$ and the $\{0\}$-shortened $\overline{G}_{24}$ are the only binary CR codes with more than two codewords and minimum distance of at least 8.

Three other intersection arrays mentioned in [8, 11.3EF] are $\{253, 210, 3; 1, 30, 231\}$, $\{276, 231; 1, 36\}$, $\{231, 160, 6; 1, 48, 210\}$. The corresponding linear CR codes also exist by [8, Theorem 11.1.10] and can also be related to the binary Golay family of CR codes.

The ternary Golay family of linear CR codes consists of the 2-perfect ternary Golay $[11, 6, 5]_3$ code G_{11}, the extended ternary Golay $[12, 6, 6]_3$ code $\overline{G}_{12}$ (actually, G_{11} is the $\{0, 1, 2\}$-shortened $\overline{G}_{12}$), and several shortened Golay codes:

- $\{22, 20; 1, 2\}$-CR 2-perfect ternary Golay code G_{11}, see [8, 11.3B];
- $\{20, 18; 1, 6\}$-CR $\{0, 1, 2\}$-shortened G_{11}, see [8, 11.3C];
- $\{24, 22, 20; 1, 2, 12\}$-CR extended ternary Golay code $\overline{G}_{12}$, see [8, 11.3A];
- $\{20, 18, 4, 1; 1, 2, 18, 20\}$-CR $\{0\}$-shortened G_{11}, see [8, 11.3H];
- $\{22, 20, 18, 2, 1; 1, 2, 9, 20, 22\}$-CR $\{0\}$-shortened $\overline{G}_{12}$, see [8, 11.3H].

7.2.6 Union of CR Codes

For an equitable partition, sometimes we can unify some groups of colors to make another, *coarser*, equitable partition. Such possibility, for given groups of colors, implies some obvious arithmetic identities between the elements of the quotient matrix, and if these identities are satisfied, then the corresponding coarser partition is equitable. A special case of this is the following straightforward proposition.

Proposition 7.3. *Assume that in a graph G we have a $\{b_0,\ldots,b_{\rho-1};c_1,\ldots,c_\rho\}$-CR code C and a $\{b'_0,\ldots,b'_{\rho-1};c'_1,\ldots,c'_\rho\}$-CR code C' at distance ρ from each other (i.e., $C' \subseteq C^{(\rho)}$ and $C \subseteq C'^{(\rho)}$) such that*

- $b_i = b'_i = c_{\rho-i} = c'_{\rho-i}$ *for all i except $i = \lfloor \rho/2 \rfloor$,*
- $b_{\lfloor \rho/2 \rfloor} + c_{\rho-\lfloor \rho/2 \rfloor} = b'_{\lfloor \rho/2 \rfloor} + c'_{\rho-\lfloor \rho/2 \rfloor}$.

(i) *If $C' \neq C^{(\rho)}$, then $C \cup C'$ a $\{b''_0,\ldots,b''_{\rho-1};c''_1,\ldots,c''_\rho\}$-CR code, where*

- $b''_i = b_i = c''_{\rho-i} = c_{\rho-i}$ *for all i except $i = \lfloor \rho/2 \rfloor$,*
- $c''_{\rho-\lfloor \rho/2 \rfloor} = c_{\rho-\lfloor \rho/2 \rfloor} + c'_{\rho-\lfloor \rho/2 \rfloor}$,
- $b''_{\lfloor \rho/2 \rfloor} = b_{\lfloor \rho/2 \rfloor} - c'_{\rho-\lfloor \rho/2 \rfloor}$.

(ii) *If $C' = C^{(\rho)}$, then $C \cup C'$ is a $\{b_0,\ldots,b_{\lfloor \rho/2 \rfloor-1};c_1,\ldots,c_{\lfloor \rho/2 \rfloor}\}$-CR code (odd ρ) or a $\{b_0,\ldots,b_{\rho/2-1};c_1,\ldots,c_{\rho/2-1},c_{\rho/2}+b_{\rho/2}\}$-CR code (even ρ).*

It is easy to observe that by induction we can unify more than two CR codes each two of which satisfy the hypothesis on C and C' from Proposition 7.3. A nice application of this fact, illustrating both odd-ρ and even-ρ cases, is CR codes from Preparata codes, discussed in the subsection below.

Here, we consider another example. Obviously, if C is a linear CR code whose intersection array meets the hypothesis of Proposition 7.3, then, for the role of C', we can consider any coset of C at distance ρ from C (then, inductively, any other coset at distance ρ from C and the first coset, and so on). Consider the binary linear $\{48, 15; 1, 48\}$-CR code C from family SU1 in [12] (the columns of the check matrix of the code are the 48 lexicographically largest binary 6-tuples). Taking cosets of C for the role of C', we can recursively construct a $\{48, 16 - s; s, 48\}$-CR code for every $s \in \{2, 3, 4, 5, 6, 7, 8\}$.

7.2.6.1 Preparata Family

Two notable infinite classes of CR codes with code distance of more than 4 are the classes of nonlinear codes with parameters of Preparata $(4^m - 1, 2^{4^m-2m}, 5)$ and extended Preparata $(4^m, 2^{4^m-2m}, 6)$ codes, $m \geq 2$. Using translations of the Preparata codes, we can construct a wide family of nonlinear CR codes with code distance 3 and 4.

The extended Preparata $(4^m, 2^{4^m-2m}, 6)$ code $\overline{P}_{4^m}$ (for $m = 2$ also known as the Nordstrom–Robinson code) is a CR $\{4^m, 2^m - 1, 4^m - 2, 1; 1, 2, 4^{-}1, 4^m\}$-CR code with covering radius 4 (this was essentially proved in [38], although CR codes were not yet defined in that paper). It is not difficult to observe from the intersection array that $\overline{P}_{4^m} \cup \overline{P}^{(4)}_{4^m}$ is a $\{4^m, 2^m-1, 4^m-2, 1; 1, 2, 4^m-1, 4^m\}$-CR code, i.e., a binary extended 1-perfect $(4^m, 2^{4^m-m-1}, 4)$ code. For the original extended Preparata code [34] and its generalizations [1,15,43], this code is the extended Hamming code, while for known $\mathbb{Z}_4$-linear "Preparata" codes with the same parameters [21,24], it is different ($m \geq 2$). As was shown in [1], the extended Hamming code of length $n = 4^m$, $m \geq 2$, can be partitioned into $n/2$ translations of the extended Praparata code. Using this partition, we can construct CR codes with different parameters.

Proposition 7.4. *If P_1, ..., $P_{n/2}$ is a partition of an extended one-perfect code of length $n = 4^m$ into codes with parameters of the extended Preparata code and $t \in \{1, \dots, n/2-1\}$, then $P_1 \cup \ldots \cup P_t$ is a $\{4^m, 4^m-1, 4^m-2t, 1; 1, 2t, 4^m-1, 4^m\}$-CR code. Moreover, puncturing (removing the last symbol from all codewords) gives a $\{4^m - 1, 4^m - 2t, 1; 1, 2t, 4^m - 1\}$-CR code.*

Proof. Consider the CR code $P = P_1$ and the corresponding distance equitable partition $(P, P^{(1)}, P^{(2)}, P^{(3)}, P^{(4)})$. Denote $H = P \cup P^{(4)}$; we know that H is an extended one-perfect code with minimum distance 4. By the hypothesis, $\{P_1, \dots, P_{n/2}\}$ is a partition of H into $\{4^m, 2^m - 1, 4^m - 2, 1; 1, 2, 4^m - 1, 4^m\}$-CR codes. Denote by R_i, $i = 1, \dots, n/2$, the set of vertices at distance 1 from P_i. Since the codes P_i are at distance 4 from each other, the sets R_i do not intersect and partition $P^{(1)} \cup P^{(3)}$. It is easy to see that $(P_1, \dots, P_{n/2}, R_1, \dots, R_{n/2}, P^{(2)})$ is an equitable partition: all n neighbors of P_i are in R_i; each vertex from R_i has one neighbor in P_i and the other $n - 1$ neighbors in $P^{(2)}$; each vertex from R_i has 2 neighbors from each R_i, $i = 1, \dots, n/2$. It is now straightforward that by merging t codes P_i, $i = 1, \dots, t$, we get a CR code with required intersection array.

The last claim is straightforward to check (alternatively, the similar arguments as above can be repeated for distance-5 Preparata codes). □

Remark 7.3. *In [4], it is proved that the linear span of any (extended) Preparata-like code P is the (extended) one-perfect code $P \cup P^{\rho}$. It immediately follows that the CR code from Proposition 7.4 is not linear. For power-of-two $t \leq \sqrt{n}$, there is a linear CR code with the same intersection array [7] (see also family (F.12.1") in Chapter 2 of the current book). If P is $\mathbb{Z}_4$-linear (like in [21,24]), then $P \cup P^{\rho}$ is also $\mathbb{Z}_4$-linear, and for each power-of-two $t < n/2$ there is an "intermediate" $\mathbb{Z}_4$-linear CR code.*

7.2.7 Splitting Construction

We will first consider the construction of equitable 2-partitions developed in [17]. Then, we will show how to modify it for constructing CR codes with covering radius 2.

Assume that we have an equitable 2-partition (C_0, C_1) of $H(n, 2)$ with quotient matrix

$$\begin{pmatrix} a & b \\ c & d \end{pmatrix}.$$

We also assume that there is $s \geq 0$ such that $a + s \leq c$. The simplest case is $s = 0$ (we still need $a \leq c$); if $0 < a < c$, then by Kőnig's theorem we can take $s = 1$; for some partitions, we can take $s > 1$, but there are no known cases when new parameters of CR codes are obtained with s larger than 1. We start with construction in Lemma 7.14, $t = 2$, by defining

$$D_i = \{(x, y) \in \{0,1\}^n \times \{0,1\}^n : x + y \in C_i\}, \qquad i = 0, 1.$$

The partition (D_0, D_1) of $H(2n, 2)$ is equitable with quotient matrix

$$\begin{pmatrix} 2a & 2b \\ 2c & 2d \end{pmatrix}.$$

We can split D_0 into two subsets, $D_{0;0}$ and $D_{0;1}$, according to the parity $p(x, y) = p(x) = x_1 + \cdots + x_n$ of x:

$$\text{if } (x, y) \in D_0, \text{ then } (x, y) \in D_{0;p(x,y)}. \tag{7.5}$$

After this splitting, each vertex has equal number of neighbors in $D_{0;0}$ and in $D_{0;1}$ (indeed, for a vertex (x, y), if $x + y + e_i \in C_0$, then $(x + e_i, y), (x, y + e_i) \in D_0$ and the parities of $x+e_i$ and x are different, so $(x+e_i, y) \in D_{0;0}$ and $(x, y+e_i) \in D_{0;1}$ or vice versa). It follows that the partition $(D_{0;0}, D_{0;1}, D_1)$ is equitable with quotient matrix

$$\begin{pmatrix} a & a & 2b \\ a & a & 2b \\ c & c & 2d \end{pmatrix}.$$

We modify the last step using the partition $\overline{S}$ of C_0 into s-faces as follows. For a vertex (x, y) such that $x + y \in C_0$, let $i_1, \ldots, i_s$ be the free directions of the s-face in $\overline{S}$ that contains $x + y$. We redefine

$$p(x, y) = x_1 + \cdots + x_n + y_{i_1} + \cdots + y_{i_s}$$

for $x = (x_1, \ldots, x_n)$ and $y = (y_1, \ldots, y_n)$. With this new $p(x, y)$, we again use (7.5) to define $D_{0;0}$ and $D_{0;1}$. With this modification, each vertex (x, y) from $D_{0;0}$ (similarly, $D_{0;1}$) has s less neighbors in $D_{0;0}$ and s more neighbors in $D_{0;1}$. On the other hand, each vertex (x, y) from D_1 still has the same number of neighbors in $D_{0;0}$ and in $D_{0;1}$ (indeed, if $(x, y) \in D_1$ and $(x + e_i, y), (x, y + e_i) \in D_0$, then i is not a free direction of the s-face containing $x + y + e_i$, so $p(x + e_i, y) \neq p(x, y + e_i)$). It follows that the quotient matrix of the new partition $(D_{0;0}, D_{0;1}, D_1)$ is

$$\begin{pmatrix} a-s & a+s & 2b \\ a+s & a-s & 2b \\ c & c & 2d \end{pmatrix}. \tag{7.6}$$

Next, we define $t = c - a - s$ (by the hypothesis, it is nonnegative) and repeat t times the following operation. From $(D_{0;0}, D_{0;1}, D_1)$, we construct the partition $(D'_{0;0}, D'_{0;1}, D'_1)$ of $H(2n+1, 2)$ as follows:

$$\begin{aligned} D'_{0;j} &= \{(z,0) : z \in D_{0,j}\} \cup \{(z,1) : z \in D_{0;1-j}\} \\ D'_1 &= \{(z,0) : z \in D_1\} \cup \{(z,1) : z \in D_1\}. \end{aligned}$$

The quotient matrix of $(D'_{0;0}, D'_{0;1}, D'_1)$ is

$$\begin{pmatrix} a{-}s & a{+}s{+}1 & 2b \\ a{+}s{+}1 & a{-}s & 2b \\ c & c & 2d{+}1 \end{pmatrix},$$

and repeating the last step t times (or skipping it if $t = 0$) we get an equitable partition with quotient matrix

$$\begin{pmatrix} a{-}s & a{+}s{+}t & 2b \\ a{+}s{+}t & a{-}s & 2b \\ c & c & 2d{+}t \end{pmatrix} = \begin{pmatrix} a{-}s & c & 2b \\ c & a{-}s & 2b \\ c & c & 2d{+}t \end{pmatrix}.$$

Finally, the same elements c in the first column allow to join the last two cells to obtain an equitable partition with quotient matrix

$$\begin{pmatrix} a-s & c+2b \\ c & a-s+2b \end{pmatrix}.$$

Proposition 7.5 ([17]). *If there in a $\{b;c\}$-CR code in $H(n,2)$ that can be partitioned into s-faces (which is always true if $s = 0$ or $s = 1$, $b < n$) and $n+s \le b+c$, then there exists a $\{2b+c;c\}$-CR code in $H(n+c+b-s, 2)$.*

As we see, the partition into s-faces is used to reduce the minimum dimension of the hypercube for which we can construct a $\{2b+c;c\}$-CR code.

Next, we are interested in constructing a CR code with covering radius 2. We see that in the construction above, we get an equitable 3-partition at an intermediate step. If there is a zero in a non-diagonal element of the quotient matrix (7.6), then two cells are CR codes. This happens only if $a = s = 0$ (which already gives nontrivial examples, e.g., for $\{b;c\} = \{13;3\}$). However, even for $a = s > 0$, we can redefine $p(\cdot)$ as follows:

$$p(x,y) = x_1 + \cdots + x_n - x_{i_1} - \cdots - x_{i_s},$$

which results in swapping pluses and minuses in (7.6) and makes the first cell a CR code. As a result, we have the following construction.

Proposition 7.6. *If there is an $[[a,b],[c,d]]$-CR code in $H(n,2)$ that can be partitioned into a-faces (which is always true if $a \in \{0,1\}$), then in $H(2n,2)$ there is a CR code with quotient matrix*

$$\begin{pmatrix} 2a & 2b & 0 \\ c & 2d & c \\ 0 & 2b & 2a \end{pmatrix}.$$

We are now going to generalize Proposition 7.6 as follows.

Theorem 7.2. *If there is an $[[a,b],[c,d]]$-CR code in $H(n,2)$ that can be partitioned into a-faces (which is always true if $a \in \{0,1\}$), then for every integer $\tau > 0$ and any $\gamma \in \{1,\dots,2^\tau - 1\}$ in $H(2^\tau n, 2)$ there is a CR code with quotient matrix*

$$\begin{pmatrix} 2^\tau a & 2^\tau b & 0 \\ s\gamma & 2^\tau d & (2^\tau - s)\gamma \\ 0 & 2^\tau b & 2^\tau a \end{pmatrix}. \tag{7.7}$$

Proof. Assume that we have an equitable partition (C_0, C_1) with quotient matrix $[[a,b],[c,d]]$ and C_0 is the union of disjoint s-faces from the collection $\overline{S}$ (at the very last step, we will use that $s = a$). Generalizing the previous construction (corresponding to $\tau = 1$), we start with the construction in Lemma 7.14, $t = 2^\tau$. We illustrate the generalization by the case $\tau = 2$. Define

$$D_i = \{(x,y,z,w) \in \{0,1\}^n \times \{0,1\}^n \times \{0,1\}^n \times \{0,1\}^n : x+y+z+w \in C_i\}, \quad i = 0,1.$$

The partition (D_0, D_1) of $H(4n,2)$ is equitable with quotient matrix

$$\begin{pmatrix} 4a & 4b \\ 4c & 4d \end{pmatrix}.$$

We can split D_0 into four subsets, $D_{0;0,0}$, $D_{0;0,1}$, $D_{0;1,0}$, $D_{0;1,1}$, according to the parities $p_1(x,y,z,w) = p(x,y) = x_1 + \dots + x_n + y_1 + \dots + y_n$ and $p_2(x,y,z,w) = p(x,z) = x_1 + \dots + x_n + z_1 + \dots + z_n$ of (x,y) and (x,z), respectively:

$$\text{if } (x,y,z,w) \in D_0, \text{ then } (x,y,z,w) \in D_{0;p(x,y),p(y,z)}. \tag{7.8}$$

After this splitting, each vertex has equal number of neighbors in $D_{0;0,0}$, $D_{0;0,1}$, $D_{0;1,0}$, and $D_{0;1,1}$. (indeed, for a vertex (x,y,z,w), if $x+y+z+w+e_i \in C_0$, then $(x+e_i,y,z,w), (x,y+e_i,z,v), (x,y,z+e_i,v), (x,y,z,v+e_i) \in D_0$ and the parity pairs of these four vectors are different, so they belong to distinct four subcells). It follows that the partition $(D_{0;0,0}, D_{0;0,1}, D_{0;1,0}, D_{0;1,1}, D_1)$ is equitable with quotient matrix

$$\begin{pmatrix} a & a & a & a & 4b \\ a & a & a & a & 4b \\ a & a & a & a & 4b \\ a & a & a & a & 4b \\ c & c & c & c & 4d \end{pmatrix}.$$

We modify the last step using the partition $\overline{S}$ of C_0 into s-faces, $s = a$, as follows. For a vertex (x,y,z,w) such that $x+y+z+w \in C_0$, let $i_1, \dots, i_s$ be the free directions of the s-face in $\overline{S}$ that contains $x+y+z+w$. We redefine

$$p(x) = x_1 + \dots + x_n - x_{i_1} - \dots - x_{i_s}, \quad p(x,y) = p(x) + p(y)$$

for $x = (x_1, \ldots, x_n)$ and $y = (y_1, \ldots, y_n)$. With this new $p(x, y)$ (and $p(x, z)$), we again use (7.8) to define $D_{0;0,0}$, $D_{0;0,1}$, $D_{0;1,0}$, $D_{0;1,1}$. With this modification, each vertex (x, y, z, w) from $D_{0;0,0}$ (similarly, for each other subcell) has s less neighbors in each of $D_{0;0,1}$, $D_{0;1,0}$, $D_{0;1,1}$ and $3s$ more neighbors in $D_{0;0,0}$. On the other hand, each vertex (x, y, z, w) from D_1 still has the same number of neighbors in $D_{0;0,0}$, $D_{0;0,1}$, $D_{0;1,0}$, $D_{0;1,1}$ (indeed, if $(x, y, z, w) \in D_1$ and $(x + e_i, y, z, w), (x, y + e_i, z, w), (x, y, z + e_i, w), (x, y, z, w + e_i) \in D_0$, then i is not a free direction of the s-face containing $x + y + z + w + e_i$, so

$$p_1(x + e_i, y, z, w) = p_1(x, y + e_i, z, w) \neq p_1(x, y, z + e_i, w) = p_1(x, y, z, w + e_i)$$

and

$$p_2(x + e_i, y, z, w) = p_2(x, y, z + e_i, w) \neq p_2(x, y + e_i, z, w) = p_2(x, y, z, w + e_i)$$

and the pairs (p_1, p_2) are distinct for these four vectors). It follows that the quotient matrix of the new partition $(D_{0;0,0}, D_{0;0,1}, D_{0;1,0}, D_{0;1,1}, D_1)$ is

$$\begin{pmatrix} a{+}3s & a{-}s & a{-}s & a{-}s & 4b \\ a{-}s & a{+}3s & a{-}s & a{-}s & 4b \\ a{-}s & a{-}s & a{+}3s & a{-}s & 4b \\ a{-}s & a{-}s & a{-}s & a{+}3s & 4b \\ c & c & c & c & 4d \end{pmatrix}.$$

We can now see from the matrix that merging any cells from $D_{0;0,0}$, $D_{0;0,1}$, $D_{0;1,0}$, $D_{0;1,1}$ keeps the partition equitable; in particular, the partitions $(D_{0;0,0}, D_1, D_{0;0,1} \cup D_{0;1,0} \cup D_{0;1,1})$, $(D_{0;0,0} \cup D_{0;0,1}, D_1, D_{0;1,0} \cup D_{0;1,1})$, $(D_{0;0,0}, D_1, D_{0;0,1} \cup D_{0;1,0} \cup D_{0;1,1})$ are equitable with quotient matrices

$$\begin{pmatrix} a{+}3s & 4b & 3a{-}3s \\ c & 4d & 3c \\ a{-}s & 4b & 3a{+}s \end{pmatrix}, \quad \begin{pmatrix} 2a{+}2s & 4b & 2a{-}2s \\ 2c & 4d & 2c \\ 2a{-}2s & 4b & 2a{+}2s \end{pmatrix}, \quad \begin{pmatrix} 3a{+}s & 4b & a{-}s \\ 3c & 4d & c \\ 3a{-}3s & 4b & a{+}3s \end{pmatrix},$$

respectively. Now we recall that $s = a$ and see that these three matrices coincide with (7.7), $\tau = 2$, $\gamma = 1, 2, 3$. This completes the proof for $\tau = 2$.

The general proof is similar (instead of four coordinate groups x, y, z, w, we take 2^τ groups, and use τ parity checks to split D_0 into 2^τ subcells). □

7.2.8 Bipartite Equitable Partitions

It is well known that if, in a bipartite graph, each cell of an equitable k-partition intersects with each of the bipartite parts, then these intersections form an equitable $2k$-partition. A special case of this effect is the following easy-to-see fact.

Proposition 7.7. *Let C be a $\{b_0, b_1, \ldots, b_{\rho-1}; c_1, \ldots, c_\rho\}$-CR code in a regular bipartite graph of degree b_0, where $b_0 = b_1 + c_1 = \ldots b_{\rho-1} + c_{\rho-1} > c_\rho$. If C_0 is* *the intersection of C with one of bipartite parts, then C_0 is a $\{b_0, b_1, \dots, b_{\rho-1}, b_0 - c_\rho, c_\rho, \dots, c_1; c_1, \dots, c_\rho, b_0 - c_\rho, b_{\rho-1}, \dots, b_1, b_0\}$-CR code of covering radius $2\rho + 1$.*

A series of interesting examples comes from $\{b; c\}$-codes in $H(b, 2)$. Of special interest are such codes that cannot be constructed as linear codes, i.e., when c/b is not a power of 2. The construction from [17] allows to build an infinite number of such codes. The first example is a $\{13; 3\}$-code in $H(13, 2)$, with quotient matrix $\begin{pmatrix} 0 & 13 \\ 3 & 10 \end{pmatrix}$. By Proposition 7.7, it has a $\{13, 10, 3; 3, 10, 13\}$-CR cubcode.

7.3 TABLES

The tables of small parameters of binary CR-ρ, $\rho = 1, 2, 3, 4, 5, 6, 7, 8$ and ternary CR-ρ, $\rho = 1, 2, 3, 4, 5$ codes contain all putative intersection arrays $\{b_0, \dots, b_{\rho-1}; c_1, \dots, c_\rho\}$ with small intersection numbers that satisfy the following conditions:

- the intersection numbers $b_0, \dots, b_{\rho-1}; c_1, \dots, c_\rho$ are positive integers;
- $b_0 > \dots > b_{\rho-1}$, $c_1 < \dots < c_\rho$ (Lemma 7.1);
- integrality of the weight distribution (Lemma 7.7; this automatically includes integrality of the cardinality of a putative CR code, Lemma 7.2);
- Lloyd's condition (Lemma 7.4);
- the triangle inequality for eigenvectors (Lemmas 7.5 and 7.6(i); 7.6(ii) is not considered at this step because its condition depends on n).

Lemma 7.8 also rejects some arrays for all n, but there are quite a few such arrays in the tables, and we decided to keep them as an illustration. Other necessary conditions can increase the lower bound on n but do not prove the nonexistence for all n. The tables contain the following columns.

1. Intersection array, in the form $\{b_0, \dots, b_{\rho-1}; c_1, \dots, c_\rho\}$.
2. The density. For a code C in $H(n, q)$, the density is $|C|/q^n$. This column is absent for $\rho = 1$ because the density $c_1/(b_0 + c_1)$ is obvious from the intersection array $\{b_0; c_1\}$.
3. The eigenspectrum in the form $[i_1, \dots, i_\rho]$, where the eigenvalues of the quotient matrix are θ_0, θ_{i_1}, $\dots$, θ_{i_ρ}.
4. The lower bound L and the upper bound U for the smallest n such that equitable partitions with given intersection array exist in $H(n, q)$. The bounds are given in the form "$L - U$" or only "L" if $U = L$. The following notations are used to indicate the origin of the lower bound:

- L^{c}— (for $q = 2$) Lemma 7.6(ii) (as a special case, see Proposition 7.1); in these cases, the same bound is obtained from nonnegativeness of the interweight numbers;
- L^{b}— (for $\rho = 1$, $q = 2$) see Lemma 7.3(ii); it can also be applied for $\rho > 1$ if merging some cells of the partition results in a $\{b; c\}$-CR code;
- L^{w}— (for $\rho > 1, q = 2$) for $n < L$, some interweight numbers are negative (see Section 7.1.6), but other bounds do not explain the nonexistence;
- L^{s}— if for a CR code C, the code $C \cup C^{(\rho)}$ is also CR, then every lower bound on the length for $C \cup C^{(\rho)}$ is applicable to C as well;
- L^{o}— from the orthogonal-array property (Section 7.1.3); additionally, the nonexistence of a $\{14, 4\}$-CR code in $H(7, 3)$ follows from the nonexistence of the corresponding orthogonal array $\mathrm{OA}(2 \cdot 3^5, 7, 3, 5)$ [22];
- L^{a}— computational results, see [30];

5. The last column contains some additional information, often about the construction of a code with the corresponding parameters. We will use the following notations:

- rep. — repetition $[n, 1, n]$ code;
- Hadamard — Hadamard $(12, 24, 6)$ or $(11, 12, 6)$ code;
- H_n — 1-perfect Hamming $[n = \frac{q^m-1}{q-m}, n - m, 3]$ code of length n;
- $\overline{H}_n$ — extended 1-perfect Hamming code of length n;
- G_{23} — 3-perfect binary Golay $[23, 12, 7]$ code;
- $\overline{G}_{24}$ — extended binary Golay $[24, 12, 8]$ code;
- G_{11} — 2-perfect ternary Golay $[11, 6, 5]_3$ code;
- $\overline{G}_{12}$ — extended ternary Golay $[12, 6, 6]$ code;
- in $/C$, $/G_{11}$, $/G^*_{10}$, $\mathrm{NVO}^+_4(3)$, $\overline{\mathrm{NVO}^-_4(3)}$ — found in Cayley graphs covered by $H(n, 3)$ (see Proposition 7.2): the coset graph of G_{11} (including a $\{22; 5\}$-CR Delsarte coclique [9]), of G^*_{10} ($\{0, 1, 2\}$-shortened G_{11}), or a special linear code in $H(n, 3)$; or in graphs $\mathrm{NVO}^+_4(3)$ ($\mathrm{SRG}(81, 24, 9, 6)$ [10]), $\overline{\mathrm{NVO}^-_4(3)}$ ($\mathrm{SRG}(81, 50, 31, 30)$, the complement of $\mathrm{NVO}^-_4(3)$ [10, §10.29]); these values are obtained in [20] and update [3, Table A1];
- P_n — Preparata $(n = 2^m - 1, 2^{n-2m+1}, 5)$ code;
- $\overline{P}_n$ — Preparata $(n = 2^m, 2^{n-2m}, 6)$ code;
- tC, $t = 2, 3, \ldots$ — the union of t disjoint translations of the code C (P_n, $\overline{P}_n$, see Proposition 7.4, or a linear code) at distance ρ from each other;

- L-sh. C — (where L is a list of words) the code obtained from C by L-shortening: choosing only the codewords with suffix from L and by removing this suffix;
- ext., extended — extended code, obtained from a binary code with specified parameters by appending an all-parity-check bit;
- $\times k$ — construction that multiplies the array by a constant, Section 7.2.3;
- ~~$\times k$~~ — this mark denotes that all elements of the intersection array are divisible by k but if we divide them by k then we get an unfeasible array;
- $[b; c]^k$ — Cartesian product construction, Section 7.2.4;

  split — $\rho = 1$; for the smallest known length, the CR code is constructed by Fon-Der-Flaass construction [17] ($q = 2$, see Proposition 7.5) or its non-binary generalization [3];
- $\cup$: ... — this indicates that the array satisfies $b_i = c_{\rho-i}$, $i = 0, \dots, \lfloor \rho/2 \rfloor - 1, \lfloor \rho/2 \rfloor + 1, \dots, \rho - 1$; in this case, $C \cup C^{(\rho)}$ is a $\{b_0, \dots, b_{\lfloor \rho/2 \rfloor - 1}; c_1, \dots, c_{\lfloor \rho/2 \rfloor}\}$-CR code (odd ρ) or a $\{b_0, \dots, b_{\rho/2-1}; c_1, \dots, c_{\rho/2-1}, c_{\rho/2} + b_{\rho/2}\}$-CR code (even ρ); known constructions are Theorem 7.2 (for $\rho = 2$) and Proposition 7.7 (for $\rho = 3$);
- SU1, SU2, RT2, RT3, TF1, FE1, MDS — linear codes, $\rho = 2$, see [12]; MDS denotes distance-3 Q-ary MDS (maximum distance-separable) $[n, n-2, 3]_Q$ code, SU2 ($l = 1$) in [12];
- $\frac{1}{2}H(n, 2)$ — halved hypercube graph, see also [35] for the study of the corresponding CR codes;
- $B_q(m, n)$ — bilinear forms graph (on $m \times n$ matrices over GF(q)), see also [36] for the study of the corresponding CR codes.

The colors of rows denote the following:

white : a CR code exists in $H(n, q)$ for all $n \geq L = U$;

grey : a CR code exists in $H(n, q)$ for all $n \geq U$, the existence for $L \leq n < U$ remains open;

dark grey : the existence of a CR code for any $n \geq L$ is open (there are no such rows in the tables with $\rho = 1$, because for prime q, $\{b; c\}$-CR codes exist in $H(n, q)$ for some q if and only if $\frac{b+c}{\gcd(b,c)}$ is a power of q, see [13,17] for $q = 2$ and [3, Theorem 5.4] in general);

dark & white : a CR code does not exist in $H(n, q)$ for any n (by Lemma 7.8(ii) or Theorem 7.1).

Array	Spectr.	l.b.–u.b.	Comm.
{1; 1}	[1]	1	
{3; 1}	[2]	3	Split
{2; 2}	[2]	2	×2
{3; 3}	[3]	3	×3
{7; 1}	[4]	7	Split
{6; 2}	[4]	6	×2
{5; 3}	[4]	6[c]	Split
{4; 4}	[4]	4	×4
{5; 5}	[5]	5	×5
{9; 3}	[6]	9	×3
{6; 6}	[6]	6	×6
{7; 7}	[7]	7	×7
{15; 1}	[8]	15	Split
{14; 2}	[8]	14	×2
{13; 3}	[8]	13	Split
{12; 4}	[8]	12	×4
{11; 5}	[8]	13[b]	Split
{10; 6}	[8]	12[c]	×2
{9; 7}	[8]	12[c]	[18]
{8; 8}	[8]	8	×8
{9; 9}	[9]	9	×9
{15; 5}	[10]	15	×5
{10; 10}	[10]	10	×10
{11; 11}	[11]	11	×11
{21; 3}	[12]	21	×3
{18; 6}	[12]	18	×6
{15; 9}	[12]	18[c]	×3
{12; 12}	[12]	12	×12
{13; 13}	[13]	13	×13
{21; 7}	[14]	21	×7
{14; 14}	[14]	14	×14
{15; 15}	[15]	15	×15

Array	Spectr.	l.b.–u.b.	Comm.
{31; 1}	[16]	31	Split
{30; 2}	[16]	30	×2
{29; 3}	[16]	29	Split
{28; 4}	[16]	28	×4
{27; 5}	[16]	27 – 28	Split
{26; 6}	[16]	26	×2
{25; 7}	[16]	25 – 27	Split
{24; 8}	[16]	24	×8
{23; 9}	[16]	24[c]– 27	Split
{22; 10}	[16]	25[b]– 26	×2
{21; 11}	[16]	24[c]– 28	Split
{20; 12}	[16]	24[c]	×4
{19; 13}	[16]	25[b]– 28	Split
{18; 14}	[16]	24[c]	×2
{17; 15}	[16]	24[c]– 30	Split
{16; 16}	[16]	16	×16
{17; 17}	[17]	17	×17
{27; 9}	[18]	27	×9
{18; 18}	[18]	18	×18
{19; 19}	[19]	19	×19
{35; 5}	[20]	35	×5
{30; 10}	[20]	30	×10
{25; 15}	[20]	30[c]	×5
{20; 20}	[20]	20	×20
{21; 21}	[21]	21	×21
{33; 11}	[22]	33	×11
{22; 22}	[22]	22	×22
{23; 23}	[23]	23	×23
{45; 3}	[24]	45	×3
{42; 6}	[24]	42	×6
{39; 9}	[24]	39	×3
{36; 12}	[24]	36	×12

Array	Spectr.	l.b.–u.b.	Comm.
{33; 15}	[24]	36[c]– 39	×3
{30; 18}	[24]	36[c]	×6
{27; 21}	[24]	36[c]	×3
{24; 24}	[24]	24	×24
{25; 25}	[25]	25	×25
{39; 13}	[26]	39	×13
{26; 26}	[26]	26	×26
{27; 27}	[27]	27	×27
{49; 7}	[28]	49	×7
{42; 14}	[28]	42	×14
{35; 21}	[28]	42[c]	×7
{28; 28}	[28]	28	×28
{29; 29}	[29]	29	×29
{45; 15}	[30]	45	×15
{30; 30}	[30]	30	×30
{31; 31}	[31]	31	×31
{63; 1}	[32]	63	split
{62; 2}	[32]	62	×2
{61; 3}	[32]	61	split
{60; 4}	[32]	60	×4
{59; 5}	[32]	59	split
{58; 6}	[32]	58	×2
{57; 7}	[32]	57 – 58	split
{56; 8}	[32]	56	×8

Array	Spectr.	l.b.–u.b.	Comm.
{55; 9}	[32]	55 – 58	split
{54; 10}	[32]	54 – 56	×2
{53; 11}	[32]	53 – 59	split
{52; 12}	[32]	52	×4
{51; 13}	[32]	51 – 59	split
{50; 14}	[32]	50 – 54	×2
{49; 15}	[32]	49 – 61	split
{48; 16}	[32]	48	×16
{47; 17}	[32]	49[b]– 61	split
{46; 18}	[32]	48[c]– 54	×2
{45; 19}	[32]	48[c]– 59	split
{44; 20}	[32]	49[b]– 52	×4
{43; 21}	[32]	48[c]– 59	split
{42; 22}	[32]	48[c]– 56	×2
{41; 23}	[32]	49[b]– 58	split
{40; 24}	[32]	48[c]	×8
{39; 25}	[32]	48[c]– 58	split
{38; 26}	[32]	49[b]– 56	×2
{37; 27}	[32]	48[c]– 59	split
{36; 28}	[32]	48[c]	×4
{35; 29}	[32]	49[b]– 60	split
{34; 30}	[32]	48[c]– 60	×2
{33; 31}	[32]	48[c]– 62	split
{32; 32}	[32]	32	×32

TABLE 7.1 $q = 2$, $\rho = 2$

Array	Density	Spectrum	l.b.–u.b.	References
{2, 1; 1, 2}	$1 \cdot 2^{-2}$	[1, 2]	2	$\{1; 1\}^2$
{4, 2; 2, 4}	$1 \cdot 2^{-2}$	[2, 4]	4	×2
{4, 3; 1, 4}	$1 \cdot 2^{-3}$	[2, 4]	4	rep.
{5, 4; 1, 2}	$1 \cdot 2^{-4}$	[2, 4]	5	rep.

(Continued)

TABLE 7.1 (*Continued*) $q = 2$, $\rho = 2$

Array	Density	Spectrum	l.b.–u.b.	References
$\{6, 1; 1, 6\}$	$1 \cdot 2^{-3}$	$[3, 4]$	6	0-sh. H_7
$\{6, 3; 1, 2\}$	$1 \cdot 2^{-4}$	$[2, 4]$	6	$\{3; 1\}^2$
$\{6, 3; 3, 6\}$	$1 \cdot 2^{-2}$	$[3, 6]$	6	$\times 3$
$\{8, 4; 4, 8\}$	$1 \cdot 2^{-2}$	$[4, 8]$	8	$\times 4$
$\{8, 5; 3, 8\}$	$3 \cdot 2^{-4}$	$[4, 8]$	8	$3\overline{H}_8$
$\{8, 6; 2, 8\}$	$1 \cdot 2^{-3}$	$[4, 8]$	8	$\times 2$
$\{8, 7; 1, 8\}$	$1 \cdot 2^{-4}$	$[4, 8]$	8	$\overline{H}_8$
$\{9, 4; 1, 6\}$	$1 \cdot 2^{-4}$	$[4, 6]$	9	[38]; SU2
$\{10, 3; 1, 6\}$	$1 \cdot 2^{-4}$	$[4, 6]$	10	$\frac{1}{2}H(5, 2)$
$\{10, 3; 3, 10\}$	$3 \cdot 2^{-4}$	$[5, 8]$	12^{c}	$\cup : 2\times\{5; 3\}$
$\{10, 3; 4, 7\}$	$14 \cdot 2^{-6}$	$[4, 8]$	14^{a}– ?	??
$\{10, 5; 3, 6\}$	$9 \cdot 2^{-6}$	$[4, 8]$	12^{c}	$\{5; 3\}^2$
$\{10, 5; 5, 10\}$	$1 \cdot 2^{-2}$	$[5, 10]$	10	$\times 5$
$\{10, 8; 2, 4\}$	$1 \cdot 2^{-4}$	$[4, 8]$	10	$\times 2$
$\{11, 4; 3, 6\}$	$9 \cdot 2^{-6}$	$[4, 8]$	14^{a}– ?	??
$\{12, 2; 2, 12\}$	$1 \cdot 2^{-3}$	$[6, 8]$	12	$\times 2$
$\{12, 3; 1, 12\}$	$1 \cdot 2^{-4}$	$[6, 8]$	12	$\{000\}$-sh. H_{15}
$\{12, 6; 2, 4\}$	$1 \cdot 2^{-4}$	$[4, 8]$	12	$\times 2$
$\{12, 6; 6, 12\}$	$1 \cdot 2^{-2}$	$[6, 12]$	12	$\times 6$
$\{12, 9; 3, 12\}$	$1 \cdot 2^{-3}$	$[6, 12]$	12	$\times 3$
$\{14, 1; 1, 14\}$	$1 \cdot 2^{-4}$	$[7, 8]$	14	$\{0\}$-sh. H_{15}
$\{14, 7; 1, 2\}$	$1 \cdot 2^{-6}$	$[4, 8]$	14	$\{7; 1\}^2$
$\{14, 7; 7, 14\}$	$1 \cdot 2^{-2}$	$[7, 14]$	14	$\times 7$
$\{15, 4; 3, 10\}$	$1 \cdot 2^{-3}$	$[6, 10]$	$15 - ?$	?? nonlinear
$\{15, 12; 3, 6\}$	$1 \cdot 2^{-4}$	$[6, 12]$	15	$\times 3$
$\{16, 8; 8, 16\}$	$1 \cdot 2^{-2}$	$[8, 16]$	16	$\times 8$
$\{16, 9; 7, 16\}$	$7 \cdot 2^{-5}$	$[8, 16]$	16	$7\overline{H}_{16}$
$\{16, 10; 6, 16\}$	$3 \cdot 2^{-4}$	$[8, 16]$	16	$\times 2$
$\{16, 11; 5, 16\}$	$5 \cdot 2^{-5}$	$[8, 16]$	16	$5\overline{H}_{16}$
$\{16, 12; 4, 16\}$	$1 \cdot 2^{-3}$	$[8, 16]$	16	$\times 4$
$\{16, 13; 3, 16\}$	$3 \cdot 2^{-5}$	$[8, 16]$	16	$3\overline{H}_{16}$

(*Continued*)

TABLE 7.1 (*Continued*) $q = 2$, $\rho = 2$

Array	Density	Spectrum	l.b.–u.b.	References
$\{16,14;2,16\}$	$1 \cdot 2^{-4}$	$[8,16]$	16	$\times 2$
$\{16,15;1,16\}$	$1 \cdot 2^{-5}$	$[8,16]$	16	$\overline{H}_{16}$
$\{17,12;5,14\}$	$35 \cdot 2^{-8}$	$[8,16]$	24^{c}– ?	??
$\{17,16;3,12\}$	$9 \cdot 2^{-7}$	$[8,16]$	24^{c}– ?	??
$\{18,3;3,18\}$	$1 \cdot 2^{-3}$	$[9,12]$	18	$\times 3$
$\{18,3;4,15\}$	$9 \cdot 2^{-6}$	$[8,12]$	18 – ?	??
$\{18,5;3,14\}$	$7 \cdot 2^{-6}$	$[8,12]$	18 – ?	??
$\{18,7;7,18\}$	$7 \cdot 2^{-5}$	$[9,16]$	24^{c}– ?	?? $\cup : \{18;14\}$
$\{18,8;2,12\}$	$1 \cdot 2^{-4}$	$[8,12]$	18	$\times 2$
$\{18,9;3,6\}$	$1 \cdot 2^{-4}$	$[6,12]$	18	$\times 3$
$\{18,9;7,14\}$	$49 \cdot 2^{-8}$	$[8,16]$	24^{c}	$\{9;7\}^2$
$\{18,9;9,18\}$	$1 \cdot 2^{-2}$	$[9,18]$	18	$\times 9$
$\{18,15;1,6\}$	$1 \cdot 2^{-6}$	$[8,12]$	18	$TF1_4$
$\{18,15;4,11\}$	$22 \cdot 2^{-8}$	$[8,16]$	24^{c}– ?	??
$\{20,6;2,12\}$	$1 \cdot 2^{-4}$	$[8,12]$	20	$\times 2$
$\{20,6;6,20\}$	$3 \cdot 2^{-4}$	$[10,16]$	24^{c}	$\times 2$
$\{20,6;8,14\}$	$14 \cdot 2^{-6}$	$[8,16]$	24^{c}– ?	?? $\times 2$
$\{20,9;3,20\}$	$3 \cdot 2^{-5}$	$[10,16]$	24^{c}	$\cup : 4\times\{5;3\}$
$\{20,10;6,12\}$	$9 \cdot 2^{-6}$	$[8,16]$	24^{c}	$\times 2$
$\{20,10;10,20\}$	$1 \cdot 2^{-2}$	$[10,20]$	20	$\times 10$
$\{20,15;5,20\}$	$1 \cdot 2^{-3}$	$[10,20]$	20	$\times 5$
$\{20,16;4,8\}$	$1 \cdot 2^{-4}$	$[8,16]$	20	$\times 4$
$\{21,12;1,6\}$	$1 \cdot 2^{-6}$	$[8,12]$	21	$B_2(2,3)$, SU2
$\{21,12;5,10\}$	$25 \cdot 2^{-8}$	$[8,16]$	24^{c}– ?	??
$\{22,5;5,22\}$	$5 \cdot 2^{-5}$	$[11,16]$	25^{b}– ?	?? $\cup : \{22;10\}$
$\{22,8;6,12\}$	$9 \cdot 2^{-6}$	$[8,16]$	24^{c}– ?	?? $\times 2$
$\{22,11;5,10\}$	$25 \cdot 2^{-8}$	$[8,16]$	24^{c}– 26	? $\{11;5\}^2$
$\{22,11;11,22\}$	$1 \cdot 2^{-2}$	$[11,22]$	22	$\times 11$
$\{22,15;4,7\}$	$14 \cdot 2^{-8}$	$[8,16]$	24^{c}– ?	??
$\{23,4;7,14\}$	$46 \cdot 2^{-8}$	$[8,16]$	24^{c}– ?	??
$\{24,4;4,24\}$	$1 \cdot 2^{-3}$	$[12,16]$	24	$\times 4$

(*Continued*)

TABLE 7.1 (*Continued*) $q = 2,\ \rho = 2$

Array	Density	Spectrum	l.b.–u.b.	References
$\{24, 5; 3, 24\}$	$3 \cdot 2^{-5}$	$[12, 16]$	24	$\cup : 8\times\{3; 1\}$, 3 SU1
$\{24, 6; 2, 24\}$	$1 \cdot 2^{-4}$	$[12, 16]$	24	$\times 2$
$\{24, 7; 1, 24\}$	$1 \cdot 2^{-5}$	$[12, 16]$	24	$\cup : 8\times\{3; 1\}$, SU1
$\{24, 12; 4, 8\}$	$1 \cdot 2^{-4}$	$[8, 16]$	24	$\times 4$
$\{24, 12; 12, 24\}$	$1 \cdot 2^{-2}$	$[12, 24]$	24	$\times 12$
$\{24, 15; 9, 24\}$	$3 \cdot 2^{-4}$	$[12, 24]$	24	$\times 3$
$\{24, 18; 6, 24\}$	$1 \cdot 2^{-3}$	$[12, 24]$	24	$\times 6$
$\{24, 21; 3, 24\}$	$1 \cdot 2^{-4}$	$[12, 24]$	24	$\times 3$
$\{25, 12; 1, 18\}$	$3 \cdot 2^{-7}$	$[12, 16]$	25 – ?	??
$\{25, 16; 3, 4\}$	$3 \cdot 2^{-7}$	$[8, 16]$	26^{w}– ?	??
$\{25, 20; 5, 10\}$	$1 \cdot 2^{-4}$	$[10, 20]$	25	$\times 5$
$\{26, 3; 3, 26\}$	$3 \cdot 2^{-5}$	$[13, 16]$	26	$\cup : 2\times\{13; 3\}$
$\{26, 3; 5, 14\}$	$35 \cdot 2^{-8}$	$[8, 16]$	26 – ?	??
$\{26, 7; 4, 11\}$	$22 \cdot 2^{-8}$	$[8, 16]$	26 – ?	??
$\{26, 13; 3, 6\}$	$9 \cdot 2^{-8}$	$[8, 16]$	26	$\{13; 3\}^2$
$\{26, 13; 13, 26\}$	$1 \cdot 2^{-2}$	$[13, 26]$	26	$\times 13$
$\{27, 4; 3, 22\}$	$11 \cdot 2^{-7}$	$[12, 16]$	27 – ?	??
$\{27, 12; 3, 18\}$	$1 \cdot 2^{-4}$	$[12, 18]$	27	$\times 3$
$\{27, 16; 1, 12\}$	$1 \cdot 2^{-6}$	$[12, 16]$	27	RT2, RT3_4
$\{28, 2; 2, 28\}$	$1 \cdot 2^{-4}$	$[14, 16]$	28	$\times 2$
$\{28, 3; 1, 28\}$	$1 \cdot 2^{-5}$	$[14, 16]$	28	$\{000\}$-sh. H_{31}
$\{28, 5; 3, 12\}$	$9 \cdot 2^{-7}$	$[8, 16]$	28 – ?	??
$\{28, 14; 2, 4\}$	$1 \cdot 2^{-6}$	$[8, 16]$	28	$\times 2$
$\{28, 14; 14, 28\}$	$1 \cdot 2^{-2}$	$[14, 28]$	28	$\times 14$
$\{28, 15; 1, 12\}$	$1 \cdot 2^{-6}$	$[12, 16]$	28	$\overline{\text{RT2}}$
$\{28, 21; 7, 28\}$	$1 \cdot 2^{-3}$	$[14, 28]$	28	$\times 7$
$\{30, 1; 1, 30\}$	$1 \cdot 2^{-5}$	$[15, 16]$	30	$\{0\}$-sh. H_{31}
$\{30, 3; 5, 18\}$	$1 \cdot 2^{-3}$	$[10, 18]$	30 – ?	?? nonlinear
$\{30, 5; 5, 30\}$	$1 \cdot 2^{-3}$	$[15, 20]$	30	$\times 5$
$\{30, 7; 1, 18\}$	$3 \cdot 2^{-7}$	$[12, 16]$	30 – ?	??
$\{30, 8; 6, 20\}$	$1 \cdot 2^{-3}$	$[12, 20]$	30 – ?	?? $\times 2$

(*Continued*)

TABLE 7.1 (*Continued*) $q = 2, \rho = 2$

Array	Density	Spectrum	l.b.–u.b.	References
$\{30, 9; 3, 18\}$	$1 \cdot 2^{-4}$	$[12, 18]$	30	$\times 3$
$\{30, 9; 9, 30\}$	$3 \cdot 2^{-4}$	$[15, 24]$	36^c	$\times 3$
$\{30, 9; 12, 21\}$	$14 \cdot 2^{-6}$	$[12, 24]$	$36^c - ?$	?? $\times 3$
$\{30, 11; 5, 18\}$	$3 \cdot 2^{-5}$	$[12, 20]$	$30 - ?$	??
$\{30, 15; 1, 2\}$	$1 \cdot 2^{-8}$	$[8, 16]$	30	$\{15; 1\}^2$
$\{30, 15; 4, 15\}$	$2 \cdot 2^{-5}$	$[12, 20]$	$30 - ?$	?? nonlinear
$\{30, 15; 5, 10\}$	$1 \cdot 2^{-4}$	$[10, 20]$	30	$\times 5$
$\{30, 15; 9, 18\}$	$9 \cdot 2^{-6}$	$[12, 24]$	36^c	$\times 3$
$\{30, 15; 15, 30\}$	$1 \cdot 2^{-2}$	$[15, 30]$	30	$\times 15$
$\{30, 21; 3, 10\}$	$1 \cdot 2^{-5}$	$[12, 20]$	$30 - ?$	?? nonlinear
$\{30, 24; 6, 12\}$	$1 \cdot 2^{-4}$	$[12, 24]$	30	$\times 6$
$\{32, 16; 16, 32\}$	$1 \cdot 2^{-2}$	$[16, 32]$	32	$\times 16$
$\{32, 17; 15, 32\}$	$15 \cdot 2^{-6}$	$[16, 32]$	32	$15\overline{H}_{32}$
$\{32, 18; 14, 32\}$	$7 \cdot 2^{-5}$	$[16, 32]$	32	$\times 2$
$\{32, 19; 13, 32\}$	$13 \cdot 2^{-6}$	$[16, 32]$	32	$13\overline{H}_{32}$
$\{32, 20; 12, 32\}$	$3 \cdot 2^{-4}$	$[16, 32]$	32	$\times 4$
$\{32, 21; 11, 32\}$	$11 \cdot 2^{-6}$	$[16, 32]$	32	$11\overline{H}_{32}$
$\{32, 22; 10, 32\}$	$5 \cdot 2^{-5}$	$[16, 32]$	32	$\times 2$
$\{32, 23; 9, 32\}$	$9 \cdot 2^{-6}$	$[16, 32]$	32	$9\overline{H}_{32}$
$\{32, 24; 8, 32\}$	$1 \cdot 2^{-3}$	$[16, 32]$	32	$\times 8$
$\{32, 25; 7, 32\}$	$7 \cdot 2^{-6}$	$[16, 32]$	32	$7\overline{H}_{32}$
$\{32, 26; 6, 32\}$	$3 \cdot 2^{-5}$	$[16, 32]$	32	$\times 2$
$\{32, 27; 5, 32\}$	$5 \cdot 2^{-6}$	$[16, 32]$	32	$5\overline{H}_{32}$
$\{32, 28; 4, 32\}$	$1 \cdot 2^{-4}$	$[16, 32]$	32	$\times 4$
$\{32, 29; 3, 32\}$	$3 \cdot 2^{-6}$	$[16, 32]$	32	$3\overline{H}_{32}$
$\{32, 30; 2, 32\}$	$1 \cdot 2^{-5}$	$[16, 32]$	32	$\times 2$
$\{32, 31; 1, 32\}$	$1 \cdot 2^{-6}$	$[16, 32]$	32	$\overline{H}_{32}$
$\{33, 12; 5, 30\}$	$25 \cdot 2^{-8}$	$[16, 24]$	$36^c - ?$	??
$\{33, 12; 9, 18\}$	$9 \cdot 2^{-6}$	$[12, 24]$	$36^c - ?$	?? $\times 3$
$\{33, 16; 3, 28\}$	$7 \cdot 2^{-7}$	$[16, 24]$	$36^c - ?$	??
$\{33, 28; 1, 18\}$	$3 \cdot 2^{-8}$	$[16, 24]$	$36^c - ?$	??

(*Continued*)

TABLE 7.1 (*Continued*) $q = 2,\ \rho = 2$

Array	Density	Spectrum	l.b.–u.b.	References
$\{34, 5; 3, 30\}$	$9 \cdot 2^{-7}$	$[16, 20]$	$34 - ?$	??
$\{34, 9; 7, 30\}$	$35 \cdot 2^{-8}$	$[16, 24]$	$36^{c}-\ ?$	??
$\{34, 11; 20, 31\}$	$187 \cdot 2^{-10}$	$[16, 32]$	$48^{c}-\ ?$	??
$\{34, 15; 4, 27\}$	$18 \cdot 2^{-8}$	$[16, 24]$	$36^{c}-\ ?$	??
$\{34, 15; 15, 34\}$	$15 \cdot 2^{-6}$	$[17, 32]$	$48^{c}-\ ?$	?? $\cup : \{34; 30\}$
$\{34, 17; 15, 30\}$	$225 \cdot 2^{-10}$	$[16, 32]$	$48^{c}-\ 60$	? $\{17; 15\}^2$
$\{34, 17; 17, 34\}$	$1 \cdot 2^{-2}$	$[17, 34]$	34	$\times 17$
$\{34, 21; 12, 29\}$	$174 \cdot 2^{-10}$	$[16, 32]$	$48^{c}-\ ?$	??
$\{34, 24; 10, 28\}$	$35 \cdot 2^{-8}$	$[16, 32]$	$48^{c}-\ ?$	?? $\times 2$
$\{34, 32; 6, 24\}$	$9 \cdot 2^{-7}$	$[16, 32]$	$48^{c}-\ ?$	?? $\times 2$
$\{35, 4; 3, 30\}$	$9 \cdot 2^{-7}$	$[16, 20]$	$35 - ?$	??
$\{35, 16; 1, 20\}$	$1 \cdot 2^{-6}$	$[16, 20]$	35	RT2, alt. forms graph
$\{35, 28; 7, 14\}$	$1 \cdot 2^{-4}$	$[14, 28]$	35	$\times 7$
$\{36, 5; 3, 20\}$	$1 \cdot 2^{-4}$	$[12, 20]$	$36 - ?$	?? nonlinear
$\{36, 6; 6, 36\}$	$1 \cdot 2^{-3}$	$[18, 24]$	36	$\times 6$
$\{36, 6; 8, 30\}$	$9 \cdot 2^{-6}$	$[16, 24]$	$36 - ?$	?? $\times 2$
$\{36, 9; 3, 36\}$	$1 \cdot 2^{-4}$	$[18, 24]$	36	$\times 3$
$\{36, 10; 6, 28\}$	$7 \cdot 2^{-6}$	$[16, 24]$	$36 - ?$	?? $\times 2$
$\{36, 14; 14, 36\}$	$7 \cdot 2^{-5}$	$[18, 32]$	$48^{c}-\ ?$	?? $\times 2$
$\{36, 15; 1, 20\}$	$1 \cdot 2^{-6}$	$[16, 20]$	36	$\overline{\text{RT2}}$, $\overline{\text{RT3}_4}$
$\{36, 15; 16, 29\}$	$116 \cdot 2^{-9}$	$[16, 32]$	$48^{c}-\ ?$	??
$\{36, 16; 4, 24\}$	$1 \cdot 2^{-4}$	$[16, 24]$	36	$\times 4$
$\{36, 18; 6, 12\}$	$1 \cdot 2^{-4}$	$[12, 24]$	36	$\times 6$
$\{36, 18; 14, 28\}$	$49 \cdot 2^{-8}$	$[16, 32]$	48^{c}	$\times 2$
$\{36, 18; 18, 36\}$	$1 \cdot 2^{-2}$	$[18, 36]$	36	$\times 18$
$\{36, 21; 3, 20\}$	$5 \cdot 2^{-7}$	$[16, 24]$	$36 - ?$	??
$\{36, 21; 7, 36\}$	$7 \cdot 2^{-6}$	$[18, 32]$	$48^{c}-\ ?$	?? $\cup : \{36; 28\}$
$\{36, 27; 9, 36\}$	$1 \cdot 2^{-3}$	$[18, 36]$	36	$\times 9$
$\{36, 30; 2, 12\}$	$1 \cdot 2^{-6}$	$[16, 24]$	36	$\times 2$
$\{36, 30; 8, 22\}$	$22 \cdot 2^{-8}$	$[16, 32]$	$48^{c}-\ ?$	??
$\{36, 33; 7, 20\}$	$35 \cdot 2^{-9}$	$[16, 32]$	$48^{c}-\ ?$	??

(*Continued*)

TABLE 7.1 (*Continued*) $q = 2$, $\rho = 2$

Array	Density	Spectrum	l.b.–u.b.	References
$\{37,16;15,28\}$	$105 \cdot 2^{-9}$	$[16,32]$	$48^{c}-?$	??
$\{37,28;9,22\}$	$99 \cdot 2^{-10}$	$[16,32]$	$48^{c}-?$	??
$\{38,13;13,38\}$	$13 \cdot 2^{-6}$	$[19,32]$	$49^{b}-?$	?? $\cup : 2\times\{19;13\}$
$\{38,19;13,26\}$	$169 \cdot 2^{-10}$	$[16,32]$	$48^{c}-?$	??
$\{38,19;19,38\}$	$1 \cdot 2^{-2}$	$[19,38]$	38	$\times 19$
$\{38,21;3,18\}$	$9 \cdot 2^{-8}$	$[16,24]$	$38-?$	??
$\{38,21;12,25\}$	$150 \cdot 2^{-10}$	$[16,32]$	$48^{c}-?$	??
$\{39,4;7,30\}$	$26 \cdot 2^{-8}$	$[16,24]$	$39-?$	??
$\{39,20;3,18\}$	$9 \cdot 2^{-8}$	$[16,24]$	$39-?$	??
$\{39,36;7,14\}$	$49 \cdot 2^{-10}$	$[16,32]$	$48^{c}-?$	??
$\{40,12;4,24\}$	$1 \cdot 2^{-4}$	$[16,24]$	40	$\times 4$
$\{40,12;12,40\}$	$3 \cdot 2^{-4}$	$[20,32]$	48^{c}	$\times 4$
$\{40,12;16,28\}$	$14 \cdot 2^{-6}$	$[16,32]$	$48^{c}-?$	??
$\{40,15;9,40\}$	$9 \cdot 2^{-6}$	$[20,32]$	48^{c}	$\cup : 8\times\{5;3\}$
$\{40,18;6,40\}$	$3 \cdot 2^{-5}$	$[20,32]$	48^{c}	$\times 2$
$\{40,20;12,24\}$	$9 \cdot 2^{-6}$	$[16,32]$	48^{c}	$\times 2$
$\{40,20;20,40\}$	$1 \cdot 2^{-2}$	$[20,40]$	40	$\times 20$
$\{40,21;3,40\}$	$3 \cdot 2^{-6}$	$[20,32]$	48^{c}	$\cup : 8\times\{5;3\}$
$\{40,25;15,40\}$	$3 \cdot 2^{-4}$	$[20,40]$	40	$\times 5$
$\{40,30;10,40\}$	$1 \cdot 2^{-3}$	$[20,40]$	40	$\times 10$
$\{40,32;8,16\}$	$1 \cdot 2^{-4}$	$[16,32]$	40	$\times 8$
$\{40,35;5,40\}$	$1 \cdot 2^{-4}$	$[20,40]$	40	$\times 5$
$\{41,28;9,18\}$	$81 \cdot 2^{-10}$	$[16,32]$	$48^{c}-?$	??
$\{42,3;3,42\}$	$1 \cdot 2^{-4}$	$[21,24]$	42	$\times 3$
$\{42,3;5,30\}$	$21 \cdot 2^{-8}$	$[16,24]$	$42-?$	??
$\{42,3;20,31\}$	$63 \cdot 2^{-10}$	$[16,32]$	$48^{c}-?$	??
$\{42,7;4,27\}$	$18 \cdot 2^{-8}$	$[16,24]$	$42-?$	??
$\{42,7;7,42\}$	$1 \cdot 2^{-3}$	$[21,28]$	42	$\times 7$
$\{42,11;11,42\}$	$11 \cdot 2^{-6}$	$[21,32]$	$48^{c}-?$	?? $\cup : 2\times\{21;11\}$
$\{42,13;3,22\}$	$11 \cdot 2^{-8}$	$[16,24]$	$42-?$	??
$\{42,15;1,30\}$	$1 \cdot 2^{-6}$	$[20,24]$	42	$\mathrm{MDS}[6,4,3]_4$

(*Continued*)

TABLE 7.1 (*Continued*) $q = 2,\ \rho = 2$

Array	Density	Spectrum	l.b.–u.b.	References
$\{42,15;4,35\}$	$2\cdot 2^{-5}$	$[20,28]$	$42-?$	??
$\{42,21;3,6\}$	$1\cdot 2^{-6}$	$[12,24]$	42	$\times 3$
$\{42,21;7,14\}$	$1\cdot 2^{-4}$	$[14,28]$	42	$\times 7$
$\{42,21;11,22\}$	$121\cdot 2^{-10}$	$[16,32]$	$48^{c}-56$	? $\{21;11\}^2$
$\{42,21;21,42\}$	$1\cdot 2^{-2}$	$[21,42]$	42	$\times 21$
$\{42,24;2,12\}$	$1\cdot 2^{-6}$	$[16,24]$	42	$\times 2$
$\{42,24;10,20\}$	$25\cdot 2^{-8}$	$[16,32]$	$48^{c}-?$	?? $\times 2$
$\{43,20;11,22\}$	$121\cdot 2^{-10}$	$[16,32]$	$48^{c}-?$	??
$\{43,36;7,10\}$	$35\cdot 2^{-10}$	$[16,32]$	$48^{c}-?$	??
$\{44,7;16,29\}$	$77\cdot 2^{-9}$	$[16,32]$	$48^{c}-?$	??
$\{44,9;15,28\}$	$99\cdot 2^{-9}$	$[16,32]$	$48^{c}-?$	??
$\{44,10;10,44\}$	$5\cdot 2^{-5}$	$[22,32]$	$49^{b}-?$	?? $\cup: 4\times\{11;5\}$
$\{44,15;5,44\}$	$5\cdot 2^{-6}$	$[22,32]$	$49^{b}-?$	?? $\cup: 4\times\{11;5\}$
$\{44,16;12,24\}$	$9\cdot 2^{-6}$	$[16,32]$	$48^{c}-?$	?? $\times 4$
$\{44,22;10,20\}$	$25\cdot 2^{-8}$	$[16,32]$	$48^{c}-52$	? $\times 2$
$\{44,22;22,44\}$	$1\cdot 2^{-2}$	$[22,44]$	44	$\times 22$
$\{44,30;8,14\}$	$14\cdot 2^{-8}$	$[16,32]$	$48^{c}-?$	?? $\times 2$
$\{44,33;11,44\}$	$1\cdot 2^{-3}$	$[22,44]$	44	$\times 11$
$\{45,12;1,30\}$	$1\cdot 2^{-6}$	$[20,24]$	45	$\overline{\mathrm{TF1}_4}$
$\{45,12;9,30\}$	$1\cdot 2^{-3}$	$[18,30]$	$45-?$	?? $\times 3$
$\{45,12;13,26\}$	$169\cdot 2^{-10}$	$[16,32]$	$48^{c}-?$	??
$\{45,16;15,36\}$	$3\cdot 2^{-4}$	$[20,36]$	$54^{c}-?$	??
$\{45,20;5,30\}$	$1\cdot 2^{-4}$	$[20,30]$	45	$\times 5$
$\{45,28;1,6\}$	$1\cdot 2^{-8}$	$[16,24]$	45	$B_q(2,4)$; MDS$[3,1,3]_{16}$
$\{45,28;9,30\}$	$3\cdot 2^{-5}$	$[20,36]$	$54^{c}-?$	??
$\{45,36;9,18\}$	$1\cdot 2^{-4}$	$[18,36]$	45	$\times 9$
$\{46,8;14,28\}$	$46\cdot 2^{-8}$	$[16,32]$	$48^{c}-?$	?? $\times 2$
$\{46,9;9,46\}$	$9\cdot 2^{-6}$	$[23,32]$	$48^{c}-?$	?? $\cup: \{23;9\}$
$\{46,13;12,25\}$	$150\cdot 2^{-10}$	$[16,32]$	$48^{c}-?$	??
$\{46,15;1,18\}$	$3\cdot 2^{-8}$	$[16,24]$	$46-?$	??
$\{46,23;9,18\}$	$81\cdot 2^{-10}$	$[16,32]$	$48^{c}-?$	??

(*Continued*)

TABLE 7.1 (*Continued*) $q = 2,\ \rho = 2$

Array	Density	Spectrum	l.b.–u.b.	References
$\{46, 23; 23, 46\}$	$1 \cdot 2^{-2}$	$[23, 46]$	46	$\times 23$
$\{46, 33; 7, 10\}$	$35 \cdot 2^{-10}$	$[16, 32]$	48^{c}– ?	??
$\{46, 45; 3, 10\}$	$3 \cdot 2^{-8}$	$[20, 32]$	$49^{\underline{w}}$– ?	??
$\{47, 4; 15, 30\}$	$94 \cdot 2^{-10}$	$[16, 32]$	48^{c}– ?	??
$\{48, 8; 8, 48\}$	$1 \cdot 2^{-3}$	$[24, 32]$	48	$\times 8$
$\{48, 9; 7, 48\}$	$7 \cdot 2^{-6}$	$[24, 32]$	48	$\cup : 16 \times \{3; 1\}$; 7 SU1
$\{48, 10; 6, 48\}$	$3 \cdot 2^{-5}$	$[24, 32]$	48	$\times 2$
$\{48, 11; 5, 48\}$	$5 \cdot 2^{-6}$	$[24, 32]$	48	$\cup : 16 \times \{3; 1\}$; 5 SU1
$\{48, 12; 4, 48\}$	$1 \cdot 2^{-4}$	$[24, 32]$	48	$\times 4$
$\{48, 13; 3, 48\}$	$3 \cdot 2^{-6}$	$[24, 32]$	48	$\cup : 16 \times \{3; 1\}$; 3 SU1
$\{48, 14; 2, 48\}$	$1 \cdot 2^{-5}$	$[24, 32]$	48	$\times 2$
$\{48, 15; 1, 48\}$	$1 \cdot 2^{-6}$	$[24, 32]$	48	SU1
$\{48, 24; 8, 16\}$	$1 \cdot 2^{-4}$	$[16, 32]$	48	$\times 8$
$\{48, 24; 24, 48\}$	$1 \cdot 2^{-2}$	$[24, 48]$	48	$\times 24$
$\{48, 27; 21, 48\}$	$7 \cdot 2^{-5}$	$[24, 48]$	48	$\times 3$
$\{48, 30; 18, 48\}$	$3 \cdot 2^{-4}$	$[24, 48]$	48	$\times 6$
$\{48, 33; 15, 48\}$	$5 \cdot 2^{-5}$	$[24, 48]$	48	$\times 3$
$\{48, 36; 12, 48\}$	$1 \cdot 2^{-3}$	$[24, 48]$	48	$\times 12$
$\{48, 39; 9, 48\}$	$3 \cdot 2^{-5}$	$[24, 48]$	48	$\times 3$
$\{48, 42; 6, 48\}$	$1 \cdot 2^{-4}$	$[24, 48]$	48	$\times 6$
$\{48, 45; 3, 48\}$	$1 \cdot 2^{-5}$	$[24, 48]$	48	$\times 3$

TABLE 7.2 $q = 2,\ \rho = 3$

Array	Density	Spectrum	l.b.–u.b.	Reference
$\{3, 2, 1; 1, 2, 3\}$	$1 \cdot 2^{-3}$	$[1, 2, 3]$	3	$\{1; 1\}^3$
$\{6, 5, 4; 1, 2, 6\}$	$1 \cdot 2^{-5}$	$[2, 4, 6]$	6	rep.
$\{7, 6, 1; 1, 6, 7\}$	$1 \cdot 2^{-4}$	$[3, 4, 7]$	7	$\{0\}$-sh. $\overline{H}_8$
$\{7, 6, 5; 1, 2, 3\}$	$1 \cdot 2^{-6}$	$[2, 4, 6]$	7	rep.
$\{9, 6, 3; 1, 2, 3\}$	$1 \cdot 2^{-6}$	$[2, 4, 6]$	9	$\{3; 1\}^3$
$\{9, 6, 3; 3, 6, 9\}$	$1 \cdot 2^{-3}$	$[3, 6, 9]$	9	$\times 3$
$\{10, 9, 4; 1, 6, 10\}$	$1 \cdot 2^{-5}$	$[4, 6, 10]$	10	linear [35]

(*Continued*)

TABLE 7.2 (*Continued*) $q = 2,\ \rho = 3$

Array	Density	Spectrum	l.b.–u.b.	Reference
$\{11, 10, 3; 1, 2, 9\}$	$3 \cdot 2^{-8}$	$[4, 6, 8]$	11	Hadamard
$\{12, 8, 4; 4, 8, 12\}$	$1 \cdot 2^{-3}$	$[4, 8, 12]$	12	$\times 4$
$\{12, 10, 8; 2, 4, 12\}$	$1 \cdot 2^{-5}$	$[4, 8, 12]$	12	$\times 2$
$\{13, 6, 1; 1, 6, 9\}$	$9 \cdot 2^{-8}$	$[4, 6, 8]$	13	new [30]
$\{13, 10, 3; 3, 10, 13\}$	$3 \cdot 2^{-5}$	$[5, 8, 13]$	13	bipartite $\{13; 3\}$
$\{14, 12, 2; 2, 12, 14\}$	$1 \cdot 2^{-4}$	$[6, 8, 14]$	14	$\times 2$
$\{14, 12, 10; 2, 4, 6\}$	$1 \cdot 2^{-6}$	$[4, 8, 12]$	14	$\times 2$
$\{15, 6, 1; 1, 6, 15\}$	$1 \cdot 2^{-5}$	$[5, 8, 9]$	15	$\frac{1}{2}H(6, 2)$, [5]
$\{15, 8, 1; 1, 8, 15\}$	$1 \cdot 2^{-5}$	$[6, 8, 10]$	15	$4P_{15}$; linear [35], [5]
$\{15, 10, 1; 1, 6, 15\}$	$3 \cdot 2^{-7}$	$[6, 8, 10]$	15	$3P_{15}$
$\{15, 10, 5; 3, 6, 9\}$	$27 \cdot 2^{-9}$	$[4, 8, 12]$	18^{c}	$\{5; 3\}^3$
$\{15, 10, 5; 5, 10, 15\}$	$1 \cdot 2^{-3}$	$[5, 10, 15]$	15	$\times 5$
$\{15, 12, 1; 1, 4, 15\}$	$1 \cdot 2^{-6}$	$[6, 8, 10]$	15	$2P_{15}$
$\{15, 14, 1; 1, 2, 15\}$	$1 \cdot 2^{-7}$	$[6, 8, 10]$	15	P_{15}
$\{15, 14, 1; 1, 14, 15\}$	$1 \cdot 2^{-5}$	$[7, 8, 15]$	15	$\{0\}$-sh. $\overline{H}_{16}$
$\{18, 12, 6; 2, 4, 6\}$	$1 \cdot 2^{-6}$	$[4, 8, 12]$	18	$\times 2$
$\{18, 12, 6; 6, 12, 18\}$	$1 \cdot 2^{-3}$	$[6, 12, 18]$	18	$\times 6$
$\{18, 15, 12; 3, 6, 18\}$	$1 \cdot 2^{-5}$	$[6, 12, 18]$	18	$\times 3$
$\{20, 18, 8; 2, 12, 20\}$	$1 \cdot 2^{-5}$	$[8, 12, 20]$	20	$\times 2$
$\{21, 8, 3; 3, 8, 21\}$	$1 \cdot 2^{-4}$	$[6, 12, 14]$	$21 - ?$	?? nonlin.; $\cup$:$\{21, 3\}$
$\{21, 10, 3; 1, 6, 15\}$	$1 \cdot 2^{-6}$	$[6, 10, 12]$	21	$\frac{1}{2}H(7, 2)$
$\{21, 10, 3; 3, 10, 21\}$	$1 \cdot 2^{-4}$	$[7, 12, 15]$	$21 - ?$	?? nonlin.; $\cup$:$\{21, 3\}$
$\{21, 14, 7; 1, 2, 3\}$	$1 \cdot 2^{-9}$	$[4, 8, 12]$	21	$[7, 1]^3$
$\{21, 14, 7; 7, 14, 21\}$	$1 \cdot 2^{-3}$	$[7, 14, 21]$	21	$\times 7$
$\{21, 18, 3; 3, 18, 21\}$	$1 \cdot 2^{-4}$	$[9, 12, 21]$	21	$\times 3$
$\{21, 18, 15; 3, 6, 9\}$	$1 \cdot 2^{-6}$	$[6, 12, 18]$	21	$\times 3$
$\{21, 20, 16; 1, 2, 12\}$	$1 \cdot 2^{-9}$	$[8, 12, 16]$	21	$\{00,01,10,11\}$-sh. G_{23}
$\{22, 8, 3; 2, 16, 21\}$	$11 \cdot 2^{-8}$	$[8, 12, 16]$	$22 - ?$	??
$\{22, 20, 6; 2, 4, 18\}$	$3 \cdot 2^{-8}$	$[8, 12, 16]$	22	$\times 2$
$\{22, 21, 20; 1, 2, 6\}$	$1 \cdot 2^{-10}$	$[8, 12, 16]$	22	$\{0, 1\}$-sh. G_{23}
$\{23, 4, 3; 1, 20, 21\}$	$23 \cdot 2^{-10}$	$[8, 12, 16]$	$27^{\mathrm{w}} - ?$	??

(*Continued*)

TABLE 7.2 (*Continued*) $q = 2,\ \rho = 3$

Array	Density	Spectrum	l.b.–u.b.	Reference
$\{23,16,6;1,8,18\}$	$3 \cdot 2^{-8}$	$[8,12,16]$	$23 - ?$	??
$\{23,22,21;1,2,3\}$	$1 \cdot 2^{-11}$	$[8,12,16]$	23	G_{23}
$\{24,10,2;2,16,18\}$	$5 \cdot 2^{-7}$	$[8,12,16]$	$24 - ?$	?? ~~×2~~
$\{24,11,2;3,12,20\}$	$11 \cdot 2^{-8}$	$[8,12,16]$	$24 - ?$	??
$\{24,16,8;8,16,24\}$	$1 \cdot 2^{-3}$	$[8,16,24]$	24	×8
$\{24,20,7;2,3,16\}$	$1 \cdot 2^{-7}$	$[8,12,16]$	$27^{\text{w}} - ?$	?? nonlin.
$\{24,20,16;4,8,24\}$	$1 \cdot 2^{-5}$	$[8,16,24]$	24	×4
$\{24,21,10;1,4,12\}$	$1 \cdot 2^{-8}$	$[8,12,16]$	$24 - ?$	?? nonlin.
$\{25,18,7;7,18,25\}$	$7 \cdot 2^{-6}$	$[9,16,25]$	$25 - ?$	?? $\cup : \{25;7\}$
$\{25,24,5;3,8,15\}$	$3 \cdot 2^{-7}$	$[8,12,20]$	$30^{\text{c}} - ?$	??
$\{26,12,2;2,12,18\}$	$9 \cdot 2^{-8}$	$[8,12,16]$	$26 - ?$	?? ×2
$\{26,20,6;6,20,26\}$	$3 \cdot 2^{-5}$	$[10,16,26]$	26	×2
$\{27,18,9;3,6,9\}$	$1 \cdot 2^{-6}$	$[6,12,18]$	27	×3
$\{27,18,9;7,14,21\}$	$343 \cdot 2^{-12}$	$[8,16,24]$	36^{c}	$\{9;7\}^3$
$\{27,18,9;9,18,27\}$	$1 \cdot 2^{-3}$	$[9,18,27]$	27	×9
$\{27,22,5;5,22,27\}$	$5 \cdot 2^{-6}$	$[11,16,27]$	$27 - ?$	?? $\cup : \{27;5\}$
$\{28,24,4;4,24,28\}$	$1 \cdot 2^{-4}$	$[12,16,28]$	28	×4
$\{28,24,20;4,8,12\}$	$1 \cdot 2^{-6}$	$[8,16,24]$	28	×4
$\{28,27,16;1,12,28\}$	$1 \cdot 2^{-7}$	$[12,16,28]$	28	ext. $\{27,16;1,12\}$
$\{29,26,3;3,26,29\}$	$3 \cdot 2^{-6}$	$[13,16,29]$	29	$\cup : \{29;3\}$
$\{30,12,2;2,12,30\}$	$1 \cdot 2^{-5}$	$[10,16,18]$	30	×2, $\cup$:$\{30;2\}$
$\{30,15,2;2,9,30\}$	$3 \cdot 2^{-7}$	$[10,16,18]$	$30 - ?$	?? $\cup$:$\{30;2\}$
$\{30,16,2;2,16,30\}$	$1 \cdot 2^{-5}$	$[12,16,20]$	30	×2, $\cup$:$\{30;2\}$
$\{30,17,2;2,15,30\}$	$15 \cdot 2^{-9}$	$[12,16,20]$	$30 - ?$	?? $\cup$:$\{30;2\}$
$\{30,18,2;2,6,30\}$	$1 \cdot 2^{-6}$	$[10,16,18]$	$30 - ?$	?? ~~×2~~, $\cup$:$\{30;2\}$
$\{30,18,2;2,14,30\}$	$7 \cdot 2^{-8}$	$[12,16,20]$	$30 - ?$	?? ~~×2~~, $\cup$:$\{30;2\}$
$\{30,19,2;2,13,30\}$	$13 \cdot 2^{-9}$	$[12,16,20]$	$30 - ?$	?? $\cup$:$\{30;2\}$
$\{30,20,2;2,12,30\}$	$3 \cdot 2^{-7}$	$[12,16,20]$	30	×2, $\cup$:$\{30;2\}$
$\{30,20,10;6,12,18\}$	$27 \cdot 2^{-9}$	$[8,16,24]$	36^{c}	×2
$\{30,20,10;10,20,30\}$	$1 \cdot 2^{-3}$	$[10,20,30]$	30	×10
$\{30,21,2;2,3,30\}$	$1 \cdot 2^{-7}$	$[10,16,18]$	$31^{\text{w}} - ?$	?? nonlin. $\cup$:$\{30;2\}$
$\{30,21,2;2,11,30\}$	$11 \cdot 2^{-9}$	$[12,16,20]$	$30 - ?$	?? $\cup$:$\{30;2\}$

(*Continued*)

TABLE 7.2 (*Continued*) $q = 2,\ \rho = 3$

Array	Density	Spectrum	l.b.–u.b.	Reference
$\{30, 21, 4; 1, 10, 30\}$	$5 \cdot 2^{-9}$	$[12, 16, 20]$	$30 - ?$	??
$\{30, 22, 2; 2, 10, 30\}$	$5 \cdot 2^{-8}$	$[12, 16, 20]$	$30 - ?$	?? ~~×2~~, ∪:$\{30; 2\}$
$\{30, 23, 2; 2, 9, 30\}$	$9 \cdot 2^{-9}$	$[12, 16, 20]$	$30 - ?$	?? ∪:$\{30; 2\}$
$\{30, 24, 2; 2, 8, 30\}$	$1 \cdot 2^{-6}$	$[12, 16, 20]$	30	×2, ∪:$\{30; 2\}$
$\{30, 25, 2; 2, 7, 30\}$	$7 \cdot 2^{-9}$	$[12, 16, 20]$	$30 - ?$	?? ∪:$\{30; 2\}$
$\{30, 25, 20; 5, 10, 30\}$	$1 \cdot 2^{-5}$	$[10, 20, 30]$	30	×5
$\{30, 26, 2; 2, 6, 30\}$	$3 \cdot 2^{-8}$	$[12, 16, 20]$	$31^{\underline{w}}- ?$	?? ~~×2~~, ∪:$\{30; 2\}$
$\{30, 27, 2; 2, 5, 30\}$	$5 \cdot 2^{-9}$	$[12, 16, 20]$	$31^{\underline{w}}- ?$	?? ∪:$\{30; 2\}$
$\{30, 27, 12; 3, 18, 30\}$	$1 \cdot 2^{-5}$	$[12, 18, 30]$	30	×3
$\{30, 28, 2; 2, 4, 30\}$	$1 \cdot 2^{-7}$	$[12, 16, 20]$	30	×2, ∪:$\{30; 2\}$
$\{30, 28, 2; 2, 28, 30\}$	$1 \cdot 2^{-5}$	$[14, 16, 30]$	30	×2
$\{30, 29, 2; 2, 3, 30\}$	$3 \cdot 2^{-9}$	$[12, 16, 20]$	$36^{\underline{w}}- ?$	?? ∪:$\{30; 2\}$
$\{31, 30, 1; 1, 30, 31\}$	$1 \cdot 2^{-6}$	$[15, 16, 31]$	31	$\{0\}$-sh. $\overline{H}_{32}$
$\{31, 30, 17; 1, 2, 15\}$	$1 \cdot 2^{-10}$	$[12, 16, 20]$	31	BCH_{31} [2]
$\{33, 22, 11; 5, 10, 15\}$	$125 \cdot 2^{-12}$	$[8, 16, 24]$	$36^{c}- 39$	$\{11; 5\}^2$ for $n \geq 13$
$\{33, 22, 11; 11, 22, 33\}$	$1 \cdot 2^{-3}$	$[11, 22, 33]$	33	×11
$\{33, 30, 9; 3, 6, 27\}$	$3 \cdot 2^{-8}$	$[12, 18, 24]$	33	×3
$\{33, 30, 15; 1, 2, 15\}$	$1 \cdot 2^{-10}$	$[12, 16, 20]$	33	see [39]
$\{33, 32, 6; 3, 6, 24\}$	$3 \cdot 2^{-8}$	$[12, 16, 24]$	$36^{\underline{w}}- ?$	??
$\{35, 24, 5; 5, 24, 35\}$	$1 \cdot 2^{-4}$	$[14, 20, 30]$	$40^{c}- ?$	??
$\{35, 30, 5; 5, 30, 35\}$	$1 \cdot 2^{-4}$	$[15, 20, 35]$	35	×5
$\{35, 30, 25; 5, 10, 15\}$	$1 \cdot 2^{-6}$	$[10, 20, 30]$	35	×5
$\{36, 24, 12; 4, 8, 12\}$	$1 \cdot 2^{-6}$	$[8, 16, 24]$	36	×4
$\{36, 24, 12; 12, 24, 36\}$	$1 \cdot 2^{-3}$	$[12, 24, 36]$	36	×12
$\{36, 30, 24; 6, 12, 36\}$	$1 \cdot 2^{-5}$	$[12, 24, 36]$	36	×6
$\{36, 35, 16; 1, 20, 36\}$	$1 \cdot 2^{-7}$	$[16, 20, 36]$	36	ext. $\{35, 16; 1, 20\}$
$\{39, 10, 1; 3, 30, 37\}$	$13 \cdot 2^{-11}$	$[16, 20, 24]$	$39 - ?$	??
$\{39, 18, 3; 3, 18, 27\}$	$9 \cdot 2^{-8}$	$[12, 18, 24]$	$39 - ?$	?? ×3
$\{39, 26, 13; 3, 6, 9\}$	$27 \cdot 2^{-12}$	$[8, 16, 24]$	39	$\{13; 3\}^3$
$\{39, 26, 13; 13, 26, 39\}$	$1 \cdot 2^{-3}$	$[13, 26, 39]$	39	×13
$\{39, 30, 9; 9, 30, 39\}$	$3 \cdot 2^{-5}$	$[15, 24, 39]$	39	×3
$\{39, 34, 5; 1, 6, 35\}$	$7 \cdot 2^{-11}$	$[16, 20, 24]$	$39 - ?$	??

(*Continued*)

TABLE 7.2 (*Continued*) $q = 2,\ \rho = 3$

Array	Density	Spectrum	l.b.–u.b.	Reference
$\{40,21,2;1,20,36\}$	$3\cdot2^{-8}$	$[16,20,24]$	$40-?$	??
$\{40,33,10;6,15,24\}$	$9\cdot2^{-8}$	$[12,20,32]$	$48^{c}-?$	??
$\{40,36,16;4,24,40\}$	$1\cdot2^{-5}$	$[16,24,40]$	40	$\times4$
$\{42,16,6;6,16,42\}$	$1\cdot2^{-4}$	$[12,24,28]$	$42-?$	?? $\times2$
$\{42,20,6;2,12,30\}$	$1\cdot2^{-6}$	$[12,20,24]$	42	$\times2$
$\{42,20,6;6,20,42\}$	$1\cdot2^{-4}$	$[14,24,30]$	$42-?$	?? $\times2$
$\{42,22,9;4,9,42\}$	$6\cdot2^{-8}$	$[12,24,28]$	$42-?$	??
$\{42,28,14;2,4,6\}$	$1\cdot2^{-9}$	$[8,16,24]$	42	$\times2$
$\{42,28,14;14,28,42\}$	$1\cdot2^{-3}$	$[14,28,42]$	42	$\times14$
$\{42,35,28;7,14,42\}$	$1\cdot2^{-5}$	$[14,28,42]$	42	$\times7$
$\{42,36,6;6,36,42\}$	$1\cdot2^{-4}$	$[18,24,42]$	42	$\times6$
$\{42,36,30;6,12,18\}$	$1\cdot2^{-6}$	$[12,24,36]$	42	$\times6$
$\{42,40,32;2,4,24\}$	$1\cdot2^{-9}$	$[16,24,32]$	42	$\times2$
$\{43,32,6;5,16,42\}$	$35\cdot2^{-10}$	$[16,24,32]$	$44^{c}-?$	??
$\{43,42,21;3,6,29\}$	$87\cdot2^{-14}$	$[16,24,32]$	$48^{c}-?$	??
$\{44,16,6;4,32,42\}$	$11\cdot2^{-8}$	$[16,24,32]$	$44-?$	?? $\times2$
$\{44,40,12;4,8,36\}$	$3\cdot2^{-8}$	$[16,24,32]$	44	$\times4$
$\{44,42,40;2,4,12\}$	$1\cdot2^{-10}$	$[16,24,32]$	44	$\times2$
$\{45,18,3;3,18,45\}$	$1\cdot2^{-5}$	$[15,24,27]$	45	$\times3$
$\{45,20,7;3,28,41\}$	$287\cdot2^{-13}$	$[16,24,32]$	$45-?$	??
$\{45,24,3;3,24,45\}$	$1\cdot2^{-5}$	$[18,24,30]$	45	$\times3$
$\{45,30,3;3,18,45\}$	$3\cdot2^{-7}$	$[18,24,30]$	45	$\times3$
$\{45,30,15;1,2,3\}$	$1\cdot2^{-12}$	$[8,16,24]$	45	$H(3,16);\ \{15;1\}^3$
$\{45,30,15;5,10,15\}$	$1\cdot2^{-6}$	$[10,20,30]$	45	$\times5$
$\{45,30,15;9,18,27\}$	$27\cdot2^{-9}$	$[12,24,36]$	54^{c}	$\times3$
$\{45,30,15;15,30,45\}$	$1\cdot2^{-3}$	$[15,30,45]$	45	$\times15$
$\{45,32,3;3,32,45\}$	$1\cdot2^{-5}$	$[20,24,36]$	$48^{c}-?$	?? $\cup:\{45;3\}$
$\{45,32,10;3,16,38\}$	$19\cdot2^{-10}$	$[16,24,32]$	$45-?$	??
$\{45,34,3;3,30,45\}$	$15\cdot2^{-9}$	$[20,24,36]$	$54^{c}-?$	?? $\cup:\{45;3\}$
$\{45,34,11;3,14,37\}$	$259\cdot2^{-14}$	$[16,24,32]$	$45-?$	??
$\{45,36,3;3,12,45\}$	$1\cdot2^{-6}$	$[18,24,30]$	45	$\times3$
$\{45,36,3;3,28,45\}$	$7\cdot2^{-8}$	$[20,24,36]$	$54^{c}-?$	?? $\cup:\{45;3\}$

(*Continued*)

TABLE 7.2 (*Continued*) $q = 2,\ \rho = 3$

Array	Density	Spectrum	l.b.–u.b.	Reference
$\{45, 38, 3; 3, 26, 45\}$	$13 \cdot 2^{-9}$	$[20, 24, 36]$	54^{c}– ?	?? $\cup : \{45; 3\}$
$\{45, 40, 3; 3, 24, 45\}$	$3 \cdot 2^{-7}$	$[20, 24, 36]$	54^{c}– ?	?? $\cup : \{45; 3\}$
$\{45, 40, 9; 7, 24, 35\}$	$49 \cdot 2^{-10}$	$[16, 24, 40]$	60^{c}– ?	??
$\{45, 40, 17; 3, 8, 31\}$	$31 \cdot 2^{-12}$	$[16, 24, 32]$	46^{w}– ?	??
$\{45, 42, 3; 3, 6, 45\}$	$1 \cdot 2^{-7}$	$[18, 24, 30]$	45	$\times 3$
$\{45, 42, 3; 3, 22, 45\}$	$11 \cdot 2^{-9}$	$[20, 24, 36]$	54^{c}– ?	?? $\cup : \{45; 3\}$
$\{45, 42, 3; 3, 42, 45\}$	$1 \cdot 2^{-5}$	$[21, 24, 45]$	45	$\times 3$
$\{45, 44, 3; 3, 20, 45\}$	$5 \cdot 2^{-8}$	$[20, 24, 36]$	54^{c}– ?	?? $\cup : \{45; 3\}$
$\{45, 44, 31; 3, 4, 17\}$	$17 \cdot 2^{-13}$	$[16, 24, 32]$	51^{w}– ?	??
$\{46, 8, 6; 2, 40, 42\}$	$23 \cdot 2^{-10}$	$[16, 24, 32]$	46 – ?	?? $\times 2$
$\{46, 18, 5; 4, 33, 38\}$	$345 \cdot 2^{-13}$	$[16, 24, 32]$	48^{c}– ?	??
$\{46, 32, 12; 2, 16, 36\}$	$3 \cdot 2^{-8}$	$[16, 24, 32]$	46 – ?	?? $\times 2$
$\{46, 38, 18; 2, 10, 30\}$	$25 \cdot 2^{-12}$	$[16, 24, 32]$	46 – ?	?? ~~×2~~
$\{46, 42, 17; 4, 5, 30\}$	$50 \cdot 2^{-13}$	$[16, 24, 32]$	49^{w}– ?	??
$\{46, 44, 42; 2, 4, 6\}$	$1 \cdot 2^{-11}$	$[16, 24, 32]$	46	$\times 2$
$\{47, 22, 9; 1, 26, 39\}$	$169 \cdot 2^{-14}$	$[16, 24, 32]$	47 – ?	??
$\{47, 24, 5; 5, 24, 39\}$	$195 \cdot 2^{-12}$	$[16, 24, 32]$	48^{c}– ?	??
$\{47, 40, 27; 1, 8, 21\}$	$7 \cdot 2^{-12}$	$[16, 24, 32]$	47 – ?	??
$\{47, 42, 21; 3, 6, 25\}$	$75 \cdot 2^{-14}$	$[16, 24, 32]$	48^{c}– ?	??
$\{48, 9, 2; 11, 32, 42\}$	$9 \cdot 2^{-10}$	$[16, 24, 32]$	48 – ?	??
$\{48, 20, 4; 4, 32, 36\}$	$5 \cdot 2^{-7}$	$[16, 24, 32]$	48 – ?	?? $\times 2$
$\{48, 22, 4; 6, 24, 40\}$	$11 \cdot 2^{-8}$	$[16, 24, 32]$	48 – ?	?? $\times 2$
$\{48, 32, 16; 16, 32, 48\}$	$1 \cdot 2^{-3}$	$[16, 32, 48]$	48	$\times 16$
$\{48, 33, 6; 5, 12, 40\}$	$25 \cdot 2^{-10}$	$[16, 24, 32]$	48 – ?	??
$\{48, 36, 11; 2, 15, 32\}$	$5 \cdot 2^{-9}$	$[16, 24, 32]$	48 – ?	??
$\{48, 40, 14; 4, 6, 32\}$	$1 \cdot 2^{-7}$	$[16, 24, 32]$	48 – ?	?? $\times 2$
$\{48, 40, 32; 8, 16, 48\}$	$1 \cdot 2^{-5}$	$[16, 32, 48]$	48	$\times 8$
$\{48, 42, 20; 2, 8, 24\}$	$1 \cdot 2^{-8}$	$[16, 24, 32]$	48 – ?	?? $\times 2$
$\{48, 45, 6; 1, 12, 40\}$	$1 \cdot 2^{-8}$	$[20, 24, 32]$	48 – ?	?? nonlin.
…	…	…	…	
$\{70, 69, 20; 1, 2, 30\}$	$1 \cdot 2^{-12}$	$[24, 32, 40]$	70 – ?	?? nonlin. [11]
…	…	…	…	

TABLE 7.3 $q = 2,\ \rho = 4$

Array	Density	Spectrum	l.b.–u.b.	References
$\{4,3,2,1;1,2,3,4\}$	$1{\cdot}2^{-4}$	$[1,2,3,4]$	4	$\{1;1\}^4$
$\{8,6,4,2;2,4,6,8\}$	$1{\cdot}2^{-4}$	$[2,4,6,8]$	8	$\times 2$
$\{8,7,6,5;1,2,3,8\}$	$1{\cdot}2^{-7}$	$[2,4,6,8]$	8	rep.
$\{9,8,7,6;1,2,3,4\}$	$1{\cdot}2^{-8}$	$[2,4,6,8]$	9	rep.
$\{12,9,6,3;1,2,3,4\}$	$1{\cdot}2^{-8}$	$[2,4,6,8]$	12	$\{1;3\}^4$
$\{12,9,6,3;3,6,9,12\}$	$1{\cdot}2^{-4}$	$[3,6,9,12]$	12	$\times 3$
$\{12,11,10,3;1,2,9,12\}$	$3{\cdot}2^{-9}$	$[4,6,8,12]$	12	Hadamard
$\{16,12,8,4;4,8,12,16\}$	$1{\cdot}2^{-4}$	$[4,8,12,16]$	16	$\times 4$
$\{16,14,12,10;2,4,6,16\}$	$1{\cdot}2^{-7}$	$[4,8,12,16]$	16	$\times 2$
$\{16,15,8,1;1,8,15,16\}$	$1{\cdot}2^{-6}$	$[6,8,10,16]$	16	$4\overline{P}_{16}$; linear [5]
$\{16,15,10,1;1,6,15,16\}$	$3{\cdot}2^{-8}$	$[6,8,10,16]$	16	$3\overline{P}_{16}$
$\{16,15,12,1;1,4,15,16\}$	$1{\cdot}2^{-7}$	$[6,8,10,16]$	16	$2\overline{P}_{16}$
$\{16,15,14,1;1,2,15,16\}$	$1{\cdot}2^{-8}$	$[6,8,10,16]$	16	$\overline{P}_{16}$
$\{18,16,14,12;2,4,6,8\}$	$1{\cdot}2^{-8}$	$[4,8,12,16]$	18	$\times 2$
$\{20,15,10,5;3,6,9,12\}$	$3^4{\cdot}2^{-12}$	$[4,8,12,16]$	24^{c}	$\{3;5\}^4$
$\{20,15,10,5;5,10,15,20\}$	$1{\cdot}2^{-4}$	$[5,10,15,20]$	20	$\times 5$
$\{24,18,12,6;2,4,6,8\}$	$1{\cdot}2^{-8}$	$[4,8,12,16]$	24	$\times 2$
$\{24,18,12,6;6,12,18,24\}$	$1{\cdot}2^{-4}$	$[6,12,18,24]$	24	$\times 6$
$\{24,21,18,15;3,6,9,24\}$	$1{\cdot}2^{-7}$	$[6,12,18,24]$	24	$\times 3$
$\{24,22,8,3;2,16,21,24\}$	$11{\cdot}2^{-9}$	$[8,12,16,24]$	24 – ?	??
$\{24,22,20,6;2,4,18,24\}$	$3{\cdot}2^{-9}$	$[8,12,16,24]$	24	$\times 2$
$\{24,23,4,3;1,20,21,24\}$	$23{\cdot}2^{-11}$	$[8,12,16,24]$	29^{w}– ?	??
$\{24,23,16,6;1,8,18,24\}$	$3{\cdot}2^{-9}$	$[8,12,16,24]$	24 – ?	??
$\{24,23,22,21;1,2,3,24\}$	$1{\cdot}2^{-12}$	$[8,12,16,24]$	24	$\overline{G}_{24}$
$\{27,24,21,18;3,6,9,12\}$	$1{\cdot}2^{-8}$	$[6,12,18,24]$	27	$\times 3$
$\{28,15,6,1;1,6,15,28\}$	$1{\cdot}2^{-7}$	$[7,12,15,16]$	28	$\frac{1}{2}H(8,2)$
$\{28,21,14,7;1,2,3,4\}$	$1{\cdot}2^{-12}$	$[4,8,12,16]$	28	$\{7;1\}^4$
$\{28,21,14,7;7,14,21,28\}$	$1{\cdot}2^{-4}$	$[7,14,21,28]$	28	$\times 7$
$\{32,24,16,8;8,16,24,32\}$	$1{\cdot}2^{-4}$	$[8,16,24,32]$	32	$\times 8$

(*Continued*)

TABLE 7.3 (*Continued*) $q = 2,\ \rho = 4$

Array	Density	Spectrum	l.b.–u.b.	References
$\{32, 28, 24, 20; 4, 8, 12, 32\}$	$1{\cdot}2^{-7}$	$[8, 16, 24, 32]$	32	$\times 4$
$\{32, 30, 16, 2; 2, 16, 30, 32\}$	$1{\cdot}2^{-6}$	$[12, 16, 20, 32]$	32	$\times 2$; linear
$\{32, 30, 17, 2; 2, 15, 30, 32\}$	$15{\cdot}2^{-10}$	$[12, 16, 20, 32]$	32 – ?	$\cup : \{32, 2; 30, 32\}$
$\{32, 30, 18, 2; 2, 14, 30, 32\}$	$7{\cdot}2^{-9}$	$[12, 16, 20, 32]$	32 – ?	$\cup : \{32, 2; 30, 32\}$
$\{32, 30, 19, 2; 2, 13, 30, 32\}$	$13{\cdot}2^{-10}$	$[12, 16, 20, 32]$	32 – ?	$\cup : \{32, 2; 30, 32\}$
$\{32, 30, 20, 2; 2, 12, 30, 32\}$	$3{\cdot}2^{-8}$	$[12, 16, 20, 32]$	32	$\times 2$, $\cup$:$\{32, 2; 30, 32\}$
$\{32, 30, 21, 2; 2, 11, 30, 32\}$	$11{\cdot}2^{-10}$	$[12, 16, 20, 32]$	32 – ?	$\cup : \{32, 2; 30, 32\}$
$\{32, 30, 22, 2; 2, 10, 30, 32\}$	$5{\cdot}2^{-9}$	$[12, 16, 20, 32]$	32 – ?	$\cup : \{32, 2; 30, 32\}$
$\{32, 30, 23, 2; 2, 9, 30, 32\}$	$9{\cdot}2^{-10}$	$[12, 16, 20, 32]$	32 – ?	$\cup : \{32, 2; 30, 32\}$
$\{32, 30, 24, 2; 2, 8, 30, 32\}$	$1{\cdot}2^{-7}$	$[12, 16, 20, 32]$	32	$\times 2$, $\cup$:$\{32, 2; 30, 32\}$
$\{32, 30, 25, 2; 2, 7, 30, 32\}$	$7{\cdot}2^{-10}$	$[12, 16, 20, 32]$	32 – ?	$\cup : \{32, 2; 30, 32\}$
$\{32, 30, 26, 2; 2, 6, 30, 32\}$	$3{\cdot}2^{-9}$	$[12, 16, 20, 32]$	32 – ?	$\cup : \{32, 2; 30, 32\}$
$\{32, 30, 27, 2; 2, 5, 30, 32\}$	$5{\cdot}2^{-10}$	$[12, 16, 20, 32]$	32 – ?	$\cup : \{32, 2; 30, 32\}$
$\{32, 30, 28, 2; 2, 4, 30, 32\}$	$1{\cdot}2^{-8}$	$[12, 16, 20, 32]$	32	$\times 2$, $\cup$:$\{32, 2; 30, 32\}$
$\{32, 30, 29, 2; 2, 3, 30, 32\}$	$3{\cdot}2^{-10}$	$[12, 16, 20, 32]$	40^{w}– ?	$\cup : \{32, 2; 30, 32\}$
$\{32, 31, 30, 17; 1, 2, 15, 32\}$	$1{\cdot}2^{-11}$	$[12, 16, 20, 32]$	32	BCH_{32}
$\{36, 21, 10, 3; 1, 6, 15, 28\}$	$1{\cdot}2^{-8}$	$[8, 14, 18, 20]$	36 – ?	??
$\{36, 21, 10, 3; 3, 10, 21, 36\}$	$5{\cdot}2^{-8}$	$[9, 16, 21, 24]$	36 – ?	?? $\cup : \{20, 3; 21, 36\}$
$\{36, 27, 18, 9; 3, 6, 9, 12\}$	$1{\cdot}2^{-8}$	$[6, 12, 18, 24]$	36	$\times 3$
$\{36, 27, 18, 9; 7, 14, 21, 28\}$	$7^4{\cdot}2^{-16}$	$[8, 16, 24, 32]$	48^{c}	$\{9; 7\}^4$
$\{36, 27, 18, 9; 9, 18, 27, 36\}$	$1{\cdot}2^{-4}$	$[9, 18, 27, 36]$	36	$\times 9$
$\{36, 32, 28, 24; 4, 8, 12, 16\}$	$1{\cdot}2^{-8}$	$[8, 16, 24, 32]$	36	$\times 4$
$\{36, 33, 30, 9; 3, 6, 27, 36\}$	$3{\cdot}2^{-9}$	$[12, 18, 24, 36]$	36	$\times 3$
$\{40, 30, 20, 10; 6, 12, 18, 24\}$	$81{\cdot}2^{-12}$	$[8, 16, 24, 32]$	48^{c}	$\times 2$
$\{40, 30, 20, 10; 10, 20, 30, 40\}$	$1{\cdot}2^{-4}$	$[10, 20, 30, 40]$	40	$\times 10$
$\{40, 35, 30, 25; 5, 10, 15, 40\}$	$1{\cdot}2^{-7}$	$[10, 20, 30, 40]$	40	$\times 5$
$\{40, 39, 34, 5; 1, 6, 35, 40\}$	$7{\cdot}2^{-12}$	$[16, 20, 24, 40]$	40 – ?	?? ext. {39,34,5;1,6,35}
$\{44, 33, 22, 11; 5, 10, 15, 20\}$	$5^4{\cdot}2^{-16}$	$[8, 16, 24, 32]$	48^{c}– 52	? $\{11; 5\}^4$ for $n \geq 52$
$\{44, 33, 22, 11; 11, 22, 33, 44\}$	$1{\cdot}2^{-4}$	$[11, 22, 33, 44]$	44	$\times 11$

(*Continued*)

TABLE 7.3 (*Continued*) $q = 2,\ \rho = 4$

Array	Density	Spectrum	l.b.–u.b.	References
$\{45, 40, 35, 30; 5, 10, 15, 20\}$	$1{\cdot}2^{-8}$	$[10, 20, 30, 40]$	45	×5
$\{48, 36, 24, 12; 4, 8, 12, 16\}$	$1{\cdot}2^{-8}$	$[8, 16, 24, 32]$	48	×4
$\{48, 36, 24, 12; 12, 24, 36, 48\}$	$1{\cdot}2^{-4}$	$[12, 24, 36, 48]$	48	×12
$\{48, 42, 36, 30; 6, 12, 18, 48\}$	$1{\cdot}2^{-7}$	$[12, 24, 36, 48]$	48	×6
$\{48, 43, 32, 6; 5, 16, 42, 48\}$	$35{\cdot}2^{-11}$	$[16, 24, 32, 48]$	48 – ?	??
$\{48, 44, 16, 6; 4, 32, 42, 48\}$	$11{\cdot}2^{-9}$	$[16, 24, 32, 48]$	48 – ?	?? ×2
$\{48, 44, 40, 12; 4, 8, 36, 48\}$	$3{\cdot}2^{-9}$	$[16, 24, 32, 48]$	48	×4
$\{48, 45, 20, 7; 3, 28, 41, 48\}$	$287{\cdot}2^{-14}$	$[16, 24, 32, 48]$	48 – ?	??
$\{48, 45, 24, 3; 3, 24, 45, 48\}$	$1{\cdot}2^{-6}$	$[18, 24, 30, 48]$	48	×3
$\{48, 45, 30, 3; 3, 18, 45, 48\}$	$3{\cdot}2^{-8}$	$[18, 24, 30, 48]$	48	×3
$\{48, 45, 32, 10; 3, 16, 38, 48\}$	$19{\cdot}2^{-11}$	$[16, 24, 32, 48]$	48 – ?	??
$\{48, 45, 34, 11; 3, 14, 37, 48\}$	$259{\cdot}2^{-15}$	$[16, 24, 32, 48]$	48 – ?	??
$\{48, 45, 36, 3; 3, 12, 45, 48\}$	$1{\cdot}2^{-7}$	$[18, 24, 30, 48]$	48	×3
$\{48, 45, 40, 17; 3, 8, 31, 48\}$	$31{\cdot}2^{-13}$	$[16, 24, 32, 48]$	48 – ?	??
$\{48, 45, 42, 3; 3, 6, 45, 48\}$	$1{\cdot}2^{-8}$	$[18, 24, 30, 48]$	48	×3
$\{48, 45, 44, 31; 3, 4, 17, 48\}$	$17{\cdot}2^{-14}$	$[16, 24, 32, 48]$	53^{w} – ?	??
$\{48, 46, 8, 6; 2, 40, 42, 48\}$	$23{\cdot}2^{-11}$	$[16, 24, 32, 48]$	48 – ?	?? ×2
$\{48, 46, 32, 12; 2, 16, 36, 48\}$	$3{\cdot}2^{-9}$	$[16, 24, 32, 48]$	48 – ?	?? ×2
$\{48, 46, 38, 18; 2, 10, 30, 48\}$	$25{\cdot}2^{-13}$	$[16, 24, 32, 48]$	48 – ?	?? ~~×2~~
$\{48, 46, 44, 42; 2, 4, 6, 48\}$	$1{\cdot}2^{-12}$	$[16, 24, 32, 48]$	48	×2
$\{48, 47, 22, 9; 1, 26, 39, 48\}$	$169{\cdot}2^{-15}$	$[16, 24, 32, 48]$	48 – ?	??
$\{48, 47, 40, 27; 1, 8, 21, 48\}$	$7{\cdot}2^{-13}$	$[16, 24, 32, 48]$	48 – ?	??

TABLE 7.4 $q = 2, \rho = 5$

Array	Density	Spectrum	l.b.–u.b.	Comm.
$\{5,4,3,2,1;1,2,3,4,5\}$	$1 \cdot 2^{-5}$	$[1,2,3,4,5]$	5	$\{1;1\}^5$
$\{10,8,6,4,2;2,4,6,8,10\}$	$1 \cdot 2^{-5}$	$[2,4,6,8,10]$	10	$\times 2$
$\{10,9,8,7,6;1,2,3,4,10\}$	$1 \cdot 2^{-9}$	$[2,4,6,8,10]$	10	rep.
$\{11,10,9,8,7;1,2,3,4,5\}$	$1 \cdot 2^{-10}$	$[2,4,6,8,10]$	11	rep.
$\{15,12,9,6,3;1,2,3,4,5\}$	$1 \cdot 2^{-10}$	$[2,4,6,8,10]$	15	$\{3;1\}^5$
$\{15,12,9,6,3;3,6,9,12,15\}$	$1 \cdot 2^{-5}$	$[3,6,9,12,15]$	15	$\times 3$
$\{20,16,12,8,4;4,8,12,16,20\}$	$1 \cdot 2^{-5}$	$[4,8,12,16,20]$	20	$\times 2$
$\{20,18,16,14,12;2,4,6,8,20\}$	$1 \cdot 2^{-9}$	$[4,8,12,16,20]$	20	$\times 2$
$\{22,20,18,16,14;2,4,6,8,10\}$	$1 \cdot 2^{-10}$	$[4,8,12,16,20]$	22	$\times 2$
$\{25,20,15,10,5;3,6,9,12,15\}$	$3^5 \cdot 2^{-15}$	$[4,8,12,16,20]$	30	$\{5;3\}^5$
$\{25,20,15,10,5;5,10,15,20,25\}$	$1 \cdot 2^{-5}$	$[5,10,15,20,25]$	25	$\times 5$
$\{30,24,18,12,6;2,4,6,8,10\}$	$1 \cdot 2^{-10}$	$[4,8,12,16,20]$	30	$\times 2$
$\{30,24,18,12,6;6,12,18,24,30\}$	$1 \cdot 2^{-5}$	$[6,12,18,24,30]$	30	$\times 6$
$\{30,27,24,21,18;3,6,9,12,30\}$	$1 \cdot 2^{-9}$	$[6,12,18,24,30]$	30	$\times 3$
$\{33,30,27,24,21;3,6,9,12,15\}$	$1 \cdot 2^{-10}$	$[6,12,18,24,30]$	33	$\times 3$
$\{35,28,21,14,7;1,2,3,4,5\}$	$1 \cdot 2^{-15}$	$[4,8,12,16,20]$	35	$\{7;1\}^5$
$\{35,28,21,14,7;7,14,21,28,35\}$	$1 \cdot 2^{-5}$	$[7,14,21,28,35]$	35	$\times 7$
$\{40,32,24,16,8;8,16,24,32,40\}$	$1 \cdot 2^{-5}$	$[8,16,24,32,40]$	40	$\times 8$
$\{40,36,32,28,24;4,8,12,16,40\}$	$1 \cdot 2^{-9}$	$[8,16,24,32,40]$	40	$\times 4$
$\{44,40,36,32,28;4,8,12,16,20\}$	$1 \cdot 2^{-10}$	$[8,16,24,32,40]$	44	$\times 4$
$\{45,28,15,6,1;1,6,15,28,45\}$	$1 \cdot 2^{-9}$	$[9,16,21,24,25]$	45	$\frac{1}{2}H(10,2)$
$\{45,36,27,18,9;3,6,9,12,15\}$	$1 \cdot 2^{-10}$	$[6,12,18,24,30]$	45	$\times 3$
$\{45,36,27,18,9;7,14,21,28,35\}$	$7^5 \cdot 2^{-20}$	$[8,16,24,32,40]$	60^{c}	$\{9;7\}^5$
$\{45,36,27,18,9;9,18,27,36,45\}$	$1 \cdot 2^{-5}$	$[9,18,27,36,45]$	45	$\times 9$

TABLE 7.5 $q = 2,\ \rho = 6$

Array	Density	Spectrum	l.b.–u.b.	References
$\{6,5,4,3,2,1;1,2,3,4,5,6\}$	$1 \cdot 2^{-6}$	$[1,2,3,4,5,6]$	6	$\{1;1\}^6$
$\{12,10,8,6,4,2;2,4,6,8,10,12\}$	$1 \cdot 2^{-6}$	$[2,4,6,8,10,12]$	12	$\times 2$
$\{12,11,10,9,8,7;1,2,3,4,5,12\}$	$1 \cdot 2^{-11}$	$[2,4,6,8,10,12]$	12	rep.
$\{13,12,11,10,9,8;1,2,3,4,5,6\}$	$1 \cdot 2^{-12}$	$[2,4,6,8,10,12]$	13	rep.
$\{18,15,12,9,6,3;1,2,3,4,5,6\}$	$1 \cdot 2^{-12}$	$[2,4,6,8,10,12]$	18	$\{3;1\}^6$
$\{18,15,12,9,6,3;3,6,9,12,15,18\}$	$1 \cdot 2^{-6}$	$[3,6,9,12,15,18]$	18	$\times 3$
$\{21,20,16,6,2,1;1,2,6,16,20,21\}$	$1 \cdot 2^{-10}$	$[6,8,10,12,14,16]$	21	$\{00,11\}$-sh. G_{23}
$\{21,20,16,9,2,1;1,2,3,16,20,21\}$	$1 \cdot 2^{-11}$	$[6,8,10,12,14,16]$	21	$\{000,111\}$-sh. $\overline{G}_{24}$
$\{22,21,20,3,2,1;1,2,3,20,21,22\}$	$1 \cdot 2^{-11}$	$[7,8,11,12,15,16]$	22	$\{0\}$-sh. G_{23}
$\{24,20,16,12,8,4;4,8,12,16,20,24\}$	$1 \cdot 2^{-6}$	$[4,8,12,16,20,24]$	24	$\times 4$
$\{24,22,20,18,16,14;2,4,6,8,10,24\}$	$1 \cdot 2^{-11}$	$[4,8,12,16,20,24]$	24	$\times 2$
$\{26,24,22,20,18,16;2,4,6,8,10,12\}$	$1 \cdot 2^{-12}$	$[4,8,12,16,20,24]$	26	$\times 2$
$\{30,25,20,15,10,5;3,6,9,12,15,18\}$	$3^6 \cdot 2^{-18}$	$[4,8,12,16,20,24]$	36	$\{5;3\}^6$
$\{30,25,20,15,10,5;5,10,15,20,25,30\}$	$1 \cdot 2^{-6}$	$[5,10,15,20,25,30]$	30	$\times 5$
$\{36,30,24,18,12,6;2,4,6,8,10,12\}$	$1 \cdot 2^{-12}$	$[4,8,12,16,20,24]$	36	$\times 2$
$\{36,30,24,18,12,6;6,12,18,24,30,36\}$	$1 \cdot 2^{-6}$	$[6,12,18,24,30,36]$	36	$\times 6$
$\{36,33,30,27,24,21;3,6,9,12,15,18\}$	$1 \cdot 2^{-11}$	$[6,12,18,24,30,36]$	36	$\times 3$

TABLE 7.6 $q = 2, \rho = 7$

Array	**Density**	**Spectrum**	**l.b.–u.b.**	**References**
$\{7, 6, 5, 4, 3, 2, 1; 1, 2, 3, 4, 5, 6, 7\}$	$1 \cdot 2^{-7}$	$[1, 2, 3, 4, 5, 6, 7]$	7	$\{1; 1\}^7$
$\{14, 12, 10, 8, 6, 4, 2; 2, 4, 6, 8, 10, 12, 14\}$	$1 \cdot 2^{-7}$	$[2, 4, 6, 8, 10, 12, 14]$	14	$\times 2$
$\{14, 13, 12, 11, 10, 9, 8; 1, 2, 3, 4, 5, 6, 14\}$	$1 \cdot 2^{-13}$	$[2, 4, 6, 8, 10, 12, 14]$	14	rep.
$\{15, 14, 13, 12, 11, 10, 9; 1, 2, 3, 4, 5, 6, 7\}$	$1 \cdot 2^{-14}$	$[2, 4, 6, 8, 10, 12, 14]$	15	rep.
$\{21, 18, 15, 12, 9, 6, 3; 1, 2, 3, 4, 5, 6, 7\}$	$1 \cdot 2^{-14}$	$[2, 4, 6, 8, 10, 12, 14]$	21	$\{3; 1\}^7$
$\{21, 18, 15, 12, 9, 6, 3; 3, 6, 9, 12, 15, 18, 21\}$	$1 \cdot 2^{-7}$	$[3, 6, 9, 12, 15, 18, 21]$	21	$\times 3$
$\{22, 21, 20, 16, 6, 2, 1; 1, 2, 6, 16, 20, 21, 22\}$	$1 \cdot 2^{-11}$	$[6, 8, 10, 12, 14, 16, 22]$	22	$\{00, 11\}$-sh. $\overline{G}_{24}$
$\{23, 22, 21, 20, 3, 2, 1; 1, 2, 3, 20, 21, 22, 23\}$	$1 \cdot 2^{-12}$	$[7, 8, 11, 12, 15, 16, 23]$	23	$\{0\}$-sh. $\overline{G}_{24}$
$\{28, 24, 20, 16, 12, 8, 4; 4, 8, 12, 16, 20, 24, 28\}$	$1 \cdot 2^{-7}$	$[4, 8, 12, 16, 20, 24, 28]$	28	$\times 4$
$\{28, 26, 24, 22, 20, 18, 16; 2, 4, 6, 8, 10, 12, 28\}$	$1 \cdot 2^{-13}$	$[4, 8, 12, 16, 20, 24, 28]$	28	$\times 2$
$\{30, 28, 26, 24, 22, 20, 18; 2, 4, 6, 8, 10, 12, 14\}$	$1 \cdot 2^{-14}$	$[4, 8, 12, 16, 20, 24, 28]$	30	$\times 2$

TABLE 7.7 $q = 2, \rho = 8$

Array	**Density**	**Spectrum**	**l.b.–u.b.**	**Comm.**
$\{8, 7, 6, 5, 4, 3, 2, 1; 1, 2, 3, 4, 5, 6, 7, 8\}$	$1 \cdot 2^{-8}$	$[1, 2, 3, 4, 5, 6, 7, 8]$	8	$\{1; 1\}^8$
$\{16, 14, 12, 10, 8, 6, 4, 2; 2, 4, 6, 8, 10, 12, 14, 16\}$	$1 \cdot 2^{-8}$	$[2, 4, 6, 8, 10, 12, 14, 16]$	16	$\times 2$
$\{16, 15, 14, 13, 12, 11, 10, 9; 1, 2, 3, 4, 5, 6, 7, 16\}$	$1 \cdot 2^{-15}$	$[2, 4, 6, 8, 10, 12, 14, 16]$	16	rep.
$\{17, 16, 15, 14, 13, 12, 11, 10; 1, 2, 3, 4, 5, 6, 7, 8\}$	$1 \cdot 2^{-16}$	$[2, 4, 6, 8, 10, 12, 14, 16]$	17	rep.
$\{24, 21, 18, 15, 12, 9, 6, 3; 1, 2, 3, 4, 5, 6, 7, 8\}$	$1 \cdot 2^{-16}$	$[2, 4, 6, 8, 10, 12, 14, 16]$	24	$\{3; 1\}^8$
$\{24, 21, 18, 15, 12, 9, 6, 3; 3, 6, 9, 12, 15, 18, 21, 24\}$	$1 \cdot 2^{-8}$	$[3, 6, 9, 12, 15, 18, 21, 24]$	24	$\times 3$
$\{32, 30, 28, 26, 24, 22, 20, 18; 2, 4, 6, 8, 10, 12, 14, 32\}$	$1 \cdot 2^{-15}$	$[4, 8, 12, 16, 20, 24, 28, 32]$	32	$\times 2$
$\{32, 28, 24, 20, 16, 12, 8, 4; 4, 8, 12, 16, 20, 24, 28, 32\}$	$1 \cdot 2^{-8}$	$[4, 8, 12, 16, 20, 24, 28, 32]$	32	$\times 4$

TABLE 7.8 $q = 3$, $\rho = 1$

Array	$b_0 + c_1$	$\frac{b_0+c_1}{\gcd(b_0,c_1)}$	Spectrum	l.b.–u.b.	References
$\{2;1\}$	3	3	[1]	1	
$\{4;2\}$	6	3	[2]	2	$\times 2$
$\{5;4\}$	9	9	[3]	4	$4H_4$
$\{6;3\}$	9	3	[3]	3	$\times 3$
$\{7;2\}$	9	9	[3]	4	$2H_4$
$\{8;1\}$	9	9	[3]	4	H_4
$\{8;4\}$	12	3	[4]	4	$\times 4$
$\{10;5\}$	15	3	[5]	5	$\times 5$
$\{10;8\}$	18	9	[6]	7	in $/C$
$\{12;6\}$	18	3	[6]	6	$\times 6$
$\{14;4\}$	18	9	[6]	8°	$\times 2$
$\{16;2\}$	18	9	[6]	8	$\times 2$
$\{14;7\}$	21	3	[7]	7	$\times 7$
$\{16;8\}$	24	3	[8]	8	$\times 8$
$\{14;13\}$	27	27	[9]	11	in $/G_{11}$
$\{15;12\}$	27	9	[9]	10	in $/G_{10}^*$
$\{16;11\}$	27	27	[9]	11	in $/G_{11}$
$\{17;10\}$	27	27	[9]	11	in $/G_{11}$
$\{18;9\}$	27	3	[9]	9	$\times 9$
$\{19;8\}$	27	27	[9]	11	in $/G_{11}$
$\{20;7\}$	27	27	[9]	11	in $/G_{11}$
$\{21;6\}$	27	9	[9]	11	in $/G_{11}$
$\{22;5\}$	27	27	[9]	11	in $/G_{11}$
$\{23;4\}$	27	27	[9]	12	in $\mathrm{NVO}_4^+(3)$
$\{24;3\}$	27	9	[9]	12	$\times 3$
$\{25;2\}$	27	27	[9]	13	$2H_{13}$
$\{26;1\}$	27	27	[9]	13	H_{13}
$\{20;10\}$	30	3	[10]	10	$\times 10$
$\{22;11\}$	33	3	[11]	11	$\times 11$
$\{20;16\}$	36	9	[12]	$13-14$	$\times 2$

(Continued)

TABLE 7.8 (*Continued*) $q = 3$, $\rho = 1$

Array	$b_0 + c_1$	$\frac{b_0+c_1}{\gcd(b_0,c_1)}$	Spectrum	l.b.–u.b.	References
$\{24; 12\}$	36	3	[12]	12	$\times 12$
$\{28; 8\}$	36	9	[12]	$14 - 15$	in $/C$
$\{32; 4\}$	36	9	[12]	16	$\times 4$
$\{26; 13\}$	39	3	[13]	13	$\times 13$
$\{28; 14\}$	42	3	[14]	14	$\times 14$
$\{25; 20\}$	45	9	[15]	$16 - 19$	in $/C$
$\{30; 15\}$	45	3	[15]	15	$\times 15$
$\{35; 10\}$	45	9	[15]	$18 - 19$	in $/C$
$\{40; 5\}$	45	9	[15]	20	$\times 5$
$\{32; 16\}$	48	3	[16]	16	$\times 16$
$\{34; 17\}$	51	3	[17]	17	$\times 17$
$\{28; 26\}$	54	27	[18]	$20 - 22$	$\times 2$
$\{30; 24\}$	54	9	[18]	$19 - 20$	$\times 2$
$\{32; 22\}$	54	27	[18]	$20 - 22$	$\times 2$
$\{34; 20\}$	54	27	[18]	$20 - 22$	$\times 2$
$\{36; 18\}$	54	3	[18]	18	$\times 18$
$\{38; 16\}$	54	27	[18]	$20 - 22$	$\times 2$
$\{40; 14\}$	54	27	[18]	$20 - 22$	$\times 2$
$\{42; 12\}$	54	9	[18]	$21 - 22$	$\times 6$
$\{44; 10\}$	54	27	[18]	22	$\times 2$
$\{46; 8\}$	54	27	[18]	$23 - 24$	$\times 2$
$\{48; 6\}$	54	9	[18]	24	$\times 6$
$\{50; 4\}$	54	27	[18]	25	in $\overline{\mathrm{NVO}_4^-(3)}$
$\{52; 2\}$	54	27	[18]	26	$\times 2$

TABLE 7.9 $q = 3$, $\rho = 2$

Array	**Density**	**Spectrum**	**l.b.–u.b.**	**References**
$\{4,2;1,2\}$	$1 \cdot 3^{-2}$	$[1,2]$	2	$\{2;1\}^2$
$\{6,2;1,6\}$	$1 \cdot 3^{-2}$	$[2,3]$	3	$\mathrm{MDS}[3,1,3]_3$
$\{8,4;2,4\}$	$1 \cdot 3^{-2}$	$[2,4]$	4	$\times 2$
$\{10,2;1,8\}$	$2 \cdot 3^{-3}$	$[3,4]$	5	optimal $(5,18,3)_3$
$\{10,5;4,8\}$	$16 \cdot 3^{-4}$	$[3,6]$	$6-8$	? $\{5;4\}^2$
$\{10,9;2,6\}$	$2 \cdot 3^{-3}$	$[3,6]$	$\nexists$	Lemma 7.8
$\{11,10;2,4\}$	$4 \cdot 3^{-4}$	$[3,6]$	$\nexists$	Lemma 7.8
$\{12,4;2,12\}$	$1 \cdot 3^{-2}$	$[4,6]$	6	$\times 2$
$\{12,6;3,6\}$	$1 \cdot 3^{-2}$	$[3,6]$	6	$\times 3$
$\{12,10;2,3\}$	$1 \cdot 3^{-3}$	$[3,6]$	$\nexists$	Lemma 7.8
$\{13,2;4,8\}$	$13 \cdot 3^{-4}$	$[3,6]$	$7-?$	??
$\{14,7;2,4\}$	$4 \cdot 3^{-4}$	$[3,6]$	$7-8$	? $\{7;2\}^2$
$\{15,4;2,6\}$	$2 \cdot 3^{-3}$	$[3,6]$	$8-?$	??
$\{15,8;4,15\}$	$4 \cdot 3^{-3}$	$[5,9]$	$9-?$	?? $\cup : 3\times\{5;4\}$
$\{16,8;1,2\}$	$1 \cdot 3^{-4}$	$[3,6]$	8	$\{8;1\}^2$
$\{16,8;4,8\}$	$1 \cdot 3^{-2}$	$[4,8]$	8	$\times 4$
$\{18,5;4,18\}$	$4 \cdot 3^{-3}$	$[6,9]$	9	4SU1
$\{18,6;3,18\}$	$1 \cdot 3^{-2}$	$[6,9]$	9	$\times 3$
$\{18,7;2,18\}$	$2 \cdot 3^{-3}$	$[6,9]$	9	2SU1
$\{18,8;1,18\}$	$1 \cdot 3^{-3}$	$[6,9]$	9	SU1
$\{20,4;2,16\}$	$2 \cdot 3^{-3}$	$[6,8]$	10	$\times 2$
$\{20,10;2,13\}$	$13 \cdot 3^{-5}$	$[6,9]$	$10-?$	??
$\{20,10;5,10\}$	$1 \cdot 3^{-2}$	$[5,10]$	10	$\times 5$
$\{20,10;8,16\}$	$16 \cdot 3^{-4}$	$[6,12]$	$12^{\mathrm{w}}-16$	? $\times 2$
$\{20,18;1,6\}$	$1 \cdot 3^{-4}$	$[6,9]$	$10-10$	$\{0,1,2\}$-sh. G_{11}
$\{20,18;4,12\}$	$2 \cdot 3^{-3}$	$[6,12]$	$12-?$	?? ~~×2~~
$\{21,4;2,21\}$	$2 \cdot 3^{-3}$	$[7,9]$	11	in $/C$; $\cup$:$\{21;6\}$
$\{21,10;2,12\}$	$4 \cdot 3^{-4}$	$[6,9]$	$11-?$	??
$\{22,2;4,17\}$	$22 \cdot 3^{-5}$	$[6,9]$	$11-?$	??
$\{22,9;2,12\}$	$4 \cdot 3^{-4}$	$[6,9]$	$11-?$	??
$\{22,20;1,2\}$	$1 \cdot 3^{-5}$	$[6,9]$	11	G_{11}

(*Continued*)

TABLE 7.9 (*Continued*) $q = 3,\ \rho = 2$

Array	Density	Spectrum	l.b.–u.b.	References
$\{22, 20; 4, 8\}$	$4 \cdot 3^{-4}$	$[6, 12]$	$12 - ?$	?? ~~×2~~
$\{24, 2; 1, 24\}$	$1 \cdot 3^{-3}$	$[8, 9]$	12	SU1
$\{24, 4; 2, 15\}$	$5 \cdot 3^{-4}$	$[6, 9]$	$12 - ?$	??
$\{24, 8; 4, 24\}$	$1 \cdot 3^{-2}$	$[8, 12]$	12	×4
$\{24, 12; 6, 12\}$	$1 \cdot 3^{-2}$	$[6, 12]$	12	×6
$\{24, 14; 1, 6\}$	$1 \cdot 3^{-4}$	$[6, 9]$	12	$\mathrm{MDS}[3, 1, 3]_9$
$\{24, 20; 4, 6\}$	$1 \cdot 3^{-3}$	$[6, 12]$	$13^{\mathrm{w}} - ?$	?? ~~×2~~
$\{26, 4; 8, 16\}$	$13 \cdot 3^{-4}$	$[6, 12]$	$13 - ?$	?? ×2
$\{28, 5; 4, 26\}$	$26 \cdot 3^{-5}$	$[9, 12]$	$14 - ?$	??
$\{28, 9; 2, 24\}$	$4 \cdot 3^{-4}$	$[9, 12]$	$14 - ?$	??
$\{28, 14; 1, 20\}$	$5 \cdot 3^{-5}$	$[9, 12]$	$14 - ?$	??
$\{28, 14; 4, 8\}$	$4 \cdot 3^{-4}$	$[6, 12]$	$14 - 16$	? ×2
$\{28, 14; 7, 14\}$	$1 \cdot 3^{-2}$	$[7, 14]$	14	×7
$\{28, 14; 13, 26\}$	$169 \cdot 3^{-6}$	$[9, 18]$	$18 - 22$	? $\{14; 13\}^2$
$\{29, 10; 8, 25\}$	$40 \cdot 3^{-5}$	$[9, 15]$	$15 - ?$	??
$\{29, 18; 10, 24\}$	$40 \cdot 3^{-5}$	$[9, 18]$	$18 - ?$	??
$\{30, 6; 3, 24\}$	$2 \cdot 3^{-3}$	$[9, 12]$	15	×3
$\{30, 7; 18, 26\}$	$35 \cdot 3^{-5}$	$[9, 18]$	$18 - ?$	??
$\{30, 8; 4, 12\}$	$2 \cdot 3^{-3}$	$[6, 12]$	$15 - ?$	?? ×2
$\{30, 10; 5, 30\}$	$1 \cdot 3^{-2}$	$[10, 15]$	15	×5
$\{30, 15; 12, 24\}$	$16 \cdot 3^{-4}$	$[9, 18]$	$18 - 20$	? $\{15; 12\}^2$
$\{30, 16; 8, 30\}$	$4 \cdot 3^{-3}$	$[10, 18]$	$18 - ?$	?? ×2
$\{30, 20; 1, 12\}$	$1 \cdot 3^{-4}$	$[9, 12]$	15	FE1
$\{30, 20; 9, 22\}$	$33 \cdot 3^{-5}$	$[9, 18]$	$18 - ?$	??
$\{30, 22; 8, 21\}$	$28 \cdot 3^{-5}$	$[9, 18]$	$18 - ?$	??
$\{30, 27; 6, 18\}$	$2 \cdot 3^{-3}$	$[9, 18]$	$18 - ?$	?? ~~×3~~
$\{32, 7; 2, 22\}$	$11 \cdot 3^{-5}$	$[9, 12]$	$16 - ?$	??
$\{32, 16; 2, 4\}$	$1 \cdot 3^{-4}$	$[6, 12]$	16	×2
$\{32, 16; 8, 16\}$	$1 \cdot 3^{-2}$	$[8, 16]$	16	×8
$\{32, 16; 11, 22\}$	$121 \cdot 3^{-6}$	$[9, 18]$	$18 - 22$	? $\{16; 11\}^2$
$\{32, 18; 1, 12\}$	$1 \cdot 3^{-4}$	$[9, 12]$	16	$\mathrm{MDS}[4, 2, 3]_9$

(*Continued*)

TABLE 7.9 (*Continued*) $q = 3$, $\rho = 2$

Array	Density	Spectrum	l.b.–u.b.	References
$\{32, 18; 10, 21\}$	$35 \cdot 3^{-5}$	[9, 18]	18 – ?	??
$\{33, 4; 2, 24\}$	$4 \cdot 3^{-4}$	[9, 12]	17 – ?	??
$\{33, 4; 18, 26\}$	$22 \cdot 3^{-5}$	[9, 18]	18 – ?	??
$\{33, 10; 14, 24\}$	$55 \cdot 3^{-5}$	[9, 18]	18 – ?	??
$\{33, 20; 4, 15\}$	$4 \cdot 3^{-4}$	[9, 15]	17 – ?	??
$\{33, 30; 6, 12\}$	$4 \cdot 3^{-4}$	[9, 18]	18 – ?	?? ~~×3~~
$\{34, 8; 1, 20\}$	$5 \cdot 3^{-5}$	[9, 12]	17 – ?	??
$\{34, 9; 14, 24\}$	$51 \cdot 3^{-5}$	[9, 18]	18^{w} – ?	??
$\{34, 17; 10, 20\}$	$100 \cdot 3^{-6}$	[9, 18]	18^{w} – 22	? $\{17; 10\}^2$
$\{34, 26; 7, 14\}$	$49 \cdot 3^{-6}$	[9, 18]	18^{w} – ?	??
$\{35, 4; 8, 25\}$	$28 \cdot 3^{-5}$	[9, 15]	18 – ?	??
$\{35, 18; 4, 15\}$	$4 \cdot 3^{-4}$	[9, 15]	18 – ?	??
$\{35, 22; 8, 16\}$	$64 \cdot 3^{-6}$	[9, 18]	18 – ?	??
$\{36, 10; 8, 36\}$	$4 \cdot 3^{-3}$	[12, 18]	18	×2
$\{36, 12; 6, 36\}$	$1 \cdot 3^{-2}$	[12, 18]	18	×6
$\{36, 14; 4, 36\}$	$2 \cdot 3^{-3}$	[12, 18]	18	×2
$\{36, 16; 2, 36\}$	$1 \cdot 3^{-3}$	[12, 18]	18	×2
$\{36, 18; 9, 18\}$	$1 \cdot 3^{-2}$	[9, 18]	18	×9
$\{36, 30; 6, 9\}$	$1 \cdot 3^{-3}$	[9, 18]	19^{w} – ?	?? ~~×3~~
$\{37, 14; 10, 20\}$	$100 \cdot 3^{-6}$	[9, 18]	19 – ?	??
$\{38, 10; 11, 22\}$	$121 \cdot 3^{-6}$	[9, 18]	19 – ?	??
$\{38, 19; 8, 16\}$	$64 \cdot 3^{-6}$	[9, 18]	19 – 22	? $\{19; 8\}^2$
$\{38, 28; 5, 28\}$	$5 \cdot 3^{-4}$	[12, 21]	21 – ?	??
$\{39, 6; 12, 24\}$	$13 \cdot 3^{-4}$	[9, 18]	20 – ?	?? ×3
$\{40, 2; 4, 35\}$	$4 \cdot 3^{-4}$	[12, 15]	20 – ?	??
$\{40, 2; 13, 26\}$	$40 \cdot 3^{-6}$	[9, 18]	20 – ?	??
$\{40, 8; 4, 32\}$	$2 \cdot 3^{-3}$	[12, 16]	20 – ?	?? ×4
$\{40, 9; 2, 30\}$	$1 \cdot 3^{-3}$	[12, 15]	20 – ?	??
$\{40, 20; 1, 20\}$	$1 \cdot 3^{-4}$	[12, 15]	20	MDS$[5, 3, 3]_9$
$\{40, 20; 426\}$	$13 \cdot 3^{-5}$	[12, 18]	20 – ?	?? ×2
$\{40, 20; 7, 14\}$	$49 \cdot 3^{-6}$	[9, 18]	20 – 22	? $\{20; 7\}^2$

(*Continued*)

TABLE 7.9 (*Continued*) $q = 3,\ \rho = 2$

Array	Density	Spectrum	l.b.–u.b.	References
$\{40,20;10,20\}$	$1 \cdot 3^{-2}$	$[10,20]$	20	$\times 10$
$\{40,20;16,32\}$	$16 \cdot 3^{-4}$	$[12,24]$	$24-32$	? $\times 4$
$\{40,36;2,12\}$	$1 \cdot 3^{-4}$	$[12,18]$	20	$\times 2$
$\{40,36;8,24\}$	$2 \cdot 3^{-3}$	$[12,24]$	$24-?$	?? $\times 2$
$\{42,5;10,24\}$	$35 \cdot 3^{-5}$	$[9,18]$	$21-?$	??
$\{42,8;4,42\}$	$2 \cdot 3^{-3}$	$[14,18]$	$21-?$	?? $\times 2$
$\{42,8;9,22\}$	$33 \cdot 3^{-5}$	$[9,18]$	$21-?$	??
$\{42,14;7,42\}$	$1 \cdot 3^{-2}$	$[14,21]$	21	$\times 7$
$\{42,20;4,24\}$	$4 \cdot 3^{-4}$	$[12,18]$	$21-?$	?? $\times 2$
$\{42,20;9,28\}$	$3 \cdot 3^{-3}$	$[12,21]$	$21-?$	??
$\{42,21;6,12\}$	$4 \cdot 3^{-4}$	$[9,18]$	$21-22$	? $\{21;6\}^2$
$\{42,26;7,24\}$	$2 \cdot 3^{-3}$	$[12,21]$	$21-?$	??
$\{42,26;13,42\}$	$13 \cdot 3^{-4}$	$[14,27]$	$27-?$	?? $\cup: 3\times\{14;13\}$
$\{42,28;5,6\}$	$5 \cdot 3^{-5}$	$[9,18]$	$21-?$	??
$\{44,4;8,34\}$	$22 \cdot 3^{-5}$	$[12,18]$	$22-?$	?? $\times 2$
$\{44,18;4,24\}$	$4 \cdot 3^{-4}$	$[12,18]$	$22-?$	?? $\times 2$
$\{44,22;5,10\}$	$25 \cdot 3^{-6}$	$[9,18]$	22	$\{22;5\}^2$
$\{44,22;11,22\}$	$1 \cdot 3^{-2}$	$[11,22]$	22	$\times 11$
$\{44,40;2,4\}$	$1 \cdot 3^{-5}$	$[12,18]$	22	$\times 2$
$\{44,40;8,16\}$	$4 \cdot 3^{-4}$	$[12,24]$	$24-?$	?? $\times 2$
$\{45,12;6,18\}$	$2 \cdot 3^{-3}$	$[9,18]$	$23-?$	?? $\times 3$
$\{45,20;16,45\}$	$16 \cdot 3^{-4}$	$[15,27]$	$27-?$	?? $\cup: 9\times\{5;4\}$
$\{45,24;12,45\}$	$4 \cdot 3^{-3}$	$[15,27]$	$27-?$	?? $\times 3$
$\{45,28;8,45\}$	$8 \cdot 3^{-4}$	$[15,27]$	$27-?$	?? $\cup: 9\times\{5;4\}$
$\{45,32;4,45\}$	$4 \cdot 3^{-4}$	$[15,27]$	$27-?$	?? $\cup: 9\times\{5;4\}$
$\{46,23;4,8\}$	$16 \cdot 3^{-6}$	$[9,18]$	$23-26$	? $\{23;4\}^2$
$\{47,10;2,40\}$	$8 \cdot 3^{-5}$	$[15,18]$	$24-?$	??
$\{47,18;4,12\}$	$8 \cdot 3^{-5}$	$[9,18]$	$24-?$	??
$\{48,4;2,48\}$	$1 \cdot 3^{-3}$	$[16,18]$	24	$\times 2$
$\{48,8;4,30\}$	$5 \cdot 3^{-4}$	$[12,18]$	$24-?$	?? $\times 2$
$\{48,14;9,28\}$	$3 \cdot 3^{-3}$	$[12,21]$	$24-?$	??

(*Continued*)

TABLE 7.9 (*Continued*) $q = 3,\ \rho = 2$

Array	Density	Spectrum	l.b.–u.b.	References
$\{48, 16; 8, 48\}$	$1 \cdot 3^{-2}$	$[16, 24]$	24	$\times 8$
$\{48, 20; 1, 30\}$	$1 \cdot 3^{-4}$	$[15, 18]$	24	$\mathrm{MDS}[6, 4, 3]_9$
$\{48, 20; 9, 40\}$	$3 \cdot 3^{-3}$	$[15, 24]$	$24 - ?$	??
$\{48, 22; 11, 48\}$	$11 \cdot 3^{-4}$	$[16, 27]$	$27 - ?$	$?? \cup : 3\times\{16; 11\}$
$\{48, 24; 3, 6\}$	$1 \cdot 3^{-4}$	$[9, 18]$	24	$\times 3$
$\{48, 24; 12, 24\}$	$1 \cdot 3^{-2}$	$[12, 24]$	24	$\times 12$
$\{48, 28; 2, 12\}$	$1 \cdot 3^{-4}$	$[12, 18]$	24	$\times 2$
$\{48, 35; 4, 30\}$	$1 \cdot 3^{-3}$	$[15, 24]$	$24 - ?$	??
$\{48, 40; 8, 12\}$	$1 \cdot 3^{-3}$	$[12, 24]$	$25^{w} - ?$	$?? \times 2$

TABLE 7.10 $q = 3,\ \rho = 3$

Array	Density	Spectrum	l.b.–u.b.	References
$\{6, 4, 2; 1, 2, 3\}$	$1 \cdot 3^{-3}$	$[1, 2, 3]$	3	$H(3, 3)$
$\{12, 8, 4; 2, 4, 6\}$	$1 \cdot 3^{-3}$	$[2, 4, 6]$	6	$\times 2$
$\{15, 2, 1; 2, 10, 12\}$	$1 \cdot 3^{-4}$	$[3, 5, 6]$	$\nexists$	Lemma 7.11
$\{15, 10, 5; 4, 8, 12\}$	$64 \cdot 3^{-6}$	$[3, 6, 9]$	$9 - 12$	$?\ \{5; 4\}^3$
$\{18, 12, 6; 3, 6, 9\}$	$1 \cdot 3^{-3}$	$[3, 6, 9]$	9	$\times 3$
$\{21, 14, 7; 2, 4, 6\}$	$8 \cdot 3^{-6}$	$[3, 6, 9]$	$11 - 12$	$?\ \{7; 2\}^3$
$\{24, 16, 8; 1, 2, 3\}$	$1 \cdot 3^{-6}$	$[3, 6, 9]$	12	$\{8; 1\}^3$
$\{24, 16, 8; 4, 8, 12\}$	$1 \cdot 3^{-3}$	$[4, 8, 12]$	12	$\times 4$
$\{24, 22, 20; 1, 2, 12\}$	$1 \cdot 3^{-6}$	$[6, 9, 12]$	12	$\overline{G}_{12}$
$\{28, 18, 4; 2, 9, 20\}$	$5 \cdot 3^{-5}$	$[6, 9, 12]$	$14 - ?$	??
$\{30, 4, 2; 4, 20, 24\}$	$1 \cdot 3^{-4}$	$[6, 10, 12]$	$15 - ?$	$?? \times 2$
$\{30, 20, 10; 5, 10, 15\}$	$1 \cdot 3^{-3}$	$[5, 10, 15]$	15	$\times 5$
$\{30, 20, 10; 8, 16, 24\}$	$64 \cdot 3^{-6}$	$[6, 12, 18]$	$18 - 24$	$? \times 2$
$\{30, 20, 12; 3, 4, 30\}$	$1 \cdot 3^{-4}$	$[6, 12, 15]$	$15 - ?$	??
$\{34, 33, 8; 2, 12, 28\}$	$28 \cdot 3^{-7}$	$[9, 12, 18]$	$18 - ?$	??
$\{36, 20, 3; 1, 18, 30\}$	$1 \cdot 3^{-4}$	$[9, 12, 15]$	$18 - ?$	??
$\{36, 22, 2; 2, 27, 28\}$	$21 \cdot 3^{-6}$	$[9, 12, 18]$	$18 - ?$	??
$\{36, 24, 12; 6, 12, 18\}$	$1 \cdot 3^{-3}$	$[6, 12, 18]$	18	$\times 6$
$\{36, 26, 3; 4, 18, 30\}$	$10 \cdot 3^{-5}$	$[9, 12, 18]$	$18 - ?$	??

(*Continued*)

TABLE 7.10 (*Continued*) $q = 3,\ \rho = 3$

Array	Density	Spectrum	l.b.–u.b.	References
$\{40, 18, 10; 2, 9, 20\}$	$1 \cdot 3^{-4}$	$[6, 12, 15]$	$20 - ?$	??
$\{41, 18, 5; 4, 18, 40\}$	$32 \cdot 3^{-6}$	$[9, 15, 18]$	$21 - ?$	??
$\{42, 28, 5; 1, 2, 30\}$	$1 \cdot 3^{-6}$	$[9, 12, 15]$	$21 - ?$	?? nonlin.
$\{42, 28, 14; 4, 8, 12\}$	$8 \cdot 3^{-6}$	$[6, 12, 18]$	$21 - ?$	?? $\times 2$
$\{42, 28, 14; 7, 14, 21\}$	$1 \cdot 3^{-3}$	$[7, 14, 21]$	21	$\times 7$
$\{42, 28, 14; 13, 26, 39\}$	$2197 \cdot 3^{-9}$	$[9, 18, 27]$	$27 - 33$	? $\{14; 13\}^3$
$\{43, 6, 5; 2, 30, 40\}$	$43 \cdot 3^{-7}$	$[9, 15, 18]$	$22 - ?$	??
$\{44, 42, 10; 1, 12, 35\}$	$1 \cdot 3^{-5}$	$[12, 15, 21]$	$22 - ?$	?? nonlin.
$\{45, 6, 3; 6, 30, 36\}$	$1 \cdot 3^{-4}$	$[9, 15, 18]$	$23 - ?$	?? $\times 2$
$\{45, 28, 12; 2, 9, 30\}$	$2 \cdot 3^{-5}$	$[9, 15, 18]$	$23 - ?$	??
$\{45, 30, 15; 12, 24, 36\}$	$64 \cdot 3^{-6}$	$[9, 18, 27]$	$27 - 30$	? $\{15; 12\}^3$
$\{48, 30, 7; 9, 20, 30\}$	$15 \cdot 3^{-5}$	$[9, 15, 24]$	$24 - ?$	??
$\{48, 32, 16; 2, 4, 6\}$	$1 \cdot 3^{-6}$	$[6, 12, 18]$	24	$\times 2$
$\{48, 32, 16; 8, 16, 24\}$	$1 \cdot 3^{-3}$	$[8, 16, 24]$	24	$\times 8$
$\{48, 32, 16; 11, 22, 33\}$	$1331 \cdot 3^{-9}$	$[9, 18, 27]$	$27 - 33$	? $\{16; 11\}^3$
$\{48, 44, 40; 2, 4, 24\}$	$1 \cdot 3^{-6}$	$[12, 18, 24]$	24	$\times 2$

TABLE 7.11 $q = 3,\ \rho = 4$

Array	Density	Spectrum	l.b.–u.b.	References
$\{8, 6, 4, 2; 1, 2, 3, 4\}$	$1 \cdot 3^{-4}$	$[1, 2, 3, 4]$	4	$\{2; 1\}^4$
$\{16, 12, 8, 4; 2, 4, 6, 8\}$	$1 \cdot 3^{-4}$	$[2, 4, 6, 8]$	8	$\times 2$
$\{20, 15, 10, 5; 4, 8, 12, 16\}$	$256 \cdot 3^{-8}$	$[3, 6, 9, 12]$	$12 - 16$	? $\{5; 4\}^4$
$\{20, 18, 4, 1; 1, 2, 18, 20\}$	$1 \cdot 3^{-5}$	$[5, 6, 8, 9]$	10	$\{0\}$-sh. G_{11}
$\{24, 18, 12, 6; 3, 6, 9, 12\}$	$1 \cdot 3^{-4}$	$[3, 6, 9, 12]$	12	$\times 3$
$\{28, 21, 14, 7; 2, 4, 6, 8\}$	$16 \cdot 3^{-8}$	$[3, 6, 9, 12]$	$14 - 16$	? $\{7; 2\}^4$
$\{32, 24, 16, 8; 1, 2, 3, 4\}$	$1 \cdot 3^{-8}$	$[3, 6, 9, 12]$	16	$\{8; 1\}^4$
$\{32, 24, 16, 8; 4, 8, 12, 16\}$	$1 \cdot 3^{-4}$	$[4, 8, 12, 16]$	16	$\times 4$
$\{40, 30, 20, 10; 5, 10, 15, 20\}$	$1 \cdot 3^{-4}$	$[5, 10, 15, 20]$	$20 - 20$	$\times 5$
$\{40, 30, 20, 10; 8, 16, 24, 32\}$	$256 \cdot 3^{-8}$	$[6, 12, 18, 24]$	$24 - 36$	? $\times 2$
$\{40, 36, 8, 2; 2, 4, 36, 40\}$	$1 \cdot 3^{-5}$	$[10, 12, 16, 18]$	$20 - 20$	$\times 2$
$\{48, 36, 24, 12; 6, 12, 18, 24\}$	$1 \cdot 3^{-4}$	$[6, 12, 18, 24]$	$24 - 24$	$\times 6$

TABLE 7.12 $q = 3,\ \rho = 5$

Array	Density	Spectrum	l.b.–u.b.	References
$\{10, 8, 6, 4, 2; 1, 2, 3, 4, 5\}$	$1 \cdot 3^{-5}$	$[1, 2, 3, 4, 5]$	5	$\{2; 1\}^5$, $H(5, 3)$
$\{20, 16, 12, 8, 4; 2, 4, 6, 8, 10\}$	$1 \cdot 3^{-5}$	$[2, 4, 6, 8, 10]$	10	$\times 2$
$\{22, 20, 15, 2, 1; 1, 2, 12, 20, 22\}$	$4 \cdot 3^{-7}$	$[5, 6, 8, 9, 11]$	$\nexists$	Theorem 7.15
$\{22, 20, 18, 2, 1; 1, 2, 9, 20, 22\}$	$1 \cdot 3^{-6}$	$[5, 6, 8, 9, 11]$	11	$\{0\}$-sh. $\overline{G}_{12}$
$\{25, 20, 15, 10, 5; 4, 8, 12, 16, 20\}$	$4^5 \cdot 3^{-10}$	$[3, 6, 9, 12, 15]$	15–20	? $\{5; 4\}^5$
$\{30, 24, 18, 12, 6; 3, 6, 9, 12, 15\}$	$1 \cdot 3^{-5}$	$[3, 6, 9, 12, 15]$	15	$\times 3$
$\{35, 28, 21, 14, 7; 2, 4, 6, 8, 10\}$	$2^5 \cdot 3^{-10}$	$[3, 6, 9, 12, 15]$	18–20	? $\{7; 2\}^5$
$\{40, 32, 24, 16, 8; 1, 2, 3, 4, 5\}$	$1 \cdot 3^{-10}$	$[3, 6, 9, 12, 15]$	20	$\{8; 1\}^5$
$\{40, 32, 24, 16, 8; 4, 8, 12, 16, 20\}$	$1 \cdot 3^{-5}$	$[4, 8, 12, 16, 20]$	20	$\times 4$
$\{44, 40, 30, 4, 2; 2, 4, 24, 40, 44\}$	$4 \cdot 3^{-7}$	$[10, 12, 16, 18, 22]$	22–?	?? ∪:$\{44, 40; 2, 4\}$
$\{44, 40, 36, 4, 2; 2, 4, 18, 40, 44\}$	$1 \cdot 3^{-6}$	$[10, 12, 16, 18, 22]$	22	$\times 2$
$\{50, 40, 30, 20, 10; 5, 10, 15, 20, 25\}$	$1 \cdot 3^{-5}$	$[5, 10, 15, 20, 25]$	25	$\times 5$
$\{50, 40, 30, 20, 10; 8, 16, 24, 32, 40\}$	$4^5 \cdot 3^{-10}$	$[6, 12, 18, 24, 30]$	30–40	? $\times 2$

For $q = 3$, $\rho = 6$, $b_0 \leq 48$, there are only arithmetic intersection arrays $\{2;1\}^6$, $\{4;2\}^6$, $\{5;4\}^6$, $\{6;3\}^6$, $\{7;2\}^6$, $\{8;1\}^6$, $\{8;4\}^6$.

ACKNOWLEDGMENT

The authors thank Vladimir Potapov for fruitful discussions. The work of D. S. Krotov was carried out within the framework of the state contract of the Sobolev Institute of Mathematics (Project FWNF-2022-0017). The work of J. H. Koolen was partially supported by the National Natural Science Foundation of China (No. 12471335), and the Anhui Initiative in Quantum Information Technologies (No. AHY150000). The work of W. J. Martin was supported in part by the US National Science Foundation (NSF DMS Award #1808376).

REFERENCES

[1] R. D. Baker, J. H. van Lint, and R. M. Wilson. On the Preparata and Goethals codes. *IEEE Trans. Inf. Theory*, 29(3):342–345, 1983. https://doi.org/10.1109/TIT.1983.1056675.

[2] L. A. Bassalygo, G. V. Zaitsev, and V. A. Zinoviev. On uniformly packed codes. *Probl. Inf. Transm.*, 10(1):6–9, 1974. Translated from *Probl. Peredachi Inf.*, 10(1):9–14, 1974.

[3] E. A. Bespalov, D. S. Krotov, A. A. Matiushev, A. A. Taranenko, and K. V. Vorob'ev. Perfect 2-colorings of Hamming graphs. *J. Comb. Des.*, 29(6):367–396, 2021. https://doi.org/10.1002/jcd.21771.

[4] J. Borges, K. T. Phelps, J. Rifà, and V. A. Zinoviev. On Z_4-linear Preparata-like and Kerdock-like codes. *IEEE Trans. Inf. Theory*, 49(11):3834–3843, 2003. https://doi.org/10.1109/TIT.2003.819329.

[5] J. Borges, J. Rifà, and V. Zinoviev. New families of completely regular codes and their corresponding distance regular coset graphs. *Des. Codes Cryptogr.*, 70(1–2):139–148, 2014. https://doi.org/10.1007/s10623-012-9713-3.

[6] J. Borges, J. Rifà, and V. A. Zinoviev. On non-antipodal binary completely regular codes. *Dis. Math.*, 308(16):3508–3525, 2008. https://doi.org/10.1016/j.disc.2007.07.008.

[7] J. Borges, J. Rifà, and V. A. Zinoviev. Families of nested completely regular codes and distance-regular graphs. *Adv. Math. Commun.*, 9(2):233–246, 2015. https://doi.org/10.3934/amc.2015.9.233.

[8] A. E. Brouwer, A. M. Cohen, and A. Neumaier. *Distance-Regular Graphs*. Springer-Verlag, Berlin, 1989. https://doi.org/10.1007/978-3-642-74341-2.

[9] A. E. Brouwer, J. H. Koolen, and R. J. Riebeek. A new distance-regular graph associated to the Mathieu group M_{10}. *J. Algebr. Comb.*, 8(2):153–156, 1998. https://doi.org/10.1023/A:1008685726957.

[10] A. E. Brouwer and H. Van Maldeghem. *Strongly Regular Graphs*, volume 182 of Encyclopedia of Mathematics and Its Applications. Cambridge University Press, Cambridge, 2022.

[11] A. R. Calderbank. Nonexistence of a uniformly packed [70,58,5] code. *IEEE Trans. Inf. Theory*, 32(6):828–833, 1986. https://doi.org/10.1109/TIT.1986.1057232.

[12] A. R. Calderbank and W. M. Kantor. The geometry of two-weight codes. *Bull. Lond. Math. Soc.*, 18(2):97–122, 1986. https://doi.org/10.1112/blms/18.2.97.

[13] R. Canogar. Partition designs with two subsets of the Hamming space. In *6th International Conference on Graph Theory*, volume 5 of *Electron. Notes Discrete Math.*, pages 53–55. Elsevier, July 2000. https://doi.org/10.1016/S1571-0653(05)80123-4.

[14] P. Delsarte. *An Algebraic Approach to Association Schemes of Coding Theory*, volume 10 of Philips Research Reports: Supplement. N.V. Philips' Gloeilampenfabrieken, Eindhoven, 1973.

[15] I. I. Dumer. Some new uniformly packed codes. In *Proceedings of Moscow Institute of Physics and Technology Series "Radiotekhnika i Elektronika"*, pages 72–78. MITP, Moscow, 1976. In Russian.

[16] D. G. Fon-Der-Flaass. A bound on correlation immunity. *Sib. Èlektron. Mat. Izv.*, 4:133–135, 2007. Online: http://mi.mathnet.ru/eng/semr149.

[17] D. G. Fon-Der-Flaass. Perfect 2-colorings of a hypercube. *Sib. Math. J.*, 48(4):740–745, 2007. https://doi.org/10.1007/s11202-007-0075-4 translated from *Sib. Mat. Zh.* 48(4):923–930, 2007.

[18] D. G. Fon-Der-Flaass. Perfect colorings of the 12-cube that attain the bound on correlation immunity. *Sib. Èlektron. Mat. Izv.*, 4:292–295, 2007. In Russian. English translation: https://arxiv.org/abs/1403.8091.

[19] C. D. Godsil. *Algebraic Combinatorics.* Chapman and Hall, New York, 1993.

[20] S. Goryainov and D. Krotov. Completely regular codes in graphs covered by a Hamming graph. arXiv preprint 2411.09698, 2024. https://arxiv.org/abs/2411.09698.

[21] A. R. Hammons, Jr, P. V. Kumar, A. R. Calderbank, N. J. A. Sloane, and P. Solé. The Z_4-linearity of Kerdock, Preparata, Goethals, and related codes. *IEEE Trans. Inf. Theory*, 40(2):301–319, 1994. https://doi.org/10.1109/18.312154.

[22] A. Hedayat, J. Stufken, and G. Su. On the construction and existence of orthogonal arrays with three levels and indexes 1 and 2. *Ann. Stat.*, 25(5):2044–2053, 1997. https://doi.org/10.1214/aos/1069362386.

[23] O. Heden and C. Roos. The non-existence of some perfect codes over non-prime power alphabets. *Dis. Math.*, 311(14):1344–1348, 2011. https://doi.org/10.1016/j.disc.2011.03.024.

[24] W. M. Kantor and M. E. Williams. Symplectic semifield planes and Z_4-linear codes. *Trans. Am. Math. Soc.*, 356(3):895–938, 2004. https://doi.org/10.1090/S0002-9947-03-03401-9.

[25] A. V. Khalyavin. Estimates of the capacity of orthogonal arrays of large strength. *Mosc. Univ. Math. Bull.*, 65(3):130–131, 2010. https://doi.org/10.3103/S0027132210030101.

[26] J. H. Koolen, W. S. Lee, and W. J. Martin. Characterizing completely regular codes from an algebraic viewpoint. In R. A. Brualdi, S. Hedayat, H. Kharaghani, G. B. Khosrovshahi, and S. Shahriari, editors, *Combinatorics and Graphs: The Twentieth Anniversary Conference of IPM Combinatorics, May 15–21, 2009, Tehran, Iran*, volume 531 of Contemporary Mathematics, pages 223–242. AMS, Providence, RI, 2010.

[27] J. H. Koolen, W. S. Lee, W. J. Martin, and H. Tanaka. Arithmetic completely regular codes. *Dis. Math. Theor. Comput. Sci.*, 17(3):59–76, 2016. https://doi.org/10.46298/dmtcs.2150.

[28] D. S. Krotov. On weight distributions of perfect colorings and completely regular codes. *Des. Codes Cryptogr.*, 61(3):315–329, 2011. https://doi.org/10.1007/s10623-010-9479-4.

[29] D. S. Krotov. On calculation of the interweight distribution of an equitable partition. *J. Algebr. Comb.*, 40(2):373–386, 2014. https://doi.org/10.1007/s10801-013-0492-3.

[30] D. S. Krotov. On the existence of some completely regular codes in Hamming graphs. In *IEEE International Symposium on Information Theory, Athens, Greece*, July 7–12, 2024, pages 121–126. IEEE, 2024. https://doi.org/10.1109/ISIT57864.2024.10619341.

[31] D. S. Krotov and K. V. Vorob'ev. On unbalanced Boolean functions with best correlation immunity. *Electr. J. Comb.*, 27(1):#P1.45(1–24), 2020. https://doi.org/10.37236/8557.

[32] S. P. Lloyd. Binary block coding. *Bell Syst. Tech. J.*, 36(2):517–535, 1957. https://doi.org/10.1002/j.1538-7305.1957.tb02410.x.

[33] F. J. MacWilliams and N. J. A. Sloane. *The Theory of Error-Correcting Codes*. Amsterdam, Netherlands, North Holland, 1977.

[34] F. P. Preparata. A class of optimum nonlinear double-error correcting codes. *Inf. Control*, 13(4):378–400, 1968. https://doi.org/10.1016/S0019-9958(68)90874-7.

[35] J. Rifà and V. A. Zinoviev. On a class of binary linear completely transitive codes with arbitrary covering radius. *Dis. Math.*, 309(16):5011–5016, 2009. https://doi.org/10.1016/j.disc.2009.03.004.

[36] J. Rifà and V. A. Zinoviev. New completely regular q-ary codes based on kronecker products. *IEEE Trans. Inf. Theory*, 56(1):266–272, 2010.

[37] J. Rifà and V. A. Zinoviev. On a family of binary completely transitive codes with growing covering radius. *Dis. Math.*, 318:48–52, 2014. https://doi.org/10.1016/j.disc.2013.11.009.

[38] N. V. Semakov, V. A. Zinoviev, and G. V. Zaitsev. Uniformly packed codes. *Probl. Inf. Trans.*, 7(1):30–39, 1971. Translated from *Probl. Peredachi Inf.*, 7(1):38–50, 1971.

[39] M. Shi, D. S. Krotov, and P. Solé. A new distance-regular graph of diameter 3 on 1024 vertices. *Des. Codes Cryptogr.*, 87(9):2091–2101, 2019. https://doi.org/10.1007/s10623-019-00609-w.

[40] M. Shi, D. S. Krotov, and P. Solé. On the coset graph construction of distance-regular graphs. *Dis. Math.*, 345(11):113037(1–6), 2022. https://doi.org/10.1016/j.disc.2022.113037.

[41] P. Solé. Completely regular codes and completely transitive codes. *Dis. Math.*, 81(2):193–201, 1990. https://doi.org/10.1016/0012-365X(90)90152-8.

[42] Yu. Tarannikov. On resilient Boolean functions with maximal possible nonlinearity. Cryptology ePrint Archive 2000/005, 2000. https://eprint.iacr.org/2000/005.

[43] E. R. van Dam and D. Fon-Der-Flaass. Uniformly packed codes and more distance regular graphs from crooked functions. *J. Algebr. Comb.*, 12(2):115–121, 2000. https://doi.org/10.1023/A:1026583725202.

[44] A. Yu. Vasil'eva. Local and interweight spectra of completely regular codes and of perfect colorings. *Probl. Inf. Trans.*, 45(2):151–157, 2009. https://doi.org/10.1134/S0032946009020069, translated from *Probl. Peredachi Inf.*, 45(2):84–90, 2009.

[45] A. Yu. Vasil'eva. Local distributions for eigenfunctions and perfect colorings of q-ary Hamming graph. *Des. Codes Cryptogr.*, 87(2–3):509–516, 2019. https://doi.org/10.1007/s10623-018-0559-1.

Index

Note: **Bold** page numbers refer to tables and *italic* page numbers refer to figures.